Amorphous and Polycrystalline
Thin-Film Silicon Science
and Technology—2007

MATERIALS RESEARCH SOCIETY
SYMPOSIUM PROCEEDINGS VOLUME 989

Amorphous and Polycrystalline Thin-Film Silicon Science and Technology—2007

Symposium held April 9–13, 2007, San Francisco, California, U.S.A.

EDITORS:

Virginia Chu
INESC MN
Lisboa, Portugal

Seiichi Miyazaki
Hiroshima University
Higashi-Hiroshima, Japan

Arokia Nathan
London Centre for Nanotechnology – UCL
London, United Kingdom

Jeffrey Yang
United Solar Ovonic LLC
Troy, Michigan, U.S.A.

Hsiao-Wen Zan
National Chiao Tung University
Hsinchu, Taiwan

Materials Research Society
Warrendale, Pennsylvania

CAMBRIDGE UNIVERSITY PRESS
Cambridge, New York, Melbourne, Madrid, Cape Town,
Singapore, São Paulo, Delhi, Mexico City

Cambridge University Press
32 Avenue of the Americas, New York NY 10013-2473, USA

Published in the United States of America by Cambridge University Press, New York

www.cambridge.org
Information on this title: www.cambridge.org/9781107408722

Materials Research Society
506 Keystone Drive, Warrendale, PA 15086
http://www.mrs.org

First published 2007
First paperback edition 2012

Single article reprints from this publication are available through
University Microfilms Inc., 300 North Zeeb Road, Ann Arbor, MI 48106

CODEN: MRSPDH

ISBN 978-1-107-40872-2 Paperback

CONTENTS

Preface xix

Materials Research Society Symposium Proceedings...................................xxi

DEFECTS AND METASTABILITY

Spin Density in Thin-Film Silicon Before and After
Electron Bombardment..3
    Oleksandr Astakhov, Reinhard Carius,
    Yuri Petrusenko, Valeriy Borysenko,
    Dmitry Barankov, and Friedhelm Finger

Metastable Defects in Tritiated Amorphous Silicon 9
    Tong Ju, Janica Whitaker, Stefan Zukotynski,
    Nazir Kherani, P. Craig Taylor, and Paul Stradins

Light Soaking and Thermal Annealing Effects on
the Micro-Electrical Properties of Amorphous and
Nanocrystalline Mixed-Phase Silicon Solar Cells......................................15
    C.-S. Jiang, B. Yan, H.R. Moutinho, M.M. Al-Jassim,
    J. Yang, and S. Guha

SOLAR CELLS I

* Electrical and Optical Modeling of Thin-Film Silicon Solar
Cells..23
    Miro Zeman and Janez Krc

Enhancing Light-Trapping and Efficiency of Solar Cells
with Photonic Crystals...35
    Rana Biswas and Dayu Zhou

High Open-Circuit Voltage in Silicon Heterojunction Solar
Cells..41
    Qi Wang, Matt R. Page, Eugene Iwancizko, Yueqin Xu,
    Lorenzo Roybal, Russell Bauer, Dean Levi, Yanfa Yan,
    Tihu Wang, and Howard M. Branz

*Invited Paper

ALLOYS

Effects of Nitrogen Addition on the Properties of a-SiCN:H
Films Using Hexamethyldisilazane..49
    Amornrat Limmanee, Michio Otsubo, Tsutomu Sugiura,
    Takehiko Sato, Shinsuke Miyajima, Akira Yamada, and
    Makoto Konagai

Dependence of the Electronic Properties of Hot-Wire CVD
Amorphous Silicon-Germanium Alloys on Oxygen Impurity
Levels 55
    Shouvik Datta, J. David Cohen, Yueqin Xu, A.H. Mahan,
    and Howard M. Branz

High Quality a-Ge:H Films and Devices Through Enhanced
Plasma Chemistry 61
    Erik V. Johnson and Pere Roca i Cabarrocas

Dielectric Properties of Ultra Dense (3 g/cm^3) Silicon Nitride
Deposited by Hot Wire CVD at Industrially Relevant High
Deposition Rates...67
    Zomer Silvester Houweling, Vasco Verlaan,
    Karine van der Werf, Hanno D. Goldbach, and
    Ruud E.I. Schropp

POSTER SESSION:
ALLOYS

Crystalline Silicon Surface Passivation by PECV-Deposited
Hydrogenated Amorphous Silicon Oxide Films [a-SiO$_x$:H]....................75
    Thomas Mueller, Wolfgang Duengen, Reinhart Job,
    Maximilian Scherff, and Wolfgang Fahrner

Effect of Boron Doping on Microcrystalline Germanium
Carbon Thin Films 81
    Yasutoshi Yashiki, Seiichi Kouketsu, Shinsuke Miyajima,
    Akira Yamada, and Makoto Konagai

Characterization of Silicon Carbide Thin Films Obtained
via Sublimation of a Solid Polymer Source Using Polymer-
Source CVD Process 87
    El Hassane Oulachgar, Cetin Aktik, Starr Dostie,
    Subhash Gujrathi, and Mihai Scarlete

Low Temperature High Quality Growth of Silicon-Dioxide Using Oxygenation of Hydrogenation-Assisted Nano-Structured Silicon Thin Films 95
Nima Rouhi, Behzad Esfandyarpour, Shams Mohajerzadeh,
Pouya Hashemi, Bahman Hekmat-Shoar, and
Michael D. Robertson

Channel Reliability in MOSFETs with Gate Oxide Grown Using ECR Plasma of O_2/He 101
Vishwas Jaju and Vikram Dalal

POSTER SESSION:
CRYSTALLIZATION

A Novel Bonding Technique Using Metal-Induced Crystallization of Amorphous Silicon 109
Markus D. Ong and Reinhold H. Dauskardt

Effects of Power Density and Thickness on Aluminum-Induced Crystallization of PECVD Amorphous Silicon........................115
Kendrick S. Hsu, Jeremy Ou-Yang, Li P. Ren, and
Grant Z. Pan

Comparative Study of Hot-Wire Chemical Vapor Deposition Onto (100) Si Near 600°C: Epitaxial and Polycrystalline Silicon Films 121
Charles W. Teplin, Howard M. Branz, Kim M. Jones,
Bobby To, Eugene Iwaniczko, and Paul Stradins

ESR Study of Crystallization of Hydrogenated Amorphous Silicon Thin Films 127
Tining Su, Tong Ju, P. Craig Taylor, Paul Stradins,
Yueqin Xu, Falah Hasoon, Qi Wang, and
Walter A. Harrison

Hot-Wire Chemical Vapor Deposition Epitaxy on Polycrystalline Silicon Seeds on Glass...133
Charles W. Teplin, Howard M. Branz, Kim M. Jones,
Manuel J. Romero, Paul Stradins, and Stefan Gall

Grain Size Control by Means of Solid Phase Crystallization of Amorphous Silicon 139
Jordi Farjas, Pere Roura, and Pere Roca i Cabarrocas

Microwave Activation of Dopants and Solid Phase Epitaxy in Silicon 145
    D.C. Thompson, J. Decker, T.L. Alford, J.W. Mayer, and
    N. David Theodore

Low Temperature Epitaxy of n-Doped Silicon Thin Films Using Plasma Enhanced Chemical Vapor Deposition 151
    Mahdi Farrokh Baroughi, Hassan G. El-Gohary,
    Cherry Y. Cheng, and Siva Sivoththaman

POSTER SESSION: MICRO- AND NANOCRYSTALLINE SILICON

Relation Between Electronic Properties and Density of Crystalline Agglomerates in Microcrystalline Silicon 159
    Paula C.P. Bronsveld, Arjan Verkerk, Tomas Mates,
    Antonin Fejfar, Jatindra K. Rath, and Ruud E.I. Schropp

Optimization of Microcrystaline Silicon Deposited by Expanding Thermal Plasma Chemical Vapor Deposition for Solar-Cell Application 165
    R. Jimenez Zambrano, R.A.C.M.M. van Swaaij, and
    M.C.M. van de Sanden

Boron Doping Effects in Microcrystalline Silicon 171
    Wolfhard Beyer, Lars Niessen, and Frank Pennartz

POSTER SESSION: THIN FILM GROWTH

Phase Control and Stability of Thin Silicon Films Deposited From Silane Diluted with Hydrogen 179
    Gijs van Elzakker, Pavol Šutta, Frans D. Tichelaar,
    and Miro Zeman

Reliability of Silicon Nitride Gate Dielectric in Vertical Thin-Film Transistors 185
    M. Moradi, D. Striakhilev, I. Chan, A. Nathan,
    N.I. Cho, and H.G. Nam

**Deposition Uniformity Control in a Commercial Scale
HTO-CVD Reactor** 191
    Shigeru Sakai, Masaaki Ogino, Ryosuke Shimizu,
    and Yukihiro Shimogaki

ELECTRONICS ON FLEXIBLE SUBSTRATES

**Flexible a-Si:H-Based Image Sensors Fabricated by Digital
Lithography** 199
    William S. Wong, TseNga Ng, Michael L. Chabinyc,
    Rene A. Lujan, Raj B. Apte, Sanjiv Sambandan,
    Scott Limb, and Robert A. Street

**Stability of Amorphous Silicon Thin-Film Transistors
Under Prolonged High Compressive Strain** 205
    Jian-Zhang Chen, I-Chun Cheng, Sigurd Wagner,
    Warren Jackson, Craig Perlov, and Carl Taussig

**Single Grain Si TFTs Fabricated at 100°C for Microelectronics
on a Plastic Substrate** 211
    Ming He, R. Ishihara, T. Chen, J.W. Metselaar, and
    C.I.M. Beenakker

NOVEL APPLICATIONS

**Performance of Thin-Film Silicon MEMS on Flexible
Plastic Substrates** 219
    Samadhan Patil, Virginia Chu, and Joao Pedro Conde

**Amorphous Silicon Based TFT and MIS Nonvolatile
Memories** 225
    Yue Kuo and Helinda Nominanda

**Monolithic Integrated a-Si:H Based pin-Diodes with
Orthogonal Liquid Light Guidance Structures for
Lab-on-Microchip Applications** 231
    Heiko Schäfer, Konstantin Seibel, Lars Schöler, and
    Markus Böhm

**μ-Watt Enhanced Electroluminescent Power of Silicon
Nanocrystal Light-Emitting Diodes Made on Nano-Scale
Silicon-Tip-Array Substrate** 237
    Gong-Ru Lin and Chun-Jung Lin

THIN FILM TRANSISTORS I

* High Mobility Nanocrystalline Silicon TFTs for Display
Application 245
    Min-Koo Han and Sang-Myeon Han

Self-Aligned Nanocrystalline Silicon Thin-Film Transistor
with Deposited n^+ Source/Drain Layer 255
    I-Chun Cheng and Sigurd Wagner

Contact Effects in High Mobility Microcrystalline Silicon
Thin-Film Transistors 261
    Kah Yoong Chan, Eerke Bunte, Helmut Stiebig, and
    Dietmar Knipp

IMAGES AND SENSORS

PECVD Grown p-i-n Si and Si,Ge Thin-Film
Photodetectors for Integrated Oxygen Sensors.......................................269
    Debju Ghosh, Ruth Shinar, Vikram L. Dalal,
    Zhaoqun Zhou, and Joseph Shinar

Segmented Amorphous Silicon n-i-p Photodiodes on
Stainless-Steel Foils for Flexible Imaging Arrays 275
    Y. Vygranenko, R. Kerr, K.H. Kim, J.H. Chang,
    D. Striakhilev, A. Nathan, G. Heiler, and T. Tredwell

Transient Current Behavior of Vertically Integrated
Amorphous Silicon Diodes 281
    Gregory Choong, Nicolas Wyrsch, Christophe Ballif,
    Rolf Kaufmann, and Felix Lustenberger

Optical Readout in pinpi'n and pini'p Imagers: A Comparison 287
    P. Louro, M. Vieira, Y. Vygranenko, M. Fernandes,
    and A. Garção

Germanium-Silicon Separate Absorption and Multiplication
Avalanche Photodetectors Fabricated with Low Temperature
High Density Plasma Chemical Vapor Deposited Germanium 293
    Malcolm Carroll, Kent Childs, Darwin Serkland,
    Robert Jarecki, Todd Bauer, and Kevin Saiz

*Invited Paper

Modeling and Characterization of the Hydrogenated
Amorphous Silicon Metal Insulator Semiconductor
Photosensors for Digital Radiography 299
    N. Safavian, Y. Vygranenko, J. Chang,
    Kyung Ho Kim, J. Lai, D. Striakhilev,
    A. Nathan, G. Heiler, T. Tredwell, and
    M. Fernandes

CRYSTALLIZATION TECHNIQUES I

An Approach to Obtain Single Layer of Nanostructured
Si by Laser Irradiation on Ultrathin Amorphous Si Films 307
    Jun Xu, Zanhong Cen, Jiang Zhou, Wei Li,
    Xinfan Huang, and Kunji Chen

Rapid Crystallization of Amorphous Silicon Utilizing
the Plasma Annealing at Atmospheric Pressure 313
    Hajime Shirai, Yusuke Sakurai, Mina Ye, Koji Haruta,
    Tomohiro Kobayashi, and Yu-ichiro Takemura

THIN FILM TRANSISTORS II

Drain Bias Dependent Threshold Voltage Shift of a-Si:H
TFT Due to the Pulsed Stress..321
    Sang-Geun Park, Jae-Hoon Lee, Sang-Myeon Han,
    Sun-Jae Kim, and Min-Koo Han

Noise Performance of High Fill Factor Pixel Architectures
for Robust Large-Area Image Sensors Using Amorphous
Silicon Technology 327
    Jackson Lai, Yuri Vygranenko, Gregory Heiler,
    Nader Safavian, Denis Striakhilev, Arokia Nathan,
    and Timothy Tredwell

SOLAR CELLS II

* Status of nc-Si:H Solar Cells at United Solar and Roadmap
  for Manufacturing a-Si:H and nc-Si:H Based Solar Panels....................................335
    Baojie Yan, Guozhen Yue, and Subhendu Guha

*Invited Paper

Advanced Deposition Phase Diagrams for Guiding Si:H-Based Multijunction Solar Cells 347
Jason A. Stoke, Nikolas J. Podraza, Jian Li,
Xinmin Cao, Xunming Deng, and Robert W. Collins

Triple Junction n-i-p Solar Cells with Hot-Wire Deposited Protocrystalline and Microcrystalline Silicon 353
Ruud E.I. Schropp, Hongbo Li, Ronald H.J. Franken,
Jatindra K. Rath, Karine van der Werf,
Jan Willem Schüttauf, and Robert L. Stolk

High Rate Deposition of Amorphous Silicon Based Solar Cells Using Modified Very High Frequency Glow Discharge 359
Guozhen Yue, Baojie Yan, Jeffrey Yang, and
Subhendu Guha

Fabrication and Optimization of a-Si:H n-i-p Single-Junction Solar Cells with 8 Å/s Intrinsic Layers of Protocrystalline Si:H Materials 365
Xinmin Cao, Wenhui Du, Y. Ishikawa, Xianbo Liao,
Robert W. Collins, and Xunming Deng

CRYSTALLIZATION TECHNIQUES II

* **Laser Crystallization of Silicon for Large Area Electronics** 373
Toshiyuki Sameshima

Defect Characterization of Polycrystalline Silicon Layers Obtained by Aluminum-Induced Crystallization and Epitaxy 385
Dries Van Gestel, Ivan Gordon, Lodewijk Carnel,
Guy Beaucarne, and Jef Poortmans

Comparative Study of Solid-Phase Crystallization of Amorphous Silicon Deposited by Hot-Wire CVD, Plasma-Enhanced CVD, and Electron-Beam Evaporation 391
Paul Stradins, Oliver Kunz, David L. Young, Yanfa Yan,
Kim M. Jones, Yueqin Xu, Robert C. Reedy,
Howard M. Branz, Armin G. Aberle, and Qi Wang

*Invited Paper

POSTER SESSION:
THIN FILM TRANSISTORS

**Manufacturing of TFTs with High Deposition Rated
Microcrystalline Silicon Using Plasma Enhanced
Chemical Vapor Deposition** 399
    Kyung-Bae Park, Ji-Sim Jung, Jong-Man Kim,
    Myung-kwan Ryu, Sang-Yoon Lee, and
    Jang-Yeon Kwon

**Effect of a Channel Length and Drain Bias on the Threshold
Voltage of Field Enhanced Solid Phase Crystallization
Polycrystalline Thin-Film Transistor on the Glass Substrate**.....................405
    Won-Kyu Lee, Sang-Myeon Han, Sang-Geun Park,
    Young-Jin Chang, Kee-Chan Park, Chi-Woo Kim,
    and Min-Koo Han

POSTER SESSION:
SOLAR CELLS

**Influence of Amorphous Layers on Performance of
Nanocrystalline/Amorphous Superlattice Si Solar
Cells**..413
    Atul Madhavan, Debju Ghosh, Max Noack, and
    Vikram Dalal

**Improved Efficiency of Single Junction Microcrystalline
Silicon n-i-p Solar Cells with an i-Layer Made by Hot-
Wire CVD** 419
    Hongbo Li, Ronald H.J. Franken, Robert L. Stolk,
    C.H.M. van der Werf, Jan-Willem A. Schuttauf,
    Jatin K. Rath, and Ruud E.I. Schropp

**Resistive Losses at c-Si/a-Si:H/ZnO Contacts for
Heterojunction Solar Cells**...425
    Florian Einsele, Phillip Johannes Rostan, and
    Uwe Rau

**Hydrofluoric Acid Treatment of Amorphous Silicon
Films for Photovoltaic Processing** 431
    M. Burrows, U. Das, M. Lu, S. Bowden, R. Opila,
    and R. Birkmire

High-Performance, Tandem-Type Amorphous
Silicon Solar Cell..437
    Porponth Sichanugrist, Nirut Pingate, and
    Channarong Piromjit

Hydrogen Passivation of Thin-Film Polysilicon
Solar Cells...443
    Lode Carnel, Ivan Gordon, Dries Van Gestel,
    Guy Beaucarne, and Jef Poortmans

Photocurrent Profile in a-SiC:H Monolithic Tandem
pinpin and pinip Photodiodes 449
    Alessandro Fantoni, Manuela Vieira, and
    Yuri Vygranenko

POSTER SESSION: IMAGERS, SENSORS AND NOVEL APPLICATIONS

Study of a Fabrication Process and Characterization
of One Dimensional Array of Un-Cooled Micro-
Bolometers Based on Germanium Films Deposited
by Plasma...457
    Mario Moreno, Andrey Kosarev, Alfonso Torres,
    and Roberto Ambrosio

Preliminary Results on Large Area X-ray a-SiC:H
Multilayer Detectors with Optically Addressed Readout.................463
    Manuela Vieira, Yuri Vygranenko, Miguel Fernandes,
    Paula Louro, Pedro Sanguino, Alessandro Fantoni,
    and Reinhard Schwarz

Modeling the Laser Scanned Photodiode S-Shaped J-V
Characteristic..469
    Miguel Fernandes, Manuela Vieira, and Rodrigo Martins

Temperature Dependence of Leakage Current in Segmented
a-Si:H n-i-p Photodiodes 475
    Jeff Hsin Chang, Tsu Chiang Chuang, Yuri Vygranenko,
    Denis Striakhilev, Kyung Ho Kim, Arokia Nathan,
    Gregory Heiler, and Timothy Tredwell

High Performance Hydrogenated Amorphous Silicon
n-i-p Photo-Diodes on Glass and Plastic Substrates by
Low-Temperature Fabrication Process 481
    Kyung Ho Kim, Yuriy Vygranenko, Mark Bedzyk,
    Jeff Hsin Chang, Tsu Chiang Chuang, Denis Striakhilev,
    Arokia Nathan, Gregory Heiler, and Timothy Tredwell

POSTER SESSION:
ELECTRONICS AND
FLEXIBLE SUBSTRATES

Hot-Wire Deposited Nanocrystalline Silicon TFTs on
Plastic Substrates 489
    Farhad Taghibakhsh, Michael M. Adachi, and
    Karim S. Karim

POSTER SESSION:
ELECTRONIC PROPERTIES
AND METASTABILITY

Growth and Electronic Properties in Hot Wire Deposited
Nanocrystalline Si Solar Cells...497
    Kamal Muthukrishnan, Vikram Dalal, and Max Noack

Evolution of Structural and Electronic Properties in
Boron-Doped Nanocrystalline Silicon Thin Films 503
    Hyun Jung Lee, Andrei Sazonov, and Arokia Nathan

POSTER SESSION:
STRUCTURAL PROPERTIES

Elastic Properties of Several Silicon Nitride Films 511
    Xiao Liu, Thomas H. Metcalf, Qi Wang, and
    Douglas M. Photiadis

Structural Analysis of Nanocrystalline Silicon
Prepared by Hot-Wire Chemical Vapor Deposition
on Polymer Substrates...517
    Michael M. Adachi, Farhad Taghibakhsh,
    Karen L. Kavanagh, and Karim S. Karim

Manipulating the Hydrogen-Bonding Configuration in ETP-CVD a-Si:H 523
M.A. Wank, R.A.C.M.M. van Swaaij, and
M.C.M. van de Sanden

POSTER SESSION:
NANOCRYSTALS, NANOCLUSTERS
AND NANOWIRES

Tailored Deposition by LPCVD of Non-Stoichiometric Si Oxides and Their Application in the Formation of Si Nanocrystals Embedded in SiO$_2$ by Thermal Annealing 531
Bruno Morana, Juan Carlos G. de Sande,
Andrés Rodríguez, Jesús Sangrador,
Tomás Rodríguez, Manuel Avella,
Ángel Carmelo Prieto, and Juan Jiménez

Silicon Nanowires: Growth Studies Using Pulsed PECVD 537
David Parlevliet and John C.L. Cornish

SOLAR CELLS III

* **Recent Progress in Up-Scaling of Amorphous and Micromorph Thin-Film Silicon Solar Cells to 1.4 m^2 Modules**..545
Johannes Meier, Ulrich Kroll, Stefano Benagli,
Tobias Roschek, Andreas Huegli, Joel Spitznagel,
Oliver Kluth, Daniel Borello, Michael Mohr,
Dmitri Zimin, Giovanni Monteduro, Jiri Springer,
Christoph Ellert, Girogios Androutsopoulos,
Gerold Buechel, Arno Zindel, Franz Baumgartner,
and Detlev Koch-Ospelt

Temperature Dependence of Silicon-Based Thin-Film Solar Cells on Their Intrinsic Absorber 557
Kobsak Sriprapha, Ihsanul Afdi Yunaz, Shuichi Hiza,
Kun Ho Ahn, Seung Yeop Myong, Akira Yamada,
and Makoto Konagai

*Invited Paper

**Efficient Thin-Film Polycrystalline-Silicon Solar Cells
Based on Aluminum-Induced Crystallization** 563
    Ivan Gordon, Lode Carnel, Dries Van Gestel,
    Guy Beaucarne, and Jef Poortmans

**Materials Optimization for Silicon Heterojunction Solar
Cells Using Spectroscopic Ellipsometry** 569
    Dean Levi, Eugene Iwanizcko, Steve Johnston,
    Qi Wang, and Howard M. Branz

**Interdigitated Back Contact Silicon Heterojunction (IBC-SHJ)
Solar Cell** 575
    Meijun Lu, Stuart Bowden, Ujjwal Das, Michael Burrows,
    and Robert Birkmire

FILM GROWTH

**Time-Resolved Cavity Ringdown Spectroscopy as a Monitoring
Technique of Nanoparticles in Pulsed VHF Plasmas** 583
    Takehiko Nagai, Arno H.M. Smets, and Michio Kondo

Author Index 589

Subject Index 593

PREFACE

Hydrogenated amorphous silicon has grown into a multi-billion dollar industry encompassing commercial products such as active-matrix liquid crystal displays, solar cells, digital imagers and scanners. Advances in technology have enabled deposition on large area glass substrates with dimensions of 2.16 m × 2.46 m for the 8th generation TFT LCD displays and 1 m × 1.4 m for solar cells, as well as utilization of roll-to-roll processing on metallic and polymeric substrates allowing the development of new applications and the reduction of production costs. For active-matrix liquid crystal displays, significant progress has been made by using amorphous, nanocrystalline, and polycrystalline silicon and alloys. Current challenges lie in on-panel integration of the perpherical electronics for large area displays. With active-matrix organic light emitting diode displays, choice of appropriate and scalable materials technology still remains a question. Here, key technology requirements are large area uniformity of transistor parameters, threshold-voltage stability, and a carrier mobility that is higher than amorphous silicon. Nano-crystalline silicon is viewed with great promise for this application. For polycrystalline silicon, novel grain control technologies have been proposed and improved to realize large grain size and to diminish the defects inside the grain leading to the realization of single-crystal transistor on glass substrates. Issues on large-area and flexible substrates are addressed by both invited and contributed papers. As we witness the rapid growth of the photovoltaic market in recent years, thin-film technology using amorphous silicon based alloys has also gained momentum; large manufacturing plants are being built around the world. In the meantime, tandem structures incorporating both amorphous and nanocrystalline silicon materials have been further developed and show promise for production. In addition to the two major applications, amorphous and nanocrystalline silicon based large-area electronics have found new applications in the biomedical arena. The success of amorphous silicon and polysilicon materials in commercial products is the driving force for Symposium A being one of the longest-running symposia of the Materials Research Society, providing an excellent forum for reporting research results, exchanging ideas, and discussing scientific and technological issues.

Symposium A, "Amorphous and Polycrystalline Thin-Film Silicon Science and Technology—2007," was held April 9–13 at the 2007 MRS Spring Meeting in San Francisco, California. The symposium started off on April 9 with an extremely well-attended full-day tutorial aimed at young researchers and people new to the field, organized by Friedhelm Finger and Michio Kondo. There were 13 invited talks, 58 contributed oral presentations and 83 poster presentations spread over the next three and one-half days. Renewed interest in photovoltaics resulted in three very well-attended oral sessions on this topic. The continued technological importance of thin-film transistors also figured prominently in the symposium, with sessions on both amorphous and polycrystalline silicon TFTs. Studies of deposition, especially scaling to large area substrates and crystallization techniques, also drew special attention. Fundamental studies on long-outstanding issues in amorphous silicon have continued. Material characterizations for new materials such as nanocrystalline and polycrystalline materials and alloys are hot topics in the symposium. Carrier transport and defect structures in nanocrystalline silicon have been widely studied, providing important information for material optimization, device simulation, and device design.

The success of Symposium A is a direct consequence of the high-quality invited and contributed presentations and the energetic scientific exchange that takes place during oral and poster sessions. So we first acknowledge the invaluable contribution of the authors of oral and poster presentations as well as the written contributions in this volume.

Many people worked behind the scenes to make things run smoothly before, during and after the conference. As in past years, Mary Ann Woolf supervised and managed the manuscript reviewing process, and her experience allowed for the smooth and timely production of this

xix

proceedings volume. Many people were asked to review the papers published here, and we thank them for the time that they took to give valuable feedback to the authors. The MRS staff provided friendly and professional support throughout the organization of the symposium and proceedings.

Finally, we gratefully acknowledge the generous financial support of our corporate sponsors: AKT/Applied Materials, Asahi Glass, AU Optronics, Fuji Electric, NREL, United Solar Ovonic LLC, and ULVAC.

Virginia Chu
Seiichi Miyazaki
Arokia Nathan
Jeffrey Yang
Hsiao-Wen Zan

July 2007

MATERIALS RESEARCH SOCIETY SYMPOSIUM PROCEEDINGS

Volume 989— Amorphous and Polycrystalline Thin-Film Silicon Science and Technology—2007, V. Chu, S. Miyazaki, A. Nathan, J. Yang, H.W. Zan, 2007, ISBN 978-1-55899-949-7

Volume 990— Materials, Processes, Integration and Reliability in Advanced Interconnects for Micro- and Nanoelectronics, Q. Lin, E.T. Ryan, W-L. Wu, D.Y. Yoon, 2007, ISBN 978-1-55899-950-3

Volume 991— Advances and Challenges in Chemical Mechanical Planarization, C. Borst, L. Economikos, A. Philipossian, G. Zwicker, 2007, ISBN 978-1-55899-951-0

Volume 992E— Deposition on Nonplanar Substrates, D. Josell, M. Brett, C. Witt, M. Ritala, 2007, ISBN 978-1-55899-952-7

Volume 993E— Pb-Free and RoHS-Compliant Materials and Processes for Microelectronics, E. Chason, 2007, ISBN 978-1-55899-953-4

Volume 994— Semiconductor Defect Engineering—Materials, Synthetic Structures and Devices II, S. Ashok, P. Kiesel, J. Chevallier, T. Ogino, 2007, ISBN 978-1-55899-954-1

Volume 995E— Extending Moore's Law with Advanced Channel Materials, S. Chakravarthi, R. Arghavani, G. Klimeck, 2007, ISBN 978-1-55899-955-8

Volume 996E— Characterization of Oxide/Semiconductor Interfaces for CMOS Technologies, Y. Chabal, A. Estève, N. Richard, G. Wilk, 2007, ISBN 978-1-55899-956-5

Volume 997— Materials and Processes for Nonvolatile Memories II, T. Li, Y. Fujisaki, J. Slaughter, D. Tsoukalas, 2007, ISBN 978-1-55899-957-2

Volume 998E —Nanoscale Magnetics and Device Applications, S.S. Xue, 2007, ISBN 978-1-55899-958-9

Volume 999E —Novel Semiconductor Materials for Room-Temperature Ferromagnetism, C.R. Abernathy, S. Bedair, P. Ruterana, R. Frazier, 2007, ISBN 978-1-55899-959-6

Volume 1000E—Functional Interfaces in Oxides, 2007, ISBN 978-1-55899-960-2

Volume 1001E—Progress in High-Temperature Superconductors, P. Barnes, D. Lee, C. Park, N. Amemiya, J. Reeves, 2007, ISBN 978-1-55899-961-9

Volume 1002E—Printing Methods for Electronics, Photonics, and Biomaterials, G. Gigli, 2007, ISBN 978-1-55899-962-6

Volume 1003E—Organic Thin-Film Electronics—Materials, Processes, and Applications, A.C. Arias, J.D. MacKenzie, A. Salleo, N. Tessler, 2007, ISBN 978-1-55899-963-3

Volume 1004E—Materials and Strategies for Lab-on-a-Chip—Biological Analysis, Microfactories, and Fluidic Assembly of Nanostructures, S. Grego, J.M. Ramsey, O. Velev, S. Verpoorte, 2007, ISBN 978-1-55899-964-0

Volume 1005E—Advances in Photo-Initiated Polymer Processes and Materials, A. Guymon, C. Hoyle, M. Shirai, E. Nelson, 2007, ISBN 978-1-55899-965-7

Volume 1006E—Transport Behavior in Heterogeneous Polymeric Materials and Composites, J. Grunlan, D. Bhattacharyya, E. Marand, O. Regev, A. Balazs, 2007, ISBN 978-1-55899-966-4

MATERIALS RESEARCH SOCIETY SYMPOSIUM PROCEEDINGS

Volume 1007— Synthesis, Processing, and Properties of Organic/Inorganic Hybrid Materials, R.M. Laine, C. Sanchez, C. Barbé, U. Schubert, 2007, ISBN 978-1-55899-967-1

Volume 1008E—The Nature of Design—Utilizing Biology's Portfolio, R.R. Naik, C.C. Perry, K. Shiba, R. Ulijn, 2007, ISBN 978-1-55899-968-8

Volume 1009E—Advanced Materials for Neuroprosthetic Interfaces, D.R. Kipke, S.P. Lacour, B. Morrison III, D. Tyler, 2007, ISBN 978-1-55899-969-5

Volume 1010E—Functional Materials for Chemical and Biochemical Sensors, E. Comini, P-I. Gouma, V. Guidi, X-D. Zhang, 2007, ISBN 978-1-55899-970-1

Volume 1011E—Materials for Architecture, D.S. Ginley, K.E. Uhrich, B.J. Faircloth, J-J. Kim, 2007, ISBN 978-1-55899-971-8

Volume 1012— Thin-Film Compound Semiconductor Photovoltaics—2007, T. Gessert, K. Durose, S. Marsillac, T. Wada, C. Heske, 2007, ISBN 978-1-55899-972-5

Volume 1013E—Organic and Nanoparticle Hybrid Photovoltaic Devices, 2007, ISBN 978-1-55899-973-2

Volume 1014E—Three-Dimensional Nano- and Microphotonics, P.V. Braun, S. Fan, A.J. Turberfield, S-Y. Lin, 2007, ISBN 978-1-55899-974-9

Volume 1015E—Hybrid Functional Materials for Optical Application, A. Cartwright, T.M. Cooper, A. Koehler, K.S. Schanze, 2007, ISBN 978-1-55899-975-6

Volume 1016E—Materials and Material Structures Enabling Terahertz Technology, E. Stutz, I. Wilke, K. Kreischer, Q. Hu, 2007, ISBN 978-1-55899-976-3

Volume 1017E—Low-Dimensional Materials—Synthesis, Assembly, Property Scaling, and Modeling, M. Shim, M. Kuno, X-M. Lin, R. Pachter, S. Kumar, 2007, ISBN 978-1-55899-977-0

Volume 1018E—Applications of Nanotubes and Nanowires, L. Chen, M. Hersam, 2007, ISBN 978-1-55899-978-7

Volume 1019E—Engineered Nanoscale Materials for the Diagnosis and Treatment of Disease, V.A. Hackley, A.K. Patri, J. Stein, B.M. Moudgil, 2007, ISBN 978-1-55899-979-4

Volume 1020— Ion-Beam-Based Nanofabrication, D. ILA, J. Baglin, N. Kishimoto, P.K. Chu, 2007, ISBN 978-1-55899-980-0

Volume 1021E—Surface and Interfacial Nanomechanics, R.F. Cook, W. Ducker, I. Szlufarska, R.F. Antrim, 2007, ISBN 978-1-55899-981-7

Volume 1022E—Nanoscale Heat Transport—From Fundamentals to Devices, S.R. Phillpot, S. Volz, T. Borca-Tasciuc, M. Choi, 2007, ISBN 978-1-55899-982-4

Volume 1023E—Functional Nanoscale Ceramics for Energy Systems, E. Ivers-Tiffee, S. Barnett, 2007, ISBN 978-1-55899-983-1

Prior Materials Research Society Symposium Proceedings available by contacting Materials Research Society

Defects and Metastability

Mater. Res. Soc. Symp. Proc. Vol. 989 © 2007 Materials Research Society   0989-A02-03

Spin Density in Thin Film Silicon Before and After Electron Bombardment

Oleksandr Astakhov[1,2], Reinhard Carius[1], Yuri Petrusenko[2], Valeriy Borysenko[2], Dmitry Barankov[2], and Friedhelm Finger[1]

[1]IEF-5, Forschungszentrum Jülich, Jülich, 52425, Germany

[2]National Science Centre Kharkov Institute of Physics & Technology, Akademichna 1, Kharkov, 61108, Ukraine

ABSTRACT

The defect density in thin film silicon was increased using low temperature 2MeV electron irradiation up to a factor of 1000. More than 30 samples of different structure from highly crystalline to amorphous were prepared with PECVD and irradiated to study the dynamics of defect accumulation and role of the material structure in this process.

INTRODUCTION

The *Electron Spin Resonance* (ESR) technique is an established method for investigation of paramagnetic electrons in the defect related states in thin film silicon [1, 2]. The technique being sensitive to the nearest neighbourhood of the electron in paramagnetic state in principle might give information on the nature and configuration of the given defect [3] and not only their density. But structural disorder of the investigated material smears out fine structure of the spectrum due to inhomogeneous line broadening. Slightly asymmetric ESR lines at g-values in the range of 2.0045-2.0056 with the width of 6-8 gauss characterize the intrinsic thin film silicon [1, 2, 4, 5, 6]. The resonance is commonly assigned to the silicon dangling bonds (db) in different environments but more specific information on the defects configuration is missing. Also the role of defects in the transport properties of the material separated from the role of the microstructure is of interest.

We apply 2MeV electron irradiation with successive annealing to the samples of thin film silicon prepared over the whole range of structural compositions in order to (i) gain additional information on the nature of defects from analysis of ESR spectra at different steps of the experiment (ii) study the role of defects in the transport properties of the material.

The bombardment of thin film silicon with MeV electrons was a subject for earlier investigations [7-10] but in each case the experiment was limited to the certain material structure (a-Si:H or μc-Si:H) leaving space for a comprehensive study.

Other group of studies utilizes 1-50 keV electrons for investigation of metastability of a-Si:H [11, 12] and μc-Si:H [13]. These results however should be compared to results of MeV irradiation with great caution. Since keV electrons do not produce atom displacements whereas several displacements are possible per incident 2MeV electron [14] and therefore the defect creation mechanisms are different in these two cases.

In our previous publications we reported appearance of new lines in the ESR spectra after irradiation and reversibility of radiation induced change upon annealing [15, 16].

This paper is concentrated on the spin density (N_S) in the material before and after irradiation which is discussed with respect to the material structure.

Beyond the scope of our study the results presented here are of interest from the radiation stability point of view for devices made of different types of thin film silicon.

EXPERIMENT

Samples were deposited with *Plasma Enhanced Chemical Vapour Deposition* (PECVD) technique [17, 18] from the gas mixture of silane (SiH_4) and hydrogen. The silane to hydrogen ratio (or silane concentration - $SC=SiH_4/(SiH_4+H_2)$ was varied in order to obtain samples with different structure from microcrystalline (2%<SC<7%) to amorphous (SC>8%). Other parameters of the deposition listed in Table 1 were constant for all samples.

Table 1. Parameters of the deposition.

$SiH_4/(SiH_4+H_2)$	Discharge power	Discharge frequency	Substrate temperature	Pressure
2-100%	$0.1 W/cm^2$	95MHz	200°C	40 Pa

The samples were deposited on a 50μm thick Mo foil. The thickness of the films was in range of typically 1-3μm. The foil was bent and the flakes of the material were weighed and sealed in Wilmad 710-SQ-250M quartz tubes in atmosphere of 500hPa He for the ESR measurements.

The crystalline volume fraction of the samples was semi-quantitatively determined from Raman spectroscopy as a ratio of the integrated intensities of the Raman signal at $520cm^{-1}$ and $500cm^{-1}$ (attributed to the crystalline phase) and $480cm^{-1}$ (attributed to the disordered phase), i.e. $I_C^{RS} = (I_{500} + I_{520})/(I_{480} + I_{500} + I_{520})$ [19]. The measurements were performed on the samples on the metal foil using laser with wavelength of 647nm.

The irradiation with 2MeV electrons was performed at 100K with the electron beam current density of $5\mu A*cm^{-2}$. The scheme of the irradiation procedure is presented in Fig. 1.

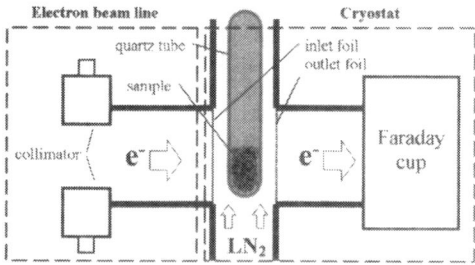

Fig. 1 Schematic view of the irradiation chamber.

The samples assembly was irradiated in flow of liquid nitrogen (LN_2) during 10 hours to acquire the dose of $1.1*10^{18}cm^{-2}$. The high vacuum volume of the electron beam line was separated from the volume of the irradiation chamber with an 80μm thick stainless steel foil. The beam of electrons which passed through the samples was measured with a Faraday cup for the relative control of the dose. The absolute value of the dose was determined from the current and the cross section of the electron beam. The accelerated electrons loose energy in each medium they pass but losses in the outlet foil and nitrogen are negligible (less then 1keV) and the most attenuation takes place in the quartz tube wall where the beam looses around 200keV [20] therefore the samples indeed were exposed to the electrons of approximately 1.8MeV.

Samples were handled, transferred and stored at LN_2 temperature after irradiation prior to measurements in order to prevent spontaneous annealing at room temperature.

ESR measurements are performed with conventional continuous wave technique in X band (9.3 GHz) at 40K. As one can see in Fig. 1 only the section of the quartz tube containing the sample was irradiated with electrons. Nevertheless scattered γ-rays created E´-centres [21] in the shielded part of the tube. E´-centre has a sharp and intensive ESR line that overlaps with the resonance from the sample. In order to reduce E` concentration one end of the irradiated tube was annealed while the part of the tube containing the sample was kept in $LN_2$ [15, 22] and shielded with Al foil. The samples were measured in the annealed part of the tube thereafter. The procedure was sufficient to reduce $N_S$ of E` to a negligible level.

RESULTS AND DISCUSSION

The variation of SC results in a change of the material structure as shown in Fig. 2 (a)

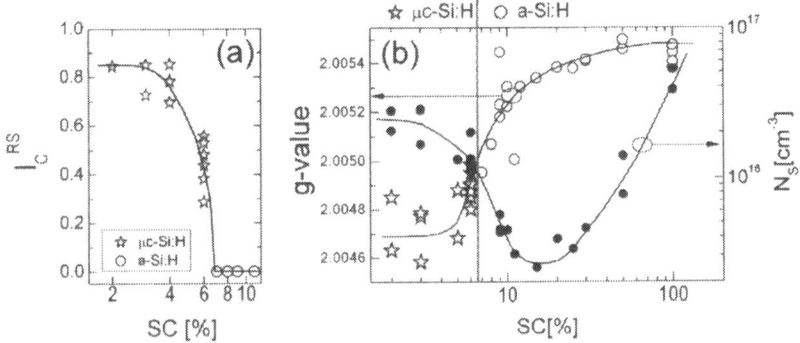

Fig. 2 (a) I_C^{RS} in the material versus SC. (b) N_S and g-value of the samples prepared over the whole range of SC. The lines are guides for the eye.

The increase of SC from 2% to 7% results in a strong reduction of I_C^{RS}. At SC of about 7% no crystalline contribution was detected.

In Fig. 2 (b) the g-value and N_S are plotted versus SC. The low N_S of as-deposited samples in Fig. 2 (b) demonstrates the good quality of the material. μc-Si:H has $N_S \approx 2 * 10^{16} cm^{-3}$ that decreases with decrease of crystalline volume fraction (increase of SC in the graph). The data agree well with earlier reports on device quality μc-Si:H [23]. a-Si:H after the transition (SC=9-25%) shows a minimum of N_S and corresponds to good quality a-Si:H with $N_S = 2-5 * 10^{15} cm^{-3}$. Further increase of SC results in higher deposition rate (from 3 Å/s at SC=10% to 25Å/s at SC=100%) and consequently poorer quality of the material.

g-value of the ESR spectra changes through the whole range of SC from 2.0047 for highly crystalline μc-Si:H to 2.0055 for a-Si:H prepared at SC=100%. The transition between these extremes is of interest. In particular the region where according to Raman spectroscopy the material is amorphous but g-value changes from 2.0050 to 2.0054. The region corresponds to the minimum of N_S that underlines the difference between samples of a-Si:H prepared with different SC.

The electron bombardment results in considerable changes of the ESR lineshape as shown in earlier reports [15, 16]. At the same time the position of the resonance does not change with irradiation and db-resonance dominates the spectra also after irradiation [15, 16]. Therefore we will compare N_S in as-deposited material (N_{SDep}) to N_S after irradiation (N_{SIrr}) not considering the details of the spectral shape. In Fig. 3 (a) N_{SDep} and N_{SIrr} are presented together with the ratio of these values versus SC.

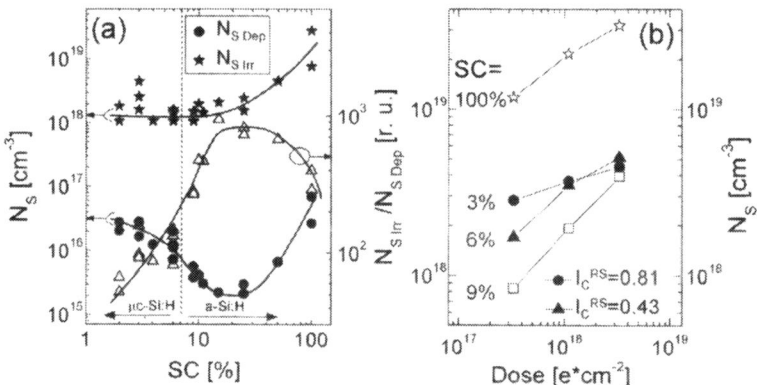

Fig. 3 (a) N_S in as-deposited material (black circles) and N_S after irradiation (black stars) as a function of SC. The ratio $N_{S\ Irr}/N_{S\ Dep}$ is shown with triangles. The lines are guides for the eye. (b) N_S in the irradiated material as a function of irradiation dose. Empty symbols correspond to a-Si:H with different SC, black symbols represent highly crystalline and transition μc-Si:H.

N_S increases by a factor of 50 to a factor of 1000 after irradiation depending on the material structure. In Fig. 3 (a) triangles represent the ratio N_{SIrr}/N_{SDep}. The smallest relative change of N_S is observed for the μc-Si:H with highest I_C^{RS}=0.8. The highest N_{SIrr}/N_{SDep} ratio is found in a-Si:H with SC of 10%...50% caused by the initially low N_S in this material. a-Si:H prepared with SC=100% has a smaller relative change although the absolute value of N_S is the highest after irradiation for this material type.

After irradiation one can see approximately equal values of N_S in a wide range of SC i.e. for all μc-Si:H samples and a-Si:H up to SC=30%. At SC>30% N_S increases with increase of SC. This equity of N_S for material with different structure composition should be considered a coincidence resulting from different dynamics of defects accumulation in the respective materials. This is shown in Fig. 3 (b) where N_{SIrr} is plotted as a function of the irradiation dose. At low dose N_S of highly crystalline μc-Si:H is significantly higher than the value of good quality a-Si:H (SC=9%) while the dose dependence of μc-Si:H itself is more flat than for a-Si:H. Therefore at the highest dose these materials have by coincidence very similar values of N_S. This is the case for the samples shown in Fig. 3(a) that were irradiated with dose of $1.1*10^{18}e*cm^{-2}$ (corresponds to medium dose in Fig. 3 (b)).

At any dose a-Si:H (SC=100%) prepared with high deposition rate shows almost an order of magnitude higher N_S in comparison to any other material. As we already mentioned in description of Fig. 2(a) there is a considerable difference between a-Si:H prepared with SC of 9-30% and material made of pure silane. The dose dependence of N_S for this material underlines this difference. High defect stability seems to be a characteristic feature of this material separating it from the good quality a-Si:H both before and after irradiation.

The additional defects created by electron bombardment can be annealed at 160°C back to the initial value. For the details of N_S annealing see our reports [15, 16].

CONCLUSIONS

We investigated the density of paramagnetic centres in thin film silicon covering a wide range of material structure before and after low temperature electron bombardment.

The smallest relative change of N_S after electron irradiation was found in highly crystalline μc-Si:H (factor 50), whereas the biggest relative change of N_S took place in good quality a-Si:H prepared with SC of 10…30% (factor 1000).

a-Si:H prepared at high deposition rate with pure silane showed the highest N_S both before and after irradiation – $10^{17}e*cm^{-3}$ and $3*10^{19}e*cm^{-3}$, respectively. The value of N_S after irradiation was an order of magnitude higher than in all other samples. This and the dose dependence in the range of $3*10^{17}…3*10^{18}e*cm^{-3}$ indicates that generated defects are more effectively stabilized in this type of material while they might more easily reconstruct in the other investigated materials.

ACKNOWLEDGEMENTS
The work was supported in part by STCU project #655 A. We are grateful to A. Lambertz for assistance with the samples deposition, M. Hülsbeck for Raman measurements and Prof. I. Nekliudov for his support of the irradiation experiments in Ukraine.

REFERENCE

1. R. A. Street, Hydrogenated amorphous silicon, Cambridge University Press (1991) ISBN 0-521-37156-2, pp. 307-320
2. R.A.Street, D.K.Biegelsen, Topics in applied physics 56 (1984) pp.198-208.
3. C. P. Poole, Electron spin resonance : a comprehensive treatise on experimental techniques, New York : Wiley (1983) ISBN 0-471-04678-7.
4. J. K. Rath, Solar Energy Materials & Solar Cells, 76 (2003) pp. 431-487.
5. K. Morigaki, H. Hikita, M. Yamaguchi, Y. Fujita, Materials Science & Engineering B, 103 (2003) pp. 37-44.
6. F. Finger, J. Müller, C. Malten, H. Wagner, Phil. Mag. B, 77 3 (1998) pp. 805-830.
7. R. Street, D. Biegelsen, J. Stuke, Philosophical Magazine B, 40 6 (1979) pp. 451-464.
8. H. Dersch, A. Skumanich, N. M. Amer, Phys. Rev. B, 31 10 (1985) p. 6913.
9. R. Brüggemann, W. Bronner, M. Mehring, Sol. State Comm. 119 (2001) pp. 23-27.
10. W. Bronner, M. Mehring, R. Brüggemann, Phys. Rew. B, 65 (2002) p. 165212.

11. F. Diehl, W. Herbst, S. Bauer, B. Schröder, H. Oechsner, J. Non-Cryst. Sol. **198-200** (1996) pp. 436-440.
12. A. Yelon, H. Fritzsche, H. M. Branz, J. Non-Cryst. Solids, **266-269** (2000) pp.437-443.
13. M. V. Chukichev, P. A. Forsh, W. Fuhs, A. G. Kazanskii, J. Non-Cryst. Solids, **338-340** (2004) pp. 378-381.
14. V. S. Vavilov, N. A. Ukhin, Radiation effects in semiconductors and semiconductor devices, New-York, (1977).
15. O. Astakhov, R. Carius, Yu. Petrusenko, V. Borysenko, D. Barankov and F. Finger, Phys. Stat. Sol. (RRL) **1** 2 (2007) pp. R77-R79.
16. O. Astakhov, F. Finger, R. Carius, A. Lambertz, Yu. Petrusenko, V. Borysenko and D. Barankov, Thin Solid Films, in press (avalible online).
17. W. Luft, Y. S. Tsuo. Hydrogenated amorphous silicon alloy deposition processes, New York, NY : Dekker (1993) ISBN 0-8247-9146-0.
18. G. Bruno, P. Capezzuto, A. Madan, Plasma deposition of amorphous silicon-based materials, Boston, MA : Academic Pr. (1995) ISBN 0-12-137940-X.
19. L. Houben, M. Luysberg, P. Hapke, R. Carius, F. Finger, H. Wagner, Philos. Mag. A, **77** (1998) p. 1447.
20. Stopping powers for electrons and positrons, ICRU Report, 37 (1984).
21. K. Awazu, K. Watanabe, H. Kawazoe, J. Appl. Phys. **73** 12 (1993) p. 8519.
22. W. M. Pontuschka, W. W. Carlos, P. C. Taylor, Phys. Rew. B, **25** 7 (1982) p. 4362.
23. L. Baia Neto, A. Lambertz, R. Carius, F. Finger, J. Non-Cryst. Sol. **299-302** (2002) pp. 274-279.

Mater. Res. Soc. Symp. Proc. Vol. 989 © 2007 Materials Research Society   0989-A02-04

Metastable Defects in Tritiated Amorphous Silicon

Tong Ju[1], Janica Whitaker[2], Stefan Zukotynski[3], Nazir Kherani[3], P. Craig Taylor[4], and Paul Stradins[5]

[1]Physics, University of Utah, Salt Lake City, UT, 84112
[2]ATK Thiokol Hazard Analysis, Brigham City, UT, 84302
[3]University of Toronto, Toronto, M5S 3G4, Canada
[4]Colorado School of Mines, Golden, CO, 80401
[5]National Renewable Energy Laboratory, Golden, CO, 80401

ABSTRACT

We have observed the growth of defects caused by tritium decay in tritiated a-Si: H instead of inducing defects optically. We kept the samples in liquid nitrogen for two years. After two years the ESR signal reached $\sim 10^{19}$ cm^{-3} with no evidence of saturation. However, the density is still less than the density of tritium that has decayed. We step-wise annealed (isochronally annealed) one sample up to 200 °C, where all of the defects were annealed out. Another sample was isothermally annealed at 300 K for several months. At this temperature, the defects anneal slowly.

INTRODUCTION

The appearance of optically or electrically induced defects in hydrogenated amorphous silicon (a-Si: H), especially those that contribute to the Staebler-Wronski effect [1], has been the topic of numerous studies, yet the mechanism of defect creation and annealing is far from clarified. This paper presents another method to induce silicon dangling-bond defects by replacing some of the hydrogen, ^{1}H, with tritium, ^{3}H. Tritium decays to ^{3}He, emitting a beta particle (average energy of 5.7 keV) and an antineutrino. This reaction has a half –life of 12.5 years. The samples discussed in this paper contain approximately 7 and 10.4 at. % tritium. In these tritium-doped a-Si: H samples each beta decay will create a defect by converting a tritium, which is bonded to silicon, to an interstitial helium, leaving behind a silicon dangling bond.

We have tracked these defects through electron spin resonance (ESR) and photothermal deflection spectroscopy (PDS) [2]. The densities we measured at room temperature were smaller by orders of magnitude – only about 5×10^{17}cm^{-3}. Therefore, there should exist a mechanism of defect annealing that is capable of healing $\sim 10^{20}$cm^{-3} defects at room temperature. In the present work, we extend these studies to 77K, in order to establish the saturation behavior at this temperature and the thermal stability of the Si dangling bond defects introduced by tritium decay.

EXPERIMENTAL

Both samples studied were made at the University of Toronto in 1996. The tritium gas was mixed with SiH$_4$, and samples were deposited using a DC glow discharge deposition system at various substrate temperatures. The samples used in this experiment were deposited on glass substrates at temperatures of 423 K (150 °C) (further referred to as G181) and 498 K (225 °C) (referred to as G83). The G181 sample used in this study was 1.5 um thick and the G83 sample used in this study was 0.26 um thick. Shortly after deposition, high temperature tritium effusion experiments determined the tritium concentration to be approximately 7.0 and 10.4 at. % in samples G181 and G83, respectively.

All ESR measurements were made using a Bruker ESR spectrometer operating at approximately 9.5 GHz with 4 gauss magnetic field modulation amplitude. ESR measures only the paramagnetic defects, such as neutral silicon dangling bonds. ESR does not measure the charged dangling bonds. Therefore, photo thermal deflection spectroscopy (PDS) was used to measure both the charged and uncharged defects in these samples at room temperature.

We expect the tritium decay to accumulate Si dangling bond defects because of the silicon-tritium bonds. The density of these defects is related to the number of tritium atoms that have decayed per unit volume. We first measured the samples 7 years after deposition, where the density of tritium atoms that had decayed since making the films was about $6x10^{20}$ cm^{-3}[2]. However, both ESR and PDS measurements of the defect densities were lower by about 3 orders of magnitude because the defect density saturates at room temperature [2]. Next, we annealed the samples near the deposition temperature and kept the two samples at liquid nitrogen temperature for almost two years. During this time we used ESR to track the defect densities. After two years, the defect densities were about 10^{19} cm^{-3} for both samples

After two years in liquid nitrogen, we annealed the two samples. We step-wise annealed the G83 sample at successive temperatures up to 473 K isochronally while the G181 sample was annealed isothermally at 300 K.

RESULTS AND DISCUSSION

After annealing the films, the defects began to accumulate in the dark at 77 K. The spin densities as functions of time stored at 77K are shown in Figs. 1 and 2 for samples G181 and G83, respectively. The data at the shortest times are the defect densities just after annealing. These densities are about 10^{16} cm^{-3} and 10^{17} cm^{-3} for G181 and G83, respectively. In both cases, the densities are probably due to surface or interface defects and not representative of residual densities in the bulk. The spin densities increase linearly with time. The final data points are the defect densities after about two years. In Fig.1, the final density is about $3x10^{19}$ cm^{-3}, which is about 4 times lower than $1.4x10^{20}$ cm^{-3}, the density of tritium atoms, which have decayed since the sample was annealed. In Fig. 2, the final density is about $2x10^{19}$ cm^{-3}. In both samples there is no saturation in the growth as is the case at 300 K [2].

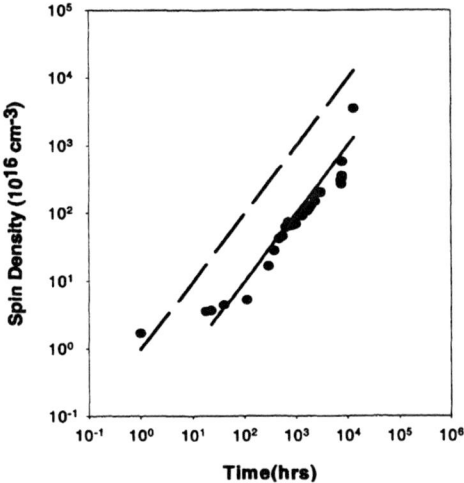

Figure 1. Increase of the defect densities of GJ81 as a function of time at 77 K.

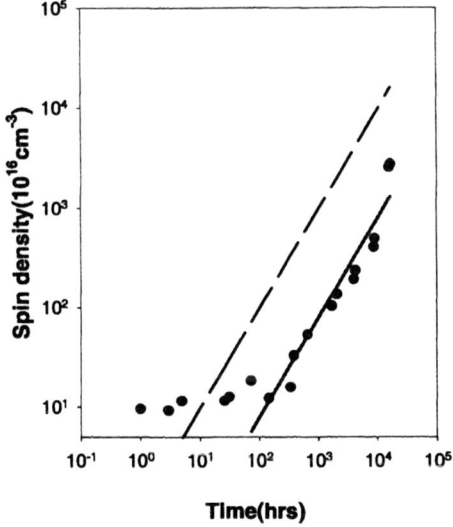

Figure 2. Increase of the defect density of G83 as a function of time at 77 K.

11

Figure 3 shows the decrease in defect densities after sample G181 was heated to room temperature and then kept at room temperature for several months. The defect densities were tracked by ESR. At 300 K, the defects created at 77 K anneal slowly. Even after several months, the density is still higher than the saturation density for defects created at room temperature, which is 6×10^{17} cm^{-3}; only after about one year does the defect density reach the saturated value at 300 K[2]. Because PDS measures both charged and uncharged defects in the sample, we used this method to check that the ESR was measuring all of the defects. The densities of defects as measured by ESR in these samples of a-Si: H are the same as those measured by PDS within a factor of two.

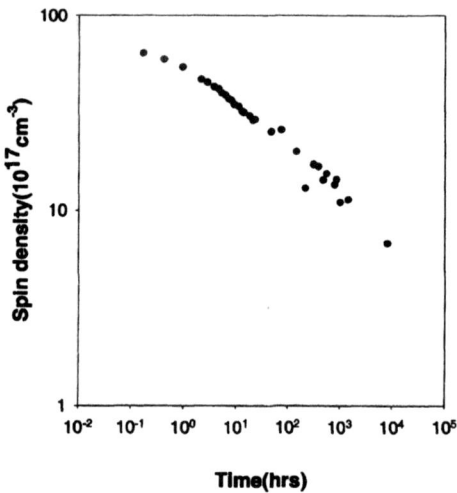

Figure 3. Decrease in the defect density after sample G181 was heated to room temperature and then kept at room temperature for several months of isothermal annealing.

Figure 4 shows data for stepwise annealing of the G83 sample up to 380 K. The sample was kept for 30 minutes at each annealing temperature, T_a. We continued annealing step by step up to 380 K, and after each annealing step we measured spin densities at 20 K. The data show that the defects are fully annealed at 473 K. The final density of approximately 10^{17} cm^{-3} matches the number after we annealed the sample at the start of the 77 K experiments. As mentioned above, this final "density" is due to surface defects and is not indicative of the remaining bulk spin density.

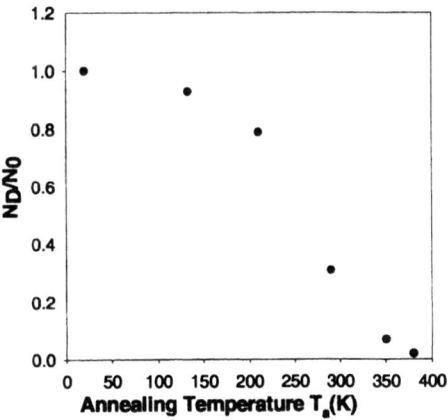

Figure 4. Relative decreases in defect density in sample G83 after stepwise annealing for 30 minutes at each successive annealing temperature, T_a.

At 77 K the defects accumulate almost linearly with the time. The spin density after two years is smaller than the density of tritium atoms that have decayed. This result possibly suggests that the tritium decay in a clustered hydrogen environment does not produce a dangling bond due to bond reconstruction as a result of emission of hydrogen from a nearby Si atom. This suggestion is similar to that proposed previously for thermal emission of hydrogen from the clustered phase [3, 7]. In addition, some tritium decays do not produce a silicon dangling bond because the tritium occurs in molecular form (trapped ^3H-^1H or ^3H-^3H). The sample G83 was annealed step-by-step up to 473K. These annealing results are similar to those that are observed when defects are created with light (Staebler-Wronski effect) at low temperatures. In particular, two thirds of the defects decay near 300 K in both experiments [4, 5].

We checked that there is no effect of the energetic electrons that are emitted during the decay of tritium by electron irradiating a sample with an order of magnitude greater dose that that received in the tritium decay experiments. No increase in spin density was observed. There are two or three hydrogen-related signals that appear in the glass substrates, presumably after tritium decay. These signals are clearly in the glass substrates because their lineshapes correspond to centers commonly observed in oxide glasses on irradiation.

SUMMARY

We have shown that the defect densities of two tritiated amorphous silicon samples at 77 K increase linearly in time up to 10^{19} cm^{-3}. The final densities, however, are factors of 4 to 8 smaller than the density of tritium atoms that decayed. From NMR experiments, we know that 3 at. % of the total atoms in the film (Si plus ^1H plus ^3H) consist of hydrogen and tritium atoms that reside in a dilute phase and the rest of the hydrogen and tritium atoms exist in a clustered phase [9]. Therefore, some of the tritium atoms in the clustered phase probably do not produce silicon dangling bonds at 77 K due to reconstruction. There is no evidence of saturation at 77 K. In our tritium experiments, hydrogen plays an important role in limiting the creation of the dangling bonds, even at 77 K. The annealing isochronal process of G83 is similar to that previously observed for Staebler-Wronski defects created at low temperature. This result provides further hints for the roles of hydrogen in creating and annealing defects in light soaked samples at low temperature, such as those suggested in [6-8].

ACKNOWLEDGEMENTS

Research at the University of Utah was supported by NREL under subcontract No. XXL-5-44205-09 and by NSF under grant No. DMR 0073004. Work at NREL was supported by DOE contract #DE-AC36-99G010337. Research at the University of Toronto was supported by the Natural Sciences and Engineering Research Council of Canada and by the Ontario Centres of Excellence.

REFERENCES

1. D. L. Staebler and C.R.Wronski, Appl. Physics. Lett.31, 292 (1977).
2. J. Whitaker, J. Viner, P. C. Taylor, S. Zukotynski, N. P. Kherani, E. Johnson, and P. Strandins, Mat. Res. Soc. Proc. 808,153 (2004)
3. S. Zafar and E. A. Schiff, Phys. Rev. Lett. 66, 1493 (1991).
4. P. Stradins and H. Fritzsche, Philosophical Magazine B 69, 1 (1994)
5. N. A. Schultz and P. C. Taylor, Phys. Rev. B 65, 235207 (2002)
6. R. A Street, Hytrogenerated Amorphous Silicon (Cambridge Univ. Press, Cambridge, 1991)
7. H. M. Branz, Phys. Rev. B 59, 5498 (1999)
8. M. Stutzmann, W. B. Jackson and C. C. Tsai, Phys. Rev. B 32, 23 (1985)
9. J. A. Reimer, R. W. Vaughn, J. C. Knights, Phys. Rev. B 24 (1981)

Mater. Res. Soc. Symp. Proc. Vol. 989 © 2007 Materials Research Society   0989-A02-05

Light Soaking and Thermal Annealing Effects on the Micro-Electrical Properties of Amorphous and Nanocrystalline Mixed-Phase Silicon Solar Cells

C.-S. Jiang[1], B. Yan[2], H. R. Moutinho[1], M. M. Al-Jassim[1], J. Yang[2], and S. Guha[2]
[1]National Renewable Energy Laboratory, Golden, CO, 80401
[2]United Solar Ovonic LLC, Troy, MI, 48084

ABSTRACT

We report on the measurement of local current flow in hydrogenated amorphous and nanocrystalline mixed-phase n-i-p silicon solar cells in the initial, light-soaked, and annealed states using conductive atomic force microscopy (C-AFM). The C-AFM measurement shows that the nanometer-size grains aggregate and the local current densities in the nanocrystalline aggregation areas significantly decreased after light soaking and recovered to similar values as the initial state after annealing at a high temperature in vacuum. This result supports the two-parallel-connected-diode model for explaining the light-induced open-circuit voltage increase in the mixed-phase solar cells.

INTRODUCTION

Amorphous and nanocrystalline mixed-phase hydrogenated silicon (Si:H) n-i-p solar cells have been fabricated using glow-discharge chemical vapor deposition [1, 2]. This was achieved by optimizing the deposition parameters in the amorphous to nanocrystalline transition regime. Measurement of micro-electrical properties of the mixed-phase cell is not only useful for the device optimization of both amorphous silicon (a-Si:H) and nanocrystalline silicon (nc-Si:H) solar cells, but also helpful for the understanding of device physics with complex structures. The mixed-phase solar cells show an open-circuit voltage (V_{oc}) ranging from 0.5 to 1.0 V, which is between the V_{oc} values of typical a-Si:H and nc-Si:H cells. One interesting phenomenon for the mixed-phase solar cells is that the V_{oc} increases after light soaking [1,2], opposite to commonly observed for a-Si:H and nc-Si:H solar cells where the V_{oc} decreases after light soaking due to the Staebler-Wronski effect [3]. The original explanation for the light-induced V_{oc} increase in the mixed-phase solar cell was based on light-induced structural changes from crystalline to amorphous phase [1, 2]. Subsequently, a complementary model of two parallel-connected diodes (two-diode model) was proposed [4]. There are two key points made in the two-diode model. First, the amorphous phase and nanocrystalline phase can be considered as two separate diodes with significantly different characteristics. Second, the current versus voltage (I-V) characteristics of nc-Si:H diode in the mixed-phase cells degrade by light soaking. The degradation of nc-Si:H causes a decrease in the electric current, when a forward voltage larger than the V_{oc} of the nc-Si:H cell is applied, resulting in an increase in the V_{oc} of the mixed-phase solar cells [4]. However, the size of the nanocrystallites observed by X-ray diffraction and Raman spectroscopy is very small, ranging from a few nm to 30 nm. It is difficult to believe that such small grains can form complete diodes through the entire thickness of the intrinsic layer. Recently, we used a conductive atomic force microscopy (C-AFM) to measure the local current flow and found that the nanocrystallites aggregated and the aggregation regions showed a significantly higher forward current than the amorphous regions [5]. More recently, using a scanning Kelvin probe microscopy (SKPM), we found that the V_{oc} of the mixed-phase solar cell shows localization features on the p-layer of the n-i-p structure [6]. The local V_{oc} showed significantly smaller values for the nc-Si:H aggregates than for the surrounding a-Si:H areas. These findings support the first key point of the two-diode model that the mixed-phase solar cell can be considered as tow separate diodes.

In this paper, we report new results of light-soaking and high-temperature annealing effects on the local current flow in the nc-Si:H aggregate areas. Indeed, we found that the forward electric current density in the aggregates is significantly reduced after light soaking, and recovered to values similar to the initial value after annealing in vacuum. These results provide further evidence for the second key point of the two-diode model that the I-V characteristics of the nanocrystalline phase degrade after light-soaking.

EXPERIMENT

Mixed-phase Si:H n-i-p solar cells were deposited onto a 4×4 cm^2 stainless-steel substrate. The n layer is an a-Si:H layer doped with phosphorus and the p layer is an nc-Si:H layer doped with boron. The i layer was deposited under the condition in the amorphous to nanocrystalline transition regime. Due to the high sensitivity of the transition region to the plasma properties, we can make mixed-phase solar cells showing different characteristics at different locations on the same substrate. Detailed description of the sample preparation was reported in ref. [5]. The sample was then cut to 4 equal pieces along the center line in both the horizontal and vertical directions. Two pieces were subjected to light soaking, and the other two were kept in the initial state. Light soaking was carried out under a white light illumination with an intensity of 100 mW/cm^2 for over 1000 h at 50°C. The V_{oc} values were reduced in both the amorphous and nanocrystalline regions, but were increased in the mixed-phase region as previously reported [1, 2]. After the light soaking, one piece of the light-soaked samples was annealed at 150°C for 2 h in a vacuum, and the V_{oc} values were recovered to the initial values. Then, the local current maps of the three pieces at the initial, light-soaked, and annealed states were measured using C-AFM, along with conventional AFM topographic images

The C-AFM measurements were performed using a Digital Instrument's Dimension 3100 System with diamond-like-carbon coated conducting tips. The C-AFM tip makes a direct contact with the p layer, which significantly reduces the contact resistance as compared to the contact made directly to the intrinsic a-Si:H or nc-Si:H films [7, 8]. At given locations, we measured the I-V characteristics through the tip to confirm that the current is indeed limited by the device instead of by the contact resistance. The scan rate for one line (back and forth) was 0.5 Hz with 256 pixels. The AFM tips were replaced frequently to ensure the same measurement conditions are used for the various scans.

RESULTS AND DISCUSSION

In order to investigate the light-soaking and thermal annealing effects on the current flow through the nanocrystalline phase, we applied a fixed bias voltage of 0.75 V to the sample during the C-AFM measurements. The value was chosen to be between the V_{oc} of a-Si:H and nc-Si:H solar cells. As previously reported [5], the nanocrystalline phase is not uniformly distributed in the amorphous matrix, but aggregates to form clusters. Each aggregate is substantially nanocrystalline and contains many small grains in the nanometer scale. The local current on the nanocrystalline aggregates is much larger than on the surrounding amorphous phase as shown in

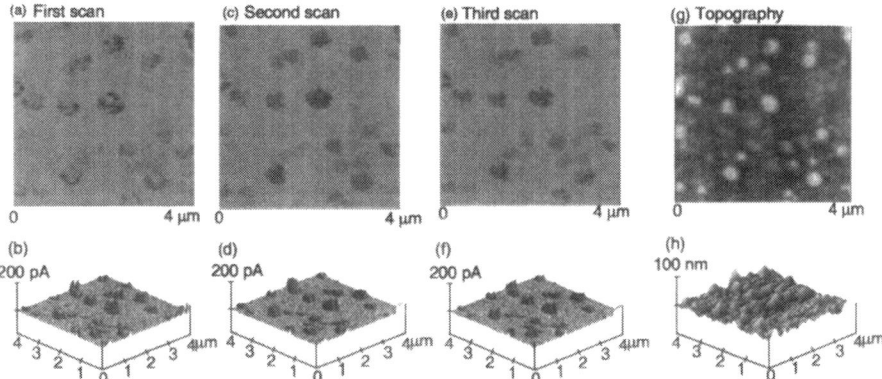

Fig. 1. 2-D and corresponding 3-D C-AFM images on a mixed-phase silicon solar cell from the first C-AFM scan (a) and (b), the second scan (c) and (d), and the third scan (e) and (f). The AFM topographic images of the same area are shown in (g) and (h).

Figs. 1(a)-(f), where the high current regions are nanocrystalline aggregates, corresponding to the hill-like structures on the surface morphology images shown in Figs. 1(g) and (h). It should be pointed out that for multiple scans over the same area, the current image changes, as illustrated in Figs. 1(a) - (f). This is probably resulting from the surface modification by the AFM tip. In the first scan, the current flow is very nonuniform in each individual aggregate. The nonuniformity was reduced in the second scan. We suspect that the improvement in the current uniformity could be caused by a mechanical modification of the p layer when the AFM tip contacted the sample surface, because the current image does not significantly depend on the bias voltage. However, in the following scans, the current value became smaller and smaller. The reduction of current from the previous scans could result from both re-oxidation of the sample surface caused by the high local current and by the wearing off the tip. In this case, the change in the current image depends on the bias voltage. In fact, the tip-induced oxidation process has been reported by other groups previously [9, 10].

The C-AFM image changes from different scans resulted in difficulties for *in-situ* light soaking experiment, because we could not distinguish whether the changes in the C-AFM images were caused by the light soaking or by an artifact of sample surface modification during the measurements. To ensure a proper comparison, the same measurement condition has to be used. Therefore, we pretreated the samples to reach the initial, light-soaked, and annealed states as described above. Then, we took the C-AFM images on the fresh sample surface areas and use relatively fresh tips. In most cases, the C-AFM images in the second scan were the best, where the current values were relatively uniform in an individual aggregate and the average value was the largest among the multiple scans. Therefore, the data presented below were from the second scans. We took a large number of C-AFM images at different areas of the samples in the different states for statistical comparisons. Although the current values showed a large scattering, we found that a larger aggregate exhibits a higher average current value over the aggregate area. Figure 2 shows an example of the C-AFM images in the three states, with a slightly larger average current values in the annealed state than in the initial states, but

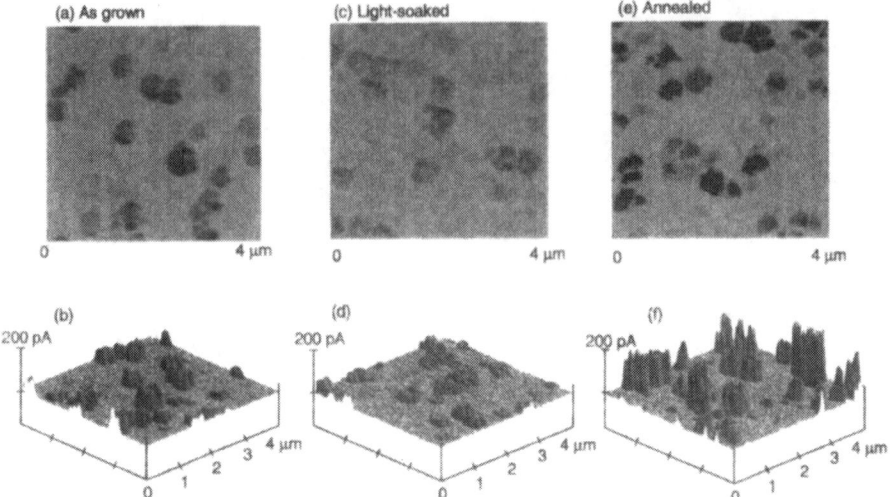

Fig. 2. Examples of 2-D and 3-D C-AFM current images measured on an a-Si:H and nc-Si:H mixed-phase solar cell in the initial state (a) and (b), light-soaked-state (c) and (d), and thermal-annealed state (e) and (f).

significantly smaller average current values in the light-soaked state. Figure 3 plots the average current value versus aggregate area. Although the data are very scattered, the average current values with the same size of aggregates at the initial and annealed states are similar and significantly larger than that at the light-soaked state. We further linearly fitted the current data versus aggregate area, and got the current densities of $(9.4\pm0.7)\times10^{-1}$, $(2.8\pm0.3)\times10^{-1}$, and $(10\pm1)\times10^{-1}$ $(nA/\mu m^2)$ for the initial, light-soaked, and annealed states, respectively. The similar current density on aggregates with different sizes reveals that the current flow in individual aggregates is limited by the size of the aggregates, but not by the tip-sample contact. This means that the current route spreads out in the nanocrystalline aggregates, and was not confined in small areas under the tip. The large scattering in the current data reflects structure fluctuations in the aggregates. Each aggregate contains a large number of small grains. The grain boundaries and amorphous component between the nanometer-size grains significantly affected the electron transport and resulted in the current density fluctuations.

The current density measurements by C-AFM above and other micro-electrical measurements as previously reported [6] reveal that the amorphous and nanocrystalline phases in the mixed-phase solar cells can be distinguished, and thus support the model that the V_{oc} of the mixed-phase solar cells can be considered from the two parallel-connected diodes. The light-induced decrease in average current density in the nanocrystalline aggregates further supports this model. The V_{oc} of a mixed-phase cell is the forward bias voltage, at which the reverse photocurrent flowing through the amorphous phase equals to the forward current through the nanocrystalline aggregates, as schematically shown in Fig. 4(a). The V_{oc} values of mixed-phase solar cells should be between those of amorphous cells (~ 1.0 V) and nanocrystalline cells (~ 0.5 V). If the current through the nanocrystalline phase around this bias voltage decreases, the

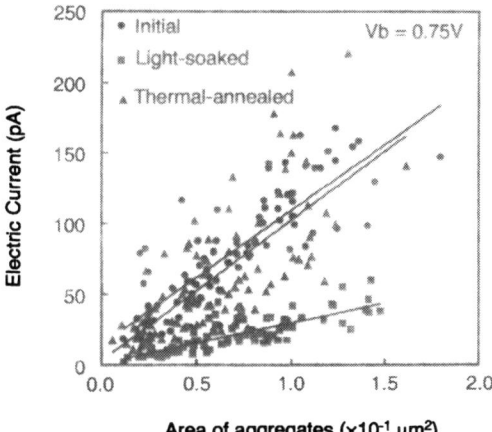

Fig. 3. Electric current as a function of the aggregation area in the initial, light-soaked, and annealed states. Straight lines are linear fitting of the data and represent the current densities.

compensating reverse current through the amorphous phase also decreases, resulting in an increase in V_{oc} of the mixed-phase cell, as shown in Fig. 4(b). It is well known that a-Si:H solar cells degrade after light soaking, as shown in Fig. 4(c), resulting in a corresponding decrease in the reverse current for a given bias voltage. As a result, the V_{oc} of the mixed-phase cell would decrease, which is opposite to what was observed. If the currents through the amorphous and nanocrystalline phases both decrease after light soaking, the V_{oc} could increase or decrease, depending on the net effect. Experimentally, we observe that the light-induced V_{oc} increase occur in most mixed-phase cells, therefore, we believe that the current density decrease in the nanocrystalline aggregates is the dominant factor for the observed light-induced V_{oc} increase. We should point out that the nanocrystalline aggregates in the mixed-phase contain many nanometer-size grains and grain boundaries as observed by transmission electron microscopy images. Although the mechanism of the light-induced current decrease in the nanocrystalline aggregate is not clear, defect generation in the grain boundaries inside the aggregates is a logical speculation. In addition, we observed that the fill factor (FF) also degraded after light soaking in the mixed-phase solar cells and the light-induced FF degradation was fully recovered by high temperature annealing [1,2]. Base on the analysis in Fig. 4, we can see that no mater the amorphous or nanocrystalline diodes degraded, the overall solar cell FF will degrade.

SUMMARY

Using C-AFM measurements, we have investigated the local current flow through amorphous and nanocrystalline mixed-phase silicon solar cells. The local current flows mainly through the nanocrystalline aggregates. Moreover, the local current density decreased significantly after light soaking, and recovered to a value similar to the initial state after thermal annealing. This effect of light-soaking on the local current flow provides further support for the two-diode model of the light-induced V_{oc} increase in the mixed-phase cells.

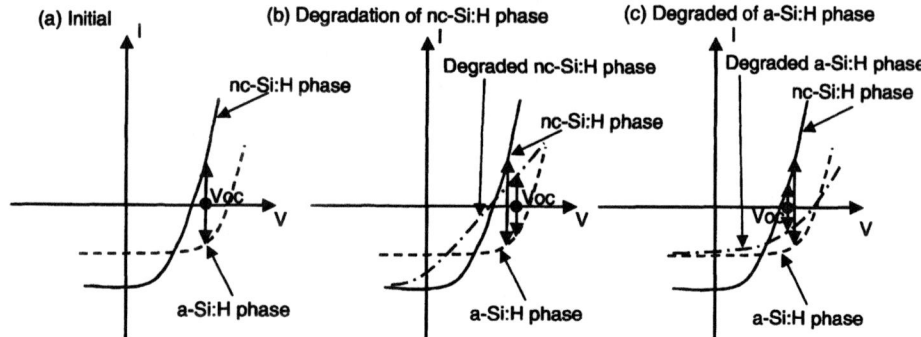

Fig. 4. Schematics of I-V curves and Voc positions (a) in the initial state, and in the case of (b) nc-Si:H phase degradation, and (c) a-Si:H phase degradation.

ACKNOWLEDGMENTS

This work at NREL was supported by DOE under Contract No. DE-AC36-99GO10337, at United Solar by NREL under the Thin Film Partnership Program Subcontract No. ZXL-6-44205-14.

REFERENCES

[1]   K. Lord, B. Yan, J. Yang, and S. Guha, Appl. Phys. Lett. **79**, 3800 (2001).
[2]   J. Yang, K. Lord, B. Yan, A. Banerjee, S. Guha, D. Han, and K. Wang, Mat. Res. Soc. Symp. Proc. **715**, 601 (2002).
[3]   D. L. Staebler and C. R. Wronski, Appl. Phys. Lett. **31**, 292 (1977).
[4]   B. Yan, J. Yang, and S. Guha, Proc. of 3rd World Conference on Photovoltaic Energy Conversion, Osaka, Japan, 2003, p. 1627.
[5]   B. Yan, C.-S. Jiang, H. R.Moutinho, M. M. Al-Jassim, and J. Yang, S. Guha, Mat. Res. Soc. Symp.Proc., A23.6 (2006); B. Yan, C.-S. Jiang, C. W. Teplin, H. R. Moutinho, M. M. Al-Jassim, J. Yang, and S. Guha, J. Appl. Phys. **101**, 033711 (2007).
[6]   C.-S. Jiang, H. R. Moutinho, M.M. Al-Jassim, L. L. Kazmerski, B. Yan, J. M. Owens, J. Yang, and S. Guha, Proc. of 4th World Conference on Photovoltaic Energy Conversion, Hawaii, USA, 2006, p. 1552.
[7]   B. Rezek, J. Stuchlík, A. Fejfar, and J. Kočka, J. Appl. Phys. **92**, 587 (2002).
[8]   D. Azulay, I. Balberg, V. Chu, J. P. Conde, and O. Millo, Phys. Rev. B **71**, 113304 (2005).
[9]   X.-Z Bo, L. P. Rokhinson, Y. Haizhou, D. C. Tsui, and J. C. Sturm, Appl. Phys. Lett. **81**, 3263 (2002)
[10]   S. Myhra, Appl. Phys. A76 63 (2003).

Solar Cells I

Mater. Res. Soc. Symp. Proc. Vol. 989 © 2007 Materials Research Society 0989-A03-01

Electrical and Optical Modelling of Thin-Film Silicon Solar Cells

Miro Zeman[1], and Janez Krc[2]

[1]ECTM/DIMES, Delft University of Technology, Feldmannweg 17, 2628 CT Delft, Netherlands
[2]Lab. of Photovoltaics and Optoelectronics, University of Ljubljana, Trzaska 25, 1000 Ljubljana, Slovenia

ABSTRACT

Today amorphous and microcrystalline silicon based solar cells use surface-textured substrates for enhancing the light absorption and buffer and graded layers in order to improve the overall performance of the cells. Tandem and triple-junction configurations are utilized to assure better use of the solar spectrum and, thus, achieve higher conversion efficiencies of the devices. Resulting structures of the solar cells are complex and computer modeling has become an essential tool for a detailed understanding and further optimization of their optical and electrical behavior.

The performance limits of tandem and triple-junction silicon based solar cells are studied by simulations using the optical simulator SunShine developed at Ljubljana University and the opto-electrical simulator ASA developed at Delft University of Technology. First, both simulators were calibrated with realistic optical and electrical parameters. Then, they were used to study the required scattering properties, absorption in non-active layers, antireflective coatings, the crucial role of the wavelength selective intermediate reflector, and a careful current matching in order to indicate the way for achieving a high photocurrent, more than 15 mA/cm² for a tandem a-Si:H/μc-Si:H and 11 mA/cm² for a triple-junction a-Si:H/a-SiGe:H/μc-Si:H solar cells. By optimizing electrical properties of the layers and interfaces, for example using a p-doped a-SiC layer with a larger band gap ($E_G > 2$ eV) and introducing buffer layers at p/i interfaces, the extraction of the charge carriers, the open-circuit voltage and the fill factor of the solar cells are improved. The potential for achieving the conversion efficiency over 15% for the a-Si:H/μc-Si:H and 17 % for the triple-junction a-Si:H/a-SiGe:H/μc-Si:H solar cells is demonstrated.

INTRODUCTION

Numerical simulation of optical and electrical behavior of semiconductor devices has been world-wide established as an essential tool for the improvement of existing devices, obtaining insight into their physical operation and for the development of new ones. A number of sophisticated semiconductor device simulation packages are already commercially available on the market such as Atlas from SILVACO company [1], Apsys from Crosslight company [2] and Taurus Medici from Synopsis company [3]. These programs are mostly designed for two-dimensional (2-D) modeling of a broad range of crystalline semiconductor devices, but they are gradually updated to offer possibilities to model polycrystalline and amorphous semiconductor based devices such as thin-film transistors and solar cells. The advantage of these programs is that they are modular, so the users need to acquire only the minimum set of modules to meet their needs.

However, modeling of thin-film solar cells based on amorphous and polycrystalline semiconductors, such as hydrogenated amorphous silicon (a-Si:H) and hydrogenated microcrystalline silicon (μc-Si:H), requires the use of advanced physical models which describe specific features of material electronic properties and device operation. These advanced models are not readily implemented in the commercial computer programs. An ideal thin-film solar cell simulation program should meet at least the requirements listed in Table I [4]. For this reason several research groups have developed their own computer programs for modeling thin-film solar cells. Examples are: the AMPS program developed at Penn State University [5], the ASPIN program from Ljubljana University [6], the ASA program developed at Delft University of Technology [7], the SCAPS program written and maintained at the University of Ghent [8] and the AFORS program from Hahn-Meitner-Institute in Berlin [9]. Excellent overview and comparison of these programs is presented in the article of Marc Burgelman [4].

All above mentioned computer programs are primarily limited to one dimensional (1-D) modeling. In principle 1-D modeling is well suited for thin-film silicon solar cells on flat substrates. However, there are at least two important developments in the field of thin-film silicon solar cells that point out that 2-D or 3-D modeling will be required in the future for more accurate modeling of these solar cells. The first development is the use and optimization of surface-textured substrates in solar cells, which introduce spatial variations also in lateral dimensions. The second one is the application of absorber materials such as μc-Si:H which are spatially not homogeneous. There are groups that have already extended their programs to 2-D modeling [10-12] or have used commercially available 2-D or 3-D programs for simulations of a-Si:H based solar cells [13,14]. Furthermore, 2-D modeling and 3-D modeling are becoming an important tool for simulating thin-film photovoltaic modules behavior [15].

In this article the specific issues regarding the modeling of thin-film silicon solar cells will be discussed. The performance limits of a tandem micromorph (a-Si:H/μc-Si:H) and a triple-junction solar cell in the configuration a-Si:H/amorphous silicon germanium (a-SiGe:H)/μc-Si:H are studied by simulations using the ASA program [7] and the optical simulator SunShine developed at Ljubljana University [16]. Both simulators were first calibrated with realistic optical and electrical parameters. The required scattering properties, absorption in non-active layers, antireflective coatings, the crucial role of the wavelength-selective intermediate reflector, and a careful current matching are presented in order to indicate the route towards record-high efficiencies of 15 % for the tandem and 17 % for the triple-junction solar cell.

SPECIFIC ISSUES IN MODELING OF THIN-FILM SILICON SOLAR CELLS

Modeling of thin-film silicon devices requires one to take the electronic structure of a-Si:H and μc-Si:H into account. The spatial disorder in the atomic structure of a-Si:H results in a continuous density of states (DOS) in the band gap with no well defined conduction-band (CB) and valence-band (VB) edges. When considering the transport properties of charge carriers in a-Si:H we have to distinguish between the extended states and the localized states in the DOS distribution. The localized states within the mobility gap strongly influence the trapping and recombination processes and therefore the trapped charge in the localized states cannot be ignored as is often the case in modeling of crystalline semiconductor devices. The localized states in the mobility gap of a-Si:H are represented by the CB and VB tail states and the defect states. These states are different in nature. The tail states behave as acceptor-like states

Table I. Models required for a thin film PV simulation program (adapted from [4])

- Multiple layers
- Band discontinuities in the conduction and valence bands
- Large band gap materials: $E_g > 2.0 - 3.7$ eV
- Grading of material parameters
- Recombination and charge in the localized states within the band gap
- Simulation of non-routine measurements: J(V), QE, C(V), etc., all as a function of T
- Fast and easy to use

(CB-tail states) or donor-like states (VB-tail states) and their density is described by an exponential decay into the mobility gap. The most common defect in a-Si:H is a dangling bond. A dangling bond can be in three charge states: positive (D^+), neutral (D^0) and negative (D^-). An imperfection with three possible charge states acts to a good approximation like a pair of two imperfections consisting of a donor-like state (DB$^{+/0}$) and an acceptor-like state (DB$^{0/-}$) and is therefore represented by two so called transition-energy levels $E^{+/0}$ and $E^{0/-}$ in the band gap. The continuous density of defect states is represented by two equal (Gaussian) distributions located around the mid gap. The corresponding pair of defect energy states is separated by the correlation energy. The different nature of the localized states in a-Si:H requires different approaches for the calculation of recombination-generation (R-G) statistics through these states. The models that are commonly used to describe the localized states in a-Si:H and their corresponding R-G statistics are described in detail in the book of Schropp and Zeman [17].

From the optical point of view, both the efficient use of the solar spectrum and the light management inside a solar cell are important to obtain high conversion efficiencies. In today's thin-film solar cells light management is accomplished by implementing light trapping techniques. The light trapping techniques are based on the introduction of surface-textured substrates and the use of special (back-) reflector layers. The surface-textured substrates introduce rough interfaces into the solar cell. The incident light is scattered at rough interfaces and modeling of solar cells must take into account the scattering processes at rough interfaces in order to determine accurately the generation profile of charge carriers inside the solar cell. This requires the development of optical models that take both coherent non-scattered (specular) and incoherent scattered (diffused) light propagation through a device into consideration. Examples of optical simulators that have implemented both specular and scattered light propagation are optical model developed at Ecole Polytechnique Palaiseau [18], Sunshine program from Ljubljana University [16], Genpro3 module implemented in the ASA program of Delft University of Technology [19], and the Prague optical model [20].

The efficient use of solar spectrum requires a multi-junction approach to thin-film silicon solar cells. The tunneling assisted recombination at an interface between two adjoining junctions is responsible for charge carrier transfer through a multi-junction solar cell. This interface is described as the tunnel-recombination junction (TRJ). Modeling of multi-junction solar cell has pushed for the development of models that can describe the tunnel-recombination processes at the interface between the component cells. A computer model that aims to simulate a tandem or triple thin-film silicon solar cell as a complete device should contain a model for the TRJ. There are two approaches that are used to model TRJ and are implemented in simulation programs [7]. The Delft approach is based on the trap-assisted tunneling model and enhanced carrier transport in the high-field region of the TRJ and the Pennsylvania approach is based on the introduction of

25

a highly defective layer with strongly reduced band gap at the n/p interface and grading of the mobility gap of the n-layer and p-layer in the regions adjacent to the defective layer.

In summary, the following models should be included in a thin-film silicon solar cell simulator: (i) description of density of states distribution in amorphous and microcrystalline semiconductors with the proper recombination-generation statistics, (ii) an optical model that takes both coherent (specular) and incoherent (scattered) light propagation into account, (iii) a model for the TRJ that enables to simulate multi-junction solar cells.

DESCRIPTION AND MODEL CALIBRATION OF THE DEVICE SIMULATORS

A simulation program solves the model equations in order to obtain information about device performance or properties. The effective use of the simulation results and their predictive power strongly depend on the reliability of the values of the input parameters that are required by the numerical model. Assigning the input parameters their proper values is an important step in device modeling and is referred to as calibration of model parameters. The numerical model is calibrated when realistic values of the parameters are used and a good agreement between simulated and measured layer or device characteristics is reached. Well calibrated computer model should reproduce a broad scale of experimental results.

The optical simulator SunShine and the opto-electrical simulator ASA were used to investigate the performance limits of thin-film silicon-based tandem and triple-junction solar cells. The SunShine program is used to calculate the total reflectance from the device, R_{tot}, optical absorbance, A, in the individual layers, and the charge-carrier generation-rate profile in the device. In the SunShine simulator, the specular light is analyzed in terms of electromagnetic waves, whereas the scattered light is treated using ray tracing. Scattering at each rough interface is taken into account. For the electrical analysis of the solar cells the ASA program was used. The ASA program is based on the set of semiconductor equations. Tunneling in the regions of high electric field (at n/p junctions connecting the component cells) is taken into account. As a final result of simulations with the ASA program, internal electrical parameters, such as electron and hole densities, band-diagram, electric field profile, potential, and the external electrical parameters, such as external quantum efficiency, QE, current-voltage characteristic, $J(V)$, short-circuit current, J_{SC}, open-circuit voltage, V_{OC}, fill factor, FF, and conversion efficiency, η, of a solar cell are obtained.

The calibration of the SunShine model parameters is demonstrated on simulations of a state-of-the-art tandem micromorph (a-Si:H/μc-Si:H) solar cell deposited on a surface-textured glass/ZnO:Al superstrate [21]. The thicknesses of the top and the bottom absorbers are $d_{i,top}$ = 250 nm and $d_{i,bot}$ = 2.8 μm. Undoped ZnO is used as an interlayer between the top and bottom cells. Experimentally determined complex refractive indexes of the actual layers [20], realistic scattering parameters of the surface-textured substrates [22,23] and a realistic reflectance at a rough metal back reflector [20] are used in the simulations. Considering the ideal extraction of generated charge carriers from the top and bottom absorbers (i-a-Si:H and i-μc-Si:H), external quantum efficiencies QE_{top} and QE_{bot} can be compared directly with the absorbances in the absorber layers, $A_{i,top}$ and $A_{i,bot}$. As demonstrated in figure 1 a good agreement between the calculated and measured absorbances and quantum efficiencies, as well as in R_{tot} is obtained.

RESULTS AND DISCUSSION

Tandem micromorph (a-Si:H/μc-Si:H) [24-26] and triple-junction solar cells, consisting of an a-Si:H top cell, an a-SiGe:H middle cell and a μc-Si:H bottom cell are both of interest for the production [25,26]. The difference between the tandem micromorph solar cell in the standard a-Si:H/μc-Si:H configuration and the triple-junction a-Si:H/a-SiGe:H/μc-Si:H solar cell is the middle a-SiGe:H based cell, which is introduced to gain in the open-circuit voltage of the device and, thus, raise the conversion efficiency further. Today, state-of-the-art a-Si:H/μc-Si:H solar cells reach the stabilized efficiency up to 11 % [27]. In case of a-Si:H/a-SiGe:H/μc-Si:H solar cells the highest initial conversion efficiencies of ~ 15 % have been reported [25, 26]. For indicating the possibilities and directions for increasing the efficiency of the tandem and the triple-junction silicon solar cells further, as well as to detect the realistic efficiency limits, computer modeling is utilized. Since the relatively thin amorphous absorber layers (intrinsic i-a-Si:H and i-a-SiGe:H layers) (d_i < 300 nm) were intentionally used in our simulating structures, the calculated initial conversion efficiencies can be considered as the stabilized efficiencies, because the effect of light-induced degradation in such thin amorphous layers is small [28].

Optical and electrical simulation of a-Si:H/μc-Si:H solar cell

A micromorph solar cell with the following structure was used in the simulations: glass/ ZnO:Al(500nm)/p-a-SiC:H(10nm)/i-a-Si:H($d_{i,top}$)/n-a-Si:H(15nm)/p-μc-Si:H(10nm)/i-μc-Si:H($d_{i,bot}$)/n-a-Si:H(15nm)/ZnO/Ag. The ZnO/Ag forms the back reflector (BR). The absorber layers were relatively thin ($d_{i,top}$ = 200 nm and $d_{i,bot}$ = 2.2 µm) and no interlayer was applied in this starting cell. Calibrated optical and electrical parameters of un-doped and doped a-Si:H and μc-Si:H layers were utilized in simulations [7]. The generation profile calculated with the SunShine program was exported into the ASA program, which carried out electrical analysis.

The calculated QE_{top} and QE_{bot} obtained by combined optical and electrical analysis are compared to $A_{i,top}$ and $A_{i,bot}$ in figure 2. A good agreement is obtained between A and QE. However, two main deviations are observed (effect (a) and (b)) in the spectral characteristics of the cells. A detailed electrical analysis revealed that an increase in the short-wavelength QE_{top} (effect (a)) originates from, apparently, non-negligible contribution of photo-generated charge carriers from a thin (10 nm) p-a-SiC:H layer. A slightly decreased peak of QE_{bot} (effect (b)) originates from a limited charge extraction from the bottom μc-Si:H absorber. The analysis of the electric field in the bottom cell showed that the carrier extraction becomes critical when applying the solar illumination (AM 1.5) and a forward bias-voltage as expected in a normal operation under the maximal power point conditions. Limited extraction in the bottom cell under normal working conditions should be taken into account when designing the top and bottom cell and matching their photocurrents. Similar trends are observed between the calculated A curves and measured (figure 1) and calculated (figure 2) QE curves. In addition to the effects discussed in this paragraph, there are other possible reasons for the differences observed in figure 1, such as the difference between the true optical functions of the thin p-layer and those assumed in the model.

Besides the QE, $J(V)$ characteristic and the external parameters J_{SC}, V_{OC}, FF and η were calculated for the analyzed solar cell considering standard AM 1.5 solar illumination. The results for this starting micromorph solar cell are shown in figure 3 (thin black curve). With realistic optical and electrical parameters today's state-of-the-art solar cell efficiency of 10 % is calculated, which is in good agreement with the realistic values.

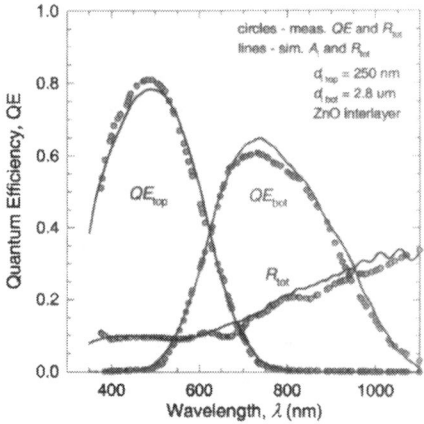

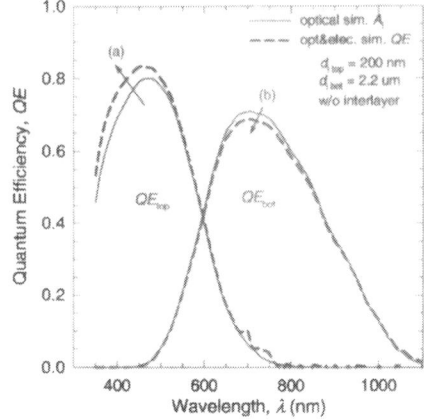

Figure 1. Calibration of the optical simulator SunShine for a micromorph silicon solar cell with ZnO interlayer.

Figure 2. Results of optical and combined optical and electrical simulation of *QE* of micromorph solar cell.

Optical and electrical simulation of a-Si:H/a-SiGe:H/μc-Si:H solar cell

The triple-junction solar cell in the following configuration was used as a starting point in the simulations: glass/ZnO:Al(500nm)/p-a-SiC:H(10nm)/i-a-Si:H($d_{i,top}$)/n-a-Si:H(15nm)/p-a-SiC:H(10nm)/i-a-SiGe:H($d_{i,mid}$)/n-a-Si:H(15nm)/p-μc-Si:H(10nm)/i-μc-Si:H($d_{i,bot}$)/n-a-Si:H(15nm)/ZnO/Ag. The thicknesses of the absorber layers in the starting cell are: $d_{i,top}$ = 180 nm, $d_{i,mid}$ = 220 nm and $d_{i,bot}$ = 2.4 μm, assuring a current matching between all three p-i-n component cells in this case. In the top cell a wide band gap i-a-Si:H absorber is used (E_G = 1.9 eV) in order to enable efficient passing of the middle- and long-wavelength light through the top cell. The band gap of the middle i-a-SiGe:H and bottom μc-Si:H absorbers are E_G = 1.5 eV and 1.3 eV, respectively. No intermediate reflectors (interlayers) are applied here between the individual component cells. The $J(V)$ characteristic of this starting cell, calculated for standard AM 1.5 illumination, is shown in figure 4 (thin black curve). Conversion efficiency of 11.4 % is calculated and the other external parameters are included in the figure.

In figure 5 the absorbances in the absorber layers, $A_{i,top}$, $A_{i,mid}$ and $A_{i,bot}$, of the starting triple-junction solar cell are presented (dashed curves). The absorbances of these individual layers can be identified with the external quantum efficiencies as confirmed by the detailed optical and electrical analysis of the micromorph solar cell.

Optimisation of a thin-film silicon solar cell for achieving the record-high efficiency

Simulations were used to evaluate the efficiency limits of tandem and triple-junction silicon solar cells and determine the optical and electrical requirements for improving the efficiency. The following improvements were included in optical simulations:

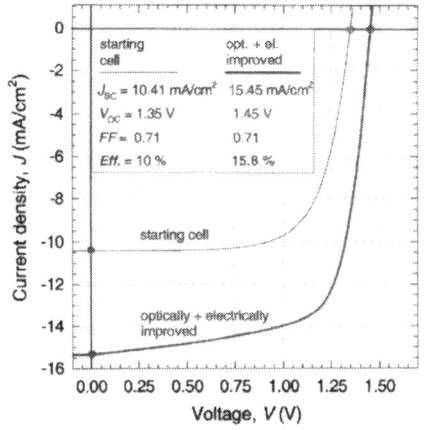

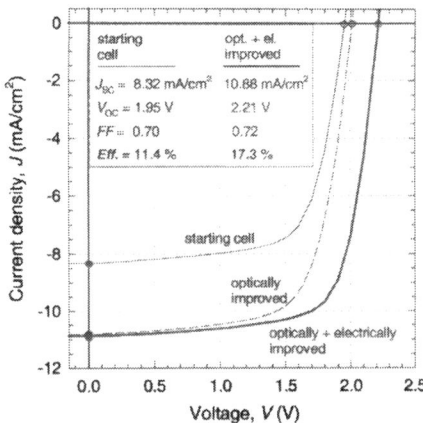

Figure 3. Calculated *J(V)* curves and the external parameters of a micromorph solar cell before and after optimization.

Figure 4. Calculated *J(V)* curves for the starting, optically improved and both optically and electrically improved triple-junction solar cell.

- *Enhanced scattering parameters to improve light trapping in the cell:*
  - (i) ideal haze parameter (scattering level) $H = 1$,
  - (ii) broad (Lambertian) angular distribution function (*ADF*) of scattered light applied to all rough interfaces,
- *Reduced absorptions in non-active layers:*
  - (iii) significant decrease of the absorption coefficient (5×) in the front TCO,
  - (iv) significant decrease of the absorption coefficient (5×) in *p*- and *n*-doped *a*-Si:H and *μc*-Si:H layers,
- *Improved back reflector:*
  - (v) back reflector with an enhanced reflectance of 98 % in the whole wavelength range,
- *Improved light in-coupling:*
  - (vi) optimized single-layer anti-reflective coating (ARC) on the top of the glass substrate with refractive index of 1.25 and thickness of 100 nm and an antireflective interlayer at ZnO:Al/*p-a*-SiC:H interface,
  - (vii) (a) an optimized single-layer intermediate reflector (interlayer) and (b) a (hypothetical) wavelength-selective interlayer [29] between the top and bottom cell in tandem configuration,
  - (viii) adjustment of the absorbers thickness for obtaining the current matching in the triple-junction cell.

High *H* and broad *ADF* can be achieved by further optimization of the morphology of surface-textured substrates and interfaces. Besides random roughness, substrates with periodical surface-texture indicate a large potential for effective light scattering [29]. In order to decrease the absorption losses TCO's with a low free-carrier concentration, resulting in a lower optical absorption in the long-wavelength region are under investigation [30]. By taking into account

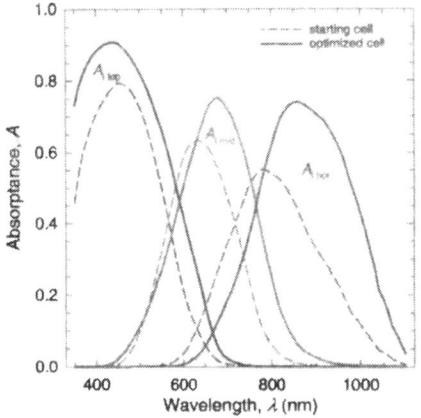

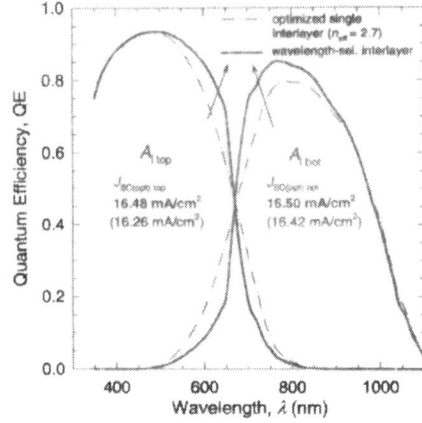

Figure 5. Absorbances in the absorbers for the starting and optimized triple-junction solar cell.

Figure 6. Absorbances in the absorbers for the optimized micromorph solar cell: (a) using a single (J_{SC} values in brackets) and (b) wavelength-selective interlayer.

that a significant reduction of absorption coefficients of p- and n-doped layers may be a critical issue, the use of a very thin efficiently doped layers are of interest. High reflectance of the back reflector may, in principle, be achieved using some specially designed dielectric reflectors [29], avoiding surface plasmon absorption in surface-textured metal layers. Glass substrates covered with a single- as well as multi-layer broadband ARC's are already available [28]. Significant improvements in light in- coupling related to anti-reflective interlayers (such as TiO_2) at TCO/p interfaces have already been reported for the single-junction silicon thin-film solar cell [31].The research on intermediate reflector is in progress nowadays. Single interlayers with different refractive indexes and low absorption are under investigation [25]. As a further improvement a wavelength-selective interlayer with a high reflectivity at the shorter wavelengths and a low reflectivity at the longer wavelengths is of interest. In this way an efficient gain in the absorption can be obtained in the top cell (shorter wavelengths) without losing the absorption in the bottom cell (longer wavelengths). The interlayer is important since it enables the use of a thin top a-Si:H absorber which is a crucial requirement for the long-term stability of the efficiency of the micromorph solar cells.

As a starting point for the optimization of the micromorph solar cell, the cell from previous section with $d_{i,top}$ = 200 nm and $d_{i,bot}$ = 2.2 μm was used. These thicknesses do not assure current-matching in the starting cell, but, as shown later, in the optimized cell. In figure 6 $A_{i,top}$ and $A_{i,bot}$ of the optically optimized micromorph solar cell (considering the improvements i-vii) are plotted for two cases: using an optimized single interlayer and the wavelength-selective interlayer. For the last one we apply a reflectance of 80 % for shorter wavelengths (λ < 670 nm), whereas for the longer wavelengths 10 % reflectance was used. Significant enhancement of $A_{i,top}$ and $A_{i,bot}$ are observed for both interlayers compared to the ones of the non-optimized cell (figure 2). Potential short-circuit currents, $J_{SC(opt),top}$ and $J_{SC(opt),bot}$, calculated from the optical absorbances are included in figure 6. Very high J_{SC} values are obtained (>16 mA/cm^2), predicting a high potential for significant improvements in the solar-cell efficiency. Simulations indicate that an optimal single interlayer, which assures the highest $J_{SC\,(opt)\,top}$ and does not decrease $J_{SC(opt),bot}$ for this solar cell structure, should have a refractive index of 2.6. Materials with such refractive index and low absorption at the same time are of interest. However, simulations with the wavelength-selective interlayer indicate, that there is still a possibility for a further gain in $J_{SC(opt),top}$ without losing in $J_{SC(opt),bot}$. Moreover, by using such a wavelength-selective interlayer, even $J_{SC(opt),bot}$ can be slightly increased, compared to the results obtained with the optimized single interlayer. The important improvements in $A_{i,top}$ and $A_{i,bot}$ that are related to the wavelength-selective interlayer are indicated by the arrows in figure 6.

The results of optical simulation of the optically improved triple-junction solar cell, including the improvements (i-vi,viii) are shown in figure 5 (full curves). The starting triple-junction cell from previous section was optimized. In order to obtain current matching in the optically improved cell the thicknesses in the optimized solar cell were adjusted to $d_{i,top} = 110$ nm, $d_{i,mid} = 280$ nm and $d_{i,bot} = 2.1$ μm. One can see that considering the improvements (i-vi) the thickness of the absorbers in the top and bottom cell can be even reduced (especially in the top cell), assuring high $J_{SC} = \sim 11$ mA/cm^2 at the same time. However, to achieve sufficiently high absorbance in the i-a-SiGe:H absorber (middle cell) the increase of its thickness from $d_{i,mid} = 220$ nm (starting cell) to 280 nm was required. The introduction of an intermediate reflector between the middle and the bottom cell (such as thin ZnO layer) may help to reduce $d_{i,mid}$, however the role of an interlayer in the triple-junction solar cell is not investigated here.

In figure 7 and 8 we present the results of a comparative study of the role of each optical improvement (i-viii) regarding an increase of the J_{SC} of each component cell (top, middle, bottom) in the optimized micromorph and triple-junction solar cell, respectively. In the figures the height of each bar can be interpreted as a fraction of the total increase in J_{SC} (in %) of the optimized cell that would be missing when the particular improvement was excluded. From the all applied improvements we exclude only one and determine the effect. The total increase in J_{SC} is the difference between J_{SC} of the optimized and starting cell.

In case of the micromorph cell, the interlayer plays the most important role for the top cell, whereas the broad ADF (Lambertian distribution) is the most important for the bottom cell. However, a synergy between the improvements should be taken into account, e.g. without having a high H the effect of *ADF* is minimized. In case of the triple-junction solar cell, the broad *ADF* of scattered light plays the most important role for the top and the bottom cell. The main contribution to J_{SC} of the middle cell is the adjustment of the thickness of the middle a-SiGe:H absorber layer (from 220 nm to 280 nm) in our case.

To raise the efficiency of the investigated solar cells further, in addition to optical also electrical improvements were investigated. The following improvements were introduced in the electrical simulations:

(i)   the band gap of the p-doped a-SiC:H layer of the top cell was increased from 1.95 eV to 2.1 eV in the micromorph cell and from 1.9 eV to 2.2 eV in the triple-junction cell,

(ii) thin (5nm) i-a-Si:H buffer layers with large band gap and low defect density were introduced at the p/i interfaces (1.95 eV top cell in the micromorph cell and 1.9 eV top cell, 1.8 eV middle cell in triple-junction cell),

(iii) the band gap of the a-Si:H absorber in the micromorph cell was increased from 1.75 eV (non-optimized cell) to 1.8 eV.

31

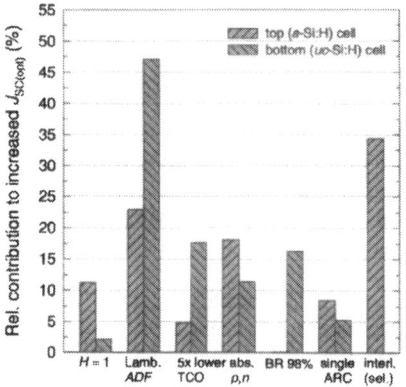

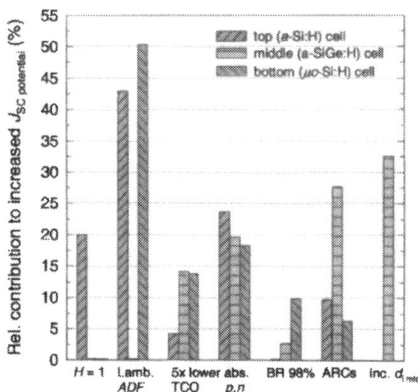

Figure 7. Relative contribution of each optical improvement to the J_{SC} of the component cells in the micromorph solar cell.

Figure 8. Relative contribution of each optical improvement to the J_{SC} of the component cells in the triple-junction solar cell.

In figure 3 the $J(V)$ results of electrical simulations of the improved (optically and electrically) micromorph solar cell are shown (thick red curve). Much larger J_{SC} (15.45 mA/cm^2) of the cell compared to the initial one (10.41 mA/cm^2) is obtained as a result of optical optimization. However, as shown by simulations non-ideal extraction of charge carriers from the bottom absorber decreases the current when increasing forward bias. Further on, V_{OC} is increased from 1.35 V to 1.45 V for the optimized solar cell due to applied electrical improvements of the cell. The FF remains unchanged (0.71). Considering these data for the optimized solar cell, the efficiency of 15.8 % is obtained. However, there still exist possibilities for further electrical improvements (such as improving extraction from the bottom cell or increasing V_{OC} and FF), which may push the efficiency of the cell up to 16 %.

In figure 4 the $J(V)$ results of electrical simulations of the optically improved triple-junction solar cell is presented (dashed blue curve). A significant increase in the J_{SC} (from 8.32 mA/cm^2 to 10.82 mA/cm^2, not included in the figure) is observed as a consequence of the optically enhanced $A_{i,top}$, $A_{i,mid}$ and $A_{i,bot}$. Due to the increased J_{SC}, the efficiency of the optically improved solar cell is improved noticeably (from 11.4 % to 15.0 % , not included in the figure). The V_{OC} and FF of the solar cell are changed slightly. The results of both optically and electrically improved triple-junction solar cell are presented by solid red curve. The electrical improvements result in an increased V_{OC} (from 2.01 V to 2.21 V) and in improved FF (from 0.69 to 0.72). From these values the efficiency of 17.3 % is calculated for the optimized cell.

It is important to note that in the electrical simulations the input parameters referring to standard device quality materials were considered (not state-of-the-art materials with advanced properties). By improving the quality of materials (e.g. lowering defect density, increasing electron and hole mobilities) there is still space for further electrical improvements. For example, the simulations indicated that by decreasing the concentration of dangling bonds in the bottom μc-Si:H absorber (from 4 x 10^{15} cm^{-3} to 1 x 10^{15} cm^{-3}) the FF of the optimized cell can be improved significantly (up to 0.80). In this way the efficiency of 17 % of the optimized a-Si:H/a-SiGe:H/μc-Si:H solar cells can be even overcome.

CONCLUSIONS

Accurate modeling of thin-film solar cells based on a-Si:H and μc-Si:H requires that models describing specific material properties of these films, such as continuous DOS in the band gap and appropriate R-G statistics involving the gap states, and specific device processes, such as light scattering at rough interfaces and tunneling assisted recombination, are implemented in device simulators.

Optical and electrical analysis of a-Si:H/μc-Si:H micromorph and a-Si:H/a-SiGe:H/μc-Si:H triple-junction silicon based solar cells were carried out by means of numerical simulations. Detailed electrical analysis of the micromorph cell showed that the small differences between simulated optical absorbances in intrinsic layers and external quantum efficiencies of the cells originate from a contribution of generated charge carriers in p-a-SiC:H layer (top cell) and a limited carrier extraction from the μc-Si:H absorber (bottom cell). Optical and electrical properties of the solar cell were optimized in order to improve the efficiency of the solar cell. The comparative study of optical improvements revealed an important role of the interlayer for the increased J_{SC} of the top cell, especially the wavelength-selective one, whereas for the bottom cell the broad ADF (Lambertian distribution) is of great importance in the optimized cell provided there is an ideal scattering at rough interfaces. By applying the improvements, the simulated efficiency above 15 % was obtained.

The broad ADF of scattered light is important for the increase of J_{SC} in the top and bottom cells of the triple-junction cell. In order to assure high J_{SC} (~ 11 mA/cm^2) in the middle cell an increase of the a-SiGe:H absorber thickness (from 220 nm to 280 nm) was required in the analyzing structure. By applying the electrical improvements to the cell (higher band gap of the top p-doped layer, high-quality buffer layers at p/i interfaces in the top and middle cells) a potential to achieve and overcome the conversion efficiency of 17 % of the triple-junction cell was demonstrated by the simulations.

ACKNOWLEDGMENT

The authors thank W. Reetz, H. Stiebig and B. Rech from Research Centre Juelich, Germany, for providing the experimental data for the micromorph silicon solar cell. Bart Pieters is acknowledged for developing the ASA simulation program. This work has been partially funded through the Dutch SenterNovem agency.

REFERENCES

1. http://www.silvaco.com/products/device_simulation/atlas.html
2. http://www.crosslight.com/Product_Overview/prod_overv.html#APSYS
3. http://www.synopsys.com/products/tcad/taurus_medici_ds.html
4. M. Burgelman, J. Verschraegen, S. Degrave and P. Nollet, Prog. Photovolt: Res. Appl. **12**, 143–153 (2004).

5. J. K. Arch, F. A. Rubinelli, J.-Y. Hou, and S. J. Fonash, J. Appl. Phys. **69**, 7674 (1991).
6. M. Topic, F. Smole, and J. Furlan, J. Appl. Phys., **79** (1996) 8537.
7. M. Zeman, J.A. Willemen, L.L.A. Vosteen, G. Tao and J.W. Metselaar, Solar Energy Materials and Solar Cells **46**, 81 (1997).
8. M. Burgelman, P. Nollet, S. Degrave, Thin Solid Films **361 – 362**, 527 (2000).
9. A. Froitzheim, R. Stangl, L. Elstner, M. Kriegel, W. Fuhs, Proc. 3rd WCPEC, Osaka, Japan, 2003, 1P-D3-34.
10. T. Sawada, H. Tarui, N. Terada, M. Tanaka, T. Takahama, S. Tsuda and S. Nakano, Proc. 23rd IEEE PVSC, Louisville, KY, 1993, p. 803.
11. A. Fantoni, M. Vieira, J. Cruz, R. Schwarz and R .Martins, J. Phys. D: Appl. Phys. **29**, 3154 (1996).
12. J. Zimmer, H. Stiebig, and H. Wagner, Mat. Res. Soc. Proc. **507**, Warrendale, PA, 1998, p. 377
13. J. Furlan, S. Amon, P.Popovič, F. Smole, Proc. 1st WCPEC-1, Hawaii, USA, (1994) p. 658.
14. Ch. Haase and H. Stiebig, Proc. 21st EU PVSEC, Dresden, Germany, 2006, p. 1712.
15. K. Brecl, D. Fischer, F. Smole, M. Topic, Proc. 21th EU PVSEC, Dresden, Germany, 2006, p. 1662
16. J. Krc, F. Smole, M. Topic, Prog. in Photovolt: Res. Appl. **11**, 15 (2003).
17. R.I.E. Schropp and M. Zeman, *Amorphous and Microcrystalline Solar Cells: Modeling, Materials, and Device Technology*, (Kluwer Academic Publishers, 1998).
18. F. Leblanc, J. Perrin, J. Schmitt, J. Appl. Phys. **75**, 1074 (1994).
19. ASA simulator, User's manual v5.0, Delft University of Technology, 2005.
20. J. Springer, A. Poruba, and M. Vanecek, J. Appl. Phys. **96**, 5329 (2004).
21. J. Mueller, B. Rech, J. Springer, and M. Vanecek, Solar Energy Materials and Solar Cells **77**, 917 (2004).
22. M. Zeman, R.A.C.M.M. van Swaaij, J.W. Metselaar, and R.E.I. Schropp, J. Appl. Phys. **88**, 6436 (2000).
23. J. Krc, M. Zeman, O. Kluth, F. Smole, M. Topic, Thin Solid Films **426**, 296 (2003).
24. D. Fischer *et al.*, Proc. 25th IEEE PVSC, Washington, DC, 1996, p. 1053.
25. K. Yamamoto *et al.*, Proc. 15th PVSEC, Shanghai, China, 2005, p. 529
26. S. Guha *et al.*, Proc. 15th PVSEC, Shanghai, China, 2005, p. 35.
27. A. Shah *et al.*, Prog. in Photovolt: Res. Appl. **12**, 113 (2004).
28. J. Meier, J. Spitznagel, U. Kroll, C. Bucher, S. Fay, T. Moriarty, A. Shah, Thin Solid Films **451-542**, 518 (2004).
29. J. Krc, M. Zeman, A. Campa, F. Smole, M. Topic, Mater. Res. Soc. Proc. **910**, Warrendale, PA, 2006, A25.1.
30. J.A. Anna Selvan, A.E. Delahoy, S. Guo, Y. Li, Proc. 14th PVSEC, Bangkok, Thailand, 2004, p. 179.
31. M. Kondo et al., Proc. 15th PVSEC, Shanghai, China (2005), 43-4.

Mater. Res. Soc. Symp. Proc. Vol. 989 © 2007 Materials Research Society 0989-A03-02

Enhancing Light-trapping and Efficiency of Solar Cells with Photonic Crystals

Rana Biswas[1], and Dayu Zhou[2]

[1]Dept of Physics, Microelectronics Res Ctr, Ames Laboratory, Iowa State University, Dept of Electrical & Computer Engineering, Ames, IA, 50011

[2]Dept. Electrical and Computer Engineering, Microelectronics Res. Center., Iowa State University, Ames, IA, 50011

ABSTRACT

A major route to improving solar cell efficiencies is by improving light trapping in solar cell absorber layers. Traditional light trapping schemes involve a textured metallic back reflector that also introduces losses at optical wavelengths. Here we develop alternative light trapping schemes for a-Si:H thin film solar cells, that do not use metallic components, thereby avoiding losses. We utilize low loss one dimensional photonic crystals as distributed Bragg reflectors (DBR) at the backside of the solar cells. The DBR is constructed with alternating layers of crystalline silicon and SiO_2. Between the DBR and the absorber layer, there is a layer of two dimensional photonic crystal composed of amorphous silicon and SiO_2. The 2D photonic crystal layer will diffract light at oblique angles, so that total internal reflections are formed inside the absorber layer. We have achieved very high optical absorption throughout optical wavelengths (400–700 nm) and enhanced light-trapping at near-infrared (IR) wavelengths (700–800 nm) for amorphous silicon solar cell. The optical modeling is performed with a rigorous three dimensional scattering matrix approach where Maxwell's equations are solved in Fourier space.

INTRODUCTION

Enhancing light-trapping is a major route to improving solar cell efficiency. Typically, enhancing light-trapping in thin film solar cells is achieved by a back reflector that confines light within the absorber layer. The back reflector is usually textured to scatter light at the interface through large reflected angles. This increases the optical path length within the cell i-layer, and it is necessary to scatter as much light as possible in oblique directions.

A typical metallic back reflector consisting of Ag coated with ZnO [1], suffers from intrinsic losses. The granularity of the interface [1] produces small metallic nanoparticles that can exhibit surface plasmon modes. Surface plasmons of free Ag nano-particles are at ultra-violet wavelengths. Ag coated with a dielectric (with refractive index n), has surface plasmon wavelengths lowered by $\sim 1/n$, and can reside at optical wavelengths. Such surface plasmon modes induce intrinsic loss with every light passage in the cell, which was measured by Springer et al [2] to be 3% to 8% at 650 nm for different surface roughness of the silver back reflector. The losses accumulate and become severe at infrared wavelengths where the absorption length of photons in a-Si:H is long and multiple optical passes are required. Even a small loss of 4% with each reflection in a metallic back-plane incurs a severe loss of $1-(0.96)^{50}$ or 87% with 50 passes. These considerations have motivated us to examine the novel use of loss-less non-metallic photonic crystal structures for light-trapping.

BACKGROUND

Photonic crystals (PCs) have been a major scientific revolution in manipulating and guiding light in novel ways [3]. It is likely that photonic crystals can be utilized to harvest solar photons in ways not possible conventionally. Heine and Dorf [4] utilized gratings to enhance absorption in solar cells. Two groups at MIT [5,6] have recently developed a novel scheme for enhancing light-trapping in c-Si solar cells, where the metallic back-reflector was replaced by a DBR with a reflective band at near-IR wavelengths ($\lambda > 0.8$ μm) where the absorption lengths of photons are large. A diffractive grid on the DBR reflected light at oblique angles to increase the path lengths of near IR photons.

In the limit of a *loss-less* random metallic scattering surface, geometrical optics predicts an increase of optical path length by at most $4n^2$ [7] corresponding to ~ 45 for Si (n is the refractive index of the absorber). However, in conventional light-trapping schemes considerably less path length enhancement (~ 10) can be achieved. The advantage of photonic crystals is to introduce diffraction, where the photon momentum (k) can be scattered away from the specular direction with ($k^{\|} = k_i^{\|} + G$), where G is a reciprocal lattice vector. As emphasized in recent work by MIT groups [5, 6], diffractive wave optics can enhance the path length of light far in excess of the classical limit of $4n^2$ from geometrical optics.

THEORETICAL MODEL

We simulate solar cell structures with a rigorous scattering matrix method [8], where Maxwell's equations are solved in Fourier space and the electric/magnetic fields are expanded in Bloch waves. The structure is divided into slices (along z). In each slice the dielectric function $\varepsilon(r)$ is a periodic function of x and y. Hence the dielectric function and its inverse are Fourier expansions with coefficients $\varepsilon(G)$ and $\varepsilon^{-1}(G)$.

In the scattering matrix method [8, 9], a transfer matrix M in each layer is calculated and diagonalized to obtain the eigenmodes within each layer. Both polarizations are included. The continuity of the parallel components of E and H at each interface leads to the scattering matrices s_i of each layer, from which we obtain the scattering matrix S for the entire structure. Using the S-matrix, we simulate the reflection, transmission and absorption [9] for incident light. The advantage of this approach is that any number of layers of differing widths can be easily simulated since a real-space grid is not necessary. Since the solutions of Maxwell's equations are independent for each frequency, the computational algorithm has been parallelized where each frequency is simulated on a separate processor. The individual layers utilize realistic frequency dependent dielectric functions that include absorption and dispersion.

We focus on optical engineering of single junction cells, which ideally have a-Si:H absorbers with band gaps ~1.6 eV. We use the frequency dependent dielectric functions (ε_1, ε_2) (Fig. 1) determined from spectroscopic ellipsometry for a-Si:H and analytically continued to the infrared by Ferlauto et al [10]. The absorption length $\zeta(\lambda)$ of photons ($\zeta = 1/\alpha = 1/4\pi n_2$) is < 1 μm below wavelengths of 0.65 μm (Fig. 2) and these photons can be absorbed effectively in a thin (~ 0.5 μm) a-Si:H layer. However, when $\lambda > 0.65$ μm, the absorption length exceeds 1 μm and in the near-IR below the band gap wavelength (0.7–0.775 μm), ζ exceeds 5 μm. It is very difficult to harvest these near-IR photons. It is necessary to have lossless multiple reflections within the absorber, for effective light-trapping.

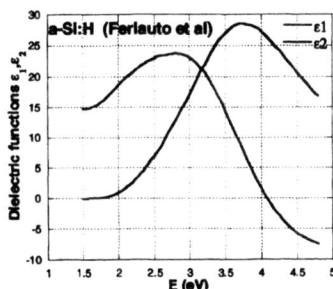

Fig. 1. Energy dependent dielectric functions for a-Si:H with Eg = 1.6 eV, determined from spectroscopic ellipsometry by Ferlauto et al (Ref [10]).

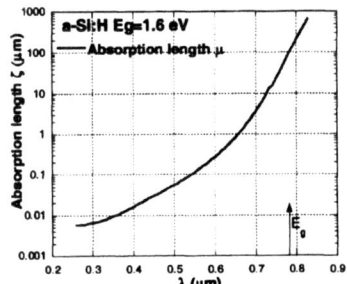

Fig. 2. Absorption length of photons as a function of wavelength for a-Si:H with E_g = 1.6 eV. The wavelength corresponding to the band edge E_g is indicated by the arrow.

LIGHT-TRAPPING SCHEME

The solar cell that we simulate (Fig. 3) consists of the following layers:

i) The top indium tin oxide (ITO) layer (n = 1.95), is an antireflective coating and top contact with a thickness (d_0) that is determined using scattering matrix simulation.

ii) The a-Si:H absorber layer has a thickness of 0.5 μm, typical for single-junction thin film solar cells. The bandgap is 1.6 eV (λ_g = 0.775 μm).

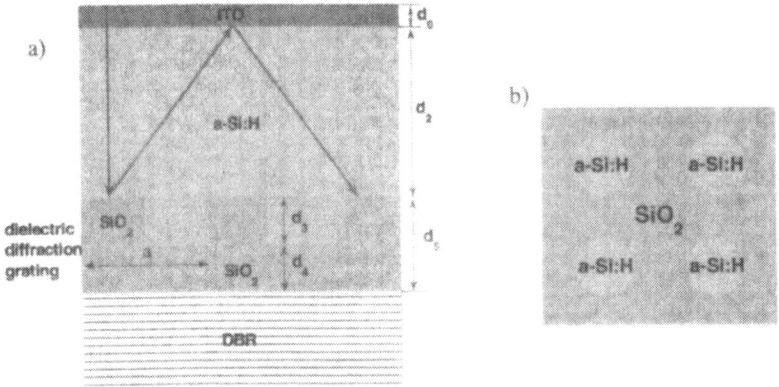

Fig. 3. (a) Schematic solar cell configuration with antireflective coating, two-dimensional photonic crystal and distributed Bragg reflector (DBR) (b) Top view of 2D photonic crystal grating layer with amorphous silicon cylinders in SiO₂ background

iii) The two dimensional (2D) photonic crystal layer is a square lattice (lattice constant a) of dielectric cylinders (radius R) within another dielectric background material. The dielectric contrast of the two materials needs to be high enough for sufficient diffraction. We choose a-Si:H (n ≈ 4) and SiO₂ (n = 1.46) as grating materials. There are two alternative designs, with a-Si:H cylinders in an SiO₂ background (Fig. 3b) or SiO₂ cylinders in an a-Si:H matrix. The PC layer thickness (grating depth d_3) and lattice constant a are varied to maximize the absorption.

iv) The DBR of 6 alternating c-Si and SiO_2 layers represents an economical method for making an omni-directional reflector, that is much simpler than fabrication of 3D PCs. c-Si is chosen because it has a high index contrast (n ~ 3.8) with SiO_2 (n = 1.46) and low absorption. Although a-Si:H has a high refractive index, it also has considerable absorption, which is not desirable for the DBR. With c-Si thickness $d_1 = 50$ nm and SiO_2 thickness $d_5 = 120$ nm, an omni-directional bandgap from 565 to 788 nm is obtained [11]. The incident light in the absorber layer in confined in a small light cone ($\alpha < 16°$). The closer the incident light is to the normal, the wider the bandgap (Fig. 4).

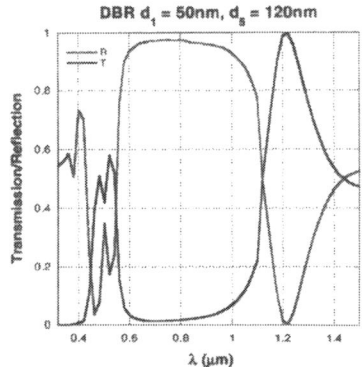

Fig. 4. Reflection and transmission through 6-layer c-Si/SiO_2 DBR for normal incidence

SIMULATION AND RESULTS:

We understand the functions of different components of the solar cell separately. As shown above, the DBR provides a wide reflective stop band. To understand the diffraction of the 2D PC within the absorber layer, we simulate just the grating consisting of SiO_2 cylinders in an a-Si background in air on a thick (~ 300 μm) c-Si substrate. We obtain reflectance for different R/a with fixed grating depth ($d_3 = 40$ nm) and lattice constant (a = 0.7 μm). By comparing the difference between specular and non-specular reflection, we can examine the effectiveness of the diffraction grating (Fig 5). Diffraction occurs when $\lambda < a$. The maximum diffraction occurs between R/a = 0.35 and 0.4, when ~ 30% of the light is diffracted. The value of R/a=0.38 also maximizes the strength of the Fourier component of grating.

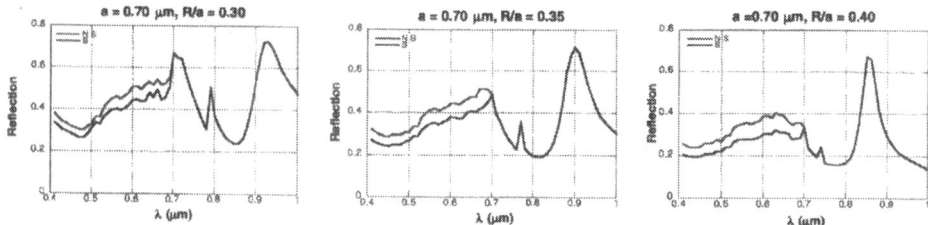

Fig. 5. Comparison of specular (S) and non-specular (NS) reflection from the 2D photonic crystal for different R/a with a = 0.70 μm, $d_3 = 0.04$ μm and $d_4 = 300$

Scattering matrix simulations are performed for the solar cell configuration of Fig. 3a and the absorption is maximized by varying one parameter at a time. An optimal structure consists of a = 600 nm, R/a = 0.38, $d_1 = 50$ nm, $d_2 = 500$ nm, $d_3 = 50$ nm and $d_5 = 120$ nm. The thickness of ITO d_0 is expected to range from 500 nm/$4n_{ITO}$ = 64 nm to 775 nm/$4n_{ITO}$ = 100 nm. By simulations we find that $d_0 = 65$ nm provides the least reflection from the top surface.

For fixed $d_0 = 65$ nm, we optimize the lattice constant. We find that the lattice constants in the range from 600 to 700 nm give the highest absorption in the absorber layer, which correspond to the third order of diffraction. With a = 680 nm, we run simulations with different d_3 for both grating configurations. We find that d_3 = 80 nm for SiO_2 layer with a-Si cylinders and d_3 = 50 nm for a-Si layer with SiO_2 cylinders give the best diffraction and highest light-trapping within the absorber layers.

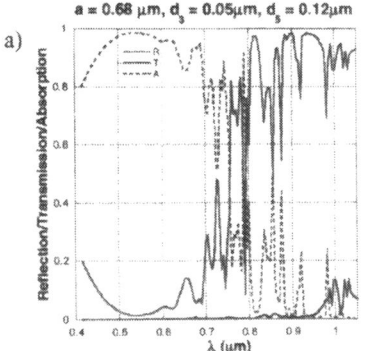
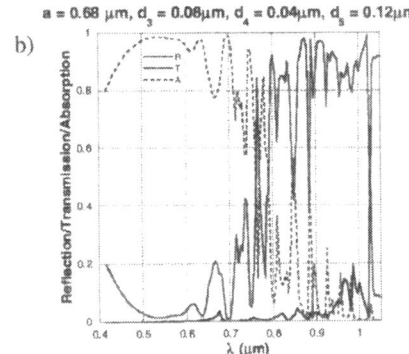

Fig. 6. Non-specular (total) reflection/transmission/absorption in the absorber layers for the two competing designs with d_0 = 65 nm, d_1 = 50 nm, d_2 = 500 nm and (a) a-Si layer with SiO_2 cylinders (b) SiO_2 layer with a-Si cylinders

The optimized solar cell with SiO_2 cylinders inside the a-Si layer (Fig. 6a) and the complementary structure of a-Si cylinders in the SiO_2 matrix (Fig. 6b) both show very high absorption (A > 0.8) through most of the optical wavelengths (0.4–0.7 μm) (Fig. 6), accompanied by small reflection. There is appreciable photon absorption between 0.7–0.775 μm, a region where the absorption length of photons exceeds 5 μ. The enhanced absorption improves the near-IR solar response. Wavelengths larger than 0.775 μm are largely reflected from the cell in both configurations – these are above the bandgap and do not contribute to photocurrent.

It is instructive to compare the light-trapping in our optimized solar cell with that of a bare a-Si:H film, an a-Si:H film with antireflective coating and an a-Si:H film with DBR and antireflective coating (Fig. 7). Below λ = 0.65 μm, the absorption length is

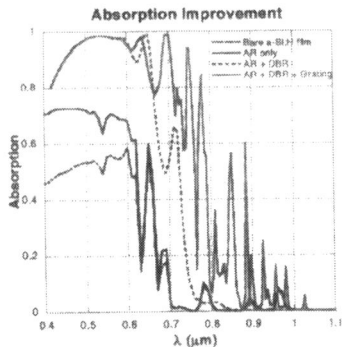

Fig. 7. Comparison of absorption of a bare a-Si:H film, a-Si:H film with AR coating only, a-Si:H film with AR coating and DBR and the optimized solar cell with photonic crystal grating.

less than 1 μm. With an antireflective coating and DBR, most of incident light can be absorbed in the absorber layer within two optical paths. Above λ = 0.65 μm, the PC enhances the light trapping efficiency up to a factor of 2 without introducing losses.

The difference in grating depths of the two competing designs can be understood by different effective dielectric constants of the grating layers. With $R/a = 0.38$, the filling ratio $f = \pi R^2 / a^2 = 0.45$. Maxwell-Garnett approximation gives $n_{eff} \sim 3$ for a-Si grating layer and $n_{eff} \sim 2.3$ for SiO$_2$ grating layer. Grating depth can be estimated by quarter wavelength inside the grating layer to reduce the specular reflection.

CONCLUSIONS

We develop a novel light trapping scheme in a-Si:H solar cells with dielectric photonic crystals. By combining a 2D photonic crystal diffraction grating and DBR, we can efficiently harvest optical photons, without losses associated with metallic reflectors. We enhance the absorption at longer wavelength ($\lambda > 0.65$ μm – where absorption lengths are long) up to the bandgap by a factor of ~ 2. Also, the absorption peak of the optimized a-Si:H solar cell matches the peak of solar radiation spectrum. For future work, we wish to place a 2D photonic crystal diffraction grating at the interface between ITO and the absorber.

ACKNOWLEDGEMENTS

We acknowledge support from the Catron Solar Foundation. The Ames Laboratory is operated for the Department of Energy by Iowa State University under contract No. W-7405-Eng-82. It is a pleasure to thank V. Dalal for many stimulating discussions and C.G. Ding for the initial scattering matrix computation code. We also acknowledge support from the NSF under grant ECS-06013177. We thank R. Collins and N. Podraza for kindly supplying the optical data for amorphous silicon films and the most helpful discussions.

REFERENCES

[1] B. Yan, J. M. Owens, C. Jiang and S. Guha, MRS Symp. Proc. 862, A23.3 (2005).
[2] J. Springer, A. Poruba, L. Mullerova, M. Vanecek, O. Kluth and B. Rech, J. Appl. Phys. 95, 1427 (2004).
[3] J. D. Joannopoulos, R. D. Meade and J. N. Winn, Photonic Crystals, Princeton, NJ: Princeton University Press, 1995.
[4] C. Heine, and R. H. Morf, Appl. Opt. 34, 2476 (1995).
[5] L. Zeng, Y. Yi, C. Hong, J. Liu, N. Feng, X. Duan, L.C. Kimmerling and B.A. Alamariu, Appl. Phys. Lett. 89, 111111 (2006); MRS Symp. Proc. 862, A12.3 (2005).
[6] P. Bermel, C. Luo and J. D. Joannopoulos, to be published in Opt. Express (2007).
[7] E. Yablonovitch and G. Cody, IEEE Transactions Electron Devices ED-29, 300 (1982).
[8] Z. Y. Li and L. L. Lin, Phys. Rev. E 67, 046607 (2003).
[9] R. Biswas, C.G. Ding, I. Puscasu, M. Pralle, M. McNeal, J. Daly, A. Greenwald and E. Johnson, Phys. Rev. B. 74, 045107 (2006).
[10] A.S. Ferlauto, G. M. Ferreira, J. M. Pearce, C. R. Wronski, R. W. Collins, X. Deng and G. Ganguly, J. Appl. Phys. 92, 2424 (2002).
[11] Y. Fink, J. N. Winn, S. Fan, C. Chen, J. Michel, J. D. Joannopoulos and E. L. Thomas, Science. 282, 1679 (1998).

Mater. Res. Soc. Symp. Proc. Vol. 989 © 2007 Materials Research Society 0989-A03-04

High Open-Circuit Voltage in Silicon Heterojunction Solar Cells

Qi Wang[1], Matt R Page[1], Eugene Iwancizko[1], Yueqin Xu[1], Lorenzo Roybal[1], Russell Bauer[1], Dean Levi[1], Yanfa Yan[1], Tihu Wang[2], and Howard M. Branz[1]
[1]EDMD, National Renewable Energy Laboratory, Golden, CO, 80410
[2]Suntech Power, Wuxi, China, People's Republic of

ABSTRACT

High open-circuit voltage (V_{oc}) silicon heterojunction (SHJ) solar cells are fabricated in double-heterojunction a-Si:H/c-Si/a-Si:H structures using low temperature (<225°C) hydrogenated amorphous silicon (a-Si:H) contacts deposited by hot-wire chemical vapor deposition (HWCVD). On p-type c-Si float-zone wafers, we used an amorphous n/i contact to the top surface and an i/p contact to the back surface to obtain a V_{oc} of 667 mV in a 1 cm^2 cell with an efficiency of 18.2%. This is the best reported p-type SHJ voltage. In our labs, it improves over the 652 mV cell obtained with a front amorphous n/i heterojunction emitter and a high-temperature alloyed Al back-surface-field contact. On n-type c-Si float-zone wafers, we used an a-Si:H (p/i) front emitter and an a-Si:H (i/n) back contact to achieve a V_{oc} of 691 mV on 1 cm^2 cell. Though not as high as the 730 mV reported by Sanyo on n-wafers, this is the highest reported V_{oc} for SHJ c-Si cells processed by the HWCVD technique. We found that effective c-Si surface cleaning and a double-heterojunction are keys to obtaining high V_{oc}. Transmission electron microscopy reveals that high V_{oc} cells require an abrupt interface from c-Si to a-Si:H. If the transition from the base wafer to the a-Si:H incorporates either microcrystalline or epitaxial Si at c-Si interface, a low V_{oc} will result. Lifetime measurement shows that the back-surface-recombination velocity (BSRV) can be reduced to ~15 cm/s through a-Si:H passivation. Amorphous silicon heterojunction layers on crystalline wafers thus combine low-surface recombination velocity with excellent carrier extraction.

INTRODUCTION

Open-circuit voltage (V_{oc}) of crystal silicon (c-Si) solar cells on high-quality wafers (minority carrier diffusion length much greater than wafer thickness) is limited by the surface recombination rates. Any dark-current path through inadequately surface passivation reduces both V_{oc} and the collection of photo-generated charge carriers. Therefore, surface passivation is the key to achieve high V_{oc} and, ultimately, high-performance solar cells. The unique combination of surface passivation and current conduction of a-Si:H on c-Si allows superior a-Si:H/c-Si emitter construction as well excellent full-area a-Si:H/c-Si back contact creation [1], all at temperatures below 250°C. The a-Si:H passivation is comparable to the dielectric surface passivation means such as SiO$_2$ and SiN$_x$ but a-Si heterojunctions provide good current conduction without the fired-glass-frit or laser-fired contacts needed on these dielectric layers. Early, we reported our single heterojunction a-Si/c-Si (p-type) solar cells with V_{oc} of 645 mV, an increase of 15 mV over a diffused junction emitter. Optimization of the i/n a-Si:H front emitter enable us to achieve 17.1%-efficient single-heterojunction solar cells with a screen-printed Al back-surface field (BSF) [2]. However, an Al-BSF limits further improvements in our device performance, because of a high back-surface-recombination velocity (BSRV), on the order of 10^3 cm/sec. This causes a high back-surface dark saturation current component that limits the open-circuit voltage. Further, an Al-BSF has to be processed at temperatures above 800ºC and this may cause a problem to keep a clean front surface before the heterojunction deposition. Replacing the conventional Al-alloyed BSF on p-type wafers or P-diffused BSF on n-type wafers by the effective a-Si:H thin passivating layers will improve open-circuit voltage significantly [3] and also avoids all high-temperature processing steps.

This paper describes our progress in achieving this objective. We illustrate our systematic development of deposited a-Si:H as front emitters and back contacts by hot wire chemical vapor deposition (HWCVD) for both p- and n-type silicon wafers. In commonly used plasma-enhanced chemical vapor deposition (PECVD), great care must be taken to avoid plasma damage to the c-Si wafer surfaces to make high-efficiency SHJ cells; we opt to use HWCVD to grow a-Si:H layers to eliminate the possibility of such ion damage. For both doping types, we obtain higher open-circuit voltages than with standard Al-alloyed or P-diffused back-surface-field contacts. Our highest V_{oc} is 691 mV on n-type c-Si.

EXPERIMENTAL

Open-circuit voltage and the interface-recombination velocity are the key indicators of the a-Si:H/c-Si heterointerface quality. Our work attempts to contribute to both the technological development of high-efficiency silicon heterojunction solar cells and the fundamental understanding of the a-Si:H/c-Si interface; therefore, we study both finished isolated solar cells of 1 cm^2 in a structure of ITO/a-Si:H/c-Si/a-Si:H/Metal for complete device characterizations and as-deposited a-Si:H/c-Si/a-Si:H structures for evaluation lifetime using a Sinton lifetime tester and estimate BSRV from the measured lifetime. High-resolution transmission electron microscopy (HRTEM) is used to characterize the structure of the a-Si/c-Si interface. Both high-quality p- and n-type float-zone silicon (FZ-Si) wafers, bulk lifetime greater than 1 ms, are used in this study. Exact details of sample preparation and a-Si:H deposition are given elsewhere [2-6].

RESULTS AND DISCUSSIONS

We focus on the improvement of V_{oc} using a-Si:H emitters to achieve high performance c-Si solar cells. With a diffused junction on p-type wafer solar cell having a V_{oc} of 630 mV as the reference, the front-only single heterojunction SHJ cell on p-type wafer having a similar Al-BSF as diffused one shows an improved V_{oc} of 652 mV. When

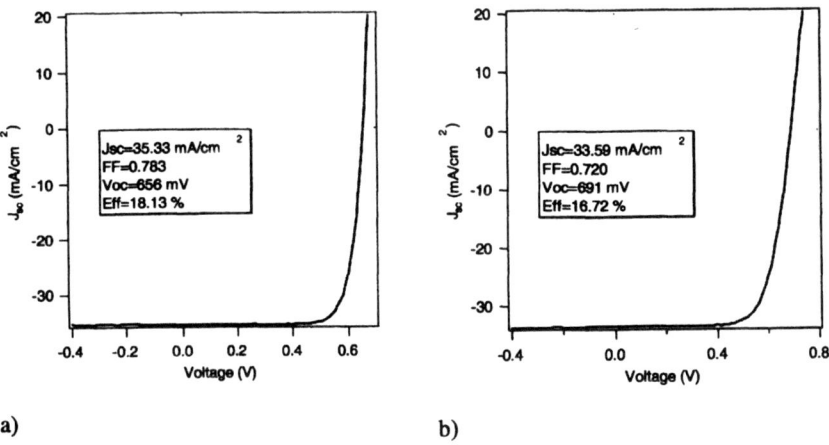

Figure 1. Solar cell performance under AM 1.5 for double-heterojunction SHJ isolated 1 cm^2 cells based on both textured p-type (a) and planar n-type (b) wafers.

we optimize the thin HWCVD a-Si:H emitter layers, we can achieve a very low surface recombination velocity about 15 cm/sec [5]. We found that it is critical to obtain and maintain a clean interface before the thin Si emitter and avoid formation of epitaxial c-Si or nanocrystalline Si at the interface. In order to increase the cell's short-circuit current, J_{sc}, we first used textured wafers with the same Al-BSF back contacts as the planar cells. An essentially identical V_{oc} of 651 mV is obtained despite the much increased surface area due to roughness, indicating the near perfect front surface passivation by the a-Si:H(i/n) emitter.

However, an Al-BSF has a back-surface-recombination velocity (BSRV) on the order of 10^3 cm/s, which causes a high back-surface saturation current component that limits further increases in V_{oc}. In addition, an Al-BSF has to be processed at temperatures above 800°C which could increase the impurity concentration at the front surface. To further improve the V_{oc}, we replace the high-temperature Al-BSF with low-temperature HWCVD-deposited i/p a-Si:H layers as the back contact. Lifetime measurement shows the BSRV is reduced to ~15 cm/sec. We obtain a higher V_{oc} of 676 mV on some devices, including even textured p-type silicon wafers in the double-heterojunction structure (front n/i heterojunction emitter and back i/p heterojunction contact).

On n-type wafers, we start with a planar a-Si:H p/i front emitter and a P-diffused BSF as the back contact. This gives an unimpressive V_{oc} of 627 mV. However, once we replace the back P-diffused BSF with a-Si(i/n) layers as the back contact, V_{oc} jumps to 691 mV. This is the highest open-circuit voltage achieved using the HWCVD technique so far, implying a great potential in reaching very high-efficiency SHJ solar cells. Again, when we texture the wafer to increase J_{sc}, a high V_{oc} of 686 mV is attained.

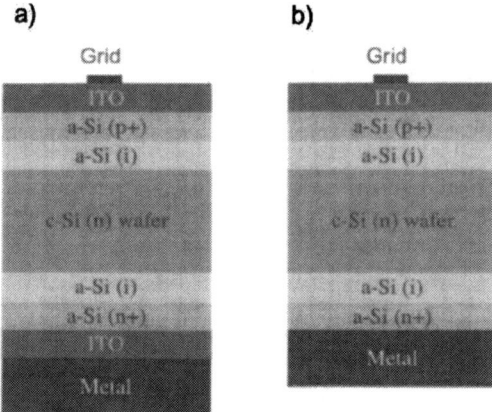

Figure 2. Cell structure for double-heterojunction SHJ cells.

Figure 1 shows an J-V curve of a high V_{oc} SHJ c-Si solar cells performance as measured by our unofficial but calibrated simulator XT-10. Figure 1a is our double heterojunction SHJ device on textured p-type wafer with high efficiency over 18%. Fill factor is another important performance parameter for back contact quality. When using a dielectric back-surface passivation by insulating materials, one must employ local contact windows [7] to obtain low-resistance and effective majority carrier collection. On the other hand, with full-area back contacts of a-Si:H, it is possible to obtain excellent hole conduction across the back c-Si(p)/a-Si:H(i/p) interface; we obtain a good fill factor of 78%. We found that micro-crystalline p-layer [8] is not necessary. High-resolution transmission electron micrographs show that our back c-Si(p)/(i/p) interface is all amorphous [9] – clearly we can make our low-resistance contact with a-Si:H (i/p).

Figure 1b is the J-V curve of our double heterojunction SHJ device on planar n-type wafer. Currently we achieved a V_{oc} of about 690 mV in the double side junction structure. This V_{oc} is higher than on p-type wafers and shows the great potential to achieve even higher efficiency using textured n-type wafer. However, this cell had a low FF of 0.72. We believe it is due to a high series resistance at the back contact. After optimization, we have improved the FF to around 0.76. Certainly, there is still room to further enhance the electron transport across the c-Si (n)/a-Si:H(i/n) back contact.

We also found the metal contact to the a-Si:H back is important to obtaining high V_{oc}. Figure 2 shows two types of back metal contacts to the n-layer: a) one is using ITO and b) the other is using a metal such as Ti, with no ITO. Figure 3 shows the importance to V_{oc} of which layer contacts the n-layer. We use an ITO contact to the n-layer (Fig. 2a) to establish the reference a V_{oc} of 680 mV. Once we use Ti direct contact to the same n-layer (Fig. 2b), the V_{oc} decreases to 56 mV. However, when we deposit a much thicker n-layer (6 times increased deposition time) and then the Ti contact, V_{oc} is restored to close to 680 mV. This effect can be explained by the hypothesis of metal diffusion

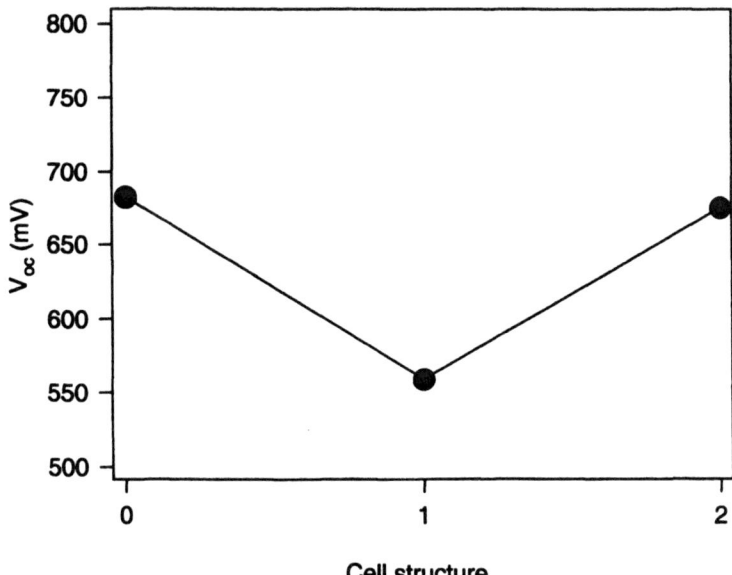

Cell structure

Figure 3. V_{oc} as a function of various back contacts. 0: n/ITO/Ti, 1: n/Ti, 2: 6x n/Ti. Lines between points are guided to eye.

through the thin BSF to the back c-Si interface which could cause a high BSRV and a decrease of V_{oc}. A thicker n-layer can stop the diffusion of metal and maintain the high V_{oc}. It is interesting to observe that ITO to n-layer has no such deleterious effect.

Finally, we discuss the effect of annealing of low temperature deposited a-Si:H on lifetime and BSRV. In recent report by De Wolf [10], the lifetime increased significantly by moderate annealing (~250°C) for about 30 min of a-Si:H grown at 105°C. Higher lifetime will normally produce a higher V_{oc}. In our SHJ cell process, we grow the i-layer at 100°C and the doped layers at higher temperature (below 225°C) to avoid epitaxy and increase the doping efficiency. It is possible that we unintentionally anneal the i-layer during the heating and deposition of the doped layers. In our early report [5], we find our best growth condition to have a high lifetime and low BSRV is at 100°C for i-layer without annealing and a high V_{oc} with an abrupt interface at c-Si surface to a-Si. Higher temperature i-layer did not yield a high lifetime in contradict to De Wolf's work. We will continue work on the understanding of the high V_{oc}.

CONCLUSIONS

We have successfully replaced the conventional high-temperature Al-BSF or P-diffused BSF with a-Si:H back contacts in double heterojunction SHJ solar cells. Excellent a-Si/c-Si heterointerfaces at both front emitter and back contact with minimal recombination losses can be obtained by using HWCVD that minimizes wafer surface damage, permits an abrupt a-Si:H deposition, and yields a flat hetero-interface to the c-Si substrate. When HWCVD a-Si:H is used as the back contact, a highly effective back-surface field is demonstrated by double-heterojunction SHJ devices with V_{oc} values of 691 mV and 676 mV on planar n-type FZ-Si and on textured p-type FZ-Si, respectively. Fill factors as high as 78% are realized on p-type wafer. The highly effective a-Si:H back contacts enable us to achieve high performance SHJ solar cells.

ACKNOWLEDGEMENTS

We would like to express sincere thanks to Anna Duda and Scott Ward at NREL for their help in device processing. This research was supported by the U.S. Department of Energy under Contract No. DE-AC39-98-GO10337.

REFERENCES

1. M. Taguchi, K. Kawamoto, S. Tsuge, T. Baba, H. Sakata, M. Morizane, K. Uchihashi, N. Nakamura, S. Kiyama, and O. Oota, *Prog. PV Res. Appl,* 8: p. 503, 2000.

2. T.H. Wang, E. Iwaniczko, M.R. Page, D.H. Levi, Y. Yan, V. Yelundur, H.M. Branz, A. Rohatgi, and Q. Wang. *IEEE.Proceedings of the 31st IEEE Photovoltaic Specialists Conference* p. 955, 2005.

3. T.H. Wang, M.R. Page, E. Iwaniczko, Y.Q. Xu, Y.F. Yan, L. Roybal, D. Levi, R. Bauer, H.M. Branz, and Q. Wang. WIP-Renewable Energies. *21st European Photovoltaic Solar Energy Conference* p. 781, 2006.

4. M.R. Page, E. Iwaniczko, Q. Wang, D.H. Levi, Y. Yan, H.M. Branz, V. Yelundur, A. Rohatgi, and T.H. Wang. National Renewable Energy Laboratory.*14th Workshop on Crystalline Silicon Solar Cells & Modules:Materials and Processes* p. 246, 2004.

5. M.R. Page, E. Iwaniczko, Y. Xu, Q. Wang, Y. Yan, L. Roybal, H.M. Branz, and T. H. Wang. the *4th World Conference on Photovoltaic Energy Conversion* (WCPEC-4) p. 6, 2006.

6. T.H. Wang, M.R. Page, E. Iwaniczko, Y.Q. Xu, Y.F. Yan, L. Roybal, D. Levi, R. Bauer, H.M. Branz, and Q. Wang. MRS. **910**: *Mat. Res. Soc. Proc.* p. 731, 2006.

7. M. Schaper, J. Schmidt, H. Plagwitz, and R. Brendel, *Prog. Photovolt: Res. Appl,* **13**: p. 381, 2005.

8. P.J. Rostan, U. Rau, V.X. Nguyen, T. Kirchartz, M.B. Schubert, and J.H. Werner. *Technical Digest of the 15th International Photovoltaic Science & Engineering Conference* p. 214, 2005.

9. Y.Yan, M. Page, T. H. Wang, M.M. Al-Jassim, H. M. Branz, and Q. Wang, *Appl. Phys. Lett.,* **88**: p.3, 2006.

10. Stefaan De Wolf * and M. Kondo, *Appl. Phys. Lett.,* **90**: p.042111, 2007.

Alloys

Mater. Res. Soc. Symp. Proc. Vol. 989 © 2007 Materials Research Society 0989-A04-02

Effects of Nitrogen Addition on the Properties of a-SiCN:H Films Using Hexamethyldisilazane

Amornrat Limmanee[1], Michio Otsubo[1], Tsutomu Sugiura[1], Takehiko Sato[2], Shinsuke Miyajima[1], Akira Yamada[3], and Makoto Konagai[1]

[1]Department of Physical Electronics, Tokyo Institute of Technology, 2-12-1-S9-9 O-okayama, Meguro-ku, Tokyo, 152-8552, Japan

[2]Material&Processing Technology, Mitsubishi Electric Corporation, 1-1-57, Miyashimo,Sagamihara, Kanagawa, 229-1195, Japan

[3]Quantum Nanoelectronics Research Center, Tokyo Institute of Technology, 2-12-1-S9-9 O-okayama, Meguro-ku, Tokyo, 152-8552, Japan

ABSTRACT

We deposited a –SiCN:H films by HWCVD using a gas mixture of hexamethyldisilazane (HMDS), H_2 and N_2, and fabricated cast polycrystalline silicon (poly c-Si) solar cells with the a-SiCN:H passivation and anti-reflection layer. N_2 addition led to the reduction of the refractive index (n) of the a-SiCN:H films due to the increase in nitrogen concentration of the films. This improved performance of the antireflection layer. The advantage of adding N_2 to the process was demonstrated by the improvement in short circuit current (J_{SC}) and efficiency (η) of cast poly c-Si silicon solar cells.

INTRODUCTION

Hydrogenated amorphous silicon carbon nitride (a-SiCN:H) films deposited by hot-wire chemical vapor deposition (HWCVD) using HMDS has been reported as a new passivation layer of silicon solar cells [1-2]. Using HMDS which is a non-explosive and low cost material enables the process to become safer and more economical. The surface recombination velocity of 1~10 Ω cm n-type c-Si with the a-SiCN:H film is lower than 50 cm/s. The η of cast poly c-Si solar cell with these a-SiCN:H films as a passivation layer reached 13.75% (2cm x 2cm) [3]. The effective reflectivity of cell surface was found to be high, indicating that optimization of the anti-reflection layer is still required. Reducing the n of the a-SiCN:H films is thought to be an effective way to lower surface reflectivity that will lead to improvements of J_{SC} and η of the cells. Using only HMDS and H_2 is found to be hard to adjust chemical composition and optical properties of a-SiCN:H films. Therefore, in this work we introduce N_2 to our process for the purpose of providing more variation of the optical properties of the films. The effects of N_2 addition on the structural, electrical, optical and interface properties of a-SiCN:H films were examined. We especially focused on the changes in the optical properties and passivation quality of the films. Optimum deposition condition was deduced from experiment results and was practically used in the fabrication of cast poly c-Si solar cells. The improvements in J_{SC} and η were achieved due to the reduction of optical losses at the surface. The benefit of adding N_2 to the process was verified by the flexibility of the n

EXPERIMENTAL DETAILS

The a-SiCN:H films were deposited on n-type CZ (001) silicon wafer (380 μm thickness) with various deposition conditions as shown in Table I. The N_2 flow rate was varied from 0 to 100 sccm. The following measurements were carried out. XPS (X-ray photoelectron spectroscopy) and FT-IR (Fourier transform infrared spectrophotometer) measurements for examining chemical composition and hydrogen bonding, optical properties characterization by spectroscopic ellipsometry, effective lifetime (τ_{eff}) measurement by μ-PCD (Microwave Photo Conductivity Decay), and C-V measurement at 1 MHz for determining the fixed charge (D_f) and interface trap densities (D_{it}).

Table I. Process parameters

Material	HMDS, H_2, N_2
HMDS flow rate (sccm)	1.75
H_2 flow rate (sccm)	160
N_2 flow rate (sccm)	0~100
Total pressure (Torr)	1.2
Filament current (A)	13.5
Filament temperature ($^{\circ}$C)	1500
Substrate temperature ($^{\circ}$C)	300

RESULTS AND DISCUSSIONS

Structural properties

Nitrogen content in the a-SiCN:H films increased from 10% to 14% as N_2 flow rate increased from 0 to 100 sccm, while carbon content decreased slightly. On the other hand, there was no obvious variation in silicon content of the films. The Si-C, Si-H and C-H bonds were detected at 800, 2100 and 2850 cm^{-1}, respectively. The N-H bond density was very small so we observed the change in hydrogen bonding from only Si-H and C-H bonds. The intensities of the peak due to Si-H and C-H bonds as a function of N$_2$ flow rate are shown in Fig.1. These results indicated that the addition of N$_2$ prevented H atoms from forming Si-H and C-H bonds. And the Si-H bond is likely to get less influence than the C-H bond.

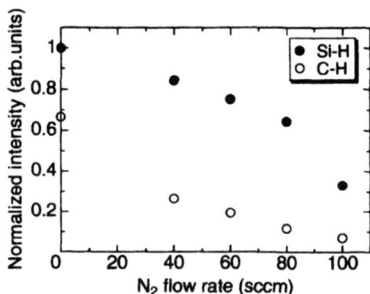

Fig.1 The normalized intensity of the hydrogen bonding as a function of N$_2$ flow rate.

Optical properties

As shown in Fig.2, the value of n decreased gradually with increasing N$_2$ flow rate. When the N$_2$ flow rate became above 60 sccm, n at the wavelength of 650 nm decreased to 2.3 and tended to become lower at higher N$_2$ flow rate. The bandgap of a-SiCN:H films increased slightly from 2.8 to 2.9 with increasing N$_2$ flow rate, where nitrogen content in the films increased from 10% to 14%. In Fig.3, we show the effective surface reflectivity of textured polycrystalline silicon (a) without passivation layer, (b) with a-SiCN:H using only HMDS and H$_2$ and (c) with a-SiCN:H using HMDS, H$_2$ and N$_2$ (60 sccm). The surface reflectivity was obviously reduced when n was lowered by the introduction of N$_2$. We can expect the reduction of optical loss at the surface of the solar cells.

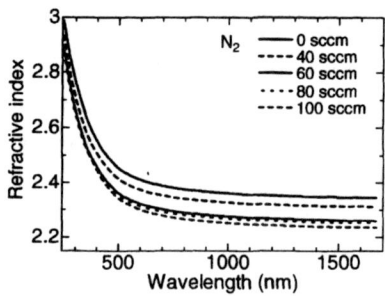

Fig.2 Effect of N$_2$ addition on the refractive index of a-SiCN:H films.

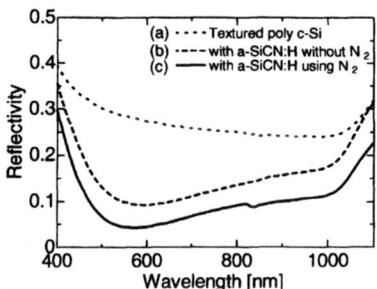

Fig.3 Surface reflectivity of (a) bared poly c-Si, (b) poly c-Si with a-SiCN:H using HMDS and H$_2$ and (c) poly c-Si with a-SiCN:H using HMDS, H$_2$ and N$_2$.

Passivation effect

In Fig.4, we show the change in the τ_{eff} with increasing N$_2$ flow rate. The τ_{eff} began to drop at the N$_2$ flow rate of 60 sccm. When the N$_2$ flow rate became higher than 80 sccm, the τ_{eff} decreased rapidly, indicating the significant degradation of passivation quality. We observed increases in D_{it} and D_f as N$_2$ flow rate became higher. C-V measurements revealed that D_{it} increased with increasing N$_2$ flow rate, indicating that the deterioration of the passivation effect was caused by the increase in the D_{it}. Increasing N$_2$ flow rate could lower n leading to the reduction of surface reflectivity, however, too high N$_2$ flow rate was likely to degrade the passivation quality of a-SiCN:H films. We can see that there is an optimum for nitrogen flow rate regarding both the optical properties and the passivation quality, which is about 60 sccm.

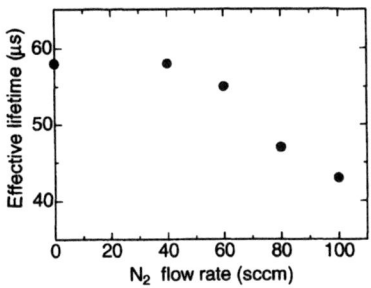

Fig.4 Effect of N_2 addition on the effective lifetime.

Cast polycrystalline silicon solar cells using a-SiCN:H films as a passivation layer

We fabricated cast poly c-Si solar cells with a-SiCN:H films deposited under the optimum condition (N_2 flow rate 60 sccm). Photovoltaic parameters of the cells are shown in Table II. Compared to our previous condition (Using only HMDS and H_2), there was an obvious improvement in the J_{SC}. The J_{SC} increased from 31.75 to 33.42 mA/cm^2 and the η reached 14.2%. Figure 5 reveals an enhancement of quantum efficiencies (QE) after N_2 addition which corresponds with the reduction of optical loss at the cell surface. Since poly c-Si wafers used for comparing cell performance were not sister-wafer, a non-enhancement of QE at the wavelength below 600 nm was probably due to the distribution of textured surface of poly c-Si wafers.

Table II. Poly c-Si solar cells using a-SiCN:H passivation layers (AM1.5, 100 mW/cm^2, 4 cm^2, 280 μm thick, p-type poly c-Si, textured surfaces, screen printed contact).

a-SiCN:H film	V_{OC}(mV)	J_{SC}(mA/cm^2)	FF	η(%)
Using HMDS and H_2	0.599	31.75	71.9	13.7
Using HMDS, H_2 and N_2	0.598	33.42	70.8	14.2

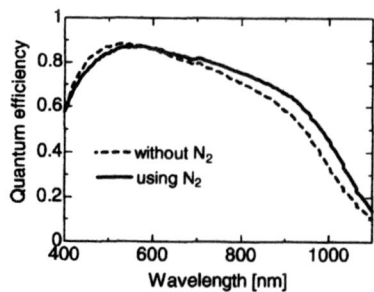

Fig.5 QE of poly c-Si solar cells using a-SiCN:H films deposited with and without N_2.

CONCLUSIONS

We have optimized the optical properties of a-SiCN:H films for the passivation and anti-reflection layer of the silicon solar cells by introducing N_2 to the process. The N_2 addition had significant effects on the optical properties and passivation quality of the films. The improvements in the J_{SC} and the η were achieved when using the optimized a-SiCN:H films as an antireflection coating and passivation layer. Further optimization by introducing NH_3 to the process probably allows a-SiCN:H films with lower n and higher hydrogen concentration that will lead to the better performance of the films.

ACKNOWLEDGMENTS

This work was supported by the New Energy and Industrial Technology Development Organization (NEDO) under the Ministry of Economy, Trade and Industry (METI) of Japan.

REFERENCES

1. A. Limmanee, M. Otsubo, T. Sato, S. Miyajima, A. Yamada, M. Konagai, Proceeding of 4th World Conference on Photovoltaic Energy Conversion, Hawai, U.S.A, May 7-12, 2006, p.1227.

2. A. Limmanee, M. Otsubo, T. Sato, S. Miyajima, A. Yamada, M. Konagai, Jpn. J. Appl. Phys **46**, 56 (2007).

3. A. Limmanee, M. Otsubo, T. Sugiura, T. Sato, S. Miyajima, A. Yamada, M. Konagai, Proceeding of 4th International Conference on Hot-Wire CVD process, Gifu, Japan, October 4-8, 2006, p.231.

Mater. Res. Soc. Symp. Proc. Vol. 989 © 2007 Materials Research Society   0989-A04-03

Dependence of the Electronic Properties of Hot-Wire CVD Amorphous Silicon-Germanium Alloys on Oxygen Impurity Levels

Shouvik Datta[1], J. David Cohen[1], Yueqin Xu[2], A. H. Mahan[2], and Howard M. Branz[2]

[1]Department of Physics, University of Oregon, 1371 E 13th Avenue, Eugene, OR, 97403
[2]National Renewable Energy Laboratory, 1617 Cole Boulevard, Golden, CO, 80401

ABSTRACT

We report the effects of intentionally introducing up to $\sim 5\times10^{20}$/cm^3 oxygen impurities into hydrogenated amorphous silicon-germanium alloys (of roughly 30at.% Ge) grown by the hot-wire chemical vapor deposition (HWCVD) method. Deep defect densities determined by drive-level capacitance profiling (DLCP) indicated a modest increase with increasing oxygen content (up to a factor of 3 at the highest oxygen level). Transient photocapacitance (TPC) spectra indicated a clear spectral signature for an optical transition between the valence band and an additional defect level which is attributed to oxygen impurities. The oxygen impurity related defect transition has an optical threshold around 1.4eV above the valence band and also results in a *negative* contribution to the TPC signal. This initially led us to believe that the bandtail for the higher oxygen samples was much narrower than it actually is. Surprisingly, this additional oxygen related defect level appears to have only a very minor effect upon the estimated minority carrier collection fraction. The effects of light-induced degradation upon some of these oxygen contaminated samples were also examined. We infer the existence of a significant thermal barrier to explain the observed spectral signatures of this oxygen impurity defect.

INTRODUCTION

Electronic properties of a-Si,Ge:H alloys deposited by the hot-wire chemical vapor deposition (HWCVD) method have now improved [1,2] to a level comparable to the best glow discharge (PECVD) a-Si,Ge:H alloy films. As was reported earlier [3-6], these were obtained by replacing the usual tungsten filament with tantalum and using a filament temperature of ~1800°C instead of ~ 2000°C during the HWCVD growth process. We recently reported studies [5] in which we carried out a systematic incorporation of oxygen impurities for a series of a-Si,Ge:H alloy samples. Here we observed that the band tail actually appeared to become narrower as the oxygen content of a-Si,Ge:H films was increased. This seemed to suggest that the oxygen impurities actually improved the quality of these alloy films! However, a more detailed study was needed to understand the root cause of such oxygen induced electronic changes. Subsequently, these alloys were characterized using a variety of junction capacitance based techniques including transient photocapacitance (TPC) [7,8], transient photocurrent (TPI) spectroscopy, and drive-level capacitance profiling (DLCP) [9,10]. After careful analysis of these results, we now have a viable alternative explanation for the oxygen induced changes in a-Si,Ge:H alloys. Specifically, we believe our experimental results provide a clear signature of the optical excitation and subsequent trapping of valence band electrons into an empty oxygen related defect state. This results in a negative contribution to the TPC signal and accounts for many of the observed electronic changes in the oxygen rich a-Si$_{0.7}$Ge$_{0.3}$:H alloy films grown by HWCVD.

Table I. Oxygen concentrations and HWCVD growth parameters of a-Si$_{0.7}$Ge$_{0.3}$:H employed in this study. The tantalum filament temperature was about 1800°C.

Sample	Growth Rate (Å/s)	Air Leak Rate (sccm)	Thickness (microns)	Initial Substrate Temp (°C)	Final Substrate Temp (°C)	Oxygen concentrations (cm^{-3})
A	2.00	~0.00	1.80	204	289	~8×10^{18}
B	2.08	~0.02	1.75	204	289	~3×10^{19}
C	2.05	~0.06	1.60	204	284	~1×10^{20}
D	2.82	~0.20	2.20	204	275	~5×10^{20}

SAMPLES

A series of a-Si,Ge:H alloys with nominal Ge fractions around 30%, as determined by Secondary Ion Mass Spectrometry (SIMS), were deposited by the HWCVD technique at the National Renewable Energy Laboratory (NREL). These HWCVD depositions were carried out simultaneously on both stainless steel (SS) substrates and also on p^{+} crystalline Si (c-Si) using a tantalum filament maintained at 1800°C. Oxygen levels were systematically varied using a controlled air leak into the growth chamber during the hot-wire CVD process. The SIMS measurements indicated a systematic increase of oxygen impurities (see Table I) as the air leak was gradually increased from sample A to sample D. At the same time, the nitrogen content was nearly identical and very low (<4×10^{16} cm^{-3}). Fourier transform infrared spectroscopy (FTIR) signatures of Si-H (2000 cm^{-1}) and Ge-H (1876 cm^{-1}) stretching vibration modes are unaffected by the sequential increase in oxygen concentration. These clearly also rule out any changes in the degree of hydrogen incorporation in these series of samples. For the 30at.% Ge samples examined, gas ratios of H$_2$/(SiH$_4$ + GeH$_4$) = 1 and GeH$_4$/(GeH$_4$+ SiH$_4$) = 0.19 were utilized for all growth runs. Substrate temperatures were always set to 204°C at the beginning of film growth, but the final temperatures were higher (~280°C) due to heating by the HWCVD filament. Detailed growth parameters of the four samples examined in our current study are listed in Table I. Films were deposited simultaneously onto both specular stainless steel and on p^{+} c-Si substrates. The silicon substrates are generally considered to be more reliable for the TPI measurements. Moreover, these samples allow us to examine the electronic properties of the a-Si,Ge:H material grown closer to the substrate where the properties were somewhat better due to the lower growth temperatures. Therefore, only the samples grown on c-silicon substrates will be discussed in this paper. A semi-transparent Pd contact was thermally evaporated on top of the intrinsic layer for junction capacitance based measurements. However, we utilized the reverse biased c-Si/a-Si,Ge:H depletion junction for all of the measurements described below.

CHARACTERIZATION METHODS

Drive Level Capacitance Profiling (DLCP) was used to determine deep defect density by fitting the non-linear capacitive response of the junction capacitance (C = C$_0$ +C$_1$ δV+ C$_2$ (δV)^2 +...) as a function of the amplitude, δV, of the applied oscillating voltage. The drive level density, N$_{DL}$, is then obtained from the coefficients C$_0$ and C$_1$ and it is directly related to an integral over the density of deep defect states near the spatial position <x> = ε A/C$_0$ from the barrier junction. Thus we are able to determine the carrier contribution from a portion of the deep defect band that responds at a particular measurement frequency and temperature. Spatial profiles of the deep defect densities were obtained by varying the dc reverse bias at each stage.

The transient photocapcitance (TPC) and transient photocurrent (TPI) techniques [7,8] were used to obtain sub-band-gap optical absorption-like spectra. Unlike true absorption measurements, however, these signals result from the charge carriers that escape the depletion region in response to the absorbed photons within certain time window. For TPC, optically excited changes in the junction capacitance indicate the underlying change in the charge within the depletion width while, for the TPI technique, differences in the detected current as a result of optical excitation represent the motion of the released charge. For these nominally n type a-Si$_{0.7}$Ge$_{0.3}$:H alloys, the TPC signal is sensitive to the *difference n-p*, where n and p are the number of electrons (majority carriers) and holes (minority carriers) collected, respectively. However, the TPI signal is proportional to the *sum n+p* because the motion of either type of carrier produces a current of the same sign. As a result, when we compare [7,8] these two types of spectra, we can estimate the relative collection fraction of the minority carriers as p/n.

EXPERIMENTAL RESULTS

In Fig. 1, we plot the observed spatial profile of deep defect densities as measured by DLCP technique. We see that the deep defect density in the annealed State A increases from a value ~7×10^{15}/cm^3 in sample A (lowest oxygen) to ~2×10^{16}/cm^3 in sample D (highest oxygen). These samples were also examined in a degraded State B produced by light soaking for 100 hours at roughly 1W/cm^2 intensity using a ELH source with a 610nm long pass filter. The deep defect densities of the samples with different oxygen contents, did increase somewhat, although less than a factor of 2 in all cases.

Figure 2 shows the TPC and TPI spectra of sample C (~1×10^{20}/cm^3 oxygen) in both State A and State B. Our initial analysis of the TPC spectra noted [5] that apparently this sample has an extremely narrow bandtail, with an Urbach energy below 40meV. However, a very different picture emerges when analyzing both the TPC and TPI spectra together. The thin solid lines shown in Fig. 2 indicate that quite a good fit can be obtained. However, these fits require that we include an additional band of transitions, from the valence band to a defect with an optical threshold centered at ~1.35 to ~1.4eV. Because this transition predominantly results in the

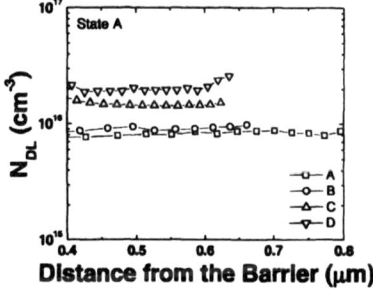 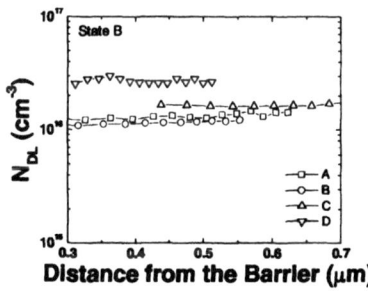

FIG. 1. Spatial Profile of deep defect density in (a) the annealed State A and (b) in the light-soaked State B. These DLCP profiles were obtained at 1.1 kHz and 370K.

release of a hole instead of an electron, they contribute *negatively* to TPC spectrum and *positively* to the TPI spectrum. It is this negative contribution to the TPC signal that makes its bandtail appear unusually narrow. In actual fact, however, the correct Urbach energy as obtained from these fits is close to 47meV. Also, from the ratio of the TPI and TPC magnitudes in the upper portion of the bandtail region we can deduce the relative hole collection fractions for this sample. It is ~96% for sample C in State A, with perhaps even a slight increase in State B. For the lowest oxygen sample A, we deduced an Urbach energy around 45meV, which is among the lowest ever measured for any a-Si,Ge:H in this alloy range. Correspondingly, the relative collection fraction of hole-to-electron was found to be around 97% for sample A. This also exceed the hole collection fraction determined for the best PECVD a-Si,Ge:H in this alloy range.

Figure 3 exhibits a comparison of the TPC spectra for three samples with increasing oxygen levels. We see that the ~1.4eV transition is basically absent from the spectrum of the sample A with no intentional oxygen incorporation during the HWCVD growth, and increases with as the oxygen level increases. *We thus attribute this 2nd defect band solely to the oxygen*

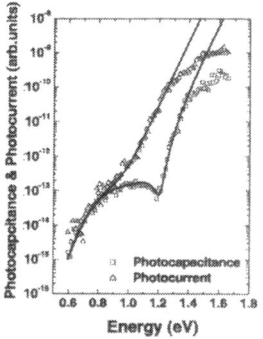 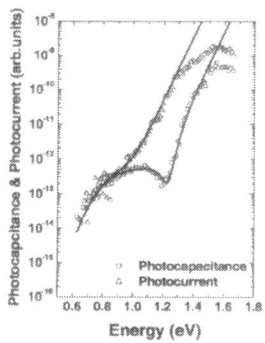

FIG. 2. Photocapacitance and Photocurrent spectra for the a-Si$_{0.7}$Ge$_{0.3}$:H sample C with ~1×10^{20}/cm^3 oxygen in (a) its annealed State, and (b) a light-soaked state. The thin solid lines are fits assuming 47meV Urbach tail plus a transition from a filled defect with an optical threshold of ~0.84eV below E$_C$, plus a second deep defect transition with an optical threshold roughly 1.4eV above E$_V$.

FIG. 3. Photocapacitance spectra for three a-Si$_{0.7}$Ge$_{0.3}$:H samples with different oxygen levels. The thin solid lines indicate fits obtained in the same manner as those in Fig. 2. In the case of fitting the spectrum for the sample without intentional oxygen contamination, the 2nd defect transition at ~1.4eV above the valence was negligible. Also, for that sample the Urbach energy was 45meV, and was 47meV for the other two samples. For the sample with highest oxygen content, the TPC signal actually became *negative* (filled triangles) in the region between 1.1 and 1.4eV.

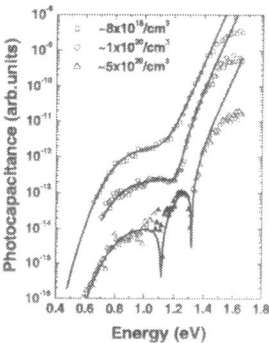

impurities present in these a-Si$_{0.7}$Ge$_{0.3}$:H alloy films. However, this increased oxygen level appears to have only a very minor effect upon the estimated minority carrier collection fractions as also seen from fig.2. Moreover, as indicated in figs. 1 and 2, these HWCVD deposited a-Si,Ge:H films appear to be much more stable with respect to light-induced degradation than the highest quality PECVD a-Si,Ge:H samples studied [5,6] previously by these methods. However, this may partially be because their initial deep defect densities are close to high 10^{15}/cm^3 level cm^{-3} as compared to the mid 10^{15}/cm^3 level of those previous PECVD samples. A recombination pathway through the 1.4 eV defect may also increase the stability by competing with the band-to-band recombination. However, more experiments are needed to examine this hypothesis.

One interesting remaining issue concerns why the transition into an empty defect band centered at ~1.4eV from the valence band should yield predominantly negative TPC signal for sample D. This can only be possible if the electron that is inserted from the valence band remains strongly trapped for the duration of the measurement time window (~ 0.4s). We attempted to study the subsequent thermal emission of this trapped electron by recording the TPC spectra for the highest oxygen doped sample over a wider range of temperatures. Three such spectra are shown in Fig. 4. Interestingly, the negative TPC signal is totally absent in the spectra measured at 330 K. It motivated us to study the temperature dependence of the TPC signal under selective photoexcitation at a photon energy of 1.2eV. This photoexcitation energy is so chosen such that the signal due to the ~1.4eV defect transition is clearly visible and is also less obscured by the positive bandtail signal. The plot in fig. 5 (circles) indicates that the negative TPC contribution due to the oxygen impurity peaks near 370K. Moreover, above ~380K, the TPC signal is reduced in magnitude due to the subsequent thermal emission of this optically trapped electron from this 2nd defect state. However, the TPC signal increases monotonically to a large positive value below ~345K. It may be due to an increasing positive contribution from the bandtail which is always found to become significantly larger at lower temperatures as shown in fig.5. However, the difference between the rate of temperature variation of TPC signal at 1.2 eV and at 1.5 eV may

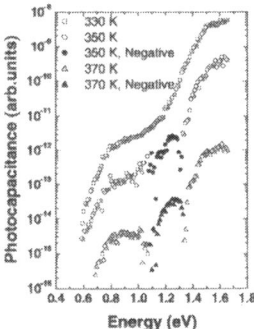

FIG. 4. TPC spectra of the highest oxygen a-Si,Ge:H sample at three different temperatures. Ploted spectra were vertically shifted from each other for greater visibility. Note the relative increase of the negative feature (filled symbols) near 1.2eV at higher temperatures, and the relative decrease in the bandtail signal.

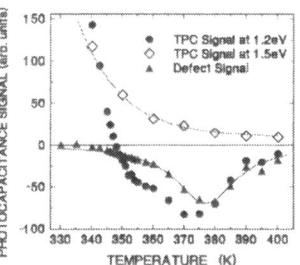

FIG. 5. Temperature variation of the TPC signal at photon energies near 1.2 eV (circles) and 1.5eV (diamonds). One may attempt to extract the underlying temperature dependence of the oxygen related defect signal (triangles) by first normalizing the 1.2eV response to the near bandgap 1.5eV response, and then subtracting off a portion of the bandtail signal to correct for its contribution at 1.2eV that becomes significant at lower temperatures. Further experiments are going on to deduce the correct picture.

59

also signify the presence of more involved thermal trapping processes as described below.

DISCUSSIONS

Based on our experimental results, we hypothesize that the observed oxygen impurity related defect state is associated with a positively charged oxygen donor level, possibly those three fold coordinated (O_3^+) type centers previously suggested in case of a-Si:H [11-14] like materials. The ~1.4eV transition energy from the valence band indicates a defect center close to the conduction band; however, to account for the observed behavior, it is also necessary that in the O_3^0 state, the trapped electron cannot be thermally emitted into the conduction band very quickly. This clearly requires the presence of a significant thermal barrier to inhibit this. We hope that further junction capacitance experiments that are currently in progress will be able to map out a more complete *configurational relaxation diagram for the optically excited electron capture* by this oxygen related defect in these HWCVD grown a-Si,Ge:H alloys.

ACKNOWLEDGEMENTS

Work at the University of Oregon was supported by NREL Subcontract ZXL-5-44205-11. Work at NREL was supported by the U.S. DOE under Contract DE-AC36-99GO10337.

REFERENCES

1. Y. Xu, B. P. Nelson, D. L. Willamson, L. M. Gedvilas, and R.C. Reedy, Mat. Res. Soc. Symp. Proc. **762**, A10.2 (2003).
2. D.L. Williamson, G. Goerigk, Y. Xu, and A.H. Mahan, Proceedings DOE Solar Energy Technologies Review Meeting, DOE/GO-102005-2067 (2005) p. 444.
3. Shouvik Datta, J. D. Cohen, Y.Xu, and A. H. Mahan, Mat. Res. Soc. Symp. Proc. **862**, A7.2 (2005).
4. Shouvik Datta, Yueqin Xu, A. H. Mahan, Howard M. Branz, and J. David Cohen, J of Non-Cryst Solid, **352**,1250 (2006).
5. Shouvik Datta, J David Cohen, Steve Golledge, Yueqin Xu, A. H. Mahan, James R. Doyle and Howard M. Branz, Mat. Res. Soc. Symp. Proc **910**, A2.5 (2006).
6. J.D. Cohen, Shouvik Datta, K. Palinginis, A.H. Mahan, E. Iwaniczko, Y. Xu, and H.M. Branz. Thin Solid Films (in press).
7. A.V. Gelatos, K.K. Mahavadi, J.D. Cohen, and J.P. Harbison, Appl. Phys. Lett. **53**, 403 (1988).
8. J. David Cohen and Avgerinos V. Gelatos, in *Amorphous Silicon and Related Materials*, Vol. A, ed. by Hellmut Fritzsche (World Scientific, Singapore, 1989), p 475.
9. C. E. Michelson, A. V. Gelatos, and J. D. Cohen, Appl. Phys. Lett. **47**, 412 (1985).
10. J. David Cohen and Avgerinos V. Gelatos, in Amorphous Silicon and Related Materials, edited by Hellmut Fritzsche Vol A (World Scientific, Singapore, 1989), p 475.
11. R. A. Street and N. F. Mott, Phys. Rev. Lett. **35**, 1293 (1975).
12. David Adler, Solar cells. **9**, 133 (1983).
13. Tatsuo Shimizu, Minoru Matsumoto, Masahiro Yoshita, Masahiko Iwami, Akiharu Morimoto, and Minoru Kumeda, J. Non-Cryst. Solids. **137&138**, 391 (1991).
14. H. Fritzsche, J. Non-Cryst. Solids. **190**, 180 (1995).

Mater. Res. Soc. Symp. Proc. Vol. 989 © 2007 Materials Research Society 0989-A04-04

High Quality a-Ge:H Films and Devices Through Enhanced Plasma Chemistry

Erik V. Johnson, and Pere Roca i Cabarrocas

LPICM (CNRS, UMR 7647), Ecole Polytechnique, Palaiseau Cedex, F-91128, France

ABSTRACT

We present a material study on RF PECVD-grown a-Ge:H showing that thin films of such material can be produced without using the conventional techniques of high power density or powered-electrode substrate placement. We demonstrate the production of material with PDS signatures superior to material produced at ten times higher power density. This is achieved through the use of Ar and H_2 dilution and by growing the films at high pressures under conditions where nanoparticles and nanocrystals formed in the gas phase contribute significantly to the growth as confirmed by HRTEM. The conditions described result in material which demonstrates activated conduction down to room temperature. Additionally, the quality of the material has been demonstrated through its application in n-i-p diodes. A spectral response at 0.9um of 0.38 and an AM1.5 efficiency of 2.1% have been demonstrated utilizing an absorber layer thickness of only 60nm.

INTRODUCTION

The goal of producing PECVD-grown, device-quality a-Ge:H for use as a low-bandgap semiconductor has lead to important discoveries about the ideal growth conditions of the material. It is generally accepted that the best quality a-Ge:H is grown under conditions that are in stark contrast with those ideal for a-Si:H, a similar tetrahedrally-bonded amorphous semiconductor. For example, the quality of a-Ge:H is greatly improved when it is grown under conditions of high H_2 dilution of germane (GeH_4) and under high power conditions, whereas the growth of a-Si:H under these conditions (without an accompanying increase in pressure) leads to low quality, porous material. High quality a-Ge:H has been produced through deposition on the powered cathode of asymmetric RF-PECVD systems [1,2] and in ECR systems [3-6]. Recently, hot-wire CVD using undiluted GeH_4 has been employed to grow high-quality films [7], demonstrating that high H_2-dilution is not a strict requirement but that the conditions for good quality films are more complicated.

The causes behind the higher quality of films grown under the above-listed conditions are still under dispute. Authors variously ascribe the high quality of the material to be due to 1) copious amounts of atomic hydrogen arriving at the growth surface, 2) ion bombardment during growth, 3) growth in the gas depletion regime for germane, and/or (4) heated precursors arriving at the surface. In this work, we promote this growth environment through the use of simultaneous Ar+H_2 dilution of GeH_4 and an elevated growth pressure. The metastable argon ions (Ar*) created through electron collision will ionize GeH_4 molecules, increase the density of atomic H, and release energy to the surface through relaxation of the metastable state. As well, under the high-pressure conditions employed, plasma-formed and heated nanoparticles will contribute to the growth, resulting in "polymorphous" (pm-) material. A low substrate temperature is also employed to encourage the incorporation of H to passivate dangling bonds. The use of conditions promoting polymorphous material has been previously shown to enhance the stability of the transport properties of pm-SiGe:H films under light-soaking [8].

It has been demonstrated in an accompanying study [9] that pm-Ge:H grown in conditions promoting nanoparticle incorporation into the film can be applied as a 60nm thick i-layer in n-i-p photodiodes demonstrating 2.1% efficiency at AM1.5 and spectral response in the near infrared (900nm) of 0.38 despite illumination through the n-layer. We investigate this device-quality pm-Ge:H through optical and electronic measurements to further examine the nature of the material and to explain the origin of these high quality films produced in non-standard conditions.

EXPERIMENT

The films were deposited in the ARCAM reactor [10], a multiplasma, monochamber RF-PECVD system operating at 13.56 MHz. The walls of the chamber were heated to the same temperature as the substrate holder. The substrates were placed on the un-powered, grounded electrode, in contrast with results from the literature showing that the best films are produced through placement on the powered electrode. This substrate holder can be rotated in and out of the plasma/deposition zone. Two pressure series of films on Corning Glass Type 1737 (CG1737) and two films on carbon-covered TEM grids were deposited under the conditions listed in Table 1.

Table 1 Growth Conditions

	Material Series A	Material Series B	HRTEM A	HRTEM B
Argon Flow Rate (FR_{Argon})	200 sccm	200 sccm	200 sccm	200 sccm
Hydrogen Flow Rate (FR_{H2})	200 sccm	200 sccm	200 sccm	200 sccm
Germane Flow Rate (FR_{GeH4})	6 sccm	6 sccm	6 sccm	5 sccm
Substrate/Chamber Temp.	175°C	175°C	175°C	175°C
Pressure (mTorr)	1000-3000	1000-3000	1560	2370
Interelectrode Distance (mm)	17	22	22	22
Power Density (mW/cm^2)	60	120	60	120

The material is characterized through HRTEM, Raman scattering, temperature-dependent conductivity, Spectroscopic Ellipsometry (SE) and photothermal deflection spectroscopy (PDS).

RESULTS AND DISCUSSION

HRTEM

In Fig. 1 are presented images of two films grown under different growth conditions. The first two images are of Sample HRTEM A (conditions listed in Table 1), grown at the lower power density and pressure. This sample was exposed to air after deposition. Apparent in Fig.1(a) is an agglomerate of larger nanoparticles 10-20nm in diameter (possibly collected during rotation out of the deposition zone) surrounded by amorphous particles sized from 2-4 nm. The greater magnification image in Fig.1 (b) shows these contrasted regions. These regions have been previously interpreted as island-like growth [11]. However, cross-sectional HRTEM results (not included in this study) suggest that they are the result of plasma-formed nanoparticles being deposited onto the growth surface and incorporated into the film. Fig. 1(c) shows the image from Sample HRTEM B, grown at higher power and pressure and with a slightly lower germane flow (5 vs 6 sccm). Clearly visible are the nanocrystalline domains, up to 10nm in diameter, despite the films being on the order of 10nm thick.

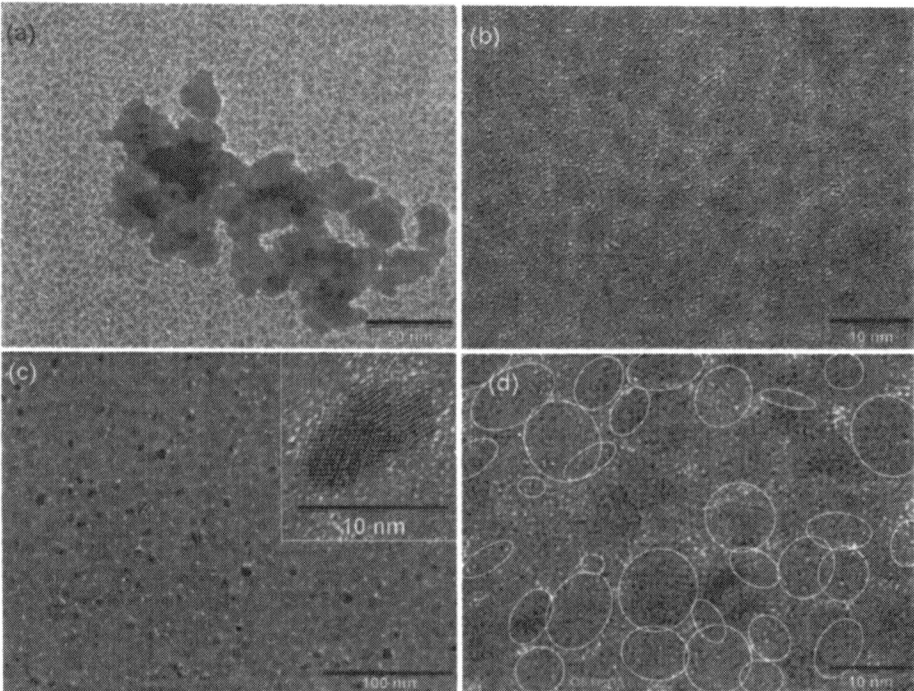

Figure 1 - HRTEM images of two pm/nc-Ge:H samples. Subfigures (a) and (b) show Sample A, consisting of amorphous nanoparticles 2-4 nm in diameter, and subfigures (c) and (d) show Sample B, consisting of nanocrystals up to 10nm in size. See text for discussion.

Temperature Dependent Conductivity

The temperature dependence of the dark conductivity was measured for the films grown in Material Series A and B. The samples were measured in vacuum, and the curves presented in Fig. 2 are measurements taken after the samples had been put through one temperature cycle (heating to 120°C, cooling to 40°C). The results for the samples grown at 60mW/cm^2 are plotted in Fig. 2(a), and these results can be fit well to activated conduction, indicating transport predominantly through extended states, at temperatures descending down to 40°C. The results for the samples grown at 120mW/cm^2 show a similar behaviour, with the exception of the sample grown at the highest pressures (2000 and 2600 mTorr). These samples show two temperature regions of conduction: a lower temperature region with smaller activation energy than the rest of the samples, and a high temperature region with larger activation energy.

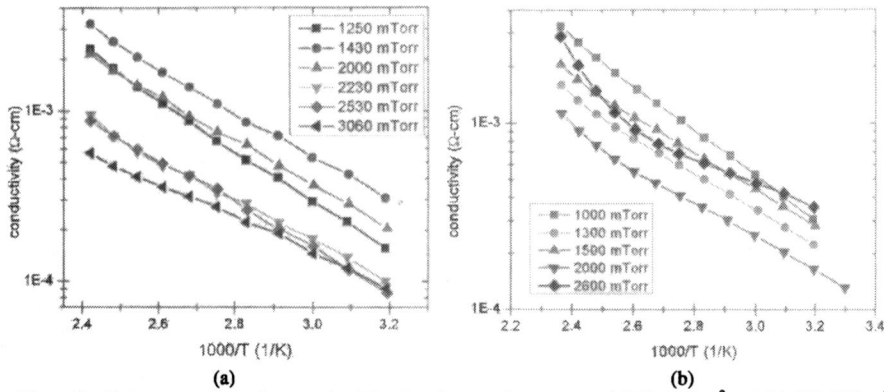

Figure 2 – Temperature dependent conductivity data for samples grown at (a) 60mW/cm^2 and (b) 120mW/cm^2

This behaviour coincides with an elevated nanocrystalline content, as determined by Raman scattering and by ex-situ spectroscopic ellipsometry. However, the samples grown at low pressure and low power - which show a small crystalline signal from Raman – nevertheless exhibit activated conduction throughout the temperature range. In general, the samples grown at 60mW/cm^2 exhibit higher activation energies than those grown at 120mW/cm^2. The activation energies for the 60mW/cm^2 samples, however, are smaller than half the gap. This result is consistent with the fact that intrinsic a-Ge:H is slightly n-type, and additionally, since no efforts were made to cap the material, some post-deposition O-contamination may have occurred. The nanoparticle-based growth confirmed by HRTEM emphasizes the large surface area available for post-deposition oxidation. The sample grown at 60mW/cm^2 and at 1250 mTorr exhibits an activation energy of 0.3eV, the largest amongst these samples.

PDS and SE

The mobility gap density of states was measured using Photothermal Deflection Spectroscopy (PDS), and the absorption coefficient spectra derived from this measurement are presented in Fig. 3. The clearest trend can be noted from the absorption coefficient value at 0.9eV, shown in Fig. 4a for the samples from Material Series A. In general, samples that show a microcrystalline character from Raman exhibit the highest absorption coefficients at 0.9eV. This result is consistent with the plasma-formation and incorporation of nanocrystals into the film, and these showing a remnant of the smaller c-Ge gap.

The films were also characterized through ex-situ spectroscopic ellipsometry (SE). A decrease in the maximum value of $<\varepsilon_2>$ - the imaginary part of the pseudo–dielectric function - with increasing pressure is noted for both sets of samples, as shown in Fig. 4(b). Comparison with results for high-quality a-Ge:H from other groups [7,12,13] indicate that the material grown at low power and low pressure is comparably dense, but at increasing power is less so.

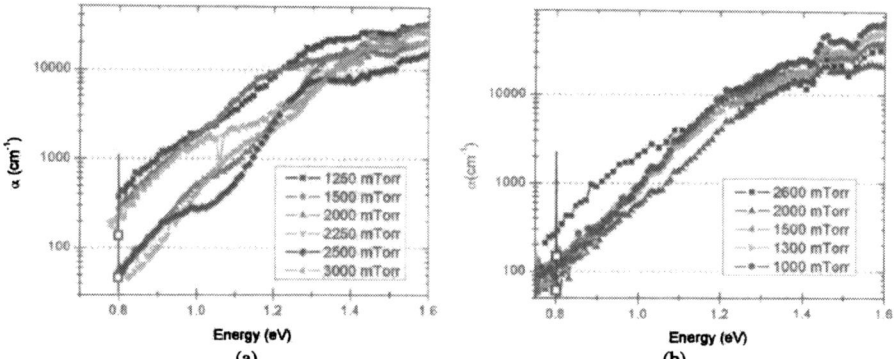

(a)   (b)

Figure 3 – Absorption spectrum from PDS for samples grown at (a) 60mW/cm^2 and (b) 120mW/cm^2. The open squares indicate points extracted from Ref. [1] for samples grown at 80mW/cm^2 (upper square) and 800mW/cm^2 (lower square).

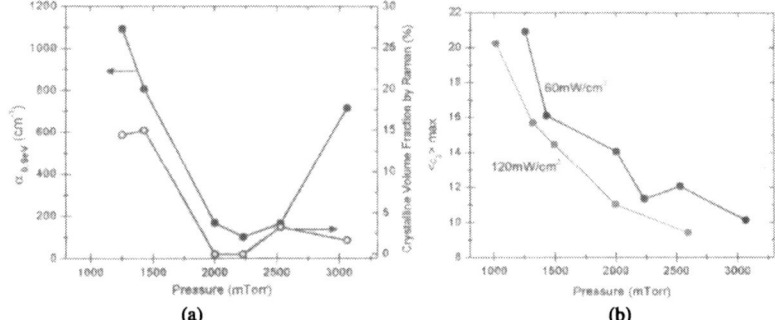

(a)   (b)

Figure 4 – (a) Absorption at 0.9eV as measured by PDS and approximate volume fraction of crystalline Ge (fractional area under Raman peak at 300cm^{-1}) for sample grown at 60 mW/cm^2, and (b) maximum value of $<\varepsilon_2>$ from the acquired SE curve.

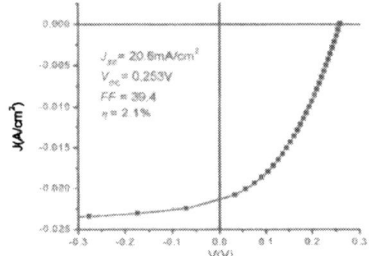

Figure 5 – J-V Characteristic for pm-Ge:H cell grown under elevated plasma chemistry conditions

Use of the material grown at 60mW/cm^2 and at lower pressure (1 Torr) in photodiodes has resulted in an elevated infrared photovoltaic response and a very high short circuit current of $J_{SC} = 20.6$mA/cm^2 despite an extremely thin i-layer (60nm) [8]. A J-V characteristic from such a device is presented in Fig. 5. These results are comparable to those achieved by growing diodes with ECR [14] and by mounting the substrate on the asymmetric, powered electrode [15].

CONCLUSIONS

This material study elucidates the promising results from the incorporation of this material in devices, as presented in Ref. [9]. Of the a-Ge:H grown for this study, only "device-quality conditions" material exhibited

- activated conduction from 40°C to 120°C (indicating a single transport mechanism)
- high material density (from SE)
- a small but existing nanocrystalline fraction (from TEM, SE, Raman, and PDS)

These properties coincided with the use of plasma conditions that encourage the formation and heating of nanoparticles in the plasma, as well as their incorporation into the film. It is proposed that under these conditions, the heating of Ge nanoparticles in the plasma encourages the formation of strong Ge-Ge bonds, while the low substrate temperature allows the incorporation of H in the film. The strong-bond formation role is typically played by ion-bombardment or by hot atoms reforming weak bonds at the surface, as proposed by Doyle *et al* [16] and as experimentally supported by results from hot-wire growth [6]. The results of this study show that it can also be enacted through enhancing plasma chemistry by simultaneous Ar+H_2 dilution.

ACKNOWLEDGMENTS

EVJ acknowledges the generous support of NSERC.

REFERENCES

[1] F.H. Karg, H. Bohm, and K. Pierz, J. Non-Cryst. Solids 114, 477 (1989).
[2] W.A. Turner, et al, J. Appl. Phys. 67, 7430 (1990).
[3] T. Aoki, S. Kato, Y. Nishikawa, and M. Hirose, J. Non-Cryst. Solids 114, 798 (1989).
[4] A. Matsuda and K. Tanaka J. Non-Cryst. Solids 97 & 98, 1367 (1987).
[5] N. Shibata, A. Tanabe, J. Hanna, S. Oda, and I. Shimizu, Jpn. J. Appl. Phys. 25, L540 (1986).
[6] Xuejin Niu, V.L. Dalal, J. Appl. Phys. 98, 096103 (2005).
[7] L. Zanzig, W. Beyer, and H. Wagner Appl. Phys. Lett. 67, 1567 (1995).
[8] M.E. Guenier, J.P. Kleider, P. Chatterjee, P. Roca i Cabarrocas, and Y. Poissant, J. Appl. Phys. 92, 4959 (2002).
[9] E.V. Johnson and P. Roca i Cabarrocas, (2007) Solar Energy Mater. and Solar Cells, doi:10.1016/j.solmat.2007.01.019.
[10] P.Roca i Cabarrocas, J.B. Chévrier, J. Huc, A. Lioret, J.Y. Parey, and J.P.M. Schmitt, J. Vac. Sci. Technol. A 9, 2331 (1991).
[11] B. Drevillon and C. Godet, J. Appl. Phys. 64, 145 (1988).
[12] J.R. Blanco, P.J. McMarr, K. Vedam, and R.C. Ross, J. Appl. Phys. 60, 3724 (1986).
[13] J.E. Yehoda, B. Yang, K. Vedam, and R. Messier, J. Vac. Sci. Technol. A 6, 1631 (1988).
[14] J. Zhu, V.L. Dalal, M.A. Ring, and J.J. Guitterez, J.D. Cohen, J. Non-Cryst. Solids 338-340, 651 (2004).
[15] W. Kusian, E. Gunzel and R.D. Plattner, Solar Energy Materials 23, 303 (1991).
[16] J.R. Doyle, D.A. Doughty, and A. Gallagher, J. Appl. Phys. 69, 4169 (1991).

Mater. Res. Soc. Symp. Proc. Vol. 989 © 2007 Materials Research Society   0989-A04-05

Dielectric Properties of Ultra Dense (3 g/cm3) Silicon Nitride Deposited by Hot Wire CVD at Industrially Relevant High Deposition Rates

Zomer Silvester Houweling, Vasco Verlaan, Karine van der Werf, Hanno D. Goldbach, and Ruud E. I. Schropp
Surfaces, Interfaces and Devices, Utrecht University, Princetonplein 5, PO box 80.000, Utrecht, NL-3508 TA, Netherlands

ABSTRACT

For silicon nitride (SiN_x) deposited at 3 nm/s using hot wire chemical vapor deposition (HWCVD), the mass-density reached an ultra high value of 3.0 g/cm^3. Etch rates in a 16BHF solution show that the lowest etch rate occurs for films with a N/Si ratio of 1.2, the ratio where also the maximum in mass density occurs. The thus found etch rate of 7 nm/min is better than that for PECVD layers, even when made at a much lower deposition rate. The root-mean-square (*rms*) roughness measured on 300 nm thick $SiN_{1.2}$ layers is only about 1 nm, which is advantageous for obtaining high field-effect mobility in thin-film transistors. $SiN_{1.2}$ films have succesfully been tested in "all hot wire" thin film transistors (TFTs). SiN_x films with various x values in the range 1.0 < x <1.5 have been incorporated in metal-insulator-semiconductor structures with n-type c-Si wafers to determine their electrical properties from C-V and I-V measurements. We analyzed the behavior of the static dielectric constant, fixed nitride charges and trapped nitride charges as function of N/Si ratio. I-V measurements show that the HW SiN_x films with N/Si $\geq$ 1.33 have high dielectric breakdown fields that exceed 5.9 MV/cm. For these films we deduce a low positive fixed nitride charge density of 6.2–7.8 x 10^{16} cm^{-3} from the flat band voltage and from the small hysteresis in the backward sweep we deduce a low fast trapped charge density of 1.3-1.7 x 10^{10} cm^{-2}. The dielectric constant ε for different compositions is seen not to change appreciably over the whole range and amounts to 6.3 $\pm$ 0.1. These high-density SiN_x films possess very low tensile stress (down to 16 MPa), which will be helpful in for instance, micro-electro-mechanical (MEMS) applications. HWCVD provides high quality a-SiN_x materials with good dielectric properties at a high deposition rate.

INTRODUCTION

Thin film amorphous hydrogenated silicon nitride (a-SiN_x:H) is a widely used insulating silicon alloy with versatile applicability. For example, it is of great importance for multi-crystalline silicon (mc-Si) solar cells where it acts as passivating anti-reflection coating [1,2,3], it serves as transparent barrier coating in plastic electronics [4] and as gate dielectric in large area electronics such as thin film transistors (TFTs) [5,6]. There are several ways to fabricate thin film SiN_x, of which plasma enhanced chemical vapor deposition (PECVD) is most commonly used. In our laboratory we investigated the use of the hot wire chemical vapor deposition (HWCVD) technique, in which the undiluted feedstock gasses silane (SiH_4) and ammonia (NH_3) are catalytically cracked at heated filaments. We employ this technique because it has some important advantages over techniques that utilize a plasma for decomposition of the source gasses.

Using HWCVD, very good decomposition efficiencies are reported in a SiH_4/NH_3 mixture at relatively low total flow rate, namely 98% and 52% for SiH_4 and NH_3, respectively [7]. The efficient catalytic decomposition of feedstock gasses has led in our laboratory to high deposition rates of up to 7.3 nm/s for device quality, transparent films [8]. Both the good gas utilization and high deposition rate are highly desirable from a cost perspective point of view. Furthermore, the technique is free of dust formation, is free from ion-bombardment issues and can easily be scaled up to offer large area capability [9]. The low H concentration of 9 at.-%, specifically the low density of Si-H bonds [10], is an advantage for the use as sidewall and liner material in ULSI p-MOS transistors [11].

In earlier work we have shown that the material properties of HWCVD SiN_x films are correlated to the internal structure of the film, which is determined by the amount of incorporated nitrogen in the film [8]. The composition is easily tunable in the HWCVD reactor due to the simplicity of the setup. The internal atomic structure is determined by the nitrogen-to-silicon ratio N/Si (the x in SiN_x). The value of x = 1.33 represents stoichiometry. Our group has previously shown that hot wire $SiN_{1.2}$ has excellent passivating properties for mc-Si solar cells [12]. At this composition the films are found to possess the maximum mass density, at a value of 3.0 g/cm^3. These films have been shown to withstand rapid thermal annealing (RTA) temperatures of 800 °C [10].

SiN_x films deposited with conventional methods like PECVD tend to have high stress values in the range of 100–1000 MPa [13]. Passivating layers for organic light emitting diodes (OLEDs) for example, require very low stress in the order of 10 MPa [14,15], but also require low deposition temperatures. Low stress in thin films is also important in micro-electro-mechanical (MEMS) applications, plastic electronics and when TFTs are deposited on thin polymer foil.

In this paper we discuss the stress and the dielectric properties of our HWCVD SiN_x deposited at high deposition rates of around 3 nm/s. The dielectric properties are determined from C-V and I-V measurements on metal-insulator-semiconductor (MIS) structures and the $SiN_{1.2}$ films have been tested in "all hot wire" thin film transistors (TFTs).

EXPERIMENT

We used the HWCVD technique to deposit thin films of SiN_x of various compositions. In our HWCVD reactor [16] pure source gasses silane (SiH_4) and ammonia (NH_3) are brought into a stainless-steel ultra high vacuum hot wire reactor through a homogeneous gas distributor. The gasses are decomposed to radicals only at resistively heated tantalum (Ta) filaments held at 2100 °C. The substrate reaches a temperature of 450 °C due to radiation from the wires only. A shutter situated between the sample and the wires controls the duration of the deposition, which occurs on a region of 5 x 5 cm^2 where homogeneous SiN_x films are deposited. The substrates consist of Corning 1737 and Eagle 2000 glass and/or on ultra-flat thin Schott D263T glass (100 μm), and single side polished c-Si wafers with <100> orientation. The c-Si wafers are cleaned in a 1.0% HF dipping procedure for one minute prior to deposition to remove native oxide layers.

Reflection/transmission measurements are performed for determination of the layer thickness [17,18] for films on glass substrates. The thin Schott glass substrates were used for measuring the mechanical stress present in the nitride-substrate stack. Elastic Recoil Detection (ERD) [19] measurements were performed to determine the atomic N/Si ratio and atomic mass density.

To investigate the dielectric properties of SiN_x films with 1.0 < x <1.5, metal-insulator-semiconductor (MIS) structures were made. We have used n-type Czochralski wafers (ε_s = 11.9) with resistivities of 2-3 Ωcm and 1-5 Ωcm. Aluminum (Al) is evaporated on the backside of the c-Si wafer as Ohmic contact and Al dots (0.16 cm^2), were evaporated as front contact. I-V measurements were performed at room temperature with a Keithley 238 high-current source measure unit, supplying voltages in the range of –110 V to +110 V to determine the electric breakdown field, which is defined as the field when the current density is $J_{break} = 10^{-6}$ A/cm^2. C-V measurements are performed in the dark, also at room temperature, with a voltage sweep speed of 0.4 - 0.9 V/s with an Agilent HP 4284A precision LCR meter. A signal frequency of 1 MHz is supplied and a bias voltage that is swept between –40 V and +40 V. The small signal capacitance is measured when a differential voltage is superimposed on the bias voltage with an amplitude of 20 mV. A bias sweep direction from inversion to accumulation is applied to ensure that the obtained C-V characteristic is independent of generation time for minority charge carriers in the semiconductor [20]. When extracting quantities from the C-V curves we follow a treatment that is analogous to Dubey *et al.* [21] and Sze [22].

For the stress measurements SiN$_x$ films with thicknesses d$_f$ between 214 and 291 nm were deposited on ultra-flat thin Schott glass substrates with a thickness d$_s$ of 100 μm. A 4-cm long line along the surface of the substrate is scanned with a DEKTAK surface profilometer before and after a SiN$_x$ deposition. Due to stress in the SiN$_x$ film, the film-substrate stack curves. To quantify the total mechanical stress we follow the approach given by Glang et al. [23]. Hooke's generalized law is applied and the deflection δ at a distance ρ from the center of the substrate is given by expression 1).

$$\sigma_i = \frac{E_s}{3(1-v_s)}\frac{d_s^2}{d_f}\frac{\rho^2}{\delta} \quad 1) \qquad \sigma_{th} = \frac{E_f}{1-v_f}\left(\alpha_s - \alpha_f\right)\cdot\left(T_{dep} - T_{meas}\right) \quad 2) \qquad \sigma_{total} = \sigma_i + \sigma_{th} \quad 3)$$

The Poisson ratio v_s, Young's modulus E_s, and thermal expansion coefficient α_s of the substrate have the values 0.208, 72.9 GPa and 7.2 x 10^{-6} K^{-1}, respectively [24,25]. When the sample is cooled from the deposition temperature T_{dep} to the temperature at which the sample is measured T_{meas}, thermal stress is built up due to the mismatch in thermal expansion coefficient of substrate and film. The thermal stress is given by expression 2). The values of the Poisson ratio v_f, Young's modulus E_f and thermal expansion coefficient α_f of the SiN film have been set to values obtained for HWCVD a-Si:H with 8-12 at.% hydrogen, which are 0.2, 140 GPa and 4 x 10^{-6} K^{-1}, resp. [24,25]. Intrinsic stress and thermal stress combined give the total stress as can be seen in expression 3).

RESULTS AND DISCUSSION

Mass density, (B)HF wet-etching and mechanical stress

The top graph in Figure 1 shows the mass density of the SiN$_x$ films deposited under various different combinations of settings of pressure, gas flow and filament temperature. It clearly shows a maximum in mass density at x = 1.2. The lower graph shows results from

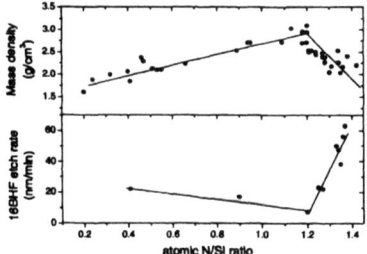

Figure 1 Mass density and 16BHF wet-etch rate as a function of atomic composition. The solid lines are guides to the eye.

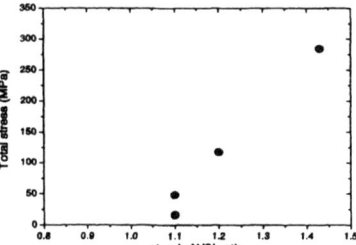

Figure 2 Tensile stress in SiN$_x$ films as a function of atomic composition.

wet-etch experiments in a buffered hydrofluoric acid 16BHF solution (5 parts 40 % NH$_4$F with 1 part 50 % HF). The minimum in etch rate (7 nm/min) at x = 1.2 coincides with the maximum in mass density for our HWCVD SiN$_x$ films. This lowest etch rate is better than that obtained for PECVD SiN$_x$ made at a much lower deposition rate, which for example in [26,27] amounts to 10 nm/min, obtained with a much weaker 10:1 BHF solution.

Intrinsic stress is generated during film growth and is strongly dependent on the stoichiometry of the SiN$_x$ film, the deposition methods and the process parameters used. The thermal stress in a SiN$_x$ film is constant with stoichiometry and amounts to -240 MPa (compressive). In Figure 2 the total stress as a result of the stress measurements is shown.

Stress measurements were performed on Schott glass substrates for SiN$_x$ depositions with $1.1 < x < 1.43$. The most N-rich samples have the highest amount of tensile stress. For layers with N/Si = 1.1 the tensile stress becomes as low as 16 MPa. Claassen *et al.* [13] and Smith *et al.* [28] discussed the origin of tensile stress and attributed it to shrinkage of the film accompanied by desorption of gas such as H$_2$ or NH$_3$.

All Hot-Wire TFTs

The root-mean-square *(rms)* roughness measured on 300 nm thick SiN$_{1.2}$ layers is only about 1 nm, which is necessary for high field-effect mobility in thin-film transistors, for more N-rich layers (x >1.4) even *rms* values of 0.5 nm are found. To test whether our fast deposited standard HW SiN$_{1.2}$ films are suitable for application in a field effect transistor (FET), trilayer structures, consisting of HWCVD SiN$_{1.2}$ (made at 3 nm/s), HWCVD a-Si:H and μc-Si:H highly doped n-layers, were deposited at Utrecht University on Corning 1737 glass with pre-patterned Cr gates as provided by Japan Institute of Advanced Science and Technology (JAIST). Analysis shows that these "all hot wire" TFTs have a V$_{th}$ = 1.7 – 2.4 V, an on/off ratio of 10^6, and a mobility of 0.4 cm^2/Vs after a forming gas anneal [29]. No pinholes have been found within the matrix made (86 TFTs). In these first experiments the SiN$_{1.2}$ gate insulation was deliberately made thick (400 nm) to avoid electrical breakdown. Although the mobility is only moderately high, the characteristics [29] of the present TFTs are already suitable for applications.

C-V and I-V characteristics of the MIS capacitor

SiN$_x$ depositions have been incorporated in MIS structures using n-doped c-Si wafers. The C-V characteristics of two MIS structures are shown in Figure 3. The left characteristic (dots) is from a 106 nm thick SiN$_x$ layer with x = 1.2. The right characteristic (open squares) is from a 100 nm thick layer with x = 1.05. For the left characteristic the sweep starts at -40 V, where the MIS structure is in deep depletion. Following the forward sweep (right curve), for increasing voltages the capacitance increases. The small plateau (bump) in the curve is caused by a hole-inversion layer, which is proportional to the nitride thickness and can vanish for high sweep rates. For higher voltages the capacitance reaches a constant level, where the MIS structure reaches the accumulation condition. In the backward sweep (left curve) a hysteresis effect is observed which can be explained by hole trapping due to shallow interface trapped states (trapped charges) [25].

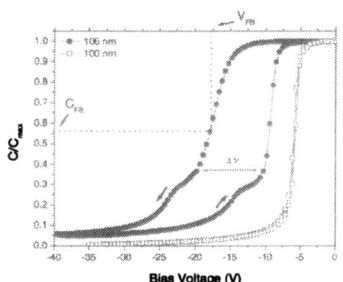

Bias Voltage (V)

Figure 3 C-V characteristics of two MIS structures. The two arrows indicate the sweep direction.

These states are easily filled when the voltage over the MIS structure is swept. The doping profile of the c-Si substrate can be extracted from the C-V characteristic if the MIS structure has gone into the deep depletion mode [30]. From the doping profile the inversion capacitance (bump) and flat band capacitance can be calculated, from the latter the fixed charges can be deduced [21,22]. When calculating charge densities, we make no distinction between fixed and mobile nitride charge and surface state charge. We call the total number of charges fixed nitride charge (N$_{fix}$). N$_{fix}$ is calculated from the flat band voltage (V$_{FB}$) with equation 4. From the hysteresis (ΔV) between the forward (right curve) and backward (left curve) sweep of the C-V characteristic we determine the number of trapped levels (N$_{traps}$) using formula 5).

$$N_{fix} = \frac{-C_{max}V_{FB}}{Aqd} \quad 4) \qquad N_{trap} = \frac{C_{max}\Delta V}{Aq} \quad 5)$$

The results for the charges for the different HWCVD SiN_x films as a function of the layer composition are shown in Figure 4. In the top graph the static dielectric constant (ε) is seen not to change appreciably over the whole range $1.05 < x < 1.43$ and it amounts to an averaged value of 6.3 ± 0.1. The middle graph shows the behavior of N_{traps} and the the bottom graph shows N_{fix}, which we assume to be homogeneously distributed in the SiN_x film. We deduce $N_{fix} = 6.2–7.8 \times 10^{16}$ cm^{-3} from the flat band voltage and from the small hysteresis in the backward sweep $N_{traps} = 1.3$-1.7×10^{10} cm^{-2}, both for N-rich samples with $x \geq 1.33$. It is striking that at $x = 1.2$ the nitride fixed charges and trapped charges are present in relatively high numbers with respect to compositions that are slightly N-richer. This ($x = 1.2$) material has been shown to have the best hydrogen passivation and also to posses the highest mass density. N-richer films have a more open structure, which is likely to allow unsaturated dangling bonds to be passivated more easily. This could explain the low fixed and trapped charges for nitrogen rich ($x > 1.33$) films [25].

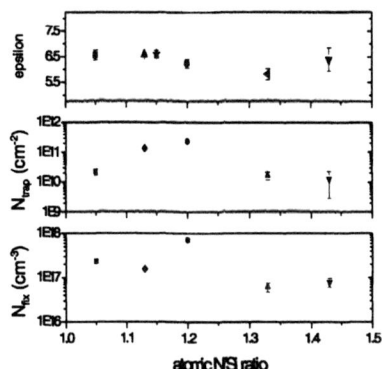

Figure 4 Static dielectric constant, number of trapped charges and total nitride charges as a function of N/Si ratio.

I-V measurements were performed on the various MIS structures. The normalized currents ($J_{break} = 10^{-6}$ A/cm^2) versus the breakdown fields are shown in Figure 5. The MIS structures show no breakdown for SiN_x films with $x \geq 1.33$ and $d > 174$ nm. The breakdown fields obtained hence exceed 5.9 MV/cm. When the SiN_x is too thin, shunting easily occurs. For $SiN_{1.15}$ no breakdown occurred when the nitride thickness was 400 nm. Other groups also report that SiN_x films with higher N/Si have higher breakdown fields but also report higher dielectric constants as the layers become more N-rich [31,32], while we report a roughly constant value for the dielectric constant. Breakdown fields for PECVD or HWCVD SiN_x of 2-6 MV/cm have been reported [31,33,34]. In order to meet the requirements for a good gate dielectric, the a-SiN_x material has to withstand electric fields of ≥ 2 MV/cm [31].

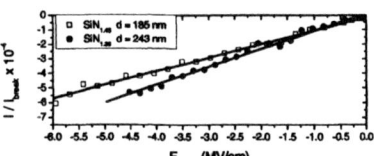

Figure 5 Normalized current versus electric breakdown field for two MIS structures.

CONCLUSIONS

Using the HWCVD method silicon nitride can be deposited possessing very low stress, which is as low as 16 MPa (tensile) for layers with N/Si = 1.1. This makes these nitrides suitable for application in MEMS. Etch rate experiments with a 16BHF solution show a minimum of 7 nm/min for SiN_x films with N/Si = 1.2. At the same N/Si ratio a maximum in mass density is found, which amounts to an ultra high value of 3.0 g/cm^3. Such high densities have been obtained at an industrially relevant high deposition rate of 3 nm/s. A dielectric constant of $\varepsilon = 6.3 \pm 0.1$ and high dielectric breakdown fields exceeding 5.9 MV/cm show the potential for nitrides to be incorporated in electronic circuitry. All Hot-Wire bottom gate staggered TFT's have successfully been made. We therefore conclude that the HWCVD provides SiN$_x$ materials with good dielectric properties.

ACKNOWLEDGEMENTS
We thank S. Nishizaki and Y. Endo from JAIST for the preparation of the TFT samples. We acknowledge SenterNovem for partial financial support.

REFERENCES
[1] W. Soppe, H. Rieffe, and A.W. Weeber. *Prog. Photovolt: res. appl.* 13 (2005) 551

[2] H.F.W. Dekker, L. Carnel, and G. Beaucarne. *Appl.Phys. Lett.* 89 (2006) 013508

[3] B. Hoex. A.J.M. van Erven, R.V.M. Bosch, W.T.M. Stals, M.D. Bijker, P.J. van den Oever, W.M.M. Kessels and M.C.M. van de Sanden *Prog. Photovolt: res. appl.* 13 (2005) 705

[4] A. S. da Silva Sobrinho, G. Czeremuszkin, M. Latre`che, and M. R. Wertheimer. *J. Vac. Sci. & Techn.* 18,1 (2000) 149-157

[5] Y. Lin, C, Chen, J. Shieh, Y. Lee, C. Pan, C. Chen, J. Peng, and C. Chao. *Appl. Phys. Lett.* 88 (2006) 233511.

[6] A. T. Hatzopoulos, N. Arpatzanis, D.H. Tassis and C.A. Dimitriadis, F. Templier, M. Oudwan, and G. Kamarinos. *J. Appl. Phys.* 100 (2006) 114311

[7] S.G. Ansari, H. Umemoto, T. Morimoto, K. Yoneyame, A.Izumi, A. Masuda, H. Matsumura *Thin solid films* 501 (2006) 31

[8] V.Verlaan, Z.S. Houweling, C.H.M. van der Werf, H. D. Goldbach, and R. E. I. Schropp. *MRS Proc.* 910 (2006) A3.3.

[9] R.E.I. Schropp. *Jpn. J. Appl. Phys.* 45 (2006) 4309

[10] V. Verlaan, C.H.M. van der Werf, W.M. Arnoldbik, H.D. Goldbach, R.E.I. Schropp. *Phys. Rev. B* 73 (2006) 195333.

[11] Y. Akasaka, *Ext. Abstr. of the 4th Conf. on Hot-Wire CVD Process*, Takayama, Jap., (2006)

[12] V. Verlaan, C.H.M. van der Werf, Z.S. Houweling, H. F. W. Dekkers, I. G. Romijn, A. W. Weeber, H. D. Goldbach, and R. E. I. Schropp. *Prog. In Photovolt. In Press.* DOI: 10.1002/pip.760

[13] W. A. P. Claassen, W. G. J. M. Valkenburg, W. M. v. d. Wijgert and M. F. C. Willemsen, *Thin Solid Films* 129, 3-4 (1985) 239-247

[14] A. Masuda, M. Totsuka, T.Oku, R. Hattori, and H. Matsumura. *Vacuum* 74 (2004) 525

[15] M. Takano, T. Niki, A. Heya, T. Osono, Y. Yonezawa, T. Minamikawa, S. Muroi, S. Minami, A. Masuda, H. Umemoto, and H. Matsumura. *Jpn. J. Appl. Phys.* 44, 6A (2005) 4098-4102.

[16] R.E.I. Schropp, K.F. Feenstra, E.C. Molenbroek, H. Meiling, and J.K. Rath, *Philos. Mag B* 76, (1997), 309.

[17] S.G. Tolmlin, *J. Phys. D* 5 (1972) 847

[18] Y. Hishikawa, N. Nakamura and Y. Kuwano. *Jpn. J. Appl. Phys.* 30 (1991) 1008

[19] W.M. Arnold Bik and F.H.P.M. Habraken. *Rep. Prog. Phys.* 56 (1993) 859

[20] B.J. Gordon, C-V plotting: Myths and Methods, *Sol. State Techn.* (1993)

[21] P.K. Dubey, V.A. Filikov and J.G. Simmons. *Thin Solid Films* 33 (1976) 49-63

[22] S.M. Sze, *Physics of Semiconductor Devices*, Wiley, London, (1969) 425-504.

[23] R. Glang, R.A. Holmwood, and R. L. Rosenfeld, *Rev. Sci. Instr.* 36, 7 (1965).

[24] R.B. Wehrspohn, S.C. Deane, I.D. French, I. Gale, J. Hewett, M. J. Powell, and J. Robertson, *J. Appl. Phys.* 87, 144 (2000).

[25] B. Stannowski, *Silicon-based thin-film transistors with a high stability*, Ph.D. thesis, Universiteit Utrecht, (2002)

[26] R. Guo, Y. Kurata, T. Inokuma, and S. Hasegawa. *J. of non-Cryst.Sol.* 351 (2005) 3006

[27] G.C. Han, P. Luo, K.B.Li, Z.Y. Liu, and Y.H. Wu. *Appl. Phys. A.* 74 (2002) 243

[28] D. L. Smith, A. S. Alimonda, C.-C. Chen, S. E. Ready and B. Wacker. *J. Electrochem. Soc.* 137 (1990) 614.

[29] R.E.I. Schropp, S. Nishizaki, Z.S. Houweling, V. Verlaan, C.H.M van der Werf , H. Matsumura. *Submitted to solid state electronics.*

[30] R.S. Muller and T.I. Kamins, *Device electronics for Integrated Circuits* 3rd ed. 145 - 148

[31] A. Sazonov, D. Stryahilev, A. Nathan, L. D. Bogomolova, *J. Non-Cryst. Sol.*, 299–302 (2002) 1360–1364

[32] L.J. Quinn, S.J.N. Mitchell, B.M. Armstrong, H.S. Gamble, *J. Non-Cryst. Sol.*, 187 (1995) 347-352

[33] B. Stannowski, J. K. Rath and R. E. I. Schropp, *Thin Solid Films* 395 (2001) 339-342.

[34] F. Liu et al. *J. Appl. Phys.* 96,5 (2004) 2973

Poster Session:
Alloys

Mater. Res. Soc. Symp. Proc. Vol. 989 © 2007 Materials Research Society   0989-A05-02

CRYSTALLINE SILICON SURFACE PASSIVATION BY PECV-DEPOSITED HYDROGENATED AMORPHOUS SILICON OXIDE FILMS [a-SiOx:H]

Thomas Mueller, Wolfgang Duengen, Reinhart Job, Maximilian Scherff, and Wolfgang Fahrner
Chair of Electronic Devices, University of Hagen, Haldener Str. 182, Hagen, 58084, Germany

ABSTRACT

In the research field of crystalline silicon (c-Si) solar cells, electronic surface passivation has been recognized as a crucial step to achieve high conversion efficiencies. The main issue of this article is to analyze the surface passivation properties of both, n-type and p-type crystalline silicon wafers by hydrogenated amorphous silicon sub oxide [a-SiOx:H] films the for use in hetero-junction (a-Si/c-Si) solar cells. A window layer is obtained with a certain fraction of oxygen in the a-SiOx:H layers.

The a-SiOx:H films were deposited by decomposition of silane, carbon dioxide and hydrogen as source gases using plasma enhanced chemical vapor deposition (PECVD). Films with varying deposition parameters such as gas flow ratio (oxygen fraction) and plasma frequency (13.56, 70.0 and 110.0 MHz) are compared.

To determine the passivation quality of the a-SiOx:H films, microwave-detected photo conductance decay (μ-PCD) provides a contactless measurement of the effective recombination lifetime of free carriers. The film compositions and also the changes in the microscopic structure of the amorphous network upon thermal annealing are studied using Raman spectroscopy and optical profiling techniques. The Raman spectra reveal the generation of Si-(OH)x and Si-O-Si bonds after thermal annealing in the layers, leading to a higher effective lifetime, as it reduces the defect absorption of the sub oxides.

For n-type FZ material, lifetime values as high as 1650 μs are obtained, resulting in a surface recombination velocity $S_{eff} < 9.5$ cm/s.

INTRODUCTION

The suppression of surface recombination by surface passivation is one of the basic prerequisites to obtain high efficiency solar cells. Thermally grown silicon oxide has shown excellent surface passivation properties, but it implies a high temperature application (~ 1050 °C). Low-temperature processing sequences are based mainly on passivation with silicon-nitride ($Si_xN_y:H_z$) [1], amorphous silicon films [2], amorphous silicon oxide [3], or amorphous silicon carbide [4] and stacks of those.

Focusing on the junction fabrication techniques of a-Si/c-Si solar cells using a low-temperature PECVD technique, passivation of the contact and surface regions of the cell avoiding high-temperature cycling becomes an important issue. Hydrogenated amorphous silicon sub oxides (a-SiOx:H) represent a material system suitable for this application. In this work, we investigate the applicability of a-SiOx:H films as passivation layers. For comparison, intrinsic amorphous silicon (a-Si(i)) layers are identically processed.

EXPERIMENTAL DETAILS

Amorphous hydrogenated silicon sub oxides are PECV-deposited by decomposition of silane (SiH_4), hydrogen (H_2) and carbon dioxide (CO_2) as oxygen source at various plasma frequencies and gas compositions (0 – 50 at. %). The intrinsic a-SiO_x:H layers are deposited at a heater temperature of 350 °C and a chamber pressure of 200 mTorr. Initially, the silicon wafers were cleaned using the standard RCA procedure followed by a dip in fluoric acid (HF). The gas flow rates for SiH_4 and CO_2 had been kept in the range of 0 – 30 sccm; the hydrogen dilution had been in the range of 0 – 20 sccm.

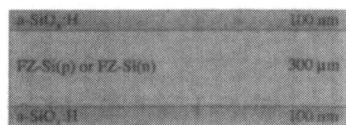

Fig. 1. Sketch of the sample structure for lifetime measurements

The oxygen content and the resulting optical band gap E_{04} of these films are controlled by varying the CO_2 partial pressure $CO_2/(CO_2+SiH_4)$. For determination of E_{04} and the optical constants, the films are deposited onto c-Si wafers with thermal SiO_2 (1000 Å). Corning 7059 glass is used for deposition for micro-Raman spectroscopy. For lifetime measurements, 0.425 – 0.575 Ωcm ($1 \cdot 10^{16} - 4 \cdot 10^{16}$ cm^{-3}) p-type FZ silicon wafers and 0.8 – 1.2 Ωcm n-type FZ silicon wafers of 300 µm thicknesses are used. The typical sample configuration to determine the passivation quality of the a-SiO_x:H is a symmetric a-SiO_x:H/c-Si(FZ)/a-SiO_x:H structure, equally treated on both sides (as seen in Fig. 1). For comparison, intrinsic amorphous silicon layers (a-Si(i)) are deposited at 13.56 MHz with the same parameters, but only 40 sccm SiH_4 for the source gas.

To study the influence of annealing on the surface passivation quality and network bonding structure, the samples are annealed in a diffusion furnace under forming gas atmosphere (10 at. % hydrogen diluted in nitrogen) at temperatures ranging from 250 °C to 500 °C. By varying the SiH_4/CO_2 gas flow ratio, different oxygen concentrations can be achieved [5, 6]. All oxygen concentrations quoted in this paper – unless stated differently – refer to the ratio [O]/([O]+[Si]). The concentration of carbon (C) and other impurities was below 1 at. % for samples investigated in this work.

DISCUSSION

A. Compositional analysis - changes due to annealing

Changes in the amorphous network of the a-SiO_x:H films depending on the plasma deposition frequency as well as compositional changes in the microscopic structure of the amorphous network upon thermal annealing are investigated by means of µ-Raman spectroscopy (µ-RS). The µ-RS setup consists of a microscope confocally coupled to a 300 mm focal length spectrometer. The excitation of the µ-RS was supplied by an Ar$^+$ ion laser using a wavelength of 488 nm and a power of 25 mW. The spot size of the focused laser light was ~1 µm. The spectra were collected at room temperature by a Peltier cooled CCD detector. The resolution was limited to ~0.7 cm^{-1}. The spectra were recorded in the wavenumber range from 200 to 2500 cm^{-1}.

In case of a-Si:H ([O]=0), the hydrogen content is roughly 12-14 at. % as reported by Janotta et al. [5, 6]. However, our silicon sub oxides contain a significant higher fraction of hydrogen. The Raman spectra around the 2000 cm^{-1} band region show typical peaks for Si-H bondings (spectra not shown) for all prepared samples. We thereby assume that no additional hydrogen from annealing in forming gas is built up in the amorphous network. Lattice distortions, such as SiO$_2$ or Si clusters, molecular hydrogen inclusions, or micro-voids with hydrogen terminated internal surfaces are likely to exist in our a-SiO$_x$:H samples. Examples are given in the following.

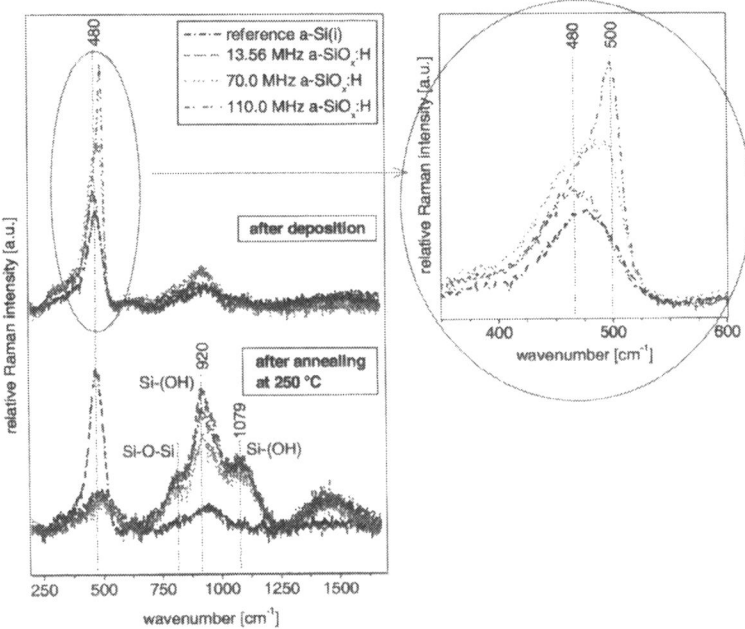

Fig. 2. Raman spectra of a-SiO$_x$:H films after deposition (upper spectra) and after annealing (lower spectra) at 250 °C for 1 h, depending on the plasma deposition frequency. On the right side, the spectra in the range from 350 – 600 cm^{-1} are enlarged for better view. For comparison the spectra of a-Si(i) are inserted.

In Fig. 2 (upper spectra) the Raman spectra of films deposited with various plasma frequencies are presented. For comparison the spectra of the reference a-Si(i) layer are overlapped. As seen from the enlarged part in Fig. 2 (right side), the amorphous character of films deposited at 13.56 MHz and 70 MHz shifts from 480 cm^{-1} towards a more micro-crystalline character at 500 cm^{-1} for films deposited with 110 MHz (cf. Fig. 2 (right)). It is also seen in Fig. 2 (cf. upper spectra and lower spectra) that after post-annealing of the films a generation of Si-(OH)$_x$ bonds appears at peaks around 920 cm^{-1} and 1079 cm^{-1} for all deposited a-SiO$_x$ films, which agrees well with the results shown in [7, 8]. Furthermore, Si-O-Si bonds occur around 820 cm^{-1} for our a-SiO$_x$:H films, whereas the reference sample depicted in Fig. 2 does not show any compositional changes due to thermal annealing in the microscopic structure. However, the generation of those bonds due to thermal annealing does not depend on the applied plasma frequency. Raman spectra for samples annealed at higher temperatures (T = 500 °C for 1 h) exhibit significantly lower Si-(OH) or Si-O-Si peaks for all samples (not shown). Effusion of hydrogen at temperatures above 400 °C might be the reason, as reported by Park et al. [9]. From Raman analysis it can be concluded that the different applied plasma frequencies only influence the amorphous/micro-crystalline character during deposition, but show no effect on the generation of bonds like Si-(OH) or Si-O-Si after thermal annealing.

The optical band gaps E$_{04}$ strongly depend on the oxygen content and are steadily rising with increasing O from 1.9 eV up to 2.4 eV for 50 at. % O. Thermal annealing at 250°C does not noticeably influence the band gap, but annealing at higher temperatures (500 °C) leads to a decrease of E$_{04}$ for the samples deposited at 13.56 and 70 MHz, which can be attributed to an effusion of hydrogen. The optical band gaps of samples deposited with 110 MHz are constant even after annealing at higher temperatures (500 °C), due to the micro-crystalline structure.

B. Surface passivation by a-SiO$_x$:H films

Investigations on PECV-deposited a-SiO$_x$:H films have been made measuring the carrier lifetime by the laser-induced microwave-detected photo conductance decay (μ-PCD) method. The basis of this measurement method is charge carrier injection in semiconductor by irradiating a spot like area with a laser pulse and determination of charge carrier concentration versus time. The wavelength of irradiation is less than that of the absorption edge. The charge carrier concentration is estimated by microwave reflection. The relaxation constant is derived from non-equilibrium charge carrier bulk-lifetime. The simplified formula for a homogeneous, defect-free wafer with a spatially uniform carrier lifetime is given by

$$\frac{1}{\tau_{eff}} = \frac{1}{\tau_{bulk}} + \frac{1}{\tau_{surf} + \tau_{diff}}$$

Equation (1)

where τ_{eff} is the effective lifetime (measured), τ_{bulk} is the bulk lifetime, τ_{surf} is the characteristic surface recombination lifetime component determined by wafer thickness, W, and surface recombination velocity, S$_{eff}$, (given by $\tau_{surf} = W/2S_{eff}$) and τ_{diff} is the characteristic time of carrier diffusion from the wafer layer center to its surface (given by $\tau_{diff} = W^2/\pi^2 D_{n,p}$, where D$_{n,p}$ is the ambipolar diffusity, with D$_n$ = 28 cm^2/s for p-type wafer and D$_p$ = 11 cm^2/s for n-type wafer as assumed for our samples) [10]. From Equation (1) it can be concluded, that a better passivation leads to a higher τ_{eff}. For calculation of S$_{eff}$ we assume that both surfaces provide a sufficiently low recombination velocity and have the same values (S$_{eff}$ = S$_{front}$ = S$_{back}$), as the sample structure is symmetric. The main error in determining S$_{eff}$ arises from the uncertainty in τ_{bulk}. Schmidt and Aberle [11] reported the lowest S$_{eff}$ with 5.5 cm/s for a 1000 Ωcm n-type wafer, passivated with a PECVD SiN$_x$ film. However, the uncertainty of S$_{eff}$ could have been as high as 10 cm/s [12] depending on the value used for τ_{bulk}. Therefore, we calculate S$_{eff}$ for the case that no Shockley-Read-Hall recombination is considered ($\tau_{bulk} \rightarrow \infty$).

The lifetime had been obtained by mapping the lifetime values on a circular area of 1 cm diameter in 1 mm steps and calculating the average value. In Fig. 4, the effective lifetimes of samples deposited on p-type FZ wafers with plasma frequencies of 13.56 MHz, 70 MHz and 110

MHz are shown - a-SiO$_x$:H films of two different compositions were applied: (a) 30 % O and (b) 50 % O.

As seen from Fig. 3, we achieve the highest lifetimes with a plasma frequency of 70.0 MHz. We found that subsequent annealing of the samples at 250 °C for 1 h in forming gas atmosphere drastically increases the effective lifetime. This might be (i) due to hydrogen saturation of Si dangling bonds at the a-SiO$_x$:H surface, or (ii) due to the generation of Si-(OH) and Si-O-Si bonds in the microscopic structure of the amorphous network (cf. Fig. 3).

Thermal annealing at higher temperature (60 min at 500 °C) leads to a drastic decrease of the effective lifetime of the samples deposited with 13.56 MHz and 70 MHz, as well as of the reference a-Si(i), due to an effusion of hydrogen [9]. Surprisingly, samples with 110 MHz deposited a-SiO$_x$:H show a further increase of the effective lifetime. This effect might be due to the micro-crystalline character preventing hydrogen effusion during annealing, as seen in the optical band gap analysis above, or due to further diffusion of hydrogen to the c-Si interface.

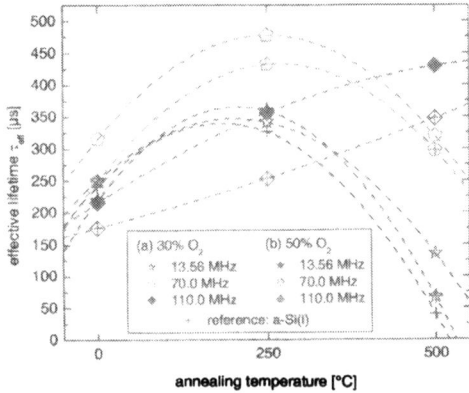

Fig. 3. Effective lifetime against annealing temperature depending on the plasma frequency for layer deposited at 13.56, 70 and 110 MHz. Two oxygen ratios are applied (a) 30% O and (b) 50% O. For comparison, the effective lifetime of intrinsic amorphous silicon without O addition is added to the graph. The lines are only a guide to the eye.

Next, the effective lifetime of passivated n-type FZ-Si is presented in Fig. 4. For this sample series a plasma frequency of 70.0 MHz is applied (highest lifetime values of samples shown in Fig. 4) and the oxygen fraction is varied from 0 at. % (a-Si(i) reference) to 50 at. % O.

Here, the samples after post-annealing are consistent to the results achieved on p-type FZ material. A high effective lifetime (τ_{eff} = 1650 µs, leading to S_{eff} < 9.5 cm/s) has been achieved for samples deposited with an oxygen content of 20 at. % after subsequent annealing at 250 °C. With higher oxygen content in the gas phase, the lifetime decreases. One possible explanation could be, that even though we assumed the content of C below 1 %, C increases with increasing oxygen content from the carbon dioxide, resulting in a higher defect density. As stated in [13], the quality of passivation on silicon material of different conduction type differs. In our case, we attribute the different lifetime values achieved on p-type FZ material and n-type FZ material mainly to the defect states in our non-stoichiometric a-SiO$_x$:H films. We assume that some of these defect states are within the band gap of the c-Si at the interface, and therefore directly induce the recombination at the interface to the c-Si material of different conduction type.

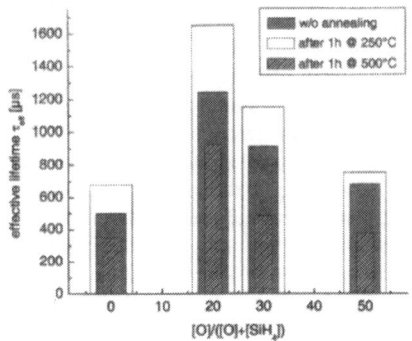

Fig. 4. Effective lifetime as a function of oxygen content. For each oxygen content, the effect of post-annealing is shown for temperatures of 250 °C and 500 °C.

CONCLUSIONS

We investigated a-SiO$_x$:H films deposited in PECVD with various plasma frequencies and gas compositions. After thermal annealing of those films at 250 °C, we achieved lifetime values on p-type and n-type FZ material as high as 500 µs and 1650 µs (S$_{eff}$ < 9.5 cm/s, note $\tau_{bulk} \to \infty$), respectively. Thermal annealing at higher temperature (above 400 °C) leads to a decreasing lifetime and to an effusion of hydrogen for samples deposited with 13.56 and 70 MHz, reducing the optical band gap, E$_G$, of the samples. Films deposited with 110 MHz show a more micro-crystalline network. We observed optical band gaps from 1.9 eV up to 2.4 eV depending on the oxygen content in the film.

REFERENCES

1. J. Schmidt, M. J. Kerr, Sol. Energy Mater. Sol. Cells 65, 585 (2001).
2. M. Taguchi et al., Prog. Photovolt. 8, 503 (2000).
3. Z. Chen, A. Rohatgi, D. Ruby, 1st WCPVEC, Hawaii, 49-56 (2001).
4. I. Martin et al., Appl. Phys. Lett. 79, 2199 (2001).
5. R. Janssen et al., Physical Review B 60, 19 (1999).
6. A. Janotta et al., Phy. Rev. B 69, 115206 (2004).
7. B. Hoex et al., J. Vac. Sci. Technol. A 24(5), 1823 (2006).
8. J. D. Kubicki, D. Sykes, American Mineralogist 78, 253-259 (1993).
9. M. Park et al., J. Appl. Phy. 89, 1130-1137 (2001).
10. P. A. Basore, B. R. Hansen, Conference Record of 21st IEEE PVSC, USA, 374 (1990).
11. J. Schmidt, A. Aberle, J. Appl. Phys. 85, 3626 (1999).
12. M. J. Kerr, A. Cuevas, Semicond. Sci. Technol. 17, 166-172 (2002).
13. S. Dauwe, J. Schmidt, R. Henzel, Proc. 29th IEEE PVSC, 1246-1249 (2002).

Mater. Res. Soc. Symp. Proc. Vol. 989 © 2007 Materials Research Society   0989-A05-04

Effect of Boron Doping on Microcrystalline Germanium Carbon Thin Films

Yasutoshi YASHIKI[1], Seiichi KOUKETSU[1], Shinsuke MIYAJIMA[1], Akira YAMADA[2], and Makoto KONAGAI[1]

[1]Department of Physical Electronics, Tokyo Institute of Technology, Tokyo, 152-8552, Japan
[2]Quantum Nanoelectronics Research Center, Tokyo Institute of Technology, Tokyo, 152-8552, Japan

ABSTRACT

Effects of boron doping on microcrystalline germanium carbon alloy (μc-Ge$_{1-x}$C$_x$:H) thin films have been investigated. We deposited boron-doped p-type μc-Ge$_{1-x}$C$_x$:H thin films by hot-wire chemical vapor deposition technique using hydrogen diluted monomethylgermane (MMG) and diborane (B$_2$H$_6$). A dark conductivity of 1.3 S/cm and carrier concentration of 1.7×10^{20} cm^{-3} were achieved with B$_2$H$_6$/MMG ratio of 0.1. Furthermore, the activation energy decreased from 0.37 to 0.037 eV with increasing B$_2$H$_6$/MMG ratio from 0 to 0.1. We also fabricated p-type μc-Ge$_{1-x}$C$_x$:H/n-type c-Si heterojunction diodes. The diodes showed rectifying characteristics. The typical ideality factor and rectifying ratio were 1.4 and 3.7×10^3 at $\pm$ 0.5 V, respectively.

INTRODUCTION

Thin film silicon-based solar cells are attracted much attention as a low cost solar cells compared to single- or multi-crystalline Si solar cells. Amorphous and microcrystalline Si are mainly used for absorber layer of thin film solar cells. In the case of silicon-based multi junction solar cells, such as tandem and triple junction solar cells, microcrystalline Si (μc-Si) is used as a bottom cell material. A thick bottom cell whose thickness is lager than 2μm is required for absorbing long-wavelength light effectively because of the low absorption coefficient of μc-Si. Therefore, the development of new material with high absorption coefficient is required.

Microcrystalline germanium carbon alloy (μc-Ge$_{1-x}$C$_x$:H) is one of the candidates of a new bottom cell material for the solar cells. μc-Ge$_{1-x}$C$_x$:H has high absorption coefficients like a Ge and its band-gap can be controlled by changing carbon concentration [1-3]. In our previous experiments, we reported that the deposition and characterization of μc-Ge$_{1-x}$C$_x$:H by hot wire chemical vapor deposition (HWCVD) technique and very high frequency plasma chemical vapor deposition technique [3-4].

Doping technique is very important for device application of μc-Ge$_{1-x}$C$_x$:H. The carbon contents of our μc-Ge$_{1-x}$C$_x$:H films deposited by HWCVD are less than 8 % [3-4]. Therefore, it is expected that doping characteristics of μc-Ge$_{1-x}$C$_x$:H is similar to pure Ge material. Chambouleyron et al. reported that the doping properties of group III impurities in a-Ge:H films and they suggested that boron is very effective for doping into a-Ge:H films compared to other group III elements[5]. Herrold et al. reported that boron and phosphorous work as p-type and n-type dopant in microcrystalline germanium carbon thin films, respectively [6]. However, effects of boron doping on structural and electrical properties of μc-Ge$_{1-x}$C$_x$:H have not been clarified yet. In this work, we report effects of boron doping on electrical and structural properties of μc-Ge$_{1-x}$C$_x$ films deposited by HWCVD.

EXPERIMENTAL DETAILS

Films were prepared by HWCVD technique using MMG (99.9999%, Tri Chemical Lab. Inc.) with H_2 dilution as a source gas. 1% H_2 diluted B_2H_6 was used as the p-type doping gas. Four rhenium (Re) wires with a diameter of 0.5 mm and a length of 6 cm were employed as hot filaments. The distance between the filaments and the substrate was 5 cm. The substrate temperature and the filament current were 300 °C and 13A, respectively. MMG flow rate was kept constant at 2 sccm. Total flow rate of H_2 and B_2H_6 was 40 sccm and B_2H_6/MMG flow ratio was varied. The deposition pressure ranged from 9.4 to 9.8 Pa. Films were deposited on Corning 7059 glass and low resistance n-type Si (001) substrate. Film structure and carrier concentration were measured by Raman scattering spectroscopy and Hall measurement. For the Hall measurement ECOPIA Hall Effect measurement system (HMS-3000) was used. For the measurement of conductivities and their activation energies, Al electrodes were evaporated on the top surface of the films in a coplanar configuration. The gap distance and the width of the electrode were 0.2 mm and 3 mm, respectively.

RESULTS AND DISCUSSION

Raman spectra of the films deposited with different B_2H_6/MMG ratio are shown in Fig. 1. Each sample has a sharp peak around 295 cm^{-1}. This peak is attributed to the Ge TO mode peak [3]. Increasing B_2H_6/MMG ratio from 0 to 0.15, the intensity of the Ge TO mode peak become small. This result revealed that all samples have a microcrystalline structure and boron doping

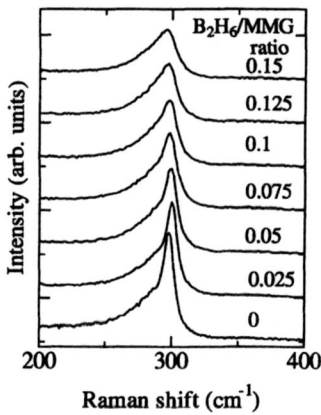

Fig. 1. Raman spectra of the films deposited with different B_2H_6/MMG ratio.

suppress the growth of microcrystalline phase.

Figure 2 (a) shows the dependence of the film conductivity and activation energy on B_2H_6/MMG ratio. Conductivity of the undoped μc-$Ge_{1-x}C_x$ films is 7.6 x 10^{-3} S/cm. Introducing B_2H_6, conductivities of the films increase rapidly. Then the conductivity of the film becomes 1.1 S/cm at B_2H_6/MMG ratio 0.05. This dark conductivity is enough for applying to doping layer of solar cells. We measured activation energy of the films from temperature dependence of the conductivity. To prevent the oxidation of the films, the measurements were carried out at a temperature from 25 to 160 °C. Annealing over 200°C in atmosphere, the Ge-O bond in the films increased in FTIR spectra. B_2H_6 flow rate dependence of the activation energy is also shown in Fig. 2 (a). The activation energy of the films deposited without B_2H_6 is 0.37 eV. However, the activation energy of the films decreased rapidly with introducing B_2H_6. With B_2H_6/MMG ratio 0.005, the activation energy of the films was 0.13 eV. Then, the activation energy has decreased down to 0.037 eV with B_2H_6/MMG ratio 0.05. This steep change in the activation energy was very similar to the conductivity change of the films.

Carrier concentration and mobility of the films measured by Hall measurement is shown in Fig. 2 (b). The carrier concentration of the films deposited with B_2H_6/MMG ratio 0.005 is 5.1 x 10^{17} cm^{-3}. Carrier concentration increases in proportion to B_2H_6/MMG ratio and reaches a 1.9 x 10^{20} cm^{-3} with B_2H_6/MMG ratio 0.15. The decreasing trend of mobility with increasing B_2H_6 flow rate is observed in Fig. 2 (b). The mobility of the films is decreased from 8.5 x 10^{-1} to 6.1 x 10^{-2} cm^2/Vs with B_2H_6/MMG ratio from 0.01 to 0.15. This is probably due to the impurity scattering and the decrease of crystalline volume fraction of the films. From Hall measurement, we found that all boron doped films were p-type conductivity. Conduction type of the undoped films could not be obtained by our Hall measurement system due to the detection limit of the system. Therefore, we checked conduction type of the undoped films by hot-probe method. This method is based on the Seebeck electro motive force. According to the hot-probe measurements, conduction type of the undoped films was found to be p-type.

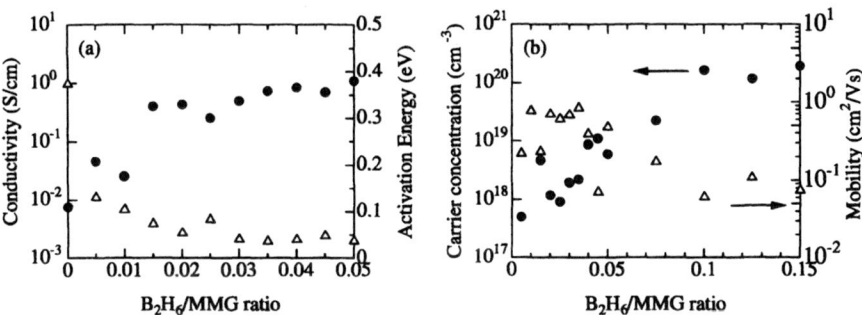

Fig. 2. (a) Dark conductivity and its activation energy of the films and (b) carrier concentration and mobility of the films deposited with different B_2H_6/MMG ratio.

We fabricated p-type μc-Ge$_{1-x}$C$_x$:H/n-type c-Si heterojunction diodes to investigate properties of the heterojunction. The structure of the device was as follows: Al / p-type μc-Ge$_{1-x}$C$_x$:H / n-type c-Si (100) / Al. The p-type μc-Ge$_{1-x}$C$_x$:H was deposited with B$_2$H$_6$/MMG ratio 0.1. The resistivity of the c-Si was 1 to 10 Ω/cm. Thickness of μc-Ge$_{1-x}$C$_x$:H and c-Si was 880 nm and 380 μm, respectively. The results of the current-voltage and capacitance-voltage measurements are shown in Fig. 3 (a) and (b). The diodes showed rectifying characteristics. The typical ideality diode factor and rectifying ratio were found to be 1.4 and 3.7 x 10^3 at ± 0.5 V, respectively.

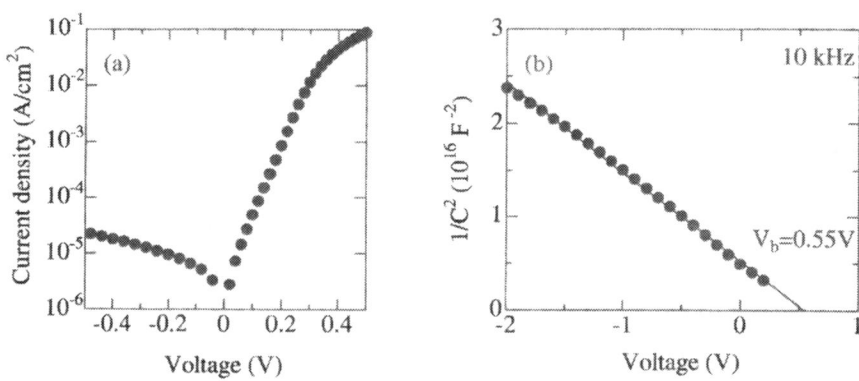

Fig. 3. (a) Current density-voltage and (b) 1/C^2-voltage curve of the p-type μc-Ge$_{1-x}$C$_x$:H/n-type c-Si heterojunction diode.

As mentioned before, carrier concentration of the p-type μc-Ge$_{1-x}$C$_x$:H layer was over than 10^{20} cm^{-3}. Therefore, depletion layer of p-n junction must be formed in only c-Si layer side, and the junction can be considered as one-sided abrupt junction. Capacitance of the p-n junction is expressed by the following fomula;

$$C = \left\{ \frac{e\varepsilon^{*}\varepsilon_0 N_D}{2(V_D - V)} \right\}^{\frac{1}{2}} \quad (1)$$

Where ε^{*} is relative permittivity, ε_0 is electrical constant, N_D is donor concentration, and V_D is a built-in potential. Assigned relative permittivity of Si (12.0) to ε^{*}, N_D of 1.2 x 10^{15} cm^{-3} is obtained. This donor concentration is in good agreement with that of the c-Si substrate (1 to 10 Ω/cm). We also obtained built in potential from the intercept (at 1/C^2 = 0). Built-in potential of our sample was found to be 0.55 V. As a result, we obtained flat band energy diagram of the hetero junction. The flat band energy diagram is shown in Fig. 4.

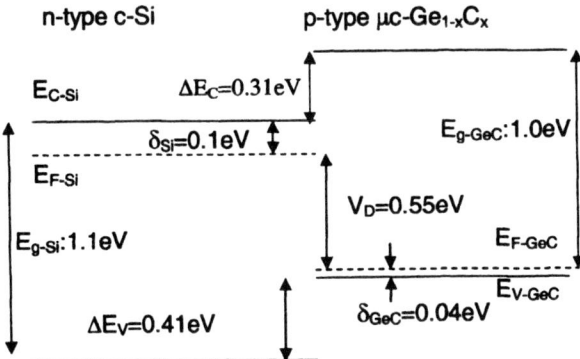

Fig. 4. Flat band energy diagram.

According to the carrier concentration, Fermi level position of n-type c-Si measured from the bottom of the conduction band ($\delta_{Si}=E_{C\text{-}Si}-E_{F\text{-}Si}$) is calculated to be 0.1 eV. The energy difference between Fermi level of c-Si and p-type μc-Ge$_{1\text{-}x}$C$_x$:H (V$_D$) was 0.55 eV. Assuming that activation energy of the dark conductivity is same as the Fermi level, Fermi level position of p-type μc-Ge$_{1\text{-}x}$C$_x$:H measured from the valence band is 0.04 eV ($\delta_{GeC}=E_{F\text{-}GeC}-E_{V\text{-}GeC}$). The energy between conduction band of n-type c-Si and valence band of p-type μc-Ge$_{1\text{-}x}$C$_x$:H is given by $\delta_{Si}+V_D+\delta_{GeC}$. The valence band offset (ΔE_V) at the p-type μc-Ge$_{1\text{-}x}$C$_x$:H/n-type c-Si heterojunction interface is expressed by the following formula;

$$\Delta E_V = E_{g-Si} - \left(\delta_{Si} + V_D + \delta_{GeC}\right) \qquad (2)$$

Using formula (2) and the values of δ_{Si} and V$_D$ and δ_{GeC}, ΔE_V is found to be 0.41 eV. The band gap of μc-Ge$_{1\text{-}x}$C$_x$:H (E$_{g\text{-}GeC}$) is about 1.0 eV[1-4]. Therefore, the conduction band offset of the heterojunction interface is expressed by the following formula;

$$\Delta E_C = E_{g-GeC} - \left(\delta_{Si} + V_D + \delta_{GeC}\right) \qquad (3)$$

From this formula, ΔE_C is found to be 0.31 eV.

CONCLUSIONS

We deposited boron doped p-type μc-Ge$_{1\text{-}x}$C$_x$ thin films by HWCVD using MMG and B$_2$H$_6$. The conductivity of 1.1 S/cm and carrier concentration of 1.9 x 10^{20} cm^{-3} were achieved. We found that the valence band offset of the p-type μc-Ge$_{1\text{-}x}$C$_x$:H/n-type c-Si heterojunction interface was 0.41 eV and boron doped μc-Ge$_{1\text{-}x}$C$_x$ is one of the promising candidates for doped layer of solar cell with μc-Ge$_{1\text{-}x}$C$_x$:H absorbing layer.

ACKNOWLEDGMENTS

This work was supported by the New Energy and Industrial Technology Development Organization (NEDO) under Ministry of Economy, Trade and Industry (METI).

REFERENCES

1. J. T. Herrold and V. L. Dalal, "Growth and properties of microcrystalline germanium-carbide alloys grown using electron cyclotron resonance plasma processing", Journal of Non-Crystalline Solids **270** (2000) 255.
2. J. Kolodzey, P. A. O'Neil, S. Zhang, B. A. Orner, K. Roe, K. M. Unruh, C. P. Swann, M. M. Waite, and S. Ismat Shah, "Growth of germanium-carbon alloys on silicon substrates by molecular beam epitaxy", Appl. Phys. Lett. **67** (1995) 1865.
3. Yasutoshi Yashiki, Shinsuke Miyajima, Akira Yamada, and Makoto Konagai, "Deposition and characterization of μc-Ge$_{1-x}$C$_x$ thin films grown by hot-wire chemical deposition using organo-germane", Thin Solid Films **501** (2006) 202.
4. Yasutoshi Yashiki, Seiichi Kouketsu, Shinsuke Miyajima, Akira Yamada, and Makoto Konagai "Growth and Characterization of Germanium Carbon Thin Films Deposited by VHF Plasma CVD Technique", Proc. 2006 IEEE 4th World Conference on Photovoltaic Energy Conversion (2006) 1608.
5. I. Chambouleyron and D. Comedi, "Column III and V elements as substitutional dopants in hydrogenated amorphous germanium", J. of Non-Cryst. Solids **227–230** (1998) 411.
6. Vikram L. Dalal and Jason T. Herrold, "Microcrystalline Germanium Carbide-A new material for PV conversion", Presented at 2001 NCPV Program Review Meeting (2001) 348.

Mater. Res. Soc. Symp. Proc. Vol. 989 © 2007 Materials Research Society   0989-A05-05

Characterization of Silicon Carbide Thin Films Obtained Via Sublimation of a Solid Polymer Source using Polymer-Source CVD Process

El Hassane Oulachgar[1], Cetin Aktik[1], Starr Dostie[2], Subhash Gujrathi[3], and Mihai Scarlete[2]

[1]Nanofabrication and Nanocharacterization Research Center (CRN2), Department of Electrical and Computer Engineering, Sherbrooke University, Sherbrooke, J1K 2R1, Canada

[2]Department of Chemistry, Bishop's University, Sherbrooke, J1M 0C8, Canada

[3]Thin Film Physics and Technology Research Center (GCM), Department of Physics, Montreal University, Montreal, H3C 3J7, Canada

ABSTRACT

Silicon carbide thin films have been deposited via sublimation of a solid organosilane polymer source using atmospheric pressure chemical vapour deposition process (PS-CVD). The advantages of this new process include high deposition rate, compatibility with batch process, hazard-free working environment and low deposition cost. The silicon carbide (SiC) thin films obtained through this process exhibit a highly uniform film thickness, highly conformal coating, and very high chemical resistance to acids and alkaline solutions. These characteristics make the SiC thin films obtained by PSCVD process very attractive as a structural material for micro-electro-mechanical systems (MEMS) and as a coating film in a wide range of other applications. These SiC thin films are also expected to be attractive as a semiconductor material provided that the defects, oxygen and nitrogen contaminations can be reduced and effectively controlled. In this work we have investigated the chemical, structural, electrical and optical properties of these films, using scanning electron microscope (SEM), Fourier transform infrared spectroscopy (FT-IR), elastic recoil detection spectroscopy (ERD), ultraviolet-visible photospectroscopy (UV-Vis), ellipsometry, and capacitance-voltage measurements (C-V).

INTRODUCTION

Thin films of silicon carbide have been intensively investigated as an alternative to crystalline silicon carbide mainly because of their low preparation cost, simple processing and easy integration with other thin film materials. Nowadays they are commonly used in a wide range of semiconductor devices including window layer for solar cells [1], rectifying diode [2], photo-detector [3], MEMS pressure sensor [4], surface coating [5], membrane for X-ray lithography [6] and a variety of other applications.

Chemical vapour deposition (CVD) process has been widely used in semiconductor industry for several decades and it is commonly used to prepare both crystalline and amorphous thin films of silicon carbide [7-12]. However, high quality single crystal silicon carbide are usually obtained with this process at very high temperature. In recent years, thin film of 3C-SiC epitaxially grown on silicon at a relatively low temperature (800 °C -1300 °C) are being widely investigated as a promising alternative to bulk crystalline silicon carbide [8, 12]. However these films can only be grown on silicon wafer and a carbonisation process is always required, to minimize film stress and to achieve epitaxial growth. Because the CVD process requires highly volatile precursors to achieve uniform and continuous vapour transport, the majority of CVD processes are designed for the utilization of gaseous or liquid sources with high vapour pressure at specific temperature. Most of these gases precursors are highly toxic, oxygen and moisture sensitive, and highly flammable, so that special handling precautions are always required.

The pyrolysis process on the other hand has received a great deal of attention as a promising low cost alternative method to deposit both bulk and thin films of silicon carbide. Amorphous, nanocrystalline and polycrystalline silicon carbide films have been deposited as the ceramic residue obtained via spin-coating or dip-coating of various liquid and solid polymeric precursors, such as polycarbosilane or organo-polysilane species [13-16]. In contrast with the CVD process, the low volatility is a crucial property for the polymeric precursors in order to achieve a high ceramic yield. These silicon carbide ceramic films are sought due to their valuable intrinsic properties such as high tensile strength, high thermal conductivity and radiation resistance, and rarely for their electrical properties. Currently, many efforts are dedicated to the synthesis of high purity polymeric precursor, and to develop new methods for the deposition of silicon carbide thin films better adapted to the chemical and structural purity required by the semiconductor industry.

A CVD process using a polymeric source appeared to be an attractive method to deposit silicon carbide thin films at a relatively lower temperature and at a lower cost [17]. This idea has been rapidly accepted by various academic research groups and industrial players, covering the whole chart between fundamental research and new emerging commercial products. Currently, a handful of important players are in competition either for the design of various polymeric sources adapted to different applications, or for the optimization of CVD-parameters adapted to the utilization of polymeric sources. In this paper we present and discuss the chemical, electrical and optical characteristics of SiC thin films obtained by this method using polydimethylsilane (PDMS) solid polymer source.

EXPERIMENT

Silicon carbide thin films were obtained through the sublimation of polydimethylsilane (PDMS) polymer powder using Lindberg horizontal three-zone furnace. The furnace is equipped with a 2.5 inch quartz tube attached to an argon tank, an oil bubbler, and a Fisher Maxima vacuum pump capable of achieving a vacuum down to 10^{-3} torr. The films were deposited on both polished N-type crystalline silicon wafer and on electronic grade quartz substrates. Prior to deposition of SiC thin films, silicon and quartz substrates were properly cleaned in an ultrasonic bath for several minutes using acetone, isopropyl alcohol (IPA) and deionized-water, and were dried using boiling isopropyl alcohol. The samples were introduced in the furnace together with a boat containing the PDMS polymer powder. In order to eliminate residual moisture and oxygen contamination, the reactor is pumped down for extended period of time and purged three times with the flow of ultra high purity argon before starting the polymer sublimation process. Argon was used as the carrier gas, with a constant flow during the deposition process. The deposition pressure is kept near atmospheric pressure while the deposition temperature is gradually increased up to 800 °C at a rate of 16 °C per minute. The samples were maintained at the maximum temperature for approximately 30 minutes, immediately after that the reactor is shut down to allow natural cooling of the samples, to avoid film cracking.

Fourier transform infrared spectroscopy (FT-IR) has been used to examine the deposited film on single crystal silicon wafers. The absorption spectra were acquired in the MID-IR range (4000 cm^{-1} to 400 cm^{-1}), using a Mattson Galaxy Series FTIR-6020 spectrophotometer. The atomic composition of the films was obtained by elastic recoil detection spectroscopy, using time-of-flight analysis (ERD-TOF). The films morphology was analyzed using high resolution scanning electron microscope (SEM). The dielectric constant was extracted from capacitance-voltage measurement obtained by Schlumberger impedance analyzer S1260 equipped with a mercury probe. The optical band gap was extracted from the ultraviolet-visible absorption spectra measured using Hewlet-Packard HP8452 UV/Vis spectrophotometer. The films thickness and the refractive index were determined by J. A. Woollam Alpha-SE ellipsometer.

RESULTS AND DISCUSSION

The thin films obtained by polymer sources CVD process using PDMS powder are found to exhibit a very uniform film thickness over a relatively large area. Figure 1-a show the cross section image of the film deposited on silicon substrate obtained by scanning electron microscope (SEM). Since the process is not conducted under continuous flow of polymer it was not possible to determine a deposition rate. Because the deposition process is conducted at near atmospheric pressure all the film obtained by this process are conformal; this is confirmed by the SEM cross section image of figure 1-c, which shows the deposition of the film on the edge of the wafer. The top view SEM image of the film obtained at high resolution shows very uniform film morphology with very small grain size, which suggests that these films are amorphous. The atomic concentration of oxygen in the samples was found to vary from sample to sample (from 2 at.% to 10 at.%) due to reactor leakage and residual contamination [18]. In this study we only focus on the samples with low oxygen concentration (around 2 at.%).

The chemical resistance to acid and alkaline solutions was investigated by dipping a silicon substrate coated with 200 nm thick films in both HNA ($C_2O_2H_4$:HNO_3:HF; 8:3:1) and molten KOH (30 %, 95 °C) solutions for 24 hours. During this experiment, the back sides of the silicon substrates were coated with polyimide and copper foil respectively. After this period of exposure, the films thickness were measured again by ellipsometry and did not show any noticeable change, which indicates that the etch rate in these solutions is extremely slow. These films are therefore highly resistant to both acids and alkaline solutions. The above characteristics make these films very attractive as a coating material and possibly as a thin film membrane for MEMS devices.

Figure 1: High-resolution SEM of SiC film on silicon; (a, c) cross section view, (b) top view

Immediately after the films deposition the chemical structure was analyzed using FT-IR spectroscopy. Figure 2 show the typical absorbance spectrum of the thin films deposited on silicon substrate under the experimental conditions described above. A preliminary look at the FT-IR spectrum shows a main peak around 780 cm^{-1} commonly associated with silicon carbide formation [16, 17, 19, 20]. The broad shoulder around 1250 cm^{-1} is generally associated with the bending mode of residual Si-CH$_3$ groups in silicon carbide thin film [16, 19, 20].

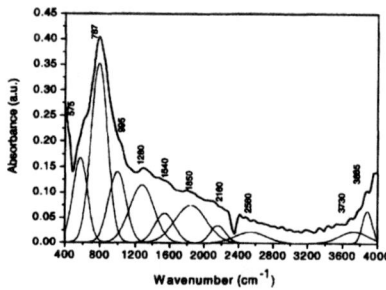

Figure 2: FTIR spectra of SiC thin film and the corresponding deconvolution peaks

In order to better understand the structural composition of these films, we further analyzed their spectral decomposition using Peakfit software. The deconvolution of the FT-IR spectra was obtained assuming a Gaussian distribution. The deconvolution reveals several hidden peaks as shown in figure 2. The association of each peak was determined by comparing the peak wavenumbers with those reported in the literature. The results are summarized in table I.

Table I: Absorption peaks assignments of SiC thin film obtained by PS-CVD

Wavenumber (/cm)	Peak Assignment and Vibration Mode	References
575	Si-H rocking and Si-Si stretching	21
787	Si-C stretching	16, 19-21
995	Si-(CH$_2$)$_n$-Si, Si-C-C bending, Si-O-Si	20, 21
1280	Si-CH$_3$ symmetric Bending of CH$_3$	16, 19, 20
1540	Si-CH$_3$ asymmetric bending of CH$_3$	16, 19, 20
1850	Unassigned	-
2160	Si-H stretching	14-16, 19-22
2560	C-H stretching	16, 19, 20
3730	O-H stretching in H-OH	16, 19
3885	O-H stretching in Si-OH	16, 19

The relatively high absorbance intensity of Si-C stretching mode observed at 787 cm^{-1} and the broad absorption band around 1280 cm^{-1} indicate that these films are very similar to amorphous silicon carbide thin films obtained by other deposition methods [20-22]. The presence of oxygen in this film was established with the weak absorption bands observed around 3770 cm^{-1} and 3885 cm^{-1} due to stretching mode of O-H bonds [16, 19]. Since the PDMS polymer source in its purified form contains only silicon, carbon and hydrogen, the presence of oxygen in these films is likely due to the reactor or polymer contamination and film oxidation.

We have further analyzed the chemical composition of this film using elastic recoil detection spectroscopy. The concentration of chemical elements was determined from time-of-flight analysis. Figure 3 show the percentage of the atomic concentration of chemical elements in the film as a function of the atomic depth. The ERD analysis shows that this film contain carbon, silicon, hydrogen, oxygen and nitrogen with very uniform depth profile except near the surface of the film. The very small and uniform atomic concentration of both oxygen and nitrogen in the bulk of these films suggest that these elements are due to polymer or reactor contamination.

However, the increase of the oxygen near the surface of the film is obviously caused by the oxidation of the film probably during the cool-down process. These results appear to be in good agreement with the FT-IR analysis. Based on the ERD results we can conclude that the thin films synthesized in this series of experiments were carbon rich, non-stoichiometric SiC thin films.

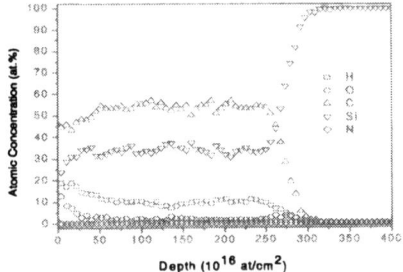

Figure 3: TOF-ERD analysis of the atomic depth profile of SiC thin film

The electrical properties of these films were investigated using capacitance-voltage measurements. Figure 4 shows the C-V characteristics of SiC/Si structure obtained by using a mercury probe as the metal contact. The silicon substrate is of an N-type with a doping concentration of $3x10^{15}$ cm^{-3}. The SiC film thickness was approximately 405 nm. The C-V curve exhibit a characteristics of a metal-insulator-semiconductor (MIS) structure, and the accumulation, depletion and inversion regions can be clearly distinguished. These results indicate that the SiC thin films are highly resistive. The film resistivity measured on quartz substrate was approximately $6x10^5$ Ω.cm (at about 2 at.% of oxygen). The shift of the C-V characteristic toward the negative voltage is the result of the high density of positive fixed charge in this film most likely due to high concentration of dangling bonds in the bulk and near the interface of silicon carbide and silicon. The slow variation of the capacitance in the depletion region also indicates a high density of traps at the interface between silicon and silicon carbide thin film. The relative dielectric constant of SiC thin film extracted from the C-V characteristics at 1 MHz was found to be approximately 7.76 (at about 2 at. % of oxygen).

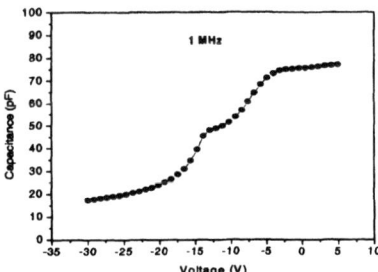

Figure 4: High frequency C-V characteristics of SiC/Si structure

The optical characteristics of SiC thin film deposited on quartz were characterized by ellipsometry and the UV-Vis photospectroscopy in order to determine the refractive index and the optical band gap respectively. Figure 5 shows the plot of the square root of the absorption coefficient as a function of photon energy, calculated from the UV-Vis absorption data. The square root of the absorption coefficient shows a nearly linear variation at high photon energy. According to Tauc's law the optical band gap can be extracted from this graph by extrapolating the linear region of the curves as indicated by the dotted line. The value of the optical bandgap obtained for SiC carbide thin film with about 2 at. % of oxygen is approximately 2.1 eV and the value of the refractive index measured by ellipsometry at a wavelength of 632.8 nm is about 2.2. The values of the optical properties are in good agreement with the results reported for carbon rich amorphous SiC thin films obtained by other methods [23].

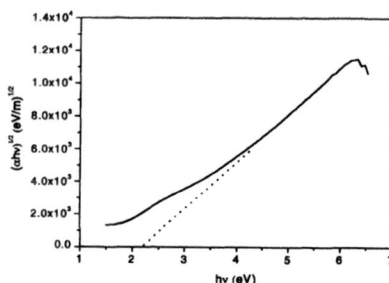

Figure 5: The square-root of the absorption coefficient of SiC thin film as a function of photon energy. The optical band gap is extracted by extrapolating the linear region of the curve as indicated by the dotted line.

CONCLUSIONS

Thin films of silicon carbide have been prepared via sublimation of polydimethylsilane polymer powder, using polymer-source CVD process. The structural analysis indicates that these films are very similar to amorphous silicon carbide thin films obtained by other deposition methods. The analysis of the bulk atomic concentration of chemical element showed that these thin films are carbon rich and non-stoichiometric with very uniform atomic depth profile.

The electrical and optical constants of these films were determined. Electrical characterization of SiC/Si junction showed a MIS structure behavior which suggests that these silicon carbide thin films are highly resistive. The film resistivity was as high as 6×10^5 Ω.cm, which is explained by the relatively high atomic concentrations of oxygen found in these films (about 2 at. %) due to both polymer and reactor contamination. By reducing the concentration of oxygen, through process optimization and polymer purification, electronic grade silicon carbide thin films could be obtained using PS-CVD process.

The strong chemical resistance to acid and alkaline solutions and the uniform thickness of these thin films make them very attractive as a conformal coating material and possibly as a structural material for MEMS devices. Some of the advantages of the polymer-source CVD process include low deposition cost, batch process compatibility, and hazard free working environment.

REFERENCES

1. G. Ambrosone, U. Coscia , S. Lettieri , P. Maddalena, S. Ferrero, Thin Solid Films, 403, 349 (2002)
2. S. Kerdiles, A. Berthelot, R. Rizk, Applied Physics Letters, 80, 3772 (2002)
3. P. Servati, Y. Vygranenko, A. Nathan, Journal of Applied Physics, 96, 7578 (2004)
4. D. J. Young, J. A. Du, C. A. Zorman, W. H. Ko, IEEE Sensors Journal, 4, 464 (2004)
5. J.-H. Boo, M. C. Kim, S.-B. Lee, S.-J. Park , J. G. Han, J. Vac. Sc. Tech. A, 18, 1713 (2000)
6. M. Shimada, T. Ono, I. Okada, S. Matsuo, J. Vac. Sc. Tech. B, 15, 736 (1997)
7. R. S. Kern, R. F. Davis, Materials Science and Engineering B, 46, 240 (1997)
8. M. B. J. Wijesundara, G. Valente, W. R. Ashurst, W. R. Ashurst, R. T. Howe, A. P. Pisano, C. Carraro, R. Maboudian, J. Electroch. Soc., 151, C210 (2004)
9. J. Chen, A. J. Steckl, M. J. Lobodab, J. Electroch. Soc., 147, 2324 (2000)
10. A. Fissel, U. Kaiser, B. Schroter, W. Richter, F. Bechstedt, App. Surf. Sc., 184, 37 (2001)
11. M. S. Lee, S. F. Bent, Journal of Applied Physics, 87, 4600 (2000)
12. M. Eickhoff, H. Moller, J. Stoemenos, S. Zappe, G. Kroetz, M. Stutzmann, J. Appl. Phys. 95, 7908 (2004)
13. P. Colombo, T. E. Paulson, C. G. Pantano, J. Amer. Cer. Soc., 80, 2333 (1997)
14. T. Iseki, M. Narisawa, Y. Katase, K. Oka, T. Dohmaru, K. Okamura, Chem. Mat., 13, 4163 (2001)
15. F. I. Hurwitz, T. A. Kacik, Xln-Ya Bu, J. Masnovl, P. J. Heimann, K. Beyene, J. Mat. Sc., 30, 3130 (1995)
16. H. Q. Ly, R. Taylor, R. J. Day, F. Heatley, J. Mat. Sc., 36, 4037 (2001)
17. M. Scarlete and C. Aktik, U.S. Patent Application Publication, 20050139966-A1 (2005)
18. Y. Awad, M.A. El Khakani, C. Aktik, J. Mouine, N. Camire, M. Lessard, M. Scarlete, Mat. Chem. Phys., 104 (2007)
19. B. Lahlouh, T. Rajagopalan, N. Biswas, and J. Sun, D. Huang, S.L. Simon, J.A. Lubguban, S. Gangopadhyay, Thin Solid Films, 497, 109 (2006)
20. A. Soum-Glaude, L. Thomas, E. Tomasella, Surf. Coat. Tech., 200, 6425 (2006)
21. A. Soum Glaude, L. Thomas, E. Tomasella, J.M. Badie, R. Berjoan, Surf. Coat. Tech., 201, 174 (2006)
22. Z. Hua, X. Liaob, H. Diaob, G. Kong, X. Zeng, Y. Xu, J. Crys. Growth, 264, 7 (2004)
23. L. Wang, J. Xu, T. Ma, W. Li, X. Huang, K. Chen, J. Alloys and Comp., 290, 273 (1999)

Mater. Res. Soc. Symp. Proc. Vol. 989 © 2007 Materials Research Society   0989-A05-08

Low Temperature High Quality Growth of Silicon-Dioxide Using Oxygenation of Hydrogenation-Assisted Nano-Structured Silicon Thin Films

Nima Rouhi[1], Behzad Esfandyarpour[1], Shams Mohajerzadeh[1], Pouya Hashemi[1], Bahman Hekmat-Shoar[1], and Michael D. Robertson[2]
[1]University of Tehran, Tehran, 14395, Iran
[2]Acadia University, Wolfville, 14395, Canada

ABSTRACT

We report a low temperature high quality oxide growth of nano-structured silicon thin films on silicon substrates obtained through a hydrogenation-assisted PECVD technique followed by a plasma enhanced oxidation process. A 100nm-thick electron-beam evaporated a-Si layer was hydrogenated at a plasma power of $2W/cm^2$ and a temperature of 250°C, followed by a post treatment at a temperature of 300°C for 30 min. with the plasma "off". Nano-crystalline silicon layers were obtained through this process with an average grain size of less than 5nm, which their grain size and porosity could be well controlled by process conditions. After the hydrogenation process, a plasma oxidation step was performed in an RF-PECVD for an extended period of two hours. The treated layers were investigated and compared with respect to their electrical, optical and stoichiometrical properties by means of Ellipsometry, transmission electron microscopy (TEM), Raman spectroscopy, Fourier transform infrared (FTIR) spectroscopy, and by current voltage and capacitance voltage measurements on metal-oxide-semiconductor (MOS) structures.

INTRODUCTION

There have been many attempts to develop a high quality low temperature oxidation process required for inter-level dielectrics, gate dielectrics in MOSFETs and thin film transistors. Low temperature oxides or LTO for short are usually realized using a LPCVD method which requires temperatures as high as 410°C. Also the deposition of silicon-oxide and silicon-nitride layers by means of plasma enhanced CVD is regularly used. On the other hand the formation of silicon oxide films by means of a growth approach requires temperatures above 850°C which is not suitable for many substrates such as glass and plastic bases that must be processed at low temperatures.

The oxidation technique reported in this paper is based on low temperature plasma assisted oxygen incorporation into high porosity nano-crystalline silicon thin films with an average grain size of less than 3nm. Micro-crystalline and nano-crystalline silicon films fabricated at low temperatures are promising materials in the areas of flexible electronics, optoelectronics, single-electron tunneling devices, solar cells and imaging sensors [1]. Several conventional approaches have been proposed for lowering the crystallization temperature of silicon, including metal-induced crystallization (MIC) [2], metal-induced lateral crystallization (MILC) [3] and excimer laser annealing (ELA) [4]. In these works, Pt and Ni-induced crystallization of amorphous silicon (a-Si) have been studied [5,6]. Recently, it is shown that RF plasma hydrogenation of evaporated amorphous Si films can lead to the formation of nano-crystalline layers suitable for thin film transistor fabrication and optical applications [7-9].

In this article, the effects of DC plasma-enhanced hydrogenation and RF PECVD oxidation on the crystallization and oxidation of e-beam evaporated a-Si are studied. The samples were prepared at various hydrogenation/annealing and oxidation/annealing temperatures and plasma-power densities, and were characterized by several techniques. The plasma hydrogenation followed by a plasma oxidation led to the formation of a dense silicon-oxide (SiOx) layer and their capacitance-voltage (C-V) and current-voltage (I-V) behaviors were characterized.

EXPERIMENT

A thin layer of amorphous silicon, about 100nm, was deposited on RCA-cleaned silicon or glass substrates by electron-beam evaporation at a temperature of 150°C and a base pressure of 2.6×10^{-4} Pa. In the case of silicon wafers we have used 1-Ωcm, P-type (100) silicon wafers. Following this step a hydrogenation-assisted PECVD technique was used to obtain a nano-structured silicon thin film. The morphology and crystallinity of the samples prepared using this approach has been examined by means of transmission electron microscopy and scanning electron microscopy tools and the results have been previously reported in Ref [1,10]. Using a DC-PECVD system, samples were hydrogenated at a temperature of 250 °C with a hydrogen gas flow of 20sccm and a plasma power of 2W/cm^2 for 15 minutes, followed by a post treatment at a temperature of 300°C for 30 min. The nano-crystalline samples were then exposed to oxygen plasma in order to create silicon oxide. The samples were placed in UNAXIS RF-plasma-enhanced chemical vapor deposition (RF-PECVD) apparatus, heated up to 350°C with RF power of 0.25W/cm^2 and chamber pressure of 20Pa for two hours. These conditions of the process were found to be suitable to achieve a nano-porous silicon layer which can turn into a silicon dioxide layer upon exposure to a proper oxygen plasma step. A schematic of the complete process is shown in Figure 1. The rate of oxide growth according to the above conditions was obtained to be about 0.42nm per minute.

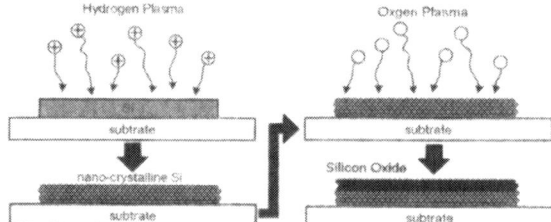

Figure 1: Schematic diagram of the process flow of the hydrogen and oxygen plasma treatments of the samples. The main steps in this method are the formation of a nano-porous layer by a hydrogenation step followed by formation of the oxide layer by oxygen plasma treatment.

DISCUSSION

The silicon-Oxygen bonding was investigated using FTIR analysis on the samples prepared for this study. The FTIR spectra clearly shows a peak at a wavenumber of about 1074cm^{-1} which is believed to be due to Si-O-Si bonds [9] and confirms the growth of an oxide layer after the RF plasma process. In Figure 2 one can see the FTIR spectra in which the Si and SiO$_2$ peaks are shown at 611(cm^{-1}) and 1074(cm^{-1}) respectively. Also the peak at about 2500(cm^{-1}) is assumed to be due to the absorption of residual CO$_2$ in the ambient environment. Inset in this figure highlights the SiO$_2$ peak.

Since the initial layer has been a thin layer of amorphous silicon deposited directly on a silicon substrate, one expects to obtain high leakage current through such a semi-conducting layer. The tunneling current and capacitance characteristics of samples were studied by electrical I-V and C-V measurements on MOS structures. For this investigation, metallic (Cr) pads of 500×500μm2 sizes were patterned using standard photolithography on the surface of the treated sample and the measurements was done by a Keithley parameter analyzer and a K82 C-V meter.

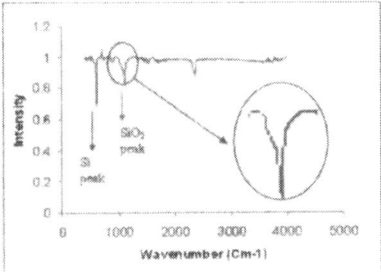

Figure 2: FTIR spectra; SiO$_2$ on Si, peak between 1050 cm^{-1} and 1100 cm^{-1}. The peak at 2500 cm-1 is believed to be due to the absorbed CO$_2$ molecules.

As it can be inferred from the curves, the leakage current is less than 1 nA. Also a threshold voltage of -0.98V can be extracted from the capacitance-voltage characteristics where a flat-band voltage of -3.5V is directly obtained from the curve. The capacitance-voltage behavior, presented in Figure 3(a), belongs to the sample which has been prepared under optimal condition as described before with the oxide thickness of about 50nm, evidencing a suitable low-high transition. Also performing the capacitance-voltage measurement at different time slots does not show any significant shift in the characteristics.

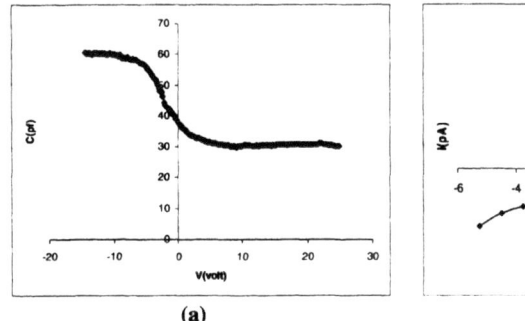 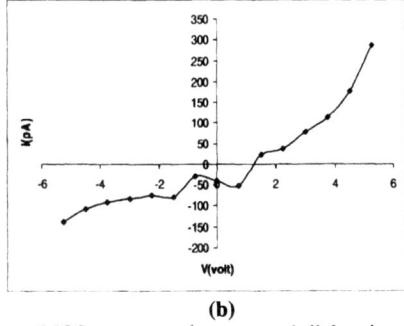

| (a) | (b) |

Figure 3: (a) C-V characteristic of the oxide sample on MOS structure; the extracted dielectric constant (ε$_r$) is about 3.8 from this data. (b) I-V characteristic of the oxide sample on MOS structure; a leakage current below 300pA is observed from this figure, further confirming the quality of the grown oxide.

According to the C-V measurements and the relation between the capacitance and the dielectric constant of the oxide layer (ε), a value of 3.8 was extracted for the sample. The extracted value for the processed oxide layer is slightly smaller than the value of the dielectric constant of the thermally-grown silicon-dioxide layers. For the samples which have been prepared under conditions away from the optimal case the difference is more dramatic. This difference in "ϵ" is due to the fact that the samples are highly porous after the DC PECVD hydrogenation. So, in the oxidation process the radicals of oxygen can go through the pores and a porous oxide will grow. Also Figure 3(b) depicts the results of the current-voltage measurement on the MOS structure evidencing a low leakage current as expected for high quality insulators.

To further investigate the physical properties of the samples, we have conducted Raman spectroscopy on various samples. The Raman spectroscopy on the processed samples showed a hump at the wavenumber of 954cm^{-1} indicating the SiO$_2$ on Si-substrate. Also a sharp peak is observed at a wavenumber of 519cm^{-1} which is the characteristic peak of the silicon substrate.

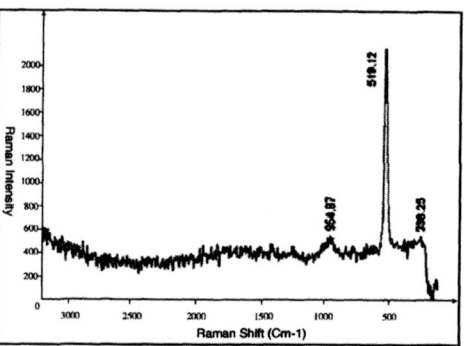

Figure 4: Raman spectroscopy of the grown SiO$_2$ on a silicon substrate. The Si peak at 519 cm^{-1} and a hump at about 954 cm^{-1} evidence the evolution of an oxide layer.

In addition to the above techniques, transmission electron microscopy (TEM) has been used to better understand the oxide formation at low temperatures. A cross-sectional view of one of the samples prepared using this approach has been presented in Figure 5. This sample has been realized on a glass substrate to show the ability of the process to create silicon-oxide layers at low processing temperatures compatible with glass bases.

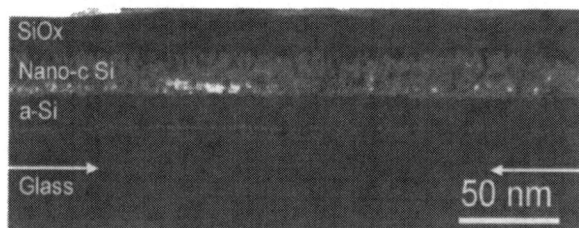

Figure 5: The transmission electron microscopy image of the silicon-oxide/glass interface. Arrows highlight the interface.

Meanwhile, Ellipsometry was used to investigate the optical constants of silicon dioxide thin films. The measurement was done by means of a PLAS MOS system with a beam wavelength of 632.8nm and an incident angle of 60°. The table below collects the optical constants for different samples prepared under various processing conditions. From the results of these measurements it is observed that the refraction index for the hydrogenated sample is 1.15 which suggests the high porosity of this sample. Besides, the refraction index of the RF-PECVD oxide sample, with the conditions stated in the previous section, is 1.3 that is close to the thermal oxide's refraction index (1.41). As we reduced the oxygen pressure in the chamber, the refraction index would decrease. Table (I) collects the results of this investigation and relates the measured refraction index to the annealing conditions of various samples. In contrast, by applying a higher partial pressure of oxygen in the chamber, a higher refraction index was achieved, close to the sample which was not bombarded by oxygen plasma (6th column).

Process	Refraction Index (n)
Only DC-Plasma hydrogenated sample	1.15
Oxygen Plasma (pressure:3.3 Pa) after DC-Plasma hydrogenation	1.2
Oxygen Plasma (pressure:20 Pa) after DC-Plasma hydrogenation	1.3
Oxygen Plasma (pressure:100 Pa) after RF-Plasma hydrogenation	1.58
Thermal Oxide	1.41
Exposed to oxygen (pressure:20 Pa) after DC-Plasma hydrogenated	1.75

Table I: A summary of the Ellipsometry results for the samples prepared under various annealing conditions.

We have also studied the C-V and I-V behavior of the samples which have been prepared under conditions away from optimal values and their electrical characteristics have been displayed in Figure 6.

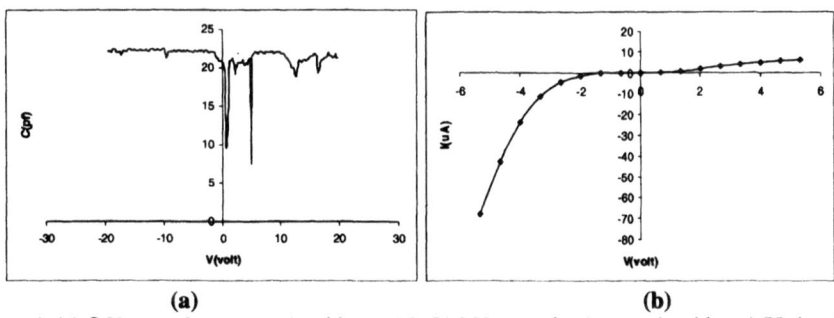

(a)   (b)

Figure 6: (a) C-V curve for the sample with n=1.75. (b) I-V curve for the sample with n=1.75 showing a high leakage current.

The C-V analysis shows that the sample do not have the expected step-like behavior for the gate oxide and also from the I-V curve we can see that the leakage current is at several micro-ampere levels not acceptable for MOS-based devices. In addition, the I-V measurement for the hydrogenated sample demonstrated a leakage current of about 1mA which is very high for a favorable insulator. The sharp drops in the capacitance-voltage characteristics are believed to be due to the breakdown of the oxide and subsequent flowing of a considerable leakage current through the SiOx layer.

CONCLUSIONS

We have studied the effects of plasma-enhanced hydrogenation on the crystallization of electron-beam evaporated a-Si at temperatures as low as 250 °C. Application of successive hydrogenation/annealing steps in the PECVD chamber resulted in the formation of nano-crystalline and porous Si. The crystalline structure and stoichiometrical properties were examined using TEM. The growth of low temperature SiO_2 (as low as 300 °C) was achieved by subsequent plasma oxidation of the nano-crystalline silicon films. The C-V and I-V characteristics show the dielectric constant of about 3.8 and leakage current of below 1nA. In addition, FTIR spectroscopy demonstrates an oxide peak and Raman shift also shows a hump of SiO_2 on Si substrate. However, decrease in the chamber pressure (or increase in the plasma power) led to the formation of porous and partially etched structures which can be inferred from the Ellipsometry results (for measurement of refraction index) and may not show acceptable electrical characteristics. Further improvement in the oxide quality and fabrication of low-temperature TFTs is presently under way.

REFERENCES

1. M. Jamei, F. Karbasian, S. Mohajerzadeh, Y. Abdi, M. D. Robertson, S. Yuill, *IEEE EDL*, V 28. Issue 3, Pages 207-210.
2. S. R. Herd, P. Chaudhari, and M. H. Brodsky, J. Non-Cryst. Solids 7, 309 (1972).
3. C. Hayzelden and J. L. Batstone, J. Appl. Phys. 73, 8279 (1993).
4. T. Sameshima and S. Usui, J. Appl. Phys. 70, 1281 (1991).
5. M. Ivanda, K. Furić, O. Gamulin, M. Peršin, and D. Gracin, J. Appl. Phys. 70, 4637 (1991).
6. S. M. Han, M. C. Lee, M. Y. Shin, J. H. Park, and M. K. Han, Proc. IEEE 93, 1297 (2005).
7. S. W. Lee, Y. C. Jeon, and S. K. Joo, Appl. Phys. Lett. 66, 1671 (1995).
8. R. A. Ditizio, G. Liu, S. J. Fonash, B. C. Hseih, and D. W. Greve, Appl. Phys. Lett. 56, 1140 (1990).
9. Steven A. MacDonald, Craig R. Schardt, David J. Masiello, Joseph H. Simmons, "Dispersion analysis of FTIR reflection measurements in silicate glasses", *Journal of Non-Crystalline Solids 275*, 72-82 (2000).
10. P. Hashemi, Y. Abdi, S. Mohajerzadeh,a_ J. Derakhshandeh, and A. Khajooeizadeh, M. D. Robertson, R. D. Thompson, and J. M. MacLachlan, JOURNAL OF APPLIED PHYSICS 100, 104320 (2006).

Mater. Res. Soc. Symp. Proc. Vol. 989 © 2007 Materials Research Society   0989-A05-09

Channel Reliability in MOSFETs with Gate Oxide Grown using ECR Plasma of O2/He

Vishwas Jaju, and Vikram Dalal

Electrical and Computer Engr., Iowa State University, Coover Hall, Ames, IA, 50011

ABSTRACT

In this paper we present the channel degradation properties of MOSFET with gate oxide grown using electron cyclotron resonance (ECR) plasma. Si was oxidized using ECR plasma of 10% O_2/He at 450°C for 60 min. To evaluate the electrical properties, MOS and MOSFETs devices were fabricated using ECR grown oxide as a gate oxide. We found that the n-MOSFET with as-grown ECR oxide shows higher hot carriers induced channel degradation. It was approximately two orders of magnitude higher compared to the channel with thermally grown (at 950°C) dry oxide. When as-grown oxide was annealed at much higher temperature (~800°C), channel resistance for the hot carriers improved and became comparable to that of the thermally grown oxide. We think that because of the lower processing temperature used during plasma oxidation there are many nonbridging O atoms and also strain is present in Si-O bonds. Annealing at higher temperatures connects these O atoms with Si to form Si-O bonds and re-arrangement of atoms could relieve strain. This assumption was found be self-consistent with the infrared absorption study presented in this paper.

INTRODUCTION

As the device dimensions are shrinking, high temperatures used during thermal oxidation and annealing are posing new challenges. Device quality thermal oxide is usually grown in the range of 900°C-1100°C. High temperature processes could redistribute the dopants profile in Si and could induce stresses between the different layers on Si wafer [1,3,9]. Thus there has been always a need of low temperature (<600°C) semiconductor processing.

Several groups have reported the electron cyclotron resonance (ECR) plasma assisted oxidation of silicon [1-8]. Si wafer was successfully oxidized from room temperature to up to 640°C using ECR plasma [1,5]. Oxide has been grown in a floating substrate and also with an applied external bias. Good electrical properties and improved growth rates have been reported when the substrate was positive biased [5]. Defect density and oxide breakdown strength has been reported to be in the range of 1.2×10^{10}- 5×10^{11} cm^{-2}eV^{-1} and 5-20 MV/cm respectively [3,7,8]. Chemical composition was studied using XPS and found to be fairly close to thermally grown silicon dioxide [7]. Actual growth mechanism still remains unknown although many groups believe that it is because of the transport of O⁻ ions through the growing oxide layer, as suggested by the higher growth rates achieved when substrate was biased positive [5,6]. Gate oxide susceptibility to hot carriers induced degradation during device operation is an important property to test if this oxide can to be used as a gate oxide. In our knowledge none of the groups have reported on the fabrication of MOSFET devices using ECR plasma oxide as gate oxide so far. For the first time, we have fabricated the MOS and MOSFET devices using ECR grown silicon dioxide as a gate oxide. Defect density and oxide breakdown strength were measured using MOS capacitors. N-channel MOSFET devices were fabricated to measure the hot electron induced degradation. These properties were compared against MOSFETs fabricated with thermally grown dry-oxide as the gate oxide.

EXPERIMENT

Reactor Design
The ECR reactor used in this study is shown in Figure 1. A microwave source (SAIREM GPM 03KSM) of 2.45 GHz frequency, 300 W power was used. Microwave power was supplied to 3 stub tuner on waveguide WR340 via a shielded coaxial cable. Reflected power was read in the power meter present in the microwave source. To reduce the reflected power tuning was done manually using a 3 stub tuner. Reflected power was always kept below 5% of the incident power. A circular quartz window was used as a coupler between reactor chamber and waveguide. Two coil based magnets were used in a mirror configuration to provide the magnetic field (>875 G) required for ECR condition. Substrate holder is made up of Inconel alloy and could be heated up to 900°C. Substrate was kept perpendicular to the magnetic field axis and was electrically isolated from rest of the reactor. Distance of sample from the source was ~20 cm. 10% O_2/He gaseous mixture was used for oxidation and it was introduced into the reactor via a shower ring present just after the quartz window.

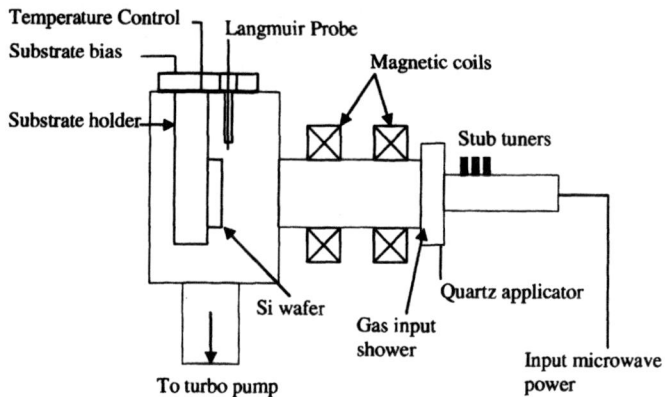

Figure 1: Schematic diagram of the ECR reactor

Sample Preparation
3-inch Si wafers of <100> orientation and p-type doping (Boron, 0.1-0.07 ohm-cm) were used. Wafers were cleaned using standard RCA process and prior to loading they were dipped in HF solution to remove any native oxide present. Pressure below 1×10^{-6} T was achieved before heating the wafer to the desired temperature. MOSFET of dimension 5 μm x 40 μm (L x W) and MOS devices 100 μm x 100 μm were fabricated using Al as the gate metal. Aluminum was thermally evaporated and deposited (~250 nm thick) on the oxide.

Thicknesses of the oxide layers were measured using an Ellipsometer. High frequency (at 1 MHz frequency) capacitance-voltage (C-V) characteristic were acquired using HP 4280 LCR meter. Keithley 595 was used to obtain Quasi C-V curve. Defect density reported is at the minimum of Quasi C-V curve. Oxide breakdown strength was measured using V-ramp method [10]. Current voltage characteristics and n-channel stressing of the MOSFET was performed using HP 4156A Semiconductor Parameter Analyzer. Threshold voltage and linear mobility were calculated from the I_{DS}-V_{GS} and I_{DS}-V_{DS} (in linear region) [10]. Lifetime of the MOSFET is defined either as a 50 mV shift in threshold voltage or a 10% change in any other device parameter [11], which in our case was the linear mobility. FTIR analysis was done using Nicolet magna IR 760 spectrometer.

RESULTS AND DISCUSSION

Electrical properties measured on MOS device
ECR oxides were grown using 125 W of microwave power at 450°C substrate temperature. Gas flow (10% O_2/He mixture) was 20 sccm and 5 mT pressure was maintained for all the runs. Oxides of thicknesses in the range of 140-170 Å were obtained after 60 min of oxidation time. Defect density (D_{it}) in the as-grown oxide was ~1×10^{12} cm^{-2}eV^{-1} and after post-metallization annealing at 450°C, D_{it} reduced to ~2×10^{11} cm^{-2}eV^{-1}. Annealing was performed in a N_2 ambient at atmospheric pressure for 30 min. MOS devices annealed at 450°C in 3% H_2/N_2 ambient had D_{it}~7×10^{10} cm^{-2}eV^{-1}. This shows that the Hydrogen environment helps in reducing the defect density.

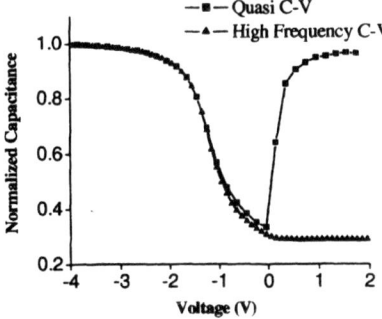

Figure 2: Typical high frequency C-V curve and Quasi C-V curve used to determine the defect density

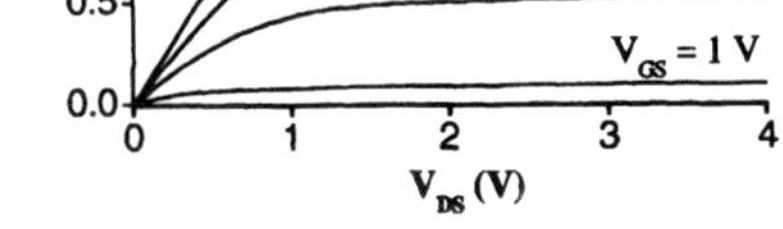

Figure 3: Typical I_{DS}-V_{DS} plot for MOSFET with gate oxide as ECR grown oxide.

Oxide breakdown field is defined as the field at which 1 µA/cm^{-2} current passes through oxide [5, 7]. Oxide breakdown strength was found to be ~8 MV/cm. This oxide strength is well above the oxide strength required for standard CMOS processing (>7 MV/cm) and comparable to the ECR oxidized oxides reported in literature elsewhere [3,7]. For ECR grown oxide electron mobility in the channel was ~350 cm^{2}V^{-1}s^{-1}, while for a comparable sample with thermally grown oxide it was ~410 cm^{2}V^{-1}s^{-1}. The lower value of mobility in the case of ECR grown oxide could be because of the high defect density present at the interface.

Hot carrier induced degradation study

N-channel MOSFETs were stressed under DC bias and change in threshold and linear mobility was observed. In order to optimize the stressing conditions, substrate current (I_{sub}) was measured by varying the gate voltage for a fixed voltage (6 V) applied to the drain. Gate voltage was chosen at the point where substrate current was maximum [11]. HP parameter analyzer was programmed to stress the channel and to record I_{DS}-V_{GS} and I_{DS}-V_{GS} sweep data after time intervals of 10, 30, 100, 300, 1000, 3000, 10000, 30000 and 50000 seconds. MOSFETs devices were prepared with gate oxides: i. ECR as-grown oxide (Sample-A), ii. ECR oxide annealed at 800°C prior to metallization (Sample-B) and iii. Thermal oxide. All samples received post-metallization annealing in N$_2$ gas at 450°C for 30 min at 1 atm. pressure. As shown in Figure 5 and 6, Sample-A showed much higher degradation of channel compared to both Sample-B and the thermal oxide.

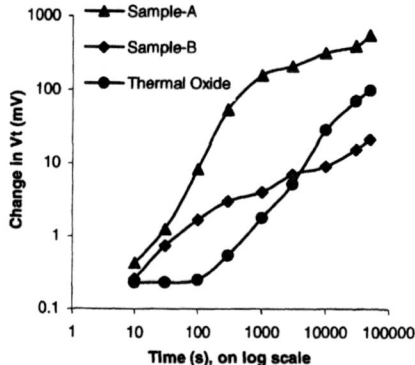

Figure 4: Change in threshold voltage with time

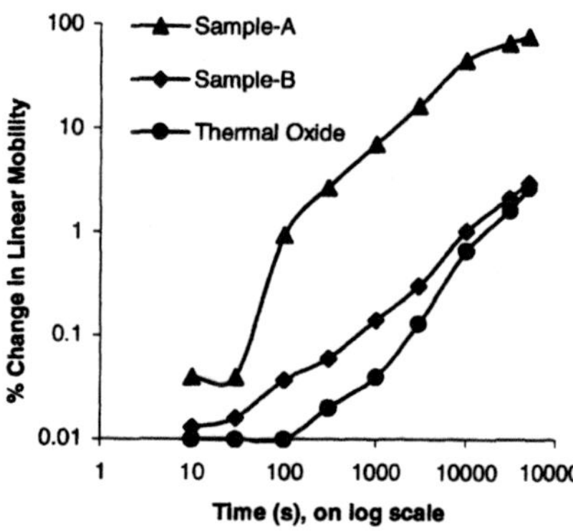

Figure 5: Percentage change in linear mobility with time. Sample-B and thermal oxide shows similar trend.

Initial degradation was faster for Sample-B (pre-metallization annealing at 800°C) due to the higher starting defect density. Later its degradation properties became comparable to that of the thermal oxide. Defect density on Sample-B was only slightly lower when compared to Sample-A, but Sample-B channel resistance to hot carriers was remarkably higher than Sample-A. Higher degradation of as-grown oxide (Sample-A) could be attributed to the strain present in Si-O bonds. Strain in the ECR grown oxide could be because of low processing temperatures (<450°C) used. Annealing at 800°C probably relieved the strain in the bonds and oxide becomes much like the thermally grown oxide.

Infrared absorption properties on SiO$_2$ films

Studies from the several groups have established that the IR absorption peaks and half-width of band would vary with the change in the physical properties of SiO$_2$ such as density of strain in the film [12]. Kimura et al compared IR absorption of ECR oxides grown at different temperatures with thermal oxide. The thickness of ECR oxides grown at low temperature (<450°C) were of low thickness and shift in the peaks could be because of the thin ECR oxide films used for measurement [1].

In our experiment we first measured IR absorption of as-grown ECR oxide and same sample was then re-measured after annealing it for 30 min at 800°C in N_2 gas ambience. IR absorption of as-grown and annealed ECR oxide was then compared against the thermal oxide (see figure 6). For as-grown ECR oxide stretching peak for Si-O bond has lower intensity when compared with annealed ECR oxide. This intensity is directly related to the number of Si-O bonds exist in the oxide. We think that there are many nonbridging O atoms present in the as-grown ECR oxide. During the annealing many of these O atoms bridge with neighboring Si atoms to form Si-O bonds and this results in the higher intensity for annealed samples. Stretching peak of Si-O bond for ECR grown oxide annealed at 800°C has the same location (1074 cm^{-1}) as that of the thermally grown dry-oxide. In as-grown ECR oxide peak for Si-O bond is shifted to ~1068 cm^{-1} and it has larger half-width compared to annealed ECR oxide. Shifting to a lower wavenumber could be because of higher Si-O bond length or change in bond angle. These results indicate that the inferior quality plasma silicon dioxide films were grown at low temperature. Because of the low temperatures used during film formation and plasma oxidation reaction mechanism, atoms may have not arranged themselves in minimum energy configuration and there are many nonbridging O atom are also present. Annealing ECR oxide at higher temperature made them physically indistinguishable from the thermally grown dry-oxide.

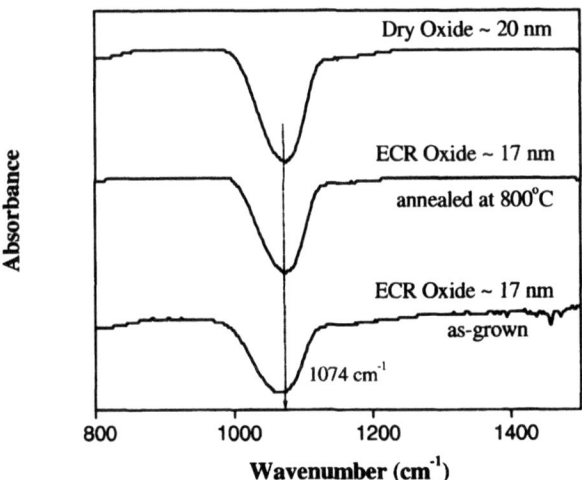

Figure 6: Infrared absorption spectrum of dry oxide and ECR oxide. Annealed ECR oxide and thermally grown dry oxide both have peaks at ~1074 cm^{-1}. As-grown ECR oxide has low intensity, larger half-width and peak shifted to ~1068 cm^{-1}.

CONCLUSIONS

In this work we have successfully oxidized Si wafers using ECR plasma of 10% O_2/He gaseous mixture. Silicon dioxide of thicknesses in the range of ~140-170 Å was grown and device electronic properties were found to be comparable with thermally grown dry oxides at 950°C. It was observed that as-grown ECR oxides showed higher channel hot electron induced degradation properties for them to be used as a gate in MOSFETs. FTIR studies showed that in ECR as-grown oxide many of O atoms are not bonded to Si atoms and strain was also present in Si-O bonds. Good degradation properties, similar to that of thermal oxides, were achieved by annealing the ECR oxides at high temperature (~800°C). Infrared absorption graph confirms that annealing ECR grown oxides at higher temperature make their physical properties almost same as that of the thermal oxide.

REFERENCES

1. Kimura, Murakami, Miyake, Warabisako, Sunami and Tokuyama, "Low Temperature Oxidation of Silicon in a Microwave-Discharged Oxygen Plasma," *J. Electrochem. Soc.*, Vol. 132, No. 6, 1460-1466 (June 1985).
2. D. A. Carl, D. W. Hess, and M. A. Liberman, "Oxidation of Silicon in an Electron Cyclotron Resonance Oxygen Plasma: Kinetics, Physiochemical, and Electrical Properties," *J. Vac. Sci. Technol. A* 8, 2924-2930 (1990).
3. G. T. Salbert, D. K. Reinhard, and J. Asmussen, "Oxide Growth on Silicon Using a Microwave Electron Cyclotron Resonance Oxygen Plasma," *J. Vac. Sci. Technol. A* 8, 2919-2923 (1990).
4. Keunjoo Kim, M. H. An, Y. G. Shin, M. S. Suh, C. J. Youn, Y. H. Lee, K. B. Lee, and H. J. Lee, "Oxide Growth on Silicon (100) in the Plasma Phase of Dry Oxygen using an Electron Cyclotron Resonance Source," *J. Vac. Sci. Technol. B* 14 (4), 2667- (1996).
5. D. A. Carl, D. W. Hess, M. A. Lieberman, T. D. Nguyen, and R. Gronsky, "Effects of dc Bias on the Kinetics and Electrical Properties of Silicon Dioxide Grown in an Electron Cyclotron Resonance Plasma," *J. Appl. Phys.* 70, 3301-3313 (1991).
6. J. Joseph, Y. Z. Hu, and E. A. Irene, "A Kinetics Study of the Electron Cyclotron Resonance Plasma Oxidation of Silicon," *J. Vac. Sci. Technol. B* 10, 611-617 (1992).
7. K. T. Sung, S. W. Pang, "Oxidation of Silicon in an Oxygen Plasma Generated by a Multipolar Electron Cyclotron Resonance Source," *J. Vac. Sci. Technol. B* 10, 2211- (1992).
8. Kunio Saito, Yoshito Jin, Toshiro Ono and Masaru Shimada, "Low-Temperature Silicon Oxidation with Very Small Activation Energy and High-Quality Interface by Electron Cyclotron Resonance Plasma Stream Irradiation," *Japanese Journal of Applied Physics,* Vol. 43, No. 6B, pp. L 765–L 767 (2004).
9. D. W. Hess, "Plasma-assisted Oxidation, Anodization and Nitridation of Silicon," *IBM J. Res. Develop.* Vol. 43, No.18, January/March. 1999.
10. Dieter K. Schroder, "Semiconductor Device and Materials Characterization," John Willey, New York, (1990).
11. JEDEC STANDARD, "Procedure for Measuring N-Channel MOSFET Hot-Carrier-Induced Degradation Under DC Stress," *JESD28-A*.
12. W. A. Pilskin, H. S. Lehman. "Structural Evolution of Silicon Oxide Films," *J. Electrochem. Soc.*, Vol. 18, No. 10, 1013-1019 (Oct. 1965).

Poster Session:
Crystallization

Mater. Res. Soc. Symp. Proc. Vol. 989 © 2007 Materials Research Society   0989-A06-01

A Novel Bonding Technique Using Metal-Induced Crystallization of Amorphous Silicon

Markus D. Ong, and Reinhold H. Dauskardt
Materials Science and Engineering, Stanford University, 416 Escondido Mall, Stanford, CA, 94305

ABSTRACT

This study investigates the use of aluminum-induced crystallization of amorphous silicon as a potential bonding mechanism for a sandwich stack of films between two silicon substrates. Similar procedures using copper diffusion bonds have been in use, but these require temperatures as high as 400 °C. Using the crystallization of amorphous silicon as the bonding mechanism has allowed the bonding temperature to be lowered by more than 100 K. Fracture experiments for a low-k material were conducted, and the results using amorphous silicon bonding was compared to epoxy bonding control experiments. Essentially identical results were obtained for the two bonding mechanisms. Low-temperature bonding techniques are of great interest to future progress in the microelectronics industry, and these results are promising advances.

INTRODUCTION

Wafer to wafer bonding is a critical step in the fabrication of complex three-dimensional (3-D) structures. Complex 3-D structures have many technologically relevant applications such as increasing transistor densities in microelectronic devices while reducing interconnect RC time delay and power consumption [1], increasing efficiencies of thin film solar cells [2], and microelectromechanical systems [3]. Current state-of-the-art bonding schemes include adhesive bonding [1,3], copper diffusion bonding [4], and oxide-oxide bonding [5]. Disadvantages and common drawbacks of these techniques include insertion of thick intermediate layers, weak bonding, high temperature requirements, and incorporation of incompatible materials [1].

In this study, we demonstrate a novel bonding technique using metal-induced crystallization (MIC) of amorphous silicon (a-Si). This novel technique creates a thin, strong bond at relatively low temperatures while incorporating materials already commonly in use in microelectronics fabrication. The integration of 3-D stacks using low temperature bonding is especially desirable, particularly when incorporating organic or hybrid organic-inorganic dielectric materials that degrade at excessively high temperatures [6]. Previous studies have shown that crystallization of a-Si can be induced by some metals such as aluminum, gold, or palladium [7] at temperatures below 200 °C, well below the 400 °C required for copper diffusion bonding [8]. Therefore, this reaction is of great interest for use as a low-temperature bonding technique. This study shows that aluminum-induced crystallization of amorphous silicon is a potential bonding mechanism as it is used as the bonding technique in thin-film fracture experiments and compared to results of a control group consisting of well-established values of fracture energies measured using an epoxy bond.

EXPERIMENT

Amorphous silicon was deposited on silicon substrates (40 mm x 40 mm squares) with native oxide using CVD. A mating substrate with a low-k film stack (40 nm SiC underlayer, 260 nm low-k, and 80 nm oxide cap) had Cr (25 nm) and Al (175 nm) evaporated on it. The films and their respective thicknesses in this stack were specifically chosen to match a previous study which measured the fracture energies of low-k films [9]. Although the entire low-k film stack is not inherently necessary for this bonding technique, reproduction of the stack allowed comparisons to previous work which incorporated an epoxy bond.

A hot press with temperature and pressure feedback control was used. A pressure of 11 MPa was applied to the substrates. Each run began with an initial hold at 60 °C followed by an hour-and-a-half ramp up to the desired temperature. In this study, the maximum temperature and the holding time were two main experimental variables. The maximum temperatures were 310 °C, 270 °C, or 250 °C (+/- 10 °C), and the holding times were 4, 10, or 20 hours. A diagram of the temperature and pressure in their progressive steps is shown in Figure 1. Four-point bend (FPB) and double cantilever beam (DCB) fracture experiments (described elsewhere) were used to measure critical fracture energies [10]. A schematic of the fracture geometries and films stacks are shown in Figure 2. The results were then compared a control group of specimens made using an epoxy bond. Characterization of the fracture path was done primarily with x-ray photoelectron spectroscopy (XPS).

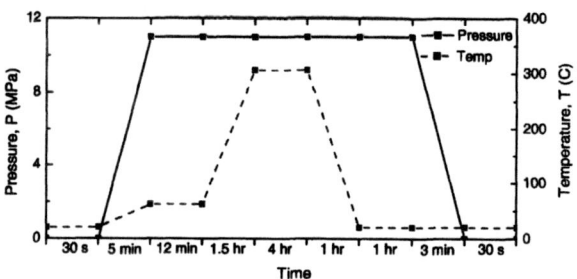

Figure 1. Load and temperature controls for each stage of the pressing procedure. The maximum temperature (shown here as 310 °C) and the holding time at this temperature (shown here as 4 hr) were the two main experimental variables in this study.

(a)   (b)   (c)

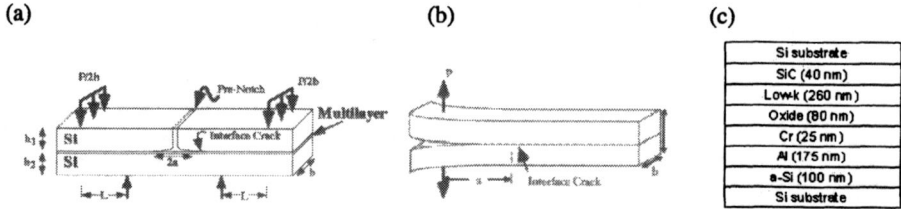

Figure 2. Schematics of the (a) FPB geometry, (b) DCB geometry, and (c) film stack.

RESULTS

Four point bend, notch on low-k side

Figure 3 shows the results for four-point bend testing with a pre-notch on the substrate with the low-k film. The bar farthest to the right is the epoxy control standard to which the amorphous silicon bonding measurements are compared. Essentially identical results were achieved for the two successful bonding conditions (310 °C for 4 hr and 270 °C for 20hr). The other two bonding conditions resulted in markedly lower fracture energies, indicating that a bond was not successfully formed. The mating fracture surfaces were characterized using x-ray photoelectron spectroscopy (XPS) to determine the fracture locations. Specimens from the epoxy control and the two successful bonding conditions fractured near the interface between the low-k and the oxide cap. The fracture path for the specimens with unsuccessful bond was in the a-Si and Al bonding layer since both Al and Si were present on both sides of the fracture path.

Four point bend, notch on blank side

Since the four-point bend fracture geometry is a mixed-mode geometry, the substrate which is notched affects both the fracture path and fracture energy along that path. The experiments in this geometry were only performed if the bonding appeared to be successful in the FPB, low-k side notch geometry. Figure 4 shows that the results agree with the epoxy control for both the condition of a 310 °C hold for 4 hours and the condition of a 270 °C hold for 20 hours. XPS characterization indicated that the fracture path was near the interface between the low-k and the SiC underlayer.

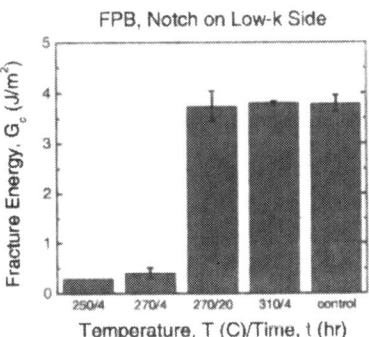

Figure 3. Fracture energy measurements for FPB, notch on low-k side.

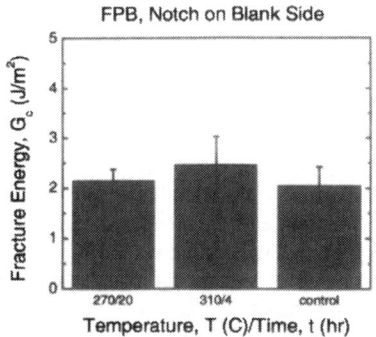

Figure 4. Fracture energy measurements for FPB, notch on blank side.

Double cantilever beam

The double cantilever beam geometry is a pure Mode I geometry, and fracture is usually cohesive within the weakest material. The experiments in this geometry were only performed only if the bonding appeared to be successful in the FPB geometry. Figure 5 shows that the results are consistent with the epoxy control for both the condition of a 310 °C hold for 4 hours and the condition of a 270 °C hold for 20 hours. XPS characterization indicated that the fracture was cohesive within the low-k material.

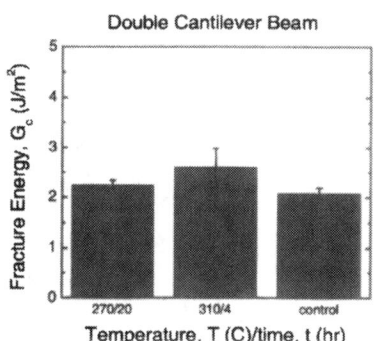

Figure 5. Fracture energy measurements for DCB.

DISCUSSION

For all fracture geometries, the results from the a-Si bonding were essentially identical to those of the epoxy bonding control. Both temperature and time were factors in determining whether a successful bond was formed. Lower temperatures can be used if the holding time is increased sufficiently. Figure 6 shows a time-temperature map that summarizes the tested conditions and essentially outlines the region in which successful bonding was achieved.

The results clearly indicate that the novel bonding technique of using metal-induced crystallization of amorphous silicon gives comparable results to the epoxy bonding control group when measuring the fracture energy of low-k films. However, future work remains to characterize the properties of the bond and to further improve the time and temperature parameters needed to create the bond. The fracture toughness of the bond itself has not been characterized since the weak low-k film and its adjacent interfaces were preferred fracture paths. When the low-k film is removed (along with its cap and underlayers), measuring a critical fracture energy for the bond itself is difficult since steady-state cracking within the bond region is difficult to achieve, and the silicon substrates typically fail instead. In addition to the fracture properties of the bond, the characterization of the structure of the interface is in progress to determine the crystallinity, grain size, and orientation of the Si region after the bonding process. Additional work to lower the required temperatures and shorten the required times is also in progress. In this study, the time between the Al deposition and the bonding during which oxidation of the Al could occur was not carefully controlled. Although there were no problems with repeatability of successful bonds observed in this study, it is possible that minimizing the oxidation time of either interface could shorten the required bonding time. Controlling and minimizing the surface roughness of the mating substrates could also help to reduce the required time and temperatures to create a successful bond.

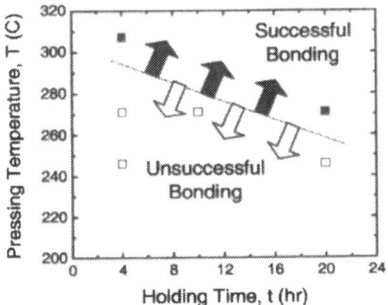

Figure 6. Time-temperature map showing relationship between required temperature and holding time for a successful bond. The dotted line is the approximate border between the conditions for successful bonding and unsuccessful bonding.

CONCLUSIONS

The use of metal-induced crystallization of amorphous silicon was used to bond two mating silicon substrates to form sandwich stack fracture test specimens. The fracture energies measured from the experiments were identical to well-established values from a control group which employed an epoxy bond. Successful bonding occurred for a temperature set point of 310 °C with a 4-hour hold, nearly 100 K lower than the temperatures needed for a comparable copper diffusion bond. The temperature could be lowered further to 270 °C if a longer 20-hour hold was used. This bonding technique is promising tool for many applications that require strong, thin bonds at lower temperatures, including fabrication of 3-D structures.

ACKNOWLEDGMENTS

This work was supported by the Semiconductor Research Corporation under Task ID 1391.001. Work was performed in part at the Stanford Nanofabrication Facility of NNIN supported by the National Science Foundation under Grant ECS-9731293.

REFERENCES

1. F. Niklaus, G. Stemme, J.Q. Lu, and R.J. Gutmann, Journal of Applied Physics, **99**, 031101 (2006).
2. H. Spanggaard and F. Krebs, Solar Energy Materials & Solar Cells, **83**, 125 (2004).
3. C.T. Pan, H. Yang, S.C. Shen, M.C. Chou, and H.P. Chou, Journal of Micromechanics and Microengineering, **12**, 611 (2002).
4. K.N. Chen, C.S. Tan, A. Fan, and R. Reif, Electrochemical and Solid-State Letters, 7 G14 (2004).
5. K. Guarini, A. Topol, M. Ieong, R. Yu, L. Shi, D. Singh, G. Cohen, H. Pogge, S. Purushothaman, and W. Haensch, in *International Symposium on Thin Film Materials, Processes, and Reliability*, PV2003-13, 390, ECS (2003).
6. W. Volksen, R. Cook, P. Furuta, C. Hawker, J.L. Hedrick, E. Liniger, R. Malek, D. Miller, R.D. Miller, C. Nguyen, D. Shon, M. Toney, D.Y. Yoon, *Dielectrics for ULSI Multilevel Interconnection Conference*, Tampa, FL (1999).
7. T.J. Konno and R. Sinclair, Materials Science and Engineering, **A179/A180**, 426 (1994).
8. P.I. Wang, T. Karabacak, H.F. Li, G.G. Pethuraja, S.H. Lee, M.Z. Liu, J.Q. Lu, and T.M. Lu: Low Temperature Copper-Nanorod Bonding for 3D Integration, in *Enabling Technologies for 3-D Integration*, edited by C.A. Bower, P. Garrou, K. Takahashi, P. Ramm, (Mater. Res. Soc. Symp. Proc. **970**, Pittsburgh, PA, 2007) 0970-Y04-07.
9. M.D. Ong, V. Jousseaume, S. Maitrejean, and R.H. Dauskardt: Fracture Properties of Porous MSSQ Films: Impact of Porogen Loading and Burnout, in Materials, Technology and Reliability of Low-k Dielectrics and Copper Interconnects, edited by Ting Y. Tsui, Young-Chang Joo, Lynne Michaelson, Michael Lane, Alex A. Volinsky (Mater. Res. Soc. Symp. Proc. **914**, Warrendale, PA, 2006), 0914-F04-07.
10. R.H. Dauskardt, M. Lane, Q. Ma, and N. Krishna, Engineering Fracture Mechanics, **61**, 141 (1998).

Mater. Res. Soc. Symp. Proc. Vol. 989 © 2007 Materials Research Society   0989-A06-03

Effects of Power Density and Thickness On Aluminum-Induced Crystallization of PECVD Amorphous Silicon

Kendrick S. Hsu[1], Jeremy Ou-Yang[1], Li P. Ren[2], and Grant Z. Pan[1]
[1]Microfabrication Laboratory, University of California at Los Angeles, Los Angeles, CA, 90095
[2]Nanoelectronics and Nanophotonics Laboratory, Global Nanosystems, Inc., Los Angeles, CA, 90025

ABSTRACT

The effect of power density and thickness on aluminum-induced crystallization (AIC) of amorphous silicon (a-Si) was studied by using N_2-protected conventional furnace reaction and optical microscopy. The a-Si with a thickness ranging from 500 to 5000 Å was formed by plasma enhanced chemical vapor deposition (PECVD) at a power density from 0.05 to 1.00 W/cm^2, followed by an E-beam evaporation of 300 Å Al. It was found that a low power density as well as a large a-Si thickness could result in a decrease of activation energy and therefore a significant reduction of the AIC reaction temperature. Scanning and transmission electron microscopy and X-ray diffraction were used to check the crystallinity and quality of the AIC thin films. Polysilicon thin films consisting of submicron-scale Si crystallites within large AIC networks of about 10 μm were achieved at an AIC reaction temperature as low as 120 °C.

INTRODUCTION

Large-area microelectronics on flexible substrates such as flexible displays on plastic substrates requires novel thin film transistors (TFTs) made of high carrier mobility materials and fabricated at temperatures below 150 °C [1]. It is well known that Al can induce the crystallization of a-Si at a temperature much lower than the eutectic temperature of silicon with Al [2-4]. This is the so-called aluminum-induced crystallization (AIC). It was recently reported that polysilicon (c-Si) could be achieved from PECVD a-Si by AIC reaction at a temperature as low as 200 °C if the Al layer is placed on top of the a-Si before the AIC reaction [5]. In order to further reduce the AIC temperature and to achieve device grade c-Si for use in large area microelectronics on flexible substrates, the effects of power density and a-Si thickness on the AIC reaction and mechanism of PECVD a-Si were investigated extensively. In this investigation, polysilicon thin films consisting of submicron crystalline silicon were achieved at an AIC temperature as low as 120 °C.

EXPERIMENTAL DETAILS

On top of a 4 inch Si or 1737 corning glass wafer coated with 3000 ° PECVD SiO_2, a-Si with a thickness ranging from 500-5000 ° was deposited by PECVD in a PlasmaTherm790 system at a power density from 0.05 to 1.00 W/cm^2, a temperature of 150 ˚C, and a vacuum of 500 mTorr with a He-diluted SiH_4 gas. This PECVD Si was confirmed to be amorphous through X-ray diffraction testing. Within a maximum of 1 h right after the a-Si deposition, the wafers were transferred into a CHA Mark40 evaporator and a thin Al layer of 300 ° was evaporated at room temperature with a predeposition vacuum of 2×10^{-7} Torr. After deposition, the wafers were first cut into 1 x 1 cm squares and stored in a N_2-protected cabinet. Within a period of days, they were then isothermally annealed in a conventional box furnace with N_2 protection at temperatures ranging from 120 to 150 °C for various times. The reacted squares were placed under an optical microscope and various pictures were taken in terms of deposition parameters and reaction conditions. These optical pictures were analyzed to characterize the microstructural evolution and AIC kinetics. SEM, TEM, and X-ray diffraction analyses were used to determine the crystallinity and quality of the AIC thin films.

RESULTS AND DISCUSSION

Microstructural evolution

Though the AIC reaction in this investigation was achieved at a temperature range significantly lower than what has previously been reported [3-6], the microstructural evolution during the AIC reaction appears pretty similar under optical microscopy, consisting of two major stages: i.e. first nucleation of Si crystallites and then growth of crystallite regions in the matrix until they meet each other, resulting in micron-scale AIC networks. It was found that the deposition power density and the thickness of a-Si strongly affect the AIC process by altering the density, growth rate and final AIC network size, which can be clearly seen by optical microscopy. Figure 1 compares the microstructures at three typical evolutional stages during an isothermal anneal at 120 ℃, i.e. nucleation, growth and the stage right before completion. Please note that the lighter areas are the remaining Al regions because of its high reflectivity and the circular darker areas are the AIC crystallite regions. It is seen that the average AIC crystallite regions as large as 10 μm can be reached at the AIC completion. Fig. 1(d-f) in the middle row shows a wafer deposited with an a-Si thickness of 1000 ° and a power density of 0.05 W/cm^2; Fig. 1(a-c) in the top row demonstrates the effect of an increased power density to 1.00 W/cm^2 from 0.05 W/cm^2 in the middle row; and Fig. 1 (g-i) in the bottom row depicts the effect of an increased a-Si thickness to 5000 Å from the 1000 Å in the middle.

When the middle row is compared with the top row under the same a-Si thickness at 1000 Å, it is seen that the decrease of the deposition power density results in a fewer number of nuclei at nucleation, a faster growth rate during growth, and a larger network at completion. Further, when the middle row is compared with the bottom row, where the power density is kept the same at 0.05 W/cm^2 but the thickness is increased from 1000 to 5000Å, it is seen that the increase of the a-Si thickness leads to a greater number of nuclei at nucleation, a slower growth rate during growth, and a smaller network at completion. Furthermore, when the top row is compared with both the middle and bottom rows of which the a-Si was deposited at a decreased power density of 0.05 W/cm^2 as opposed to the 1.00 W/cm^2 in the top row, it is interesting that the power density seems to play a more effective role in the AIC microstructural evolution than the a-Si thickness as seen that the top row has a greater number of nuclei, slower growth rate, and smaller network than both the middle and bottom rows. Please note that these observed phenomena in the microstructural evolution at 120 ℃ are consistent with other isothermal anneal temperatures.

Activation energy

To investigate the kinetics of the AIC diffusion process for the various wafers deposited at different deposition power densities and a-Si thicknesses, the AIC activation energy was calculated by using Avrami's equation as described in [5].

Shown in figure 2 is the dependence of activation energy on: (a) deposition power density ranging from 0.05 to 1.00 W/cm^2 with a fixed a-Si thickness of 1000 Å on Si wafers, and (b) a-Si thickness from 450 to 5000Å at a fixed power density of 0.05 W/cm^2 on glass wafers. It is seen that the activation energy increases with the increase of power density as well as the decrease of a-Si thickness. Furthermore, the overall result of activation energy from both power density and a-Si thickness changes from 1.11 to 1.27 eV, which is very close to the activation energy of Si diffusion in Al [7], confirming that the AIC is an Si diffusion-controlled process.

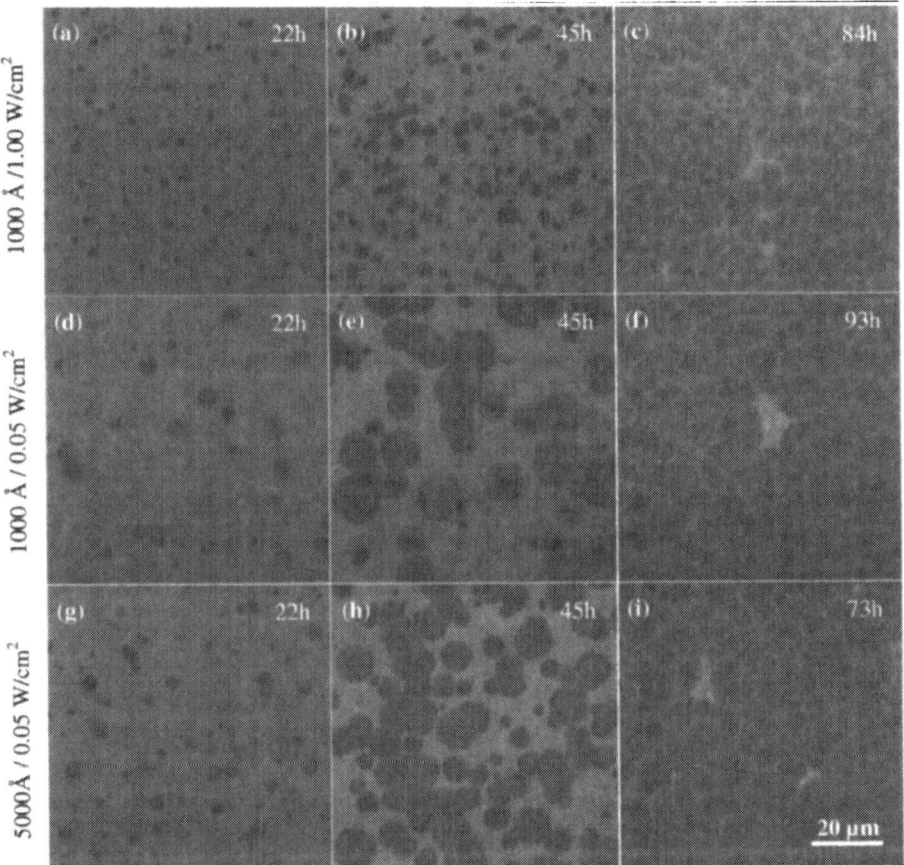

Figure 1. Optical microscopic images comparing the effects of deposition power density and a-Si thickness on the microstructural evolution at nucleation, growth and completion

Activation energy can be described as the energy barrier that diffusion species overcome to diffuse from one energy state to another. Thus, the effects of both deposition power density and a-Si thickness on activation energy should be related to the differences in the microstructural evolution and a lower activation energy will result in a faster growth. The activation energies of the AIC microstructural evolutions in the top and middle rows in Fig. 1 are also depicted in Fig. 2. They are 1.22 eV for the top row at a power density of 1.00 W/cm² and 1.11 eV for the middle row at 0.05 W/ cm². Thus, the lower activation energy from Fig. 2 is directly attributed to the faster growth of crystallite areas in the middle and bottom rows in Fig. 1.

Although nucleation and growth are both diffusion processes in a system, nucleation can occur favorably at high energy sites such as the surface, interface, boundaries, and any defects. Thus, the number of AIC nuclei not only is dependent of activation energy but existence of such high energy sites. During an AIC process, Si atoms diffuse into Al from a-Si through a thin interface SiO_2 layer formed due to the vacuum break between a-Si and Al depositions, and then precipitate in the form of crystalline Si. Because of the limited Al solubility in Si, Al atoms in the Si crystallites will be expelled out and diffuse through the interface into the a-Si. Clearly, the diffused Al atoms will continuously diffuse throughout the a-Si and/or precipitate within a-Si. This means that the AIC nucleation process not only depends on the nature of the Al layer and the interface between Al and a-Si but that of the a-Si layer. Therefore, as seen from Fig.1, there are a greater number of nuclei when a-Si thickness is increased due to the increased nucleation sites with a thick a-Si induced by Al during the AIC process, and when deposition power density is raised because of the increased defect states within the a-Si formed during its deposition.

Clearly, both a greater number of Si crystallites and a faster growth of crystallite areas can lead to early completion of the AIC process. However, the early completion of the AIC process not only is related to low activation energy for fast growth but to available nucleation sites for a great number of Si crystallites. This holds true when comparing the top row to the middle row in Fig. 1. The AIC reaction of the top row completed about 9 hours earlier than that of the middle row because of its greater number of crystallites even though its activation energy is higher.

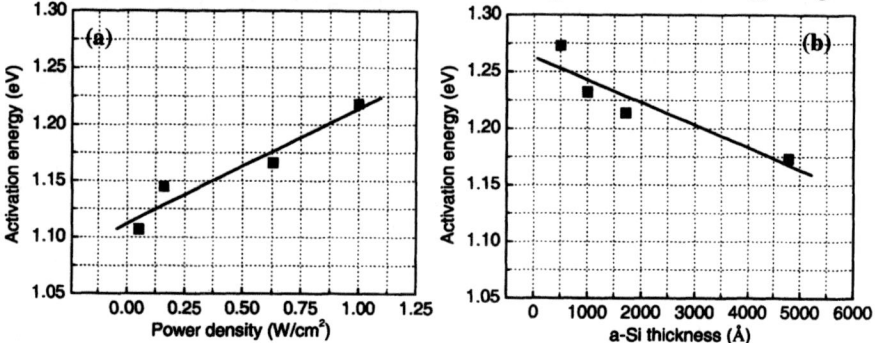

Figure 2. Dependence of activation energy on (a) deposition power density on Si wafers and (b) a-Si thickness on glass wafers

Finally it must be noted that the isothermal anneal temperature strongly affects the number of c-Si nuclei. A higher temperature is always attributed to a larger number of nuclei in this investigation, which is why a lower AIC temperature always corresponds to a larger network [5-6]. However, it is the activation energy that greatest affects the lowest temperature under which an AIC reaction can actually occur. Since the activation energy is strongly affected by the deposition power density, in order to achieve ever low AIC temperatures the deposition power density must be lowered. From the authors' knowledge, an AIC reaction at a temperature as low as 120 °C has never been previously reported.

Crystallinity and microstructural quality

To determine the microstructural quality of the low temperature AIC polysilicon films, high magnification SEM was carried out. Shown in figure 3 are two SEM pictures of a wafer of 1000 Å a-Si deposited at a power density of 0.05 W/cm^2, reacted at 120 °C for 137 hours, and then etched for 2 hours in an Al etchant: (a) high magnification and (b) higher magnification. In Fig. 3(a) the AIC network previously observed under optical microscopy can be clearly identified. The microstructures within the network as well as on the boundaries are uniform, which means that the AIC thin film is of high microstructural quality. Fig. 3(b) shows the dendratic characteristics with submicron-scale structures uniformly distributed within one of the networks.

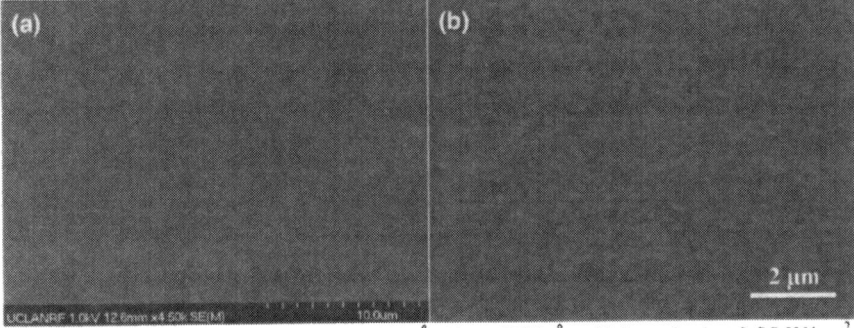

Figure 3. SEM images of a wafer with 300 Å Al on 1000 Å a-Si deposited at 0.05 W/cm^2, annealed for 137h, and etched for 2 h in an Al etchant: (a) at 4.5 K and (b) 15 K

To verify the crystallinity of the AIC thin films, X-ray diffraction pattern and TEM pictures were taken. The X-ray sample was prepared from the same wafer as of the SEM observations. However, the TEM samples were prepared by dipping a small piece of the wafer in a 50:1 HF solution, which etched away the PECVD SiO$_2$ so that the reacted AIC film was floated and picked up with a 3 mm copper grid for TEM observations.

Shown in figure 4 are (a) the typical X-ray diffraction pattern of the AIC film and (b) a bright-filed TEM diffraction contrast image of an AIC network with the inset that indicates its electron diffraction pattern. From Fig. 4 (a), the X-ray pattern shows exactly the existence of a strong {111} diffraction peak at 28.58°, which confirmed the crystallinity of the AIC film. From the TEM image in Fig. 4(b), the submicron-scale structures seen in Fig. 3 (b) can be seen clearly. The inset in Fig. 4(b), is the electron diffraction pattern of the TEM image on which three intense continuous rings are clearly seen and they are correspondent to the Si diffractions from {111}, {220} and {311} planes, respectively, from the inside to the outside. Thus, the inset diffraction pattern not only confirms that the AIC film is indeed crystalline again but also the submicron-scale structures seen by SEM in Fig. 3(b) and by TEM in Fig. 4(b) are Si crystallites. Existence of submicron-scale Si crystallites within large AIC networks has been reported before [4]. It is believed that formation of the submicron-scale crystalline Si is closely related to the nucleation and growth mechanisms of the AIC process, which will be reported in detail elsewhere.

CONCLUSION

The effects of power density and a-Si thickness on aluminum-induced crystallization of PECVD a-Si were investigated by using optical microscopy to characterize the AIC microstructural evolution and kinetics, and SEM, TEM, and X-ray diffraction analyses to determine the crystallinity and quality of the AIC films. It was found that the power density and a-Si thickness significantly influenced the activation energy, and inevitably the microstructural evolution, and the lowest AIC temperature that could be achieved. It was determined that the activation energy decreases when the power density decreases and/or when a-Si thickness increases. It is believed the AIC polysilicon films achieved by an AIC reaction at a temperature as low as 120 °C can be used for large-area microelectronics on flexible substrates.

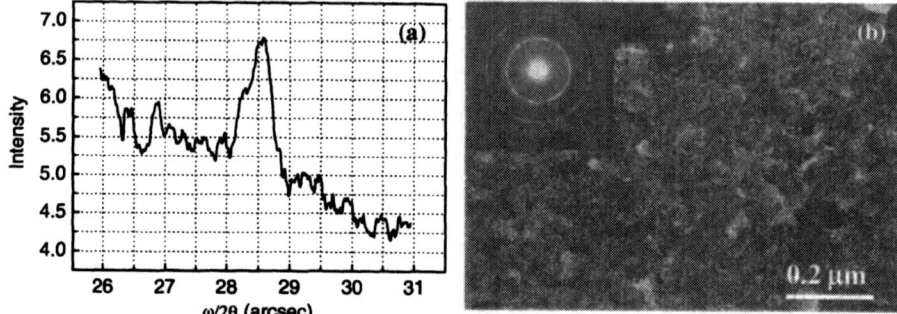

Figure 4. (a) X-ray diffraction and (b) bright-field TEM diffraction contrast image with an inset showing the electron diffraction pattern of submicron-scale crystalline Si

ACKNOWLEDGEMENTS

We are grateful to Dr. Mark Goosky for the use of his X-ray diffraction facility and Mr. George Malouf for his help taking the X-ray diffraction patterns. We would like to thank Mr. Minji Zhu for his help on taking SEM pictures.

REFERENCES

1. E.W. Forsythe, D.C. Morton, and G. L. Wood, SPIE Aerosense Proceedings, Proc. SPIE-Int. Soc. Opt. Eng. Vol. **4712** (2002) 262.
2. S.-I. Ishihara, M. Kitagawa and T. Hirao, J. Appl. Phys. **62** (1987) 837.
3. O. Nast and S. K. Wenham, J. Appl. Phys. **88** (2000) 124.
4. R. Kishore, C. Hotz, H. A. Naseem, and W. D. Brown, Electrochem. Sol.-Stat. Lett. **4** (2001) G14.
5. K. Jenq, S. S. Chang, Y.G. Lian, G.Z. Pan, and Yahmat-Samii, in *Amorphous and Nanocrystalline Silicon Science and Technology*, edited by Gautam Ganguly, Michio, Vol. **808**, Warrendale, PA, 2004), pp. 303-308.
6. Li Cai, H. Wang, W.D. Brown, and M. Zou, Electrochem. Sol.-Stat. Lett. **8** (2005) G179.
7. J. O. McCaldin and H. Sankur, Appl. Phys. Lett. **19** (1971) 524.

Mater. Res. Soc. Symp. Proc. Vol. 989 © 2007 Materials Research Society   0989-A06-12

Comparative Study of Hot-Wire Chemical Vapor Deposition onto (100) Si Near 600°C: Epitaxial and Polycrystalline Silicon Films

Charles W. Teplin, Howard M. Branz, Kim M. Jones, Bobby To, Eugene Iwaniczko, and Paul Stradins
NCPV, NREL, 1617 Cole Blvd, Golden, CO, 80401

ABSTRACT

Previously, we reported improved silicon epitaxy by hot-wire chemical vapor deposition (HWCVD) between about 600 and 650°C. Such temperatures are compatible with the thickening of large-grained Si seed layers on borosilicate glasses or other inexpensive substrates. Here, we provide detailed real-time spectroscopic ellipsometry (RTSE) and x-ray diffraction (XRD) analysis of two films grown near 600°C. A film grown at 594°C shows breakdown to a polycrystalline phase,.while a film grown at 627°C is entirely epitaxial. Transmission electron microscopy (TEM) of this epitaxial film shows dislocation defects that originate at the substrate/film interface, suggesting that an optimized surface preparation could yield lower defect densities.

INTRODUCTION

Economically produced crystal silicon on glass would enable significant cost reductions in photovoltaics. Typically, schemes for this technology include an initial step where a crystal silicon (c-Si) seed is grown onto an inexpensive substrate and a second step where this seed is epitaxially thickened [1-3]. Because low-cost substrates are unable to withstand prolonged times at high-temperature, epitaxy processes must be optimized for lower temperatures.

Our initial experiments at NREL explored epitaxy below 500°C by HWCVD [4]. Under optimized deposition conditions, we were able to grow ~0.5 μm of epitaxy at 380°C and 12 nm/min. After the initial stage of epitaxy, we observed that all of our films broke down to an amorphous silicon phase (a-Si:H) [5]. More recent experiments [6] at higher temperatures (500 - 700°C) showed that very thick epitaxy (at least 11μm) was possible above 600°C at significantly higher growth rates (>100 nm/min). Such thicknesses and temperatures would enable thickening of c-Si seeds on borosilicate glasses for PV applications.

In this report, we provide additional insight through comparison of two 0.5 μm films grown near 600°C by HWCVD onto (100)-oriented silicon.

EXPERIMENT

The films were grown in a HWCVD reactor with a base pressure before deposition near 1×10^{-6} Torr. A 0.5 mm diameter Ta filament at about 1800°C (11.5 A) is located 5 cm from the substrate and was used to crack 20 sccm of pure SiH$_4$ gas. A throttle valve was used to control the pressure at 10 mT. Each film was grown for 10 minutes and each is about 0.6 microns thick. To achieve temperatures above 500°C, we heat the substrate with second Ta filament located ~5 mm below the substrate that snakes over substrate area. To prevent silicidation of the heater wire from the process gases, we are careful to cover any openings between the substrate and heater wire. Nonetheless, reaction with process gases leaking into this zone eventually cause the heater wire to fail. (100) silicon substrates were carefully cleaned and dipped in 4% HF dip to strip the native oxide and terminate the surface with hydrogen [7].

In-situ, real-time spectroscopic ellipsometry (RTSE) data was acquired during deposition. X-ray diffraction measurements were performed using a XRD system (Bruker) with a two-dimensional X-ray detector. Substrate temperatures were determined from pyrometer measurements and ellipsometry data. Temperature information is obtained by fitting the ellipsometry data using temperature-dependent optical constants of c-Si that were previously measured at NREL under ultra-high vacuum conditions

RESULTS AND DISCUSSION

Figures 1 and 2 show RTSE data for films grown at 594°C and 627°C, respectively. In the bottom panel of each figure, the psuedodielectric function $<\varepsilon_2>$ spectrum are shown for different deposition times. In the top panels, the evolution of $<\varepsilon_2>$ at 1.5 eV is plotted vs growth time. $<\varepsilon_2>$ and $<\varepsilon_1>$ are related to the customary ellipsometry parameters Ψ and Δ through a simple algebraic expression and correspond to the complex dielectric function, $\varepsilon = \varepsilon_1 + i\varepsilon_2$, if the material being measured is accurately described by a uniform semi-infinite slab [8]. The $<\varepsilon_2>$ spectrum before deposition (solid lines in the bottom panels of Figs. 1 and 2) are close to the true ε_2 spectrum for heated crystal silicon.

During deposition just below 600°C, the $<\varepsilon_2>$ spectrum broadens and is reduced in intensity (Fig. 1). This is consistent with a loss of crystalline order and the development of surface texturing. The relatively slow change (occurring over ~300 nm of growth) at photon energies above 3 eV, where the penetration depth in silicon is very small (<10 nm), is consistent with

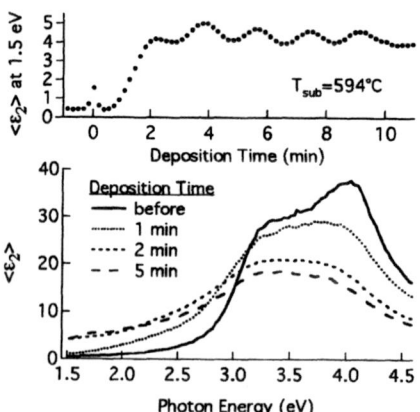

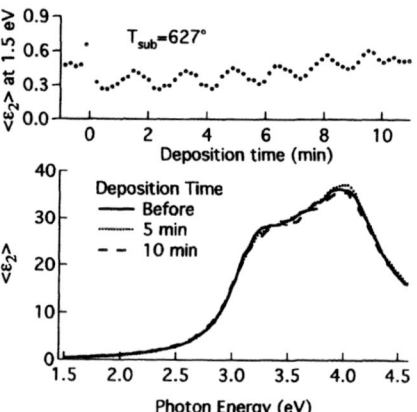

Figure 1. RTSE data for film grown at 594°C. In the bottom panel, spectra of the complex part of the pseudodielectric function, $<\varepsilon_2>$, are shown at various deposition times. At top, the temporal evolution of $<\varepsilon_2>$ at 1.5 eV during growth is shown.

Figure 2. RTSE data for film grown at 627°C. In the bottom panel, spectra of the complex part of the pseudodielectric function, $<\varepsilon_2>$, are shown at various deposition times. At top, the temporal evolution of $<\varepsilon_2>$ at 1.5 eV during growth is shown.

initial epitaxy that eventually fails. Atomic force microscopy, shown in Figure 3, does reveal that the film has developed a rough surface morphology. The sharp edges at the surface appear characteristic of small ~100 nm crystalline grains.

During deposition at 627°C, however, the $<\varepsilon_2>$ spectrum remains essentially unchanged (Fig. 2). This is what is expected for phase-pure epitaxial growth of silicon on silicon – the sample still responds like a semi-infinite slab.

In both samples, there is evidence of the substrate/film interface in the RTSE signal at low photon energies. Thickness fringes are visible in the data at 1.5 eV (top panels of Figs. 1 and 2) where the light penetration depth in silicon is thicker than in the sample. The existence of fringes indicates that there is optical

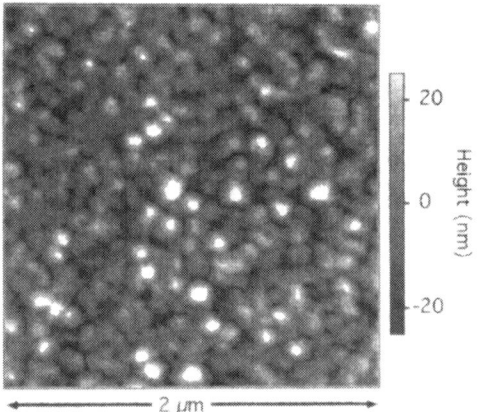

Figure 3. Atomic force microscopy image of the sample grown at 594°C.

contrast at the substrate/film interface that leads to optical reflection [5]. Additionally, a brief peak in $<\varepsilon_2>$ at 1.5 eV is observable (less than 3 data points) at the moment deposition is started. A similar "dip" in $<\varepsilon_2>$ is seen at higher photon energies (not shown). This effect is consistent with an initial increase in roughness before the initial layers of growth coalesce into a uniform film.

The thickness fringes at low energy allow us to determine the film growth rate. A simple 3-layer model of [c-Si substrate/EMA/c-Si film] can be used to fit the RTSE data at low photon energies. Here, EMA indicates an effective medium of c-Si with a small amount of void that represents the substrate/film interface. Using this model, we measure growth rates of 56 and 61 nm/min for the films grown at 594°C and 627°C, respectively.

Figures 4 and 5 show XRD data acquired on the two films. In Fig. 4A, the appearance of x-rays at angles (2θ) other than those corresponding to (100)-oriented silicon indicate that epitaxy has failed with the growth of new crystalline orientations. The non-uniform "smile" distribution in 2θ and χ shown in Fig. 4A suggest that the nucleation of specific crystalline orientations are favored during breakdown (if breakdown to random poly-crystalline silicon had occurred, rings of uniform intensity in χ would appear at each Si 2θ reflection). Similar XRD patterns have been observed in other films, suggesting that this breakdown crystallite orientation pattern is typical and not a chance observation of a small number of grains in this particular sample.

In Fig. 5, only a (400) peak is visible on the χ=0 midline, consistent with growth of an epitaxial film. They faint "eyes" observed at high power near 2θ=28.4° correspond to (111) reflections at non-zero χ angles and are consistent with an entirely (100)-oriented film.

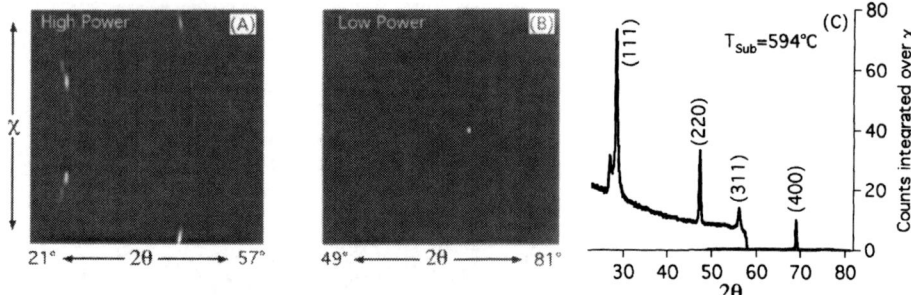

Figure 4. XRD data for the sample grown at 594°C. In panels A and B, detector images are shown for measurements taken at low 2θ values at high x-ray intensity (A) and at high 2θ values at low x-ray intensity. The horizontal midline of these images correspond to χ=0 measurements that are typically performed in θ-2θ measurements with a single detector. In C, the data in A and B is shown after being integrated over a range of χ values. Each peak is labeled with the silicon crystallographic plane corresponding to the observed 2θ position.

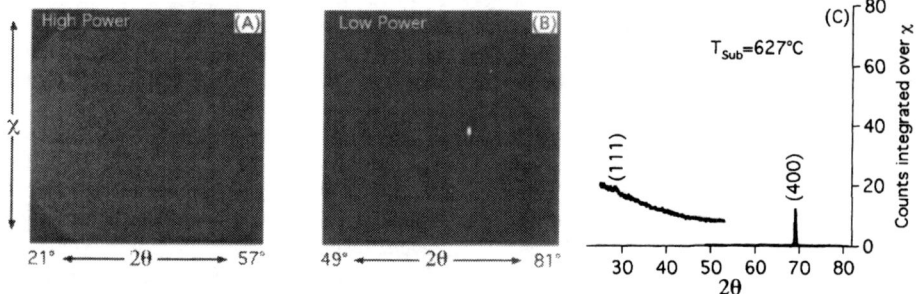

Figure 5. XRD data for the sample grown at 627°C, in the same format as in Fig. 4.

A TEM image of the film grown at 627°C is shown in Figure 6. The image shows that the film is entirely epitaxial. The image also reveals numerous dislocations in the film, though fewer than in our previously-published results [6]. Importantly, the defect density is highest near the substrate/film interface, with numerous dislocations originating from the interface itself. We also note that there is excellent agreement between the thickness measured by TEM (620 nm) and by fitting the low photon energy ellipsometry data. We note that in films grown at even higher temperatures (>640°C), another 10X reduction in defect density (compared with Figure 6) was obtained and will be published elsewhere in detail.

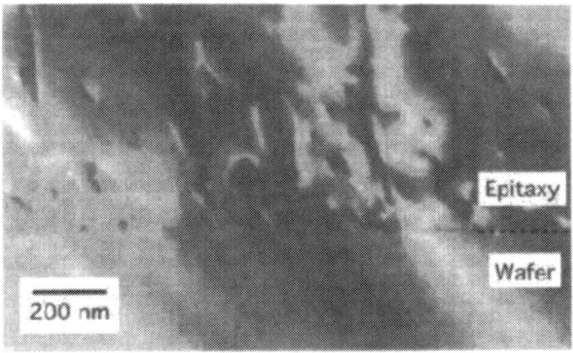

Figure 6. TEM image of the film grown at 627°C.

CONCLUSIONS

The results shown above confirm our early report that thick HWCVD epitaxy is only possible above 600°C [6]. Importantly, we note that the films reported here have lower defect densities than the similar films we reported previously. The TEM data strongly suggest that the defect densities are controlled by the quality of the substrate/film interface. We believe that these films are improved over previous results because of more careful surface cleaning and a lower system base pressure. The films reported here were cleaned using a modified RCA procedure [7], whereas our previous results above 500°C used substrates cleaned with only organic solvents and an HF dip to remove the surface oxide layer. Still further improvements have been made to reduce defect densities, and will be reported elsewhere.

ACKNOWLEDGEMENTS

This work was supported by the U.S. Dept. of Energy under Contract DE-AC36-99GO10337. The authors thank Qi Wang for helpful discussions and Matthew Page for substrate preparation.

REFERENCES

[1] R.B. Bergmann and J.H. Werner, Thin Solid Films 403, 162 (2002).
[2] W. Fuhs, S. Gall, B. Rau, M. Schmidt and J. Schneider, Sol. Energy 77, 961 (2004).
[3] C. Teplin, D.S. Ginley and H. Branz, J. Non-Cryst. Sol. 352, 984 (2006).
[4] C.W. Teplin, Q. Wang, E. Iwaniczko, K.M. Jones, M. Al-Jassim, R.C. Reedy and H.M. Branz, J. Cryst. Growth 287, 414 (2006).
[5] C.W. Teplin, D.H. Levi, E. Iwaniczko, K.M. Jones, J.D. Perkins and H.M. Branz, J. Appl. Phys. 97, (2005).
[6] Q. Wang, C.W. Teplin, P. Stradins, B. To, K.M. Jones and H.M. Branz, J. Appl. Phys. 100, 93520 (2006).
[7] C.W. Teplin, M.R. Page, E. Iwaniczko, K.M. Jones, R. Readey, B. To, H. Moutinho, Q. Wang and H.M. Branz, Amorphous and Nanocrystalline Silicon Science and Technology-2006 (Materials Research Symposium Proceedings Vol. 910) A15 (2006).
[8] H. Yao, J. Woollam, P. Wang, M. Teiwani and S. Alterovitz, App. Surf. Sci. 63, 52 (1993).

Mater. Res. Soc. Symp. Proc. Vol. 989 © 2007 Materials Research Society   0989-A06-13

ESR Study of Crystallization of Hydrogenated Amorphous Silicon Thin Films

Tining Su[1], Tong Ju[2], P. Craig Taylor[1], Pauls Stradins[3], Yueqin Xu[3], Falah Hasoon[3], Qi Wang[3], and Walter A. Harrison[4]

[1]Department of Physics, Colorado School of Mines, Golden, CO, 80401
[2]Department of Physics, University of Utah, Salt Lake City, UT, 84112
[3]National Renewable Energy Laboratory, Golden, CO, 80401
[4]Department of Applied Physics, Stanford University, Stanford, CA, 94305

ABSTRACT

Electron-spin-resonance (ESR) is used to investigate the evolution of the local order surrounding the dangling bonds produced by hydrogen effusion in a-Si:H thin films prepared by both plasma-enhanced-chemical-vapor-deposition (PECVD) and hot-wire CVD (HWCVD). At 560° C, the HWCVD sample fully crystallizes after ~ 800 min, while the sample made by PECVD does not crystallize. The PECVD sample crystallizes at a higher temperature (580° C) and after a much longer annealing time ($\Delta t = 1300$ min). The ESR signal of the defects in both samples remains at about 5×10^{18} cm^{-3} as long as the sample remains amorphous during the grain nucleation period. In both HWCVD and PECVD samples, as the crystallites appear, the defect densities gradually decrease and saturate at about 3×10^{17} cm^{-3} as the crystallization is completed.

INTRODUCTION

Solid-phase crystallization and the subsequent re-hydrogenation of the amorphous silicon thin films provides a low cost approach for thin-film crystalline Si:H-based photovoltaic devices [1,2]. During the hydrogen effusion, significant lattice reconstruction occurs, as hydrogen is driven out of the film. This process is accompanied by creation and migration of a large number of dangling bonds. Optical techniques, such as measurements of the reflectance and transmittance, provide *in situ* information of the kinetics of the crystallization [1,2]. However, these techniques do not provide information of local structural changes during crystallization. Experimentally, samples prepared by HWCVD crystallize faster and at lower temperatures, compared to samples made by PECVD [1,2]. This phenomenon is not well understood. Although most of the hydrogen is driven out of the film during effusion, there exists a rather persistent residual hydrogen concentration of about 10^{19} cm^{-3}. It is not clear if this residual hydrogen plays any role during the crystallization. Another issue of interest is the effect of rehydrogenation after the film is crystallized. The local order of the hydrogen atoms is crucial in understanding the effect of hydrogen passivation of the excessive defects.

We have investigated the evolution of the defects created during hydrogen effusion and the crystallization process, using magnetic resonance techniques such as nuclear magnetic resonance (NMR) and electron spin resonance (ESR). In this article, we report the results from ESR during the annealing and crystallization and present a simple model of the exchange interaction in the disordered *a*-Si.

EXPERIMENTAL DETAILS

Amorphous silicon samples were deposited by both HWCVD and PECVD at a facility at the National Renewable Energy Laboratory (NREL). Information of the samples is listed in Table I. One HWCVD sample (H2029) and one PECVD sample (L1477) were annealed at the same time at $T = 560°C$ in nitrogen gas. A second PECVD sample was annealed at $T = 580°C$ for 1300 min, and reflectance measurements showes that the film crystallized after $t \sim 1050$ min. The annealing is monitored by a n&k reflectance and transmittance spectrometer.

Table I. Growth conditions of the HWCVD and PECVD a-Si:H samples.

Sample #	Deposition method	Substrate temperature (°C)	Deposition rate (Å/s)	Thickness (μm)
L1447	PECVD	250	1-2	~1
H2029	HWCVD	400	7	~1

ESR measurements of the defect densities were carried out on a Bruker Elex 500 Spectrometer at room temperature. The defect densities were obtained by comparing the double integrated intensity in the samples to that in a standard weak pitch sample measured under the same conditions.

RESULTS

Dependence of Defect density on annealing time

Figure 1 shows the evolution of the defect densities in H2029 and L1477 as a function of the annealing time at $T = 560°C$. In the HWCVD sample (H2029), the defect density begins to decrease after about 60 min, and the film is fully crystallized after about 750 min. Reflectance measurements show that crystallization occurs during the same time period. After the film is crystallized, the defect density deceases to about 4×10^{17} cm^{-3}, and remains at this value through the rest of the annealing.

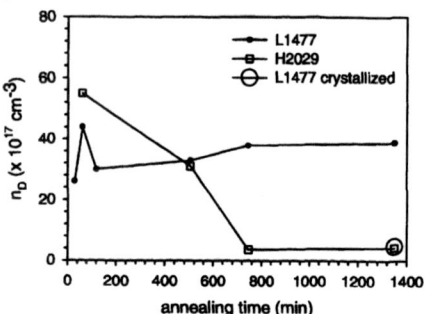

Figure 1   The defect densities in HWCVD (H2029) and PECVD (L1477) samples as a function of annealing time at $T = 560°C$. Open squares and solid circles indicate data for

HWCVD and PECVD samples, respectively. The open triangle indicates the defect density of the PECVD sample after annealing at $T = 580°C$ for $t\sim1300$ min.

On the other hand, the defect density in PECVD sample (L1477) shows an initial sharp increase to about 4×10^{18} cm^{-3}, followed by a decrease to 3×10^{18} cm^{-3}, and a gradual increase over the remainder of the annealing. This sample does not show any evidence of crystallization, which is consistent with the reflectance measurements. Due to the lack of detailed data on the HWCVD sample, we will focus our discussion in the PECVD sample.

Comparison of the lineshapes

Figure 2 shows the comparison of the lineshapes of the ESR signals in H-effused and crystallized samples to that in a typical a-Si:H sample. Trace (a) is from a typical a-Si:H sample, the peak-to-peak width is about 7 G. Trace (b) is from the sample in the H-effused state before crystallization, the line-width is about 5 G. The line-width of the crystallized sample (6.5 G) is much smaller than previously reported values in micro-crystalline silicon (μc-Si:H) samples [3]. The lineshape changes can be attributed to exchange-narrowing that is well known for high defect density films [4]. This narrowing effect will be discussed in detail later. Figure 2 also shows a slight difference in the g-values as calculated from the field of zero crossing, which could be due to the difference in ESR resonance frequency or to a slight change in g-value of the defect signal.

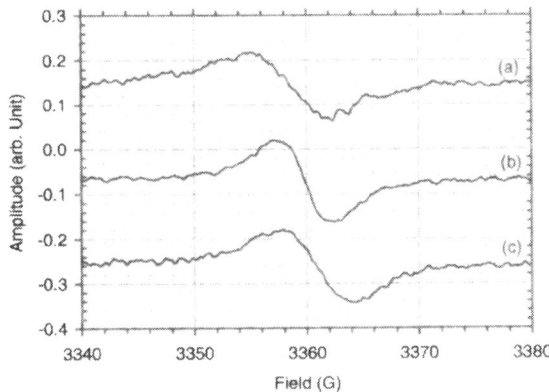

Figure 2. Comparison of the ESR lineshapes in H-effused and crystallized samples to a typical a-Si:H sample. (a) typical a-Si:H, (b) H-effused, before crystallization ($n_D = 3.6\times10^{18}$ cm^{-3}), and (c) after crystallization ($n_D = 4.3\times10^{17}$ cm^{-3}). The amplitudes are scaled to show the difference in line-width.

Evidence of clustered defects

Since only the total number of spins is measured in typical ESR spin count measurements, these measurements do not determine whether or not the spins are clustered. The bulk defect density is only accurate based on the assumption that the defects are randomly distributed within the sample. This is unlikely during the crystallization, as well as right before and after crystallization. To investigate the potential clustering of the spins, we compared the lineshape in the hydrogen effused PECVD sample to that in a tritiated amorphous silicon thin film (a-Si:HT) sample. In a-Si:HT, the defects are crested by decay of the tritium toms that are bonded to silicon atoms,

$$^{3}H \rightarrow {}^{3}He + \beta^{-} + \text{anti-neutrino} + 5.7 \text{ keV} \qquad (1)$$

The resulting ^{3}He atom will detach from the silicon atom and leave a dangling bond on the Si atom. The energy of the β particle is ~ 5.7 keV. Irradiating the sample with an electron beam with the same energy and dosage as the tritium decay in the a-Si:HT sample shows that no significant amount of defects are created [5, 6]. Therefore the main mechanism for defect creation is due to the reaction in Eq. (1). This process does not involve large-scale lattice reconstruction, due to the slow decay rate of the tritium nuclei, and therefore this process presumably creates defects that are more likely to be randomly distributed. Figure 3 shows the dependence of the line-width on the defect densities in one a-Si:HT sample. For comparison, the line-width of the H-effused PECVD sample with a bulk defect density of $n_D = 3.6 \times 10^{18} \text{ cm}^{-3}$ and the line-width in an a-Si sample prepared by RF sputtering [4] are also shown.

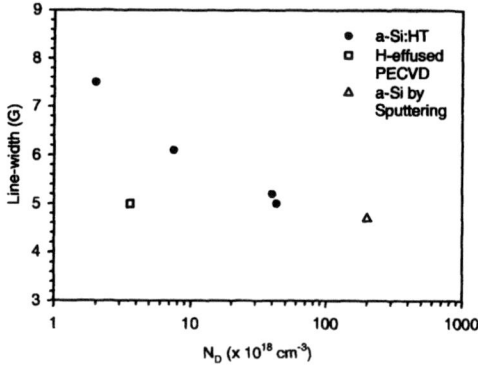

Figure 3   The dependence of the ESR line-width on the defect densities in a-Si:HT sample and in H-effused a-Si:H sample prepared by PECVD. Solid circles represent the data from the a-Si:HT sample, the open square represents the line-width for the defect density of $n_D = 3.6 \times 10^{18} \text{ cm}^{-3}$ in the H-effused PECVD sample. The open triangle is the line-width from Ref. [4] in a-Si sample prepared by RF sputtering.

Figure 3 clearly shows the discrepancy in line-width between the H-effused PECVD sample and the a-Si:HT sample with the same bulk defect density. The defect density in a-Si:HT is about one order of magnitude higher with the same line-width (~5 G). If we assume the defects are randomly distributed in a-Si:HT, then there must be significant clustering in the H-effused sample to have a comparable line-width, with a much lower bulk defect density. This result will be further discussed below.

DISCUSSION

It is well known that at high spin densities, the observed ESR line-width is narrower due to the so-called exchange-narrowing effect [7]. This effect arises from the exchange interaction between the spins, and has the form

$$H_{ex} = J\vec{S}_1 \cdot \vec{S}_2 , \tag{2}$$

where $\vec{S}_1$ and $\vec{S}_2$ are the spins of the two defects. J is the coupling constant. This interaction allows the spin-pair to undergo rapid mutual flipping and effectively average out the broadening due to other interactions, such as the dipole-dipole interaction and g-tensor anisotropy, between the spins. Exchange interaction deceases rapidly as the distance between the two spins increases, and therefore this interaction does not affect the line-width at low spin densities. Although this effect has been extensively studied in crystalline materials, no details to estimate the coupling constant have been reported. Here we develop a rather simple method to estimate the interaction in a-Si:H.

Consider two defects (dangling bonds) separated by n Si atoms, the exchange-interaction will be through the n-1 bonds in between. The dangling bond, which is a singly occupied sp^3 hybrid on a Si atom, interacts with another dangling bond by coupling to the Si-Si bond on the same atom, which in turn couples to another Si-Si bond, and so on. If we only consider the coupling through the bonding orbitals, the matrix of the total Hamiltonian is a $(n+1)\times(n+1)$ matrix,

$$\begin{pmatrix}
0 & \frac{V_1}{\sqrt{2}} & 0 & . & . & . & 0 \\
\frac{V_1}{\sqrt{2}} & -V_2 & \frac{V_1}{2} & . & & & . \\
0 & \frac{V_1}{2} & . & . & . & & . \\
. & . & . & . & . & \frac{V_1}{2} & 0 \\
. & & & 0 & \frac{V_1}{2} & -V_2 & \frac{V_1}{\sqrt{2}} \\
0 & . & . & . & 0 & \frac{V_1}{\sqrt{2}} & 0
\end{pmatrix} \tag{3}$$

where $\frac{V_1}{\sqrt{2}}$ is the coupling between the dangling bond and the Si-Si bond, $\frac{V_1}{2}$ is the coupling between two hybrid within a bond, and V_2 is the energy lowered by forming the bonding orbital. Diagonalization of this matrix results in $n+1$ energy levels, of which n-1 levels are close to V_2 (bonding orbitals). The other two energy levels are higher and are separated only by a small energy, which is due to the exchange interaction between the two dangling bonds.

To estimate the onset of the exchange-narrowing of the lineshape, we assume that the narrowing will be measurable when the exchange interaction is about the same as the line-width of the ESR signal. By choosing $V_1 = 1.8$ eV and $V_2 = 4.4$ eV for Si [8], we numerically diagonalized the Hamiltonian (Eq. 3), and obtained a value of the exchange interaction of $\Delta E_{ex} = 5.4 \times 10^{-8}$ eV when the two dangling bonds are separated by 12 atoms. This value of ΔE_{ex} corresponds to an ESR line-width of about 4.6 G. The defect density is calculated to be about 3×10^{19} cm^{-3}, with an average distance between two defects of 12 inter-atomic distances of Si. This is consistent with the experimentally observed relation between defect density and the ESR line-width in a-Si:HT, in which we assume the defects are randomly distributed, and not significant clustered.

We found that ΔE_{ex} decreases exponentially with an increasing number of bonds between the two dangling bonds. Therefore, the exchange interaction is effectively a short-range interaction. As a consequence, our results do not distinguish the case in which more than two defects are close to each other, and the case in which only paired defects are present.

CONCLUSIONS

In conclusion, we have perform ESR measurements in H-effused a-Si:H thin films made by both HWCVD and PECVD. The evolution of the defect density as a function of annealing time is consistent with that found in optical measurements. Analysis of the line-width shows that in H-effused states, the defects are either clustered or exist in pairs. A simple model of the exchange interaction in a-Si can account for the difference in line shapes observed in these films as compared to an a-Si:HT samples.

ACKNOWLEDGMENTS

The authors thank S. Zukotynski and N. Kherani for providing the a-Si:HT samples. This work is partially supported by NREL under subcontract number XXL-5-44205-09 and by NSF under grant number DMR-0073004.

REFERENCES

1. P. Stradins, D. Young, Y, Yan, E. Iwaniczko, Y. Xu, R. Reedy, H. M. Branz, and Q. Wang; Appl. Phys. Lett. **89**, 121921, (2006).
2. D. L. Young, P. Stradins, Y. Xu, L. Gedvilas, R. Reedy, A. H. Mahan, H. M. Branz, Q. Wang, and D. L. Williamson, Appl. Phys. Lett, **89**, 161910 (2006).
3. M. M. de Lima, Jr, P. C. Taylor, S. Morrison, A. LeGeune, and F. C. Marques, Phys. Rev. B **65**, 235324-1 (2002). (And references therein)
4. M. H. Brodsky, R. S. Title, K. Weiser, and G. D. Pettit, Phys. Rev. B **1**, 2632 (1970).
5. J. Whitaker, J. Viner, S. Zukotynski, E. Johnson, P. C. Taylor, P. Stradins, *Tritium Induced Defects in Amorphous Silicon*, in *Amorphous and Nanocrystalline Silicon Science and Technology—2004*, edited by Gautam Ganguly, Michio Kondo, Eric A. Schiff, Reinhard Carius, and Rana Biswas (Mater. Res. Soc. Symp. Proc. **808**, Warrendale, PA, 2004), A2.3.
6. T. Ju, J. Whitacker, S. Zukotynski, N. Kherani, P. C. Taylor, P. Stradins (To be published).
7. For example see, J. H. Van Vleck, Phys. Rev. **74**, 1168 (1948); P. W. Anderson and P. R. Weiss, Rev. Mod. Phys. **25**, 269 (1953).
8. *"Elementary Electronic Structures"*, W. A. Harrison, (World Scientific Publishing Co. Singapore, Singapore, 2004).

Mater. Res. Soc. Symp. Proc. Vol. 989 © 2007 Materials Research Society

Hot-Wire Chemical Vapor Deposition Epitaxy on Polycrystalline Silicon Seeds on Glass

Charles W. Teplin[1], Howard M. Branz[1], Kim M. Jones[1], Manuel J. Romero[1], Paul Stradins[1], and Stefan Gall[2]
[1]NCPV, NREL, 1617 Cole Blvd, Golden, CO, 80401
[2]Hahn-Meitner-Institut Berlin, Kekuléstr. 5, D-12489, Berlin, Germany

ABSTRACT

During the last few years, hot-wire chemical vapor deposition (HWCVD) has been explored as a low-temperature process for epitaxially thickening c-Si seeds layers on low cost substrates. Here, we demonstrate HWCVD epitaxy on thin polycrystalline silicon seed layers on borosilicate glass substrates. The crystal Si seeds are large-grained (~10 μm) polycrystalline silicon that were fabricated by Al-induced crystallization of a-Si. We report the growth of 0.5 μm of epitaxy at ~670°C.

INTRODUCTION

The high cost of the crystal silicon (c-Si) wafers in wafer-based photovoltaics has motivated recent research into cheaper forms of c-Si [1-4]. Numerous approaches have been explored to achieve c-Si of adequate quality to fabricate efficient solar cells. Perhaps the simplest approach has been to simply crystallize a-Si films on glass by annealing after deposition (solid phase crystallization – or SPC). Using SPC, mini-module efficiencies of up to 9.8% have been achieved [5]. However, SPC results in a log-normal distribution of Si grain sizes with many small grains and the largest grain size of about 1 micron [6]. The boundaries between these grains affect the electrical properties of the material. Reducing or eliminating these grain boundaries could dramatically improve the carrier lifetimes in this material and increase the achievable photovoltaic efficiency. One approach to reduce the density of grain boundaries is to carefully control the crystallization process using Al-induced crystallization (AIC) of a-Si [7]. This process yields dramatically larger grains. However, film thicknesses are limited and the Al doping level is too high for the Si to act as an effective photovoltaic absorber. Thus, a second step – in which the AIC seed layers are epitaxially thickened – is required. The initial devices based on Al induced crystallization of a-Si have shown promising open circuit voltages [7, 8].

Epitaxy on AIC seed layers has been achieved using electron cyclotron resonance CVD (ECR-CVD) [7, 9], ion beam assisted deposition (IAD) [10], and electron-beam evaporation [11]. We have explored the use of hot-wire chemical vapor deposition (HWCVD) for the epitaxy step at substrate temperatures above 600°C, because HWCVD appears straightforward to scale for large-scale production and because we recently demonstrated the possibility of thick epitaxy (at least 11 μm) at high growth rates (>100 nm/min) using HWCVD on (100) Si wafers [12]. Previously, Stradal and coworkers explored HWCVD epitaxy on poly-Si seed layers below 500°C and found only limited epitaxial thicknesses (<100 nm) [13]. Here, we report successful epitaxial thickening at ~670°C. The initial results suggest that HWCVD epitaxy is possible on all grain orientations.

EXPERIMENT

The fabrication process for Al-induced seed layers, used by the Hahn Meitner Institute Berlin, has been described in detail previously [7]. The procedure involves depositing first Al (300 nm) and then a-Si (375 nm) layers onto a borosilicate glass substrate (Borofloat 33 from Schott). During annealing, a layer exchange process occurs, resulting in a glass/poly-Si/Al(+Si) structure. The top Al(+Si) layer is then removed by chemical mechanical polishing (CMP) leaving a glass/poly-Si structure. Before epitaxy at the National Renewable Energy Laboratory (NREL), the AIC seed layers were cleaned using an organic solvent and then the surface oxide was removed by dropping 4% HF solution onto the poly-Si surface for 30 seconds.

These seed layers were epitaxially thickened in a HWCVD reactor with a base pressure near 10^{-6} Torr. A 0.51 mm diameter W filament at about 2100°C (heated by a 16 A current), located 5 cm below the inverted substrate, was used to crack 20 sccm of pure SiH_4 gas. A throttle valve was used to control the pressure to 10 mT. We report here on a film grown for 5 minutes at a growth rate of ~100 nm/min. The substrate was heated with a second Ta filament that snakes over the back of the substrate at a height of about 5 mm.

Temperature measurements on these samples are an experimental challenge. Pyrometry measurements are inaccurate because the infrared (IR) radiation from the sample includes a contribution from both the glass and the silicon film. The pyrometry measurements are further complicated by optical interference between the reflections from the front and back surfaces of the thin film of silicon. Ellipsometry measurements are also challenging because the optical constants of the poly-Si layer are not identical to those of single crystal silicon and because the surface of the seed layer is not identical to that of an ultra-high vacuum prepared c-Si wafer. After experimenting with other methods, we believe that our best estimate of the temperature arises from comparison of the ellipsometry data to that of the known temperature-dependence of single crystal silicon [14]. We estimate that the temperature of the film grown here was near to 670°C. When removed from the HWCVD reactor, the sample was slightly bowed because it was unsupported during growth.

We report here on scanning electron microscopy (SEM) and transmission electron microscopy (TEM) analysis of a successful epitaxial growth. We also performed electron-beam scattered diffraction (EBSD) mapping of the crystal grain orientations which will be reported in detail elsewhere.

RESULTS AND DISCUSSION

Figure 1 shows an SEM image of the surface of the grown film (after epitaxy) and a similar SEM image of an unthickened AIC seed layer (bare seed). The grain boundaries in the films are clearly visible in these images. The large grain sizes in the two images are very similar, suggesting that epitaxy was achieved. If epitaxy failed, one would expect much smaller poly-Si grains (see paper A6.12 in this Proceedings). EBSD measurements taken on the areas shown in Fig. 1 reveal similar grain orientation distributions. Cathode luminescence measurements of the epitaxially thickened sample suggest low densities of intra-grain defects.

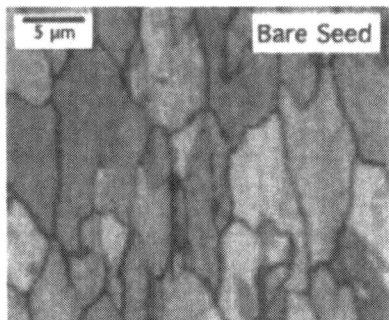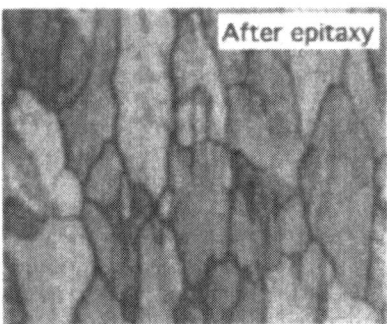

Figure 1. SEM images of an AIC seed on glass and of a thickened sample. The similar grain sizes in the two images suggest that epitaxy was successful on the majority of grain orientations.

A TEM image of epitaxy along two adjacent grains is shown in Fig. 2. The TEM suggest that epitaxy occurred on both grains, and high-resolution TEM images of the lattice planes do confirm epitaxy on all grains examined. Interestingly, the grain boundary in this film did not propagate normal to the interface during growth, suggesting that the growth rate may be different on different crystalline orientations [15]. We note that few dislocations nucleate at the seed/epitaxy interface. This is somewhat surprising, since we normally observe a high dislocation density during deposition on (100) silicon wafers [12]. We speculate that the glass softening reduces strain between the seed and epitaxial layer. In other TEM images, we observe that intra-grain defects in the seed layer (presumably formed during AIC) continue into the epitaxial film.

Figure 2. TEM images of the thickened AIC seed layer.

If HWCVD is to be used for thickening AIC seed layers, it is important to understand any additional challenges posed by growth on poly-Si, as compared to growth on single crystal wafers where most previous epitaxy has been done. An obvious challenge is achieving epitaxy on all grain orientations. In the SEM and TEM data presented here, we have not observed any indications of epitaxy failure, regardless of grain orientation. However, the TEM images suggest that different grains may thicken at different growth rates. This difference in growth rate could be beneficial if faster growing grains overtook slower growing grains, effectively increasing the grain size at the surface of the film. However, differing growth rates are also likely to increase the roughness of the film, which could be problematic for subsequent processing steps.

The bowing of the glass substrate during deposition should also be prevented, as nonplanar substrates could be problematic for subsequent processing steps. However, there are simple steps that could be taken to reduce this problem (other than lowering the temperature). First, optical heating could be implemented that would preferentially heat the silicon seed layer (and epitaxial film) over the glass substrate. Second, the substrate could be supported or held vertically during growth. Such arrangements would improve upon our current reactor configuration, where the substrate is suspended below the substrate holder with small clips at the edges.

CONCLUSIONS

We have demonstrated that HWCVD epitaxy can be used to thicken poly-Si seed layers on borosilicate glass substrates at about 100 nm/min. There is no breakdown to finecrystalline silicon observed in the first 600 nm of growth. With further optimization, it is likely that hot-wire epitaxy on seed layers produced by Al-induced crystallization of a-Si could be used to fabricate c-Si solar cells on glass.

ACKNOWLEDGEMENTS

This work was supported by the U.S. Dept. of Energy under Contract DE-AC36-99GO10337.

REFERENCES

[1] R.B. Bergmann and J.H. Werner, Thin Solid Films **403**, 162 (2002).
[2] W. Fuhs, S. Gall, B. Rau, M. Schmidt and J. Schneider, Sol. Energy **77**, 961 (2004).
[3] C. Teplin, D. Ginley and H.M. Branz, J. Non-Cryst. Sol. **352**, 984 (2006).
[4] C. Richardson, M. Mason and H. Atwater, Thin Solid Films **501**, 332 (2006).
[5] P. Basore, 21st European Photovoltaic Solar Energy Conference - Dresden, Germany Oral presentation (2006).
[6] R. Bergmann, F. Shi and J. Krinke, Phys. Rev. Lett. **80**, 1011 (1998).
[7] S. Gall, J. Schneider, J. Klein, K. Hubener, M. Muske, B. Rau, E. Conrad, I. Sieber, K. Petter, K. Lips, M. Stoger-Pollach, P. Schattschneider and W. Fuhs, Thin Solid Films **511**, 7 (2006).
[8] A. Aberle, J. Cryst. Growth **287**, 386 (2006).
[9] G. Ekanayake, T. Quinn, H.S. Reehal, B. Rau and S. Gall, J. Cryst. Growth **299**, 309 (2007).

[10] A. Straub, D. Inns, M.L. Terry, Y. Huang, P. Widenborg and A. Aberle, Thin Solid Films **511**, 41 (2006).

[11] B. Gorka, P. Dogan, I. Sieber, F. Fenske and S. Gall, Thin Solid Films in press (2007).

[12] Q. Wang, C. Teplin, P. Stradins, B. To, K.M. Jones and H.M. Branz, J. Appl. Phys. **100**, 093520 (2006).

[13] J. Stradal, G. Scholma, H. Li, C.H.M. Van Der Werf, J.K. Rath, P. Widenborg, P. Campbell, A. Aberle and R. Schropp, Thin Solid Films **501**, 335 (2006).

[14] P. Lautenschlager, M. Garriga, L. Vina and M. Cardona, Phys. Rev. B **36**, 4821 (1987).

[15] C. Teplin, E. Iwaniczko, B. To, H. Moutinho, P. Stradins and H.M. Branz, Phys. Rev. B **74**, 235428 (2006).

Mater. Res. Soc. Symp. Proc. Vol. 989 © 2007 Materials Research Society   0989-A06-17

Grain Size Control by Means of Solid Phase Crystallization of Amorphous Silicon

Jordi Farjas[1], Pere Roura[1], and Pere Roca i Cabarrocas[2]

[1]GRMT, Physics department, University of Girona, Campus Montilivi, Girona, E-17071, Spain
[2]LPICM (UMR 7647 CNRS), Ecole Polytechnique, Palaiseau Cedex, F-91128, France

ABSTRACT

The grain size of thermally crystallized a-Si films is controlled by the rates of nucleation, r_N, and growth, r_G, according to the standard Avrami theory. Despite this evidence, most papers devoted to improving the crystallized grain size analyze their results with a qualitative reference to this theory. In this paper, we will show that it is possible to identify the standard set of r_N and r_G values for a-Si and that experiments reveal that deviations from these values always result in a smaller grain size. It is also shown that no substantial improvement with non-conventional heat treatments can be expected. Finally, it is argued that a larger grain size is expected from a-Si films containing, in their as-grown state, a controlled density of embedded nanocrystals.

I. INTRODUCTION

Solid phase crystallization (SPC) of amorphous silicon (a-Si) thin films results in a material with improved transport properties and stability against Staebler-Wronski effect [1]. The present shortage in the worldwide production of silicon [2] makes the crystallized thin films an interesting alternative to monocrystalline or multicrystalline solar cells [3]. On the one hand, it takes advantage of the much lower material consumption, characteristic of thin film technology, and on the other, it has the stability of crystalline silicon. The main drawback of crystallized thin films comes from the deleterious effect of grain boundaries on the minority carrier lifetime which results in a lower photovoltaic conversion efficiency [4]. Although extensive research has been carried out for more than twenty years, no significant improvement has been achieved.

In this paper, we first review a number of significant papers that allow us to identify the 'standard' value of the kinetic parameters governing the SPC process. We will show that all experiments indicate that the observed deviations from the standard values never result in a larger grain size. Section III is devoted to analyzing if the final grain size can be improved after thermal treatments other than the conventional single isothermal annealing. Finally, in Section IV it is shown that it is possible to obtain a larger grain size through the crystallization of a-Si layers containing silicon nanocrystals embedded in the amorphous matrix.

II. REVIEW OF THE LITERATURE

The standard kinetic parameters and grain size

The crystallization of a-Si begins with the spontaneous formation of nanometric ordered regions: the nuclei. They are usually revealed by TEM. The incubation time, t_0, and the nucleation rate, r_N are obtained from the density of nuclei as a function of the annealing time.

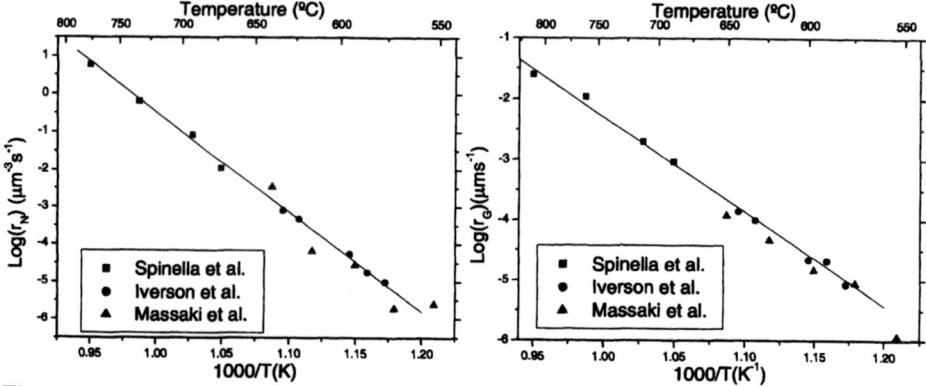

Figure 1. The standard kinetic parameters of a-Si crystallization.

The growth of these nuclei by continuous addition of atoms coming from the surrounding amorphous phase is described by the growth rate, r_G. In figure 1 the results reported by independent groups are plotted altogether. Iverson et al. [5] and Spinella et al. [1] measured them from low pressure chemical vapour deposition (LPCVD) films after complete amorphization by ion implantation whereas Masaki et al. [6] made the measurements on plasma enhanced (PE)CVD films. Despite the different microstructure, hydrogen content and deposition methods, the values of r_G and r_N almost coincide and can be described by an Arrhenius dependence:

$$r_N = (1.7 \pm 0.4)10^{26} e^{-(5.30 \pm 0.1)eV/kT} \mu m^{-3} s^{-1} \quad r_G = (2.1 \pm 0.4)10^{13} e^{-(3.10 \pm 0.1)eV/kT} \mu m \ s^{-1} \tag{1}$$

From figure 1 we see that the results of Masaki et al.[6] deliver slightly lower values of r_G. The incubation time, t_0, measured by Iverson et al. [5] and Spinela et al. [1] can be fitted to:

$$t_0 = (0.92 \pm 0.2)10^{-15} e^{3.4eV/kT} s \tag{2}$$

whereas no t_0 was observed by Massaki. This different behaviour is not a treat of PECVD materials since finite values of t_0 have been measured in these kinds of samples by other authors [7-9]. We will refer to the values of equations 1-2 as the "standard kinetic parameters".

The evolution of the crystallized fraction, α, with the annealing time, t, as well as the final grain size, d_G, can be described with reasonable accuracy by the equation of Avrami [10]:

$$\alpha(t) = 1 - \exp(-(kt)^n) \tag{3}$$

where n is the Avrami exponent (n = 4 for 3D growth) and $k = ((\pi/3)r_N r_G^3)^{1/4}$ for 3D growth of spherical grains. Within the framework of this theory one can predict the peak temperature and the FWHM of the transformation rate when crystallization takes place at a constant heating rate, $\beta \equiv dT/dt$ [11]. We have measured the transformation rate by calorimetric experiments on PECVD films [12] and the results show a nice agreement with the prediction according to the standard parameters over a large temperature range (figure 2). Additionally, this theory predicts the final grain size, which for 3D (i.e. films much thicker than d_G) is expressed as [13]:

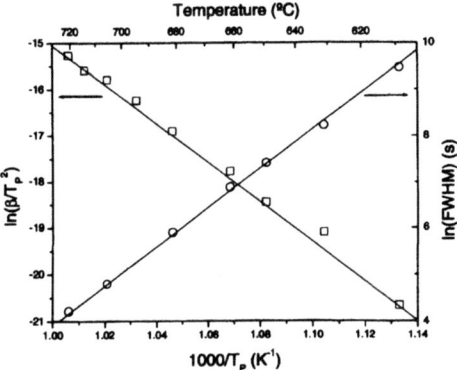

Figure 2. Peak temperature and FWHM of the crystallization rate measured by differential calorimetry at several heating rates (β). The straight lines are not fits to the data; they correspond to the prediction with the standard parameters.

$$d_G = 1.29(r_G / r_N)^{1/4} \tag{4}$$

Equation 4 predicts a monotonous decrease of the grain size the annealing temperature (figure 3). We will refer to d_G given by equation 4 with the standard r_G and r_N values as the "standard grain size", d_{ST}. At 600°C, $d_{ST}=1.10$ μm and it increases up to 1.75 μm at 550°C. This increase requires a ten-fold increase of the annealing time which, at 600°C, is already quite long ($t_{0.95} = 22.5$ h). By $t_{0.95}$ we mean the time needed to get a crystalline fraction of 0.95 (figure 3).

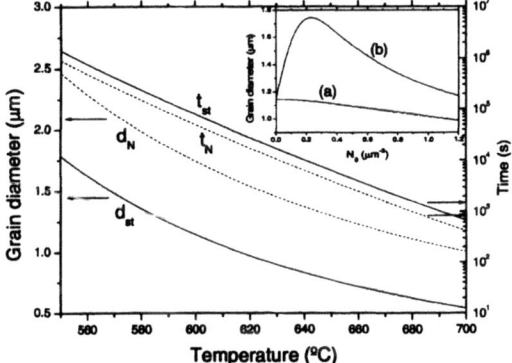

Figure 3. Grain size and crystallization time predicted by the standard parameters (d_{st}) and the optimum value when nanocrystals are embedded in the amorphous matrix (d_N). Inset, grain size at 600°C as a function of the nuclei already present in the amorphous matrix. a) Nuclei created by a thermal period at a higher temperature and b) preexisting nuclei in the as-grown state.

Comparison with experimental results

Table I.- Summary of the grain size (d), kinetic parameters (r_G, r_N, t_0 stand for growth and nucleation rate and incubation time respectively) and crystallization time (t_{crist}) found in the literature. All these values are normalized to the standard values.

	Deposition technique	Deposition temperature (°C)	$r_G/r_G(st)$	$r_N/r_N(st)$	$t_0/t_0(st)$	$t_{crist}/t_{0.95}$	d_G/d_{st}
Nakazawa et al. [7]	PECVD	50	7	6	--	0.2	0.7
		150	15	30	0.2	0.1	0.4
		320	2	110	0.5	0.4	0.3
Hatalis et al.[14]	LPCVD	545	>0.9	>60	--	--	0.2
		560	>0.6	>150	--	--	0.2
Ryu et al.[15]	LPCVD	475	>1.5	>0.5	--	--	>1
Pangal el al.[8]	PECVD	150	1.5	5-10	1.1	--	0.6
	epitaxy	--	6	--	--	--	--
Bo et al.[16]	PECVD	150	>0.15	>25	--	--	0.3
		150	>0.75	>0.5	--	--	1
Kumoni et al.[17]	LPCVD	550	1.2	25	0.4	--	--
	ii-LPCVD	--	0.9	1.5	1.5	--	--
Young et al.[9]	HWCVD	--	--	--	0.15	0.2	--
	PECVD	--	--	--	1	1	--
Yamauchi et al. [18]	ii-LPCVD	--	0.8	0.8	--	--	--
Kim et al.[19]	PECVD	150	--	--	1.4	1.3	--

In this subsection we want to address two questions. First, to what extent can the kinetic parameters of equations 1-2 really be considered standard values? Second, is it possible to alter r_G and r_N through the deposition conditions in order to obtain a grain size larger than the standard value (figure 3)? The answer to both questions will come from a review of a number of papers, most of them detailed in Table I. We have analyzed the results published there in order to extract the kinetic constants whenever possible. All the annealing experiments were done within a narrow range of temperatures between 580 and 600°C. The films' thicknesses ranged from 0.1 to 2 μm. In order to make the analysis as clear as possible, the reported values in Table I are relative to the standard in the particular conditions of the experiment. For most of the papers in which the incubation time was not known [14-16], only lower bounds of r_N and r_G could be extracted.

Without exception, we conclude from Table I that the particular values of r_G, r_N and t_0 coincide with the standard values within a narrow range or else the kinetics is clearly faster. The kinetic parameters almost coincide for the films crystallized after ionic implantation (ii-LPCVD) [17,18] and also for PECVD samples.[9,19]. For our PECVD films (figure 2) and those of Masaki et al.[6], r_G and r_N coincide but $t_0 = 0$.

Deviations of r_N from the standard values are usually due to nucleation at surfaces [1,8,15] or to an inhomogeneous amorphous structure [7], whereas the origin of the r_G variations is less clear. Nakazawa et al.[7] argue that they could be due to residual impurities. We think that it is reasonable to state that, in the absence of any extrinsic contribution (nucleation at surfaces, impurities...) and, for a homogeneous amorphous structure (such as that obtained by ionic implantation), crystallization is governed by the standard kinetic parameters.

Now, taking into account that r_N and r_G can be changed with the deposition conditions, we can hope to obtain films with a faster growth rate which, according to equation 4, would develop larger grains. Indeed this enhancement is usually surpassed by a larger increase in r_N with the overall result of a grain size that is smaller than the standard size (last column of Table I). In some cases (the first two films of Nakazawa [7]), the measured grain size clearly deviates (is smaller) from the expected value given by equation 4. This discrepancy is also found in the paper by Iverson [5], which indicates that microscopic measurements tend to underestimate the final grain size. Despite this lack of accuracy, we can conclude from the experimental results that it is very difficult to surpass the standard grain size by modifying r_N and r_G. The nucleation rate is never reduced with respect to the standard value and the growth rate does not increase enough.

III. NON-CONVENTIONAL HEAT TREATMENTS

In view of the difficulties of modifying r_G and r_N in a good direction, we will analyze now if it is possible to obtain larger grains with the standard kinetic parameters but following non-conventional heat treatments. First of all, we will discard two possibilities. Once the film has crystallized, it can 'recrystallize' into very big grains of hundreds of microns [20]. However, the process is very slow consequently the material has to heated at too high temperature (>1100°C) [20]. The second proposal consists of heating the sample at a constant rate β [12]. In this case, the growth is favoured with respect to the nucleation because of its lower activation energy and the fact that the material approaches a low temperature range before it crystallizes. A quantitative analysis confirms [21] that the grain size obtained at a given heating rate (which corresponds to a given peak temperature, figure 2) is larger than that obtained with an isothermal treatment at the peak temperature. However, the enhancement is too moderate (around 8%) to be significant.

For the rest of this section, we will consider a two-step treatment consisting of a short period at a high temperature to create a controlled density of nuclei, N_0, followed by a long period at a lower temperature at which these nuclei would grow. This proposal is similar to the two-step experiment by Fan et al.[22] who induced nucleation by laser annealing. From a formal point of view, it can be considered that after the incubation time, t_0, the amorphous phase undergoes crystallization in two parallel processes: the growth of the 'pre-existing' nuclei and the usual nucleation and growth process. It is possible to analytically predict the final grain size [13] which, as shown in the inset of figure 3, never exceeds d_{ST}. The reason for this negative result is the fact that although N_0 can be low enough to give big grains, the parallel nucleation at the low temperature period is too intense to be neglected.

IV. AMORPHOUS FILMS WITH EMBEDDED NANOCRYSTALS

The two-step treatment fails to give larger grains because the pre-existing nuclei begin to grow at the same time as the conventional nucleation and growth process. It is possible to make it better with a single isothermal treatment as follows. According to the standard kinetic parameters, a long time elapses before 'thermal' nucleation begins (i.e. the incubation time). If the as-deposited a-Si film already contains N_0 pre-existing nuclei, then they can grow from $t = 0$ without the parallel thermal nucleation. During the incubation period, the pre-existing nuclei could grow, which would led to large grains whose size would depend on N_0. In the inset of figure 3 we see that there is an optimum value of N_0 for which the grain size is at its maximum, d_N, and substantially larger than d_{ST}. In figure 3, the value of d_N is compared with d_{ST} for the practical temperature range. An increase by a factor of 2.5 is expected in the whole range. An additional benefit of this approach is the reduction in the time needed for crystallization.

It is possible to grow PECVD a-Si films with embedded nanocrystals. It has recently been shown that silicon nanocrystals formed in the plasma can be trapped in the substrate by a proper bias voltage [23].

CONCLUSIONS

In this paper, we have shown that, from the results reported in the literature, it is possible to identify the standard values of the nucleation rate, the growth rate and the incubation time governing the crystallization kinetics of a-Si. We have seen that deviations from these standard values have not led to larger grain sizes and that this improvement cannot be reached by non-conventional heat treatments. Finally, it is concluded that a substantially larger grain size could be achieved with thin films containing embedded nanocrystals in their as-grown condition.

ACKNOWLEDGMENTS

This work has been partially supported by the Spanish Programa Nacional de Materiales under contract number MAT2006-11144.

REFERENCES

1. C. Spinella, S. Lombardo and F.Priolo, *J. Appl. Phys.* **84**, 5383 (1998).
2. S.Hegedus, *Prog. Photovol. Res. & Appl.* **14**, 393 (2006).
3. M.A. Green, *Solar Energy* **74**, 181 (2003)
4. J. Nelson, *The physics of solar cells*, (Imperial College Press, London, 2003).
5. R. B. Iverson and R. Reif, *J. Appl. Phys.* **62**, 1675 (1987).
6. Y. Masaki, P. G. LeComber and A. G. Fitzgerald, *J. Appl. Phys.* **74**, 129 (1993).
7. K. Nakazawa and K. Tanaka, *J. Appl. Phys.* **68**, 1029 (1990).
8. K. Pangal, J. C. Sturm, S.Wagner and T. H. Büyüklimanli, *J. Appl. Phys.* **85**, 1900 (1999).
9. D. L. Young, P. Stradins, Y. Xu, L. Gedvilas, B. Reedy, A. H. Mahan, H. M. Branz, Q.Wang, and D. L. Williamson, *Appl. Phys. Lett.* **89**, 161910 (2006).
10. A. N. Kolmogorov, *Izv. Akad. auk. SSSR Ser.Fiz.* **1**, 355 (1937).
11. J. Farjas, P. Roura, *Acta Mater.* **54**, 5573 (2006).
12 J. Farjas, Chandana Rath, P. Roura and P.Roca i Cabarrocas, *Appl. Surf. Sci.* **238**, 165 (2004).
13. J. Farjas and P.Roura, *Phys. Rev. B.* **75** (scheduled issue: 01 May 2007).
14 M. K. Hatalis and D. W. Greve, *J. Appl. Phys.* **63**, 2260 (1988).
15 M.-K. Ryu, S.-M. Hwang, T.-H. Kim, K.-B.Kim, S.H.Min, *Appl. Phys. Lett.* **71**, 3063 (1997).
16 X.-Z. Bo, N. Yao and J. C. Sturm, *J. Appl. Phys.* **91**, 2910 (2002).
17 H. Kumonia and T. Yonehara, *J. Appl. Phys.* **75**, 2884 (1994).
18 N. Yamauchi and R. Reif, *J.Appl.Phys.* **75**, 3235 (1994).
19 H.-Y. Kim, J.-B. Choi and J.-Y. Lee, *J. Vac. Sci. Technol. A* **17**, 3240 (1999).
20 C. D. Ouwens and H. Heijligers, *Appl. Phys. Lett.* **26**, 569 (1975).
21 J. Farjas and P. Roura, *Acta Mater* (submitted).
22 Ch.-L.Fan, M.-Ch.Chen and Y.Chang, *J. Electrochem. Soc.* **150**, H178 (2003).
23 N.Chaabane, V.Suendo, H.Vach, P.Roca i Cabarrocas, *Appl.Phys.Lett.* **88**, 203111 (2006).

Mater. Res. Soc. Symp. Proc. Vol. 989 © 2007 Materials Research Society   0989-A06-18

Microwave Activation of Dopants & Solid Phase Epitaxy in Silicon

D. C. Thompson[1], J. Decker[1], T. L. Alford[1], J. W. Mayer[1], and N. David Theodore[2]
[1]School of Materials, Arizona State University, Tempe, AZ, 85287
[2]Wireless & Packaging Systems Lab., Freescale Semiconductor Inc., Tempe, AZ, 85284

ABSTRACT

Microwave heating is used to activate solid phase epitaxial re-growth of amorphous silicon layers on single crystal silicon substrates. Layers of single crystal silicon were made amorphous through ion implantation with varying doses of boron or arsenic. Microwave processing occurred inside a 2.45 GHz, 1300 W cavity applicator microwave system for time-durations of 1-120 minutes. Sample temperatures were monitored using optical pyrometery. Rutherford backscattering spectrometry, and cross-sectional transmission electron microscopy were used to monitor crystalline quality in as-implanted and annealed samples. Sheet resistance readings show dopant activation occurring in both boron and arsenic implanted samples. In samples with large doses of arsenic, the defects resulting from vacancies and/or micro cluster precipitates are seen in transmission electron micrographs. Materials properties are used to explain microwave heating of silicon and demonstrate that the damage created in the implantation process serves to enhance microwave absorption.

INTRODUCTION

Microwave heating has been presented as a viable alternative to other high temperature processing methods used to repair the damage inherent in ion implant processing [1-4]. The most common methods of repairing implant damage in silicon processing, lamp and laser rapid thermal processing (RTP), may result in uneven heating of the near surface device layers due to emissivity differences between device layers [2, 5]. Uneven heating results in vertical and lateral dopant diffusion, which degrades device performance [6]. Further complications with lamp and laser heating can arise because the depth of the heated region is determined by the depth of absorbed photons, 1 µm or less in silicon, resulting in conduction losses and higher power input needs [1]. Microwave heating of silicon allows for more even heating of a near surface volume due to the penetration depth associated with microwave processing. This volumetric heating may result in less power use and less dopant diffusion. Microwave heating has already found utility by activating solid state reactions in silicon [7, 8]. For this work microwave processing was used to study the effects of microwave heating on highly damaged layers of ion implanted silicon.

EXPERIMENTAL

To fabricate the samples used in this work Czochralski (CZ) grown, 100 mm, P-type boron doped, 50-60 Ω cm (100) orientated silicon wafers were cleaned and placed in an NV10-180 batch ion implanter. Samples for monitoring implant damage and boron activation were implanted with 33 keV B^+ ions, to doses ranging $1 \times 10^{14} - 5 \times 10^{15}$ B^+ cm^{-2}, at room temperature. Samples used to monitor implant damage and arsenic activation were implanted with 180 keV As^+ ions, to doses ranging 1×10^{14} $cm^{-2} - 5 \times 10^{15}$ As^+ cm^{-2}, at room temperature. For microwave processing, boron implanted silicon samples were placed in a 2.45 GHz, cavity applicator microwave system where they were processed for time durations ranging 0.15 - 120 minutes using a 1300 Watt magnetron source.

In arsenic implanted silicon samples the microwave reactor could not achieve a large enough microwave power input to initiate dopant activation with incident microwaves alone. To achieve the desired sample temperature in arsenic implanted silicon silicon carbide (SiC) microwave susceptors were used to supplement incident microwaves. Although this type of heating would best be characterized as microwave co-heating, previous reports have shown that co-heating does not negate the effects of microwave processing [3].

Silicon surface temperature was estimated during microwave anneal using a Raytek Compact MID series pyrometer with a spectral response of 8 – 14 μm and an estimated emissivity of 0.7 [9]. Previous reports have shown that the silicon emissivity used in pyrometric determination of temperature for low doped silicon is dependant on response wavelength and temperature. In moderately doped silicon the dependence is not as pronounced [9]. In this work we believe that the use of a spectral response range by the pyrometer, coupled with in situ microwave activation of dopants which result in a moderately doped silicon surface, resulted in an adequate estimate of surface temperature.

To monitor lattice damage as a function of microwave process time samples were analyzed using Rutherford backscattering spectrometry (RBS) with a 2.0 MeV He^{++} analyzing beam. Samples were mounted in random and [001] channeled orientations. He^{++} ions were collected using a solid state detector, positioned 13° from the incident beam. To examine crystalline quality and extended defect networks cross-sectional transmission electron microscopy (XTEM) was performed using a Philips CM200-FEG TEM operating at a voltage of 200 kV. To monitor dopant activation, samples surfaces were contacted with an in-line four-point-probe (FPP) equipped with a 100 mA Keithley 2700 digital multimeter.

RESULTS & DISCUSSION

Figure 1A displays the results of RBS analysis on silicon implanted with 33 keV, $5x10^{15}$ B$^+$ cm^{-2}, followed by microwave processing for up to 30 minutes. Spectra (3)-(5) shown in Figure 1A are the result of microwave processing involving incident microwaves only. As can be seen in figure 1A, spectra (1) corresponds to the randomly oriented RBS spectra for un-implanted silicon, spectra (2) – (4) correspond to RBS analysis with boron implanted samples oriented in a [001] channeled direction, and spectra (5) corresponds to the [001] channeled spectra for un-implanted silicon. RBS analysis of an as-implanted sample in an [001] channeled direction (spectra 2) demonstrates that a highly damaged silicon layer exists near the surface of the silicon. Comparison of the normalized yield of spectra (2) with the normalized yield of spectra (1) demonstrates that the as-implanted sample contains a layer of disorder less than that of amorphous silicon. Spectra (3, 4) in Figure 1A demonstrates that samples processed for times greater than 10 minutes resulted in a significant reduction in the lattice damage incurred during high dose ion implant. Sample temperature during microwave processing, monitored by pyrometer, ranged 450 – 500 C depending on process time. Figure 1B displays the RBS spectra that resulted after microwave processing arsenic implanted samples while using SiC microwave susceptors. Spectra (1) in Figure 1B corresponds to the randomly oriented RBS spectra for un-implanted silicon, spectra (2) and (3) correspond to $5x10^{15}$ cm^{-2} and $1x10^{14}$ cm^{-2} arsenic implanted samples as-implanted and oriented in a [001] channeled direction, respectively.

Spectra (4) and (5) correspond to the [001] channeled spectra for 5×10^{15} cm^{-2} and 1×10^{14} cm^{-2} arsenic implanted and microwave annealed silicon, respectively. As can be seen in Figure 1B spectra (3), a highly disordered surface layer remains after ion implant. Spectra (2) demonstrates that for implant doses of 5×10^{15} As$^+$ cm^{-2} a thin amorphous silicon layer resides at the surface of implanted samples. RUMP simulation software used on spectra (2) in Figure 1B determined the amorphous layer thickness to be 245 nm, approximately twice the projected range of 180 keV As$^+$ implanted into silicon [10]. Spectra (4) and (5) in Figure 1B demonstrate that after microwave processing nearly all implant induced damage in the arsenic implanted silicon has been annealed away. Comparison of spectra (4) and (5) with the channeled spectra of virgin silicon (not shown) demonstrate that all three spectra coincide. The normalized yield of arsenic implanted samples did not decrease further with increasing time of anneal. Samples temperatures during microwave processing ranged 500 – 650 C, depending on process time.

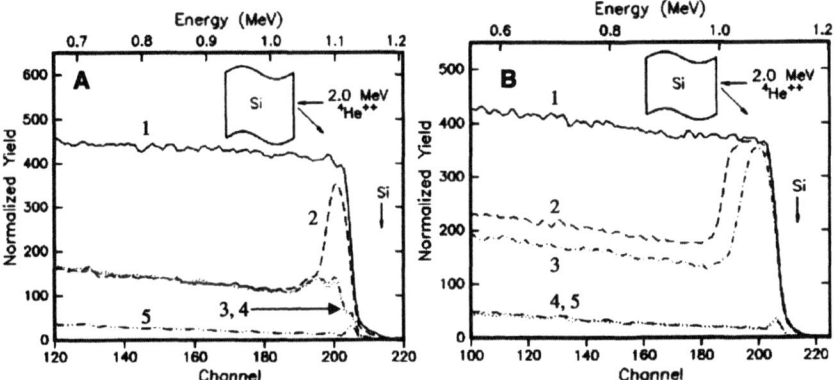

Figure 1. RBS spectra for boron and arsenic implanted silicon. 1A: 5×10^{15} B$^+$ cm^{-2} implanted Si (1) as-implanted random orientation (2) as-implanted in a channeled orientation (3, 4) post 10 and 30 minute anneals, respectively, in a channeled orientation, (5) virgin Si in channeled orientation. 1B: arsenic implanted Si, as-implanted, (1) in random orientation, (2) 5×10^{15} cm^{-2} in a channeled orientation, (3) 1×10^{14} cm^{-2} in a channeled orientation, (4, 5) 5×10^{15} cm^{-2} and 1×10^{14} cm^{-2} arsenic implanted silicon, post anneal, in a channeled orientation.

Figure 2 demonstrates the change in sheet resistance (R$_s$) during successive microwave anneals for boron and arsenic implanted silicon. Sheet resistance was monitored between microwave anneals to determine if dopant activation was occurring during microwave processing [6]. Figure 2 reports average sheet resistance, as measured from several samples within each wafer. Sheet resistance was measured at several points on each sample wafer in order to capture any non uniformity during heating. In all measured samples, sheet resistance readings differed by less than 10%. As can be seen in Figure 2A, R$_s$ of the boron implanted silicon falls dramatically within 1 minute of microwave processing. For microwave times greater than 5 minutes, R$_s$ of each implanted sample was approximately constant. Also apparent in the figure, the sheet resistance of implanted samples saturated at successively lower values for increasing implant doses. Figure 2B demonstrates the change in sheet resistance for arsenic implanted and microwave annealed silicon, as a function of microwave processing time. As can be seen in the figure, samples implanted with arsenic doses ranging $1 \times 10^{14} - 5 \times 10^{15}$ cm^{-2} realized a decrease in R_s after microwave annealing. Similar to figure 2, increasing implant dose corresponds to a decrease in the saturation R_s, an expected result if dopant activation occurs during microwave processing [8]. Also like Figure 2A, R_s in Figure 2B nearly saturates for all microwave processing longer than 5 minutes.

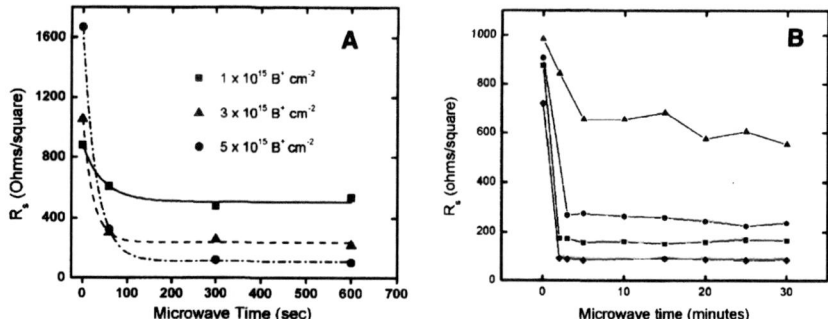

Figure 2. (A) R_s vs. microwave time for Si with (■ – 1×10^{15}, ▲ – 3×10^{15}, ● - 5×10^{15}) B$^+$ cm^{-2} followed by microwave processing. (B) R_s vs. microwave time for Si with (▲ – 1×10^{14}, ● - 3×10^{14}, ■ – 5×10^{14} and ♦ - 1×10^{15}) As$^+$ cm^{-2} followed by microwave processing.

To view the radiation effects in the ion implanted silicon post microwave processing XTEM was performed on arsenic implanted and microwave annealed silicon samples. To supplement the XTEM data SAED patterns were taken in order to demonstrate the extent of crystallinity in processed samples [11]. Figure 3 demonstrates the results of XTEM analysis. As can be seen in Figure 3, the as-implanted samples result in an amorphous layer at the surface of the sample. The amorphous region can be distinguished as the lightly shaded area in Figure 3A. The corresponding SAED pattern in Figure 3A demonstrates the characteristic opaque rings associated with highly disordered silicon [11]. Figure 3B demonstrates the effect of microwave processing the sample pictured in Figure 3A for 30 minutes. Even though the overlaid SAED pattern demonstrates great crystallinity, two small bands of defects can be seen in the XTEM micrograph. The center of the upper band of defects lies at approximately 120 nm and compares well to the TRIM simulated implant projected range (R_p) of 180 keV As$^+$ into silicon (122 nm), while the lower band of defects lie at approximately twice the implant projected range ($2R_p$) [12]. The location of these defect bands gives insight into their mechanism of formation. The lower defect band in Figure 3B can be accounted for as a coagulation of deep level interstitials indicative of ion irradiation. Using the +1 model of implant induced damage in solids, an excess of interstitials would be present after vacancy-interstitial annihilation during anneal of the damaged silicon [13, 14]. The upper band of defects seen in figure 3B may be the consequence of excess vacancies coalescing into a defect network. Another possibility is that the defects are the result of precipitates of arsenic. Arsenic implanted into silicon has been shown to form micro clusters of precipitates for implant doses as low as 2-5×10^{14} As$^+$ cm^{-2}. When the micro clusters form interstitials are released and these interstitials can condense into extended dislocation loops [15].

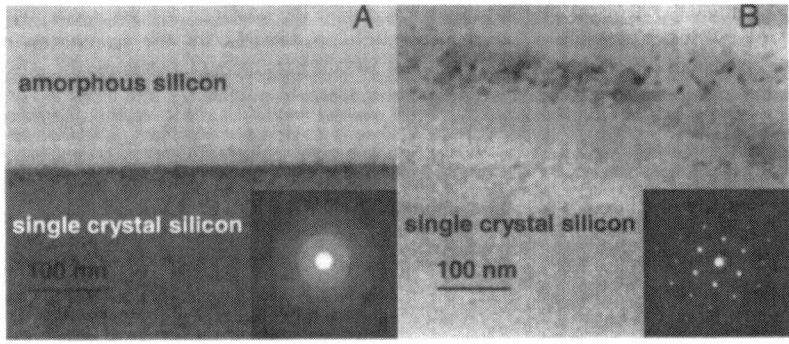

Figure 3. XTEM micrographs and SAED patterns of 5×10^{15} As$^+$ cm^{-2} implanted Si A) as-implanted, and B) post microwave processing

Due to the nature of the RBS spectra in Figure 1, the process temperatures attained during microwave heating, the defect bands seen in Figure 3B, and the absence of any dislocation networks indicative of nucleated crystal growth within an amorphous layer, we believe that the crystal re-growth mechanism during microwave anneal of arsenic implanted silicon samples is that of solid phase epitaxial re-growth (SPEG), nucleating at twice the projected implant range ($2R_p$) and proceeding toward the sample surface [13, 14]. The mechanism by which incident microwaves heated the silicon used in this work involves volumetric heating through power transfer from incident microwaves to the silicon sample. Power transfer from microwaves to a dielectric involves power loss from the microwaves, to the dielectric [16]. The significant microwave loss mechanisms in ion implanted silicon are conduction and polarization. Microwave conduction losses occur when free electrons in the extrinsic silicon move in response to an applied electric field. Microwave polarization losses occur in extrinsic silicon due to electric and dipolar polarization. Electric polarization occurs when electrons are displaced from their equilibrium positions around nuclei in response to an applied electric field. Dipolar polarization occurs in extrinsic silicon as a consequence of coupled defects such as vacancy – vacancy, vacancy – interstitial pairs, or complex dipoles interacting with applied electric fields [16]. A materials parameter that captures the effect of conduction and polarization loss mechanisms in microwave heating of extrinsic silicon is the loss factor ϵ". When the loss factor includes the effects of both conduction and polarization it is called an effective loss factor. Using an effective loss factor power absorption and penetration depth can be modeled within the silicon [7].

CONCLUSION

This work has demonstrated that microwave processing can be used for dopant activation and radiation repair in ion implanted silicon. Boron and arsenic implanted silicon samples were microwave processed to temperatures required for solid phase epitaxy in silicon. Upon reviewing surface temperature data, XTEM, SAED patterns, RBS, and sheet resistance curves we determined that the mechanism of damage repair in arsenic implanted silicon was solid phase epitaxial re-growth. Microwave loss mechanisms were responsible for the conversion of microwave power to heat in ion implanted silicon. A materials parameter that captures the effect of microwave loss mechanisms is the effective loss factor.

Acknowledgements

This work was partially supported by a grant from the NSF (DMR-0308127, L. Hess).

REFERENCES

1. A. V. Rzhanov, N. N. Gerasimenko, S. V. Vasil'ev and V. I. Obodnikov, Sov. Tech. Phys. Lett. 7, 521, (1981)
2. S. – L. Zhang, R. Buchta and D. Sigurd, Thin Solid Films, 246, 151 (1994).
3. H. Zohm, E. Kasper, P. Mehringer and G. A. Müller, Microelec. Eng. 54 247 (2003).
4. K. Thompson, J. H. Booske, R. F. Cooper and Y. B. Gianchandani, Mat. Res. Soc. Symp. Proc. 717 (2002).
5. R. Buchta, S.-L. Zhang, and D. Sigurd, Appl. Phys. Lett. 62, 3153 (1993).
6. D. K. Schroder, *Semiconductor Material and Device Characterization*, John Wiley & Sons, Hoboken, New Jersey (2006).
7. D. C. Thompson, H. C. Kim, T. L. Alford, J. W. Mayer, Appl. Phys. Lett. 83, 3918 (2003).
8. D. C. Thompson, T. L. Alford, J. W. Mayer, T. Höchbauer, M. Nastasi, S. S. Lau, N. David Theodore, K. Henttinen, Ilkka Suni and Paul K. Chu, Appl. Phys. Lett. 87, 224103 (2005).
9. T. Sato, Jpn. J. Appl. Phys. 6, 339 (1967).
10. L. R. Doolittle, Nuclear Instruments & Methods in Physics Research, B B9, 344-351 (1985).
11. B. D. Cullity and S. R. Stock, *Elements of X-Ray Diffraction*, Prentice Hall, Upper Saddle River, NJ (2001).
12. J. F. Zeigler, IBM Research, Yorktown, NY, 10598.
13. J. W. Mayer, S. S. Lau, *Electronic Materials Science: For Integrated Circuits in Si and GaAs*, Macmillan Publishing Company, New York, NY (1990).
14. J. D. Plummer, M. D. Deal, P. B. Griffin, *Silicon VLSI Technology: Fundamentals, Practice and Modeling*, Prentice Hall, Upper Saddle River, NJ (2000).
15. K. S. Jones, S. Prussin and E. R. Weber, Appl. Phys. A 45, 1 (1998).
16. A. C. Metaxas, R. J. Meredith, *Industrial Microwave Heating*, IEEE Power Engineering Series 4, (Peter Peregrinus Ltd, London, U. K. 1983).

Mater. Res. Soc. Symp. Proc. Vol. 989 © 2007 Materials Research Society   0989-A06-19

Low Temperature Epitaxy of n-Doped Silicon Thin Films using Plasma Enhanced Chemical Vapor Deposition

Mahdi Farrokh Baroughi[1,2], Hassan G. El-Gohary[1], Cherry Y. Cheng[1], and Siva Sivoththaman[1]

[1]Electrical and Computer Engineering, University of Waterloo, 200, Univ. Ave. W., Waterloo, Ontario, N2L3G1, Canada

[2]Current affiliation: Electrical Engineering and Computer Science Department, South Dakota State University, 201 Harding Hall, Brookings, SD, 57007

ABSTRACT

Highly conductive epiraxial silicon thin films, with conductivities more than 680 $\Omega^{-1}cm^{-1}$, were obtained using plasma enhanced chemical vapor deposition (PECVD) technique at 300°C. The effect of hydrogen in growth of low temperature extrinsic Si thin films was studied using conductivity, Hall, and Raman measurements, and it was shown that epitaxial growth was possible at hydrogen dilution (HD) ratios more than 85%. The epitaxial growth of the extrinsic Si thin films at high hydrogen dilution regime was confirmed by high resolution transmission electron microscopy (HRTEM).

INTRODUCTION

Highly conductive Si thin films are widely used in semiconductor devices such as emitters in solar cells and bipolar transistors, and as source/drains in MOSFETs. For these applications, the highly doped Si material is obtained either by high temperature diffusion process or by ion implantation followed by a high temperature annealing process [1]. At low temperatures (<300 °C), however, conductive Si thin films with a maximum reported conductivity value of less than 100 $\Omega^{-1}cm^{-1}$, are normally obtained by PECVD [2-5] of highly diluted silane in hydrogen (HD> 99%) on glass substrates. It has been shown that low temperature PECVD of thin intrinsic Si films on c-Si with similar process conditions of microcrystalline Si films, deposited on glass substrate, results in near epitaxial growth of Si films [6]; this causes some major differences in the structural and the electrical properties of the epitaxial Si and microcrystalline Si thin films. Unlike the development of microcrystalline Si thin films on glass substrates, much less work has been reported on epitaxial Si film growth on Si substrate using PECVD technique at low temperatures [7-9]. Further, the studies of the epitaxial Si films on c-Si substrates using PECVD were mainly limited to the structural study of thin intrinsic (undoped) Si films [7,9]. Near epitaxial films with very high doping concentrations add to the challenge of obtaining high crystallinities in the low temperature Si thin films.

Conductivity in an (n$^+$) Si film depends on electron mobility and the free electron density, which depends on active doping density. High crystallinity is the key towards obtaining highly conductive films because an improved crystallinity can improve both the carrier mobility and also the active doping density. We have developed highly conductive near epitaxial (n$^+$) Si thin films using PECVD of hydrogen diluted silane and phosphine. We have studied the electrical and structural properties of the films using conductivity, Hall, Raman and HRTEM analyses and confirmed an epitaxial growth of highly conductive (n^+) Si films. Also we have shown that the epitaxial growth highly improves the electron mobility and the doping efficiency inside the film.

EXPERIMENTAL

Films were grown by a standard RF PECVD system running at 13.56 MHz and a base pressure of 2-4×10^{-6} Torr. To prepare samples for conductivity measurement, Hall measurement and Raman analysis, (n^+) Si films of 50 nm thickness were deposited on Czochralski silicon (CZ-Si) wafers. The CZ-Si wafers were doped with boron (p-type) and the resistivity of the wafers were in the range of 1-2 Ω-cm. For HRTEM analysis, films of about 100 nm thickness were grown on multicrystalline Si substrates with substrate resistivity values in the range of 1-2 Ω-cm and random crystal orientation.

The Si substrates were cleaned by the standard RCA procedure, rinsed in DI water, and dipped in 1% HF solution for 30 sec before loading into the process chamber. The following process conditions were employed for growth (deposition) of the films using the PECVD system: pressure = 400 mT, RF power density = 47 mW/cm^2, SiH$_4$ flow = 25 sccm, PH$_3$ flow = 0.1 sccm, and T = 300°C. The hydrogen flow (HD =$100[H_2]/[H_2+SiH_4]$) was varied in a wide range of 10 sccm (28.5%) to 475 sccm (95%) to study the effect of hydrogen on the growth mechanism and the conductivity of the films.

To make metal electrodes on the films for the conductivity and Hall measurements, an Al layer with thickness of 200 nm was deposited on the (n^+) Si films using RF magnetron sputtering technique. The Al layer was then patterned using conventional lithography technique to form the Al electrodes. The isolation of the highly doped, $[PH_3/SiH_4]=0.4\%$, (n^+) Si films from the Si substrates were obtained from the n^+p junction formed between the film and the substrate. Furthermore, the high doping density of the (n^+) Si thin films assures that the depletion layer width in the (n^+) Si film is ultra thin, less than 5 nm. Therefore, the electrical measurements only reflect the electrical properties of the deposited films.

UV Raman measurements were performed on the films of 50 nm thickness. The use of a short wavelength laser (328 nm), with very short penetration depth in Si (<10 nm), in Raman measurements allows us to completely eliminate the effect of the crystalline substrate on the measured Raman signal. For HRTEM analysis films of 100 nm thickness were grown. The HRTEM was performed by a JEOL 2010F field emission electron microscope operated at 200 kV.

MEASUREMENT RESULTS AND DISCUSSION

The evolution of the film growth was studied by varying hydrogen dilution from 28.5% - 95%. Fig. 1 shows the conductivity of 50 nm films versus the hydrogen dilution ratio during the film growth. Three different regimes can be identified in the figure: (a) The films deposited with HD<80% show very low conductivities, about 0.008 $\Omega^{-1}cm^{-1}$, comparable to the conductivity of (n^+) a-Si:H films. (b) The films deposited with 80%<HD<85% show a rapid change from low conductivities to high conductivities. Such a sharp transition, close to four orders of magnitude change due to only 5% change in hydrogen dilution, suggest a rapid phase transform from amorphous phase to some form of crystalline phase for the highly phosphorous-doped thin films. (c) The films grown with HD>85% show very high film conductivities, about 680 $\Omega^{-1}cm^{-1}$, comparable to the conductivity of highly doped high temperature diffused c-Si films.

This set of experiments result in two major conclusions: (i) The conductivity of the film highly depends on the amount of hydrogen dilution and it changes up to about 5 orders of magnitude with only varying hydrogen dilution. This suggests that hydrogen plays extremely important role in the growth mechanism (amorphous phase growth or epitaxial phase growth) of the developed films. (ii) The growth mechanism and the film conductivity is extremely substrate dependent. While extremely high conductivities (about 680 $\Omega^{-1}cm^{-1}$) were obtained by growing films on CZ-Si substrate using 90% and 95% hydrogen dilution values, the grown films on glass substrates under identical conditions showed conductivity values less than 0.01 $\Omega^{-1}cm^{-1}$. Such a significant substrate dependence of the conductivity of the films suggests that the use of Si substrate significantly improves the crystallinity of the grown films. This is inline with the theory of the epitaxial growth on c-Si substrates under high hydrogen dilution conditions.

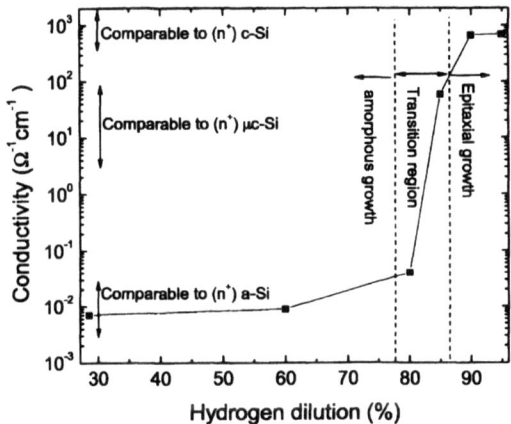

Fig.1 : Conductivity of the films grown under different hydrogen dilution ratios. About five orders of magnitude enhancement in the conductivity of the films were observed by only varying the HD from 28.5% to 95%. The experiment suggests that the phase of the developed film depends on the value of HD, amorphous growth at low HD values and epitaxial growth at high HD values.

Hall measurement was employed to characterize the mobility of the high conductivity films (HD>85%). The films developed with HD of 90% and 95% showed an electron mobility of 39 $cm^2/v.s.$ and 42 $cm^2/v.s.$ and free electron density of 1.1×10^{20} cm^{-3} and 1.05×10^{20} cm^{-3}, respectively. Such high electron mobility in a highly degenerate material with electron density of about 10^{20} cm^{-3} is very high and can be also compared with the electron mobility values in (n$^+$) c-Si material [10]. If we roughly assume an equal gas phase and solid phase phosphorous concentration of 0.4% the doping density of the film would be equal to 2×10^{20} cm^{-3}. Considering the doping density of 2×10^{20} cm^{-3} and the electron density of 10^{20} cm^{-3} and knowing the fact that the material is highly degenerate, it can be concluded that the doping efficiency of the low temperature films obtained at high hydrogen dilution values is very high, close to one. Further, very high electron mobility and a high free electron density in these films suggest a very long range order in the structure of the films.

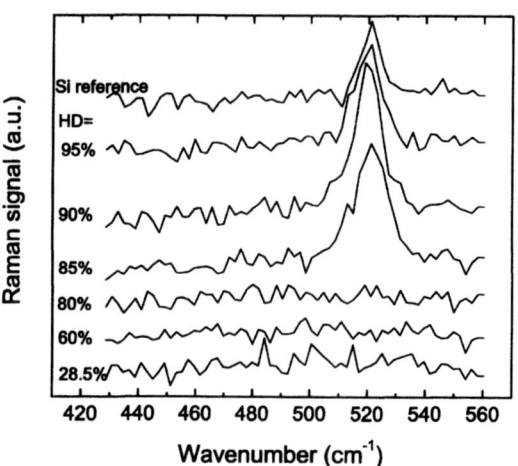

Fig.2: Raman spectra of the 50 nm Si thin films on c-Si substrate. Films were deposited using different HD values of 28.5%, 60%, 80%, 85%, 90%, and 95%. A transition from amorphous growth to crystalline growth is seen between hydrogen dilutions of 80% and 85%.

Fig. 2 shows the Raman spectra of the 50 nm (n$^+$) Si thin films grown on c-Si substrate using different HD values. No Raman peaks at 520 cm^{-1} and 480 cm^{-1}, the well known Raman peaks associated with c-Si and a-Si:H networks, were observed for the films deposited by HD values less than 80% in Fig. 2. However, the films deposited with hydrogen dilutions more than 85% showed a strong peak at 520 cm^{-1}, very similar to the Raman signal from the Si reference. Although the short penetration depth of the UV laser in the Si films has resulted in no Raman peak at 480 cm^{-1} for the films obtained by low HD values and no narrow peaks for the films obtained by high HD values, the results of the Raman measurements qualitatively show a very rapid phase transform from the amorphous phase to the crystalline phase by just varying the HD. This is consistent with the results obtained from the conductivity and Hall measurements. We also measured Raman spectra of the films deposited on glass substrate using the same process conditions and the same HD values in the range of 28.5% and 95%. We observed that all of the films on glass substrates were amorphous. This is consistent with the reported phase transition regime for amorphous to nano- (micro) crystalline Si on glass substrates [11].

Both conductivity and Raman measurements suggest that the films developed by low HD values (<80%) have amorphous structure and the films developed by higher HD values (>85%) are epitaxial. We have confirmed this proposition by HRTEM analysis of the epitaxial Si films obtained at 95% HD. Fig. 3(a) shows the HRTEM picture the interface between the (n^+) epitaxial Si thin film and the (p) mc-Si substrate. The figure clearly shows a long-range order, up to at least 40 nm away from the c-Si substrate, in the film very similar to the crystal order of the c-Si substrate. The crystallinity of the film is very high and no well defined boundary was observed between the (p) c-Si substrate and the (n^+) qEpi-Si film. Further TEM analysis from the bulk of the film showed that the epitaxial growth was obtained up to the top of the film with thickness of 100 nm. Fig. 3(b) shows the HRTEM picture of the bulk of the (n^+) Si film about 50 nm deep inside the film. The TEM picture also shows highly crystalline order in the bulk of the film. Study of the defects in the TEM pictures qualitatively show that the quality of the film in terms of the planar defect density increases gradually at higher thickness. This suggests that the quality of the film decreases gradually with increasing thickness of the film.

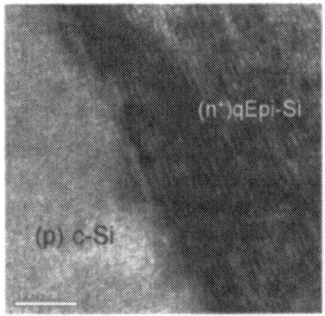
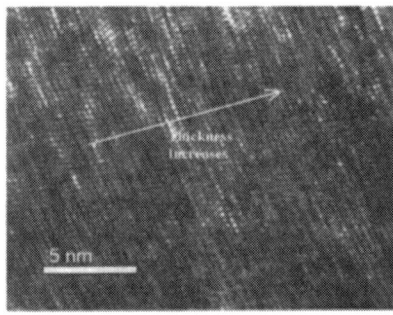

(a)   (b)

Fig.3: (a) HRTEM image of a (n^+) epitaxial Si thin film on a c-Si substrate showing the epitaxial nature of the highly doped Si film grown at 300°C using PECVD. The figure clearly shows that the crystallinity of the film, in presence of very high doping density, is very high. (b) HRTEM picture of the bulk (about 50 nm away from the (p) c-Si interface) of the (n^+) epitaxial Si film. The picture qualitatively shows that the quality of the film gradually decreases moving away from the c-Si substrate. This suggests that the film growth is not pure epitaxial but near epitaxial.

Effect of Hydrogen in Growth of (n^+) Epitaxial Si Films

It is important to discuss the role of hydrogen in the growth of the extrinsic epitaxial Si thin films at low temperatures. We believe that this is linked to the hydrogen diffusion model that was widely discussed in growth of microcrystalline Si thin films [11]. Based on this model, sufficient amount of atomic hydrogen, formed in the plasma environment, cover the film surface and result in a fully H-covered surface. The H-covered surface plays two major roles: (i) reduces the sticking factor of the precursors into the film surface, which results in the higher diffusion length of the precursors; (ii) causes a local heating through hydrogen-recombination reactions on the growth surface of the film. As a result, film precursors adsorbed on the surface can find energetically favorable (stable) sites, subsequently leading to the formation of ordered structure in the film. Also, the effect of hydrogen in breaking of weak Si-Si bonds and etching disordered tissues can justify the improved crystallinity in our films [12]. Based on the results from the conductivity and Hall measurements, we believe hydrogen dilution has a significant impact on doping efficiency. Basically, a higher hydrogen dilution results in a higher crystallinity in the film and the higher crystallinity in the film improves the doping efficiency. This is so because P atoms inside a highly crystalline Si film follow the order of the crystal and result in four-fold coordination, rather than three-fold coordination. As a result, presence of a c-Si substrate, presence of $-SiH_3$ precursors from the plasma environment, and presence of H-covered surface all together result in the (n^+) epitaxial Si films with high carrier mobility and high doping efficiency even at low temperatures.

CONCLUSIONS

The conductivity, Hall, Raman, and HRTEM measurements confirmed that high quality phosphorous doped epitaxial Si thin films were obtained using relatively high HD of silane in a standard PECVD system at low temperature. The high crystallinity of the developed epitaxial Si films has resulted in high electron mobility of about 40 $cm^2/v.s$ and high doping efficiency and free electron concentration of about 10^{20} cm^{-3}. Similar to the growth of microcrystalline silicon, hydrogen plays an extremely important role in the growth of the epitaxial Si film. In this work, we showed that HD values more than 85% can result in epitaxial growth of highly phosphorous doped Si thin films.

ACKNOWLEDGEMENT

The authors would like to thank Natural Sciences and Engineering Research Council (NSERC) of Canada for partially supporting this research. We would like to thank Mr. Fred Pearson from McMaster University for HRTEM analysis and Dr. R. Arocca and Dr. N. Pieczonka from University of Windsor for UV Raman measurements.

REFERENCES

[1] S. Wolf and R. N. Tauber, Silicon Processing for the VLSI Era, vol.1, Lattice Press, Sunset Beach, CA, 1986.

[2] R. E. Hollingsworth and P. K. Bhat, Applied Physics Letters -- January 31, 1994 -- Volume 64, Issue 5, pp. 616-618.

[3] S. Hammal and P. Roca i Cabarrocas, Solar Energy Materials and Solar Cells, Volume 69, Number 3, October 2001, pp. 217-239(23)

[4] S. C. Shah, J. K. Rath, S. T. Kshirsagar, and S. Ray, J. Phys. D: Appl. Phys. vol.30, p.p. 2686-2692, 1997

[5] Czang-Ho Lee, Denis Striakhilev, and Arokia Nathan, Journal of Vacuum Science & Technology A: Vacuum, Surfaces, and Films, Volume 22, Issue 3, p.p. 991-995

[6] C. H. Chen, T. R. Yew, Journal of Crystal Growth, vol.147, p.p.305-312, 1995

[7] W. J. Varhue, J. L. Rogers, and P. S. Andry, Appl. Phys. Lett. vol.68, p.p.349-351, 1996.

[8] T. Kitagawa, M. Kondo, A. MatsudaKondo, Applied Surface Science vol. 159-160, p.p. 30-34, 2000.

[9] E. Centurioni, D. Iencinella, R. Rizzoli, F. Zignani, IEEE Transactions on Electron Devices, vol. 51, p.p.1818-1824, 2004.

[10] G Masetti, M Severi, S Solmi, Electron Devices, IEEE Transactions on Electron Devices, vol.30, p.p. 764-769, 1983.

[11] Akihisa Matsuda, Thin Solid Films, vol. 337, p.p.1-6, 1999

[12] L Haji, P Joubert, J Stoemenos, NA Economou, Journal of Applied Physics, vol. 75 , p.p. 3944-3952, 1994

Poster Session:
Micro- and
Nanocrystalline Silicon

Mater. Res. Soc. Symp. Proc. Vol. 989 © 2007 Materials Research Society   0989-A07-01

Relation between Electronic Properties and Density of Crystalline Agglomerates in Microcrystalline Silicon

Paula C.P. Bronsveld[1], Arjan Verkerk[1], Tomas Mates[2], Antonin Fejfar[2], Jatindra K. Rath[1], and Ruud E.I. Schropp[1]
[1]Faculty of Science, Utrecht University, Princetonplein 5, Utrecht, 3508 TA, Netherlands
[2]Institute of Physics, Academy of Sciences of the Czech Republic, Cukrovarnická 10, Praha, 162 53, Czech Republic

ABSTRACT

A series of silicon thin films was made by very high frequency plasma enhanced chemical vapour deposition (VHF PECVD) at substrate temperatures below 100 °C at different hydrogen to silane dilution ratios. The electronic properties of these layers were studied as a function of the surface crystalline fraction as determined accurately from a combination of microscope images at different length scales (gathered by using different types of microscopes). The results show that the electrical conductivity increases monotonously as a function of crystalline surface coverage and no discontinuity is observed at the percolation threshold. An increase in conductivity of four orders of magnitude for layers with a high crystalline content is observed after annealing at temperatures up to 170 °C. Combined with the information that oxygen is incorporated at Si-H surface bond sites, this suggests that gas adsorption effects might be dominantly responsible for the electronic properties of mixed phase silicon.

INTRODUCTION

It is often observed that hydrogenated silicon layers deposited just over the amorphous to microcrystalline transition edge consist of cone-shaped agglomerates of crystallites, embedded in amorphous material. According to the observations of Collins et al. [1], changing the crystallinity by changing the hydrogen to silane dilution ratio can influence both the incubation height and the density of these cones. At higher dilutions, the incubation layer diminishes and the nucleation density of crystalline cones becomes higher, leading to a surface covered with microcrystalline material and an incubation layer in the order of a few tens of nanometer.

It is also observed that an increase in crystallinity in a (mixed phase) μc-Si:H layer causes an effective increase in the conductivity. For example, recent studies by Azulay et al. [2] that make use of conductive atomic force microscopy (C-AFM) to test the transport properties of the different phases, have shown that for layers with a very high crystalline content, as determined by Raman spectroscopy, a percolation network can be formed by the tissue surrounding the crystalline agglomerates. The formation of this network is associated with a large "jump" in conductivity at a crystalline volume fraction of ≈ 0.7. Another percolation threshold was found at a volume fraction of ≈ 0.3, associated with conduction through a regular percolation path formed by the more conducting crystallites embedded in a less conducting amorphous tissue. The low volume fraction threshold is most often reported for p-type or n-type doped layers (refs. in [2]), but has also been observed for undoped μc-Si [3].

Only in the case where the crystalline fraction is too low and no lateral percolation network of crystallites is completed, the current must pass through the a-Si:H matrix , bcause it cannot flow through the crystallites or their borders, which was also confirmed by observations [2].

In the substrate temperature range <100°C, the hydrogen dilution dependent transition from amorphous to microcrystalline growth goes more gradual compared to higher substrate temperatures [4]. This fact in particular facilitates a study of the evolution from amorphous material, from a non-percolation state, via a percolation state, to complete microcrystallinity at the surface. In these materials, deposited in the transition regime, at a thickness of about 1 micron the crystallites are of such a large size and sufficiently well separated that they can easily be resolved in scanning electron microscopy (SEM) and even in optical microscopy (OM) images. The latter is particularly useful to determine the presence of a percolation path on the macroscopic scale of the typical distance (~0.5 mm) between two coplanar electrodes in a lateral conductivity measurement. Microscopy can be considered to be a more reliable method for the detection of geometrical percolation than Raman spectroscopy, since the values given for the crystalline content by Raman spectroscopy are based on several assumptions, such as the ratio between the different optical cross-sections for amorphous material and crystalline grains and knowledge of the thickness of the incubation layer, while through microscopy the percolation path can be observed directly. In this paper, the electronic transport properties of at low temperature deposited transition materials are reviewed in direct connection with the existence of a geometrical percolation path. It seems that an increase in conductivity can already be observed before a percolation path is formed by the crystallites. Moreover, an even larger increase in conductivity can be achieved by post-deposition annealing of the layers at temperatures below 170 °C, where no changes to the crystalline structure are expected.

EXPERIMENT

Three hydrogen dilution series were made by VHF-PECVD in a high vacuum deposition system called ASTER at the substrate temperatures of 40 °C, 70 °C and 100 °C. The 40 °C series is studied in greatest detail, while the two other series at higher temperatures are made using fewer dilution steps. The layers were deposited on HF-dipped Corning 1737F glass. The HF dip serves to improve the adhesion of the silicon layer. The hydrogen to silane ratios that were necessary to cross the amorphous to microcrystalline transition were estimated from a previous hydrogen dilution series of thinner layers [4]. All layers had a thickness (on Corning) of about 1 µm. With the exception of the dilution ratio and the substrate temperature, all deposition parameters were kept constant in a low pressure (0.16 mbar), low power (5W) regime.

For the isochronal annealing and dark conductivity studies, two coplanar Ag contacts at a distance of 0.5 mm were deposited on top of the sample on Corning glass and the annealing experiment was performed in vacuum for a fixed period of 30 minutes at successive temperatures of 50°C, 70°C, 90°C, 110°C, 130°C, 150°C, 170° C. OM and SEM images of the area between the Ag contacts were taken using an Olympus confocal reflection microscope and a FEI NanoSEM 200, respectively. Details on the AFM and Cross-Sectional Transmission Electron Microscopy (X-TEM) measurements are described in [5].

DISCUSSION

Microscopy

In figure 1 a series of microscope images on different scales and three different techniques (AFM, SEM and OM) for a typical mixed phase layer, with no percolation network but a reasonable amount of crystallites, is presented. The spherical caps of cone-shaped agglomerates stick out from the surface [5] which makes it possible to estimate their diameter and count the number of crystallites. These spherical caps form a field of randomly distributed crystalline circles. Where the circles overlap, the circular shape has been disturbed by the collision of one or more cones and in between a straight boundary, the intergrain tissue, is formed, which makes the total surface coverage of the two caps equal to that of two overlapping circles. Only for layers deposited within a narrow range of hydrogen dilution values, the amorphous to microcrystalline transition range, the surface is not yet covered with clusters of crystallites and it is still possible to distinguish a crystalline and an amorphous part of the surface.

Figure 1. Different scale micrographs of the same mixed phase layer (below the percolation threshold) deposited at 40 °C. From left to right: AFM $(5 \times 5 \ \mu m^2)$, SEM $(11.8 \times 10.8 \ \mu m^2)$ and OM $(29.2 \times 22.3 \ \mu m^2$ cut out of larger, $132.9 \times 106.1 \ \mu m^2$ image)

For the substrate temperatures of 70 °C and 100 °C, also AFM, SEM, OM and X-TEM images were taken. Here, the caps get broader and smoother and more spherical, but the general behavior is similar.

The threshold for a geometrical percolation path through the crystallites is crossed at the point where, by increasing the density of crystallites, a path is formed by a connected line of crystallites from one side to the other side of the image (and the scale is large enough such that statistical homogeneity can be assumed). Whether such a path is present can be determined by making use of a known relationship between the formation of a percolation path by overlapping of filled circles and the surface coverage of all circles. A percolation probability of 1.0 is expected for a surface coverage of 67%, with only a small deviation for slightly non-circular shapes [6].

The surface coverage of the crystalline circles in the OM and SEM images was determined using a four-step method, which was necessary to overcome the issue of determining the 'edge' of the crystallites in our micrographs. As a first step, the image is converted to gray-scale. Then, a threshold grayscale value is set, to separate the gray-scale values attributed to the crystallites from those attributed to the background, and is varied until the radius of one, previously selected, isolated crystallite, remains practically constant for several consecutive threshold-steps. Subsequently, the surface coverage is determined in this 'constant radius range', and this coverage is converted to a number of counted cones according to the relationship [7]:

$$\varphi = 1 - e^{-\rho A}, \tag{1}$$

where φ is the surface coverage of the circles, ρ is the number of cones per area, and A is the area of one circle. After determination of the cone density, the 'real' surface coverage is calculated using the same equation, but then with the cone diameter measured at high magnifications in SEM. By comparing the original images to synthetic images with a known crystalline surface fraction, generated by randomly drawing spatially distributed spherical caps with a single radius, it was found that the four-step method has the tendency to underestimate the surface coverage slightly, but not by more than 5%. Moreover, for the low densities, the values could be checked by a Fourier transform based method, of which the results also agreed well. This method first applies a cut-off filter in the spatial frequency domain to reduce the effect of noise and particles in the image. The maxima in the resulting image are either accepted and counted or rejected as a possible crystalline cone by calculating the average intensity of the surrounding pixels.

Conductivity at 300K and higher

In figure 2a the dark conductivities at 300 K of the samples of the hydrogen dilution series after annealing at 50 °C in vacuum are presented as a function of hydrogen dilution. In figure 2b the same measurements are given as a function of crystalline surface coverage, as determined by the above described method. Several conclusions can be based upon these observations. First of all, the range of hydrogen dilution values where the transition from amorphous material to mixed phase material with percolation, visible as a steep increase in conductivity in figure 2a, shifts to lower dilution values as a function of substrate temperature. However, the general behavior seems to be rather independent of substrate temperature. Defect rich amorphous material has a conductivity of about 10^{-11} which decreases to $\sim 10^{-12}$ at the point where the first, separated, crystallites become visible (at surface coverage values up to about 0.2). Increasing the crystallite density even further results in a large increase in conductivity, which is very steep as a function of dilution, but which is surprisingly monotonous as a function of surface coverage, with no apparent discontinuities at the percolation threshold (surface coverage of 0.67). At full surface coverage, the rise in conductivity seems to saturate, showing only a minor increase with a further increase in dilution. Apparently, the conductivity is mainly determined by the crystalline ratio at the surface and the presence of a percolation path has a secondary role for the transport mechanism.

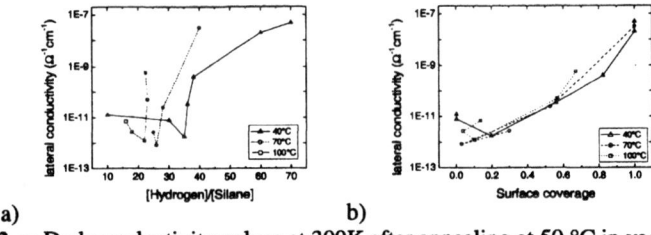

a)   b)

Figure 2. a: Dark conductivity values at 300K after annealing at 50 °C in vacuum. b: The same conductivity measurements as presented in (a), but as a function of crystalline surface coverage.

More information is provided by a second experiment, in which the conductivity of the layers was determined after isochronal annealing at different temperatures up to 170 °C. In figure 3, the results of this experiment for the series deposited at 40 °C are shown. For all layers with a crystalline fraction, and for some without, the conductivity increases with increasing temperature when annealed for equal periods at temperatures up to 170 °C. This effect increases strongly with crystalline fraction, bridging about four orders of magnitude between 50 °C and 170 °C at a surface coverage above 0.8. This is remarkably, as no changes to the total amount of bulk material or drastic changes inside the bulk are expected at such low annealing temperatures.

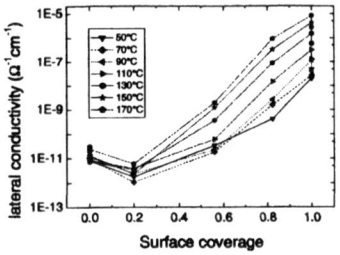

Figure 3. Conductivity at 300K for layers deposited at 40 °C after annealing in vacuum at different temperatures below 170 °C. Similar graphs can be made for 70 °C and 100 °C samples.

Azulay et al. [2] suggested that unintentional doping of the intergrain tissue by air exposure can cause the conductivity to increase, resulting in a main conductive path through the grain boundaries. Indeed, for a similar dilution series at 40 °C [8] at similar low annealing temperatures ≤ 170°C, oxygen incorporation was observed in infrared spectra, which might explain the higher conductivity due to annealing (and due to higher crystallinity because of increased inter-crystallite area). However, it is unclear how an increase of the grain boundary conductivity could affect the total conductivity below the percolation threshold and also the possibility of oxygen donor activation below 170 °C is questionable. An alternative explanation [9] is that adsorbed oxygen or water, acting as an electron acceptor, traps the free charge carriers at the grain boundaries and is actually removed at low annealing temperatures in vacuum, therewith increasing the conductivity. The processes of incorporation and adsorption of water seem to be related [9] and would both intensify at an increase of grain boundary area. This would, however, not explain why the main conductive paths at room temperature and high crystallinity follow the grain boundaries. More detailed research is necessary to discriminate between the different options.

CONCLUSIONS

From a combination of X-TEM, AFM, SEM and OM microscope images the percolation threshold and crystalline surface coverage for mixed phase amorphous/microcrystalline materials can be determined within an error of approximately 5%. Comparison of the results of dark conductivity measurements with the corresponding surface coverage values reveals a monotonous increase of conductivity with crystalline surface fraction and no steep changes around the percolation threshold. Combining this information with low-temperature annealing studies at temperatures below 170 °C, a large influence on the conductivity of gas adsorption effects due to air exposure is suspected, either causing the conductivity to be artificially low or high.

ACKNOWLEDGMENTS

The authors gratefully acknowledge Dirk Knoesen (UWC, Cape town, South Africa), Vasco Verlaan, Hans Meeldijk, Pim van Maurik and Ingo Gestmann (FEI Company, Eindhoven, the Netherlands) for their help with the acquisition of the micrographs.

REFERENCES

1. R.W. Collins, A.S. Ferlauto, G.M. Ferreira, Joohyun Koh, Chi Chen, R.J. Koval, J. M. Pearce, C.R. Wronski, M.M. Al-Jassim and K.M. Jones, Mat. Res. Soc. Symp. Proc., **762**, 2003, A10.1.1
2. D. Azulay, I. Balberg, V. Chu, J.P. Conde and O. Millo, Phys. Rev. B, **71**, 2005, 113304
3. S. Koynov, S. Grebner, P. Radojkovic, E. Hartmann, R. Schwarz, L. Vasilev, R. Krankenhagen, I. Sieber, W. Henrion, M. Schmidt, J. Non-Cryst. Sol. **198**, 1996, 1012
4. P.C.P. Bronsveld, J.K. Rath, R.E.I. Schropp, Proceedings of the 20th EUPVSEC Barcelona, 2005, 1675
5. P.C.P. Bronsveld, T. Mates, A. Fejfar, B. Rezek, J. Kocka, J.K. Rath, R.E.I. Schropp, Appl. Phys. Lett. **89**, 2006, 051922
6. Y.-B. Yi and A.M. Sastry, Phys. Rev. E, **66**, 2002, 066130
7. J. Quintanilla and S. Torquato, Phys. Rev. E, **54**, 1996, 5331
8. P.C.P. Bronsveld, H.J. van der Wagt, J.K. Rath, R.E.I. Schropp, and W. Beyer, "Post-deposition thermal annealing studies of hydrogenated microcrystalline silicon deposited at 40°C", Thin Solid Films (2007), DOI: 10.1016/j.tsf.2006.11.158
9. S. Veprek, Z. Iqbal, R.O. Kühne, P. Capezzuto, F.-A. Sarott and J.K. Gimzewski, J. Phys. C **16**, 1983, 6241

Mater. Res. Soc. Symp. Proc. Vol. 989 © 2007 Materials Research Society   0989-A07-02

Optimisation of Microcrystalline Silicon Deposited by Expanding Thermal Plasma Chemical Vapor Deposition for Solar-Cell Application

R. Jimenez Zambrano[1], R.A.C.M.M. van Swaaij[1], and M.C.M. van de Sanden[2]

[1]Department of Micro-electronics, DIMES-ECTM, Delft University of Technology, P.O. Box 5053, Delft, NL-2600 GB, Netherlands
[2]Department of Applied Physics, Eindhoven University of Technology, P.O. Box 513, Eindhoven, NL-5600 MB, Netherlands

ABSTRACT

Microcrystalline silicon (μc-Si:H) deposited with ETP-CVD shows a high degree of porosity. In this paper the underlying reasons for this porosity are investigated, using infrared absorption measurements. The results suggest that this porosity is due to the relatively low surface mobility of radicals on the surface and the incorporation of clusters in the material. The porosity of the material could be influenced by applying RF biasing on the substrate. In that case material with a higher density is obtained, which is ascribed to a higher surface mobility of radicals on the film surface during growth as a result of ion bombardment and to modifications in the mechanism of cluster incorporation..

INTRODUCTION

Amorphous and microcrystalline silicon (a-Si:H and μc-Si:H) are materials with demonstrated properties for photovoltaic application. These materials are expected to be used for mass production of cheap and high efficient solar cells. Due to the low absorption of μc-Si:H a thick layer (> 1 μm) is required, which renders the deposition rate a crucial parameter. Several techniques are used to grow μc-Si:H at high deposition rates. Promising results have been achieved by very high frequency (VHF) plasma-enhanced chemical vapor deposition (PECVD) in the high-pressure depletion (HPD) regime [1,2] and VHF-PECVD in a triode-reactor configuration [1]. The combination of VHF-PECVD and HPD achieves high deposition rates due to an efficient dissociation of feed gases together with a decrease in ion energy and an increase in ion flux at higher frequencies. With this technique μc-Si:H solar cells with 9.9% efficiency at 0.45 nm/s and 6.4% at 4.5 nm/s have been obtained [2]. Using the triode reactor configuration μc-Si:H solar cells with 3.4% efficiency at 6 nm/s have been deposited [3].

Expanding thermal plasma chemical vapor deposition (ETP-CVD) is a remote plasma technique characterized by high deposition rates and low-energy ion bombardment. In this technique a plasma is created in a cascaded arc at high pressure and this plasma then expands into the reactor that is maintained at lower pressure. It was reported earlier that with this technique μc-Si:H could be grown at 3.7 nm/s [4]. Also it was shown that the material properties of μc-Si:H could be influenced significantly by moving the injection ring towards the substrate [5], though the results suggested that the mix of depositing radicals did not change when moving the injection ring [6]. The initial implementation of μc-Si:H in solar cells shows that porous material with crack-like voids is deposited with ETP-CVD [7]. So far, the porosity of the μc-Si:H is a characteristic systematically observed when using ETP-CVD.

In this paper we will study the causes of the porosity and we will study the changes in porosity with bias and substrate temperature. First, the ETP-CVD is described briefly. Then, the different experimental results are presented and discussed. The paper will conclude with remarks.

EXPERIMENT

The films were deposited by ETP-CVD, a remote plasma technique in which a plasma is created in the cascaded arc at high pressure (0.2 to 0.5 bar). In our vertical configuration, the arc is located above the substrate. Via a nozzle the plasma expands to a reactor at lower pressure ($\sim$ 0.2 mbar), where the atomic hydrogen emanating from the arc reacts with SiH_4 injected further down into the expanding plasma. Due to the limited pumping capacity the reactor pressure is determined by the total gas flow. The depositions were performed at pressures between 0.12 and 0.28 mbar. The current through the cascaded arc plasma was 35 A. The hydrogen can be injected in the arc and/or the nozzle. The hydrogen flow injected into the arc was kept constant to maintain the same plasma conditions. The hydrogen flow injected in the nozzle was varied to control the dilution. The substrate temperature was 200°C, whereas during the temperature series it was varied between 170 and 420°C. Further, the substrate can be biased independently to control the ion bombardment by applying a 13.5-MHz RF signal to the substrate with the walls of the reactor as ground. The power varied between 0 and 30 W. When an RF voltage is applied to the substrate in combination with ETP-CVD, the substrate potential consists of two parts: a DC voltage, V_{dc}, and an AC voltage. With the frequency used, the DC voltage determines the energy of the impinging ions. Films of 500 nm were grown on Corning 1737 glass and on one-side-polished monocrystalline Si wafer with native oxide.

The IR-absorption measurements between 400 and 6000 cm^{-1} wave numbers were performed with a Nicolet 6700 FT-IR spectrometer on films deposited on crystalline Si wafers. The wave-number range from 2000-6000 cm^{-1} was used to determine the thickness of the sample accurately, which in turn was used to calculate the absorption coefficient and the hydrogen concentration. To analyze the FTIR spectra in the region 800-950 cm^{-1} we have followed the assignment of these peaks reported in Ref. [8]. For the region 1800-2200 cm^{-1} we have followed Ref. [9], as summarized in Table 1. In this table the bulk and surface modes are listed separately. Surface modes originate from dangling bonds terminated by hydrogen at the inner surfaces of voids. From the IR-absorption, we determine the microstructure factor, R*, which is defined as the ratio of the areas of the multihydrides in the bulk to monohydrides and multihydrides in the bulk. To determine the crystal fraction of the films deposited on glass we used a Renishaw InVia Raman microscope with a 514-nm Ar laser. In order to obtain the crystal fraction, fc, first three peaks are used to fit the amorphous part of the Raman spectrum (not including from 470 to 540 cm^{-1}). Afterwards, the amorphous part is subtracted from the measurement. The remaining spectrum is associated with the microcrystalline part of the material and is fitted with two modes at 510 and 520 cm^{-1}. The crystal fraction is then calculated as
fc = $I_{510}+I_{520}$ / (0.8*$I_{480}+I_{510}+I_{520}$).

Table 1. Assignment of the stretching-mode peak positions and their FWHM from Ref [9].

Species	Position (cm^{-1})	FWHM (cm^{-1})
Bulk		
SiH	1995 – 2020	80 – 110
SiH$_x$D$_{2-x}$ (x=1,2)	2095	55 – 70
SiH$_x$D$_{3-x}$ (x=1,2,3)	2135 – 2140	40 – 50
Surface		
SiH	2069 – 2084	10 – 20
SiH$_x$D$_{2-x}$ (x=1,2)	2105 – 2112	29 – 35
SiH$_x$D$_{3-x}$ (x=1,2,3)	2130 – 2140	10 – 30

RESULTS AND DISCUSSION

External RF biasing

Figure 1 compares the IR absorption spectra of two dilution series, the first *without* RF biasing (Fig. 1(a)) and the second, similar to the first one, but now *with* RF biasing (Fig. 1(b)). The change in dilution is achieved by increasing the hydrogen flow through the nozzle. This change in dilution (increasing from samples a1 to a4 and from b1 to b4 in Fig. 1) results in higher crystalline fractions and deposition pressure (as explained above). Both the crystalline fraction and the deposition pressure are given in Fig. 1.

With the dilution several remarkable changes happen. First, the mode at 2000 cm^{-1}, assigned to the stretching mode of monohydride (SiH) bonding configuration in the bulk of the material, is reduced and the peak position shifts to higher frequencies indicating changes in the structure even before the crystallization occurs (compare a1 vs a2 and a3, and b1 vs b2 and b3 in Fig. 1) . Second, for biased material (see b2, b3 and b4 Fig. 1(b)) two small peaks at 2080 and 2100 cm^{-1} are visible. These peaks have been assigned to hydrogen bonded on crystal boundaries. For unbiased material these two peaks are also present for microcrystalline material (see a3 and a4 in Fig. 1), although the mode at 2100 cm^{-1} can be observed as well for amorphous material (see a1 and a2 in Fig. 1). Third, the mode at 2140 cm^{-1} (associated with SiH$_3$) is visible when there is a crystalline phase in the material (see a3, a4, b3, and b4 in Fig. 1). The structural changes modify the absorption spectra in the region 800-950 cm^{-1} as well, which is more obvious for the samples deposited with RF biasing (compare b1 vs b4 in Fig. 1).

For RF biased a-Si:H, the integrated absoption of the monohydride mode at 2000 cm^{-1} is higher than that of the dihydride mode at 2095 cm^{-1}, giving a low microstructure factor (R* ~ 0.24 for sample b1). The opposite is found for unbiased amorphous material (R* ~ 0.55 for sample a1). We conclude that the material deposited with no RF bias has a higher degree of porosity than the material deposited with RF bias. Further, the narrow infrared absorption peak at 2105-2112 cm^{-1} observed for unbiased amorphous material (see a1 and a2 in Fig. 1(a)) is not observed for biased material. Following Ref. 9, we assign this mode to high hydride configurations on void surfaces. Therefore, the density of these configurations is large in unbiased material.

In the FTIR spectra of ETP-CVD μc-Si:H a mode is observed at 2140 cm^{-1}. This mode is associated with high hydrides on void surfaces and in the material bulk. When the substrate temperature is increased, the intensity of this mode is reduced. We ascribe this reduction at higher substrate temperature to higher surface diffusion, similar to what has been observed for the deposition of ETP-CVD a-Si:H [10].

The substrate temperature is a crucial parameter when depositing solar cells. To avoid reaching temperatures that will influence the properties of a previously deposited layer, we have decided to increase the surface mobility of radicals by applying external RF biasing. The electric field created accelerates the ions to an energy controlled by the power of the RF bias. The amorphous material presented in Fig. 1 (sample b1) shows an optimum with respect to R* for a DC voltage on the substrate of 40 V. This self-bias voltage is sufficient to increase the surface mobility of the radicals, making the growth surface reactive, without damaging the structure. Higher voltages (higher ion energies) result in sputtering or displacement of Si atoms from the network and the material becoming more porous.

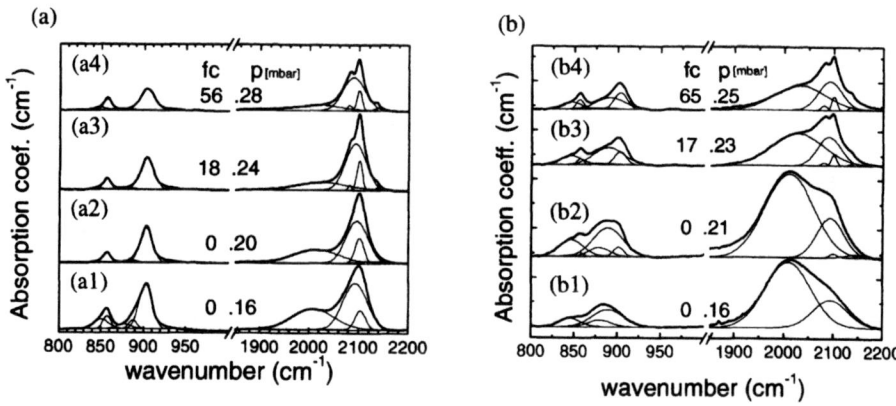

Figure 1. FT-IR absorption spectra of μc-Si films prepared at several H_2 flow conditions: (a) without RF bias (b) with RF bias. These flows result in different crystalline factors (fc) and deposition pressure (see text), given in the graphs.

The effect of biasing is also observable for the microcrystalline material. The integrated absorption of the SiH stretching mode at 2000 cm^{-1} increases with biasing (from 11 cm^{-1} to 25 cm^{-1}, for samples a4 and b4, respectively) , while the stretching mode at 2095 cm^{-1} decreases (from 26 cm^{-1} to 13 cm^{-1}, for samples a4 and b4, respectively), indicating a more compact structure when external RF biasing is used. The increase in density is also observed in the refractive index at 2 eV, which increases from 2.8 to 3.6 when external RF biasing is used (samples a4 and b4, respectably). The integrated absorption of the peak doublet at 2080 and 2100 cm^{-1} is reduced (from 4 cm^{-1} to 2 cm^{-1}, for samples a4 and b4, respectively), possibly an indication of better intergrain and interphase boundaries between the amorphous and the crystals.

Deposition pressure

With the use of biasing the structure of the material has become more compact, although the microstructure factor is still relatively high (R* ~ 0.4 for sample b4 in Fig. 1). We think that this high microstructure factor is due to the lack of control on the deposition pressure for the deposition of μc-Si:H. For the deposition of μc-Si:H the hydrogen flow is increased and due to the limited pumping capacity this extra flow increases the deposition pressure. As shown in Fig. 1 the pressure nearly doubles when increasing the dilution in order to obtain microcrystalline material.

To study the influence of the pressure on the FTIR spectra, we have deposited the low-pressure a-Si:H (sample b1 in Fig. 1) at a similar pressure used for the deposition of μc-Si:H (sample b4 in Fig. 2). The pressure increase was achieved by closing the valve to the pump. The FTIR spectra of the low-pressure a-Si:H (sample b1), the high-pressure a-Si:H (sample c1), and the μc-Si:H (sample b4) are shown in Fig. 2. Note that sample b1 and b4 in Fig. 2 are also presented in Fig. 1(b). For these materials an external RF biasing has been used.

The spectrum of the high-pressure a-Si:H (c1) in the region 800-950 cm^{-1} is very similar to that of μc-Si:H (b4). The doublet at 860 and 905 cm^{-1} is visible in both the materials. It has been assigned to high hydrides (SiH$_3$), in combination with the mode at 2140 cm^{-1}, which is also visible.

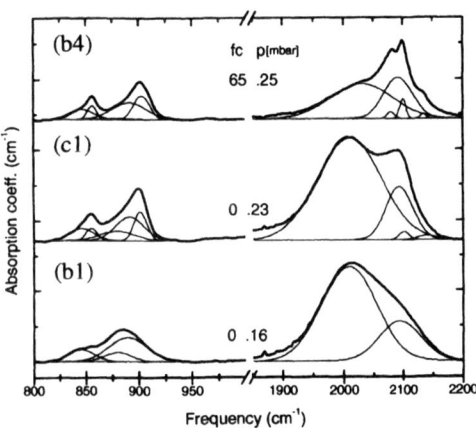

Figure 2. FT-IR absorption spectra of a-Si at low pressure, a-Si at high pressure and μc-Si films (from bottom to top). First and third from Fig. 1b.

Compared to the low-pressure a-Si:H (b1), the pressure increase leads to a less compact structure, as is deduced from the increase of the stretching mode at 2095 cm^{-1} and the appearance of the mode at 2100 cm^{-1}.

When depositing at higher pressures gas phase polymerization reactions start to become dominant, resulting in ionic cluster formation. The incorporation of these clusters into the material changes the structure from vacancy to void dominated. Similar results have been observed by Petit [11] for a-Si:H.

In conclusion, an increase of the deposition pressure results in more porous material, although more research is required to confirm this suggestion. This work is being carried out.

CONCLUSIONS

We have identified two reasons for the high degree of porosity of microcrystalline silicon deposited by the ETP-CVD. First, the surface mobility of radicals on the surface is controlled by the energy of the ion bombardment. Due to intrinsic characteristics of this technique the ion energy is low (2 eV). Second, the deposition of microcrystalline material takes place in a regime with high polymerization reactions. This regime renders the void-dominated material.

We have shown that the use of external RF bias during deposition results in a material with more compact structure. We suggest that the energy of the ion bombardment is transferred to the radicals, increasing the surface mobility. In addition, RF biasing might influence the cluster formation also and this can contribute to the densification of the material as well.

We have shown the influence of the deposition pressure on the porosity of the material for amorphous material and we have related these results to microcrystalline material. More work is been carried out to study the structure of microcrystalline material at lower pressure.

ACKNOWLEDGEMENTS

Martijn Tijssen and Kasper Zwetsloot are thanked for their technical support. This work was financially supported by SenterNovem.

REFERENCES

1. A. Matsuda, J. Non-Cryst. Solids **338–340**, 1 (2004).
2. A. Gordijn, L. Hodakova, J. K. Rath, and R. E. I. Schropp, J. Non-Cryst. Solids **352**, 1868 (2006).
3. M. Kondo, T. Nishimoto, M. Takai, S. Suzuki, Y. Nasuno, and A. Matsuda, Tech. Dig. Intern. PVSEC-12, Jeju, Korea, 41 (2001).
4. C. Smit, E. A. G. Hamers, B. A. Korevaar, R. A. C. M. M. van Swaaij, and M. C. M. van de Sanden, J. Non-Cryst. Solids **299-302**, 98 (2002).
5. C. Smit, R. A. C. M. M. van Swaaij, E. A. G. Hamers, and M. C. M. van de Sanden, J. Appl. Phys. **96**, 4076 (2004).
6. R. A. C. M. M. van Swaaij, R. Jiménez Zambrano, C. Smit, and M. C. M. van de Sanden, J. Non-Cryst. Solids **352**, 933 (2006).
7. C. Smit, A. Klaver, B. A. Korevaar, A. M. H. N. Petit, D. L. Williamson, R. A. C. M. M. van Swaaij, and M. C. M. van de Sanden, Thin Solid Films **491**, 280 (2005).
8. P. J. Zanzucchi, in Semiconductors and Semimetals, 21, chapter 4 (1984).
9. S. Agarwal, A. Takano, M. C. M. van de Sanden, D. Maroudas, and E. S. Aydil, J. Chem. Phys. **117**, 10805 (2002).
10. A. H. M. Smets, W. M. M. Kessels, and M. C. M. van de Sanden, Appl. Phys. Lett. **86**, 041909 (2005).
11. A. M. H. N. Petit, Ph. D. thesis, Delft University of Technology, 2006.

Mater. Res. Soc. Symp. Proc. Vol. 989 © 2007 Materials Research Society   0989-A07-07

Boron Doping Effects in Microcrystalline Silicon

Wolfhard Beyer, Lars Niessen, and Frank Pennartz

IEF-5 Photovoltaik, Forschungszentrum Jülich GmbH, Leo Brandt Strasse, Jülich, 52425, Germany

ABSTRACT

Conditions leading to high conductivities (up to 300 S/cm) in chlorosilane-based boron-doped microcrystalline Si:Cl:H films are investigated. It is found that the high conductivity originates primarily from the growth of highly crystalline material with a high concentration of boron. Furthermore, these films grow with relatively low chlorine and hydrogen concentrations of a few percent and, according to effusion measurements of hydrogen and implanted helium, in a relatively compact structure. At a boron doping level of 1%, admixture of 10% silane to the tetrachlorosilane results in the growth of amorphous material of low conductivity while for admixture of up to 90% of silicontetrafluoride, microcrystalline Si films with high conductivities can be grown.

INTRODUCTION

Highly conductive microcrystalline silicon films are of interest for application as contact layers in microcrystalline silicon thin film solar cells. However, in the standard silane (SiH_4) - based technology, an asymmetry is observed for p- and n-type doping. Highest conductivities reached by boron doping are found to be more than a factor of 10 lower than those obtained by phosphorus doping, where maximum conductivities lie in the range of 100 $(\Omega cm)^{-1}$ [1,2]. Recently we found high conductivity values for boron-doped microcrystalline Si films prepared by plasma-enhanced chemical vapour deposition (PECVD) from tetrachlorosilane-hydrogen-diborane gas mixtures [3, 4, 5]. Room temperature conductivities of nearly 300 $(\Omega cm)^{-1}$ were obtained [4]. In this article we investigate the peculiarities of this B-doped chlorinated material and explore to what degree such high conductivities can be reached by using gas mixtures of chlorosilane with silane or silicontetrafluoride.

EXPERIMENTAL DETAILS

The films were prepared by plasma deposition (PECVD) employing the same regime as applied previously [3-5], namely a high dilution of $SiCl_4$ /SiH_4 /SiF_4 in hydrogen, a rather high pressure of 8 mbar (measured by a baratron) and a high rf (13.56 MHz) power of 40 -60 W (0.7 - 1W/cm^2). A hydrogen flow of 100 to 300 sccm and a flow of $SiCl_4$ /SiH_4 /SiF_4 of 1 to 8 sccm were used. For doping, flows of diborane (B_2H_6) were added. Typical substrate temperature was $T_S = 250°C$. As substrates, crystalline silicon and quartz were used. The film thickness ranged between 0.24 and 1.2 µm with typical values near 0.8 µm. For structural characterization (crystallinity), Raman measurements were employed (using the 488 nm line of an argon laser) [6]. The chlorine, fluorine and boron content were analyzed by secondary ion mass spectrometry

(SIMS) while the hydrogen concentration was determined by hydrogen effusion (using a heating rate of 20K/min) measurements. Effusion measurements of implanted He were used for structural characterization. For electrical characterization, dark conductivity (using coplanar geometry) was measured at room temperature.

RESULTS

Parameters leading to high conductivity

In Fig. 1, some properties of Si:Cl:H films with highest conductivities > 90 $(\Omega cm)^{-1}$ are plotted as a function of the room temperature conductivity σ. These films were grown with a nominal diborane doping level of 2-10 % from $SiCl_4$ –H_2 gas mixtures [4]. The Raman crystallinity X shown in Fig. 1a reveals a weak increase with rising σ. This is in agreement with the previously observed trend [4] that for high conductivities high Raman crystallinities are required. The concentrations c of chlorine, hydrogen and boron shown in Fig. 1b do not vary (within measurement accuracy) as the conductivity σ of the films changes from 100 to 300 $(\Omega cm)^{-1}$. Note, however, that rather low chlorine concentrations (near 1at.%) seem to be prerequisites to obtain high conductivities according to previous work [4]. For phosphorus-doped Si:Cl:H films which grow under our conditions with a much higher chlorine concentration than the boron doped material, the maximum conductivity reached was 4 orders of magnitude below the highest values of B-doped material [4]. Fig. 1b also shows the parameter $\sigma/e\ N_B$, where e is

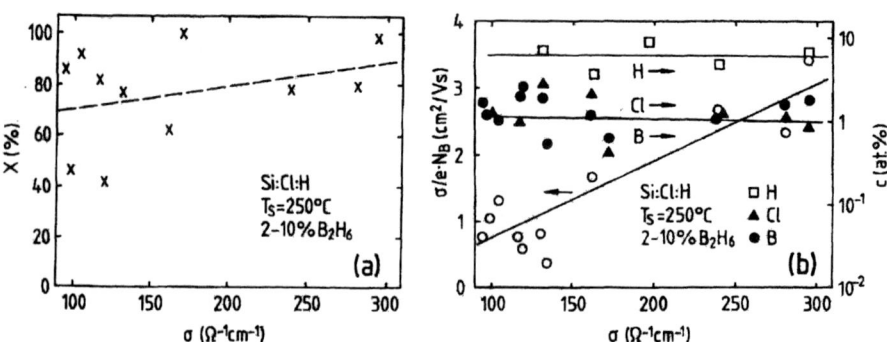

Figure 1. (a) Crystalline fraction X and (b) $\sigma/e\ N_B$ and hydrogen, boron and chlorine concentrations c as a function of electrical conductivity σ of highly conducting boron-doped Si:Cl:H films.

the electronic charge and N_B is the boron density in the films measured by SIMS depth profiles. This quantity $\sigma/e\ N_B$ is found to increase strongly with rising conductivity. Defining the doping efficiency γ ($\gamma \leq 1$) of incorporated boron atoms by $p = N_B\ \gamma$ (p is the boron-induced carrier density in the valence band), $\sigma/e\ N_B = \gamma\ \mu$ is valid with μ the carrier mobility. Accordingly, the increase of σ/eN_B with rising conductivity can be due to an increase in carrier mobility or in boron doping efficiency (or in both). Calculations of the mobility in microcrystalline Si using a percolation model give an increase in mobility by less than a factor of two if the crystalline volume fraction rises from 70 to 90 % [2]. Since the quantity $\sigma/e\ N_B$ in Fig. 1b varies for the same range of crystallinity by more than a factor of 3, a variation of the doping efficiency cannot be ruled out at the present stage. Note that the doping efficiency of boron and phosphorus is quite low for amorphous silicon while for microcrystalline Si doping efficiencies close to unity have been reported [2].

Another parameter which may affect electrical properties of a given film is the material microstructure. Previously we have shown that a characterization method of microstructure in Si:Cl:H films is the effusion of implanted He [7]. Our results for the samples discussed in Fig. 1 show the presence of rather compact material with a temperature T_M of the He effusion maximum between 300 and 350°C. No clear correlation of T_M with conductivity is observed. These values of T_M are higher than those for typical undoped μc-Si:Cl:H films but lower than those for good quality μc-Si:H films. In the latter case, T_M values > 400°C are observed for the substrate temperature of 250°C [7].

Films grown from SiCl$_4$ – SiH$_4$ and SiCl$_4$ – SiF$_4$ gas mixtures

Since the use of SiCl$_4$ in thin film silicon technology leads to some corrosion of the deposition apparatus, the possibility to grow highly conductive B-doped Si films was also studied for mixtures of SiCl$_4$ with silane and silicontetrafluoride. Results for films deposited

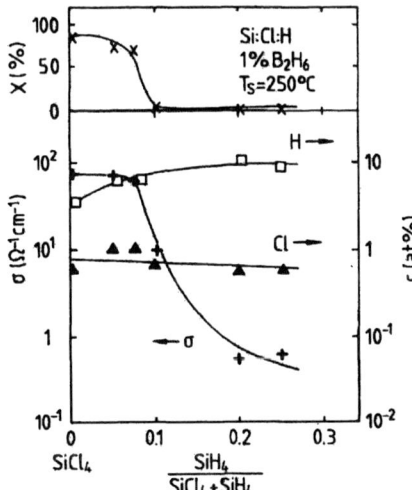

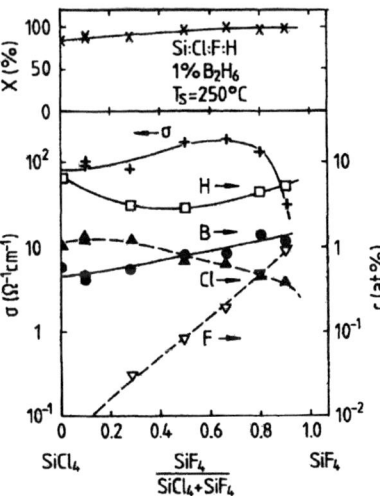

Figure 2. Raman crystallinity X, conductivity σ and Cl and H concentrations c as a function of silane fraction in the SiCl$_4$ –SiH$_4$ mixture.

Figure 3. Raman crystallinity X, conductivity σ and Cl, F, H and B concentrations c as a function of SiF$_4$ fraction in the SiCl$_4$ –SiF$_4$ mixture.

with 1 % diborane doping are shown in Figs. 2 and 3. In Fig. 2 it is seen that the admixture of only 10% of SiH_4 to the process gas leads (under the present deposition conditions) to the growth of amorphous material. Clearly, this transition is related to the presence of the diborane in the gas mixture since undoped films remained crystalline in the whole mixture range. It is interesting to note that the transition from the growth of crystalline to that of amorphous material has no influence on the chlorine concentration and only little influence on the hydrogen concentration. The latter concentration increases somewhat when microcrystalline growth changes to amorphous growth in agreement with earlier observations [3]. However, this increase in H cioncentration could be partially due also to an increase in deposition rate which changed from about 1.4 Å/s (0% SiH_4) to 3.2 Å/s (25 % SiH_4).

As seen in Fig. 3, films grown from $SiCl_4$ –SiF_4 gas mixtures remain crystalline in the whole mixture range investigated. Starting from Raman crystallinity values of $X \approx 90\%$ for purely $SiCl_4$ - based material, the crystallinity is found to increase with rising SiF_4 admixture. This may be related to a slight decrease in deposition rate from about 1.5 Å/s (0 % SiF_4) to about 0.7 Å/s (90 % SiF_4). The results of the concentration measurements show that the fluorine/chlorine concentration ratio in the solid differs considerably from that in the gas phase. E.g., for a gas mixture with 50% SiF_4, the fluorine to chlorine ratio in the solid is about 0.1. As the SiF_4/$SiCl_4$ gas phase ratio rises, the boron incorporation increases too. We relate this effect to a decrease in boron incorporation in the presence of chlorine as reported earlier [3]. When the concentration of chlorine in the plasma decreases, the boron incorporation in the material apparently approaches the gas phase ratio. Highest conductivities (exceeding 10^2 $(\Omega cm)^{-1}$) are obtained in this gas mixture system for about 60% SiF_4. It is interesting to note that in the gas mixture range where conductivity is high, the hydrogen concentration shows low values. The

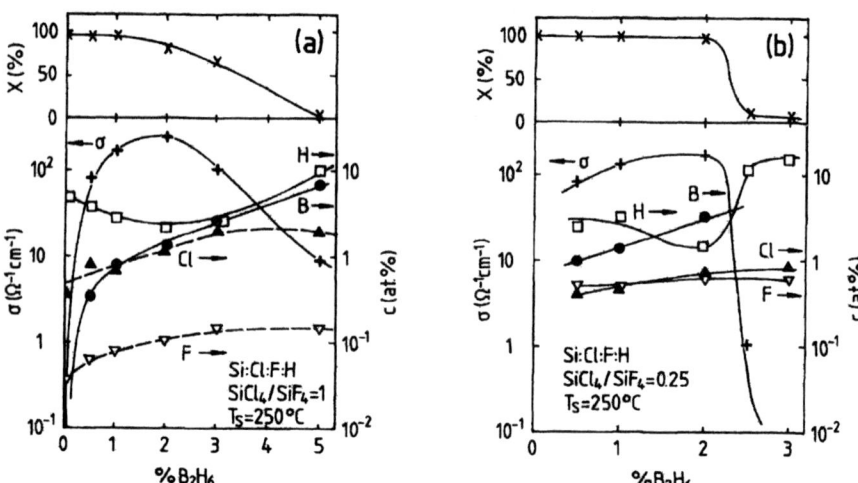

Figure 4. Raman crystallinity X, conductivity σ and Cl, F, H and B concentrations c in Si:Cl:F:H for the gas mixtures $SiCl_4$/SiF_4 = 1 (a) and $SiCl_4$/SiF_4 = 0.25 (b).

nature of the decrease in conductivity at high $SiF_4/SiCl_4$ gas phase ratio is not completely clear. Tentatively we attribute it to an increased porosity and post-deposition oxidation of these films of high Raman crystallinity, i.e. of poor grain boundary passivation.

The influence of boron doping on the composition and the properties of Si:Cl:F films prepared with the gas mixtures of $SiCl_4/SiF_4 = 1$ and 0.25 is shown in Figs. 4a and 4b, respectively. In both cases high conductivities exceeding 10^2 $(\Omega cm)^{-1}$ are observed for gas phase doping levels of 1-2%. Similar as in Fig. 3, the hydrogen concentration shows a minimum for samples with maximum conductivity. For both $SiCl_4/SiF_4$ gas mixtures, a high boron doping level eventually leads to the growth of amorphous material. In case of $SiCl_4/SiF_4 = 1$ (Fig. 4a), this transition is rather smooth so that only at a gas phase boron doping level of 5% completely amorphous material grows. For the gas mixture of $SiCl_4/SiF_4 = 0.25$ (Fig. 4b), the transition occurs rather abrupt near 2% of diborane suggesting that the tendency for the growth of crystalline material in the presence of diborane is higher for $SiCl_4$ than for SiF_4. An interesting feature of Fig. 4a is, furthermore, that a rising boron incorporation increases also the incorporation of Cl and F in the solid, demonstrating an influence of boron on the plasma/growth chemistry. Note that the increase in chlorine and fluorine incorporation is not due to a rising deposition rate, as the deposition rate for the films of Fig. 4a was almost constant at about 1.5 Å/s.

DISCUSSION

The results demonstrate that for the purpose to obtain boron-doped material with high electrical conductivity, $SiCl_4$ and SiF_4 are much superior process gases than silane. For silane based gas mixtures, the effect of gas phase diborane doping favoring the growth of amorphous material is well known [8] and has been attributed primarily to a boron-induced release of surface hydrogen, thus increasing the number of growth sites and the deposition rate [8,9]. As a possible mechanism for surface hydrogen depletion, the scavenging action of BH_3 radicals was suggested [9] as proposed by Perrin et al. [10] to explain an increase in growth rate of amorphous silicon in the presence of diborane. The present results do not contradict to this explanation since, qualitatively, the concentration of BH_3 radicals is expected to get reduced when the concentration of Cl or F in the plasma rises. We note, however, that boron may also induce a depletion of surface hydrogen via the Fermi level dependence of hydrogen diffusion and desorption in amorphous and microcrystalline Si materials [11]. A Fermi level close to the valence band is known to shift hydrogen surface desorption to lower temperatures so that this process becomes highly active at temperatures below 200°C. Thus, the transition from microcrystalline to amorphous growth with rising boron doping can be explained by a boron-induced enhanced H surface desorption. While this hydrogen depletion is presumably active also in the presence of fluorine and chlorine, the presence of rather stable Si-Cl and Si-F bonds at the growth surface could result in a similar flexibility and mobility of silicon surface species as in the presence of a high concentration of hydrogen, and thus in microcrystalline growth. The result that at very high boron concentrations a transition to amorphous material always occurs would suggest that some minimum hydrogen concentration is required for crystalline growth. Differences between chlorine and fluorine surface chemistry and bonding on silicon surfaces could be responsible for the differences in the boron-induced crystalline to amorphous transitions in Figs 4a and b.

A general result of the present work is that a high conductivity of boron-doped material is usually paired with rather low hydrogen content. This result agrees with the concept that a high conductivity implies a high p-type doping effect, i.e. a low Fermi level. Thus, the low hydrogen content would reflect the boron doping-induced instability of hydrogen incorporation. However, an alternative explanation for the low H content is also conceivable. In this second model, high conductivities require the presence of highly crystalline material where the hydrogen content (due to the low hydrogen solubility in crystalline silicon) is low. Further work appears necessary to clarify this point. Finally we note that from our experience the corrosion of stainless steel apparatus by chlorinated gases is smaller when SiF_4 is present.

CONCLUSIONS

The results demonstrate that high conductivities exceeding 100 $(\Omega cm)^{-1}$ can be achieved for boron-doped microcrystalline Si by using $SiCl_4$ and SiF_4 precursor gases, as in both cases highly crystalline material can be grown at rather high boron concentration. These films may open new options for contact layers of thin film silicon solar cells.

ACKNOWLEDGMENTS

The authors would like to thank M. Hülsbeck and R. Carius for the Raman crystallinity measurements, D. Lennartz and W. Hilgers for technical assistance, U. Zastrow for support with the SIMS measurements and A. Dahmen for the ion implantations.

REFERENCES

1. K. Prasad, U. Kroll, F. Finger, A. Shah, J. Dorier, A. Howling, J. Baumann, M. Schubert, *MRS Symp. Proc.* **219**, 469 (1991).
2. R. Carius, F. Finger, U. Backhausen, M. Luysberg, P. Hapke, L. Houben, M. Otte, H. Overhof, *MRS Symp. Proc.* **467**, 283 (1997).
3. W. Beyer, B. Rech, R. Carius, M. Albert, R. Terasa, in *Proceedings PV in Europe Conference, Rome, 7.-11.Oct 2002* (WIP Munich and ETA Florence, 2002) p. 75.
4. W. Beyer, R. Carius, M. Lejeune, U. Zastrow, *MRS Symp. Proc.* **808**, 389 (2004).
5. W. Beyer, R. Carius, and U. Zastrow, *MRS Symp. Proc.* **862**,139 (2005).
6. L. Houben, M. Luysberg, P. Hapke, R. Carius, H. Wagner, *Philos. Mag.* A**77**, 1447 (1998).
7. W. Beyer, R. Carius, U. Zastrow, *J. Non-Cryst. Solids* **352**, 1402 (2006).
8. R. Flückinger, J. Meier, A. Shah, A. Catana, M. Brunel, H.V. Nguyen, R.W. Collins, R. Carius, *MRS Symp. Proc.* **336**, 551 (1994).
9. S. Ghosh, A. De, S. Ray, A.K. Barua, *J. App. Phys.* **71**, 5205 (1992).
10. J. Perrin, Y. Takeda, N. Hirano, Y. Takeuchi, A. Matsuda, *Surface Science* **210**, 114 (1989).
11. W. Beyer, *Solar Energy Materials and Solar Cells* **78**, 235 (2003).

Poster Session:
Thin Film Growth

Mater. Res. Soc. Symp. Proc. Vol. 989 © 2007 Materials Research Society

Phase Control and Stability of Thin Silicon Films Deposited from Silane Diluted with Hydrogen

Gijs van Elzakker[1], Pavol Šutta[2], Frans D. Tichelaar[3], and Miro Zeman[1]
[1]DIMES, Delft University of Technology, Feldmannweg 17, Delft, 2628 CT, Netherlands
[2]NT-RC, West Bohemian University, Univerzitní 8, Plzen, 306 14, Czech Republic
[3]NCHREM, Delft University of Technology, Lorentzweg 1, Delft, 2628 CJ, Netherlands

ABSTRACT

Hydrogen dilution of silane during the rf-PECVD growth of a-Si:H absorber layers is used to suppress light-induced degradation of a-Si:H solar cells. The increased stability of cells and films deposited using hydrogen dilution is verified in an accelerated degradation experiment. At higher hydrogen dilutions the early phase transition to the microcrystalline phase complicates the growth of fully amorphous films as absorbers with a sufficient thickness. In a systematic study on the influence of various deposition conditions on the material properties the pressure is identified as an important factor for controlling the structural phase evolution of the films.

INTRODUCTION

Hydrogenated amorphous silicon (a-Si:H) is a promising semiconductor material for low-cost solar cells. Unfortunately a-Si:H suffers from light-induced degradation known as Staebler-Wronski effect [1]. It has been demonstrated that solar cells with a-Si:H absorber layers prepared from silane source gas diluted with hydrogen in plasma enhanced chemical vapour deposition (PECVD) showed less degradation during light exposure than their conventional undiluted counterparts [2,3]. This a-Si:H absorber material is generally referred to as protocrystalline silicon due the fact that it will eventually evolve from the amorphous to microcrystalline phase when grown to a sufficient thickness [4]. The thickness of the amorphous 'protocrystalline' phase decreases strongly with increasing hydrogen dilution [5]. This means that the minimum required thickness (in solar cells this is typically 300 nm for the a-Si:H absorber layer) limits the amount of hydrogen dilution that can be used, unless phase control techniques such as dilution profiling [6] or layer-by-layer deposition [7] are applied. In this work the latter technique was applied to fabricate an amorphous silicon solar cell at a relatively high dilution ratio of $R=[H_2]/[SiH_4]=30$. However, the inferior performance of the resulting solar cell demonstrates that a high hydrogen dilution is not the only requisite for a good absorber material. A systematic study was carried out to investigate the influence of deposition parameters such as rf-power, pressure and substrate temperature on the structural and optoelectronic properties of materials deposited at $R=20$. The parameters' range was determined for which the phase transition in the deposited films was not crossed. The increased stability of a solar cell with an absorber layer from hydrogen diluted silane ($R=20$) is verified in an accelerated degradation experiment. The evolution of the fill factor during illumination is compared to the evolution of the defect density in the light-soaked individual films deposited under equivalent deposition conditions.

EXPERIMENTAL DETAILS

Deposition of films and solar cells

Several silicon films were deposited on Corning glass substrates using different conditions during the rf-PECVD growth. A film was deposited using the layer-by-layer (LBL) approach [7] and consists of a stack of 5 layers deposited at R=30 with a thickness of 50 nm, interrupted by 4 interlayers with a thickness of 10 nm. The deposition conditions of the layers are described in [7]. A second film with the same thickness as the previous stack was deposited in a continuous deposition using a dilution ratio of R=30. Further, three series of films were deposited at a (continuous) hydrogen dilution ratio of R=20 in which the rf-power, pressure and substrate temperature were varied. The rf-power was varied between 4 and 16 W, the pressure was varied between 1.4 and 2.6 mbar and the substrate temperature was varied between 150°C and 200°C. The parameters that were not varied within a specific series were kept constant at an rf-power of 4 W, a pressure of 2.0 mbar and a temperature of 180°C. A reference film deposited at these conditions appears in all three series. A second reference film grown without hydrogen dilution (R=0) was deposited using an rf-power of 4 W, a pressure of 0.7 mbar and a temperature of 180°C. The thickness of all films is approximately 300 nm.

Single junction p-i-n solar cells were deposited on Asahi U-type substrates using the above described films as the absorber layers. The solar cells have the following structure: p-type a-SiC:H layer (10 nm)/a-SiC:H buffer layer (5 nm)/intrinsic absorber layer (300 nm)/n-type a-Si:H layer (20 nm). The back contact consists of 100 nm silver and 200 nm aluminum.

Characterization of films and solar cells

Absorption coefficient spectra were determined from Reflection and Transmission and Dual Beam Photoconductivity (DBP) measurements. The absorption coefficient, $\alpha(1.2$ eV), is linked to the defect density, N_d, in amorphous silicon via a factor of 2.4×10^{16} cm^{-2} [8]. The Reflection and Transmission data were also used to determine the E_{04} and E_{klazes} optical gap [9] as well as the film thickness, which was used to determine the deposition rate r_d. From infrared absorption spectroscopy measurements the microstructure factor R* of the films was determined [10]. Infrared spectra were recorded using a Thermo Electron Nicolet 5700 spectrometer with H-ATR accessory. The H-ATR method was used instead of transmission mode measurements to reduce absorption in the glass substrates. The microstructure factor was estimated by fitting the spectra with two Gaussian distributions at ~ 2000 cm^{-1} and ~ 2100 cm^{-1} (after background signal removal). Raman spectra were recorded using a Renishaw InVia type spectrometer with a laser wavelength of 514 nm. The external parameters of the solar cells (efficiency η, fill factor ff, short-circuit current I_{SC}, open-circuit voltage V_{OC}) were determined from I-V measurements carried out under standard illumination conditions using an Oriel Corporation solar simulator.

Degradation experiment

An accelerated degradation experiment was performed on the R=20 and R=0 reference films. The absorption coefficient spectra were measured in the initial state, after 64 minutes and after 8000 minutes of light-induced degradation. The corresponding solar cells were also light soaked and their external parameters were monitored in time. The light soaking was performed at room temperature using a semiconductor laser with a wavelength of 670 nm and a power density of 300 mW/cm^2.

RESULTS AND DISCUSSION

Phase control in Si:H films

At higher hydrogen dilution ratios the inhomogeneous growth of Si:H has to be manipulated in order to avoid the amorphous to microcrystalline phase transition. The most common approach is to gradually decrease the hydrogen dilution during the deposition process [6]. We presented an alternative approach using a layer-by-layer technique [7], in which thin interlayers from pure silane are applied to interrupt the hydrogen diluted growth and thereby suppress the development of the microcrystalline phase. In order to investigate the suitability of these layer-by-layer stacks as absorber layers p-i-n solar cells were fabricated. The absorber layer consists of a stack of a total thickness of ~300 nm that is formed by 50 nm thick diluted layers using R=30 separated by 10 nm thick interlayers. Table 1 shows the external parameters of the cell with the layer-by-layer (LBL) absorber and a cell with an absorber deposited at R=30, and the properties of corresponding individual films. When comparing these two cells, the LBL cell has a higher fill factor and efficiency. However, the fill factor remains relatively low at 0.65. Apparently, the use of a high dilution ratio of R=30 is not the prerequisite for obtaining a good solar cell.

To further improve the material properties a systematic study on the influence of the rf-power, pressure and substrate temperature was conducted. For this study a reduced hydrogen dilution of R=20 was chosen to increase the range of deposition conditions for which the phase transition is not crossed, allowing for a material study without the need for a layer-by-layer scheme. Trends in material properties found at R=20 will later be applicable to the growth at higher hydrogen dilution ratios. Prior to this systematic study the increased stability of the R=20 reference solar cell was verified. The properties of reference layers deposited at R=20 and at R=0 and the external parameters of the corresponding solar cells are also presented in Table 1.

Degradation characteristics of solar cells and films

The stability of the reference solar cells with the absorber layer deposited at R=20 and R=0 was investigated using an accelerated degradation experiment. Figure 1 shows the normalized fill factor and efficiency as a function of illumination time. The figures demonstrate that the cell with the absorber layer deposited at R=20 is more stable than the conventional cell. The efficiency of the R=20 cell stabilizes at 80% of the initial value, whereas the conventional cell degrades below 65% of the initial value. Table 2 shows the absolute value of the external parameters of the cells; initial, after 64 minutes and after 8000 minutes of light soaking. For the R=0 cell the V_{OC} decreases during degradation, whereas for the R=20 cell the V_{OC} is approximately constant. The table also shows $\alpha(1.2 \text{ eV})$ and N_d for the corresponding individual films. It is interesting to note that the defect density shows a similar behaviour as the fill factor, and increases less for the diluted film.

Table 1. Comparison of film properties and solar cell external parameters

	E_{klazes}	E_{04}	R^*	$\alpha_{1.2\,eV}$	N_d	r_d	η	ff	I_{SC}	V_{OC}
	eV	eV	1	cm^{-1}	$\times 10^{16}$ cm^{-3}	Å/s	%	1	mA/cm^2	mV
R0	1.57	1.90	0.06	0.33	0.79	2.0	9.5	0.72	158	0.84
R20	1.65	1.95	0.08	0.38	0.91	0.60	8.1	0.69	135	0.87
R30	1.58	1.99	0.11	33		1.1	6.4	0.60	128	0.84
LBL	1.64	1.96	0.13	0.72	1.7	1.2	6.7	0.65	122	0.85

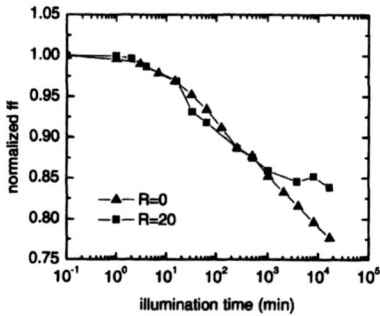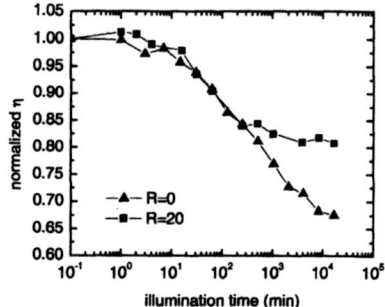

Figure 1. Normalized fill factor and efficiency as a function of illumination time of solar cells with absorber layers deposited at R=20 and R=0.

Dependence of material properties on deposition conditions

In a further effort to improve the stabilized efficiency of solar cells a systematic study of the effect of the deposition conditions at a reduced hydrogen dilution of R=20 was carried out. The effect of the rf-power, pressure and substrate temperature was investigated to investigate the parameter range for which the phase transition in a 300 nm thick layer does not occur. Figure 2 shows r_d, $\alpha(1.2\ eV)$, E_{klazes} and the R* values of the three series of layers deposited at R=20. The deposition rate is clearly the most influenced by the rf-power. The pressure and temperature have little effect on r_d within the investigated parameter range. To interpret the series data correctly, it was investigated how the structural phase of the material is influenced by the deposition parameters. The crystallinity of the films was determined using Raman spectroscopy. Figure 3a shows the Raman spectra of the pressure series and the development of the peak at 520 cm^{-1} for films deposited at pressures ≤ 1.8 mbar is observed. This peak is an indication of the crystalline phase in these films. Raman spectra measured on the films of the power and temperature series show that these samples are amorphous, except for the film in the temperature series deposited at 200°C which contains a crystalline fraction.

Table 2. External parameters of the solar cells and $\alpha(1.2\ eV)$ and N_d of the corresponding films as a function of illumination time t.

	t	η	ff	I_{SC}	V_{OC}	$\alpha_{1.2eV}$	N_d
	min	%	1	mA/cm^2	mV	cm^{-1}	$\times 10^{16}$ cm^{-3}
R0	0	9.5	0.72	158	0.84	0.33	0.79
R20	0	8.1	0.69	135	0.87	0.38	0.91
R0	128	8.0	0.65	151	0.83	0.49	1.2
R20	128	7.0	0.62	128	0.88	0.42	1.0
R0	8000	6.5	0.57	139	0.82	1.6	3.9
R20	8000	6.6	0.59	128	0.88	0.86	2.1

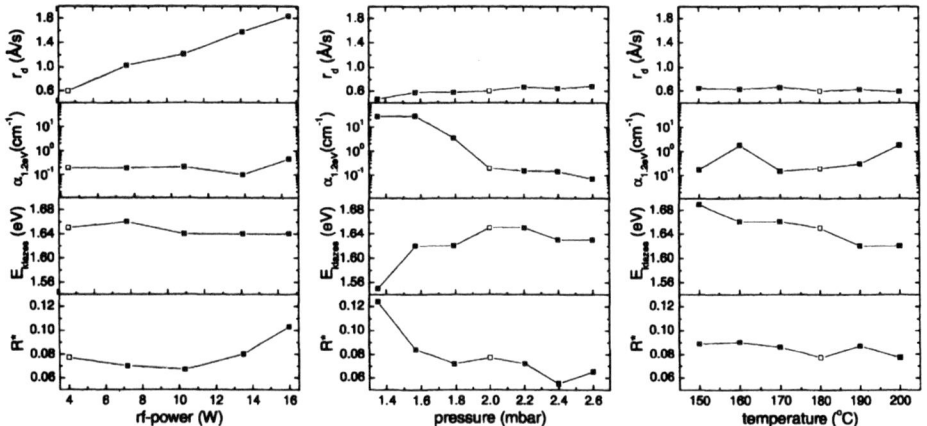

Figure 2. Deposition rate r_d, microstructure factor R*, optical gap E_{klazes} and absorption coefficient $\alpha(1.2\ eV)$ of films from the rf-power, pressure and temperature series with R=20.

The power series does not show much variation in $\alpha(1.2\ eV)$, indicating a constant defect density with varying power. The optical gap becomes slightly lower at higher powers and the microstructure factor shows an increase at higher powers. The pressure series shows a strong increase in $\alpha(1.2\ eV)$ at lower pressures. However, this cannot be interpreted as an increase of the defect density. Instead, the high value of $\alpha(1.2\ eV)$ in the films at pressures ≤ 1.8 mbar is caused by the microcrystalline phase in the material. This is reflected in the shape of the absorption coefficient spectra as shown in figure 3b. For the amorphous films grown at pressures of ≥ 2.0 mbar the defect density decreases slightly with pressure. The optical gap decreases with pressure in the amorphous range, and increases with pressure in the microcrystalline range. The microstructure factor and $\alpha(1.2\ eV)$ show a comparable trend, with a strong increase at lower

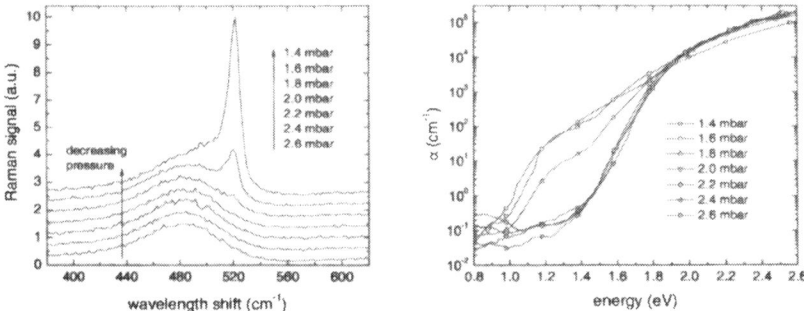

Figure 3. (a) Raman spectra and (b) absorption coefficient spectra of the pressure series.

pressures and a slight decrease at higher pressures. The temperature series shows a minor dependence of $\alpha(1.2$ eV) and the microstructure factor. Only for the sample at $200°$C there is a strong increase in $\alpha(1.2$ eV), again due to the contribution of a crystalline phase. The observation that the microcrystalline fraction increases at elevated temperatures and lower pressures is in accordance with the stress-induced nucleation mechanism [11], as those conditions are expected to result in films with increased stress [11, 12].

CONCLUSIONS

At high hydrogen dilutions phase-control techniques can be applied to stay in the amorphous growth regime. However, a high hydrogen dilution of the absorber layer is not the only requisite for a good solar cell. Accelerated degradation experiments verified that a solar cell with an absorber layer deposited at R=20 is more stable against light-induced degradation than a cell with an absorber layer from pure silane. The study on the influence of various deposition conditions on the material properties has identified the pressure as an important factor for controlling the structural phase evolution within the investigated parameter range. For the amorphous films, the data on R^* and N_d indicate that the material quality improves with decreasing rf-power and increasing pressure. As the differences between the amorphous samples of all series are small further experiments in which these films are included as absorber layers in solar cells are required to obtain conclusive data which could guide the improvement of material and cells deposited at higher hydrogen dilution using the layer-by-layer technique.

ACKNOWLEDGEMENTS

This work was carried out as part of the SENECU project of Delft University of Technology.

REFERENCES

1. D. L. Staebler and C. R. Wronski, Appl.Phys. Lett. **31** (4), 292 (1977).
2. L. Yang and L. F. Chen, Mater. Res. Soc. Proc. **336**, Pittsburg, PA, 1994 p. 669
3. J. Yang, X. Xu and S. Guha, Mater. Res. Soc. Proc. **336**, Pittsburg, PA, 1994 p. 687.
4. C.R. Wronski, J.M. Pearce, R.J. Koval et al., Mater. Res. Soc. Proc. **715**, Pittsburg, PA, 2002 A13.4.
5. G. van Elzakker, V. Nadazdy, F. D. Tichelaar, J.W. Metselaar and M. Zeman, Thin Solid Films **511-512**, 252 (2006).
6. J. Koh, Y. Lee, H. Fujiwara et al., Appl. Phys. Lett. **73** (11), 1526 (1998).
7. G. van Elzakker, F. D. Tichelaar, and M. Zeman, Thin Solid Films (2007), (in press).
8. N. Wyrsch, F. Finger, T. J. McMahon et al., J. Non-Cryst. Solids **137-138**, 347 (1991).
9. R. H. Klazes, M. H. L. M. van den Broek, J. Bezemer et al., Phil. Mag. B **45**, 377 (1982).
10. M. Zeman, "Advanced Amorphous Silicon Solar Cell Technologies", *Thin Film Solar Cells*, ed. by J. Poortmans and V. Arkhipov (John Wiley & Sons, Chichester, 2006).
11. H. Fujiwara, M. Kondo, and A. Matsuda, Jpn. J. Appl Phys **41**, 2821 (2002).
12. C.L. Garrido Alzar, Mater. Sci. Eng. B **65**, 123 (1999).

Mater. Res. Soc. Symp. Proc. Vol. 989 © 2007 Materials Research Society 0989-A08-05

Reliability of Silicon Nitride Gate Dielectric in Vertical Thin-Film Transistors

M. Moradi[1], D. Striakhilev[1], I. Chan[1], A. Nathan[2], N. I. Cho[3], and H. G. Nam[3]

[1]Electrical and Computer Eng., University of Waterloo, 200 University Ave. West, Waterloo, N2L3G1, Canada

[2]London Centre for Nanotechnology, University College London, 17-19 Gordon Street, London, WC1H 0AH, United Kingdom

[3]Electronic Eng., Sun Moon university, Asan-Si, Chungnam, 336-708, Korea, Republic of

ABSTRACT

This paper presents results of a systematic investigation of the impact of film thickness on leakage current and electrical breakdown of plasma enhanced chemical vapor deposited (PECVD) silicon nitride (SiN_x). We consider SiN_x films of various thicknesses, in the range 50 to 300 nm, deposited on both planar and vertical sidewalls in resemblance to the structural topology of the vertical thin film transistor (VTFT). The electrical breakdown strength for 150–300 nm thick films was approximately 7 MV/cm, while the value dropped to ~3 MV/cm for 50 nm thick films deposited under the same process conditions. In all cases, failure is inevitably accompanied by an increase in pinhole density. The results show that the reliability and leakage current of the gate dielectric in vertical thin film transistors depends on the step coverage of the SiN_x on the vertical sidewall.

INTRODUCTION

Thanks to its inherent structural attributes of short channel length and small device area, the VTFT offers an excellent platform for a new generation of low-cost, high-speed, and high-resolution flat-panel electronics [1]. While the device holds great promise, there are issues related to its electrical performance and reliability when scaled to smaller geometries. In particular, the thickness of the gate dielectric has a significant impact on the short channel effects in the device as channel lengths approach the nano-scale regime. An aggressive scaling of the gate dielectric is necessary to improve the short channel performance and sustain high ON/OFF current ratio [2, 3]. VTFTs with channel lengths of 1μm and below require gate dielectrics as thin as 50nm to comply with scaling rules. However, and as is well known, scaling down the gate dielectric thickness adversely impacts the electrical characteristics of the device. High gate leakage and, in particular, early breakdown are the major issues (see Fig. 1).

EXPERIMENT

To examine the impact of thickness on the physical properties and electrical characteristics of SiN_x, we considered films deposited on both planar and vertical surfaces. The latter resembles the topology of the vertical thin film transistor. Fig. 2 shows a schematic diagram of the planar and vertical structures. For the electrical properties, current-voltage characteristics of the nitride with different thicknesses were measured to study the leakage current and the breakdown voltage. In both structures the silicon nitride film was sandwiched between two metal electrodes with different areas. The dependence of the leakage current on thickness was investigated with SiN_x films of different thicknesses in the range 50-300 nm. The SiN_x films were deposited by conventional 13.56MHz PECVD using silane and ammonia gas precursors at a substrate temperature of 300°C.

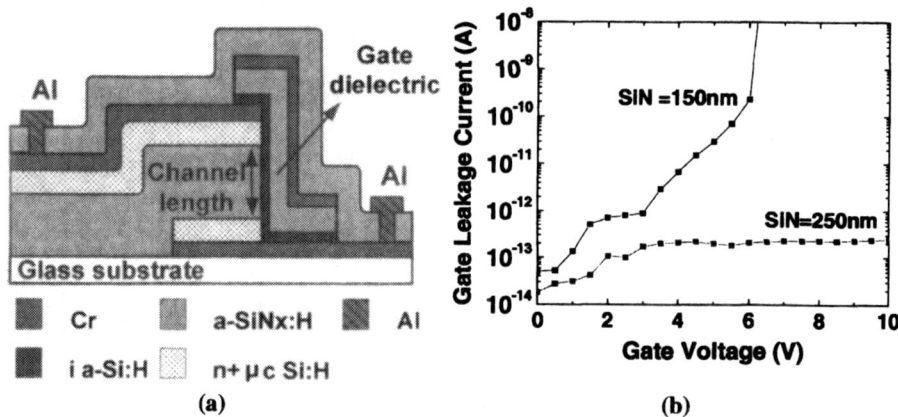

(a)   (b)

FIG. 1 (a) Vertical thin film transistor structure and (b) its gate leakage illustrating the effect of silicon nitride dielectric scaling.

In order to study the physical properties of the nitride films with different thicknesses, we used the selective etching method, in which an etchant enters the pinholes and/or cracks of the nitride to etch the underlying metal layer. Using this method we obtain a rough estimate of the pinhole density for the different nitride thicknesses.

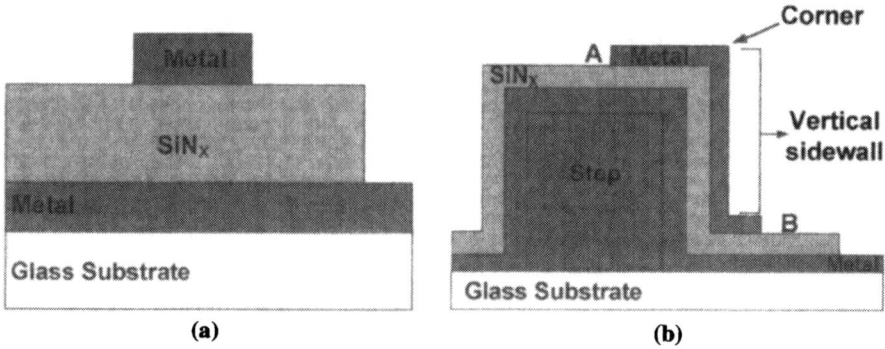

(a)   (b)

FIG. 2: Cross-section schematic of the (a) planar and (b) vertical structures used in this study (not to scale).

RESULTS AND DISCUSSION

To study the source of the high gate leakage in the VTFT, we fabricated both lateral and vertical structures with different areas. We found that nitride films deposited on a metallized planar substrate (Fig. 2(a)) have a lower leakage current and higher breakdown strength compared to those deposited on vertical sidewalls (Fig. 2(b)) for the same deposition conditions (Fig. 3). The data reveals that for VTFTs whose gate has both planar and vertical areas, the latter is responsible for the high leakage current whose magnitude is three orders higher.

Intuitively there may be two possible causes for the high leakage and the low breakdown in vertical structures. The first could be the high electric field at the corners of the vertical structure. Fig. 4 shows the electric field distribution of the vertical structure, numerically simulated using the Medici software. Here, the step height is 500nm and the thickness of the nitride is 200nm. The horizontal axis on the plot corresponds to the distance from A to B shown in Fig. 2(b). The two noticeable areas on Fig. 4 which show the highest electric field are related to the corners of the vertical structure. The electric field here is ~ 40% higher compared to the middle of the sidewall. We also tested structures which featured a variable number of corners to verify if the number of corners has an effect on the leakage and breakdown behavior. The associated current-voltage characteristics are illustrated in Fig. 5. Contrary to intuitive predictions, the current-voltage characteristics show no clear certainty in the breakdown voltage values, i.e. there is no systematic effect of the number of corners on the breakdown voltage.

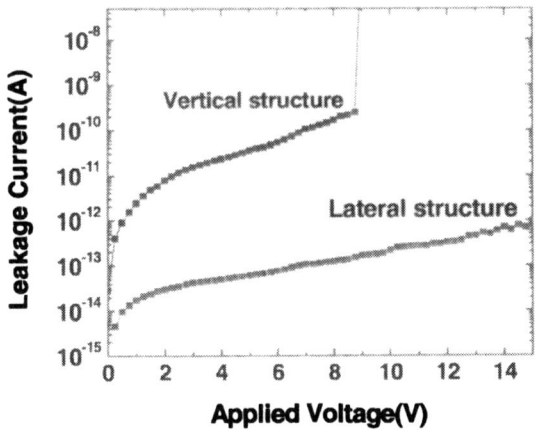

FIG. 3: Current-voltage characteristics of vertical and lateral structures.

The other possible cause for high leakage current and early breakdown may be related to material differences or inhomogeneities in the material deposited on the vertical side walls. It has been reported that the thickness of PECVD films on vertical sidewalls are approximately half that on the horizontal surface due to the nature of plasma kinetics, which governs step coverage behavior [1]. We therefore examined the dependence of the leakage current and breakdown voltage on nitride thickness.

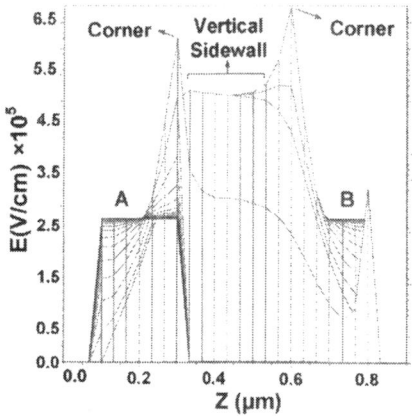

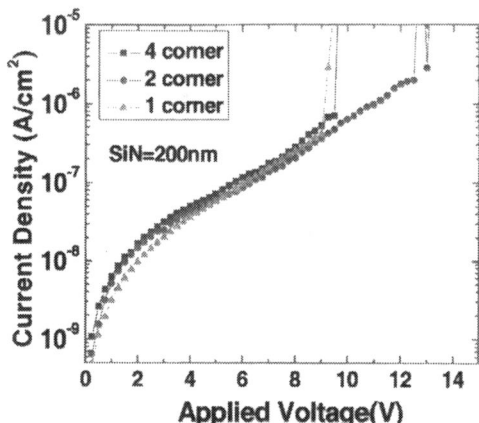

FIG. 4: Simulated electric field distribution of the vertical structure. The horizontal axis corresponds to the distance from A to B shown in Fig. 2(b)

FIG. 5: Current density of vertical structures with different number of corners.

Planar structures with nitride thicknesses in the range 50-300nm were fabricated. The associated current-voltage characteristics are shown in Fig. 6. The results show that the electrical breakdown strength for 150-300nm thick films was approximately 7 MV/cm, whereas the value drops to ~3MV/cm for 50nm thick films deposited under the same process conditions. Fig. 7 shows the critical electric field leading to breakdown, which is lower for the thinner nitride. This behavior led us to look at the possible presence of defects in the nitride which may be responsible for the early failure in thinner films. Nitride films with different thicknesses were deposited onto metallized substrates and subject to a chemical solution which enters cracks or pinholes to etch the underlying metal. The spots where the metal is etched are easy to observe and provide a rough estimate of the defect pinhole density [4], which is obtained by counting the number of metal-etched spots in a 500μm-radius area. In order to increase the accuracy of pinhole count, the observation field was magnified by a factor of 20. It is observed that while the pinhole density of the silicon nitride with 50nm thickness is $8.6 \times 10^3/cm^2$, it is $3 \times 10^2/cm^2$ for the 200 nm thickness film. Experimental data show that the lower breakdown strength is accompanied by a pinhole density that is larger by a factor of 20. High pinhole density of the thin nitride can be explained by the nature of thickness evolution during the deposition. At the initial stages of deposition there is formation of islands or nuclei, which subsequently connect to form a continuous film [5]. Hence at the vicinity of the substrate, the nitride is more defective than in the bulk [6].

The thickness of the film in which the adjacent nuclei meet to produce a continuous layer, can be reduced by changing the deposition conditions [5]. In fact, there are speculations in the published literature regarding the role of inert gases like helium in the chemistry of plasma assisted deposition. It is reported that the addition of He to the deposition chamber reduces the incorporation of impurities and produces a high-quality film structure [7]. Work along these lines is currently in progress to gauge its possible impact on dielectric leakage and breakdown.

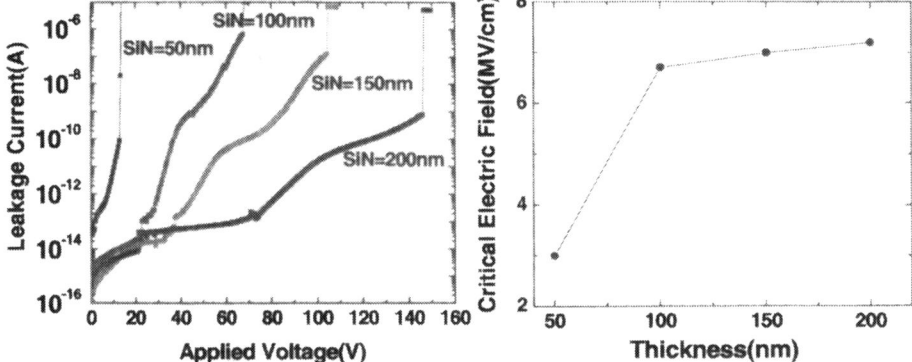

FIG. 6: Current-voltage characteristics of silicon nitride with different thicknesses.

FIG. 7: Critical electric field for breakdown as a function of silicon nitride thickness.

The above results have direct implications in the VTFT, in which silicon nitride is used as a gate dielectric. Here, because of poor step coverage, its thickness on the vertical sidewall is half of that on the horizontal area, inevitably leading to material inhomogeneities. We believe this is the primary cause of the high gate leakage and early breakdown. The breakdown usually occurs in a small area of the device [8] and is related to the concentration of pin-hole defects. In other words, VTFTs that break down at low electric filed are due to the high density of pinholes which are dominant in the thin silicon nitride that is on the vertical sidewall. Thus modification of the deposition conditions aimed towards achieving good dielectric characteristics of thin SiN films is crucial for future advancement in VTFT structures and short channel a-Si TFTs in general.

CONCLUSIONS

Current-voltage measurements reveal that the leakage current in vertical sidewall SiN films, typical to vertical transistor structures, is at least 3 orders of magnitude higher than that in films deposited on planar geometries, along with a significantly lower breakdown voltage. We explored the two possible causes for the high leakage current and low breakdown voltage of thin nitride films, i.e. the high electric field at corners and the poor step coverage, Results show that the electric field singularity at corners in the VTFT is not the reason for the high leakage current and poor device reliability. Rather it is the poor step coverage due to the plasma kinetics in PECVD, which makes the thickness of the deposited film on the vertical sidewall half that on the planar substrate. Thinner silicon nitride films are observed to contain a higher density of pinholes leading to higher leakage and lower breakdown voltage.

ACKNOWLEDGMENTS

This research is funded by the Natural Sciences and Engineering Research Council of Canada (NSERC), the Ontario Graduate Scholarship, the Korean Ministry of Commence, Industry and Energy, and Grant No. RTI04-01-02 from the Regional Technology Innovation Program of the Ministry of Commerce, Industry, and Energy (MOCIE). The authors would like to thank S. Fathololoumi and A.A. Fomani for technical assistance.

REFERENCES

1. I. Chan, S. Fathololoumi, and A. Nathan, "Nanoscale channel and small area amorphous silicon vertical thin film transistor," *Journal of Vacuum Science & Technology A*, vol. 24, no. 3, pp. 869-874, 2006.
2. D. Matsushita, K. Muraoka, K. Kato, Y. Nakasaki, S. Inumiya1, K. Eguchi1, and M. Takayanagi1, "Novel fabrication process to realize Ultra-thin (EOT=0.7 nm) and ultra-low-leakage SiON gate dielectric," *Microelectronic Engineering*, vol. 80, pp. 424-431, 2005.
3. T. P. Ma, "Making Silicon Nitride Film a Viable Gate Dielectric," *IEEE Transaction on Electron Devices*, vol. 45, no. 3, pp. 680-690, 1998.
4. C. G. Shirley, S. C. Maston, "Electrical measurements of moisture penetration through passivation," *28th Annual Proceedings. Reliability Physics*, pp. 72-80, 1990.
5. K. Joohyun, J. S. Burnham, L. Yeeheng, L. Hongyue, C. Ing-Shin; , L. J. Pilione, C. R. Wronski., and R.W Collins, "Structural evolution of top-junction a-Si:C:H:B and mixed-phase (microcrystalline Si)-(a-Si$_{1-x}$C$_x$:H) p-layers in a-Si:H n-i-p solar cells," *Amorphous Silicon Technology Symposium*, pp. 69-74, 1996.
6. H. Fritzsche, *Amorphous Silicon and Related Materials*, Vol. 1, Singapore: World Scientific, 1989.
7. T. Karabacak, Y.-P. Zhao, G.-C. Wang, and T.-M. Lu, "Growth front roughening in silicon nitride films by plasma-enhanced chemical vapor deposition," *Physical Review B*, vol. 66, pp.075329/1-10, 2002.
8. D. K. Schroder, *Semiconductor Material and Device Characterization*, New York: Wiley, 1998.

Mater. Res. Soc. Symp. Proc. Vol. 989 © 2007 Materials Research Society 0989-A08-08

Deposition Uniformity Control in a Commercial Scale HTO-CVD Reactor

Shigeru Sakai[1], Masaaki Ogino[1], Ryosuke Shimizu[1], and Yukihiro Shimogaki[2]
[1]Fujielectric Advanced Technology Co.,Ltd., 4-18-1,Tsukama,Matsumoto, Nagano, 390-0821, Japan
[2]School of Engineering, University of Tokyo, 7-3-1,Hongo,Bunkyo-ku, Tokyo, 113-8656, Japan

ABSTRACT

High-temperature silicon dioxide (HTO) chemical vapor deposition (CVD), using SiH_2Cl_2 and N_2O, can realize dense and conformal oxide film, not only on large size silicon wafers, but even inside of microscopic silicon trenches, at high-temperature around 800°C.

In this work, we investigated the kinetics of HTO-CVD using a commercial scale low pressure (LP) CVD reactor, focusing on the correlation between deposition rate and surface-to-volume ratio (S/V ratio), which is a specific surface area of substrate wafer divided by the space volume between two adjacent wafers. We also investigated the deposition rate profile on wafers, and along the axial direction of the reactor near the region where one, two or three substrate wafers are extracted from the quartz holder. The deposition rate profiles on wafer characteristically change from skillet-like to pancake-like, according to the increase of wafer spacing. The influence of wafer spacing on the deposition rate spreads to ranges not only downstream, but also upstream in the gas flow. These experimental results strongly suggest that in the HTO-CVD gas-phase reactions through intermediate states of active species contribute to deposition reaction as well as direct deposition reaction of source gases on Si surface.

INTRODUCTION

HTO film is deposited on a higher temperature around 800°C [1-3], compared with the deposition temperature, 350°C of LTO (Low-temperature silicon dioxide) film, usually applied as passivation films of semiconductor devices. Because of high-temperature deposition, dense and conformal oxide film can be obtained even inside microscopic silicon trenches. HTO is promising as gate oxides of gallium nitride (GaN) MOSFET [4] and silicon trench-type MOSFET. We found that the deposition rate profile of HTO shows unique characteristics, greatly different from the profile of conventional films, such as poly-Si films obtained by commercial scale batch type reactors as in fig.1. From the background mentioned above, in this work, to improve the thickness uniformity, we investigated the kinetics of HTO-CVD using a commercial scale LPCVD reactor, through an analysis on deposition rate profiles both in the radial direction of 6inch silicon wafers and along the axial direction of the reactor.

The deposition rate profile in the radius direction of the wafer gave us the insight of SiH_2Cl_2 (Dichlorosilane; DCS) and N_2O based reaction chemistry. The experimental results of actual deposition behavior revealed the chemical species that controls the uniformity in HTO-CVD. Thus we could obtain the simple reaction model in a manufacturing scale reactor.

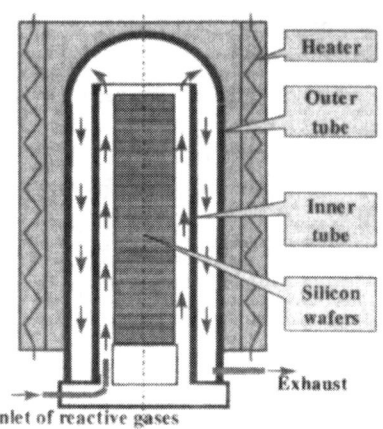

Figure.1 Schematic diagram of a longitudinal type CVD reactor of $\phi 6$" manufacturing scale applied in this work. $\phi 6$" silicon wafers are set up successively with a distance of w (5.35mm). Reactive gases flow upwards from the bottom of the reactor and diffuse between two wafers together with generated precursors to the center of the wafers.

EXPERIMENT

We used a longitudinal type CVD reactor of $\phi 6$" manufacturing scale, which is drawn schematically in fig.1. Sample wafers were fully charged on the quartz holder in horizontal manner and the 100% DCS and 100% N_2O gases were introduced from the bottom of the reactor. The deposition temperature is kept about 800°C and the total pressure is 60Pa in the reactor. Gas flow rates are 150sccm of DCS and 75sccm of N_2O. Each wafer is usually charged on the quartz holder with equal interval w (5.35mm).

In this work, we paid attention to wafer interval space volume V and surface area S on a sample wafer that deposition reaction occurs.

1. A correlative investigation of surface area S and wafer interval space volume V

We set a sample wafer that had surface area S at center of quartz holder. By removing 1, 2, or 3 wafers just above the sample wafer, we made the wafer interval space volume V changed, where the wafers were usually charged with equal interval (5.35mm) as shown in fig.2. In this way, we investigated the change of deposition rate and rate profile when S and wafer interval space volume V were changed. We changed S by producing trenchs with 2 types of mask patterns and 3 types of trench depths on the sample wafer. In total, we prepared 6 types of sample wafers that have different surface area S's compared with non patterned bare Si wafer. In case of this samples, trench patterned sample wafers have 1.35~3.3 times larger S, compared with bare Si wafer (177cm²).

2. Investigation of influences to neighboring wafers

To confirm the change of a deposition condition in the area where S and V were changed, we investigated deposition rate characteristics of neighboring wafers near the sample wafer that was set at center of quartz holder. In this experiment, these neighboring wafers were all non-patterned bare Si wafers.

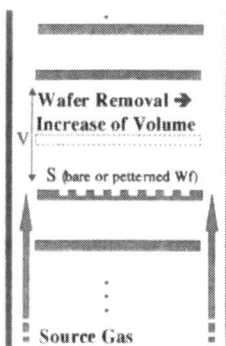

Figure.2 Sample wafer arrangement in the furnace center region. We placed one sample wafer to the furnace central region, the surface area S of which was changed by making trench structures over the surface, and we made the space volume V directly above the sample wafer change by removing 1, 2 or 3 wafers just above the sample wafer.

Experimental Results and Discussion

1. A correlative investigation of surface area S and wafer interval space volume V

We evaluated the dependence of deposition rate on S with constant V between adjacent wafers. By increasing S, deposition rate tends to decrease monotonously. With constant S and V increased, deposition rate tends to increase monotonously.

Fig. 3 shows the distribution of deposition rate ratio on sample wafer (bare Si) when V was changed. From this result, we can confirm that the increase of the deposition rate with larger V, which suggests that the more the wafer spacing, the more the quantities of source gases are supplied from the outside of the sample wafer. But, when we concentrate to the distribution on the sample wafer, in case of condition of wafer spacing roughly over 10mm, the deposition rate profiles characteristically change from skillet-like to pancake-like, according to the increase of the wafer spacing. This tendency was also observed on trench patterned wafers. These tendencies that the deposition rate is larger in the wafer center region rather than in wafer outer region cannot be explained by a mechanism that source gases come from the space outside the wafer column, chemically react, waste on and over wafer surface and formed SiO_2 film.

Figure.3 The dependence of deposition rate profiles on the sample wafer on wafer spacing w. The deposition rate increases according to the increase of wafer spacing. And the deposition rate profile characteristically changes from skillet-like to pancake-like.

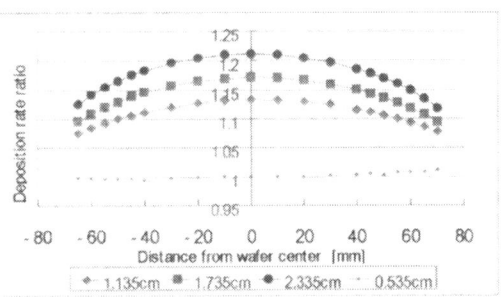

While, we also confirmed about the distribution of deposition rate ratio and wafer spacing dependence in poly-Si CVD by use of similar longitudinal type CVD reactor of ϕ6" manufacturing scale. In poly-Si CVD, we obtained a result that with the increase of wafer spacing w or wafer interval space volume V, the deposition rate increases similarly as in the case of HTO-CVD, but the deposition rate profiles did not change from skillet-like to pancake-like according to the increase of wafer spacing [5].

Based on this result, it is considered that HTO reaction process has mainly 2 types of reaction paths as shown in fig.4 [6]. One is the deposition reaction path where source gases react directly on wafer surface and deposit as SiO_2 film (Direct deposition), and the other is the deposition reaction path where source gases react to form intermediate species in wafer interval space volume V, and intermediate species reach to the wafer surface and react to deposit as SiO_2 film (Deposition of Intermediate). More precisely, there are so many paths existing, where the formation of intermediate species and the reaction through the gas phase and on wafer surface occur. But, in this work, to simplify the model of the deposition processes, we summarized these many reaction processes in gas phase and on surface to one reaction path (Non direct deposition).

Figure.4 Schematic diagram of deposition reaction model in DCS/N_2O HTO process. Deposition reaction has 2 types of reaction paths. One is the direct deposition, and the other is the deposition through the intermediate species generated from source gases.

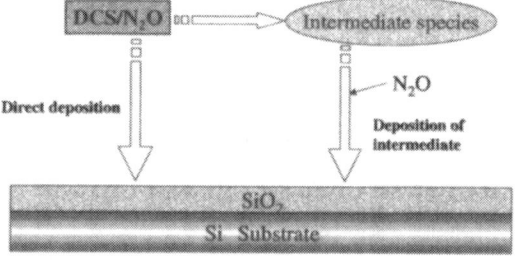

Fig. 5 shows the summary of correlation between the deposition rate and the S/V ratio based on these experimental results. The space volume where the deposition reactions occur is surrounded by the sample wafer and the adjacent wafer facing to the sample wafer. SiO_2 film is deposited not only on the sample wafer surface but also on the backside surface of the adjacent wafer. So, we summarized the deposition rate by use of parameter V/S' instead of V/S, where S' means the total surface area summing the sample wafer surface and the adjacent wafer backside surface. From fig.5, we confirmed that the deposition rate increases monotonously, as increase of V/S'.

Fig. 5 (b) shows a simple model of factors that dominate the deposition rate. As mentioned above, the deposition rate is the sum of the direct deposition rate and the deposition rate through intermediate species. The growth rate at zero point of V/S' in this figure shows the deposition rate of the direct deposition itself. From the experimental result in fig.5 (a), we can confirm that in this HTO deposition condition, the direct deposition is predominant in whole deposition reaction.

But in the region of large V/S' ratio above about 0.3cm, the increase of growth rate tends to slow down. We consider this reason is as follows. According to increase of V, the deposition rate through the intermediate species increases, but these intermediate species are easy to flow out from wafer region as the wafer spacing is becoming larger, and so the growth rate on the sample wafer decreases. From this point of view, we investigated the influence of these flown-out intermediate species by checking the deposition rate on neighborhood wafers of the sample wafer.

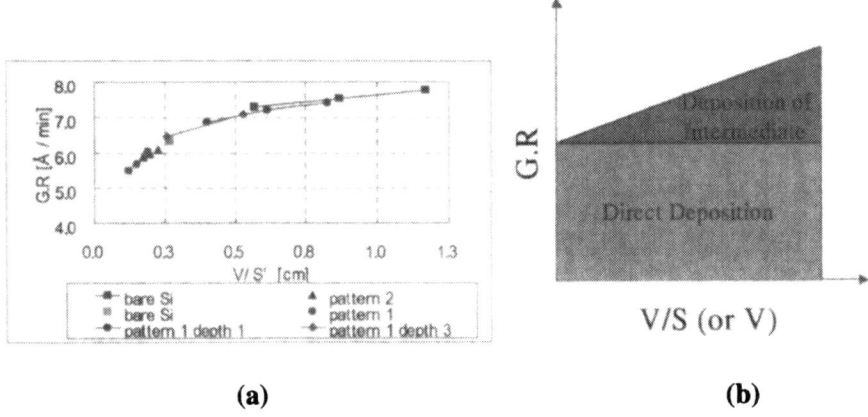

(a)   (b)

Figure.5 Experimental correlation of growth rate (G.R) and V/S' (a) , and simple growth rate model (b). From this experimental result, it turned out that the direct deposition contributes to the growth rate more than the deposition through the intermediate species.

2. Investigation of the influences to the deposition rate on neighboring wafers

Fig. 6 shows the change of deposition characteristic on neighboring wafers (bare Si) near the sample wafer when the spacing over the sample wafer is changed. In this experiment, the sample wafer was trench patterned wafer. In case the sample wafer was bare Si, we also confirmed the similar tendency as with trench patterned sample wafer (not shown in this paper). In fig. 6, horizontal axis shows wafer number counted from bottom of quartz holder shown in fig.1. Usually, the deposition rate on whole wafers in the furnace are controlled to be constant. But, in case that V is partly increased at the center region of holder, deposition rate on the sample wafer increased as shown in the figure. Not only on the sample wafer, but it was found that the deposition rate on the neighboring wafers near the sample wafer also increases. The more near neighboring wafers is from the sample wafer, the more influenced the deposition rate is. This result explains that the intermediate species, which are more formed in wafer interval space volume V of the sample wafer than in the case of the normal wafer spacing, flow out outside of the sample wafer, diffuse, and influence on the deposition reaction on not only downstream wafers but also upstream wafers against gas flow.

We also investigated the deposition rate profiles on these neighboring wafers and confirmed that the profiles remained to be pancake-like on even the nearest wafer from the sample wafer.

Figure.6 Growth rate distribution along the axial direction of furnace (sample wafer: pattern 1 depth 1 V/S'=0.61). When wafer spacing w is large, formation of the intermediate species increases and number of flown-out species to outside of the wafer region also increases. So, the flown-out intermediate species contribute to the increase of deposition reaction and growth rate on neighboring wafers.

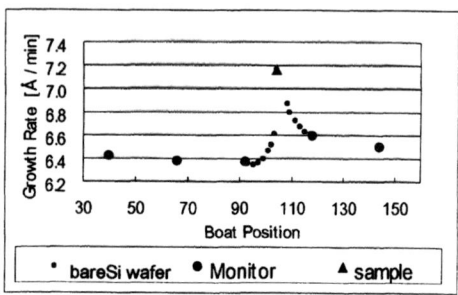

CONCLUSIONS

We investigated HTO-CVD of DCS/N_2O deposition reaction mechanism experimentally using a commercial scale LPCVD reactor. From this work, it turns out that HTO deposition process is composed of mainly 2 types of reaction paths, that is, the direct deposition and the deposition through intermediate species, and the direct deposition dominates in whole deposition reaction process. Intermediate species of source gases increase greatly according the increase of the wafer spacing and attribute to the convex deposition rate profile on wafers. And in the case of large wafer spacing, the intermediate species flow out to outside of the wafer region, diffuse to not only downstream but also upstream to the neighboring wafers, and influence the deposition reactions on those wafers.

Identifications of these intermediate species and reaction processes in gas phase and on wafer surface region are the pressing issues.

ACKNOWLEDGMENTS

The authors would like to thank Dr. Shinji Fujikake and Mr. Kunio Mochizuki of Fuji Electric Advanced Technology Co., Ltd. for their continuous encouragement to this work.

REFERENCES

1. A. Philipossian and K. van Wormer, World Congr. 3 Chem. Eng. **1**, 290 (1986).
2. A. Philipossian, Proc. Int. Conf. Chem. Vap. Depos. **10**, 530 (1987).
3. D.W. Freeman, J. Vac. Sci. Tech. A. **5**, 1554 (1987).
4. K. Matocha, T. P. Chow, and R. J. Gutmann, Proc. 2003 Int. Symp. Power Semicon. Dev. & ICs, p.54 (2003).
5. R. Shimizu, M. Ogino, M. Sugiyama, and Y. Shimogaki, J. Electrochem. Soc. **154**, No.6 (2007), in press.
6. R. Shimizu, M. Ogino, S. Sakai, and Y. Shimogaki, Meeting Abstr. 209th Meeting of the Electrochem. Soc. (2006).

Electronics on
Flexible Substrates

Mater. Res. Soc. Symp. Proc. Vol. 989 © 2007 Materials Research Society 0989-A09-03

Flexible a-Si:H-based Image Sensors Fabricated by Digital Lithography

William S. Wong[1], TseNga Ng[1], Michael L. Chabinyc[1], Rene A. Lujan[1], Raj B. Apte[1], Sanjiv Sambandan[1], Scott Limb[2], and Robert A. Street[1]

[1]Electronic Materials and Devices Lab, Palo Alto Research Center, 3333 Coyote Hill Road, Palo Alto, CA, 94304

[2]Hardware Systems Lab, Palo Alto Research Center, 3333 Coyote Hill Road, Palo Alto, CA, 94304

ABSTRACT

Amorphous silicon-based x-ray image sensor arrays were fabricated on poly-ethylene naphthalate substrates at process temperatures below 180°C. Patterning of the thin-film transistor backplane was accomplished using ink-jet printed etch masks. The sensor devices were found to be comparable to high-temperature processed devices. The integration of the sensor stack, TFT array and PEN substrate resulted in a flexible x-ray image sensor with 180×180 pixels with 75 dpi resolution.

INTRODUCTION

The processing of electronic devices on flexible platforms possesses many intrinsic difficulties. Film stress, thermal cycling, environmental conditions, and handling are all issues that have a direct impact on the successful fabrication of flexible electronics. The incorporation of multiple thin-film layers over a wide range of thicknesses adds more complexity to managing the fabrication process. For example, in x-ray image sensor array applications, the thickness of the sensor layer itself may be more than twice the thickness of the thin-film transistor (TFT) device layer.[1] The built-in stress for this layer may lead to strain-induced cracking of the sensor layer and the underlying TFT backplane.[2] In order to reduce the built-in stress of these films, thinner layers and lower process temperatures may be used to minimize the possibility of cracking and distortion of the substrate and the thin film. To address potential misalignment problems, jet-printed etch masks may be more applicable for aligning to features on a distorted surface. The mask layers can be first aligned to a local reference mark and the non-contact jet-printed mask can then be placed directly onto the distorted feature to complete the masking process. We have combined the use of a jet-printed etch-mask process, digital lithography, with 880 nm p-i-n sensor structures processed at < 200°C to integrate a-Si:H-based image sensor arrays onto flexible platforms.

EXPERIMENTAL DETAILS

a-Si:H-based p-i-n sensors were first fabricated on glass substrates at temperatures ranging from 150°C to 250°C. The layers were deposited using plasma-enhanced chemical-vapor deposition. The heterostructure consisted of a bottom-metal electrode followed by a 70 nm thick n-layer, i-layer, and a 10 nm thick p-layer. The i-layer thickness was varied between 500 nm to 1 micron. Current-voltage measurements were done using a Keithly 617 programmable electrometer. External quantum efficiency measurements were performed using irradiation from an Ar ion laser at a wavelength of 488 nm.

The digital lithography process combines jet-printed lithography with digital imaging and processing to register virtual masks for TFT device patterning. In digital lithography, layer alignment is accomplished by electronically imaging the coordinates of existing alignment marks using a camera attached to a microscope objective. Once obtained, the coordinates are used to reposition an electronic mask layer aligned to the process wafer prior to printing the mask pattern. Digital lithographically patterned TFT devices on poly-ethylene naphthalate (PEN) substrates was used in place of conventional photolithography for the backplane patterning. These devices were fabricated using conventional plasma-enhanced chemical vapor deposition of the a-Si:H and silicon nitride layers at process temperatures between of 150-180°C. The print-patterned back-channel etch structures formed a TFT array of 180×180 pixels having 75 dpi resolution.

DISCUSSION

Figure 1 shows the dark-current and light-current characteristics for sensors fabricated at 150°C on PEN substrates having an i-layer thickness between 300nm to 800 nm. For the 300nm thick i-layer device the leakage current increased with reverse bias voltages > 2 V compared to the 800 nm thick device deposited at 150°C.

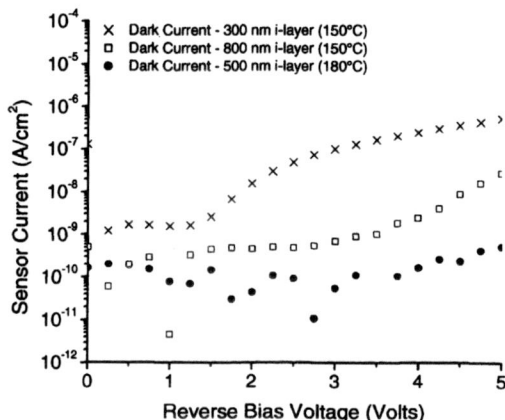

Figure 1: Dark current characteristics of low-T and high-T processed sensors under reverse bias conditions

This increased dark current is due to tunneling current leakage across the contacts through the thinner i-layer. By increasing the deposition temperature of the i-layer to 180°C, the dark current was found to be constant for reverse bias voltages up to 5 volts for a 500 nm thick i-layer. The ideality factor of the sensors at different deposition temperatures were $n_{low-T} \sim 3$ and $n_{high-T} \sim 2$ for the low-temperature ($T_{low} < 180°C$) and high-temperature ($T_{high} > 180°C$) deposition, respectively. The ideality factor did not change as a function of thickness (at similar process temperatures) suggesting that the I-V characteristics are more sensitive to process temperature changes.[3]

The external quantum efficiency of the p-i-n structures was measured to further characterize the quality of the low-temperature sensors. Figure 2 shows the external quantum efficiency for two deposition temperatures and i-layer thicknesses as a function of applied electric field for the p-i-n- structures. The sensor fabricated at high temperature having a 1 micron i-layer showed an external quantum efficiency of ~ 50% and was flat across the applied field. For the low-temperature (T = 150°C) p-i-n having a 1 micron thick i-layer, the quantum efficiency was found to decrease with decreasing electric field. For the same deposition temperature, the thinner i-layer device (600 nm thick i-layer) possessed a characteristic comparable to the high-temperature sensor. The results suggest the carrier mean free path length in the low-temperature device is much lower than that of the high-temperature material. In the case of the thinner i-layer processed at 150°C, the trapping probability is reduced, resulting in the improved external quantum efficiency.

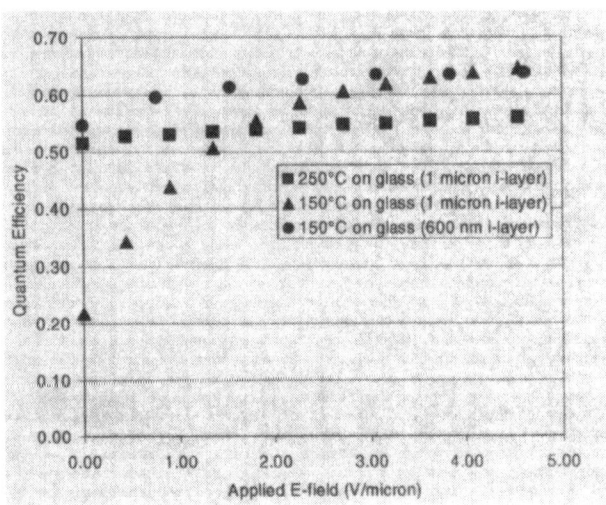

Figure 2: External quantum efficiency as a function of applied electric field.

BACKPLANE AND SENSOR INTEGRATION

The device performance for the low-temperature processed TFT backplane by digital lithography on PEN was reported earlier.[4,5] In summary, the devices were comparable to a conventionally processed TFT on glass having a field-effect mobility of ~ 0.9 cm^2/V·s, and on/off ratio of 10^7 and a threshold voltage of ~ 3V for a device with W/L of ~ 1.5. Once the backplane was fabricated, an encapsulation layer consisting of 800 nm of oxynitride was deposited by PECVD. Next, via contacts were etched onto the pixel and the sensor layer was deposited consisting of a 70 nm thick n-layer, 800 nm thick i-layer, and a 10 nm thick p-layer. Figure 3 shows the successful integration of an a-Si:H TFT array with a low-temperature a-Si:H p-i-n sensor to create an x-ray imager array. An image of the steel screws and bolts was created from light captured from an x-ray illuminated phosphor onto the flexible sensor array. The image shows some line and point defects that were due to process handling of the free standing flexible PEN. The flexible sensor array does demonstrate the integration of high-quality electronics on flexible platforms at low-temperatures using an ink-jet print-patterning process.

6 cm

6 cm

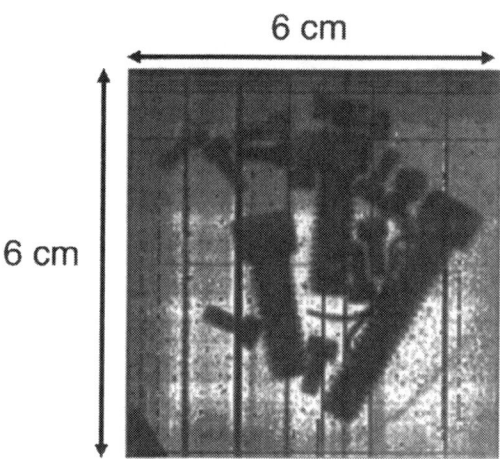

Figure 3: Demonstration of a flexible x-ray image sensor array. The sensor and TFT backplane were fabricated using low-T processed a-Si:H patterned by digital lithography on PEN.

CONCLUSIONS

Amorphous silicon-based p-i-n sensor diodes arrays were fabricated on PEN substrates at process temperatures below 180°C. The dark current of the sensors were found to increase with decreasing process temperature while the efficiency of the devices were found to improve with decreasing i-layer thickness. A digital lithographic patterning process was used to fabricate the TFT array backplane directly onto the PEN substrates.

The p-i-n sensor was then directly deposited onto the backplane to create a flexible sensor array. A 180×180 pixel x-ray image sensor array, with 75 dpi resolution was demonstrated.

ACKNOWLEDGEMENTS

The authors gratefully acknowledge M. Young, B. Russo, M.A. Rosenthal, and Steve Ready of PARC for assistance with the experiments and for helpful discussions. We also wish to acknowledge W.A. MacDonald of DuPont-Teijin Films for supplying the polyethylene naphthalate substrates. This work is partially supported by the Advanced Technology Program of the National Institute of Standards and Technology (contract #: 70NANB7H3007 and 70NANB3H3029).

REFERENCES

1. R. A. Street, Large Area Image Sensor Arrays, *Technology and Applications of Amorphous Silicon*, R.A.Street (Ed.), pp. 147-221 (Springer-Verlag, 2000).
2. Z. Suo, E.Y. Ma, H. Gleskova, and S. Wagner, Appl. Phys. Lett. **74**, 1177 (1999).
3. M. A. Kroon and R. A. C. M. M. van Swaaija, J. Appl Phys. **90**, 994 (2001).
4. W.S. Wong, S. Ready, R. Matusiak, S.D. White, J.-P. Lu, J. Ho, R.A. Street, Appl. Phys. Lett. **80**, 610-12 (2002).
5. W.S. Wong, S.E. Ready, J.-P. Lu, R.A. Street, IEEE Electron Dev. Lett. **24**, 577 (2003).

Mater. Res. Soc. Symp. Proc. Vol. 989 © 2007 Materials Research Society   0989-A09-04

Stability of Amorphous Silicon Thin Film Transistors under Prolonged High Compressive Strain

Jian-Zhang Chen[1,2], I-Chun Cheng[1,3], Sigurd Wagner[1], Warren Jackson[4], Craig Perlov[4], and Carl Taussig[4]

[1]Department of Electrical Engineering, Princeton University, Princeton, NJ, 08544
[2]Photovoltaics Technology Center, Industrial Technology Research Institute, HsinChu, 310, Taiwan
[3]Graduate Institute of Electro-Optical Engineering, National Taiwan University, Taipei, 100, Taiwan
[4]Hewlett Packard Laboratories, Palo Alto, CA, 94304

ABSTRACT

We studied the effect of prolonged mechanical strain on the electrical characteristics of thin-film transistors of hydrogenated amorphous silicon made at a process temperature of 150°C on 51-μm thick Kapton polyimide foil substrates. Effects are observed only at very high compressive strain of 1.8%. Tensile strain up to fracture at 0.3% to 0.5% does not show any effect, nor does compressive strain substantially less than 1.8%. The TFTs were stressed for times up to 23 days by bending around a tube with axis perpendicular to the channel length, and were evaluated in the flattened state. The changes observed are small. The threshold voltage is increased, the "on" current and the field effect mobility remain essentially constant, and the subthreshold slope, "off" current and gate leakage current drop somewhat. Overall, the observed changes are small. We conclude that mechanical strain caused by roll-to-roll processing and permanent shaping will have negligible effects on TFT performance.

INTRODUCTION

Thin film transistors of hydrogenated amorphous silicon (a-Si:H TFTs) are the pixel switches of the mainstream technology for active matrix liquid crystal displays. Recent research interest has focused on fabricating a-Si:H TFTs on plastic substrate to obtain nonbreakable, comformable, and elastic flexible electronics [1-4]. A major application of on-plastic a-Si:H TFTs is the backplane for active matrix organic light emitting displays (AMOLEDs). In AMOLEDs a-Si:H TFTs function not only as electronic switches but also as analog current sources, for which long-term stability is a critical issue [5-7].

The stability of a-Si:H TFT under constant electrical gate bias has been studied widely. Two mechanisms of instability have been identified – charge trapping in the gate dielectric and defect creation in the a-Si:H channel layer [8-14]. The former mechanism dominates at high electrical field in the gate; the latter dominates at low gate bias voltage, which is the typical operating condition [8,9,12]. On-plastic electronics must keep working during and after mechanical flexing. The electro-mechanical stability of a-Si:H TFTs becomes more of a concern if the TFTs are fabricated at low process temperature (100-200°C), which may render them more susceptible to defect generation than TFTs made in the standard range of process temperature of 250-350°C) [1]. If the electric field of the gate can create defects by breaking Si-Si bonds [8-14], how do a-Si:H TFTs respond to the prolonged mechanical strain that might be applied in permanently bent flexible displays?

Gleskova *et al.* studied the effect of short-term mechanical strain on TFTs processed on Kapton at 150°C [3-4,15]. The only changes observed were reversible with strain: primarily a rise of the electron field effect mobility in tension and a drop in compression, and a very small change in subthreshold slope in the opposite direction [3,4,15]. In the study reported here we tested the effects of long-term mechanical strain, short of fracture, on the performance of TFTs. We found small but measurable changes in the electrical characteristics when a very large strain was applied for long times.

EXPERIMENTS

Figure 1(a) shows the cross-section of the TFT, fabricated for this study at 150°C process temperature on 51-μm thick Kapton®E polyimide foil. We used the masks and followed the fabrication procedure described earlier [3-4,15], except that the gate electrode was made as a sandwich of 10nm Cr/100nm Aℓ/10nm Cr. The Aℓ is used for its low built-in stress and ductility, and the Cr for adhesion and chemical protection. The TFTs' current-voltage characteristics were measured with an HP 4155A parameter analyzer and probes that had been re-fitted for low-noise, high-current sensitivity measurements.

Experiments conducted at the beginning of this study had shown no effect of mechanical strain when added on top of electrical gate bias, in either tension or compression. Therefore we dropped biasing the TFTs electrically and only tested under mechanical stress. The TFTs were evaluated before and after bending around cylinders of decreasing radii to set defined values of compressive or tensile strain. We applied strains up to TFT fracture. In tension a-Si:H TFTs break at about 0.3% strain, and in compression at about 2% strain [3,4]. Lengthy experimentation also showed that only very large strains produce measurable changes in TFT characteristics. Therefore we concentrated on applying the largest mechanical strain that can be applied without breaking the TFTs, 1.8% in compression, in the source-drain direction by bending the TFT/Kapton around a cylinder with axis perpendicular to the channel direction. For electrical evaluation the TFT was taken off the cylinder and then was mounted again for more bending, up to a total bending time of 23 days.

Figure 1(b) shows the transfer characteristic of an a-Si:H TFT as-fabricated on ~ 51μm thick Kapton E foil. For the strain study we used device parameters extracted from the characteristics in linear operation (drain-source voltage $V_{ds} = 0.1V$), and also measured the gate leakage current. The points for "on" current I_{on} and gate leakage current I_g are taken at $V_g = 20V$, off current I_{off} at $V_g = -5V$, and the threshold voltage V_t at $I_{ds} = 10^{-8}A$, as indicated in Fig. 1(b). The subthreshold slope S, $dV_g/dlog(I_{ds})$ is obtained from fitting an exponential function to the subthreshold region of the transfer curve. The field effect mobility μ was calculated from a least square fit to data points at $I_{ds} > 10^{-8}$ A. I_{off} is $\sim 10^{-12}$A (~ 25 fA/μm gate width), and the ratio I_{on} in saturation ($V_{ds} = 10V$) to I_{off} is $\sim 10^7$. Other characteristics are $V_t \cong 3.6V$, $\mu = 0.75$ cm^{2}V^{-1}s^{-1}, and S = 0.45 V/decade. The gate leakage current I_g is very low at 0.3pA (~ 1 fA/μm gate width).

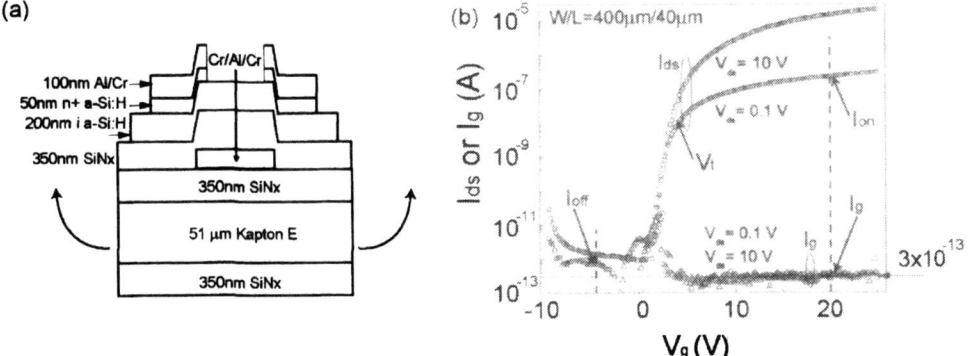

Figure 1 (a) Cross-section of the a-Si:H TFTs used in this study. The arrows designate compressive bending. (b) Transfer characteristics of the TFT. The source is grounded. Data points for I_{on}, I_{off}, I_g and V_t were used for evaluation of the effects of mechanical compression.

RESULTS AND DISCUSSION

Figure 2 shows the effects of prolonged compression on the TFT parameters. All data were normalized to their initial, as-fabricated, values. Figure 2(a) shows that V_t initially drops and then increases with $t^{0.10}$, while I_g decreases slightly. Figure 2(b) indicates little effect of strain on I_{on} and some reduction of I_{off}. Figure 2(c) shows that μ is essentially constant, in agreement with the nearly constant I_{on}, while S decreases slightly.

We mentioned earlier that we were unable to distinguish any effect of mechanical stress when superimposed on gate bias electrical stress. Figure 3, which shows the responses of V_t and I_g to gate bias stress, gives the reason. Note that the unit of time in Figure 2 is minutes, while in Figure 3 it is seconds. We conclude that, within the boundaries of practicable gate electrical bias stress and compressive mechanical stress, electrical stress changes V_t and I_g about 100 times faster than mechanical stress.

The similar dependences on time, $V_t \propto t^x$ with x = 0.10 (mechanical) and x = 0.11 (electrical), suggest that a search for a mechanistic link between the microscopic effects of mechanical and electrical stress might produce interesting results. Wehrspohn *et al.* suggested that the activation energy of defect creation in a-Si:H is reduced under intrinsic compressive stress [14].

CONCLUSIONS

Our main conclusion is that mechanical stress short of values that cause fracture has little permanent effect on a-Si:H TFT performance. This means that the roll-to-roll processing and permanent shaping of a-Si:H TFT backplanes will not produce irreversible electrical effects. Pursuing the correlation between effects on the threshold voltage of electrical gate bias stress and mechanical compressive stress would be interesting. However, our experience shows that the mechanical portion of any required experiments will have to be conducted with great care to extract meaningful information.

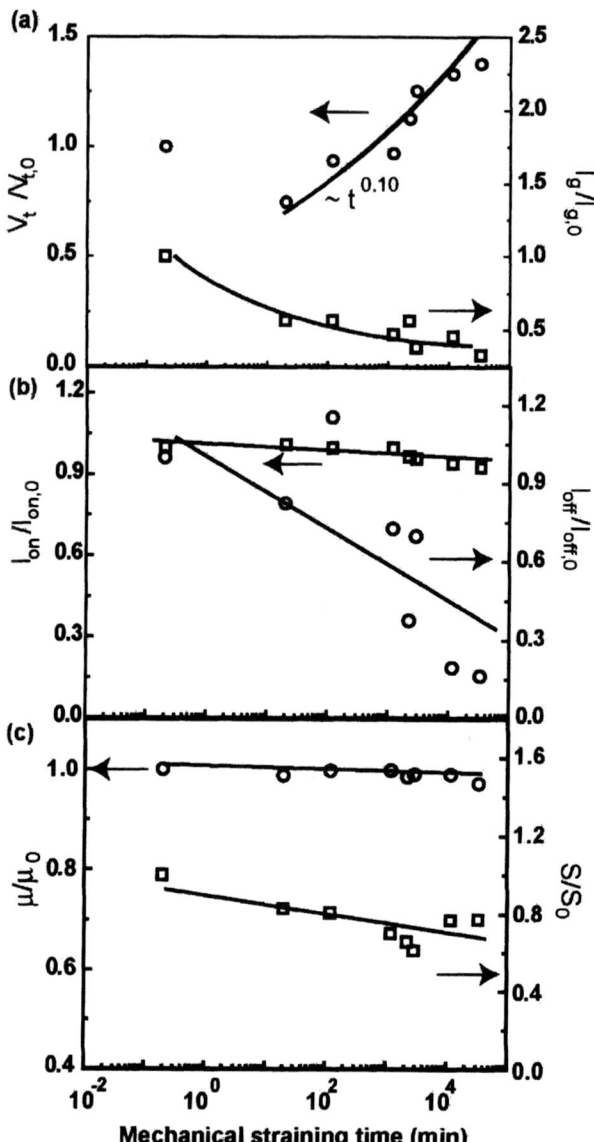

Figure 2 TFT performance after mechanical compression for up to 33,000 minutes. (a) Normalized threshold voltage $V_t/V_{t,0}$ and gate leakage current $I_g/I_{g,0}$. (b) Normalized "on" current $I_{on}/I_{on,0}$ and "off" current $I_{off}/I_{off,0}$. (c) Normalized field effect mobility μ/μ_0 and subthreshold slope S/S_0. All data are normalized to the values (subscript "0") for the as-fabricated TFT, entered at t = 0.2 minutes.

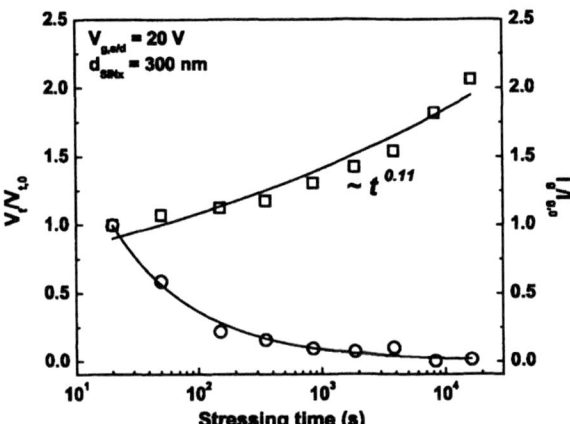

Figure 3 Normalized threshold voltage $V_t/V_{t,0}$ (squares) and gate leakage current $I_g/I_{g,0}$ (circles) against electrical gate bias stress time, at $V_g = 20$ V (S/D grounded). The gate SiN$_x$ is 300nm thick.

REFERENCES

1. C. –S. Yang, L. L. Smith, C. B. Arthur, and G. N. Parsons, J. Vac. Sci. Technol. B **18**, 683 (2000).
2. P. I. Hsu, R. Bhattacharya, H. Gleskova, M. Huang, Z. Xi, Z. Suo, S. Wagner, and J. C. Sturm, Appl. Phys. Lett. **81**, 1723 (2002).
3. H. Gleskova, S. Wagner, W. Soboyejo, and Z. Suo, J. Appl. Phys. **92**, 6224 (2002).
4. H. Gleskova, S. Wagner, and Z. Suo, Appl. Phys. Lett. **75**, 3011 (1999).
5. K. S. Karim, A. Nathan, M. Hack, and W. I. Milne IEEE Electron Device Lett. **25**, 188 (2004).
6. S. M. Jahinuzzaman, A. Sultana, K. Sakariya, P. Servati, and A. Nathan, Appl. Phys. Lett. **87**, 023502 (2005).
7. K. Long, A. Z. Kattamis, I-C. Cheng, H. Gleskova, S. Wagner, and J. C. Sturm, IEEE Electron Device Lett. **27**, 111 (2006).
8. C. van Berkel and M. J. Powell, Appl. Phys. Lett. **51**, 1094 (1987).
9. M. J. Powell, C. van Berkel, and J. R. Hughes, Appl. Phys. Lett. **54**, 1323 (1989).
10. W. B. Jackson and M. D. Moyer, Phys. Rev. B **36**, 6217 (1987).
11. W. B. Jackson, J. M. Marshall, and M. D. Moyer, Phys. Rev. B **39**, 1164 (1989).
12. Y. Kaneko, A. Sasano, and T. Tsukada, J. Appl. Phys. **69**, 7301 (1991).
13. S. C. Deane, R. B. Wehrspohn, and M. J. Powell, Phys. Rev. B **58**, 12625 (1998).
14. R. B. Wehrspohn, S. C. Deane, I. D. French, I. Gale, and M. J. Powell, J. Appl. Phys. **87**, 144 (2000).
15. H. Gleskova and S. Wagner, Appl. Phys. Lett. **79**, 3347 (2001).

Mater. Res. Soc. Symp. Proc. Vol. 989 © 2007 Materials Research Society   0989-A09-05

Single Grain Si TFTs Fabricated at 100°C for Microelectronics on a Plastic Substrate

Ming He, R. Ishihara, T. Chen, J.W. Metselaar, and C.I.M. Beenakker
Delft University of Technology, Delft Institute of Microelectronics and Submicrontechnology
(DIMES), Delft, 2628CT, Netherlands

ABSTRACT

Single grain TFTs are fabricated at a maximum temperature of 100°C for macroelectronics on a plastic substrate, as Si channels are fabricated at 100°C by combination of excimer laser crystallization and sputtering. The gate oxide is formed at 80°C by inductively coupled plasma enhanced chemical vapor deposition. These TFTs have shown a smaller threshold swing of 0.49 V/dec. and a higher field-effect mobility of 290 cm^2/V·s, which can be used to directly fabricate system circuits or a high quality display on a plastic substrate.

INTRODUCTION

Macroelectronics [1,2] is an emerging field in which a remarkable progress has made in microelectronics on a flexible, large-area substrate with size much bigger than a microelectronic fab can handle. Macroelectronics have begun to show great promise. The research in this trend will open up new applications, which doesn't exist today and reshape the global large-area microelectronics.

One desirable feature of macroelectronics is to realize system circuits (e.g., memory, CPU, ect.) on a flexible plastic substrate, which can be built directly on or within an interesting foreign object, to enable it to sense, control, compute, and communicate, ect.[2] These applications that require computation, control or communication functions cannot be addressed by today's TFTs. To achieve above abilities, it is pressing to improve TFT performance and to develop cost-efficient processes, which are compatible with a flexible substrate.

Single grain TFT (SG-TFT), which is fabricated inside a location-controlled single grain, exhibits similar characteristics as SOI-FETs [3]. SG-TFTs are fabricated within location-controlled grains with μ-Czochraski process (grain filter) [4] at a low process temperature. However the maximum process temperature is 550°C, which is at deposition of α-Si by low-pressure chemical vapor deposition (LPCVD) and is much higher than resistant temperature of a plastic substrate.

In this paper, IC-quality SG-TFT is fabricated at a maximum process temperature of 100°C, which can be applied to a direct formation of system circuits on plastics. First, single grain channel are formed at 100°C by patterning location-controlled grains, which are prepared by excimer laser crystallization of sputtered α-Si on top of grain-filters; Then, gate SiO$_2$ is deposited at 80°C by inductively coupled plasma enhanced chemical vapor deposition (ICPECVD). SG-TFTs fabricated with the above gate oxide within location-controlled Si grains have shown high performance such as a field-effect mobility of 290 cm^2/V·s, a subthreshold slope of 0.49 V/decade, an on/off-current ratio larger than 10^6.

EXPRIMENTAL DETAILS

Location-controlled Si grains

The requirement of high-quality Si materials with an ultra-low process temperature can be met by combination of sputtering and excimer laser crystallization of sputtered α-Si film [5, 6]. By sputtering, a good quality α-Si film can be deposited in a wide process temperature range, down to room temperature, which enables compatibility with a plastic substrate. Eximer laser crystallization of α-Si film has no thermal damage on a plastic substrate if there is a buffer layer between α-Si film and plastics [7]. Unfortunately, sputtered α-Si is easily ablated by explosive evolution of entrapped ambient gas (such as Ar) during excimer laser crystallization.

(a)   (b)

Figure 1: SEM images of poly-Si obtained with excimer laser at (a) 400 and (b) 500 mJ/cm^2 from a 140 nm-thick α-Si precursor film, which is sputtered with zeros substrate bias, 6 kW plasma power, at 100°C substrate temperature.

Therefore, optimization of the sputtering parameters is carried out to suppress ablation in subsequent excimer laser crystallization. α-Si is sputtered on a non-structured oxidized Si wafer with various sputtering conditions. α-Si film is crystallized with a pulsed XeCl excimer laser (with pulse duration of 50 ns) at room temperature. It is found that, when a bias is applied during sputtering, α-Si film is easily ablated during laser crystallization, even at an energy density below the super lateral growth (SLG) region [8]. α-Si film deposited without bias can endure well beyond the SLG region without ablation. For a 140 nm-thick α-Si film, the grain has a diameter of 1.8 μm with a petal-like shape after excimer laser crystallization (Fig.1). These grains can be obtained in a wide energy density window (from 300 to 600 mJ/cm^2).

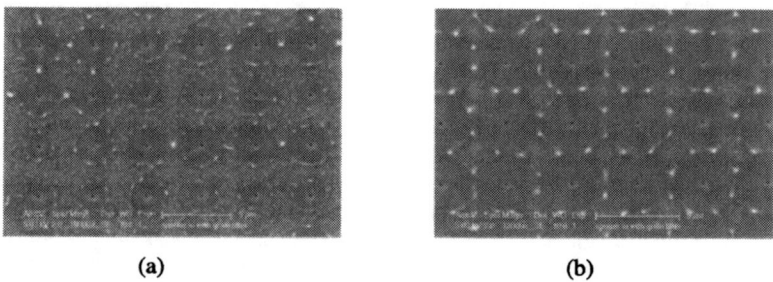

(a)   (b)

Figure 2: SEM images of location-controlled grains from 250 nm-thick α-Si crystalized at (a) 775 and 900 mJ/cm^2.

Then, μ-Czochralski process (grain filter)[4] is combined with sputtered α-Si to prepare location-controlled grains. A 250 nm-thick α-Si is sputtered on oxidized Si wafers with grain filters having a diameter of 100 nm in the SiO_2. Single shot of excimer-laser light is irradiated to the structure. Despite of a poor step-coverage of the sputtered film, grains with a maximum diameter of 4 μm can be obtained at the predetermined positions (Fig.2). With this low temperature process, these location-controlled Si grains are ideal active materials for circuits on a plastic substrate.

Low-temperature oxide

Recently, ICPECVD has been developed for low-temperature gate-oxide deposition [9]. ICPECVD, as a remote plasma deposition method, attracted a lot of attentions because ion energy and ion current density of the plasma can be controlled independently. ICPECVD can achieve a low plasma damage while have a high density of plasma. Compared to conventional PECVD, ICPECVD results in minimum plasma damage caused by direct ion-surface interaction. Electric properties of low-temperature ICPECVD SiO_2 are investigated by C-V and I-V measurements of an Al/SiO_2/p-type Si MOS capacitor.

MOS capacitors used in this study are fabricated with p-type (100) silicon wafers with a resistivity of 2~5Ω·cm. Right after dipping the wafer into 0.55% HF for 2 min to remove native oxide, then, a 100 nm-thick gate SiO_2 is deposited at a temperature of 80°C by ICPECVD. An ICP source with a frequency of 13.56 MHz and power of 500 W is used. SiH_4 and O_2 are used as precursor gases.

Figure 3 (a) plots current density (I) as function of applied electric field (E) for ICPECVD SiO_2. The breakdown field, defined as the field at which the leakage current density are 1 μA/cm^2, is 6.34×10^6 V/cm. Resistivity (ρ), defined at electric field (E) value of 1 MV/cm, is 5.20×10^{15} Ω×cm.

The SiO_2/Si interface characteristic is investigated by C-V measurements of MOS capacitor using a LCR meter of Agilent 4282A. The frequencies used for low-frequency and high-frequency are 100 Hz and 100 kHz, respectively. The densities of interface trap states D_{it}, evaluated from high- and low-frequency C-V characteristics of MOS capacitor [10] (Fig.3 (b)) are estimated to be as low as 3.21×10^{10} $cm^{-2}eV^{-1}$ (at V_{gs}=-0.7 V).

These electric properties are comparable to those of thermal grown SiO_2.

SG-TFTs

Top-gate n-channel TFTs are fabricated with above-described gate oxide within location-controlled grains. The single grains are patterned as active channel, which is made inside grains and positioned outside grain filter. Both channel length and width are 1 μm.

After patterning Si grains into channel islands, a 100 nm-thick gate SiO_2 is deposited at 80°C by ICPECVD. Subsequently, Al is sputtered at a room temperature and patterned as a gate. Subsequently, gate oxide is patterned in a low-power fluorine-based plasma with Al gate as a mask.

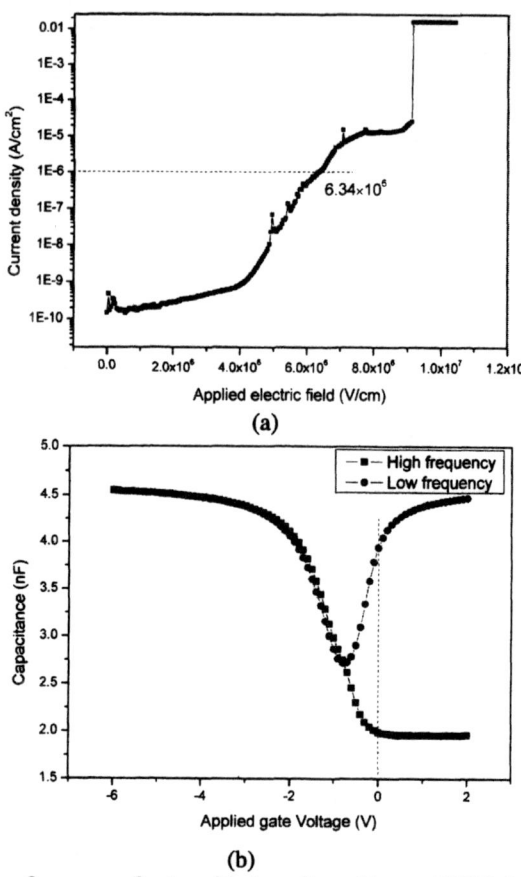

(a)

(b)

Figure 3: Electrical performance of gate oxide investigated by an Al/SiO₂/p-type Si MOS capacitor. Gate oxide is deposited by ICPECVD at 80°C: (a) Current density vs. applied electric filed (I-V) curve; (b) Capacitance vs. applied gate voltage (C-V) curve, the low- and high-frequencies used in C-V measurement are 100 Hz and 100 KHz, respectively.

Source and drain regions are doped with phosphorus with concentration of 10^{16} ion/cm² by ion-implantation at 30 keV. The implanted is performed in a self-aligned manner with the Al-gate-oxide stack as a mask. The source and drain implatation are activated with the excimer-laser irradiation at room temperature. The energy density of activation is 300 mJ/cm² with an 80% overlap scanning mode. Finally, source and drain electrodes are made by Al sputtering at room temperature.

Figure 4 shows transfer and output characteristics for SG-TFTs. Owing to the low interface trap density of gate SiO₂, SG-TFTs have shown both a smaller a subthreshold slope of 0.49 V/dec and a higher field-effect mobility for electrons of 290 cm²/V·s. The off-current is low as 6.8 pA with an on/off-current ratio larger than 10^6.

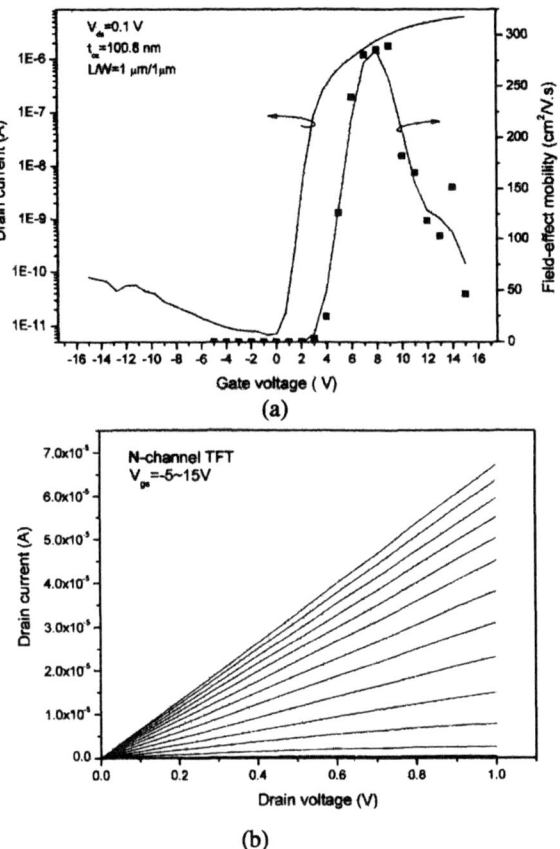
(a)

(b)

Figure 4: Transfer characteristic (a) and output characteristic (b) of single grain TFT fabricated with gate oxide deposited at 80°C. The subthreshold slope is 0.49 V/dec., the maximum field-effect mobility is 290 cm^2/V·s.

This paper only demonstrates that high-performance n-type TFTs can be fabricated at an ultra-low temperature. P-type TFTs, TFT-based CMOS, and further complicated system circuits should be fabricated within location-controlled grains with this ultra-low process temperature as well.

CONCLUSIONS

High-performance SG-TFTs can be fabricated at a maximum process temperature of 100°C inside a single, location-controlled grains with gate SiO$_2$ deposited at 80°C by ICPECVD. α-Si can be excimer-laser crystallized at a maximum process temperature of 100°C, without ablation.

The position of large grains is controlled by μ-Czochraski process, to get location-controlled silicon grains. High-quality gate oxide can be deposited at 80°C by ICPECVD. TFTs fabricated with above Si grains and gate oxide has shown a smaller threshold swing of 0.49 V/dec. and a higher field-effect mobility of 290 cm^2/V·s. These high performance TFTs fabricated at an ultra-low temperature are pressing needed devices for emerging macroelectronics on a plastic substrate.

ACKNOWLEDGMENTS

The authors would like to thank Y. van Andel, E.J.J.Neihof and H. Schellevis for technical help during excimer laser crystallization and sputtering. The author also would like thanks SENTECH instruments GmbH for the ICPECVD gate oxide demo deposition.

REFERENCES

1. R.H. Reuss, B.R. Chalamala, A. Moussessian, M.G. Kane, A. Kumar, D.C. Zhang, J.A. Rogers, M. Hatalis, D. Temple, G. Moddel, B. J. Eliasson, M. J. Estes, J. Kunze, E.S. Handy, E.S. Harmon, D.B. Salzman, J.M. Woodall, M.A. Alam, J.Y. Murthy, S.C. Jacobsen, M. Olivier, D. Markus, P.M. Campbell and E. Snow, *proc. of the IEEE*, Vol. 93, 2005, 1239-1256.
2. R.H.Reuss, D. G. Hopper and J. -G.Park, *Mater Res Bull*, Vol. 31, 2006, 447-450.
3. R. Ishihara, Y. Hiroshima, D. Abe, B. D. van Dijk, P. C. van der Wilt, S. Higashi, S. Inoue, T. Shimoda, J. W. Metselaar and C. I. M. Beenakker, *IEEE T Electr Dev*, vol 51, 2004, 500-502.
4. P. C. van der. Wilt, B. D. van Dijk, G. J. Bertens, R. Ishihara and C. I. M. Beenakker, *Appl.Phys.Lett.*,Vol.79, 2001, 1819-1821.
5. P. M. Smith, P. G. Carey and T. W. Sigmon, *Appl. Phys. Lett.*, Vol. 70 1997, 342-344.
6. D. P. Gosain and T. Noguchi and Usui, S., *Jpn. J. Appl.Phys.*, Vol. 39, 2000, L179-L181.
7. A. Burtsev, M. Apel and R. Ishihara and C. I. M. Beenakker, *Thin Solid Film*, Vol. 427, 2003, 309-313.
8. J.S.Im and H.J.Kim, *Appl.Phys.Lett.*, Vol. 64, 1994, 2303-2305.
9. R.Ishihara, T.Chen, M. He, D.Deosarran, Y.Andel, J.W.Metselaar and C.I.M. Beenakker, *Thin Solid Film*, submitted.
10. E. H. Nicollian and J. R. Brews, MOS (Metal Oxide Semiconductor) physics and technology, 1982, John Wiley & Sons.

Novel Applications

Mater. Res. Soc. Symp. Proc. Vol. 989 © 2007 Materials Research Society   0989-A10-02

Performance of Thin Film Silicon MEMS on Flexible Plastic Substrates

Samadhan Patil[1], Virginia Chu[1], and Joao Pedro Conde[1,2]

[1]INESC MN, Rua Alves Redol, 9, Lisbon, 1000-029, Portugal

[2]Dept. of Chemical and Biological Engineering, Instituto Superior Tecnico, Av. Rovisco Pais, 1, Lisbon, 1049-001, Portugal

ABSTRACT

Microresonators based on thin film hydrogenated amorphous silicon microbridges were fabricated by surface micromachining on flexible polyethylene terephthalate (PET) substrates with a maximum processing temperature of 110°C. An aluminum sacrificial layer is used which is patterned by either wet etching or lift-off. Resonance in the MHz range was observed using electrostatic actuation. Processing of the microbridges on PET with sacrificial layer patterned by lift-off has higher yield than by etching. Bending measurements show that the thin film silicon microbridges on PET can withstand a higher compressive strain (-2.5%) than tensile strain (1.25%).

INTRODUCTION

Thin film technology used to deposit amorphous silicon by Radio Frequency Plasma Enhanced Chemical Vapor Deposition (RF-PECVD) allows semiconductor device processing at temperatures ~100-250°C [1]. The combination of thin film deposition technology with microelectromechanical systems (MEMS) processing opens the possibility of realizing thin film MEMS devices on large area, low cost, and transparent substrates. Standard MEMS are fabricated on silicon wafers using processing temperatures at or above 600°C [2,3]. The low temperature processing of thin film MEMS fabrication technology has the advantage of being CMOS compatible. Hydrogenated amorphous silicon (a-Si:H) based MEMS and bolometers have been previously demonstrated on glass substrates [4-6].

Microelectronics on large area, flexible substrates is being pursued because of their low cost, low weight, robustness, mechanical flexibility and their simple processing. Devices on flexible substrate are valuable in medical electronics; they are biocompatible and can be used in disposable devices. The field of flexible electronics has already produced devices such as thin film transistors (TFTs) [7,8], image sensors [9], organic TFTs [10] and LEDs [11] fabricated on flexible plastic substrates.

In this study, thin-film silicon MEMS microbridges are fabricated on a 250 μm-thick flexible transparent polyethylene terephthalate (PET), and their performance in the as-fabricated state, as well as after substrate bending, is evaluated.

EXPERIMENTAL PROCEDURES

The device fabrication process sequence is shown in figure 1. The fabrication sequence depends on the way the sacrificial layer is patterned. First, n$^+$-a-Si:H (1000 Å) is deposited on the PET substrate using RF-PECVD at 100°C. This n$^+$-a-Si:H acts as an anchor layer for the TiW (500 Å) gate electrode, which is deposited on top of it by DC magnetron sputtering. This bilayer is patterned to form the gate electrode. Next, the Al sacrificial layer (7500 Å) is patterned either by etching or by lift-off as shown in branches (a) and (b) in fig. 1, respectively. Al is deposited by DC magnetron sputtering. Thin film TiW (150 Å) is deposited on top of the Al sacrificial layer. TiW acts as an anti-diffusion barrier between the sacrificial (Al) and structural layers (n$^+$-

a-Si:H) to be deposited on top. The structural material of the microbridge consists of a bilayer of amorphous silicon (n$^+$-a-Si:H) (4000 Å) deposited by RF-PECVD and TiW (500 Å). Low DC power has been used during deposition of all metal layers by DC magnetron sputtering to avoid excessive heating of the sample thus minimizing thermal stresses and warpage of the flexible substrate. The TiW /n$^+$-a-Si:H/TiW stack is patterned and etched by RIE. After dicing the sample, the sacrificial layer (Al) is removed with the etchant H$_3$PO$_4$: H$_2$O in a

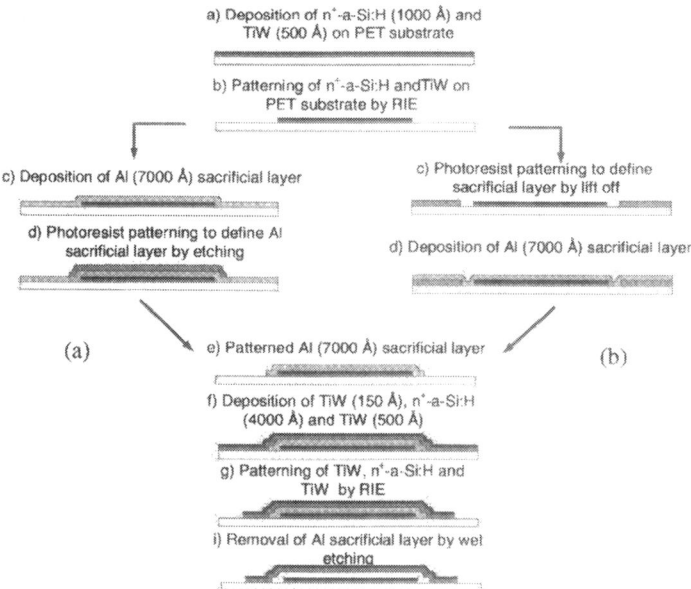

Figure 1. Schematic diagram of the fabrication process sequence for the bilayer thin film silicon microbridges on a 250 μm thick flexible PET substrate: a) the Al sacrificial layer is patterned by etching; b) the Al sacrificial layer is patterned by lift-off.

2:1 proportion. Use of a commercial Al etchant, which contains nitric acid, was found to attack the PET substrate. Special care was required during processing to ensure that the substrate remained flat. The same process sequence as described above was also used to fabricate thin film MEMS microbridges on glass substrates for comparison. In addition, another set of microbridges was fabricated on a glass substrate with photoresist (PR) as the sacrificial layer. In these microresonators, Al is used as the top contact instead of TiW as in the process described above for PET substrates. The fabrication process is similar to the one described above except that the PR sacrificial layer is patterned directly by lithography. This is the standard fabrication process for thin film silicon MEMS fabrication [4,12,13], but it cannot be adapted to PET substrates because the photoresist stripper solutions used for sacrificial layer removal chemically attacks the PET substrate. The width, w, of the microbridges is 15 μm and the length, L, varies from 25 to 70 μm. The target thickness of the silicon thin film structural layer is 0.4 μm.

Figure 2 shows SEM micrographs of microbridges fabricated on PET and glass substrates. Figures 2c) and 2d) show a protruding profile along the edge of the sacrificial layer in microbridges where the sacrificial layer is defined by lift-off for both PET and glass substrates, which is absent for microbridges fabricated using etching to pattern the sacrificial layer. Despite this, the fabrication process by lift-off has a higher yield because the even the modified Al-etching solution tends to attack the substrate.

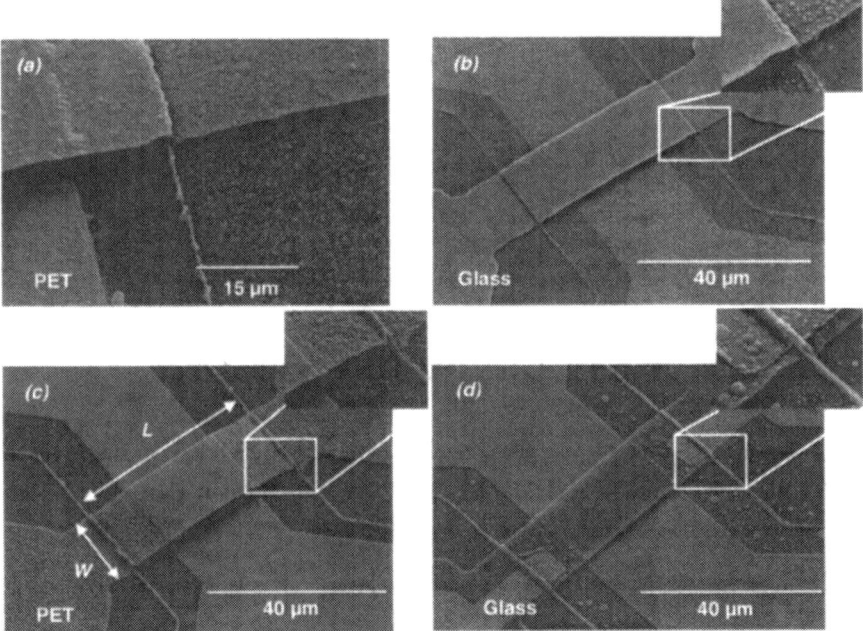

Figure 2. SEM micrographs of thin film silicon microbridges on PET and glass substrates: a) with sacrificial layer patterned by etching on PET; b) with sacrificial layer patterned by etching on glass; c) with sacrificial layer patterned by lift-off on PET; and d) with sacrificial layer patterned by lift-off on glass. Insets show an edge profile for the respective bridge.

The microbridges are electrostatically actuated by applying a voltage, V_G, between the gate electrode and metal top contact on the bridge producing an electrostatic force. The gate voltage, V_G, has DC and AC components, $V_G = V_{DC} + V_{AC}\sin(2\pi f t)$, where $V_{DC} \gg V_{AC}$, and f is the excitation frequency. The resonance frequency (f_r) is monitored in vacuum at a pressure of $\sim 10^{-6}$ Torr by means of an optical setup coupled to a spectrum analyzer, described elsewhere [12]. Quasi-DC deflection is measured at atmospheric pressure using a similar optical setup coupled to a lock-in amplifier. For these measurements, a voltage signal $V_G = V_{AC}\sin(2\pi f t)$ is applied to the gate electrode and bridge is grounded. For these measurements, $V_{DC} = 0$ V and the applied voltage frequency is kept low at around 31 Hz.

RESULTS AND DISCUSSION

Figure 3 shows the f_r of the bridges as a function of the bridge length (L).

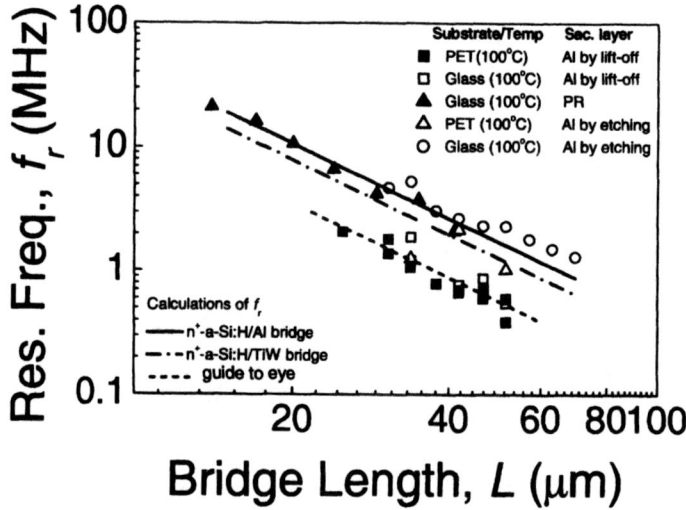

Figure 3. Resonance frequencies (f_r) of the bridges as a function of the bridge length (L) for (■) n^+-a-Si:H /TiW on PET, (□) n^+-a-Si:H /TiW on glass, (▲) n^+-a-Si:H /Al on glass, (△) n^+-a-Si:H /TiW on PET, and (○) n^+-a-Si:H /TiW bridges on glass.

The f_r of the fundamental flexural vibration mode for bilayer bridges is approximately given by [14]

$$f_r = \frac{1}{2\pi}\sqrt{\pi^4\left(a_n+\frac{1}{2}\right)^4\frac{(EI)_{\mathit{eff}}}{\mu_{\mathit{eff}}L^4}+\pi^2\left(a_n+\frac{1}{2}\right)^2\sigma_0\frac{wt_{\mathit{eff}}}{\mu_{\mathit{eff}}L^2}} \quad (1)$$

where $a_n = a_1 = 1.00562$ for the first flexural mode for clamped-clamped bridge [13] (a clamped-clamped beam is defined as a structure for which both the displacement at the end points where the beam is held, and its first derivative along the direction of the beam, is zero [14]). μ_{eff} is the mass per unit length of the bilayer suspended structure, $(EI)_{\mathrm{eff}}$ is the effective rigidity of the bridge, t_{eff} is the total thickness of the bilayer bridge, and σ_0 is the residual axial stress of the structure.

The values of f_r for n^+-a-Si:H/Al bridges on glass substrate using PR as the sacrificial layer follow a $1/L^2$ dependence, indicating that it is the rigidity of the beam that controls f_r (first term under the square root in eq. (1)), and fall within the range of the expected frequencies calculated from eq. (1) for all lengths (solid line in fig. 3). Bridges using Al sacrificial layer patterned by lift-off, both on PET and on glass substrate also exhibit $1/L^2$ dependence (dashed line in fig. 3) but lower values of f_r. A possible explanation for these lower values of f_r is the presence of the previously-mentioned protruding profiles (figs. 2c) and 2d)) along the edges of the sacrificial layer (Al) formed during the lift-off of the Al sacrificial layer, which probably alter the clamping conditions of the bridge. In the case of sacrificial layers patterned by etching, a step profile at the edge of the n^+-a-Si:H/TiW fabricated on PET and glass substrate is observed (figs. 2a) and 2b)). In this case the expected values of f_r are obtained.

Figure 4 shows the quality factors (Q) for different values of L. Average Q values for n^+-a-Si:H/Al microbridges fabricated with a PR sacrificial layer on glass substrates vary from 650 to 1000. Average Q values for the n^+-a-Si:H/TiW bridges fabricated with Al sacrificial layer patterned by lift-off vary from 100 to 450 on both PET and glass substrates. Lower values of Q for the n^+-a-Si:H/TiW bridges compared to the n^+-a-Si:H/Al microresonators on glass substrate with PR as the sacrificial material are tentatively attributed to presence of the "ears" at the boundaries of the bridges caused by the Al sacrificial layer liftoff process. Q values for a-Si:H/TiW bridges fabricated with Al sacrificial layer patterned by etching on glass substrate are comparable to those obtained using PR as sacrificial layer. Smaller Q values (~100) were observed with the same process on PET substrates, possibly due either to dissipation caused by the compliant substrate or from damage resulting from the wet etching process.

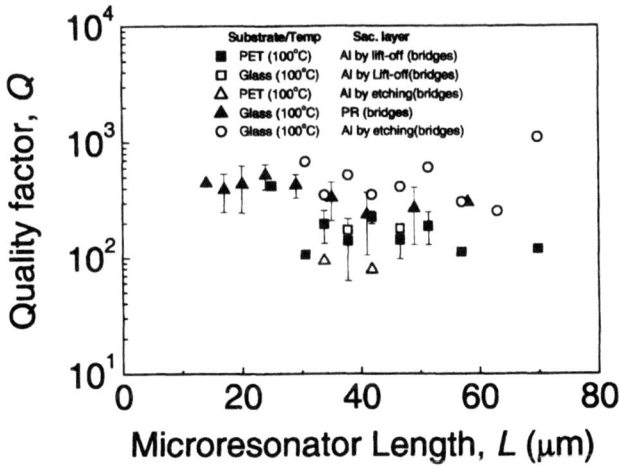

Figure 4. Quality factors (Q) of the bridges as a function of the bridge length (L) for (■) n^+-a-Si:H /TiW on PET, (□) n^+-a-Si:H /TiW on glass, (▲) n^+-a-Si:H /Al on glass, (△) n^+-a-Si:H /TiW on PET, and (○) n^+-a-Si:H /TiW bridges on glass.

Finally, the performance of two separate sets of thin-film silicon microbridges fabricated on PET after bending is evaluated by subjecting these devices to tensile and compressive strain and measuring its quasi-DC deflection in air before and after bending. The number of working devices was measured before and after every bending step. The radius of curvature (r) of the PET substrate was gradually increased until all devices failed (i.e., no longer showed electrostatically actuated quasi-DC deflection). During tensile strain bending, the relative number of working devices dropped to below 20% for $r=1$ cm (equivalent to a tensile strain of 1.25%). For compressive strain, it was possible to reduce r down to -0.5 cm (corresponding to a strain of -2.5%) with 50% of the devices still functioning.

CONCLUSIONS

Thin film silicon microbridges are fabricated on a PET substrate using Al as the sacrificial layer. Electrostatically actuated quasi-DC deflection and resonance frequency in the MHz range were measured for these structures. Resonance frequencies and quality factors for microbridges fabricated on PET substrate are process dependent and can be made comparable to those fabricated on glass. Thin silicon MEMS on PET substrates can withstand compressive strain down to a radius of curvature of -0.5 cm with 50% of the devices still functioning, but can resist far less tensile strain.

ACKNOWLEDGMENTS

The authors gratefully acknowledge J Bernardo, V. Soares, F. Silva, J. Faustino and J. Borme for help in sample/cleanroom processing and SEM imaging. S. Patil thanks Fundação para a Ciência e Tecnologia (FCT) for a post-doctoral grant (SFRH/BPD/17040/2004). FCT supported the research. J.P. Conde thanks FLAD for a travel grant. V. Chu thanks Fundação Calouste Gulbenkian for a travel grant. This work was funded by FCT through POCI projects.

REFERENCES

1. See, for example, R A Street, Hydrogenated Amorphous Silicon, Cambridge University Press, Cambridge, 1991.
2. K E Petersen, C R Guarnieri, J. Appl. Phys. **50**, 6761 (1979).
3. See, for example, M Elwenspoeck and R Wiegerink, Mechanical Microsensors, Springer, Berlin, 2001.
4. M Boucinha, P Brogueira, V Chu, and J P Conde, Appl. Phys. Lett. **77**, 907 (2000).
5. J Gaspar, V Chu, N Louro, R Cabeca, and. J P Conde, J. Non-Cryst. Solids, **299–302**, 1224 (2002).
6. A J Syllaios, T R Schimert, R W Gooch, W L McCarde, B A Ritchey and J H Tregilgas, Mater. Res. Soc. Symp. Proc. **609**, A14.4.1 (2000).
7. H Gleskova, S Wagner and Z Suo, Mat. Res. Soc. Symp. Proc. **557**, 653 (1999).
8. P. Servati, A. Nathan, Appl. Phys. Lett. **86**, 033504 (2005)
9. P Louro, M Vieira, M Fernandes and M Schubert, Optical Materials, **27 no.5**, 1069-1073 (2005).
10. H Klauk, G Schmid, W Radlik, W Weber, L Zhou, C D Sheraw, J A Nichols and T N Jackson, Solid State Electronics, **47** (2), 297(2003).
11. Z Taehyoung, S H Kim, H Y Chu, J H Lee, S C Lim, L Jeong-Ik, O Jiyoung, Flexible electronics technology, Part I: Systems & applications, **93** (7), 1265-1272 (2005).
12. J Gaspar, V Chu and J P Conde, J. Appl. Phys. **93**, 10018 (2003).
13. J Gaspar, V Chu and J P Conde, J. Appl. Phys. **97**, 094501 (2005).
14. A N Cleland, Foundations of Nanomechanics- From Solid State Theory to Device Applications, Springer, Berlin, 2003, p. 209

Mater. Res. Soc. Symp. Proc. Vol. 989 © 2007 Materials Research Society

Amorphous Silicon Based TFT and MIS Nonvolatile Memories

Yue Kuo, and Helinda Nominanda
Texas A&M University, 235 J. E. Brown Engineering Bldg., MS 3122, College Station, TX, 77843-3122

ABSTRACT

The amorphous silicon (a-Si:H) TFT and MIS capacitor, which include an a-Si:H layer embedded in the silicon nitride gate dielectric layer, have been prepared and characterized for memory functions. Large shifts of the threshold voltage and flat band voltage were detected in the current-voltage and capacitance-voltage hysteresis measurements. The embedded a-Si:H film functioned as a charge retention medium that stores and releases injected carriers. The device's memory capacity varied with the thickness of the embedded a-Si:H layer and the sweep voltage. These low-cost memory devices can be used in many low-temperature prepared circuits.

INTRODUCTION

Due to its versatility in functions, e.g., as a pixel switching, reading or controlling device, the a-Si:H TFT has been used in both LCD and non-LCD products [1]. Recently, the non-volatile memory application of a-Si:H TFT has been demonstrated [2, 3]. The device utilizes a thin layer of hydrogenated amorphous silicon (a-Si:H) layer embedded inside the TFT's SiN_x gate dielectric layer. The embedded a-Si:H layer serves as floating gate for charge trapping. The one-pump down deposition of the gate dielectric, channel and passivation layers by PECVD method promotes clean interfaces as well as provided a simpler fabrication compared to other floating gate memory structure that utilizes separate patterning steps for the embedded layer [4]. The TFT memory can be used in LCD applications during the idle state that requires no data update. For non-LCD applications, TFT memory can act as an alternative built-in memory.

Here, authors further investigated the charge trapping mechanism of the TFT-based memory device and examined the interface and bulk dielectric properties of the floating gate MIS capacitor memory device. For instance, influences of the embedded a-Si:H film thickness and the operating gate voltage (V_g) on interface density of states (D_{it}), flat-band voltages (V_{FB}), flat-band voltage shifts (ΔV_{FB}), leakage current (I_{off}), fixed charge density (Q_{ot}) were investigated. The relationship between the characteristics of the embedded a-Si:H MIS capacitor and TFT are discussed.

EXPERIMENT

The complete TFT for memory was fabricated following a self-aligned, inverted staggered, tri-layer TFT, two-photomask process on Corning 1737 glass [5]. Figure 1a shows the cross-section structure of the TFT–based floating gate memory [2]. The gate, and later, the source and drain electrodes were made of 100-nm thick DC-sputtered molybdenum (Mo). The Mo layers were etched with a solution of $CH_3COOH:H_3PO_4:HNO_3:H_2O$. The gate dielectric was composed of 150 nm SiN_x/4, 7, or 9 nm thick intrinsic a-Si:H/150 nm SiN_x layers. For comparison, the TFT with no embedded a-Si:H layer was also fabricated, as shown in Fig. 1b. A

50-nm thick undoped a-Si:H layer for channel and a 250-nm thick SiN_x for top channel passivation layers were deposited subsequently. All layers were deposited by PECVD method (Applied Materials AMP Plasma II) without breaking the vacuum at substrate temperature of 300°C and an rf-frequency of 13.56 MHz. After the "five-layer" deposition, the SiN_x channel passivation layer was defined using the gate as the self-aligned mask using a backlight exposure method. A PECVD microcrystalline n^+ a-Si:H layer was deposited as the source/drain ohmic contact layer. The n^+ layer was reactive ion etched at 13.56 MHz using a mixture of CF_4/Cl_2 (2/8 sccm) at 100 mTorr and 300 W for 2 minutes. Details of the plasma conditions of the PECVD and RIE processes can be found in references [5,6]. The finished TFT was annealed in air at 250°C for 1 hour to repair the plasma etch induced damages [6]. The TFTs were characterized with Agilent 4155C semiconductor parameter analyzer.

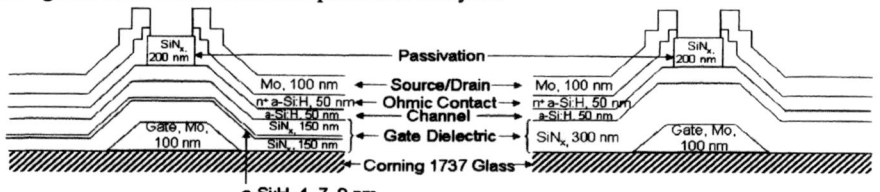

Figure 1. Cross-section schematic of a) non-volatile memory and b) "control" TFT. [2]

The fabrication steps of the floating-gate a-Si:H MIS capacitor followed closely those of the floating-gate a-Si:H TFT. Fig. 2a shows the cross-section schematic of an MIS capacitor fabricated on the Corning 1737 glass. DC-sputtered Mo (100-nm thick) was used as the bottom gate and top contact electrodes. The gate dielectric structure, channel and ohmic contact layers with the same thickness and condition as the ones for TFT samples were deposited with PECVD. For comparison purpose, a "control" MIS capacitor that contained only a SiN_x (300 nm) dielectric without the embedded a-Si:H layer was fabricated, as shown in Fig. 2b. Both types of capacitors, i.e., 65 μm diameter, had the same channel and ohmic contact layers deposited by PECVD method. The top Mo contact was wet-etched. The n^+ μc-Si:H, the a-Si:H channel and floating gate layer were RIE-defined at the same condition as the TFT fabrication described above. The finished MIS capacitors then were annealed at 250°C for 1 hour in air. The V_{FB}, ΔV_{FB} and D_{it} were extracted from the current-voltage (C-V) and conductance-voltage (G-V) curves at room temperature with Agilent 4284A LCR Meter. The V_{FB} value of the C-V curves was extracted from the intersection of the reciprocal of capacitance squared ($1/C^2$) vs. gate bias voltage [7], where the measured capacitances were corrected for series resistance [8].

a)   b)

Figure 2. Cross-section schematic of a) non-volatile and b) "control" MIS capacitor.

DISCUSSION

Figure 3 shows the hysteresis of the transfer characteristic of a TFT with a 9-nm embedded a-Si:H and a TFT without any embedded layer ("control"). The a-Si:H embedded TFT had a larger hysteresis than the "control" TFT has. The memory window can be measured from the difference in threshold voltage, ΔV_t, of the forward direction threshold voltage, $V_t^{Forward}$, in which V_g was swept from -20 V to 20 V, and the backward direction threshold voltage, $V_t^{Backward}$, in which V_g was swept from 20 V to -20 V. ΔV_t's of 7 V vs. -1 V were observed for the embedded TFT and the "control" TFT in Fig. 3, respectively. The memory window increased with the increase of V_g sweep range. For instance, the Fig. 3 TFT showed a ΔV_t of 0.98 V in the sweep range of (-5 V, 5 V) and 5.6 V in the sweep range of (-10 V, 10 V). On the other hand, the "control" TFT of Fig. 3 had a ΔV_t of -0.5 V in the sweep range of (-10 V, 10 V).

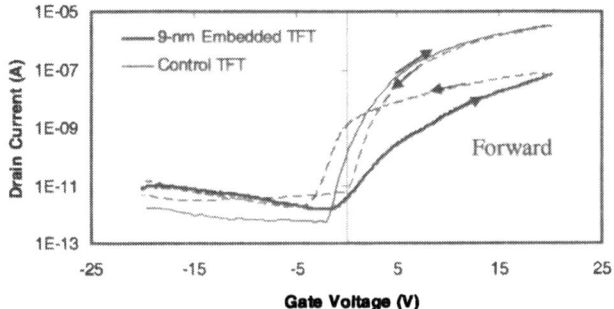

Figure 3. Transfer characteristics hystereses of a 9-nm embedded TFT and a "control" TFT.

Figure 4 shows the C-V hysteresis curves of a 9-nm embedded and a "control" MIS capacitors. The forward sweep was defined as from the positive voltage toward the negative voltage, while the backward sweep was defined as the opposite sweep direction. ΔV_{FB} was obtained from the difference in the V_{FB} in the forward sweep, $V_{FB}^{Forward}$, and the V_{FB} in the backward sweep, $V_{FB}^{Backward}$. ΔV_{FB}'s of 2.6 V and 0.3 V in (20 V, -20 V) range were detected for the embedded and the "control" capacitors, respectively. As in the TFT memory operation, the V_g sweep range effect on memory capacity was also observed in the capacitor memory device. For instance, an increase of ΔV_{FB} was observed with the increase of V_g sweep range, e.g., of 0.2 V in the sweep range of (5 V, -5 V) and 0.8 V in the sweep range of (15 V, -15 V). The control capacitor did not demonstrate such effect on the hysteresis of its C-V curves, e.g., ΔV_{FB} was 0.4 V in the sweep range of (15 V, -15 V).

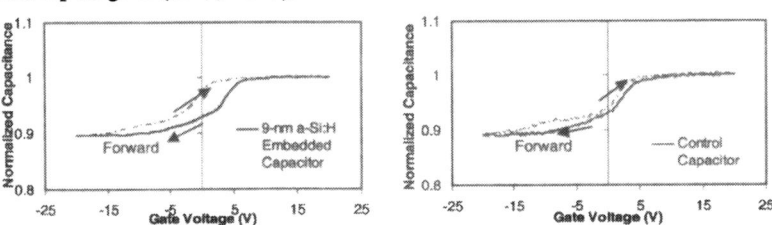

Figure 4. C-V curve hystereses of a 9-nm embedded and "control" MIS capacitors.

The larger hysteresis on the transfer characteristics of the TFT with embedded a-Si:H layer compared to the "control" TFT was due to the more effective trapping of electrons or holes at the embedded a-Si:H sites as confirmed by the hystereses behavior on the C-V curves of the capacitors with the embedded a-Si:H when compared to the "control" capacitor. For instance, at the forward direction sweep from 20 V to -20 V, the 9-nm a-Si:H embedded capacitor had a more positive V_{FB} compared to the "control" capacitor, i.e., 5 V vs. 4.5 V. The embedded layer trapped electrons and caused a larger V_g required for the formation of the inversion layer. At the backward direction sweep from -20 V to 20 V, electrons detrapped from the embedded layer contributed to the more negative V_{FB} of the embedded layer capacitor compared to the "control" capacitor, i.e., 2.4 V vs. 4.2 V.

From the above observation, simple energy-band diagrams of the capacitors at different states can be constructed, as shown in Figure 5.

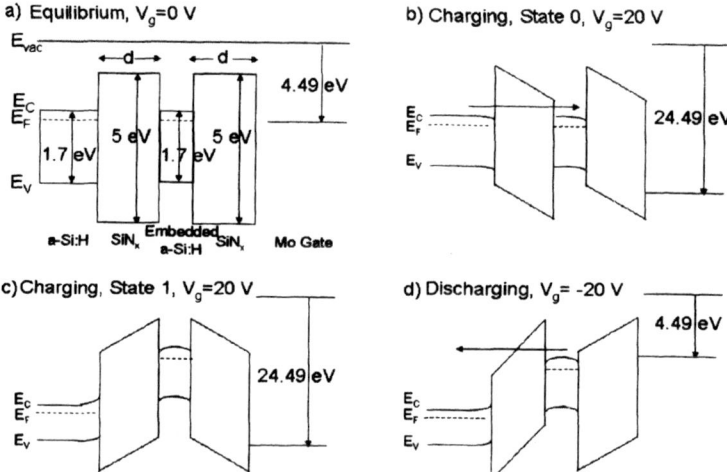

Figure 5. Energy-band diagrams for MIS capacitor with embedded a-Si:H at a) equilibrium, b) charging, state 0, c) charging, state 1, and d) discharging.

When V_g of 20 V was applied to the bottom Mo gate, electrons were injected into the a-Si:H embedded dielectric layer (Fig. 5b) and the negative charges were stored in the embedded layer (Fig. 5c), similar to the one observed in the poly-Si floating gate of the SAMOS structure [9]. As V_g = -20 V was applied at the gate, the trapped charges were released into the semiconductor layer (Fig. 5d) and the capacitor would return to equilibrium state (Fig. 5a). Even though both the control and tunneling SiN$_x$ dielectric layers had the same thickness, the 9 nm embedded a-Si:H was thin enough to ensure the needed voltage drop across the tunneling SiN$_x$ interface layer.

Separately, it was observed that the ΔV_{FB} increased with the increase of embedded a-Si:H layer thickness. The embedded layer thickness did not affect the backward $V_{FB}^{Backward}$ significantly. However, it had a more pronounced effect on the $V_{FB}^{Forward}$. For example, capacitors with 4 and 9 nm embedded a-Si:H have a $V_{FB}^{Backward}$ of 2.0 V and 2.3 V, respectively, and a $V_{FB}^{Forward}$ of 3.0 V and 4.4 V, respectively. As the gate was biased from depletion region towards accumulation region, i.e., during the backward sweep, positive charges increasingly replaced the negative charges stored at the floating gate. The limited supply of holes in the inversion region probably contributed to the slight change on the $V_{FB}^{Backward}$. In other words, the thickness of the embedded layer probably facilitates the charge trapping mechanism of the dielectric structure through regulating the amount of charge stored at the floating gate or at the interface similar to the Coulombic scattering effect [10]. For instance, the decrease of dielectric layer thickness in a conventional MIS capacitor, i.e., without an embedded layer, has been observed to reduce the screening of charge scattering centers [10].

Finally, the effective field effect mobility and the sub-threshold slope of the 9-nm a-Si:H embedded TFT in Fig. 2 was lower and higher, respectively, than the ones for the "control" TFT, e.g, 0.17 cm^2/V-s and 1.83 V/decade vs. 0.4 cm^2/V-s and 0.25 V/decade at the forward direction sweep. This can be explained by the stress mismatch between the a-Si:H and the SiN_x films. The stress in PECVD SiN_x films is related to factors such as the Si/N ratio or the hydrogen content. The SiN_x used on these capacitors was N-rich with a refractive index of 1.874. Nitrogen-rich SiN_x has been observed to be more tensile compared to stoichiometric Si_3N_4 [11] while a-Si:H film has a compressive stress [12].

The interface phenomenon, i.e., between the SiN_x and the a-Si:H layers, was also observed at the capacitors. The capacitor with an a-Si:H embedded layer had an interface trap density, D_{it}, value in the range 10^{11} cm^{-3} while the "control" capacitor has a D_{it} value in the 10^{10} cm^{-3} range. Since stress mismatch directly affects the interface state through modifying its interfacial fixed charges and density of state [13], the extra two SiN_x/a-Si:H interfaces in the embedded gate dielectric structure could enhance the charge trapping capacity. Furthermore, the a-Si:H embedded layer thickness had a tremendous influence on the interface state of the TFT [3] and on the D_{it} values of the capacitors. The intrinsic stress of a-Si:H is also a function of many factors such as the hydrogen content, the deposition temperature, and the film thickness [14]. The compressive stress increases with a-Si:H thickness. The D_{it} increases with the embedded a-Si:H layer thickness, which facilitates the mismatch between SiN_x and a-Si:H.

To observe the frequency-dependent behavior of the interface states, the C-V and G-V curves of the a-Si:H embedded and the "control" capacitors were measured at 100 kHz, 500 kHz and 1 MHz from accumulation, i.e., 20 V, toward inversion, i.e., -20 V, with a 0.1 V step. The additional interface states from the embedded a-Si:H layer significantly affected the capacitance. For instance, the depletion capacitance of a 9-nm a-Si:H embedded capacitor decreased with the increase of frequency, e.g., the capacitance at -3 V were 1.64 x 10^{-4} F/cm^2, 1.62 x 10^{-4} F/cm^2, and 1.61 x 10^{-4} F/cm^2 for 100 kHz, 500 kHz, and 1 MHz, respectively. While the depletion capacitance at -3 V of the "control" capacitor was 1.682 x10^{-4} F/cm^2, 1.668 x10^{-4} F/cm^2, and 1.655 x10^{-4} F/cm^2 at 100 kHz, 500 kHz, and 1 MHz, respectively. At the depletion region, the interface states were more responsive to the measurement at lower frequency due to the change of distribution of the minority and majority carriers at this region that magnified the frequency effect. This change in carrier distribution was also observed from the G-V curves at different frequency. For instance, the maximum conductance values for all three frequencies were at the depletion region.

The inversion capacitance for the same 9-nm embedded a-Si:H capacitor at 500 kHz and the 1 MHz showed negligible frequency dispersion, however, the inversion capacitance at 100 kHz was higher than at 500 kHz and 1 MHz. At lower frequency measurement, the generation and combination of minority carriers during depletion can follow the AC signal and contributed to the higher inversion capacitance at 100 kHz. The "control" capacitor at the inversion region showed a smaller frequency dispersion effect compared to the embedded capacitor, which probably was related to the less generation-recombination of the minority carriers from the absence of the embedded layer. The saturating capacitance of both embedded and "control" capacitors at the accumulation region showed negligible frequency dispersion which is expected due to the conduction from the majority carriers at this region.

CONCLUSIONS

Both TFT and MIS capacitor devices showed that memory capacity was governed among others by the applied V_g as well as by the thickness of the embedded a-Si:H layer. The applied V_g facilitated charge trapping by injecting proportional amount of charges into the floating gate, while the embedded layer thickness facilitated the charge trapping by regulating the amount of available sites for storage. The extra intrinsic a-Si:H/SiN$_x$ interfaces affected the charge trapping mechanism due to the creation of additional interface states. This was confirmed by the significant frequency dispersion of the depletion capacitance. This kind of memory device has low cost potential for low-temperature prepared circuits.

ACKNOWLEDGMENTS

The authors would like to thank Jiong Yan for the metal sputtering, Guojun Liu for the plasma etching work and Jiang Lu for technical discussion.

REFERENCES

1. Y. Kuo in "Non-LCD Applications of a-Si:H TFTs," *Thin Film Transistors, Materials and Processes, Volume 1: Amorphous Silicon Thin Film Transistors,* ed. Y. Kuo (Kluwer, 2004) pp. 485-503.
2. Y. Kuo and H. Nominanda, Appl. Phys. Lett. **89**, 173503 (2006).
3. H. Nominanda and Y. Kuo in *Non-volatile Amorphous Silicon Thin Film Transistor Memories with the a-Si:H Embedded Gate Dielectric Structure,* ed. Y. Kuo (Electrochem. Soc. Trans., TFT Tech. 8, Vol. 13(8), Pennington, NJ, 2006, pp. 333-339.
4. S. G. Burns, H. R. Shanks, A. P. Constant, C. Grubber, D. Schmidt, A. Landin, C. Thielen, F. Olympie, T. Schumacher, and J. Cobbs, *Design and Fabrication of α-Si:H-Based EEPROM Cells,* ed. Y. Kuo (Electrom. Soc. Proc. **PV 1994-35** 1994) pp. 370-380.
5. Y. Kuo, J. Electrochem. Soc. **138**, 637 (1991).
6. Y. Kuo and S. Lee, Appl. Phys. Lett. **78**, 1002 (2001).
7. R. J. Hillard, J. M. Heddleson, D. A. Zier, P. Rai-Choudhury, and D. K. Schroder, Diagnosis Techniques for Semiconductor Materials and Devices, J. L. Benton, G. N. Maracas, and P. Rai-Choudhury, Eds, Pennington, NJ: Electrochem. Soc., p. 261, 1992.
8. E. H Nicollian and J. R. Brews, MOS (Metal Oxide Semiconductor), Physics and Technology, Ch. 5 & 10, New Jersey: Wiley-Interscience, 2003.
9. S. M. Sze, Physics Semiconductor Devices, p.4.98, New York: John Wiley & Sons, 1981.
10. K. J. Yang, T.-J. King, C. Hu, S. Levy, H. N. Al-Shareef, Solid-Sta. Elec. **47**, 149 (2003).
11. K. Hiranaka, T. Yoshimura, and T. Yamaguchi, J. Appl. Phys. **62**, 2129 (1987).
12. J. P. Harbison, A. J. Williams, and D.V. Lang, J. Appl. Phys. **55**, 946 (1984).
13. K. Hiranaka and T. Yamaguchi, Jpn. J. Appl. Phys. **29**, 229 (1990).
14. P. Danesh and B. Pantchev, Semicond. Sci. Technol. **15**, 971 (2000).

Mater. Res. Soc. Symp. Proc. Vol. 989 © 2007 Materials Research Society 0989-A10-04

Monolithic Integrated a-Si:H Based Pin-Diodes with Orthogonal Liquid Light Guidance Structures for Lab-on-Microchip Applications

Heiko Schäfer, Konstantin Seibel, Lars Schöler, and Markus Böhm
Institute for Microsystem Technologies, University of Siegen, Hölderlinstr. 3, Siegen, 57076, Germany

ABSTRACT

We report the fabrication of an amorphous silicon based fluorescence sensor for miniaturized total analysis systems along with experimental results on optical excitation and detection elements. The pin-photodiode exhibits a dynamic range of 110dB and a room temperature dark current of less than 3000 charge carriers per ms according to a detector area of $0.1256mm^2$. The spectral response is ranging from 320nm to 780nm with a maximum at 600nm @ 80% quantum efficiency. To provide high sensitivity, the excitation light irradiates the fluid orthogonally to the active sensor detection direction by means of specifically designed microfluidic capillaries filled with e.g. methylene iodide or 1,2-o-dibrombenzene. The liquid core, which is enclosed by solid cladding materials, has been calculated to dimensions of a width of 16.75μm or 59.67μm with a height from 15μm to 50μm according to a number of propagating modes inside of 16 or 57, respectively.

INTRODUCTION

Since the early 90's, several groups have started to develop optics for microchips used in total analysis systems [1,2,3,4,5]. However, there is still a tremendous need for the development of miniaturized fluorescence detection systems [6,7,8]. Most notably, due to the higher absorption coefficient for visible light and the low dark current, an amorphous silicon detector is more suitable for the detection of fluorescence light than a crystalline silicon detector. In particular, most applied labeling dyes used for chemical and biological analysis emit in the visible part of the light spectrum.

The Application specific Lab-on-Microchip (ALM) is a monolithically integrated technology platform, combining microfluidic networks, microoptical components and microelectronic circuits [9]. With its full-fledged monolithic integration onto a standard Application Specific Integrated Circuit (ASIC), the ALM allows for dramatic reductions of sample volumes and analysis times, thus offering vastly improved speed, reliability and efficiency. As a consequence, the ALM is well suited to perform fast and mobile environmental analysis without the need of large and expensive diagnostic instrumentation. In contrast to most of the commonly used architectures that are manufactured by micromechanical procedures on unpatterned silicon or glass substrates combined with polymer materials like poly(dimethylsiloxane) (PDMS) or polymethylmethacrylate (PMMA) the ALM offers a wide range of functional components. Integrated optoelectronic devices and microoptical waveguides can provide a portable, parallel and inexpensive solution for on-chip fluorescence sensing. With regard to cost minimization during the early stages of ALM process development we are using thermal oxidized silicon wafers or glass substrates instead of externally fabricated ASICs. The development of on chip electronics, e.g. for mass flow regulated osmotic micropumps, is initially carried out with inexpensive multiproject wafers which are combined with the labchip on a hybrid substrate.

The first liquid-core light guides have been used for applications like Raman-Cells, telecommunications and other applications in the ultraviolet and visible spectral frequency region [10,11,12,13]. Integrated liquid-core light guides are a highly promising alternative to conventional integrated solid-core fibers or to hollow waveguides, because they are variable in dimension (< 1 - 500.0µm) and refractive index difference. In combination with the ALM concept they allow the integration of a mechanically stable, multiple use, and inexpensive way to guide light compatible with the technology platform.

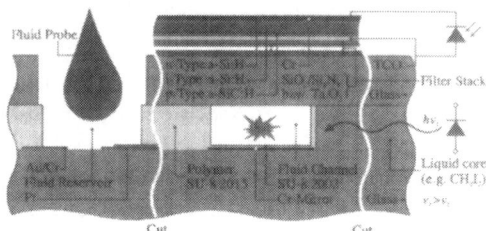

Figure 1. Cross section of the ALM device architecture with a glass as substrate, liquid light guidance and an additional optical filter stack.

FABRICATION

a-Si:H based pin-diodes

The aim of this work is to realize a novel integrated fluorescence sensor for Micro Total Analysis Systems (µTAS). The ALM architecture with glass as substrate is depicted in figure 1. It consists of a substrate plate and a thinner top plate which are sandwiched together by application of temperature, pressure and oxygen plasma to these preprocessed glass plates. The thicker bottom plate (1.2mm) carries the microfluidic network which is deposited on top of a metallization system to contact the fluid. Platinum, gold or chromium are used as materials for the metal electrodes, in some cases also electrically isolated from the fluid. It is in the fluidic lab layer that the chemical processes take place. This layer consists of a polymer, e.g. SU-8 2015, in which channels and compartments are created by lithography. The thinner top plate (150µm) supports the semiconductor optoelectronic devices. It is bonded onto the bottom substrate while the amorphous silicon (a-Si) based pin-diodes are located on top.

The pin-diode fabrication process is shown in detail in figure 2. These diodes are manufactured by plasma enhanced chemical vapor deposition from silane and dopant gases at temperatures around 200°C. The first step is the optional deposition of an interference filter to block the excitation wavelength and, moreover, to protect the diodes from high electric fields in the fluid

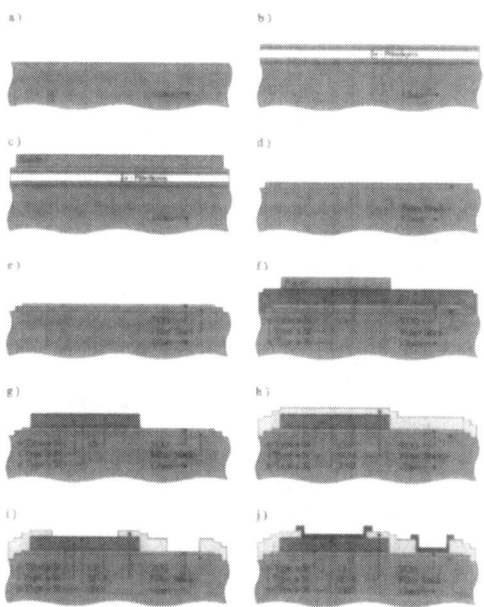

Figure 2. Process flow for the a-Si:H based thin film pin-diodes at the top plate sealing the microfluidic channel network.

channels, e.g. caused by micro capillary zone electrophoresis. The next isolated semitransparent layer acts as a common diode contact to the following a-Si:H layers depicted in step g). A Transparent Conductive Oxide (TCO), usually sputtered ZnO:Al, is used as a semitransparent front contact. For ZnO:Al the resistance is on the order of $30\Omega/\square$ with a transmission of approximately 90% for visible and near IR light and a thickness of 220nm. Cr acts as a single rear contact for each individual diode.

The layer system is patterned using 4 masks depicted in the steps c), f), i) and j). Immediately after metal etching the unmasked a-Si:H layers are removed by plasma treatment with a sulphur hexafluoride (SF_6) gas mixture. The TCO layer is wet etched by HCl in about 20 seconds. The thickness of the complete device is approximately 1.5μm to 2.8μm. The polymer SU-8 2002 is used to protect the structure against environmental influences during experiments with chemical reagents. The SU-8 2002 film thickness is below 2μm for a spin coating speed of 6000rpm, 2 minutes softbake at 95°C and 45 seconds development time.

Liquid light guidance structures

Liquids, which are primarily intended to be used as reagents in the ALM, are also able to guide light. Due to this fact the integration of liquid optics technology in ALMs does not require additional process steps and seems to be a technique with good prospects.

Integrated optical sensors and systems nowadays are mostly made of III-V semiconductors such as gallium arsenide, indium phosphite or they utilize the photo- and electroluminescence of porous-Si and silicon rich oxides which can be obtained with silicon technology and are currently being integrated with waveguides and photodetectors. For the ALM concept these solutions are not an option. Instead, we make use of liquid light guidance structures, combined with a glass fiber, to interface to an external light source.

The light guidance within the microcapillaries is based on the principle of total reflection presuming that the refractive index of the liquid is larger than that of the coating. In consideration of its application the numeric aperture (NA) of such a system should be within $0.15 \leq NA \leq 0.7$. With NA < 0.15 the integrated liquid-core fiber reacts very sensitively to coupling and structuring errors, therefore the losses would be large. Values of NA < 0.7 supply likewise no advantage, since light under such an angle has a large absorption due the higher count of reflections inside the guidance structure. As a consequence, it is very difficult to refocus the radiation at the analysis capillary.

In the concept presented the excitation light is guided towards the analysis capillary by means of a specifically designed microfluidic network, filled e.g. with a methylene iodide mixture (CH_2I_2: CH_2Cl_2). To increase the sensor signal of a pin-diode caused by the emitted fluorescent light of the analysis probe, the excitation light irradiates the fluid orthogonally to the active sensor detection direction as shown in figure 1. In order to use the liquid optical light core even in the UV wavelength range it

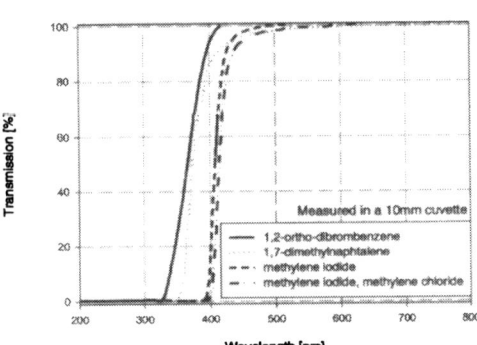

Figure 3. Transmission behavior of applicable liquid mixtures with a refractive index of 1.61, except methylene iodide with a refractive index of 1.76.

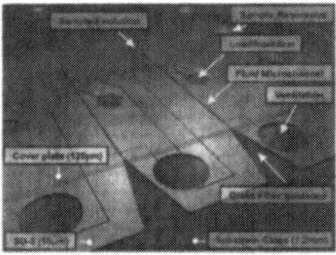 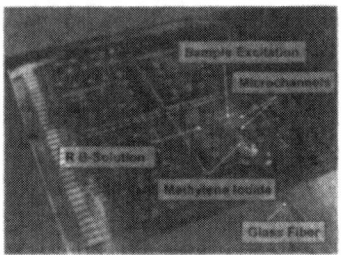

Figure 4. Left: Concept for integrated liquid light guidance with CH_2I_2, glass and a SU-8 layer. Right: Picture of the experimental ALM setup to determine the resulting fluorescence shapes in the microchannel.

needs to consist of a liquid with a suitable cut-off wavelength combined with a high index of refraction. For that purpose data for applicable liquid mixtures are presented in figure 3. The diagram shows the transmission behavior of a variety of liquid mixtures with a refractive index of 1.61, except methylene iodide with a refractive index of 1.76. The values are based on spectrometric measurements in a 10mm wide cuvette.

Within integrated optics, waveguides are classified as singlemode or multimode. Typically, on chip integrated planar singlemode waveguides are of the size of a few micrometers. Which of these two generic types should be used in an ALM analysis concept depends on the specific application. For example, to illuminate a specific part of a microfluidic channel only to cause fluorescent light, a multimode liquid core waveguide with tens of microns in size should be sufficient. Calculations for the wavelength of 532nm have resulted that a refractive index of 1.61 to 1.76 for a liquid core with a width from 16.75μm to 59.67μm enclosed by a glass substrate and sealing plate and a SU-8 layer from 15μm to 50μm height is sufficient for a multimode excitation with a number of modes of 16 or 57, respectively. The left image in of figure 4 depicts the concept for integrated liquid light guidance with a CH_2I_2 mixture as core and the combination of glass and SU-8 as solid cladding materials. The image on the right of figure 4 shows the experimental ALM setup to determine the fluorescence illumination shape in the fluidic microchannel.

CHARACTERISTICS

a-Si:H based pin-diode characteristics

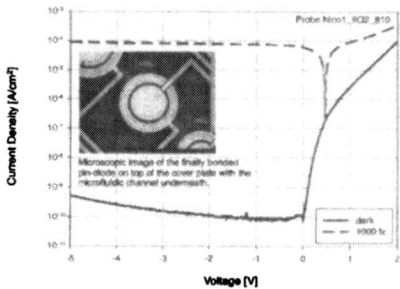 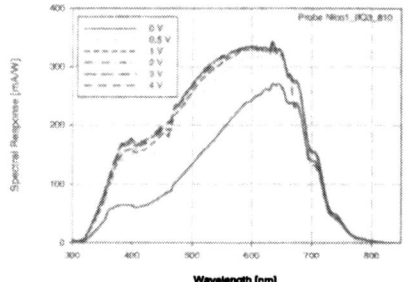

Figure 5. Logarithmic photo- and dark current density, referred to 1000lx illumination, for an a-Si:H pin-diode as function of bias voltage.

Figure 6. Spectral response for an a-Si:H pin-diode as function of wavelength and bias voltage as parameter.

Figure 5 depicts the logarithmic photo and dark current density characteristics of the ALM thin film sensor. The dynamic range, with respect to 1000lx illumination, is around 110dB in the relevant bias voltage range from 0V to 5V. In particular, the room temperature dark current density corresponds to less than 3000 charge carriers per ms for a detector area of $0.1256mm^2$. The dark current rises exponentially with temperature and the dark current doubling temperature is in the order of 6K. This is comparable with the usually reported value of 5K for crystalline devices, and it ensures stable operation in applications where temperature control is not desired or not applicable. Figure 6 depicts the spectral response ranging from 320nm to 780nm with a maximum at 600nm. The quantum efficiency is 80% at 600nm. The maximum of the spectral response can be shifted by appropriate diode design and band gap engineering between approximately 440nm to 630nm [14] at the expense of a more complicated deposition process.

Illumination shapes for liquid light guidance

The laser beam from a 532nm laser diode passes through a focus lens and is then coupled into a multimode glass fiber which is aligned with the center of the liquid core. The core, as shown in figure 4, is filled with CH_2I_2 diluted with dichloromethane. A focal lens made of SU-8 terminates the liquid core, 80µm wide and 20µm high. The analysis channel, 100µm wide and 20µm high, is connected to a syringe pump filled with rhodamine B solution (38.6µM in EtOH), applying a constant flow rate of 1µl/min. A fluorescence filter, longpass at 550nm, is placed above and perpendicular to the interface of the liquid core with the analysis channel. For observation a beam splitter is used to support the output CCD camera. Figure 7 shows the resulting illumination shape for a tapered liquid core structure, from 300µm down to 80µm width at the interface point. The upper picture was recorded without the use of the longpass fluorescence filter while the lower picture was taken with the excitation light suppressed using the blocking filter. The laser power was kept at a few microwatts to minimize photobleaching.

Figure 8 shows photomicrographs of various illumination shapes generated in an analysis capillary filled with rhodamine B solution at an excitation wavelength of 532nm.

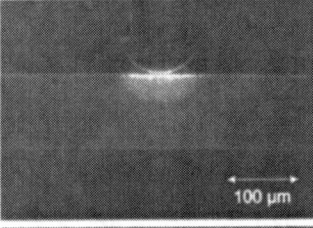
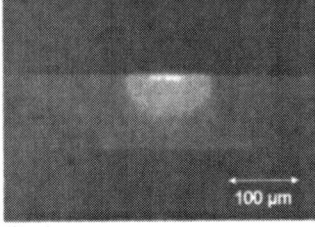

Figure 7. Resulting illumination shape for a tapered liquid core structure at the top without longpass fluorescence filter and below with filtered excitation light.

This figure shows that the output of the liquid core waveguide is divided into many single beams which are in summation asymmetric. This result indicates that several optical paths exist that enable guidance of light through the fluid filled channel, and that therefore this waveguide is multimode. The resulting illumination shapes can be optimized by variation of the alignment between the glass fiber core and the focused laser beam. The optical losses of the liquid core

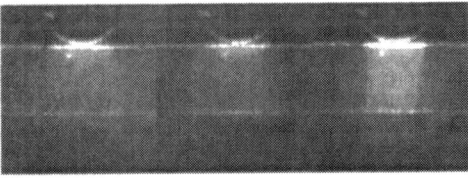

Figure 8. Various illumination shapes for a tapered liquid core structure generated in an analysis capillary.

waveguide are composed of transmission losses and coupling losses. The transmission losses of the waveguides are primarily due to three factors, namely the absorption in the fluid, the surface roughness of the SU-8 and the glass cladding, and the losses through the cladding. The coupling losses are due to scattering, reflections, mode mismatch, and misalignment. In our experiments, the observed losses were dominated by coupling losses.

CONCLUSION

In conclusion, we have demonstrated optical functionalities integrated in the ALM concept. Optical components, such as photodetectors, light guides and lens structures have been fabricated lithographically. The fabrication process for integrated a-Si:H pin-diodes has been outlined and measurements at a multipurpose test chip have been reported, leading to a compact three dimensional ALM on a glass substrate. A convenient method, in which light is guided through a liquid phase of high refractive index which is confined in channels of microscopic dimensions defined in SU-8 of lower index, has been shown. Most notably, these optical components, themselves in a liquid state, can easily be integrated to microfluidic systems possibly allowing new schemes for the optical detection required in μTAS.

ACKNOWLEDGEMENT

The authors would like to thank their colleagues at the "Research Center for Micro- and Nanochemistry and -Engineering" (Cμ) in Siegen and the "Deutsche Forschungsgemeinschaft" (DFG) under contract "DFG BO 772".

REFERENCES

1. J. M. Ng, I. Gitlin, A. D. Stroock, G. M. Whitesides, *Electrophoresis*, 23(20), 3461-3473 (2002).
2. E. Verpoorte, *Lab Chip,* 3, 42N-52N (2003).
3. K. B. Mogensen, H. Klank, J. P. Kutter, *Electrophoresis,* 25, 3498–3512 (2004).
4. R. A. Yotter, D. M. Wilson, *IEEE Sensors Journal*, 3(3), 288-303 (2003).
5. A. Brecht, G. Gauglitz, *Sensors and Actuators B*, 38-39, 1-7 (1997).
6. T. Kamei, B. M. Paegel, J. R. Scherer, A. M. Skelley, R. A. Street and R. A. Mathies, *Anal. Chem.*, 75, 5300-5305 (2003).
7. H. Schäfer, K. Seibel, M. Walder, L. Schöler, T. Pletzer, M. Waidelich, H. Ihmels, M. Schmittel, D. Ehrhardt, M. Böhm, *Proc. Micro Total Analysis Systems 2004*, vol. 2, 443-445 (2004).
8. H. Schäfer, K. Seibel, M. Walder, L. Schöler, T. Pletzer, M. Waidelich, H. Ihmels, D. Ehrhardt, M. Böhm, *Proc. Micro Electro Mechanical Systems 2005*, 758-761 (2005).
9. H. Schäfer, S. Chemnitz, V. Koizy, A. Fischer, D. Ehrhardt, M. Böhm, *Nano-Micro-Interface Conference 2003*, Berlin, H.J. Fecht, M. Werner (eds), *The Nano-Micro Interface*, Wiley-VCM, Weinheim, 119-137 (2004).
10. J. Stone, *IEEE J. Quantum Electron.*, QE-8, 386-388 (1972).
11. J. Stone, *Appl. Phys. Lett.*, 20, 239-240 (1972).
12. G. J. Ogilvie, R. J. Esdaile, *Electron. Lett.*, 8, 533-534 (1972).
13. D. N. Payne, W. A. Gambling, *Electron. Lett.*, 8, 374-376 (1972).
14. P. Rieve, M. Sommer, M. Wagner, K. Seibel, M. Böhm, *Journal of Non-Crystalline Solids*, 266-269, 1168-1172 (2000).

Mater. Res. Soc. Symp. Proc. Vol. 989 © 2007 Materials Research Society   0989-A10-05

μ-Watt Enhanced Electroluminescent Power of Silicon Nanocrystal Light-Emitting Diodes Made on Nano-Scale Silicon-Tip-Array Substrate

Gong-Ru Lin[1], and Chun-Jung Lin[2]

[1]Graduate Institute of Photonics and Optoelectronics, and Department of Electrical Engineering, National Taiwan University, No. 1, Roosevelt Road, Sec. 4, Taipei, 106, Taiwan

[2]Department of Photonics and Institute of Electro-Optical Engineering, National Chiao Tung University, No. 1001, Ta Hsueh Road, Hsinchu, 300, Taiwan

ABSTRACT

A Si nanocrystal based metal-oxide-semiconductor light-emitting diode (MOSLED) on Si nano-pillar array is preliminarily demonstrated. Rapid self-aggregation of Ni nanodots on Si substrate covered with a thin SiO_2 buffered layer is employed as the etching mask for obtaining Si nano-pillar array. Dense Ni nanodots with size and density of 30 nm and 2.8×10^{10} cm^{-2}, respectively, can be formatted after rapid thermal annealing at 850°C for 22 s. The nano-roughened Si surface contributes to both the relaxation of total-internal reflection at device-air interface and the Fowler-Nordheim tunneling enhanced turn-on characteristics, providing the MOSLED a maximum optical power of 0.7 μW obtained at biased current of 375 μA. The optical intensity, turn-on current, maximum power slope and external quantum efficiency of the MOSLED are 140 μW/cm², 5 μA, 2±0.8 mW/A and 1×10^{-3}, respectively, which is almost one order of magnitude larger than that of a same device made on smooth Si substrate.

INTRODUCTION

Silicon nanocrystal (nc-Si) based metal-oxide-semiconductor light emitting diodes (MOSLEDs) represent potential candidates for the next generation of optoelectronic applications such as optical interconnect and optical communication. [1-6] The advantages of nc-Si based MOSLEDs include full-color emission, complementary metal oxide semiconductor compatibility, system feasibility, and low cost of fabrication. Electroluminescence (EL) from nc-Si in Si-rich SiO_2 (SiO_x) film grown by plasma enhanced chemical vapor deposition (PECVD) has previously been observed, [6] however, the EL intensity and external quantum efficiency is still very low since the nature of indirect recombination, the insulating property of the host oxide, and the tunneling dependent carrier injection. Several methods were used to improve the efficiency of carrier injection into nc-Si including the increasing the density of nc-Si embedded in the SiO_x film, the decreasing thickness of nc-Si layer and the engineering of the tunneling barrier between contact metal and oxide. However, an approach to improve the carrier injection into nc-Si via Si nano-pillars has never reported. Typically, the fabrication of Si nano-pillars mainly relies on the electron-beam (E-beam) lithography and inductively couple-plasma reactive ion etching (ICP-RIE) process. [7] The <10-nm Si nano-pillar array can be produced under the assistance of E-beam lithography. [8] Nowadays, the self-assembled metallic nano-dots have emerged to produce functional nano-sensor or nano-mask. The large discrepancy between two quantum efficiencies is originated from the fact that the light extraction efficiency of most conventional LEDs is limited by the total internal reflection (TIR) of the generated light in the active region of the LED, which occurs at the semiconductor-air interface. This is due to the large difference in the refractive index between the semiconductor and air. In this letter, we demonstrate the great enhancement on the turn-on characteristics and light-emitting efficiency of nc-Si MOSLED made on a dense Si nano-pillar array based substrate. The improved injecting efficiency of carriers and the released TIR effect of emitting light through such a nano-roughened Si pillar array are elucidated.

EXPERIMENT

By rapid thermal annealing a 5-nm thick Ni film evaporated on a 20-nm thick SiO_2 layer covered Si substrate, we have successfully demonstrated the self-aggregation of two-dimensional randomized Ni nano-dots on a p-type (100) Si wafer (see Fig. 1).

 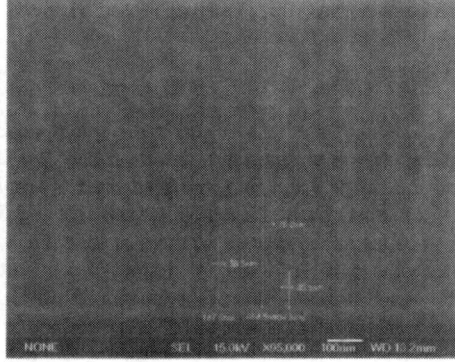

Figure 1. SEM image of Ni nano-dots precipitated from the evaporated Ni film on SiO_2/Si.

Figure 2. SEM image of Si nano-pillars formed after ICP-RIE with RF/Bias power recipe of 50/100W.

The Ni nano-dots can hardly self-aggregate on highly heat-dissipated Si substrate with a thermal conductivity of 148 W/Km. Adding a 200-Å-thick SiO_2 buffer with an ultra-low thermal conductivity of 1.35 W/Km prevents the formation of $NiSi_2$ compounds, enhances the heat accumulation, and releases the adhesion at Ni/Si interface, which greatly accelerates the self-assembly of Ni nanodots. With such a self-assemble Ni/SiO_2 nano-dots based nano-mask, a large-area Si nano-pillar array with pillar diameter of 30 nm and height of 350 nm can be formatted on Si substrate through the ICP-RIE procedure (see Fig. 2). The SiO_x films were grown on Si nano-pillar substrate by using PECVD with SiH_4/N_2O flow rate ratio of 1/4.5, chamber pressure at 60 mtorr, and plasma power of 30 W and substrate temperature of 400°C. After deposition, the Si-rich SiO_x film with thickness of 240 nm was annealed in a quartz furnace with flowing N_2 at 1100°C for 60 min to precipitate nc-Si. The device structure of a nc-Si based MOSLED on silicon nano-pillar array is shown in Fig. 3. The top view of a Cu-wire-bonded ITO/SiO_x/Si/Al with circular contact diameter of 0.8 mm is shown in Fig. 4.

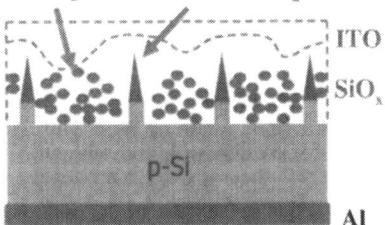

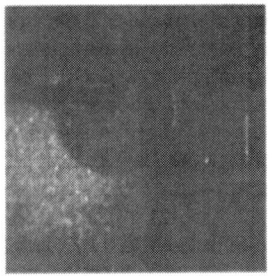

Figure 3. Device structure of a nc-Si based MOSLED on silicon nanopillar array.

Figure 4. An ITO/SiO$_x$/Si/Al diode with a circular contact diameter of 0.8 mm.

DISCUSSION

After annealing from 15 to 120 min, two PL peaks found at 550 and 770 nm are attributed to the emissions of different-size nc-Si. The maximum PL intensity was observed from the 60 min-annealed sample, as shown in Fig. 5. After annealing from 15 to 60 min, the PL intensity at 770 nm rapidly increases by 5 times, which is attributed to the precipitation of nc-Si. However, the PL intensity at 770 nm decreases after the annealing time lengths from 60 to 120 min, which is due to the re-oxidation of nc-Si.

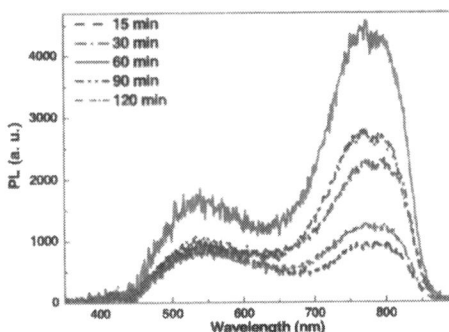

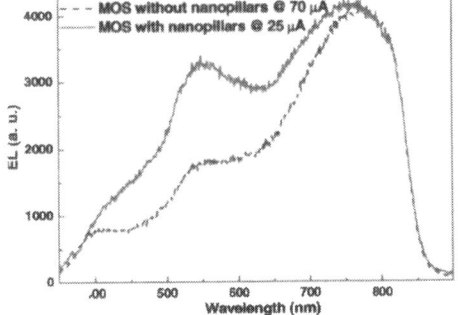

Figure 5. PL of nc-Si grown on Si nano-pillar substrate after annealing.

Figure 6. EL spectra of nc-Si based MOSLEDs with/without Si nano-pillars under the same bias.

In comparison with the SiO$_x$ film without Si nano-pillar array, the PL peak at 550 nm was enhanced in the sample with Si nano-pillar array, which is attributed to the emission from small size nc-Si in the Si nano-pillar array. Since the space of Si nano-pillar is less than 50 nm, Si nano-pillar array also facilitates the local confinement to precipitate small-size nc-Si. The localized deposition of Si-rich SiO$_x$ with in the space region of Si nano-pillars facilitates the confinement on nc-Si size after annealing. The 410-nm peak is attributed to the recombination through the neutral oxygen vacancies (NOV), while the two peaks at 550 nm and 760 nm are through the nc-Si dots with different sizes were observed in the 60 min-annealed MOSLED with Si nano-pillar array, as shown in Fig. 6. EL powers at 410 and 550 nm were significantly enhanced due to the tunneling of carriers into NOV defects and small-size nc-Sis. Furthermore,

EL intensities at 410 and 550 nm from the nc-Si based MOSLED with Si nano-pillar array were also greatly enhanced in comparison with the EL intensity from the nc-Si based MOSLED without Si nano-pillar under the same bias (see Fig. 6), which is also corresponding to the carrier-injection facilitation via the Si nano-pillar.

The turn-on voltage of nc-Si based MOSLED with Si nano-pillars is around 30 V, which is much lower than the same device made on smooth Si wafer ($V_{\text{turn-on}}$ ~42 volts), as shown in Fig. 7. The current-voltage (I-V) slopes of the nc-Si based MOSLEDs with and without Si nano-pillars are 62.5 and 21.4 $\mu A/V$, respectively, which also corresponds to the larger EL intensity from the nc-Si based MOSLED with Si nano-pillars and is due to the carrier-injection assistance of Si nano-pillars. The <u>light</u> intensity, turn-on current and <u>maximum</u> power slope of the nc-Si based MOSLED with Si nano-pillars are 140 $\mu W/cm^2$, 5 μA and 2±0.8 mW/A, respectively, as shown in Fig. 8. For the nc-Si based MOSLED without Si nano-pillars, the <u>light</u> intensity, turn-on current and <u>maximum</u> power slope are 18.5 $\mu W/cm^2$, 25 μA and 1.1±0.5 mW/A. Highest optical power of 0.7 μW is obtained at biased current of 375 μA, which is almost one order of magnitude larger than that of a same device made on smooth Si substrate, as shown in Fig. 8.

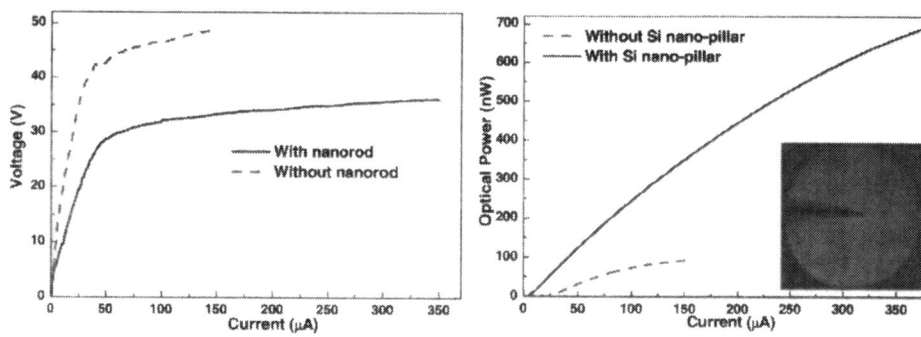

Fig. 7. I-V curves of nc-Si based MOSLEDs with/without Si nano-pillars.

Fig. 8. I-P curves of nc-Si based MOSLEDs with/without Si nano-pillars.

Electrical analysis has concluded that the Fowler-Nordheim tunneling [9] based carrier transport through the Si nano-pillar array is more pronounced than the smooth Si wafer. Such Si nano-pillars functions like a field-emission tips to facilitate the carrier tunneling into nc-Si within the SiO_x film. The external quantum efficiency of a bulk-Si LED usually limits to 10^{-5} if we define the external quantum efficiency as the rate of the output photon number and input electron number described by,

$$\eta_{ext} \cong \int P_{opt} \lambda d\lambda \Big/ I_{bias} \, hc ,$$ (1)

where λ is wavelength, <u>h is planck constant, c is light speed</u>, P_{opt} and I_{bias} are the optical spectral power and biased current, respectively. The external quantum efficiency of the nc-Si based MOSLED with Si nano-pillars is up to 0.1% under a power conversion ratio of 5×10^{-5}. This is a relatively high quantum efficiency ever reported for the nc-Si based MOSLEDs. Far field pattern of an nc-Si based MOSLED with Si nano-pillars with a brighter color is shown in the inset of Fig. 8. Growth of Si-rich SiO_2 layer on the Si nano-pillar substrate greatly enhanced the roughness of the top surface and bottom SiO_2/Si interface, which not only releases the total-internal reflection happened in device on a smooth substrate but also enhanced the Fowler-Nordheim tunneling based carrier transport through the nano-pillar structure. The Si nano-pillar further functions like a tip electrode to facilitate the carrier injection from substrate to nc-Si. In particular, the refractive indices of the top ITO contact layer and the Si nano-pillar array are only 1.8 and as small as 1.3, respectively, essentially providing a weak optical confinement for the luminescent SiO_x layer.

Roughening the top surface of an LED is also one of the efficient methods for increasing the light extraction rate to improve the external quantum efficiency of the LED. The roughened top surface reduces internal light reflection and scatters the light outward. For nc-Si based MOSLEDs, the refractive indexes of nc-Si (n_{nc-Si}) and the air (n_{air}) are 1.8 and 1, respectively. In this case, the critical angle ($\theta c = \sin^{-1} n_{air}/n_{nc-Si}$) for the light generated in the active region to escape is about 33.7°. Assuming the light emission from the active region of a nc-Si based MOSLED is a isotropic and the light can escape from the chip if the angle of incidence to the chip wall is less than the critical angle, a small fraction of light generated in the active region of the nc-Si based MOSLED can escape to the surrounding air. Assuming the output power is uniformly distributed over the whole half-spherical surface, the maximum solid angle of the MOSLED (Ω_{half}) is 2π. The solid angle of the total internal reflection (Ω_{TIR}) is 0.34π. Only a small fraction of light (about 17%) can escape from the nc-Si based MOSLED even though the refractive index of the top ITO contact layer is nearly the same with the Si-rich SiO_x film to prevent additional reflection. As compared with the nc-Si based MOSLED with and without Si nano-pillar array under based current of 150 μA, this roughening treatment resulted in an increase of output power by a factor of 3.8 from the top surface. Fujii *et al.* [10] have reported the output power of an optimally roughened surface GaN-based LED shows a twofold to threefold increase compared to that of a GaN-based LED before surface roughening. Huang *et al.* [11] have also demonstrated the light output of the InGaN–GaN LED with a nano-roughened top p-GaN surface is 1.4 times that of a conventional LED, and wall-plug efficiency is 45% higher.

The light-extraction efficiency in the nc-Si based MOSLED is limited mainly due to the large difference in the refractive index between the nc-Si film and surrounding air. The critical angle determined by Snell's law, that is, the angle that the photons can escape from the nc-Si layer to air, is crucially important to improve the light-extraction efficiency of the nc-Si based MOSLED. The key to enhance the escape probability is to give the photons generated in the active layer of the nc-Si based MOSLED structure multiple opportunities to find the escape cone. This requires angular randomization or scrambling of the photons. The maximum optical output power, biased current and biased voltage are 0.7 μW, 0.375 μA and 36 V. The peak wavelength and external quantum efficiency of the nc-Si based MOSLED with Si nano-pillars are 0.76 μm, and 1×10^{-3}.

CONCLUSIONS

In conclusion, the nc-Si based MOSLED on Si nano-pillar array is demonstrated. Rapid self-aggregation of Ni nanodots on Si substrate covered with a thin SiO_2 buffered layer is employed as the etching mask for obtaining Si nano-pillar array. Dense Ni nanodots with size and density of 30 nm and 2.8×10^{10} cm^{-2}, respectively, can be formatted after rapid thermal annealing at 850°C for 22 s. EL spectrum of Si nanocrystals grown on high-aspect-ratio Si nano-pillars is greatly enhanced. The light intensity, turn-on current, and maximum power slope of the MOSLED are 140 μW/cm^2, 5 μA and 2±0.8 mW/A, respectively. The external quantum efficiency of up to 0.1% can be obtained under a power conversion ratio of 5×10^{-5}. The Si nanopillar array with the pillar size smaller than 10 nm greatly enhances the roughness on top surface and bottom SiO_2/Si interface of the nc-Si MOSLED, which not only releases the total-internal reflection effect but also strengthens the Fowler-Nordheim tunneling effect. The maximum optical power of 0.7 μW is obtained at biased current of 375 μA. Such an output power is almost one order of magnitude larger than that of a same device made on smooth Si substrate.

ACKNOWLEDGMENTS

This work was supported in part by the National Science Council (NSC) of the Republic of China under grants NSC95-2221-E-002-448 and NSC95-2120-M-009-006.

REFERENCES

1. T. S. Iwayama, N. Kurumado, D. E. Hole and P. D. Townsend, *J. Appl. Phys.* **83**, 6018 (1998).
2. B. Garrido, M. López, O. González, A. Pérez-Rodríguez and J. R. Morante, *Appl. Phys. Lett.* **77**, 3143 (2000).
3. F. Iacona, C. Bongiorno, C. Spinella, S. Boninelli and F. Priolo, *J. Appl. Phys.* **95**, 3723 (2004).
4. Kwan Sik Cho, Nae-Man Park, Tae-Youb Kim, Kyung-Hyun Kim, Gun Yong Sung and Jung H. Shin, *Appl. Phys. Lett.* **86**, 071909 (2005).
5. K. Luterová, I. Pelant, J. Valenta, J. L. Rehspringer, D. Muller, J. J. Grob, J. Dian and B. Hönerlange, *Appl. Phys. Lett.* **77**, 2952 (2000).
6. G.-R. Lin, C. J. Lin, C. K. Lin, L. J. Chou and Y. L. Chueh, *J. Appl. Phys.* **97**, 094306 (2005).
7. M. Francois, J. Danlot, B. Grimbert, P. Mounaix, M. Muller, O. Vanbesien and D. Lippens, *Microelectron. Eng.* **61**, 537 (2002).
8. P. B. Fischer, K. Dai, E. Chen and S. Y. Chou, *J. Vac. Sci. Technol. B* **11**, 2524 (1993).
9. R. H. Fowler and L. W. Nordheim, *Proc. R. Soc. London, Ser. A* **119**, 173 (1928).
10. T. Fujii, Y. Gao, R. Sharma, E. L. Hu, S. P. DenBaars and S. Nakamura, *Appl. Phys. Lett.* **84**, 855 (2004).
11. H. W. Huang, C. C. Kao, J. T. Chu, C. C. Yu, H. C. Kuo and S. C. Wang, *IEEE Photonics Technol. Lett.* **17**, 5 (2005).

Thin Film Transistors I

Mater. Res. Soc. Symp. Proc. Vol. 989 © 2007 Materials Research Society   0989-A11-01

High Mobility Nanocrystalline Silicon TFTs for Display Application

Min-Koo Han, and Sang-Myeon Han
School of Electrical Engineering and Computer Science (#50), Seoul National University,
Seoul, 151-742, Korea, Republic of

ABSTRACT

Nanocrystalline silicon (nc-Si) thin film transistors (TFTs) of which active layer thickness was 100nm were fabricated using inductively coupled plasma chemical vapor deposition (ICP-CVD) at 150°C. The fabricated nc-Si TFT exhibits rather high field effect mobility exceeding 22 cm^2/Vs and excellent sub-threshold slope of 0.45 V/dec. The nc-Si film deposited 150°C as an active layer of the TFT shows high crystallinity more than 50% and very thin incubation layer less than 20 nm. ICP-CVD provides high density plasma with reduced ion bombardment during the deposition on nc-Si and He dilution can enhance the decomposition of SiH$_4$ into Si, SiH$_X$ radicals and atomic H, so that high quality nc-Si film can be fabricated. The SiO$_2$ film deposited by ICP-CVD at 150°C shows good electrical characteristics such as flat band voltage of -1.8 V and breakdown voltage of 6.2 MV/cm, which was used as a gate insulator.

INTRODUCTION

Recently, flexible displays have attracted considerable attention due to the robustness, lightweight, flexibility and low cost [1]. Various devices such as a-Si thin film transistors (TFTs), poly-Si TFTs and organic TFTs for flexible displays have been reported [2]. The conventional processes for low-temperature poly-Si (LTPS) TFT (< 450°C) on glass substrate may not be suitable for flexible substrates such as plastic because of a relatively high process temperature exceeding 200°C [3]. Therefore, complicated techniques and careful processes are required to fabricate poly-Si TFT under 200°C. On the other hand, peripheral driving circuits and pixel circuits are required in active matrix liquid crystal displays (AMLCDs) and active matrix organic light emitting displays (AMOLEDs). However, conventional a-Si TFTs and organic TFTs have difficulties in order to fabricate peripheral driving circuits, due to their inherent inferior electrical characteristics and their electrical instability [4,5]. Directly deposited nanocrystalline silicon (nc-Si) TFTs may be promising devices to fabricate various flat displays, due to a rather simple process and good uniformity compared with poly-Si TFTs, and superior performance and stability compared with a-Si and organic TFTs [6]. Inductively coupled plasma chemical vapor deposition (ICP-CVD) employs remote plasma [7-9] , which reduces problematic ion bombardment problems issues [10]. It may also be noted that ICP-CVD generates high density plasma [7,8]. ICP-CVD may also provide certain advantages such as a high deposition rate and improved film quality over plasma enhanced chemical vapor deposition (PECVD), which has been widely used to deposit Si film and SiO$_2$ film at low temperature.

The purpose of our work is to report the characteristics of nc-Si film deposited by ICP-CVD (13.56 MHz) and the electrical characteristics of nc-Si TFT fabricated at 150°C. The nc-Si TFT a high field effective mobility exceeding 20 cm^2/Vs and a low subthreshold slope less than 0.5 V/dec, so that the perpformance of the nc-Si TFT is enough to compose integrated driving circuits.

EXPERIMENT & RESULTS

nc-Si film deposition by ICP-CVD

The active layer of the nc-Si TFT is prepared by ICP-CVD, at the substrate temperature of 150°C using SiH_4 diluted with He. The pressure was kept at 20 mTorr and the He/SiH_4 ratio was 20/3 [sccm]. To verify the crystallinity of the ICP-CVD Si film, 100 nm thick Si films on glass were analyzed using the Raman spectrum. Figure 1 demonstrates the Raman spectrum of Si deposited by a He dilution. The crystalline peak at 520 cm^{-1} and amorphous peak near 480 cm^{-1} were observed [11] and the intensity of the crystalline peak is shown as much higher, indicating that the nc-Si film was deposited by ICP-CVD. The crystalline volume fraction is over 50%. Generally, the dilution ratio is related with deposition ratio in CVD process. To fabricate nc-Si film employing conventional PECVD, the dilution ratio is kept to be high value over 95%~99% [12], so that the very low deposition rate is obtained. Although the dilution ratio was considerably not high, the nc-Si film was deposited in this experiment using ICP-CVD. The deposition ratio of the deposited nc-Si film was 2.4 Å/sec, which can be regarded as a higher value than that of the nc-Si film fabricated by PECVD.

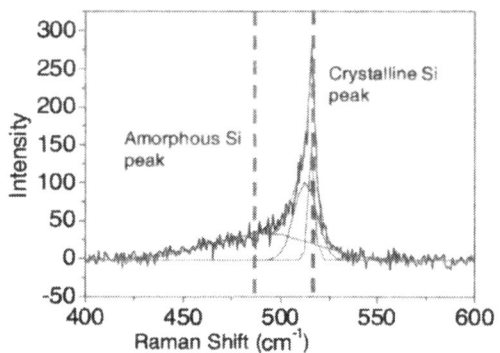

Fig. 1 The Raman spectrum of the nc-Si film. The crystalline volume fraction is 53%.

Figure 2 shows the cross sectional high resolution transmission electron microscopy (HR-TEM) image of the 100nm thick nc-Si Si film deposited by ICP-CVD on the oxidized Si wafer. The crystalline growth structure and thickness of the incubation layer of the nc-Si film was observed by cross-sectional HR-TEM. The nc-Si film consists of crystalline structures of columnar type from the bottom to the top of the nc-Si film. As shown in HR-TEM image, the ICP-CVD nc-Si film has a very thin incubation layer of 20 nm. The incubation layer of nc-Si film, which is a variety of a-Si phase mixed layer, is formed on the surface of substrate prior to crystalline phase Si deposition. It has been reported that nc-Si TFT fabricated with nc-Si film of high crystallinity is vulnerable to the V_{TH} shift problem when high gate bias is applied [6]. Thin incubation layer indicates the early nucleation of crystalline components during the Si film deposition. This thin incubation layer enables high crystallinity over 50% in the thin nc-Si film of 100 nm because the portion of amorphous phase in nc-Si thin film is decreased according to the diminishment of an incubation layer. Therefore, decrease of incubation layer thickness may improve the stability of nc-Si TFT. This thin incubation layer may be also helpful for forming large grains near the surface of nc-Si film, because the grains grow vertically in a columnar structure and the grain size can increase depending on the film thickness. Therefore, a thin incubation layer is also important for improvement of mobility. In the high resolution TEM image, the Si lattice patterns are observed clearly in the nc-Si film and those many crystalline components have their own Si lattice patterns compose the nc-Si film.

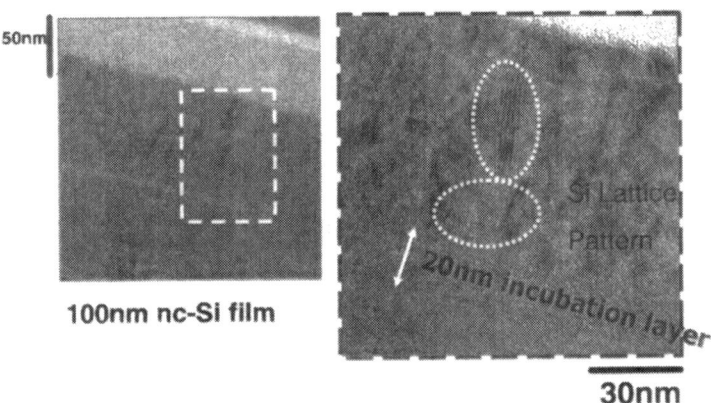

Fig. 2 The HR-TEM image of the nc-Si film fabricated at 150°C. The thickness of the nc-Si film is 100nm.

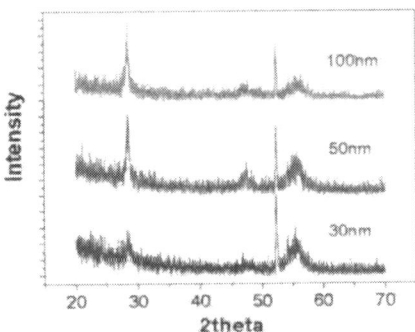

Fig.3 The XRD data of nc-Si films with various thickness. The <111> (2θ=28.442°) and <220> (2θ=47.302°) peaks were observed at thickness of 30nm and the peaks increased highly at thickness of 50nm, 100nm.

The formation of nc-Si film can be also observed by X-ray diffraction measurements (XRD) as shown in Figure 3. The <111> and <220> peaks were observed at thickness of 30nm and the peaks increased at thickness of 50 nm, 100 nm. This means the crystalline component were already formed in the Si film when the thickness is only 30 nm. And the crystalline factor became increased as the nc-Si film was deposited to the thickness of 50nm and 100nm. This

XRD result of nc-Si film's early nucleation is corresponding to previous HR-TEM results with thin incubation layer. The distinguished formation of nc-Si film may be attributed to the ICP-CVD. The ICP-CVD generates remote plasma of high density [7-9]. Therefore, ICP-CVD helps the decomposition of reacting gases and may lead the formation of nc-Si film with a thin incubation layer.

Figure 4 presents the SEM image of the Secco-etched ICP-CVD Si film. The grain size with a good uniformity presented in Figure 4 is approximately 40 nm, which can be regarded as a high value in grain size of the directly deposited nc-Si film. The large grain size of nc-Si film can contribute to the improvement of TFT mobility and the uniform grain size can contribute to the uniformity of TFTs in a display panel.

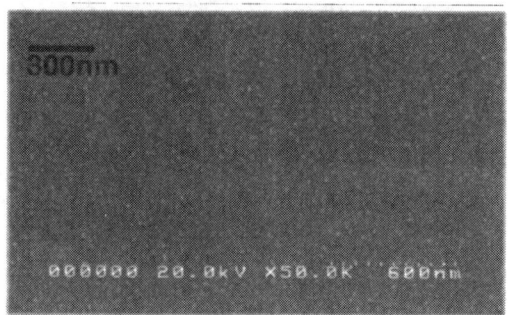

Fig. 4 The SEM images of the secco-etched nc-Si film. The grain size is about 40nm.

The RMS of the nc-Si film surface is 2.1nm, which is moderate to be used as an active layer of TFT. Interface morphology can affect the quality of gate insulator film in top gate TFT structure, so that the inspection of roughness was performed by AFM.

The formation of nc-Si film may be attributed to the ICP mode and He dilution. Figure 5 shows the Raman spectrum of the nc-Si film deposited by ICP-CVD with various substrate RF bias and constant ICP power of 400W. When the substrate RF bias is not applied, the peak of Raman spectrum is located near 510 cm^{-1} as shown in Figure 5 (a). However, as the substrate RF bias increases, the peak of Raman spectrum is shifted to near 480 cm^{-1} , which corresponds amorphous phase. ICP-CVD can generate high density plasma without acceleration of ions toward the substrate unlike PECVD because plasma generation power is delivered by coil not by parallel plates which accelerate ions toward substrate inevitably. In our experiment, we can know ions generated by plasma of ICP-CVD was accelerated by additional substrate RF bias and the crystallinity of the nc-Si film was decreased due to ion bombardment which can prevent the crystal formation [13]. Thus, ICP-CVD of no substrate RF bias with reduced ion bombardment is advantageous over conventional PECVD to fabricate high crystallinity nc-Si film. The He dilution process leads to Penning ionization observed at the addition of inert gas to plasma [14]. He plasma can increase electron concentration and electron decay time, to enhance effective decomposition of SiH$_4$ to produce abundant atomic hydrogen and energetic Si radicals, leading nc-Si formation [15].

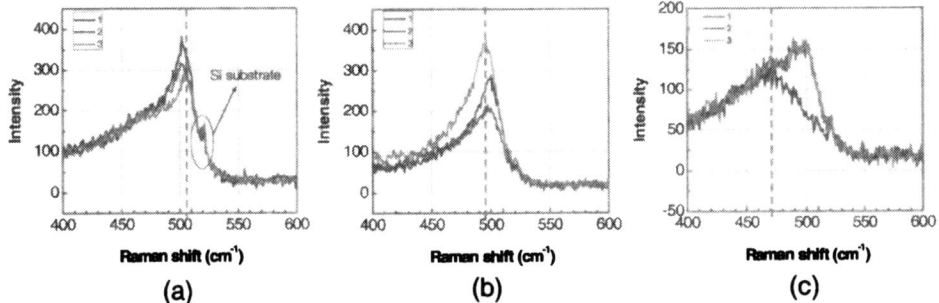

Fig. 5 The Raman spectrum of the nc-Si films deposited with various substrate RF bias
(a) Substrate RF bias : 0W, (b) Substrate RF bias : 75W, (b) Substrate RF bias : 150W

SiO₂ film deposited by ICP-CVD

The 100 nm thick silicon dioxide film is deposited by ICP-CVD at 150°C, using N_2O and SiH_4 gas. The breakdown field is 6.2 MV/cm [16]. The C-V characteristics of SiO_2 film deposited by ICP-CVD at 150°C were evaluated by modifying the flow rate of N_2O gas. The flow rate of N_2O gas varied from 30 sccm to 70 sccm while He gas and SiH_4 was fixed at 100 sccm and 5 sccm, respectively. Figure 6 demonstrates the C-V characteristics of the SiO_2 films. The more the flow of N_2O gas increased, the more the flat band voltage shifted in a negative direction. The C-V characteristics of SiO_2 film deposited with various process pressures are shown in Figure 7. According to increasing of pressure, flat band voltage become close to 0V and the gradient of the graph also increases. The increase of gradient means a decrease of the interface defects density and the flat-band voltage shift toward a negative direction means fixed oxide charge exist at the interface between Si and SiO_2 [17]. The reduction of flat-band voltage and effective oxide charge density may be attributed to the effective decomposition of reacting gases assisting the formation of SiO_2 structure at the interface at low temperature. According to above results, SiO_2 film fabricated under the condition of He:N_2O:SiH_4 = 100:30:5, ICP power $\geq$ 400W, pressure=70 mTorr was adopted as a gate insulator.

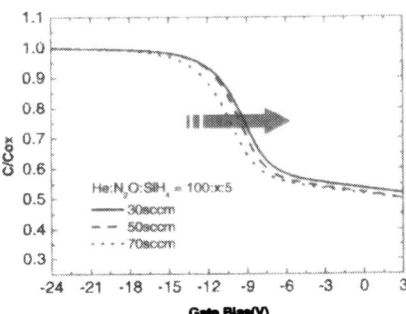

Fig. 6 The C-V characteristics of SiO_2 films concerning with the flows of N_2O (ICP RF power: 400W)

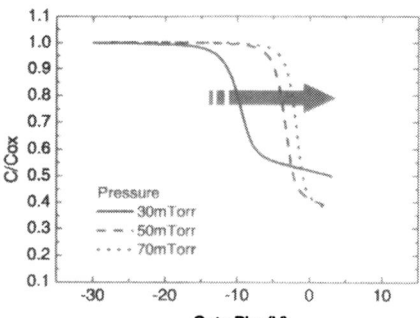

Fig. 7 The C-V characteristics of SiO₂ films concerning with the process pressures (He: N₂O: SiH₄=100:30:5 [sccm], ICP RF power: 400W)

nc-Si TFT fabrication and characteristics

We have fabricated n-type top-gate self-align nc-Si TFTs at 150°C. The process sequence is presented in Figure 8. Figure 9 shows the cross sectional schematic image of the fabricated nc-Si TFT. Source/ drain doping was achieved by ion implantation and activated by excimer laser annealing (ELA, XeCl, λ=308 nm) of which energy density is 100 mJ/cm². In the process at 150°C, there are only a few methods and techniques to enable doping and activation. Various conventional doping and activation processes used in LTPS TFTs or a-Si TFTs process may not be adopted for several reasons, including the maximum temperature limit or doping quality. Although ELA activation is relatively complicated compared with n+ Si direct deposition process, the activation efficiency of ELA is quite enough and advantageous to lead a good contact quality. In this experiment, dopants were activated by irradiating 2 ELA shots of equal energy. The channel region was located under the 300 nm thick gate metal, therefore the channel region was protected from excimer laser irradiation and still remained as a nc-Si film.

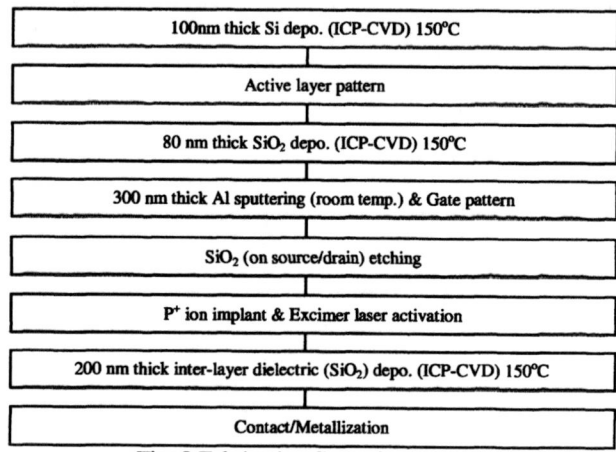

Fig. 8 Fabrication flow of nc-Si TFT

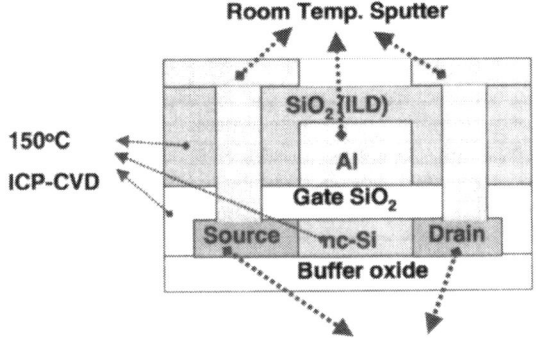

Room Temp. Sputter

SiO₂ (ILD)

Al

Gate SiO₂

150°C
ICP-CVD

Source   nc-Si   Drain

Buffer oxide

Ion implantation Phosphorous (Self-align)

Fig. 9 Cross sectional schematic image of the fabricated nc-Si TFT

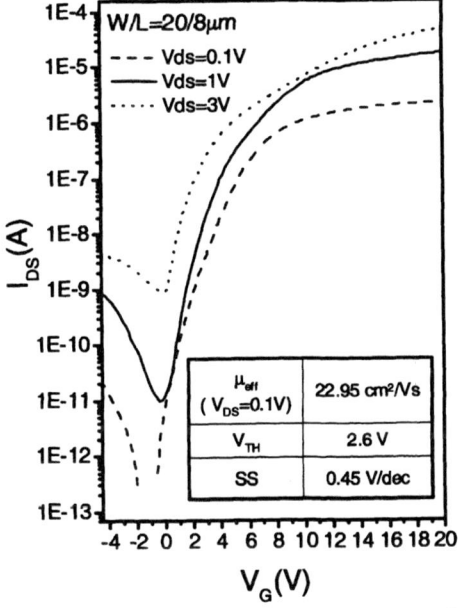

μ_{eff} (V_{DS}=0.1V)	22.95 cm²/Vs
V_{TH}	2.6 V
SS	0.45 V/dec

Fig. 10 The transfer characteristics of the fabricated nc-Si TFT. The width/length of gate is 20μm/8μm.

Figure 10 and Figure 11 present the nc-Si TFT characteristics. As shown in Figure 10, although all processes were performed below the temperature of 150°C and no post-annealing was carried out, the distinguished characteristics such as high mobility 22.95 cm²/Vs, threshold voltage (V_{TH}) of 2.6 V and a low sub-threshold slope of 0.45 V/dec were obtained [16]. The mobility was calculated from the transfer characteristic of nc-Si TFT at V_{DS}=0.1 V. These parameters show the good current driving ability of the fabricated nc-Si TFT. However, to reduce the leakage current at the reverse gate bias, post-treatment such as hydrogenation and additional Si/SiO$_2$ interface treatment at very low temperature below 150°C such as N$_2$O plasma pre-treatment may be required [18]. The transfer characteristics of the fabricated nc-Si TFT are shown in Figure 11. The saturation characteristics can be observed at the various gate voltages. In this graph the current crowding is not observed in the low V_{DS} regime. Thus, we can know the contact is well-formed in spite of very low process temperature of 150°C.

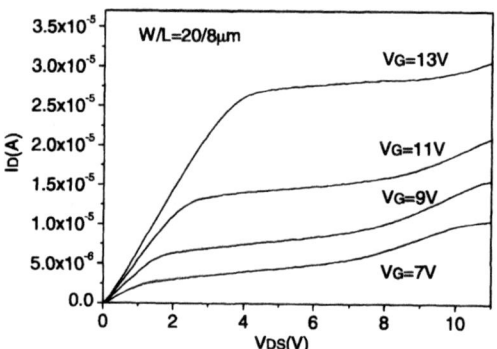

Fig. 11 The output characteristics of the fabricated nc-Si TFT

Figure 12 demonstrates the gate leakage current characteristics of the nc-Si TFT under operation conditions. The gate current level was lower than 250 pA at a high gate voltage of 13 V. These results prove that the SiO$_2$ film used as a gate insulator has a good insulating quality where the SiO$_2$ film is fabricated at 150°C on the nc-Si film in a TFT as well as on a bare Si wafer [19, 20]. This can be attributed to the high quality of ICP-CVD SiO$_2$ film and the non-rough surface of the nc-Si film formed the interface with SiO$_2$. The roughness of the nc-Si film was characterized by AFM in the previous analysis of the nc-Si film.

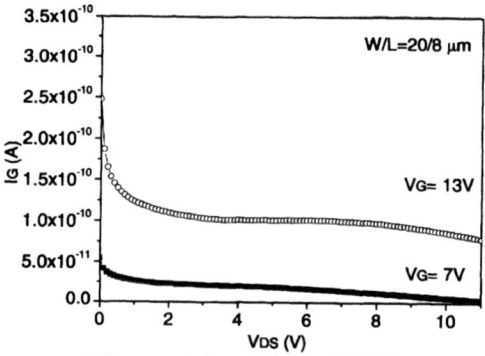

Fig. 12 The gate leakage current characteristics of the nc-Si TFT. V_{DS}: 0~11V, V_G=7 V and 13 V.

CONCLUSIONS

We have successfully fabricated nc-Si TFTs employing ICP-CVD for plastic substrate. The maximum process temperature was 150°C. High mobility of 22.95 cm^2/Vs with a low sub-threshold slope of 0.45 V/dec, V_{TH} of 2.6 V and on/off current ratio larger than 10^6 was obtained without any post-annealing. These characteristics can be attributed the nc-Si film of high crystallinity and SiO$_2$ film with good C-V characteristic. For all that the TFTs were fabricated at 150 °C, the gate leakage current was suppressed. It may be due to the ICP-CVD SiO$_2$ of high breakdown field and good interface roughness between nc-Si/SiO$_2$. The high mobility nc-Si TFT can contribute to the integration of driving circuits in the display panel and fabrication of uniform and stable TFTs in pixel circuits.

REFERENCES

1. P. M. Smith, P. G. Carey, and T. W. Sigmon, *Appl.Phys.Lett.*, Vol.70, No3, pp.342-344,1997.
2. Sang-Myeon Han, Min-Cheol Lee, Moon-Young Shin, Joong-Hyun Park and Min-Koo Han, *Proc. IEEE*, Vol. 93, No. 7, pp. 1297-1305, 2005.
3. N. D. Young, I. D. French, M. J. Trainor, D. T. Murley, D. J. McCulloch, R. W. Wilks, *Int. Display Workshop (IDW) 99*, Sendai, Japan, pp.219-222, 1999.
4. R. G. Stewart, J. Dresner, S. Weisbrod, R. I. Huq and D. Plus, *SID 95 Digest.*, pp.89-92, 1995.
5. A. Salleo and R. A. Street, *J. Appl. Phys.*, 94, pp.471-479, 2003.
6. P. Roca i Cabarrocas, R. Brenot, P. Bulkin, R. Vanderhaghen, and B. Drevillon, *J. Appl. Phys.*, Vol. 86, 12, pp.7079-7081, 1999.
7. Masashi Goto, Hirotaka Toyoda, Masatoshi Kitagawa, Takashi Hirao and Hideo Sugai, *Jpn. J. Appl. Phys.* Vol.36, pp.3714-3720, 1997.
8. J. Hopwood, *Plasma Sources Sci. Technol.* Vol. 1, pp.109-116, 1992.
9. J. H. Keller, *Plasma Sources Sci. Technol.* Vol. 5, pp. 166-172, 1996.
10. Michio Kondo, Hiroyuki Fujiwara, Akihisa Matsuda, *Thin Solid Films*, 430, pp.130-134, 2003.
11. A. T. Voutsasa, M. K. Hatalis J. Boyce and A. Chiang, *J. Appl. Phys.*78 (12), pp.6999-7005, 1995.
12. U. Kroll, J. Meier, and A. Shah, S. Mikhailov and J. Weber, *J. Appl. Phys.* 80 (9), pp. 4971-4975, 1996.
13. Michio Kondo, Makoto Fukawa, Lihui Guo, Akihisa Matsuda, *J, Non-Cryst. Solids* 266-269, pp. 84-89, 2000.
14. John L. Vossen, Werner Kern, *Thin Film Processes II, Academic Press*, 1991.
15. C. Mukherjee, C. Anandan, T. Seth, P. N. Dixit, and R. Bhattacharyya, *J. Vac. Sci. Technol.*, A 17(6), pp. 3202-3208, 1999.
16. Sang-Myeon Han, Joong-Hyun Park, Hee-Sun Shin, Young-Hwan Choi and Min-Koo Han, *Tech. Digest. of International Electron Devices Meeting (IEDM) 2005*, pp. 125-128, 2005.
17. Dieter K. Schroder, *Semiconductor Material and Device Characterization, 2nd edition*, John Wiley and Sons, pp. 337-368, 1998.
18. Sang-Myeon Han, Moon-Young Shin, Min-Cheol Lee, Joong-Hyun Park and Min-Koo Han, *Electrochem. and Solid-State Letters*, 9 (2), H5-H7, 2006.
19. W. K. Choi, K. K. Han, C. K. Choo, W. K. Chim and Y. F. Lu, *J. Appl. Phys.*83 (9), pp.4810-4815, 1998.
20. I-C. Cheng and S. Wagner, *IEE Proc.-Circuits Devices Syst.*, Vol. 150, No. 4, pp339-344, 2003.

Mater. Res. Soc. Symp. Proc. Vol. 989 © 2007 Materials Research Society   0989-A11-02

Self-Aligned Nanocrystalline Silicon Thin-Film Transistor With Deposited n+ Source/Drain Layer

I-Chun Cheng[1,2], and Sigurd Wagner[1]
[1]Department of Electrical Engineering and PRISM, Princeton University, Princeton, NJ, 08544
[2]Graduate Institute of Electro-Optical Engineering and Department of Electrical Engineering, National Taiwan University, Taipei, 10617, Taiwan

ABSTRACT

We demonstrated self-aligned nanocrystalline silicon (nc-Si:H) n-channel thin film transistors (TFTs) with directly deposited n+ layer. The silicon layers were deposited by plasma-enhanced chemical vapor deposition at a substrate temperature of 150°C. The TFTs were made in a staggered top-gate, bottom-source/drain geometry with a seed layer underneath. The self-alignment of top-gate to the bottom-source/drain was achieved by backside exposure photolithography through the glass substrate and the silicon layers, followed by a lift-off process. An extent of gate to source/drain overlap of 1.5 μm was obtained. The self-aligned TFTs have similar characteristics to their non-self-aligned counterpart. This result represents an important step toward directly deposited nc-Si:H TFT backplanes on plastic substrates.

INTRODUCTION

Large-area electronics on clear plastic substrates are under intense research and development for their light weight and flexibility. These attributes are desirable for applications in flexible transmissive or bottom-emitting displays. However, commercially available clear plastic substrates limit the fabrication of thin-film electronic circuits to process temperatures of less than ~ 200°C. Therefore, electronic materials compatible with ultra low process temperatures are needed. Nanocrystalline silicon (nc-Si:H) is a candidate, because it (a) has higher electron and hole field effect mobilities than a-Si:H, (b) is capable of CMOS operation [1], (c) is compatible with ultra-low process temperature [2], and (d) can be made by a method that is fully compatible with industrial a-Si:H technology.

Plastic substrates are restricted to low tolerable process temperatures, but also have higher coefficients of thermal expansion, lower elastic moduli, and lower dimensional stability than conventional glass substrates. These characteristics pose challenges during the fabrication of electronics on plastic. A particularly serious challenge for large area fabrication is poor overlay registration [3], which could be overcome by self-aligning the appropriate layers of the thin-film circuit. Self-aligning the source and drain to the gate also would minimize the feedthrough capacitance, and would facilitate the fabrication of the short-channel TFTs that will be needed for high-speed circuits. Self-aligned *amorphous-silicon* TFTs with bottom gate geometry have been previously reported [4-7]. Here we demonstrate that self-alignment is possible also in the fabrication of *nanocrystalline* TFTs with top-gate geometry, and can be combined with lift-off patterning of the top metal contacts. This approach is particularly interesting because it is fully compatible with present a-Si:H technology.

EXPERIMENT

All experiments described in this paper were conducted on 1.1-mm thick Corning 1737 glass substrates to develop the standard process, at a substrate temperature that is realistically close to what most clear plastic substrates can tolerate.

A. Deposition of nanocrystalline silicon films

The intrinsic nc-Si:H films, including both seed and channel layers, are deposited by plasma enhanced chemical vapor deposition (PECVD) at 80MHz VHF excitation frequency from a gas mixture of hydrogen, silane and dichlorosilane (DCS) at a substrate temperature of 150°C. The deposition parameters are a gas pressure of 500 mTorr, an absorbed power of 86 mW/cm^2, and hydrogen dilution ratios $R = [H_2]/[SiH_4+SiCl_2H_2]$ of 20 for the seed layer and 30 for the channel layer. The high hydrogen dilution ratio and the VHF excitation frequency ensure high film crystallinity [8-9]. The addition of a small amount of DCS compensates the film from n-type to intrinsic [10].

Phosphine is the dopant gases for n$^+$ source/drain (S/D) layer deposition. The S/D deposition parameters are a pressure of 500 mTorr, an absorbed power of 103 mW/cm^2, and hydrogen dilution ratio of $R = [H_2]/[SiH_4+PH_3] = \sim 100$. Because high concentrations of dopants are detrimental to film crystallinity, especially at low substrate temperature, large hydrogen dilution is needed to obtain the nanocrystalline phase [11]. A conductivity of ~ 20 S·cm^{-1} is obtained for n$^+$ films deposited at 150°C.

B. TFT geometries and fabrication sequences

The nc-Si:H TFTs are fabricated in a staggered top-gate, bottom-source/drain geometry on top of a nc-Si:H seed layer. We adopted the top-gate geometry to take advantage of the high carrier mobilities available at the top of nc-Si:H films, a consequence of their structural evolution in the growth direction. Bottom source/drains are introduced to circumvent the over-etch problem that is usually encountered in the conventional coplanar top-source/drain geometry, a result of the small etching selectivity between the doped source/drain layer and the intrinsic channel layer. A ~50-nm thick intrinsic nc-Si:H seed layer is deposited first, beneath the whole TFT structure, to promote rapid nucleation of the channel layer.

The fabrication sequence is shown in Figure 1. First the ~50-nm thick intrinsic nc-Si:H seed layer is deposited, followed by ~80-nm thick evaporated Cr for bottom S/D electrodes, and a ~50-nm thick PECVD n$^+$ nc-Si:H S/D contact layer. The source/drain patterns are defined by first, dry etch of the n$^+$ nc-Si:H layer and second, wet etch of the Cr layer. A ~50-nm thick intrinsic nc-Si:H and a ~300-nm thick SiO$_2$ layer are deposited as channel and gate dielectric layers, respectively. The thickness of the channel layer is kept small to hold the OFF current and the channel surface roughness to low values (to reduce carrier scattering), but is sufficient to let the crystallites coalesce to form a contiguous crystalline layer [12]. After the contact holes are opened by wet chemical etch and dry plasma etch, the self-aligned photoresist pattern is defined by exposure from the back-side of the sample, with the Cr source/drain contacts functioning as exposure masks. Then a ~150-nm Al layer is thermally evaporated and lifted-off in an ultrasonic acetone bath. This process aligns the gate metal with the doped nc-Si:H source/drain.

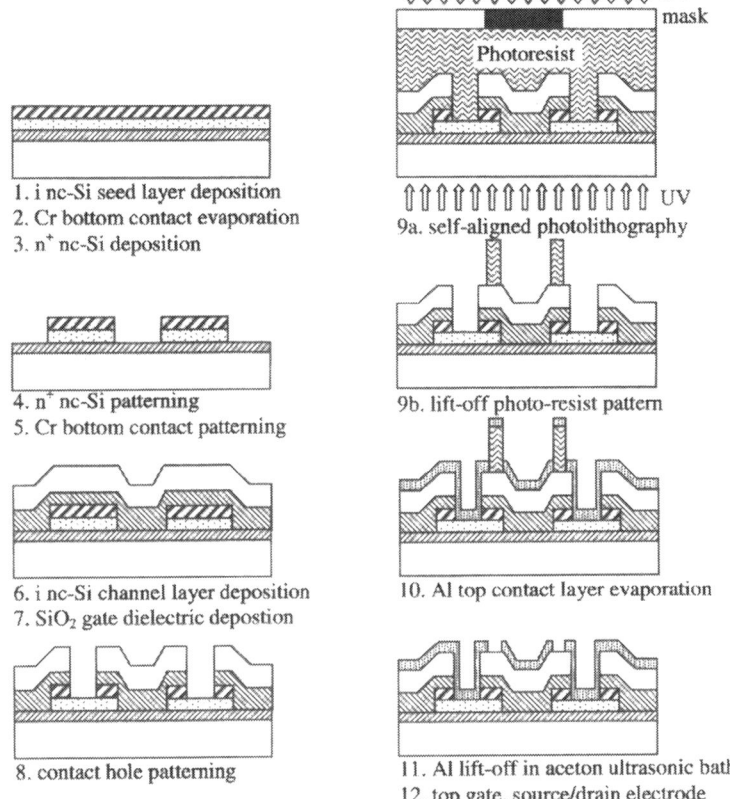

1. i nc-Si seed layer deposition
2. Cr bottom contact evaporation
3. n^+ nc-Si deposition

4. n^+ nc-Si patterning
5. Cr bottom contact patterning

6. i nc-Si channel layer deposition
7. SiO_2 gate dielectric depostion

8. contact hole patterning

9a. self-aligned photolithography

9b. lift-off photo-resist pattern

10. Al top contact layer evaporation

11. Al lift-off in aceton ultrasonic bath
12. top gate, source/drain electrode

Figure 1. Fabrication sequence for bottom-gate, top-source/drain nc-Si:H TFTs with the gate self-aligned to the n^+ nc-Si:H source/drain.

RESULTS AND DISCUSSION

A. Extent of gate-source/drain overlap

The extent of gate-source/drain overlap is determined by (a) the over-etch of the Cr bottom-source/drain contacts, (2) the over-exposure time and over-development of the photoresist. We intentionally created a large overlap for easier study. The backside exposure photolithography in step 9 of the process sequence of Figure 1 requires sufficient optical transmission through the substrate and the nc-Si:H seed / nc-Si:H channel / SiO_2 gate dielectric stack at a wavelength available for photoresist exposure. In our particular design with 50-nm thick nc-Si:H seed and 50-nm thick nc-Si:H channel, the exposure time was chosen to be ~ 8 minutes at a UV-light intensity of 2 mW/cm^2. By using commercially available high-intensity UV sources this time could be reduced to a few seconds. The resulting gate-source/drain overlap is ~ 1.5-μm out of 20, 40, and 80-μm channel lengths, as shown in the atomic force micrograph of Figure 2(b).

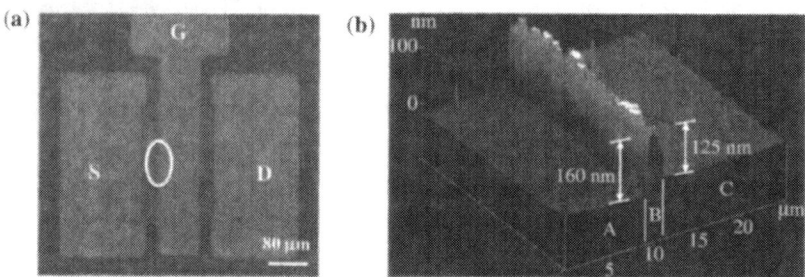

Figure 2. (a) Optical micrograph of a self-aligned nc-Si:H TFT and (b) atomic force micrograph of the source/gate edge, circled in (a). In (b), region B is the source/gate overlap region. The layer sequence in region A is: seed nc-Si + Cr + n$^+$ nc-Si + channel nc-Si + SiO$_2$, in region B: seed nc-Si + space due to Cr undercut + n$^+$ nc-Si + channel nc-Si + SiO$_2$ + Al, and in region C: seed nc-Si + channel nc-Si + SiO$_2$ + Al. The gate-source/drain overlap is ~ 1.5-μm and the thicknesses of Cr, n$^+$ nc-Si, and Al are 85 nm, 40 nm and 160 nm, respectively.

B. TFT characteristics

The TFT characteristics are evaluated with an HP 4155A parameter analyzer. The transfer and output characteristics of nc-Si:H TFTs without and with self-alignment process are shown in Figure 3 (a) and (b), respectively. The electron field-effect mobilities are calculated from the slopes of the transfer characteristics. The self-aligned nc-Si:H TFT has an electron mobility of ~ 41 cm^{2}V^{-1}s^{-1} and ON/OFF ratio of ~ 10^6, while its non-self-aligned counterpart has an electron mobility of ~ 28 cm^{2}V^{-1}s^{-1} and ON/OFF ratio of ~ 10^5. The lower carrier mobility and higher off current observed in the non-self-aligned TFTs may due to the variation from run to run of the experimental fabrication process. The high threshold voltage of ~ 17V to 20V, the non-ideal turn-on regime, and the shift observed in the OFF range are associated with the non-optimized gate SiO$_2$ deposited at ultralow temperature.

Overall the self-aligned nc-Si:H TFTs have similar characteristics to that of their non-self-aligned counterpart, which indicates that the UV exposure through the nc-Si:H channel does not cause significant degradation.

SUMMARY

We fabricated the top-gate nc-Si:H n-channel TFTs with the gate self-aligned to the source and drain. The self-alignment is obtained by combining backside exposure photolithography with lift-off patterning of the top metal contacts. Over-etch of the Cr bottom-source/drain contacts combined with over-exposure and over-development of the photoresist pattern determines the extent of the gate-source/drain overlap. Electron mobilities of

~ 40 cm^{2}V^{-1}s^{-1} are obtained in TFTs processed at 150°C. This result presents a step toward a high-performance TFT technology based on directly-deposited (in contrast to crystallized) silicon films on low-temperature substrates.

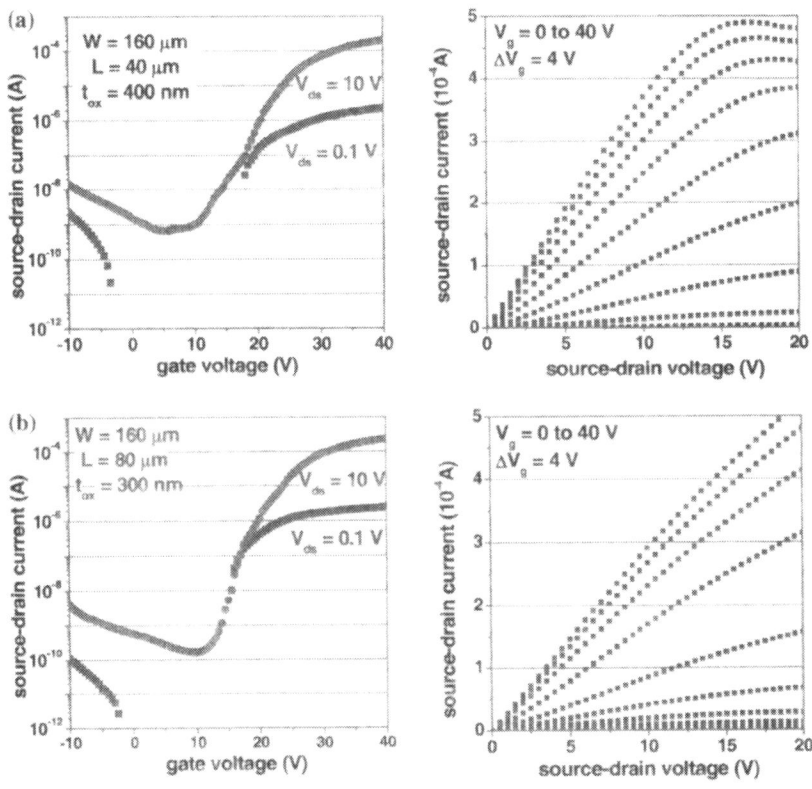

Figure 3. Transfer characteristics of a staggered top-gate, bottom source/drain nc-Si TFT (a) without self-alignment, and (b) with gate self-aligned to source and drain.

ACKNOWLEDGMENTS

We gratefully acknowledge early support from DARPA's High Definition Systems Program, from the New Jersey Commission on Science and Technology, and from the Princeton Plasma Physics Laboratory under a PPST Fellowship.

REFERENCES

1. Y. Chen and S. Wagner, Appl. Phys. Lett., 75, 1125 (1999).
2. I-C. Cheng and S. Wagner, Appl. Phys. Lett., 80, 440 (2002).
3. I-C. Cheng, A. Kattamis, K. Long, J. C. Sturm, and S. Wagner, J. Soc. Info. Disp., vol. 13/7, 563 (2005).
4. Y. Kuo, J. Electrochem. Soc, 139, 1199 (1992).
5. D. B. Thomasson and T. N. Jackson, IEEE Electron Device Lett., 19, 124 (1998).
6. C. S. Yang, W. W. Read, C. Arthur, E. Srinivasan, and G. N. Parsons, IEEE Electron Device Lett., 19, 180 (1998).
7. I-C. Cheng, A. Z. Kattamis, K. Long, J. C. Sturm, and S. Wagner, IEEE Electron Device Lett., 27, 166 (2006).
8. M. Tzolov, F. Finger, R. Carius and P. Hapke, J. Appl. Phys., 81 (11), 7376 (1997).
9. M. Mulato, Y. Chen, S. Wagner and A. R. Zanatta, J.Non-Cryst. Solids, 266-269, 1260 (2000).
10. R. Platz and S. Wagner, Appl. Phys. Lett., 73 (9), 1236 (1998).
11. P. Alpuim, V. Chu, and J. P. Conde, J. Vac. Sci. Technol. A, 19, 2328 (2001).
12. I-C. Cheng, S. Allen, and S. Wagner, J. Non-Crystal. Solids, 338-340, 720 (2004).

Mater. Res. Soc. Symp. Proc. Vol. 989 © 2007 Materials Research Society 0989-A11-03

Contact Effects in High Mobility Microcrystalline Silicon Thin-Film Transistors

Kah Yoong Chan[1,2], Eerke Bunte[2], Helmut Stiebig[2], and Dietmar Knipp[1]
[1]School of Engineering and Science, Jacobs University Bremen, Campus Ring 1, Bremen, 28759, Germany
[2]Institute of Photovoltaics, Research Center Jülich, Jülich, 52425, Germany

ABSTRACT

Microcrystalline silicon (μc-Si:H) has recently been proven to be a promising material for thin-film transistors (TFTs). We present μc-Si:H TFTs fabricated by plasma-enhanced chemical vapor deposition at temperatures below 200 °C in a condition similar to the fabrication of amorphous silicon TFTs. The μc-Si:H TFTs exhibit device mobilities exceeding 30 cm^2/Vs and threshold voltages in the range of 2.5V. Such high mobilities are observed for long channel devices (50-200 μm). For short channel device (2 μm), the mobility reduces to 7 cm^2/Vs. Furthermore the threshold voltage of the TFTs decreases with decreasing channel length. A simple model is developed, which explains the observed reduction of the device mobility and threshold voltage with decreasing channel length by the influence of drain and source contacts.

INTRODUCTION

With the advance of flat panel display technologies, an increasing demand occurred for thin-film transistors (TFTs). To date, amorphous silicon (a-Si:H) has established itself as an inexpensive and reliable technology for display backplanes. A-Si:H TFTs are mainly applied as pixel switches in active-matrix liquid crystal displays (LCDs) even though they exhibit relatively low (electron) mobility of 0.5 to 1 cm^2/Vs. However, their performance is not good enough to operate electronic circuits at frequencies well above video rate, for examples column and row drivers for LCDs or radio frequency identification tags (RFID tags). Furthermore, the hole mobility of a-Si:H TFTs is much lower than the electron mobility. Therefore, complimentary metal oxide semiconductor circuits cannot be realized. Finally, biasing of a-Si:H TFTs over a longer period of time leads to a shift of the threshold voltage. This effect complicates the realization of circuits like pixel drivers for organic light-emitting diode (oLED) displays or column multiplexer and row shift register circuits [2]. Polycrystalline silicon (poly-Si) TFTs exhibit high electron and hole mobilities above 20 cm^2/Vs and stable device performance, which recommend the technology for advanced applications like electronic circuits on glass, column and row drivers for displays, pixel switches for projector displays and pixel engines for oLED technology. However, poly-Si technology is expensive due to the need of high deposition temperature and/or crystallization steps.

In recent years a lot of alternative TFT technologies including organic and polymeric TFTs have been developed and are of interest for large area electronic applications. The performance of small molecule based TFTs is comparable to the a-Si:H TFTs. In particular pentacene has proven to be a very promising material [3,4]. However, the materials are sensitive to environmental conditions. For example the exposure of the devices to oxygen and moisture leads to an unintentional doping of the films, which affects the device characteristic [5].

A promising material for applications in thin-film electronics is hydrogenated microcrystalline or nanocrystalline silicon (μc-Si:H or nc-Si:H). μc-Si:H consists of both amorphous and crystalline silicon phase. μc-Si:H TFTs might be able to combine the two worlds of low temperature a-Si:H processes with the superior device properties of poly-Si transistors [6-8]. A summary of the different TFT technologies in terms of carrier mobility, stability, cost and maturity of the technology is shown in Table I.

Table I Summary of different TFT technologies.

Technology	Organic TFTs	A-Si:H TFTs	Poly-Si TFTs	μc-Si:H TFTs
Carrier Mobility	Low	Low	High	High
Stability	Low	Low	High	High (?)
Cost	Very low	Low	High	Low
Maturity of technology	Low (Research)	Very high (standard technology)	High	Low (Research)

EXPERIMENT

A schematic cross-section of the μc-Si:H TFT is shown in figure 1. The drain and source contacts were realized by chromium with 30 nm thickness. Afterwards, a highly doped n-type μc-Si:H film is deposited by plasma enhanced chemical vapor deposition (PECVD) at substrate temperature of 190 °C using a gas mixture of silane (SiH_4), phosphine (PH_3) and hydrogen (H_2). The n-layer exhibits a dark conductivity of 10 S/cm and an activation energy of 17 meV. Subsequently a 100 nm thick μc-Si:H intrinsic (i) layer was deposited at substrate temperature of 160 °C using a gas mixture of SiH_4 and H_2. The film was prepared using a silane concentration ($SiH_4/(H_2+SiH_4)$) of 0.75%. The n- and i-layers were prepared at radio frequency of 13.56 MHz. The i-layer was grown in the high pressure and high power regime, which facilitates high deposition rates [9,10]. The deposition pressure was 1330 Pa and the power density was 0.3 W/cm^2. For these parameters a deposition rate of 0.3 nm/s was achieved. Raman measurement of the i-layer shows that the material exhibits a crystalline volume fraction of 55%. Under this condition, the highest carrier lifetime of μc-Si:H as a function of crystalline volume fraction was observed [11]. More details concerning the material properties are given elsewhere [12].

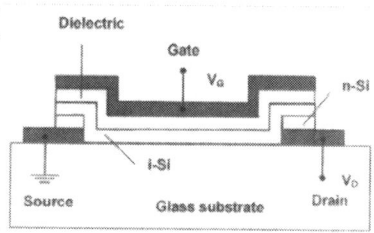

Figure 1. Schematic cross-section of a top-gate staggered μc-Si:H TFT.

Prior to the deposition of gate dielectric the sample was subjected to a hydrofluoric acid (HF) dip to clean and hydrogenate the i-layer surface. Improved device performance has been observed for μc-Si:H TFTs subjected to the HF treatment. Figure 2 shows the measured transfer characteristics for as-deposited TFTs, which were prepared with HF (closed symbols) or without HF treatment (open symbols). For positive gate voltages the drain current of the HF treated transistor is approximately one order of magnitude higher than the transistor without HF treatment. Furthermore, the threshold voltage of the HF treated transistor is considerably lower than the threshold voltage of the transistor without HF treatment.

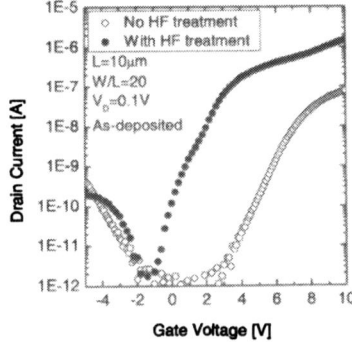

Figure 2. Transfer characteristics of as-deposited (unoptimized) μc-Si:H TFTs with channel width (W) of 200 μm and channel length (L) of 10 μm. Solid (open) circles denote the transfer curves measured for TFT subjected to (no) HF treatment prior to the deposition of gate dielectric.

Silicon oxide (SiO_2) was used as a gate dielectric to minimize the defect density at the channel/dielectric interface [13,14]. The SiO_2 was prepared by PECVD at 150 °C using a gas mixture of SiH_4, nitride oxide and helium. Finally the gate electrode was formed by an aluminium film. The devices were fabricated on Corning 1737 glass. Transistors with channel length and width ranging from 2 μm to 200 μm and 200 μm to 1000 μm were fabricated, respectively using a simple two-mask photolithography process. In order to improve the device behavior all transistors were annealed at an elevated temperature of 150 °C for 30 minutes under ambient conditions. A detailed discussion of the influence of thermal annealing on the device characteristics is given elsewhere [15].

RESULTS AND DISCUSSION

In the following the device behaviour of the μc-Si:H TFTs will be discussed. Figure 3 shows the transfer characteristics of a TFT with a channel length (L) of 10 μm and a channel width (W) of 1000 μm. The transfer curves were measured for drain voltage, V_D, of 0.1V and 1V. The applied gate voltage, V_G, leads to an exponential increase of the drain current, I_D, in the below threshold region followed by a linear increase of the I_D at higher gate voltages. The exponential increase of the I_D in the below threshold region can be described by:

$$I_D \propto \frac{W}{L} \cdot \mu \cdot \exp\left(\frac{C_G \cdot V_G}{q \cdot N_T \cdot d_s \cdot k_B \cdot T}\right), \tag{1}$$

where W and L is the channel width and channel length, respectively. μ is the device mobility, C_G is the gate capacitance, and q, N_T, d_S, k_B and T is the electron charge, the defect density in the μc-Si:H, the μc-Si:H channel layer thickness, the Boltzmann constant and the absolute temperature, respectively. The I_D in the above threshold region is described by:

$$I_D = \mu \cdot C_G \cdot \frac{W}{L}\left(V_G - V_T - \frac{V_D}{2}\right) \cdot V_D, \tag{2}$$

where V_T is the threshold voltage of the device. A device mobility of 13 cm²/Vs and a threshold voltage of 2.2V were extracted from the transfer characteristic measured for V_D=1V. The on/off ratio of the TFT for low drain voltages is larger than 10^6. The gate leakage current of the transistor is 4 orders of magnitude lower than the corresponding drain current at high gate voltages, which demonstrates the applicability of the low-temperature (150 °C) deposited SiO_2. The determined device mobility is significantly higher than the mobility of conventional a-Si:H TFTs (≤ 1 cm²/Vs) prepared by PECVD.

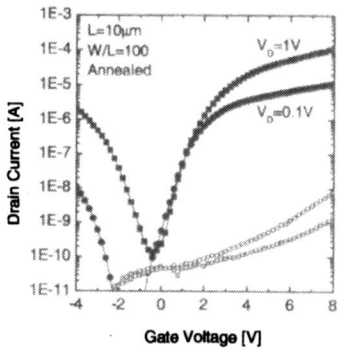
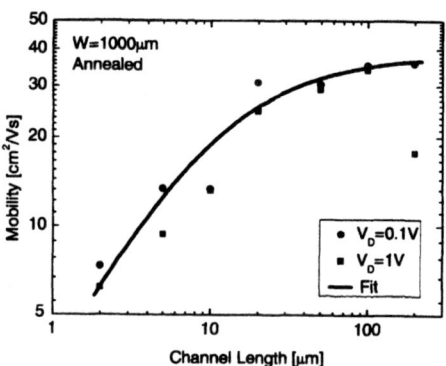

Figure 3. Transfer characteristics of an annealed μc-Si:H TFT with channel width (W) of 1000 μm and channel length (L) of 10 μm. The transfer curves were measured for V_D=0.1V and 1V. Open circles and squares denote the corresponding gate leakage currents.

Figure 4. Measured device mobility as a function of channel length.

Further measurements of devices with different geometries show that the extracted mobility strongly depends on the channel length. The device mobility as a function of the channel length is depicted in figure 4. The extracted device mobility is plotted for V_D=0.1V and V_D=1V. For long channel transistors (L>50 μm) a device mobility exceeding 30 cm²/Vs is obtained, whereas the mobility of short channel transistor (L=2 μm) decreases to 7 cm²/Vs. The reduction of the device mobility for short channel transistors can be explained by the influence of the drain and source contacts on the electric field distribution across the channel of the transistor. The non-ideal contact behaviour causes a voltage drop at the drain and source contacts. The influence of the contacts on the device behaviour is more pronounced for short channel transistors, since the drain current is proportional to V_D/L. In the following it is assumed that the contacts can be described by ohmic contacts. Under this condition, the equivalent circuit of the TFTs in linear regime is illustrated in figure 5, which shows that the drain and source contact resistance is connected in series with channel resistance. Accordingly the drain voltage, V_D, in equation 2 can be replaced by 'V_D-$I_D R_C$', where 'V_D-$I_D R_C$'=V_{D0}, which is the actual voltage drop across the channel. R_C is the resistance of the drain and source contacts. Considering the non-ideal contact behaviour, the following expression for the mobility can be derived:

$$\mu \approx \mu_0 \cdot \frac{L}{L + \mu_0 \cdot W \cdot C_G \cdot R_C (V_G - V_T)}, \tag{3}$$

where μ_0 is the electron mobility of the microcrystalline channel material which is not affected by the contact resistance. μ represents the device mobility which is extracted from the measured transfer characteristics and affected by the influence of the contact resistance. Equation 3 allows for the fit of the experimental data in figure 4. A good agreement between the experimental data and the fit was achieved, nearly independent from V_D $(0.1V<V_D<1V)$. We extracted an electron mobility, μ_0, of 38 cm^2/Vs and a contact resistance, R_C, of 6.7 kΩ from the fit. The R_C extracted using the developed model is in good agreement with the R_C determined by Transmission Line Model (TLM), which describes the contact resistance as a gate voltage dependent parameter [16]. Using TLM, we obtained a R_C ranging from 9.8 kΩ to 1.5 KΩ with increasing gate voltage from 4V to 8V.

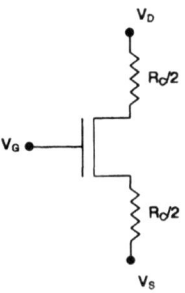

Figure 5. A simplified equivalent circuit of TFTs in the linear regime.

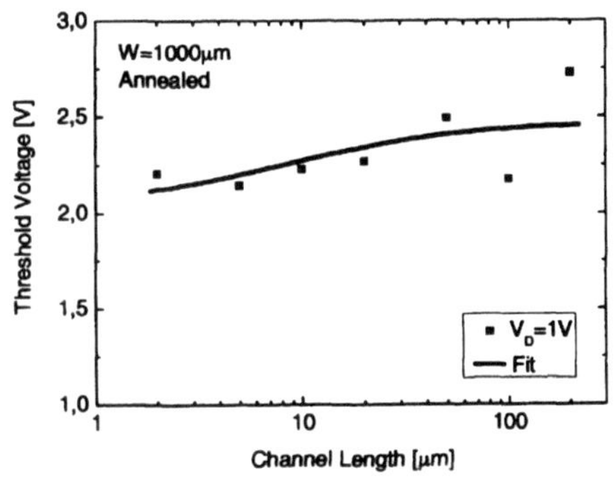

Figure 6. Measured device threshold voltage as a function of channel length for annealed μc-Si:H TFT.

The device threshold voltage as a function of the channel length is given in Figure 6. The data were extracted from the transfer curves measured at $V_D=1V$. The device threshold voltage apparently decreases with decreasing channel length. The change of the device threshold voltage as a function of the channel length can be explained by the voltage drop at the contacts. Taking the iinfluence of the contacts into account the following expression for the device threshold voltage, V_T, can be derived from equation 2 by substituting 'V_D-$I_D R_C$' for V_D:

$$V_T = V_{T0} - \frac{I_D \cdot R_C}{2}, \tag{4}$$

where V_{T0} is the threshold voltage of the microcrystalline transistor which is not affected by the contact resistance. V_T denotes the device threshold voltage which is extracted from the measured transfer characteristics and affected by the influence of the drain and source contact resistance. Equation 4 shows that the V_{T0} is reduced by '$I_D R_C/2$'. The term '$I_D R_C/2$' represents the voltage drop at the drain or source contact. The experimental data in figure 6 was fitted according to equation 4. A good agreement between the experimental data and fit was achieved. Therefore, the channel length dependence of the device threshold voltage, V_T, can also be explained by the drain and source contact effects. Depending on the contact resistance and the channel length the device threshold voltage, V_T, converges towards the actual threshold voltage, V_{T0}, of the transistor. From the fit in figure 6 we extracted a V_{T0} value of 2.5V. The contact resistance, R_C, extracted in the fit is similar to the R_C extracted from the fit of the device mobility.

CONCLUSIONS

The realized μc-Si:H TFTs exhibit high mobilities exceeding 30 cm²/Vs and threshold voltages in the range of 2.5V. The experimental results reveal that the device parameters like mobility and threshold voltage strongly depend on transistor channel length. The device mobility and threshold voltage decrease for short channel transistors. A detailed analysis reveals that the channel length dependence of the device parameters can be attributed to the influence of the drain and source contact resistance. A simple model is developed to account for the contact effects. The extracted series resistance is 6.7 kΩ and its impact on the device operation is responsible for the shift of the directly deconvoluted (device) mobility and directly deconvoluted (device) threshold voltage to smaller values.

ACKNOWLEDGMENTS

The authors like to acknowledge S. Bunte and J. Mohr (IBN-PT) for preparation of the PECVD SiO₂ and performing the HF dip, respectively, P. Foucart, M. Hülsbeck, J. Kirchhoff, T. Melle, S. Michel, and R. Schmitz for technical assistances and R. Carius, M. v. d. Donker, A. Gordijn, D. Hrunski, and B. Rech for helpful discussions.

REFERENCES

1. T. Tsukada, *Technology and Applications of Amorphous Silicon, Springer Series in Material Science, 37*, edited by R. A. Street (Springer-Verlag, Berlin, Germany, 2000).
2. I. D. French, S. C. Deane and P. R. I. Cabarrocas, *Asia Display, IDW'01*, 367 (2001).
3. C. D. Dimitrakopoulos and P. R. L. Malenfant, *Adv. Mater. (Weinheim, Ger.)* 14, 99 (2002).
4. T. W. Kelley, P. F. Baude, C. Gerlach, D. E. Ender, D. Muyres, M. A. Haase, D. E. Vogel and S. D. Theiss, *Chem. Mater.* 16, 4413 (2004).
5. D. Knipp, T. Muck, A. Benor and V. Wagner, *J. of Non-Cryst. Solids* 352, 1774 (2006).
6. I. -C. Cheng and S. Wagner, *Appl. Phys. Lett.* 80, 440 (2002).
7. C. -H. Lee, A. Sazonov and A. Nathan, *Appl. Phys. Lett.* 86, 222106 (2005).
8. A. Saboundji, N. Coulon, A. Gorin, H. Lhermite, T. Mohammed-Brahim, M. Fonrodona, J. Bertomeu and J. Andreu, *Thin Solid Films* 487, 227 (2005).
9. L. Guo, M. Kondo, M. Fukawa, K. Saitoh and A. Matsuda, *Jpn. J. Appl. Phys.* 37, L1116 (1998).
10. B. Rech, T. Roschek, T. Repmann, J. Müller, R. Schmitz and W. Appenzeller, *Thin Solid Films* 427, 157 (2003).
11. T. Brammer and H. Stiebig, *J. Appl. Phys.* 94, 1035 (2003).
12. K. -H. Jun, R. Carius and H. Stiebig, *Physical Review B* 66, 1153011 (2002).
13. Y. Ma, T. Yasuda and G. Lucovsky, *J. Vac. Sci. Technol.* A 11, 952 (1993).
14. S. W. Hsieh, C. Y. Chang and S. C. Hsu, *J. Appl. Phys.* 74, 2638 (1993).
15. K.-Y. Chan, E. Bunte, H. Stiebig and D. Knipp, *J. Appl. Phys.* 101, 074503 (2007).
16. S. Luan and G. W. Neudeck, *J. Appl. Phys.* 72, 766 (1992).

Images and Sensors

Mater. Res. Soc. Symp. Proc. Vol. 989 © 2007 Materials Research Society   0989-A12-01

PECVD GROWN p-i-n Si AND Si,Ge THIN FILM PHOTODETECTORS FOR INTEGRATED OXYGEN SENSORS

Debju Ghosh[1,2], Ruth Shinar[2], Vikram L. Dalal[1,2], Zhaoqun Zhou[3], and Joseph Shinar[1,3]

[1]Electrical and Computer Engineering, Iowa State University, Ames, IA, 50011
[2]Microelectronics Research Center, Iowa State University, Ames, IA, 50011
[3]Ames Laboratory-USDOE and Department of Physics, Iowa State University, Ames, IA, 50011

ABSTRACT

Recent efforts to advance photoluminescence (PL)-based oxygen sensors have focused on developing compact, field-deployable devices. This has led to organic light emitting device (OLED)-based sensors with a structurally integrated [OLED excitation source]/[sensing film] module. To additionally integrate a photodetector (PD), PECVD for fabrication of thin-film p-i-n and n-i-p Si- and Si,Ge-based PDs was employed. O_2 concentrations are advantageously determined by monitoring the effect of O_2 on shortening the PL decay time τ of an oxygen-sensitive dye, rather than on quenching its PL intensity. This approach, which employs pulsed OLEDs, eliminates the need for frequent sensor calibration, minimizes issues associated with background light, and eliminates the need for optical filters, which lead to bulkier sensors. However, it requires PDs with response times significantly shorter than τ. Therefore, the development of thin-film PDs focused on decreasing their response time, and understanding the factors affecting it. In this paper we show that boron diffusion during growth from the p^+ to the i layer increases the response time of PECVD grown p-i-n PDs. Incorporating a SiC buffer layer and fabricating superstrate structures, where the $p+$ layer is grown last, decrease it. Additionally, ECR fabricated PDs show a slower response in comparison to VHF PECVD-grown PDs.

INTRODUCTION

Photoluminescence (PL)-based oxygen sensors are attractive due to attributes such as high sensitivity, specificity, fast response, long-term stability, and low-maintenance. Such sensors are based on quenching of the PL intensity I and shortening of the PL decay time τ of an oxygen-sensitive dye, typically embedded in a thin sol-gel or polymeric film, due to collisions with O_2 [1-4]. Oxygen-sensors also serve as a basis for monitoring other analytes such as glucose, lactate, ethanol, and cholesterol, by monitoring oxygen consumption during the oxidation reactions of the above-mentioned analytes in the presence of an appropriate specific oxidase enzyme [5-7]. Monitoring O_2 via its effect on τ rather than I is advantageous, as it eliminates the need for frequent sensor calibration, and is unaffected by minor changes in the background light, sensor film, or excitation source [1-4].

Generally, field-deployability of PL-based chemical and biological sensors is limited due to issues related to size, ease of fabrication, and consequently cost, and calibration/maintenance. Additionally, the sensors are often restricted to monitoring a single analyte. The organic light emitting device (OLED)-based sensing platform presents an opportunity to alleviate these issues, as the compact OLED pixel array, which serves as the excitation source, can be structurally integrated with the sensing film in a uniquely simple approach, resulting in an inexpensive module that is ~2 mm thick [3]. Moreover, OLEDs are flexible in design, and their pixel outline together with the ability to individually address the pixels enable the use of the OLED platform for simultaneous detection of multiple analytes in mixtures [7].

In this paper we describe an approach to further advance the sensor's structural integration, by additionally integrating the photodetector (PD), which is based on amorphous thin films of Si and Si,Ge. In a previous report, initial results regarding the three component OLED/sensor film/thin film p-i-n PD integration were presented [8]. It was shown that the a-Si,Ge:H PD tuned to the detection wavelength is usable for monitoring the effect of the O_2 concentration on the PL intensity. The detection sensitivity, however, was very low due to the high dark current of the PD and the long-wavelength tail of the electroluminescence (EL) of the OLED. In this work, improved PDs resulted in a significantly improved detection sensitivity.

As mentioned, monitoring the O_2 concentration via its effect on τ is advantageous over I. In particular, as the OLED is pulsed when using this mode of operation and τ is monitored when the OLED is off, the issue of the background due to the long-wavelength OLED EL tail is eliminated. To this end, the development of thin-film PDs presented in this paper focused on shortening their response time, and understanding the factors affecting it

EXPERIMENTAL

The oxygen-sensitive dye used in this study was Pt octaethylporphyrin (PtOEP), which has absorption and emission bands peaking at ~535 and ~635 nm, respectively. PtOEP is attractive due to its long unquenched τ value of ~100 µs, which ensures high detection sensitivity. The dye was embedded in a polystyrene film.

The OLED used for excitation of the dye was based on tris(quinolinolate) aluminum (Alq$_3$), whose EL, peaking at ~535 nm, matches the PtOEP absorption band. The OLED arrays were fabricated by thermally evaporating the organic layers on indium tin oxide (ITO)-coated glass. The organic layers consisted of hole and electron injecting and transporting layers, and an emission layer. The OLEDs were prepared as encapsulated arrays of ~2×2 mm^2 pixels resulting from mutually perpendicular stripes of the etched ITO anode and evaporated Al cathode. Continuous or pulsed OLEDs were operated in a forward bias of 15 - 30 V; pulse widths were 50 – 300 µs and repetition rates were 20 - 300 Hz. Additional details regarding OLED fabrication and operation are presented elsewhere [3,6].

The PDs were fabricated using VHF PECVD at a pressure of 50 mTorr or an ECR reactor at 10 and 15 mTorr. The p-i-n structures were fabricated on an ITO-coated glass substrate, protected with a thin ZnO layer. The composition and thickness of the p- and i-layers were tuned to match the emission band of the oxygen-sensitive dye and to reduce the OLED background.

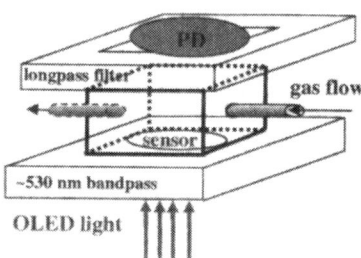

The PDs were evaluated by measuring their quantum efficiency (QE) and response time. The sensor configuration when operated in the "front detection" mode, where the PD is in front of the analyte cell, is shown in Fig. 1.

Figure 1. Sensor configuration in the "front detection" mode. The drawing is not to scale.

270

RESULTS AND DISCUSSION

The PDs were tuned to a maximal response at ~635 nm, where the dye emission peaks, and a minimal response at ~535 nm, where the OLED emission peaks. The a-Si:H-based PDs show a maximal response at ~560 nm, as seen in Fig. 2 for ECR grown devices. To shift the maximal response to longer wavelengths, the thickness of the p layer in the p-i-n device can be increased. However, by increasing the p layer thickness from 0.1 to 0.3 μm, the absolute QE decreased, as the light absorbed in the i-layer decreased. Alternatively, Ge can be incorporated in the films, or nanocrystalline-based PDs can be used. Incorporating Ge red-shifted the PD response band, as shown in Figs. 2 and 3. However, the addition of Ge, in particular in the p-layer, resulted in an increase in the dark current of the devices. The dark current was reduced by decreasing the Ge level in p layer and instead, increasing it in the i layer; the QE, however, decreased.

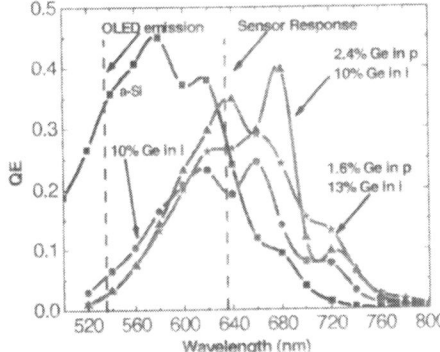

Figure 2. Effect of Ge incorporation in ECR-grown devices on the QE; the percentages are of gas-phase germane.

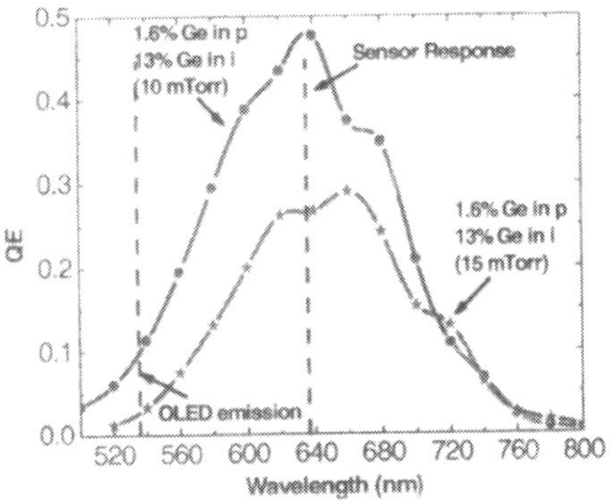

Figure 3. Effect of the deposition pressure on the QE; the percentages are of gas-phase germane.

Reducing the deposition pressure in the ECR reactor from 15 to 10 mTorr resulted in an increase of the QE up to ~700 nm (see Fig. 3) without a significant change in the dark current. It is believed that higher ion bombardment at 10 mTorr results in improved films. However, the response of the PD to the long-wavelength tail of the OLED's EL (i.e., the background) also increased.

The dark current of the PDs was also reduced by the addition of a SiC layer at the p-i interface and further by mesa etching, as shown in Fig. 4.

Fabrication of nanocrystalline Si-based PDs resulted in a broader response, and consequently an increase in the responsivity at 535 nm; therefore, studies focused on a-Si,Ge-based devices.

Using the structure shown in Fig. 1, where the PD was a VHF-grown PD (see below), the detection sensitivity, defined as the ratio of the signal intensities or the τ values at 100% nitrogen to 100% oxygen, was ~7 using the intensity mode. Using a photomultiplier tube (PMT), the sensitivity using the decay time mode is typically ~40. Though the detection sensitivity using the fully integrated sensor is smaller than that obtained when using the PMT, it still demonstrates the viability of the approach. As mentioned, the reduced sensitivity is a result of the dark current of the thin-film based PD and the long-wavelength tail of the OLED EL.

Fig. 5 shows the emission spectrum of the Alq$_3$-based OLEDs. As seen, the emission band is broad, extending to ~650 nm. One approach to eliminate the EL background issue is to operate the sensor in the τ mode, where the OLED is pulsed and the analyte is monitored from data obtained during the off period of the OLED. To operate in this mode, a fast PD is required.

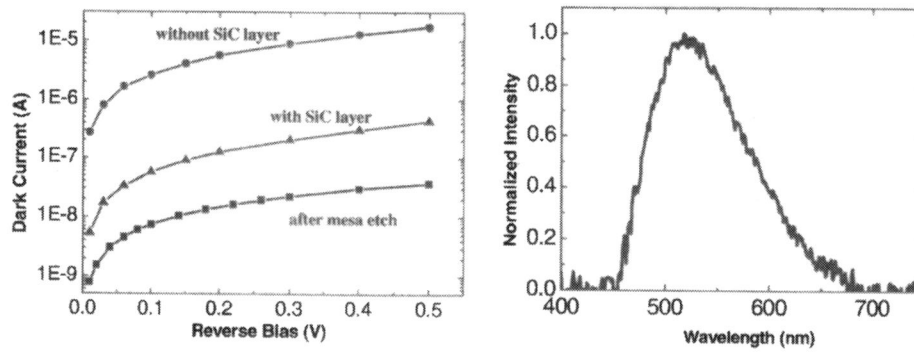

Figure 4. The dark current of an ECR-grown a-Si,Ge-based photodetector.

Figure 5. Emission spectrum of the Alq$_3$ OLED.

Frequency dependence of the PD response

The response time of the thin film PDs was evaluated using a pulsed LED at a 50% duty cycle and a lockin amplifier. As shown in the figures below, it was observed that the PD response was generally frequency dependent, more strongly at the lower frequencies. This latter behavior is not clear.

Fig. 6 shows the normalized response of comparable a-Si-based p-i-n PDs fabricated by ECR and VHF PECVD as a function of the LED pulse frequency. As seen, the response of the VHF-fabricated device was faster. Fig. 7 shows the normalized PD response as a function of the LED pulse frequency for p-i-n and n-i-p devices grown by VHF PECVD. As seen, the response of the n-i-p device is significantly faster. It is therefore suspected that boron diffusion,

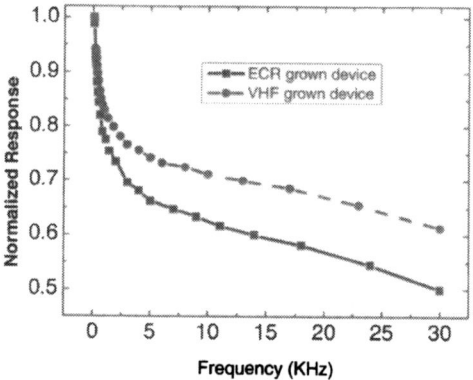

Figure 6. The response of comparable a-Si-based ECR and VHF grown devices vs the LED pulse frequency.

during growth, from the p+ layer into the intrinsic layer results in a slower response. Hence, the slower response of the ECR device may be attributed to its slower growth rate, and consequently increased boron diffusion into the i layer, in comparison to the VHF grown device.

Fig. 8 shows the improvement in the response of the VHF-grown p-i-n device achieved by incorporating a thin SiC layer, with a low boron content, at the p-i interface. The SiC layer partially blocks the diffusion of boron into the i layer during growth. However, incorporation of the SiC layer increases the series resistance of the device and hence decreases the absolute response of the photodetector. Hence, n-i-p structures show a faster response time, which may be beneficial for operating OLED-based O_2 sensors in the decay time mode. However, as shown in Fig. 9, the n-i-p configuration has a significantly lower QE for devices grown under comparable conditions, which will adversely affect the sensor performance.

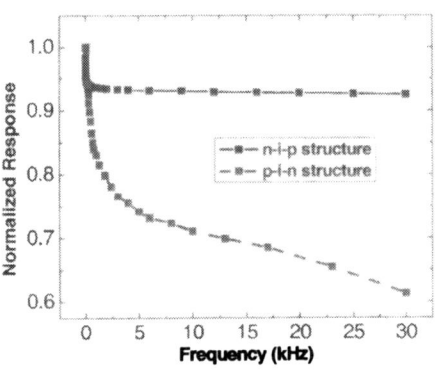

Figure 7. Comparison of the response of a-Si-based n-i-p and p-i-n VHF grown PDs.

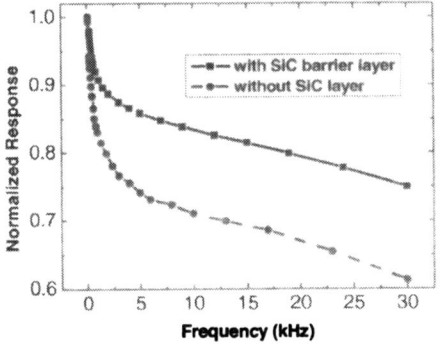

Figure 8. Frequency response improvement with a SiC layer in a VHF-grown, a-Si p-i-n PD

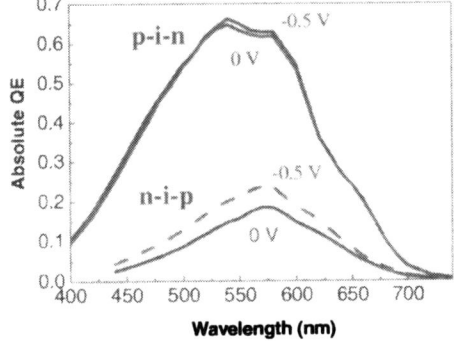

Figure 9. QE for p-i-n and n-i-p a-Si-based, VHF-grown devices.

Fig. 10 shows the effect of the frequency of the pulsed Alq_3 OLED excitation source on the signal ratios in different atmospheres using a PtOEP–doped polystyrene film and an n-i-p Si,Ge-based VHF-grown PD in the geometry shown in Fig. 1. The decrease in the ratios with increasing frequency at frequencies lower than the inverse of the longest PL decay time is unclear. At higher frequencies (> ~3 kHz), the ratio decrease may be partially attributed to the dynamics of the PtOEP PL. Specifically, at the higher frequencies, the lockin detection samples a smaller fraction of the PL rise and decay curves that vary in different atmospheres. The effect of OLED-generated, frequency-dependent noise on these ratios is currently being studied.

The sensitivity of ~7 (i.e., signal ratio at 100% N_2 and 100% O_2) obtained with this three-component integrated device at low frequencies can be seen in this figure.

Figure 10. Effect of the Alq$_3$ OLED pulse frequency on the detection sensitivity of O$_2$. The PD was an n-i-p Si,Ge-based device.

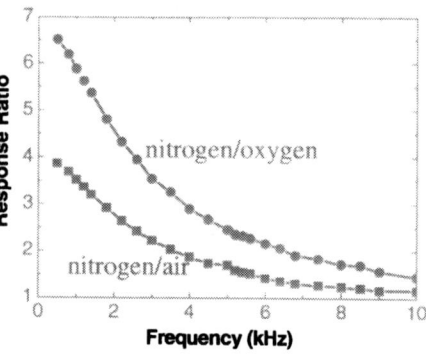

CONCLUSIONS

The viability of the integrated OLED/sensor film/a-Si,Ge-based PD compact O$_2$ sensor was demonstrated by measuring the sensor film PL in air, oxygen, and nitrogen atmospheres. The PDs were tuned by incorporating Ge in the p and i layers to enhance their sensitivity at the wavelength peak of the PtOEP emission and reduce it at shorter wavelengths. A detection sensitivity of ~7 was obtained using an Alq$_3$ OLED, a PtOEP-based sensor film, and an n-i-p VHF PECVD-grown Si,Ge PD. The sensitivity is expected to further increase for sensors operated in the PL decay time mode, where the background light is largely eliminated. However, a fast-PD is needed for this mode of operation. To that end, ECR and VHF-grown p-i-n PDs were compared. The latter were faster, possibly due to reduced boron diffusion into the i layer during the higher growth rate VHF fabrication. Indeed, partial blocking of boron diffusion by incorporating a SiC layer at the p-i interface supported this conclusion. Additionally, n-i-p structures demonstrated a faster response.

ACKNOWLEDGEMENTS

We thank Max Noack for helpful discussions. This work was partially supported by NSF and DOE. Ames Laboratory is operated by ISU for USDOE under Contract W-7405-Eng-82

REFERENCES

1. O. S. Wolfbeis, Editor, *Fiber Optic Chemical Sensors and Biosensors* (CRC Press, BocaRaton, FL, 1991).
2. Y. Amao, *Michrochim. Acta*, **143**, 1 (2003).
3. R. Shinar, Z. Zhou, B. Choudhury, and J. Shinar, *Anal. Chim. Acta* **568** 190 (2006).
4. P. Roche, R. Al-Jowder, R. Narayanaswamy, J. Young, and P. Scully, *Anal. Bioanal. Chem.* **386**,1245 (2006).
5. Z. Rosenzweig and R. Kopelman, *Sensors and Actuators* B **35-36**, 475 (1996).
6. B. Choudhury, R. Shinar, and J. Shinar, *J. Appl. Phys.* **96**, 2949 (2004).
7. R. Shinar, Chengliang Qian, Yuankun Cai, Zhaoqun Zhou, Bhaskar Choudhury, and Joseph Shinar, SPIE Conf. Proc. **6007**, 600710-1 (2005).
8. R. Shinar, D. Ghosh, B. Choudhury, M. Noack, V. L. Dalal, and J. Shinar, *J. Non Crystalline Solids* **352**, 1995 (2006).

Mater. Res. Soc. Symp. Proc. Vol. 989 © 2007 Materials Research Society   0989-A12-02

Segmented Amorphous Silicon n-i-p Photodiodes on Stainless-Steel Foils for Flexible Imaging Arrays

Y. Vygranenko[1], R. Kerr[2], K. H. Kim[1], J. H. Chang[1], D. Striakhilev[1], A. Nathan[1,3], G. Heiler[2], and T. Tredwell[2]

[1]Dept. of Electrical and Computer Engineering, University of Waterloo, 200 University Ave West, Waterloo, N2L 3G1, Canada
[2]Eastman Kodak Company, Rochester, NY, 14650-23487
[3]London Centre for Nanotechnology, University College, London, WC1H OAH, United Kingdom

ABSTRACT

This paper reports the first successful attempt to fabricate amorphous silicon (a-Si:H) n-i-p photodiodes on a thin stainless-steel foil substrate for medical X-ray imaging applications. Two architectures of the n-i-p-photosensor, where the top electrode is based on amorphous or polycrystalline ITO, have been developed and characterized. The impact of critical fabrication steps including the deposition of semiconductor layers, dry etch of the NIP stack, diode passivation and encapsulation, as well as a contact formation on the device performance is presented and discussed. The test structures comprising segmented photodiodes with a junction area ranged from 0.126×0.126 to 1×1 mm^2 have been fabricated on stainless-steel foils and on glass substrates for the purposes of process characterization. The fabricated samples are evaluated in terms of current-voltage, capacitance-voltage, and spectral response characteristics.

INTRODUCTION

Flexible stainless-steel foils present an excellent alternative to fragile glass substrates conventionally used in large area electronics. Compared to polymer films, another alternative substrate material, advantages of metal foil are in high-temperature process capability, chemical and moisture barrier properties [1]. A rapid progress in development of thin-film transistors (TFTs) on flexible substrates enables a fabrication of light, flexible and rugged displays [2-4]. It is also very attractive to apply this novel substrate technology to fabricate mechanically flexible and robust imagers for medical applications.

This paper discusses the integration issues specific for flexible substrates and reports the basic characteristics of amorphous silicon (a-Si:H) n-i-p photodiodes on a thin stainless-steel foil substrate.

EXPERIMENT

An integration issue for stainless-steel foils is a surface roughness, which should be below 20 nm. The foils available on the market are too rough to use them as substrates for the fabrication of electronic devices, and it is a technical challenge to polish metal foils with a required roughness at low cost. A simple solution of this problem is the use of stainless-steel foils of moderate surface roughness covered by a planarization layer. A process has been developed by Eastman Kodak Company. The first step is an electro-polishing of the raw stainless-steel foil to reduce the RMS roughness from 0.4 μm down to 0.15 μm. Fig. 1 includes

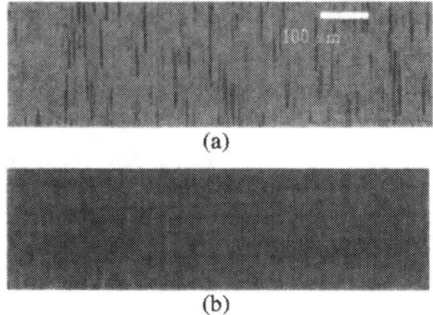

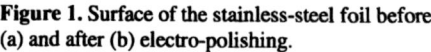

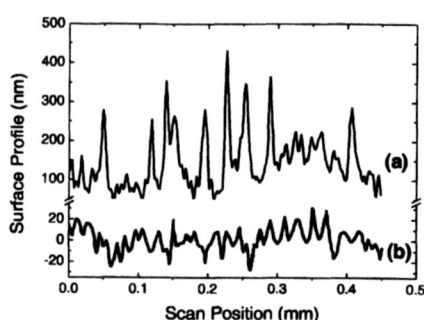

Figure 1. Surface of the stainless-steel foil before (a) and after (b) electro-polishing.

Figure 2. Surface profiles of the stainless-steel foil substrate without (a) and with (b) 3-μm-thick BCB planarization layer.

micrographs of the 132-μm-thick stainless-steel foil before (a) and after (b) the electro-polishing process. Then, the substrate is covered by a 3-μm-thick benzocyclobutene (BCB) dielectric layer; it reduces the surface roughness down to ~10 nm. Fig. 2 shows the surface profiles of the stainless-steel foil substrate without (a) and with (b) planarization layer. To enable lithography steps the 3 inch diameter substrates were attached to the 500-μm-thick stainless-steel carriers.

The segmented a-Si:H n-i-p-photodiodes of two architectures have been fabricated on the glass and stainless-steel foil substrates. Fig. 3 shows a cross-sectional view of these devices. The photodiode has a molybdenum bottom electrode, a n-i-p structure with an a-SiN$_x$ passivation layer, a top transparent electrode, and aluminium terminals. In the device shown in Fig. 3a, a combination of the a-SiN$_x$:H and amorphous ITO (a-ITO) films serves as an antireflection coating, and the spectral response characteristic can be controlled by adjusting the thicknesses of these films [5]. A new device design shown in Fig. 3b has been developed in order to simplify the fabrication process. Here, the top electrode is a polycrystalline ITO (poly-ITO) film, and a Mo buffer layer is implemented in the top metallization to enhance adhesion to the ITO. The used ~70-nm-thick poly-ITO combines high electrical conductivity (~2 × 10^3 S/cm) and good optical transmission (~85%). More importantly, poly-ITO material is stable during long-term storage and has much better resistance to process chemicals.

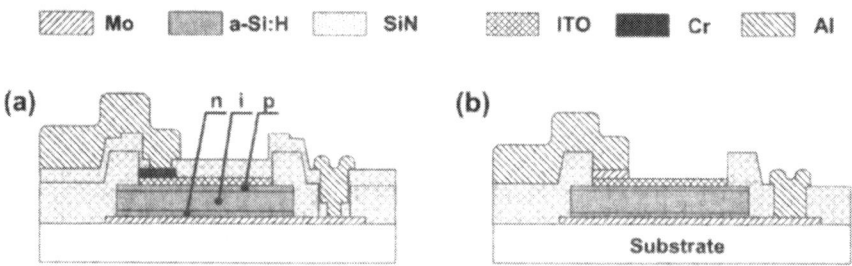

Figure 3. Cross-sectional views of the a-Si:H n-i-p photodiodes with a transparent top electrode made of amorphous (a) and polycrystalline (b) indium-tin oxide.

276

The device fabrication includes standard processing steps of metal deposition, plasma-enhanced chemical vapour deposition (PECVD), and lithography along with wet and dry etching. The a-Si:H and a-SiN$_x$:H films were deposited using a computer-controlled multi-chamber PECVD system, while the metal and ITO films were prepared using a RF sputtering system. The deposition conditions for the material used are described elsewhere [6].

The fabrication sequence of the photodiode shown in Fig. 3a includes seven lithography steps. It starts with the sputtering of Mo (~100 nm) layer on a substrate followed by deposition the a-Si:H n-i-p stack. The thicknesses of n-type and p-type a-Si:H layers were in the range of 20-40 nm, while a thickness of the undoped layer was 500 nm. Reactive ion etching (RIE) with SF$_6$ process gas is applied to etch amorphous silicon layers (*Mask #1*). The bottom metal layer is patterned by wet-etching in PAN solution (*Mask #2*). Then, an a-SiN$_x$ passivation layer is deposited at 150°C. This step can be also considered as a post-etch annealing process as it helps reduce edge components of the leakage current by annealing the defects that are located at silicon sidewalls. BHF solution is used to etch the a-SiN$_x$:H passivation layer patterned to open windows for underlying diodes (*Mask #3*). Then, a ~70-nm-thick amorphous ITO layer is deposited using RF sputtering process at a substrate temperature of 120°C followed by deposition of ~30 nm Cr film. The Cr layer is patterned and etched to form a top diode terminal (*Mask #4*). The ITO layer is patterned with *Mask #5* using wet etching in HCl:HNO$_3$:DI solution. Next, ~100-120 nm of a-SiN$_x$:H is deposited over the entire wafer at 150 °C substrate temperature to form antireflection layer and encapsulate ITO electrodes. The silicon nitride is selectively etched in BHF solution to open contact vias under the top (Cr) and bottom (Mo) contacts of the n-i-p photodiodes (*Mask #6*). Finally, a ~500-nm-thick Al layer is deposited by sputtering and patterned with *Mask #7* to form the anode metallization and contact pads.

A fabrication sequence of the photodiode with the poly-ITO top electrode is the same as described above until the formation of the top electrode. The poly-ITO film is deposited by RF sputtering at a substrate temperature of 150 °C and patterned with *Mask #4*. Then, ~50 nm Mo and Al films are sputtered and patterned using wet etched in phosphoric - acetic - nitric (PAN) solution (*Mask #5*).

The fabricated test structures include photodiodes of different sizes with a junction area in the range from 126 ^{2} 126μm^2 to 1 ^{2} 1 mm^2. To enable an accurate measurement of the leakage current, the small area diodes are connected in parallel forming arrays with up to 64 elements. Current-voltage (I-V) characteristics of photodiodes were measured using Keithely 4200 Semiconductor Characterization System and manual Cascade Summit Prober with current resolution down to <10 fA. The capacitance-voltage (C-V) characteristics of selected devices were measured at 1 kHz frequency using Agilent 4284A LCR-meter. A custom designed test setup was used for the spectral response characterization [6].

RESULTS AND DISCUSSION

Current-Voltage Characteristics

Fig. 4 shows typical quasi-static current-voltage characteristics of the a-Si:H n-i-p photodiodes with amorphous and polycrystalline ITO electrodes deposited on the glass (open symbols) and stainless-steel foil (solid symbols) substrates. The junction area is 126 × 126 μm^2. All diodes show exponential current growth over six orders of magnitude under forward biases up to 0.6 V. Under reverse bias of 5 V the dark current densities are in the range of 90-130 pA/cm^2 for devices on the glass and 180 pA/cm^2 for diodes on the stainless-steel foil substrate. The achieved level of the leakage current is similar to that in the state-of-the-art sensors with

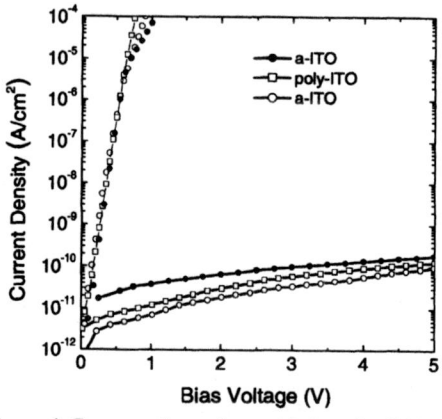

Figure 4. Current-voltage characteristics of a-Si:H n-i-p photodiodes with amorphous (circles) and polycrystalline (squares) ITO electrodes.

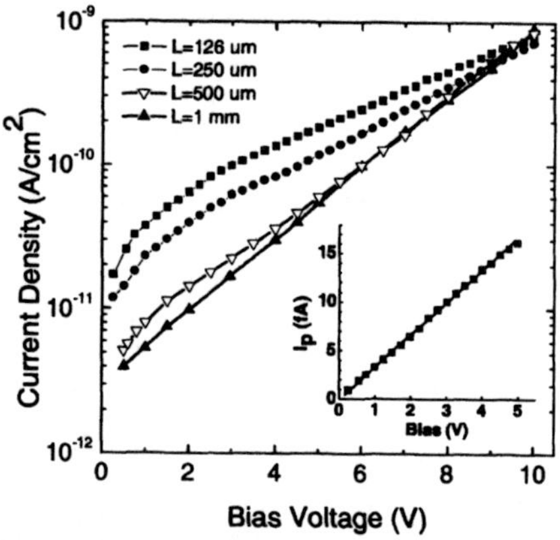

Figure 5. Reverse dark J-V plot of a-Si:H n-i-p photodiodes with various junction side lengths.

continuous i-layer and segmented diodes of comparable dimensions [7, 8]. It can be noted that diodes on the flexible substrate exhibit a noticeably higher leakage at low reverse bias voltages. In order to identify the sources of the leakage current, the reverse current-voltage characteristics of the diodes of different sizes have been measured. Fig. 5 shows $I/L^2=f(V)$ plots for square diodes with a side length L = 126, 250, 500 and 1000 μm on the stainless-steel foil substrate. A largest area diode (L = 1 mm) exhibits a lowest leakage current, which is mainly due to thermal generation of the charge carriers in the i-layer. The voltage dependence of the reverse dark current can be explained by a field-enhanced generation [9]. The relation between the bulk component J_B of the dark current and reverse bias voltage V_r is

$$J_B = J(T)e^{\beta\frac{V_r}{d_i}},\tag{1}$$

where $J(T)$ is the temperature-dependent term, d_i the i-layer thickness, β the constant. For the tested sample the β is estimated to be 2.8×10^{-5} cm/V.

An increase in current densities for diodes with L < 1 mm can be referred to a junction edge leakage component, which becomes dominant in small area diodes [10]. An additional leakage path for diodes on conductive substrate can be from top metallization through passivation and planarization layers to metal foil, then, through planarization layer to the bottom electrode. The insertion in Fig. 5 shows the reverse dark current component I_p in the 126 μm diode obtained by subtraction of the bulk component J_B from the total current. The excessive current increases linearly with the voltage yielding an equivalent shunt resistance of 300 TΩ. The similar resistance is evaluated to be due to the leakage through dielectric layers.

Capacitance-Voltage Characteristics

The boron-induced defects at the p-i interface are a potential source of an excessive leakage current in a-Si:H n-i-p structures, therefore, a doping level in the p-layer is a subject to optimization. The density of electrical active impurity in the p-layer can be estimated from C-V characteristics of the n-i-p diodes as described below.

Quasi-static capacitance C_{PIN} of the n-i-p diode under moderate reverse bias conditions is

$$C_{PIN} = \frac{\varepsilon\varepsilon_0 A}{d_p + d_I + d_N},$$ (2)

where ε is the static dielectric constant of the semiconductor material, ε_0 the permittivity of freespace, A the area of the junction, d_P the depletion width of the p-layer, d_I the thickness of the i-layer, and d_N the depletion width of the n-layer. Since under moderate reverse bias the i-layer is fully depleted, a small decrease in the diode capacitance with increasing applied voltage is due to expansion of the depleted region into the doped contacts. The doping level in the n-layer is usually much higher than that in the p-layer and for this case an acceptor density N_A can be easily estimated from C-V plot. From Poisson's equation one can conclude that a voltage variation ΔV induces a change in the depletion width of the p-layer

$$\Delta d_p = \frac{C_{PIN}(V)}{eN_A \cdot A}\Delta V,$$ (3)

where $C_{PIN}(V)$ is the diode capacitance under applied voltage V, e the electron charge. By differentiating Equation (2) and applying Equation (3), we obtain acceptor density

$$N_A = \frac{C_{PIN}^3}{e\varepsilon\varepsilon_0 A^2 (dC_{PIN}/dV)}.$$ (4)

C-V plot for the n-i-p diode on the stainless-steel foil substrate with optimized p-layer is shown in Fig. 6. The capacitance-voltage dependence is linear in the measured voltage range that suggests uniform doping across the p-layer. Estimated i-layer thickness and acceptor density are ~490 nm and 4.5×10^{17} cm^{-3}, respectively.

Spectral Response Characterization

Fig. 7 shows spectral response characteristics of a-Si n-i-p photodiodes on a stainless-steel substrate with a-ITO and poly-ITO top electrodes measured under a reverse bias of 3 V. External quantum efficiency of 84% at 570 nm has been achieved by tuning the optical properties of ITO/SiN$_x$ stack. The photodiode with the poly-ITO top electrode shows efficiency of 82% at 590 nm. At shorter wavelengths below 510 nm the sensitivity is limited by an absorption loss in the p-layer.

CONCLUSIONS

The segmented a-Si:H n-i-p-photodiodes with an top electrode based on amorphous or polycrystalline ITO have been successfully fabricated on glass substrates and on stainless-steel foils. For 126 μm diodes on the glass substrates tested at room temperature a dark current density is in the range from 90 to 130 pA/cm^2 at reverse bias of 5 V, while the diodes of the same size on stainless-steel foils shows a leakage current of 180 pA/cm^2. The analysis of the current-voltage

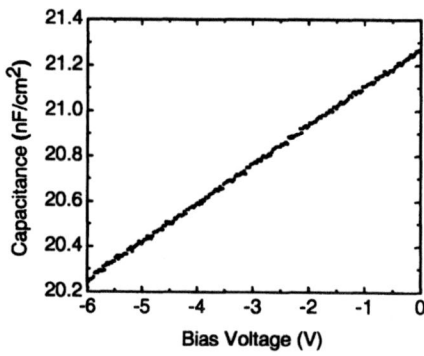
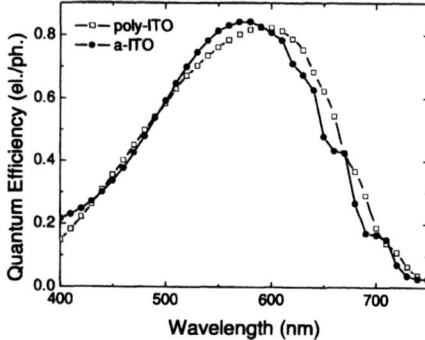

Figure 6. Capacitance-Voltage characteristic of the a-Si:H n-i-p diode with optimized doping level in the p-layer.

Figure 7. Spectral response of a-Si n-i-p photodiodes on stainless-steel substrates with a-ITO (circles) and poly-ITO (squares) top electrodes.

characteristics for diodes of different device areas reveals that a slightly increased level of the reverse dark current density in small area devices is due to the leakage through the planarization layer. Photodiodes of different architectures show external quantum efficiency above 80%. Thus, performance of the small area photodiodes on the stainless-steel foils meets the requirements for sensing elements of X-ray imagers for medical applications.

ACKNOWLEDGMENTS

Authors are grateful to Eastman Kodak Company and Natural Science and Engineering Research Council of Canada (NSERC) for financial support of their research.

REFERENCES

1. M. Wu and S. Wagner, *Mater. Res. Soc. Symp. Proc.* **664**, A17.2 (2001).
2. T. Afentakis, M. K. Hatalis, A. T. Voutsas, and J. W. Hartzell, *Mater. Res. Soc. Symp. Proc.* **769**, H2.5.1 (2003).
3. V. Cannella, M. Izu, S. Jones, S. Wagner, and I.-C. Cheng, *Info. Displ.*, 24-27 (2005).
4. A. Z. Kattamis, I-C. Cheng, Y. Hong and S. Wagner, *Mater. Res. Soc. Symp. Proc.* **910**, A16-03-L09-03 (2006).
5. Y. Vygranenko, J. Chang, and A. Nathan, *IEEE J. Quantum Electron.*, **41**, 5, 697- 703 (2005).
6. J. H. Chang, Y. Vygranenko, and A. Nathan, *J. Vac. Sci. Technol. A*, **22**, 971-974 (2004).
7. J. A. Theil, *Mater. Res. Soc. Symp. Proc.* **762**, A.21.4.1 (2003).
8. R. L. Weisefield et al. *Proc. SPIE*, **5368**, 338 (2001).
9. N. Kramer and C. van Berkel, *Appl. Phys. Lett.* **64**, 1129-1132 (1994).
10. R. A. Street (Ed.), in *Technology and Applications of Amorphous Silicon* (Springer, Berlin, 2000) pp.147-221.

Mater. Res. Soc. Symp. Proc. Vol. 989 © 2007 Materials Research Society   0989-A12-03

Transient Current Behavior of Vertically Integrated Amorphous Silicon Diodes

Gregory Choong[1], Nicolas Wyrsch[1], Christophe Ballif[1], Rolf Kaufmann[2], and Felix Lustenberger[2]

[1]Institute of Microtechnology, Breguet 2, Neuchatel, 2000, Switzerland
[2]CSEM, Badenerstrasse 569, Zurich, 8048, Switzerland

ABSTRACT

Monolithic image sensors based on Thin Film on CMOS (TFC) Technology are becoming more and more attractive as an alternative solution to conventional active pixel sensors (APS). Imager with high sensitivity, high dynamic coupled with low dark current values (10-100 pA/cm^2 @ 10^4 V/cm) have been developed. However, issues such as light-induced degradation and image lag hinder the commercial development of a-Si:H based image sensors. The problem of image lag is caused by residual current due to the release of trapped charges after the switch off of the illumination.

In this paper, we present a comprehensive study of the transient behavior of the photocurrent in a-Si:H photodiodes deposited on glass, as well as in corresponding diodes implemented in a TFC image sensor when illumination is switched off or periodically varied. The influence of the pixel architecture for two different cases is also analyzed: One setup reproduces the typical 3 transisor APS pixel architecture behavior, in which the bias voltage of the diode varies with the photogenerated charge while the second setup keeps a constant bias voltage applied to the diode by using a charge integrator.

The influence of the light-induced defect creation on the performance of the sensors is also presented and discussed.

INTRODUCTION

During the last years, the development of TFC (Thin Film on CMOS) image sensor has attracted a lot of attention [1,2,4,5]. The vertical integration of a-Si:H photodiodes on top of a dedicated CMOS integrated circuit can enhance the performance of the resulting monolithic photodiodes. As the photodiode array does not share the die area with the readout electronics, this concept greatly improve the geometrical fill factor (ratio of pixel active area to total pixel area) and leads to an improvement of the sensitivity of the sensor. The latter combined with the high quantum efficiency in the visible spectral range of a-Si:H render this technology very attractive for visible light imagers.

The dark current value J_{dark} is the key factor for high dynamic range image sensor and low light level detection and it must be as low as possible [13]. J_{dark} arises from thermal generation through bulk defect states. Optimization of a-Si:H n-i-p photodiodes (>1mm^2) deposited on glass substrates in our laboratory led to J_{dark} values as low as 1 pA/cm^2 (at −1 V for 1 µm thick detectors) while the corresponding diodes deposited on TFC readout chips exhibited J_{dark} values as low as 12 pA/cm^2 for a reverse bias voltage V=-1 V.

TFC devices (compared to c-Si CMOS sensors) usually suffer from image lag. This effect is caused by residual current due to the release of trapped charges resulting of an earlier exposition to light [11]. Moreover, light-soaking of the device results in an increase of the deep defect density in a-Si:H photodiodes (known as Staebler-Wronski Effect), and leads to a prolonged transient photocurrent emission when the illumination is switched off. Analysis of the transient behavior of the photocurrent in a-Si:H photodiodes was therefore performed on low dark current n-i-p diodes as well as TFC devices for selected cases relevant for application as image sensors.

As the dark current J_{dark} is the limiting parameter for low light level detection, the slow decrease of the trapped charges emission is expected to limit the device performance after illumination. Therefore, we first focused our analysis on the slow decay of the transient photocurrent after a light pulse on $1mm^2$ diodes deposited on glass. The same analysis was also performed on the corresponding diodes deposited on TFC chips, in order to verify that the TFC image behavior was similar. The rise time and the fall time of the signal for a light pulse were also measured in order to determine the limiting parameters at high frequency (>100 kHz). The influences of the illumination frequency on the signal amplitude were also analyzed.

Finally, the influence of the pixel architecture on the transient behavior of the photocurrent was investigated. We used a test circuit reproducing the typical 3T APS (Active Pixel Sensor) pixel architecture, in which the bias voltage varies with the photo-generated charge. The change of the bias voltage of the diode is expected to induce an additional transient current due to the thermal generation of charges from the depletion region.

EXPERIMENT

All measurements and results reported in this paper have been obtained on a-Si:H devices deposited using Very High Frequency Plasma Enhanced Chemical Vapor Deposition (VHF PECVD), details of diode fabrication have been presented elsewhere [3,5]. For the test structures on glass, 1 µm thick diodes were deposited on 200 nm chromium coated glass and the pixel areas were defined by patterning the top 65 nm thick transparent conductive oxide contact (ITO).

The CMOS readout chips used for the image sensor in TFC technology were fabricated in 0.5 µm technology of Alcatel-Mietec. The readout chip has 64x64 pixels with a pitch of 40 µm and a fill factor of 92%. Each pixel is connected to a charge integrator. Diodes similar to the ones deposited for the test structures were deposited onto the chips and covered with a common top ITO electrode. Light soaking degradation was performed at 50°C under an illumination of 1 sun in an open circuit configuration.

Various setups were used for the monitoring of the transient photocurrent caused by a light pulse from a red LED (λ=650 nm, 3.2 10^{13} photons $cm^{-2}s^{-1}$). All the measurements were made in the dark (no bias light) at room temperature. Steady-state dark current for different voltages, as well as slow photocurrent decay were measured using a Keithley 6517 electrometer and current values were measured every 1 s. A delay of 10 minutes was applied between the setup of the bias voltage and the start of the light pulse and data acquisition in order to reach a steady state for the dark current. The slow decay of the photocurrent after the switch off of the light was then measured for different bias voltage. The same procedure was used with the TFC image sensor, and the integration time of the charge integrator was varied from 20 µs to 3 s, in order to measure the illuminated state as well as the steady state dark current.

For the fast current transient monitoring, a second setup using a transimpedance amplifier (OPA627) with a transfer ratio k=V_{OUT}/I_{IN} of 10^8 V/mA was designed. In this experiment, we measured the photocurrent for single light pulse, as well as periodic light pulse. We observed the rise and the fall time of the photocurrent in a short time scale (<10 µs) and we measured the amplitude of the resulting signal for different periods of illumination while keeping a constant light pulse (10 ms – 1 s) and a constant reverse bias voltage.

In order to study the effect of the variation of the bias voltage for the typical 3T (3 transistors) APS pixel architecture, a test circuit mimicking the latter was built using a high quality IC, with a low input current (~3fA) for the source follower (INA116) as well as for the reset switch(MAX326).

RESULTS AND DISCUSSION

Analysis of the transient behavior of the photocurrent in a-Si:H photodiodes was performed on n-i-p diodes as well as TFC devices for the following selected cases .

Long time photocurrent decay

On Fig.1 one can observe, the evolution of the photocurrent after a light pulse of 10 s emitted by a red LED. We can notice that after a fast decrease of the current, first few seconds after the end of the pulse, the transient photocurrent decreases slowly by about one order of magnitude over a period of 300 s. After 300 s the current is still 2 times larger than the steady state current J_{dark}=7.2 10^{-12} A/cm^2. No clear influence of the bias voltage for the decay time can be observed, although the steady state dark current J_{dark}(-3V) increases by a factor of about 2.5.

Steady state dark current J_{dark}, identified as the bulk thermal generation current, arise from the emission of electron,

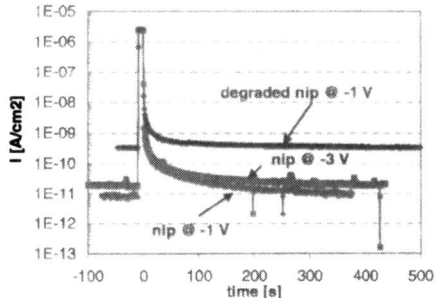

Figure 1: Comparison of transient dark current after a light pulse of 10 s, for different bias voltage, and for a diode after 1000 h of light soaking. The measures were done on a 1 μm thick n-i-p diode deposited on a glass substrate.

thermally excited, from the valence band towards the empty defect states (hole generation) in the gap, and from the emission of electron from the filled defect states to the conduction band (electron generation). The density of thermally generated current depends on the defect density and the bandgap of the semiconductor material [4, 8]:

$$J_{th} = q \cdot d_i \cdot N_{db} \cdot kT \cdot \omega_0 \exp\left[-(E_G)/2kT\right] \qquad [\text{A/cm}^2] \qquad (1.1)$$

where kT is the temperature and Boltzmann constant product, d_i the intrinsic layer thickness, N_{db} the dandling bong density, E_G the mobility band gap, and $\omega_0 \sim 10^{13}$ s^{-1} is the excitation rate prefactor (or attempt to escape frequency).

After the light pulse, carriers trapped in the band tails and dangling bonds will be released as the quasi Fermi levels return towards the thermal equilibrium Fermi level position E_F[8]. For electrons, all carriers trapped at the level of the quasi Fermi level $E_F^n(t)$ will be released at time

$$\tau_{rel}^n = \frac{1}{\omega_0}\exp\frac{E_C - E_F^n(t)}{kT} \qquad (1.2)$$

A similar expression is obtained for the hole release. The time needed to empty all traps is then given by the position of the dark Fermi level position. Assuming a position of E_C-E_F≈0.9 eV, we can estimate a decay time of >100 s in agreement with our measurements. Note that the decay is non-exponential because the quasi-Fermi energy is time dependent and the current strongly depends on the localized states distribution.

Light induced degradation increases N_{db}, hence the degraded sample has a higher dark current which is shown in Fig.1. As N_{db} is increased, the number of trapped charges increases also leading to a higher transient current.

TFC Sensor photocurrent decay

Fig. 2 shows the transient photocurrent for a single pixel of a TFC image sensor during and after a 10 s light pulse, as well as the corresponding curve for a n-i-p diode on glass substrate. The current was calculated from the change of the output voltage signal of the charge integrator implemented in each pixel of the TFC device. The transient photocurrent curves were measured for several bias voltages, but only two are represented here.

The first observation is that the steady-state dark current J_{dark} (at -1 V) for a TFC device is 5 time higher than the

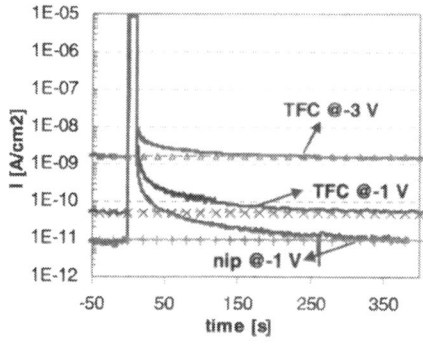
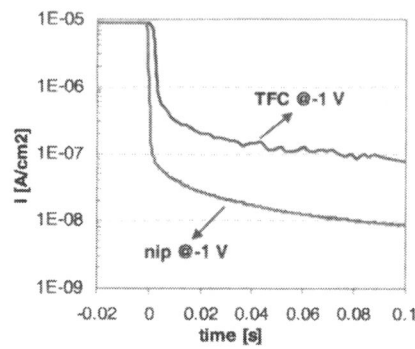

Figure 2: Dark Current Decay for 1 µm thick diode deposited on TFC readout chip and for the corresponding diode (nip) on glass substrate

Figure 3: Detail of the first 100ms after the switch off of the light for the same devices presented in Fig. 2.

corresponding diode on glass substrate; this effect is even more pronounced with reverse bias voltage of 3 V. Whereas the dark current for the n-i-p diode on glass increases by a factor 2 between -1 V and -3 V, J_{dark} increases by a factor of 33 in TFC device to a steady-state value of $1.6 \cdot 10^{-9}$ A/cm^2 at -3 V. The small pixel size (40x40 µm^2) and the non planar configuration of the pixel are responsible for this leakage increase due to a local increase of the electric field as well as an increase of the defect density at the edges and corners of the pixels [4,5]. The large increase of J_{dark} shows that this additional leakage is highly sensitive to the applied bias voltage and becomes the largest contribution to J_{dark} for reverse bias higher than 1 V.

Concerning the long time transient photocurrent decay, the same behavior is observed for both the TFC device and for the test diode; the steady state dark current is reached after 300 s, independently of the device surface. As the decay of the transient current is proportional to the probability of emission of a trapped charge and that the bias voltage is kept constant in both cases, no difference is here expected. However, a closer look at the decay during the first second shows that the decay for the TFC device is much slower (see Fig. 3) and the value of the current at 100 ms after the end of the light pulse one order of magnitude larger. This large difference is not yet well understood, but might be linked to charge release at the periphery of the pixel that contribute to a higher dark current.

Frequency response

In this experiment, we aim at determining the limiting parameters for high speed light detection when the photocurrent diode is measured using a transimpedance amplifier. In fact, the response time (for short times) can be either limited by the collection of free and shallow trapped carriers (given by their transit time) or by the response time of the complete setup (diode + measurement system). In a n-i-p structure with low defect density, the transient photocurrent decreases as the free carriers reach the electrodes and are collected. Thus, if the defect density is low, the transient photocurrent decay is determined by the transit time, which is the time that a carrier needs to travel through the device. For a 1 µm thick device under a reverse bias voltage of 1 V, we can estimate the transit time (for the free and shallow trapped carriers) to be roughly in the order of 10^{-8} s for electrons and 10^{-6} s for holes, when assuming a drift mobility of 1 cm^2/V/s for electron, resp. 0.01 cm^2/V/s for holes. Even though the observed decay time is close to the transit time of the holes, the fact that no bias dependency was observed indicates that the response time is probably limited by the circuit.

We can show that the bandwidth limit becomes in a first order approximation $f_c = 1/2\pi R_f C_f$, where $R_f = 100$ kΩ is the feedback resistor and $C_f = 18$ pF the feedback capacitance use for compensation of phase of this transimpedance amplifier stage. Thus the time fall becomes $t_f \approx 0.35/f_c \approx 5$ µs, value that is comparable to the measured one and corroborates our suggestion that the frequency response is limited by the circuit.

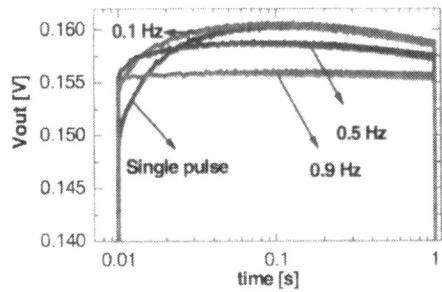

Figure 4: Magnification of the top of the output signal of the transimpedance amplifier for a nip diode @-1 V during a light pulse of 1 s, for different pulse frequency

In figure 4, we can observe the photocurrent during the light pulse for different periods of illumination (pulse frequency) while keeping a constant light pulse of 1 s. We can observe a decrease of the rise time of the signal as the frequency is increased. As a shorter time interval between two light pulses will forbid parts of the deeper traps to empty, one expects the photocurrent to reach faster its steady-state value.

The second observation is that the amplitude varies with the frequency as well. The maximum of the signal is attained after 100 ms after the switch on of the light and then the signal decrease to the steady state value under illumination. This additional transient mechanism is not yet understood (note that an effect of the measurement system can be ruled out as this overshoot is larger for slower change in photocurrent) and is still under investigation.

Photocurrent decay in 3T APS

In typical APS with small pixels, photogenerated charges are integrated in the diode itself and the voltage on the diode is monitored with a source follower. Thereby the photocurrent reduces the reverse bias voltage on the diode which induces additional carrier trapping. A dedicated setup was built in order to determine the transient behavior of n-i-p diode in this configuration. In this experiment, the photocurrent decay in the diode, after a 10 s light pulse, is monitored (using a circuit replicating a "standard" 3T APS circuit) at a 10 Hz sampling frequency.

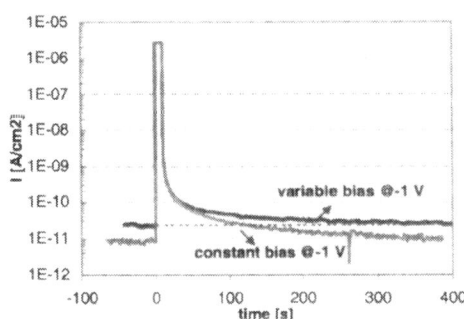

Figure 5: Comparison of the photocurrent decay after a light pulse of 10s, measured with the 3T configuration and the constant bias configuration

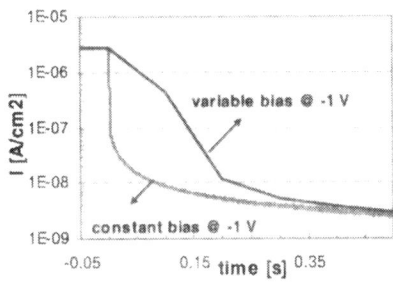

Figure 6: Detail of the first 500ms after the switch off of the light for the same devices presented in Fig. 5.

In figure 5, we can see the slow decay of the current after a 10 s light pulse, which follows almost the same time dependence as if the diode were kept under constant bias voltage. While change in the bias voltage of the diode (during one frame of 100 ms) are quite important at the maximum of the current (from -1 V to 0.2 V), effects are only visible during the first couple of measurements taken after the switch off of the light. Thereby, 100 ms after the switch off, the current is a factor of 50 larger than the one measured on a diode where the bias is kept constant, but only a factor of 2 after 200ms (see Fig.6). For longer time, the difference in photocurrent decreases very rapidly as the current decreases and induces only very small change of the diode bias. This increased lag (larger current at short times) is due to the release of the carriers trapped by the change of the diode bias.

CONCLUSIONS

In this work, we have measured and analyzed the transient current behavior of n-i-p diode for different cases. Even though a-Si:H diodes are well suited for high dynamic range imaging and low light level detection, the long photocurrent decay after an illumination limits the time before a new acquisition or can limit the dynamic.

The lowest image lag is measured on n-i-p diode on glass substrate under constant bias voltage. As soon as we leave this ideal configuration, due to light induced degradation, non-flat pixel topography or 3T pixel architecture, an increase in the image lag is observed.

We have demonstrated that a-Si:H n-i-p diode can be used for high speed light detection, as long as RC time constant of the circuit is kept short, thanks to the short collection time, that can be easily tuned with the thickness of the diode and the applied bias voltage. Finally the influence of the illumination history on the behavior of the diode under illumination has been underlined for the case of frequency variation of the illumination.

REFERENCES

[1]   T. Lulé, B. Schneider, M. Böhm, IEEE J. of Solid-State Circuits 34, 1999, p.704.
[2]   F. Blecher, S. Coors, A. Eckhardt, F. Mütze, B. Schneider, K. Seibel, J. Sterzel, M. Böhm, Int. Conf. on Mechatronics & Machine Vision in Practice, Nanjing, China (2000).
[3]   C. Miazza et al., *Mat. Res. Soc. Symp. Proc.,Vol.* 808 (2004) A4.46.1.
[4]   C. Miazza et al., *Mat. Res. Soc. Symp. Proc.,Vol.* 910 (2006) 0910-A17-03.
[5]   C. Miazza et al., *Mat. Res. Soc. Symp. Proc.,Vol.* 869 (2005) 869-D1.2.1.
[6]   J. A. Theil, *Mat. Res. Soc. Symp. Proc.,Vol.* 762 (2003) A21.4.1
[7]   J. A. Theil, *Mat. Res. Soc. Symp. Proc.,Vol.* 869 (2005) D1.3.1
[8]   R.A. Street, Appl. Phys.Lett.57 (13), pp.1334-1336, 1990.
[9]   D.S. Shen,S. Wagner, J. Appl. Phys. 79 (2) ,1996.
[10] H. Kida, KHattori,H.Okamoto,Y.Hamakawa, J. Appl. Phys. 59(12), 1986
[11] Q. Zhu, J. Sterzel,B. Schneider, S. Coors, M.Böhm, J. Appl. Phys. 83 (7) ,1998
[12] J.G. Graeme, Handbook on Photodiode Amplifer,ed. McGraw-Hill,p.31-61
[13] B. Schneider et al.,Handbook on Comp. Vis. And Appl., Ac. Press, Boston, 1999, p.237

Mater. Res. Soc. Symp. Proc. Vol. 989 © 2007 Materials Research Society   0989-A12-04

Optical Readout in Pinpi'n and Pini'p Imagers: A Comparison

P. Louro[1,2], M. Vieira[1], Y. Vygranenko[1], M. Fernandes[1], and A. Garção[2]

[1]DEETC, ISEL, Rua Conselheiro Emidio Navarro, Lisbon, 1949-014, Portugal

[2]CRI, FCT-UNL, Quinta da Torre, Caparica, 2829-516 Caparica, Portugal

ABSTRACT

Optimized a-SiC:H multilayer devices based on two different tandem configurations (TCO/pinpi'n/TCO and TCO/pini'p/TCO) are compared and tested for proper image recognition and color separation process using an optical readout technique. In both configurations the doped layers are based on a-SiC:H to increase image resolution and to prevent image blurring. To profit from the light filtering properties of the active absorbers, the intrinsic layer of the front diode (i layer) is based on a-SiC:H and the back one (i' layer) on a-Si:H.

The effect of the applied voltage on the color selectivity is discussed. Results show that the relative spectral response curves demonstrate rather good separation between the red, green and blue basic colors. Combining the information obtained under positive and negative applied bias a colour image is acquired, using the same optical technique either in pinpi'n or in the pini'p configuration, without colour filters or pixel architecture.

INTRODUCTION

The use of multilayered structures based on a-SiC:H alloys as color sensors has been an important topic of research in the field of sensing applications [1, 2, 3, 4]. In these multilayered devices the light filtering is achieved through the use of different band gap materials, namely a-$Si_{1-x}C_x$:H. In these devices the spectral sensitivity in the visible range is controlled by the external applied voltage. Thus, proper tuning of the device sensitivity along the visible spectrum allows the recognition of the absorbed light wavelength, and consequently the identification of the RGB components of a coloured image [5, 6].

In this paper color pinpi'n and pini'p sensitive detectors are tested using the laser scanned photodiode technique (LSP) [7, 8]. This technique allows a complete color analysis to be performed with a single two terminal detector element and an optically addressed readout technique. With this technique the image to acquire is optically mapped onto the sensing photodiode and a low-power light spot scans the device by the opposite side. The photocurrent generated by the moving spot is recorded as the image signal, and its magnitude depends on the light pattern localization and intensity. For image color acquisition the device is biased at different voltage values, which modulates the output image signal and allows the reconstruction of the color image [9]. Advantages to this approach are high resolution, uniformity of measurement along the sensor and the cost/simplicity of the detector. The design allows a continuous sensor without the need for pixel-level patterning, and so can take advantage of the amorphous silicon technology. It can also be integrated into a system where the signal processing can be performed by an ASIC chip underneath.

In this paper a comparison between the performance of both pinpin and pinip based image sensors with optical readout is presented. The reconstruction of color images is analyzed and the effect of threshold voltages is discussed.

EXPERIMENTAL DETAILS

Fabrication and characterization of the sensing structures

Two sensing multilayered structures (#NC10 and #Y13) in different configurations were analysed (Figure 1). Both consist of two stacked p-i-n diodes (front and back diode) deposited on a glass substrate and sandwiched between two transparent conductive electrodes (TCO). In device #NC10 (Figure 1a) both front and back diodes are stacked in series connection within a p-i-n-p-i'-n sequence. In device #Y13 (Figure 1b) the sensing structure consists of two back to back diodes (p-i-n/n-i-p) with the same n-type doped layer, which results in a p-i-n-i'-p' sequence.

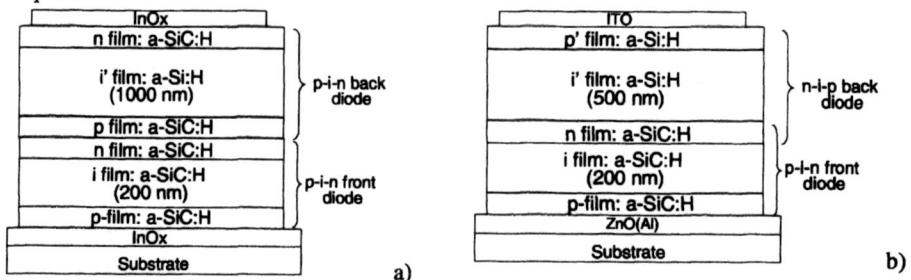

Figure 1 Schematic cross view of the: a) #NC10 (p-i-n-p-i'-n) and b) #Y13 (p-i-n-i'-p') devices.

The front diode, which is designed to face the incident illumination is based on a-SiC:H (doped and undoped layers) and the intrinsic layer is 200 nm thick, which results in an optimized structure for blue sensitivity and red transmittance [10]. The intrinsic layer of the back diode (i' layer) is based on a-Si:H and its thickness is 500 nm in #Y13 and 1000 nm in #NC10. This back diode is used for the optically addressed readout of the image mapped onto the front device. In device #NC10 the TCO contacts established for carrier collection are made of undoped InOx thin films deposited at room temperature by radio frequency plasma enhanced reactive thermal evaporation (rf-PERTE) of In [11, 12]. In device #Y13 the front contact is ZnO(Al) and the back one is ITO, deposited by sputtering.

Besides, the stacked devices, the simplified test a-SiC:H p-i-n and a-Si:H n-i-p structures have also been deposited during the same deposition process. The film layers were deposited using a parallel-plate PECVD reactor. Deposition conditions such as the RF power, partial pressure and gas flow rates are shown in Table 1. The substrate temperature was held at 260 °C.

Table 1. Deposition conditions of the a-Si:H and a-SiC:H films.

Type	RF Power (W)	Pressure (mTorr)	Gas flow(sccm)			
			SiH_4	1% TMB+99% H_2	2%PH_3+98% H_2	CH_4
p (a-SiC:H)	4	600	10	25	–	15
i (a-SiC:H)	4	500	10	–	–	15
n (a-SiC:H)	4	500	10	–	5	15
i' (a-SiC:H)	2	400	20	–	–	–
p' (a-SiC:H)	2	400	20	10	–	–

RESULTS AND DISCUSSION

Dark Current-Voltage characteristics of test structures

Figure 2 shows the dark current-voltage characteristics of test a-SiC:H p-i-n and a-Si:H n-i-p diodes, measured at room temperature.

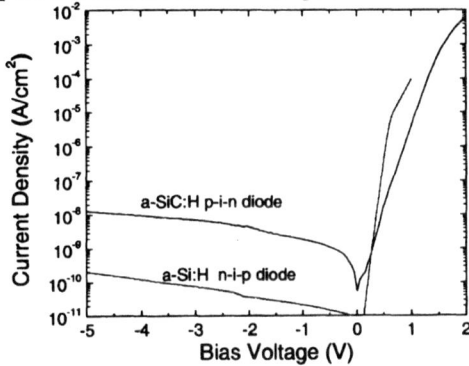

As expected both diodes exhibit different current-voltage characteristics, under forward and reverse biases. From the linear region of the forward characteristic the diode ideality factor and the saturation reverse density current were extracted. It was observed respectively, for the a-SiC:H and a-Si:H p-i-n diodes, values of 80 pA/cm^2 and 0.5 pA/cm^2 for the saturation currents, and 3.65 and 1.45 for the ideality factors. For the a-Si:H based diode the observed parameters are similar to those reported in the literature [13]. In the a-SiC:H diode both parameters exhibit larger values, which are related to the higher defect density of the a-SiC:H material and probably to higher recombination rates. The exponential growth up to 1.4 eV is due to the wider band gap of the intrinsic a-SiC:H.

Figure 2 J-V characteristics measured in dark at room temperature of a-SiC:H p-i-n (200 nm) and a-Si:H n-i-p (500 nm) diodes.

The optical band gap obtained from the Tauc's plot is 1.92 eV, which is 200 meV higher than in a-Si:H. Under reverse biases the leakage current of the a-SiC:H diode is two orders of magnitude higher than in a-Si:H. This higher current density can be attributed to the carrier generation in the intrinsic layer and also to the defects in the i-p interface. We believe that the thermal generation in the intrinsic layer is the dominant leakage mechanism in device, because the voltage dependence of the reverse dark current is not strong (the dark current exhibits saturation).

Spectral response

The structures were characterized by spectral response in the range 400 nm up to 700 nm, under different optical and electrical biasing conditions.

Figure 3 shows the spectral response of the test a-SiC:H p-i-n diode measured under reverse bias from 0 V to -4 V. No optical bias was superimposed. If compared with the standard a-Si:H p-i-n diodes, results show that the spectral response dependence on the applied voltage is enhanced and that the maximum of the spectral sensitivity is shifted to lower wavelength values. This is due to the wider band gap of a-SiC:H

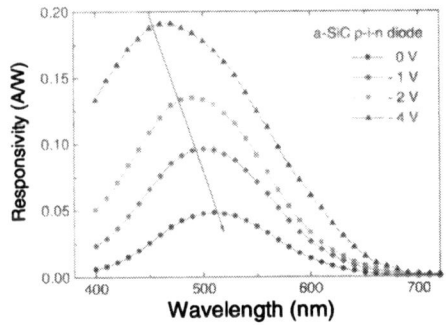

Figure 3 Spectral response of the a-SiC:H p-i-n test diode.

that makes the device transparent to the red range of the visible spectrum. Besides, the diode sensitivity is dependent on the applied reverse bias. With the increasing bias voltage the spectral response maxima shift from 509 nm (at 0 V) to 465 nm (at – 4 V), showing a voltage controlled dependence, which can be explained by lower mobility-lifetime product in comparison with a-Si:H.

Figure 4a shows the spectral response characteristics of the p-i-n-i'-p structure (#Y13). In Figure 4b it is displayed the photocurrent dependence on the light wavelength for the p-i-n-p-i'-n structure (#NC10).

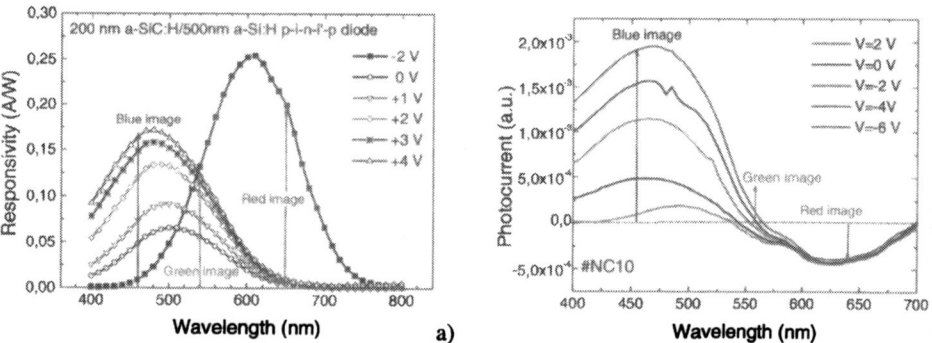

Figure 4 Wavelength dependence: a) of the responsivity in p-i-n-i'-p (#Y13) and b) of the photocurrent in the p-i-n-p-i'-n (#NC10) devices.

For each configuration the spectral response characteristics are dependent on both polarities and magnitude of the applied biases. In the p-i-n-i'-p configuration, under negative bias applied to the back electrode (Figure 1b) the a-Si:H diode is reverse biased and the incident light is filtered by the p-i-n stack. In these conditions the enhanced selectivity to the red light, as the spectral response maximum is located at 610 nm (for -2 V of applied bias). When the positive bias increases from 0 up to + 4 V an opposite behavior occurs, the spectral response monotonically increases and the spectral maxima shift from 500 nm to 480 nm. As expected, the obtained spectral response plots at positive biases are similar to that observed for the a-SiC:H diode (Figure 3). So, the color selectivity can be obtained by adjusting the biasing conditions of the structure. Defining the image signal as the difference between the photocurrent under steady-state optical bias and its value in dark (assumed to be the photocurrent at 750 nm) the arrows indicate respectively, the image signal intensity under red, green and blue illumination. Under negative bias the image signal is suppressed in the blue range while under positive bias the image signal rejection is achieved in the red part of the spectrum. Under green light the signal is the same, either under positive or negative voltages. In this configuration the leakage current of the non-absorbing diode is the restrictive factor. The high absorption coefficient of the front diode to the blue photons and its transparency to the red ones limits the current, under red irradiation and V>0, to the leakage current (dark) of front diode (red rejection), and limits the current, under blue irradiation and V<0, to the leakage current of the back diode (blue rejection). Under green irradiation the photons are absorbed in both diodes. So the generated carriers are collected either under positive or negative bias, giving always similar current signal.

290

In the p-i-n-p-i'-n configuration (Figure 4b) it is observed that the photocurrent characteristics exhibit a maximum located around 450 nm. As the wavelength light increases and shifts to the greenish and reddish parts of the visible spectrum the photocurrent decreases, reversing in sign around 550 nm. In the blue range the photocurrent is always positive, increasing as the reverse voltage increases. In the reddish part of the spectrum, the photocurrent reverses sign and its magnitude is independent on the applied voltage bias. When compared with the signal without light impinging the front diode (dark level at 750 nm) at short circuit conditions, the red and blue signals have opposite signs and the green signal is almost suppressed, allowing blue and red recognition. The green information is obtained for the voltage bias that suppresses the blue signal, which occurs at slightly positive bias. In the p-i-n-p-i'-n structure it was shown [8] that the net current depends on the level of irradiation and on the leakage current of both diodes. Taking into account the structure of the device (Figure 1a) and since the current across the structure has to be the same our interpretation points out the cause of the high collection efficiency under reverse bias and blue irradiation to a self biasing of the back cell that becomes fully depleted. An opposite behavior occur under red illumination where the a-Si:H absorber acts as a load due to the high light penetration depth of the red photons resulting in a reduced collection even under reverse bias. In the green spectral range the limiting factor is the photocurrent of the less irradiated diode.

So, combining the information obtained under positive and negative applied bias, a colour image can be acquired, using the same optical readout technique either in the p-i-n-i'-p and p-i-n-p-i'-n configurations, without colour filters or pixel architecture.

Image acquisition

Figure 5a) shows the image signal obtained with device #NC10 at − 6 V using as optical image a graded wavelength mask (rainbow) to simulate the visible spectrum in the range between 400 and 700 nm. This optical image was projected onto the a-SiC:H front diode and acquired through the a-Si:H back one with a moving red scanner. The line scan frequency was close to 500 Hz. For a readout time of 2 ms the frame time for a 40 lines image is around 80 ms. For image acquisition two applied voltages were used to sample the image signal: + 1V and -6 V. In Figure 5 b) it is displayed the 2-D colour image reconstruction using the acquired image signals. The algorithm used for image color reconstruction took into account that at -6V the positive signals correspond to the blue/green contribution and the negative ones to the red inputs. The green information was extracted from the image signal sampled at +2 V, where the blue signal is almost suppressed (Figure 4b) and the green and red signals are negative. The combined integration of this information allows

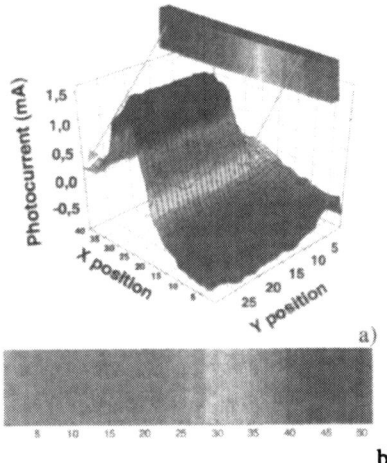

Figure 5 Image signal representation obtained with device #NC10: a) at -6 V and b) 2-D color image reconstruction. The insert at the top is the mask of the visible spectrum.

the fulfillment of the R, G and B channels .The image in Figure 5b) was created using a Matlab program that computes the digital image. A good agreement between the optical image and the electrical reconstructed color image was obtained. Further optimization of the sampling threshold voltage values seems to be needed in order to improve the R, G and B channels distributions.

CONCLUSIONS

Stacked p-i-n-p-i'-n and p-i-n-i'-p devices were analyzed using the laser scanned photodiode technique. By using a thin a-SiC:H front absorber optimized for blue collection and red transmittance and a thicker a-Si:H back absorber to spatially decouple the green/red absorption the sensors in both configurations behave themselves as a filter, giving information, at appropriated applied voltages, about the wavelength of the impinging photons.

A low level image processing algorithm was used for the reconstruction of the colour image. Further optimization of the image reconstruction is necessary, which may demand the use of a dedicated Fortran code.

ACKNOWLEDGEMENTS

This work has been financially supported by POCTI/FIS/58746/2004 and by Fundação Calouste Gulbenkian.

REFERENCES

[1] G. de Cesare, F. Irrera, F. Lemmi, F. Palma, IEEE Trans. on Electron Devices, Vol. 42, No. 5, May 1995, pp. 835-840.

[2] A. Zhu, S. Coors, B. Schneider, P. Rieve, M. Bohm, IEEE Trans. on Electron Devices, Vol. 45, No. 7, July 1998, pp. 1393-1398.

[3] M. Topic, H. Stiebig, D. Knipp, F. Smole, J. Furlan, H. Wagner, J. Non Cryst. Solids 266-269 (2000) 1178-1182.

[4] M. Mulato, F. Lemmi, J. Ho, R. Lau, J. P. Lu, R. A. Street, J. of Appl. Phys., Vol. 90, No. 3 (2001), pp. 1589-1599.

[5] H.K. Tsai, S.C. Lee, IEEE electron device letters, EDL-8, (1987) pp.365-367. .

[6] H. Stiebig, J. Gield, D. Knipp, P. Rieve, M. Bohm, Mat. Res. Soc. Symp. Proc 337 (1995) 815-821.

[7] M. Vieira, M. Fernandes, P. Louro, R. Schwarz, M. Schubert, J. Non Cryst. Solids 299-302 (2002) pp.1245-1249.

[8] M. Vieira, M. Fernandes, P. Louro, A. Fantoni, Y. Vygranenko, G. Lavareda, C. Nunes de Carvalho, Mat. Res. Soc. Symp. Proc., 862 (2005) A13.4.

[9] P. Louro, M. Vieira, A. Fantoni, M. Fernandes, C. Nunes Carvalho, G. Lavareda, Sensors and Actuators A 123-124 (2005) 326-330.

[10] P. Louro, M. Fernandes, A. Fantoni, G. Lavareda,, C. N. Carvalho, R. Schwarz, M. Vieira, Thin Solid Films 511-512 (2006) 167-171.

[11] C. Nunes de Carvalho, G. Lavareda, E. Fortunato, A. Amaral, Thin Solid Films 427 (1-2) (2003), p. 215.

[12] C. Nunes de Carvalho, G. Lavareda, A. Amaral, O. Conde and A. R. Ramos, J. Non-Cryst. Solids 352, Issues 23-25, (2006), p. 2315

[13] E. A. Schiff, R. A. Street, R. L. Weisfield, J. Non-Cryst. Solids 198–200 (1996) 1155.

Mater. Res. Soc. Symp. Proc. Vol. 989 © 2007 Materials Research Society   0989-A12-05

Germanium-Silicon Separate Absorption and Multiplication Avalanche Photodetectors
Fabricated with Low Temperature High Density Plasma Chemical Vapor Deposited Germanium

Malcolm Carroll[1], Kent Childs[2], Darwin Serkland[2], Robert Jarecki[2], Todd Bauer[2], and Kevin Saiz[2]

[1]Photonic Microsystems Technology, Sandia National Laboratories, P.O. Box 5800, M.S. 1082, Albuquerque, NM, 87185
[2]Sandia National Laboratories, Albuquerque, NM, 87185

ABSTRACT

In this paper, we evaluate a commercially available high density plasma chemical vapor deposition (HDP-CVD) process to grow low temperature (i.e., $T_{in\text{-}situ}$ & $T_{epitaxy}$ < ~460°C) germanium epitaxy for a p^+-Ge/p-Si/n^+-Si NIR separate absorption and multiplication avalanche photodetectors (SAM-APD). A primary concern for SAM-APDs in this material system is that high fields will not be sustainable across a highly defective Ge/Si interface. We show Ge-Si SAM-APDs that show avalanche multiplication and avalanche breakdown. A dark current of ~0.1 mA/cm^2 and a 3.2×10^{-4} A/W responsivity at 1310 nm were measured at punch-through. An over 400x photocurrent multiplication was demonstrated at room temperature. These results indicate that high avalanche multiplication gain is achievable in these Ge/Si heterostructures despite the highly defective interface and therefore trap assisted tunneling through the defective Ge/Si interface is not dominant at high fields.

INTRODUCTION

A desire to monolithically integrate near infrared (NIR) detectors with silicon complementary metal oxide semiconductor (CMOS) technology has motivated many investigations of single crystal germanium on silicon (Ge/Si) diodes [1-3]. Reduction of the epitaxy thermal budget below the typical chemical vapor deposition (CVD) in-situ clean temperature ($T_{in\text{-}situ\ clean}$ > 780°C) is also increasingly desired to reduce integration complexity. Reduced temperature growth approaches have included p^+-Ge/n-Si detectors formed with low temperature poly-Ge (e-beam evaporation) or heavily dislocated single crystal germanium (molecular beam epitaxy, T ~ 450°C), which have had dark currents of ~5 mA/cm^2 and responsivities of ~15 mA/W at 1310 nm, despite the large number of defects in and at the Ge/Si interface. Responsivities in these materials are however low and believed to be limited by a small diffusion length (i.e., 5-30 nm [2, 4]) due to fast electron recombination in the defect rich germanium. A silicon based separate absorption and multiplication avalanche photodiode structure (SAM-APD) is of interest to increase responsivity as well as potentially offering an alternative to III-V NIR Geiger mode (GM) single photon avalanche photodiodes (SPAD). A silicon avalanche region would be highly desirable for NIR SPAD to reduce after-pulsing effects, which are related to trap densities that are much higher in InP [5] than would be expected in the high field region of a Ge/Si APD.

EXPERIMENT

The Ge-Si vertical avalanche photodiode structure is an epitaxial stack nominally 8 μm n^+-Si ($\sim 10^{18}$ cm^{-3}) / 400 nm p$^-$ -Si ($\sim 5 \times 10^{14}$ cm^{-3})/ 200 nm p$^+$ -Ge ($\sim 2 \times 10^{18}$ cm^{-3}) grown on a <100> p$^-$ (2-20 ohm-cm), Fig. 1. The Ge epitaxy is grown using a novel low temperature (i.e., ~460°C) high density plasma chemical vapor deposition process (HDP-CVD) [6]. The Ge is grown after contact and breakdown adjust implants and anneals were performed on the silicon epi-wafer. A consequence of this low temperature growth process is that the dislocation density is very high in the Ge (i.e., $\sim 5 \times 10^{10}$ cm^{-2}). To minimize the impact of the dislocations on the dark current the germanium is heavily p-type doped, which minimizes the depletion region width in the highly defective lattice mismatched Ge. Similar approaches using high p-type doping in the Ge on Si have previously reported low dark currents in both low temperature molecular beam epitaxy (MBE) and poly-crystal p$^+$-Ge/n-Si detectors [2, 4]. One important difference in this device structure compared to the previous research efforts is that it places the p-n junction entirely in the silicon shifting the high field region further away from the highly defective Ge-Si interface.

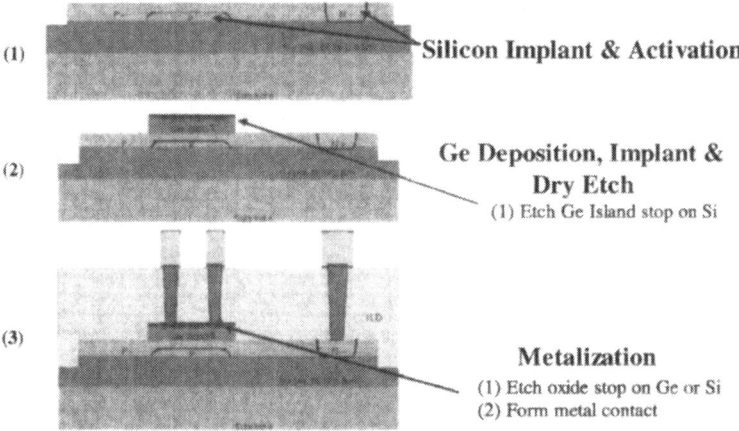

(1)   Silicon Implant & Activation

(2)   Ge Deposition, Implant &
Dry Etch
(1) Etch Ge Island stop on Si

(3)   Metalization
(1) Etch oxide stop on Ge or Si
(2) Form metal contact

Figure 1. Abbreviated process flow for Ge-Si SAM-APD

The final diode structure was formed by etching the Ge stopping on the silicon followed by an oxide cap, planarization and subsequent Ti/TiN/W plugs with Al contact pad formation that electrically connect with the buried layers, Fig. 1. We note that the maximum temperature for the Ge process flow can be kept relatively low (~460 °C) because activation of implanted dopants in germanium does not require as high temperature as in Si (350-700 °C) [7]. After the implant anneals in the silicon, the highest thermal budget step used in this work was a 700°C, 1 hour, N$_2$ anneal.

Room temperature I-V measurements, Fig. 2 (a) & (b), show several different regions of bias dependent behavior. At biases below approximately 3V the diode is not appreciably sensitive to NIR (1550 nm) photoexcitation and the dark current is relatively low. At biases above ~3V the 1550 nm photoresponse increases above the dark current by about an order of magnitude. This on-set of NIR sensitivity corresponds to the expected bias necessary to extend the depletion region through the p-type silicon into the germanium (i.e., "punch-through"). The dark current at punch-through is still relatively low, ~ 0.1 mA/cm^2 compared to previous reports of Ge-Si devices, which range from 0.1 – 10 mA/cm^2 current densities [2-4]. With increasing bias up to ~15V the photoresponse and dark current increase by about 2-3 and a factor of 10 respectively. Beyond 15V both dark current and photoresponse increase rapidly. The increase in the photoresponse is calculated in Fig. 2 (b) as: Photocurrent Multiplier = $(I - I_{dark})/I_{punch-through}$.

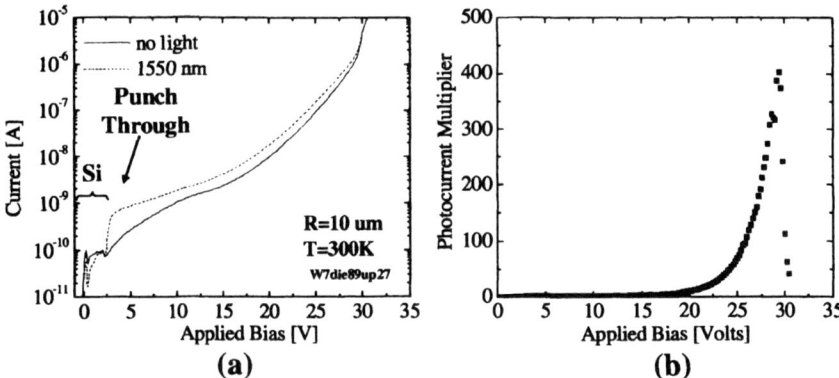

Figure 2. (a) current dependence on voltage with and without 1550 nm illumination; and (b) relative increase in photocurrent at 5 V with increasing applied bias.

The responsivity of the Ge-Si APD was measured at varying intensities and a linear response was observed over the entire range of laser powers (≤1 mW). A responsivity of 4.5x10^{-5} A/W (1550 nm) and 3.2x10^{-4} A/W (1310 nm) were measured. Several identifiable factors likely contribute to the low responsivity in this device including a thin absorption region due to the short recombination lifetime (e.g., 5-15 nm was estimated previously in highly dislocated Ge[2]) and no anti-reflection coating (R~ 0.51).

Passively gated GM detection was also used to characterize the detectors. Passive gating is a technique where the APD is biased above the breakdown voltage for short periods of time during which the APD is sensitive to single carrier injection into the junction, which is signaled by a large avalanche pulse. This technique is described in more detail elsewhere [5]. A schematic of the GM measurement set-up is shown in Fig. 3. Detection windows of 1-10 μs were used at frequencies between 0.1-20 kHz. A photon counter and time interval counter are used to discriminate Geiger pulses (i.e., counts) and measure the delay time of those counts with respect to the turn on time of the over-bias pulse. Attenuated 1310 and 1550 nm semiconductor lasers were used for the low-photon number NIR light sources. The diodes were mounted within a cryostat to enable measurements down to ~16K from room temperature.

At an over-bias of ~1V approximately 1% of the photoexcited electrons are converted into Geiger pulses. A histogram of dark counts versus time measured at 206K is shown in Fig 3 (b). A characteristic lifetime can be extracted from this dark-count time-dependence to describe the dark count probability (i.e., the exponential decay time). Assuming Poisson statistics, the probability of a dark count in a gate window of any time duration can be estimated from the characteristic lifetime. The measured lifetimes are shown in Fig. 3 (b) for three different temperatures. These lifetimes correspond to ~9x10^{-4} chance of a dark count in a 5 ns window at 206K, which is approximately one to two orders of magnitude higher than state-of-the-art InGaAs/InP APDs (note: detection efficiencies are also much higher using InGaAs/InP). The dependence of characteristic lifetime of the dark counts on gating frequency was, furthermore, examined at 208K. A weak dependence of the lifetime on frequency was observed. The lifetime ranged from 6.7 to 7.5 µs for gating frequencies between 1 to 20 kHz.

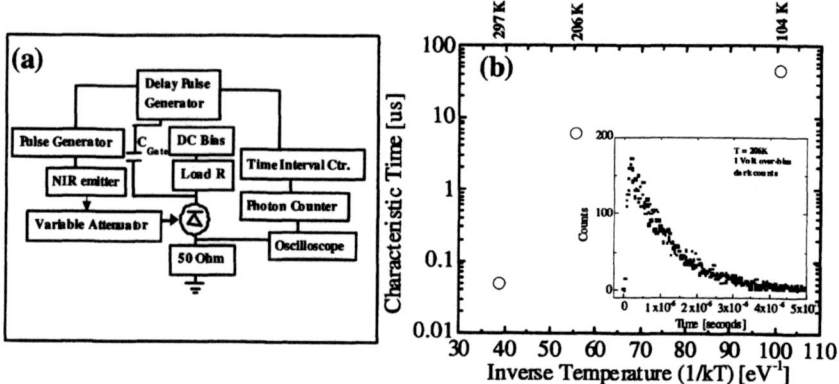

Figure 3. (a) schematic diagram of the Geiger mode experimental set-up and (b) characteristic dark count time for Ge-Si APD with 1V over-bias on the Ge-Si APD at each temperature. Inset is a histogram of dark counts of Ge-Si detector at 206K.

DISCUSSION

One-dimensional device simulations were performed using a commercial simulator (MEDICI) to better understand the dark current and photoresponse anticipated from the Ge-Si device structure, Fig. 4 (a) & (b). At low bias, 0V, the electron energy band diagram has a barrier for minority electron carrier transport from the Ge into the n-type Si. The electron barrier comes from the combination of the p-n junction and the valence bandgap off-set.

At low bias, well below the 3V, the diode behaves like a n^{+}/p^{-} Si diode leading to low dark current until the p^{-} Si is depleted out to the edge of the Ge. The rapid increase in the dark current as the p^{-} silicon is depleted is due to the increased injection of electrons from the Ge, which is greater because of the high intrinsic carrier concentrations and the shorter minority carrier lifetime in the Ge. At biases between punch-through and approximately 15V the current slowly increases. In this bias region the barrier at the Ge/Si interface is slowly decreasing as an increasing number of holes are depleted from the Ge. Because the Ge is heavily doped, the depletion region increases slowly, which is an intentional feature in the device design to limit the amount of electron-hole generation. The small increase in dark current in this bias range is due to the combined increase in depletion region generation and the continuous lowering of the barrier at the Ge/Si interface, which appears entirely lowered at 12V of applied bias. Above ~15V the simulation predicts a rapid increase in current qualitatively consistent with the experiment. The increase in current is due to impact ionization in the simulation. The Ge-Si simulation offers a qualitative explanation for the primary contributions to the dark current, which is quantitatively also within an order of magnitude for the entire bias range. We note that the simulation was done for a 1 μm p-type (5×10^{14} cm^{-3}) silicon layer, which leads primarily to a difference in the breakdown voltage, but has a small effect on the low bias current magnitudes. One insight that the simulation provides is that the observed photomultiplier could be a combination of avalanche gain and barrier lowering to electron injection. If we assume that impact ionization gain begins at ~15V as predicted in simulation, then the measured gain is of order of 200.

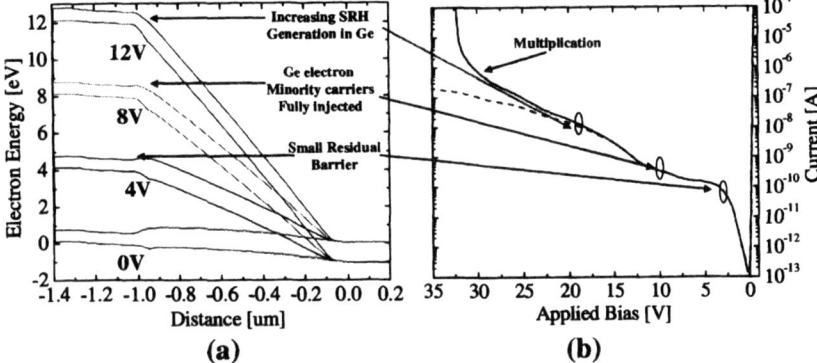

Figure 4. (a) electron energy band diagram of the simulated p$^+$Ge/p$^-$ Si/n$^+$ Si diode under different applied biases; and (b) the simulated current dependence on applied bias with and without an impact ionization model to calculate multiplication gain.

The simulated dark current is very sensitive to the minority carrier lifetime in the Ge and can be used as a fitting parameter. A lifetime of 10^{-8} seconds results in a similar dark current as experiment and is not far from that reported for lifetimes in e-beam deposited poly-Ge on silicon, ~5 ns [1]. Similar order of magnitude lifetimes would suggest similar diffusion lengths. The responsivity in the Ge-Si devices of this work are, however, relatively low compared to previous reports from poly-Ge and low temperature Ge epitaxy on Si. Assuming that high doped mobility are similar than similar diffusion lengths and absorption lengths would be expected. One difference between the poly-Ge and highly dislocated Ge-Si devices in this work is the thermal budget. A 700°C, 1 hour, N$_2$ anneal could lead to Ge and Si interdiffusion that extends beyond the diffusion length. The absorption coefficient will be significantly reduced by SiGe alloying due to the increase in bandgap if the alloy layer extends through the depletion region. The responsivity can likely be improved by increasing the electron carrier lifetime through the introduction of lower dislocation Ge on Si epitaxy techniques (e.g., cyclic annealing). Although more interdiffusion can be expected, the lifetime in low dislocation Ge will be orders of magnitude larger extending the diffusion and absorption length far beyond the alloy thickness.

Alternatives to the InGaAs/InP NIR single photon detectors have been of recent interest. A critical challenge currently to 1310 and 1550 nm GM detection is reducing the dark count rate and its dependence on traps in the InP. Relatively high numbers of charge traps in the InP, which fill during the GM pulse and de-trap after the reset, are believed responsible for this problem (i.e., after-pulsing). Silicon APDs are known to have far superior after-pulsing dependence because of the far fewer traps in the silicon starting material. The Ge-Si APD structure is, therefore, of interest for its combination of a high field silicon region and its NIR Ge absorption region. A primary concern in the Ge-Si devices structure is the highly dislocated Ge and Ge-Si interface where electrically active traps might be expected. However, the small frequency dependence up to 20 kHz on characteristic dark count lifetime is a promising result that suggests that the Ge-Si APD does not suffer from very high densities of after-pulsing traps. Because silicon is known to have very few traps, it is not unexpected that the high field avalanche region does not produce significant after-pulsing, however, this does not explain why the defective interface and germanium do not contribute large amounts of defect initiated dark counts. A possible explanation for the relatively low dark counts and very low after-pulsing dependence on operation frequency is that the electrically active threading dislocations produce acceptor defects. This hypothesis is consistent with previous deep level transient spectroscopy measurements [8]. We also note that InGaAs-Si fusion bonded APD structures that have an imperfect interface between the two materials also do not suffer unusually high after-pulsing [9].

CONCLUSIONS

Ge-on-Si APDs have been fabricated (T~700°C) using a low temperature HDP-CVD epitaxy process. The HDP-CVD Ge epitaxy is heavily dislocated and high boron doping is used to minimize dark current by minimizing the depletion width in the Ge. The Si doping profile was, furthermore, designed so that the high field avalanche region is in the silicon and not at the Ge/Si interface. When the bias is increased the photocurrent increases by as much as 400 before breakdown, which substantially compensates for the low responsivity of the low temperature Ge-Si detector. To the authors knowledge this is the first report of gain greater than ~10-20 from silicon impact ionization in a Ge-Si APD.

These Ge-Si APDs were also investigated to evaluate the promise of the combined Ge NIR absorption and Si avalanche region for GM operation. Low dark counts and dark count dependence on frequency were observed despite the large number of defects. A hypothesis to explain this observation is that the defects due to the dislocations are majority carrier traps (i.e., hole traps). These results are encouraging because much lower dislocation density Ge (on Si) can be formed and still be compatible with this Ge-Si APD process flow (e.g., cyclic annealing).

ACKNOWLEDGMENTS

The authors would like to acknowledge invaluable assistance with the cryostat set-up from T. M. Bauer. Sandia is a multiprogram laboratory operated by Sandia Corporation, a Lockheed Martin Company, for the United States Department of Energy's National Nuclear Security Administration under contract DE-AC04-94AL85000.

REFERENCES

[1] L. Colace, G. Masini, and G. Assanto, *IEEE J. of Quant. Elec.*, vol. 35, pp. 1843, 1999.
[2] P. Bandaru, S. Sahni, E. Yablonovitch, J. Liu, H.-J. Kim & Y.-H. Xie, *Mat. Sci. & Eng. B*, v 113, 79-84, 2004.
[3] G. Isella, J. Osmond, M. Kummer, R. Kaufmann, and H. v. Kanel, *Semi. Sci. & Tech.*, vol. 22, pp. S26-28, 2007.
[4] G. Masini, L. Colace, F. Galluzzi, and G. Assanto, *Mat. Sci. & Eng. B*, vol. 69, pp. 257, 2000.
[5] A. Laciata, F. Zappa, S. Cova, and P. Lovati, *Applied Optics*, vol. 35, pp. 2986, 1996.
[6] M. S. Carroll, J. Sheng, and J. C. Verley, MRS Proceedings, vol. 934, I09-02, 2006.
[7] Y. S. Suh, M. S. Carroll, R. A. Levy, G. Bisognin, D. De Salvador, M. A. Sahiner, and C. A. King, *IEEE Trans. on Elec. Dev.*, vol. 52, pp. 2416, 2005.
[8] P. M. Mooney, L. Tilly, C. P. D'Emic, J. O. Chu, F. Cardone, F. K. LeGoues, and B. S. Meyerson, *J. of Appl. Phys.*, vol. 82, pp. 688-695, 1997.
[9] Y. Kang, Y.-H. Lo, M. Bitter, S. Kristjansson, Z. Pan & A. Pauchard, *Appl. Phys. Lett.*, vol. 85, 1668, 2004.

Mater. Res. Soc. Symp. Proc. Vol. 989 © 2007 Materials Research Society 0989-A12-06

Modeling and Characterization of the Hydrogenated Amorphous Silicon Metal Insulator Semiconductor Photosensors for Digital Radiography

N. Safavian[1], Y. Vygranenko[1], J. Chang[1], Kyung Ho Kim[1], J. Lai[1], D. Striakhilev[1], A. Nathan[2], G. Heiler[3], T. Tredwell[3], and M. Fernandes[4]

[1]Electrical and Computer Engineering, University of Waterloo, Waterloo, N2L3G5, Canada
[2]London Centre for Nanotechnology, University College London, London, WC1H OAH, United Kingdom
[3]Eastman Kodak Company, Rochester, NY, 14650-23487
[4]Electronics Telecommunication and Computer Dept., ISEL, Lisboa, Portugal

ABSTRACT

Because of the inherent desired material and technological attributes such as low temperature deposition and high uniformity over large area, the amorphous silicon (a-Si:H) technology has been extended to digital X-ray diagnostic imaging applications. This paper reports on design, fabrication, and characterization of a MIS-type photosensor that is fully process-compatible with the active matrix a-Si:H TFT backplane. We discuss the device operating principles, along with measurement results of the transient dark current, linearity and spectral response.

INTRODUCTION

X-ray detection has been traditionally carried out with a phosphor screen, which converts incident x-ray photons to visible light for subsequent detection with a light-sensitive film. This system has been accepted and widely used by radiologists for more than a century. But as is well-known, it comes with huge storage requirements associated with archiving of images, and more importantly, it is digital-incompatible.

Amorphous silicon active-matrix flat-panel imagers using photosensors embedded in the thin-film transistor (TFT) backplane has been demonstrated to be a viable replacement [1]. Although the PIN or MIS structure can be employed for the photosensor [2], [3], the advantage of the latter is its fabrication-compatibility with the TFT switch, as it requires the same semiconductor and dielectric layers [4]. In addition, a continuous (semiconductor and dielectric) layer architecture can be employed for implementation of a high fill-factor array [5]. This leads to an effective reduction in the number of required masks. More importantly, since the MIS does not need a P-layer, imager fabrication can take advantage of the general purpose fabrication infrastructure used for producing large-area TFT displays.

One of the disadvantages of the MIS photosensor, however, is its high dark signal level. When photons are absorbed in the semiconductor layer, the photo-generated electrons drift to the n+ region under the influence of the externally-applied bias. Correspondingly, photo-generated holes drift to the insulator-semiconductor interface, where they become trapped at the interface. When subject to a refresh pulse periodically applied to the MIS device, electrons are injected from the n+ layer into the intrinsic amorphous silicon and drift to the dielectric interface. Some of these electrons recombine with holes at the interface while others become trapped in bulk and interface states. Upon return to reverse bias, the trapped electrons are emitted with a range of emission time constants, leading to a current transient [6]. This serves to undermine the noise performance of the imager. Thus for low dark current, the biasing conditions, and in particular, the refresh levels need to be judiciously chosen so as to minimize the possible effects of carrier trapping and recombination.

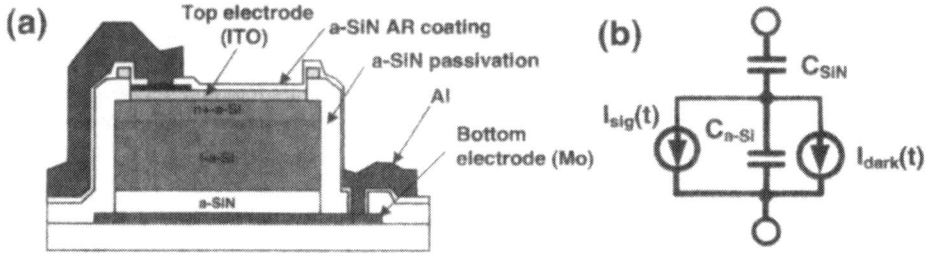

Figure 1: (a) Cross section of the fabricated MIS photosensor, and (b) its small signal model.

EXPERIMENTAL PROCEDURE

The cross-section of MIS photosensor is shown in Fig. 1(a). The MIS stack incorporates ~100-nm-thick Mo bottom electrode, 120-nm-thick a-Si:N dielectric layer, and ~1-μm-thick a-Si:H semiconductor layer. A 20-nm-thick n+-a-Si:H contact layer is provided between the a-Si:H layer and the ITO top electrode. MIS test structures were segmented using RIE and passivated with ~0.8-μm-thick a-SiN layer. A thin (~100 nm) a-Si:N layer was deposited over ITO to serve as an anti-reflection coating layer. Probing pads and interconnects were formed by ~1-μm-thick Al layer that contacted top and bottom electrodes via openings in the a-Si:N layers.

Semiconductor and dielectric layers were deposited by plasma enhanced chemical vapor deposition at 300°C. Magnetron sputtering was used for the metal- and ITO films.

The a-Si:H MIS structure can be represented by the simple model shown in Fig. 1(b). Here, the capacitances C_{SiN} and C_{a-Si} represent silicon nitride and amorphous silicon capacitances, respectively. The model also incorporates two current sources labeled as $I_{dark}(t)$ and $I_{sig}(t)$. The current course $I_{dark}(t)$ represents the dark current arising from emission of charge from the trap states at the a-SiN/a-Si interface and in the bulk of intrinsic a-Si:H layer. The current source $I_{sig}(t)$ represents the photocurrent generated in the i-layer.

The MIS samples were characterized with transient current measurements in the dark and under pulse illumination. The measurement set-up and corresponding test signal waveforms are shown in Fig. 2(a) and 2(b), respectively. The measurement set-up includes a digital pulse generator, a pulse amplifier, a light source (green LED), a charge amplifier, and a digital oscilloscope. The digital generator provides the pulses required to control bias voltage, incident light pulses, as well as the control and triggering pulses for the charge amplifier and the digital scope. The pulse amplifier enables control of the amplitude and offset voltage of the bias pulse (Fig. 2(b)). The charge amplifier integrates the output current of the MIS structure into voltage, and the digital scope captures the signal and saves the output data in the computer.

In the test measurements we have flexibility over choice of the amplitude and offset levels of the bias voltage pulse in the range from −10 V to +10 V. As it is shown in Fig. 1(b), Vpulse is the amplitude of the voltage pulse applied to the MIS structure, and Vrefresh and Vconv are the voltages applied on the cathode during refresh and conversion periods, respectively. In order to carry out consistent and reproducible measurements, the sample was fully isolated from ambient light. The output signal was analyzed as described as follows.

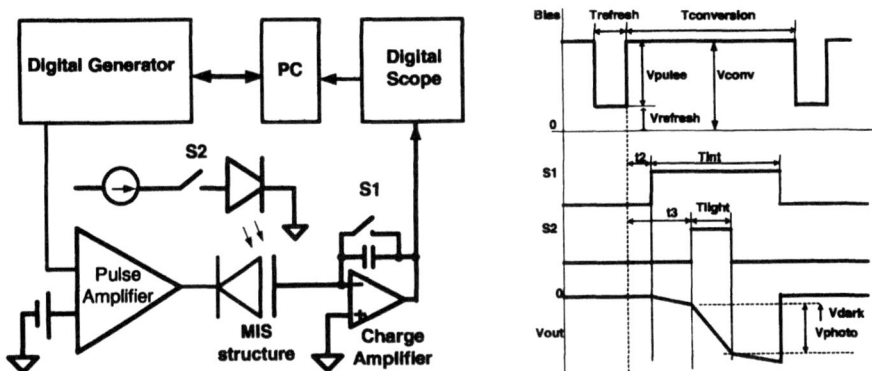

Figure 2: (a) Block diagram of the measurement setup and (b) test signal waveforms.

The relationship between the dark current I_{dark} and the output voltage V_{out} is

$$I_{dark} = C_f \cdot \frac{dV_{out}}{dt},$$ (1)

where C_f is the feedback capacitor of the charge amplifier. For the structure with an active area of 1×1 mm2, the capacitor C_f was chosen to be 490 pF. The photocurrent was calculated using the waveform parameters of the output signal under light exposure as shown in Fig. 2(b).

$$I_{photo} = \frac{C_f \cdot V_{photo}}{T_{light}} - I_{dark}.$$ (2)

RESULTS AND DISCUSSION

Fig. 3 shows the dark charge density generated by the a-Si:H MIS structure during the integration time $T_{int} = 2$ s in the conversion mode. A delay $t_2 = 20$ ms between refresh and integration pulses was introduced to avoid integration of the charge induced by the geometrical capacitance of the device. Keeping the bias voltage in conversion mode constant ($V_{conv} = 4$ V), the amplitude of the voltage pulse, V_{pulse}, was changed from 4V to 8V, and the refresh voltages varied from 0 V to -4 V in steps of 1 V. Apparently, at the negative bias voltages applied during the refresh period, the electrons injected from n+-layer fill most of the trap states in the i-layer and at the a-Si:H/SiN interface causing the large dark signal. Fig. 4 shows the dark charge density as a function of time, measured at constant pulse amplitude of 4 V and positive refresh voltages. The increase in the refresh voltage from 0 to 6 with a step of 2 V reduces the dark charge integrated during 2 s by an order of magnitude. In the case when $V_{refresh} > 0$, only some portion of the interface traps are filled by electrons during the refresh period, resulting in a significantly lower level of dark current during the conversion period.

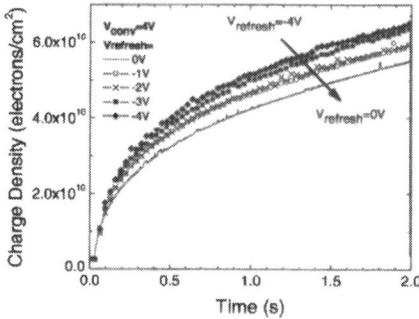 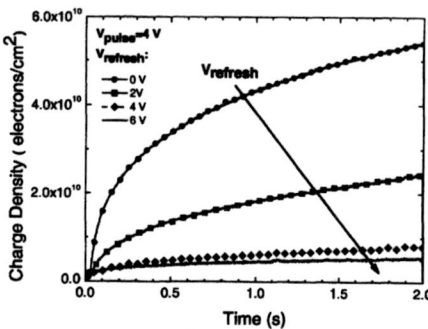

Figure 3: Dark charge as a result of trapped electron emission versus time at negative refresh bias voltages applied to the cathode of the a-Si:H MIS structure. Bias voltage in conversion mode is +4 V.

Figure 4: Dark charge versus time at positive refresh bias voltages applied to the cathode of the a-Si:H MIS structure. The refresh pulse amplitude is 4 V.

Fig. 5 shows the transient dark current density calculated according to equation (1) for the data plotted in Fig. 4. The time dependence of the dark current is

$$I_{dark}(t) \sim t^{-\beta} \ , \tag{3}$$

where β is a parameter evaluated to be in the range 0.7 to 0.85 depending on the biasing conditions. The same time dependence for the dark transient current has been observed in a-Si:H p-i-n diodes and can be explained by deep trapping and thermal emission of electrons [7], [8]. Fig. 6 shows the measured output waveforms keeping constant both the refresh pulse amplitude ($V_{pulse}= 4$ V) and intensity of the 100 ms light pulse. We observe no noticeable change in the amplitude of the photoresponse with variation of the refresh voltage $V_{refresh}$ in the range −4 V to +6 V. This experiment proves that the sensitivity of the MIS structure is about the same for positive and negative $V_{refresh}$, which determines both the electric field across the SiN layer and population of the interface traps. The refresh pulse amplitude is also a critical parameter. It determines both the sensitivity and dynamic range of the a-Si:H MIS structure, since an initial voltage drop across the i-layer in the conversion period is close to V_{pulse}. The small signal model shown in Fig. 1(b) predicts that a saturation charge capacity Q_S of the device in photodiode mode is $Q_S = C_{SiN} \cdot V_{pulse}$. For the fabricated samples, $C_{SiN}=47$ nF/cm2 and, therefore, Q_S is determined to be 1.1×10^{12} electrons/cm^2 at $V_{pulse}= 4$ V. Under some refresh conditions, the MIS device may not be completely refreshed, leading to a reduction in the saturation charge capacity.

The linearity of the MIS photosensor was examined by measuring the output charge induced by the light pulses. The amount of charge injected is varied by changing the light pulse duration. Fig. 7 shows the signal charge as a function of the light pulse duration. Obviously there is a linear relationship between the photo-injected charge and output charge demonstrating the linearity of the MIS photosensor.

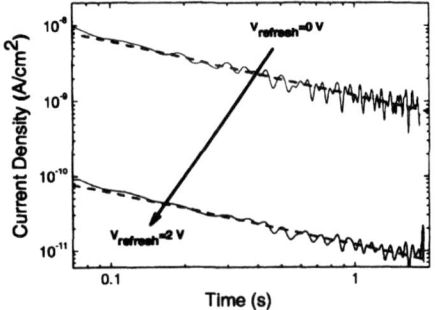

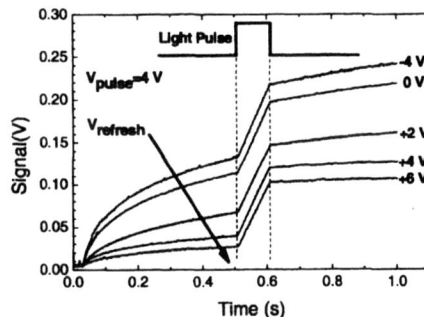

Figure 5: Transient dark current at different refresh bias voltages. The refresh pulse amplitude is 4 V.

Figure 6: Output waveforms at different refresh voltages. The light pulse intensity and refresh pulse amplitudes are constant.

The quantum efficiency (QE) can be formulated as

$$QE = \frac{I_{photo}}{q \cdot \Phi \cdot A}, \tag{4}$$

where A is the photosensitive area of the MIS structure, and Φ the photon flux. The photon flux Φ was measured using a calibrated a–Si PIN photodiode

$$\Phi = \frac{I_{photo}}{q \cdot QE_{PIN} \cdot A_{PIN}}, \tag{5}$$

where QE_{PIN} is the known quantum efficiency of the reference photodiode, and A_{PIN} is the photosensitive area.

The measured spectral response of the MIS sensor is shown in Fig. 8. The measurements were performed at V_{pulse}=4 V and $V_{refresh}$=2 V. The external quantum efficiency reaches a peak value of 67% at a wavelength of 620 nm. It is noticeably lower than the quantum efficiency of the optimized a-Si:H p-i-n photodiodes, which is typically 80-85%. The reason is that the signal charge, which is generated in the a-Si layer, cannot be fully transferred to the external circuit because of the blocking dielectric layer. According to the equivalent circuit model of the MIS structure shown in Fig. 1(b), the charge transfer efficiency (CTE) is

$$CTE = \frac{I_{out}}{I_{sig}} = \frac{C_{SiN}}{C_{SiN} + C_{a-Si}} = \frac{1}{1 + \dfrac{\varepsilon_{Si} d_{SiN}}{\varepsilon_{SiN} d_{Si}}}, \tag{6}$$

where ε_{Si}=11.8 and ε_{SiN}=6.4 are the dielectric constants of the silicon and nitride, respectively. For actual layer thicknesses d_{SiN}=120 nm and d_{Si}=1 µm a calculated CTE is 0.82.

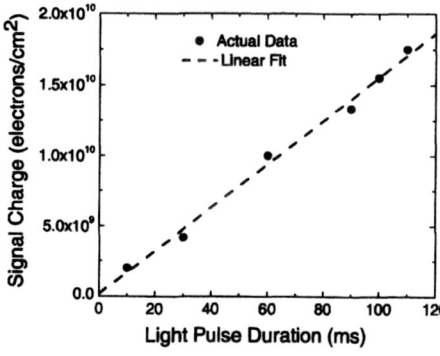

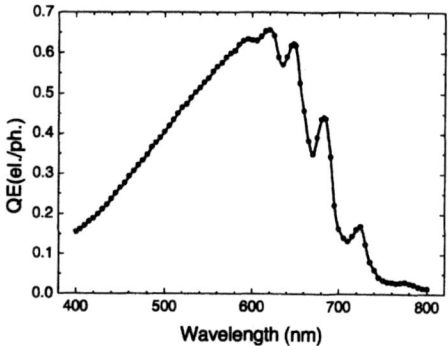

Figure 7: Signal charge versus light pulse duration.

Figure 8: Measured spectral response characteristic of the a-Si:H MIS structure.

CONCLUSION

The a-Si:H MIS-type photosensor reported here has been fabricated using a process compatible with the TFT backplane. Measurements of the transient dark current under different biasing conditions reveal that the associated noise component can be largely eliminated by adjusting the bias voltage during the refresh period. The sensor shows a linear photo-response and external quantum efficiency up to 67%.

ACKNOWLEDGMENTS

This work is supported by Eastman Kodak Company and the Natural Sciences and Engineering Research Council of Canada (NSERC).

REFERENCES

1. J. Beutel, and H. L. Kundel, "Handbook of Medical Imaging", SPIE Press, 2000.
2. R. A. Street (Ed.), in *Technology and Applications of Amorphous Silicon* (Springer, Berlin, 2000) pp.147-221.
3. M. Watanabe et al. *Proc. SPIE*, **4320**, 103 (2001).
4. C. Mochizuki, Patent US 6682960B1, Jan. 27, 2004.
5. M. D. Wright, Patent Application Publication US 2006/0001120 A1, Jan. 5, 2006.
6. Kabayashi et al. Patent US 6245601B1, Jun. 12, 2001.
7. H. Wieczorek, *J. Non-Cryst. Sol.* **164-166**, 781 (1993).
8. M.J. Powell, *Appl. Phys. Lett.* **43** (6), 15 (1983).

Crystallization Techniques I

Mater. Res. Soc. Symp. Proc. Vol. 989 © 2007 Materials Research Society
0989-A13-03

An Approach to Obtain Single Layer of Nanostructured Si by Laser Irradiation on Ultrathin Amorphous Si Films

Jun Xu, Zanhong Cen, Jiang Zhou, Wei Li, Xinfan Huang, and Kunji Chen
National Laboratory of Solid State Microstructures and Department of Physics, Nanjing University, Nanjing, 210093, China, People's Republic of

ABSTRACT

An approach to achieve single layer of nanostructured Si array on insulating layer with high density was proposed and demonstrated. It was found that a single layer of nanocrystalline Si array can be formed by using KrF pulsed excimer laser irradiation on ultrathin hydrogenated amorphous silicon films (4-30nm) followed by thermal annealing. Under the suitable fabrication conditions, the areal density of formed nanocrystalline Si is as high as $10^{11} cm^{-2}$ and the lateral size is around 10nm while the height is about 2-4 nm. Visible light emission was observed from the Si nanostructures at room temperature and the luminescence peak is varied from 660 to 725 nm with increasing the amorphous Si film thickness. The variable luminescence can be attributed to the interface state assisted radiative recombination rather than the quantum size effect.

INTRODUCTION

Recently, nanocrystalline Si (nc-Si) films have attracted much attention because of their potential applications in future optoelectronic and nano-electronic devices [1]. One of the key issues today is to find a cheap and controllable way to achieve Si nanostructures. From the viewpoint of device applications, the developed approach must be compatible with the modern Si semiconductor technology while both the size and position of the formed Si nanostructures should be well controlled. Moreover, it is also desirable to achieve the Si nanostructures on the insulating layers such as SiO_2 or SiN_x films [2].

So far, many approaches have been proposed to fabricate semiconductor nanostructures, especially nc-Si films. For example, nc-Si embedded in SiO_2 can be prepared by implantation of Si ions into SiO_2 films with subsequently high temperature annealing or by directly annealing the substoichiometric SiO_x (x<2) films. Another way is to get stacked nc-Si/ SiO_2 structures by thermal annealing the amorphous Si (a-Si)/SiO_2 multilayers. Usually, the annealing temperature is above 1000°C to induce the transition from the amorphous to nanocrystalline phase [3-5]. In our previous works, laser-induced crystallization method was used to obtain size-controllable nc-Si within nc-Si/SiN multilayers and an intense light emission was observed at room temperature under Ar^+ laser excitation [6]. In the present work, laser induced crystallization technique was used on the ultrathin a-Si films deposited on SiN layers to achieve single layer of nanostructured Si array on insulating layer. It was found that a single layer of nc-Si array can be formed by combination of KrF pulsed excimer laser irradiation and subsequent thermal annealing. The size and area density of formed nc-Si was investigated as functions of a-Si film thickness and laser fluence. Room temperature photoluminescence was observed from the obtained Si nanostructures and the origin of luminescence signals was briefly discussed.

EXPERIMENT

Conventional plasma enhanced chemical vapor deposition (PECVD) system with a radio frequency (rf) of 13.56 MHz was used to deposit ultrathin a-Si films (4-30nm) on a-SiN insulating layers. Amorphous Si thin films were prepared by decomposition of silane gas and 30nm-thick a-SiN films were prepared by using silane and ammonia gas mixtures. During the deposition, the substrate temperature and the rf power was kept at 250°C and 30W, respectively. The reaction pressure was about 20pa. Crystalline Si wafers and fused quartz glass were used as substrates for various measurements.

Laser induced crystallization was performed in air ambient by using KrF excimer laser operating at 248nm wavelength with 30ns pulse duration. Only a single pulse with laser fluence ranging from 0.4 to 1.4 J/cm^2 was employed and the corresponding laser spot area is 5×3mm^2 on the a-Si samples. After the laser irradiation, all samples were subjected to a thermal annealing at 900°C for 30min under N$_2$ ambient in order to relax the stressed film network. Raman scattering spectroscopy and transmission electron microscopy (TEM) was used to study the formation process of Si nanostructures after laser irradiation while the surface morphologies of samples were investigated by using atomic force microscopy (AFM). Room temperature photoluminescence was measured under the excitation of Ar$^+$ laser (488nm).

DISCUSSION

Laser induced crystallization of hydrogenated amorphous silicon (a-Si:H) films is widely used to obtain micro- or poly-crystalline Si for fabricating thin film transistors (TFTs) in large area electronics [7]. Usually, the film thickness of a-Si:H is in the scale of micrometer. It is interesting to investigate the laser induced crystallization of ultrathin a-Si:H films with film thickness in nanometer scale. It was found that the film surface morphologies were obviously changed under the suitable laser irradiation conditions.

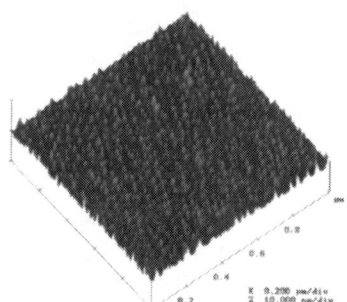

Fig. 1. AFM topographic image of irradiated a-Si:H films at a laser fluence of 0.79 J/cm^2. The initial a-Si:H film is 4 nm thick.

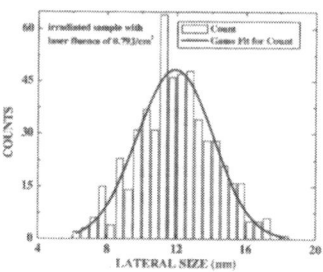

Fig. 2. Distribution of lateral size of Si nanostructures measured from AFM image for irradiated sample at laser fluence of 0.79 J/cm^2. The line is the fitting curve by using Guassian function.

Figure 1 gives AFM image of 4nm-thick a-Si:H film after laser irradiation with a single pulse at a laser fluence of $0.79J/cm^2$. The surface morphology is entirely different from that of as-deposited sample and high-density, well-defined outgrowths with nanometers size can be identified on the surface which indicated the formation of Si nanostructures was occurred under excimer laser irradiation on the ultrathin a-Si:H films.

Both the vertical and lateral size of formed Si nanostructures can be estimated from the AFM image. Figure 2 is the lateral size distribution of Si nanostructures for irradiated a-Si:H with laser fluence of $0.79J/cm^2$. It is found that the lateral size distribution can be well fitted by a Guassian Function and the average lateral size is about 12nm and the size deviation is less than 18%. It is noticeable that the vertical scale of the AFM image is 10nm and the vertical size of the Si outgrowths is estimated about 3-5nm from AFM measurements.

Fig. 3. Cross-sectional high resolution TEM micrograph of irradiated a-Si:H sample with thickness of 4nm.

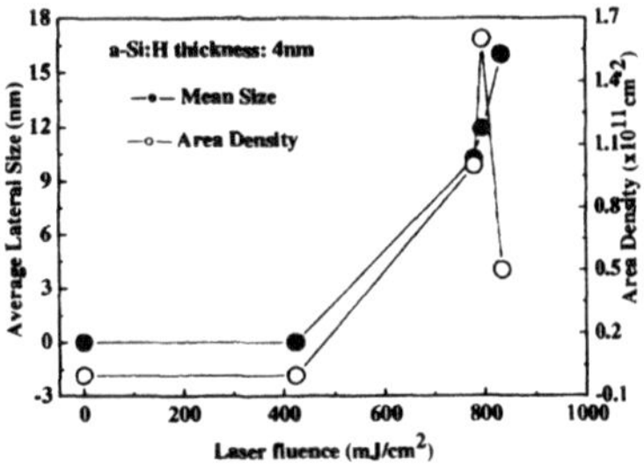

Fig. 4. Average lateral size and area density of formed Si nanostructures measured from AFM image as a function of the laser irradiation fluence for irradiated 4nm thick a-Si:H film.

The presence of nanocrystalline Si can be directly confirmed by TEM images as shown in Fig.3, which is the cross-section high resolution transmission electron microscopy (HREM) image of 4nm thick a-Si:H film irradiated at 0.79 J/cm^2. It is clearly demonstrated the formation of crystallized Si during the laser irradiation and the subsequent thermal annealing process. The Si nanocrystals show the (111) orientation faces with inter-planar distance of 3.1Å as revealed in Fig. 3. The nanocrystalline Si are not formed in spherical but rather in ellipse shape in profile with the lateral size of 6nm and the height of about 3.2nm, which is close to the initial a-Si:H layer thickness. It is worth noting that the capping layer of a-SiN shown in Fig.3 was deposited after laser irradiation for protect samples during the TEM sample preparation process.

Laser irradiation on a-Si:H films induces the melt of a-Si layer when the laser fluence exceeds the threshold value and then crystallization is occurred via the super-freezing of the silicon melt [8]. With reducing the a-Si:H film thickness to nanometer scale, the threshold laser fluence to induce the crystallization is obviously increased, which is similar with the thermal annealing where the crystallization temperature increases rapidly with decreasing a-Si:H layer thickness [9,10]. Figure 4 shows the average lateral size and area density of formed Si nanostructures esimated from AFM image as a function of the laser irradiation fluence for irradiated a-Si:H films with initial thickness of 4nm. It was found that the formation of Si nanostructures was occurred when the laser irradiation fluence exceeds 0.7 J/cm^2. The average lateral size is increased with increasing the laser fluence. It is also interesting to see the change of area density of formed Si nanostructures. As show in Fig. 4, the area density increases firstly and then decreases at high laser fluence. It can be estimated that the area density as high as 1.7×10^{11}cm^{-2} can be obtained for samples after irradiation at fluence of 0.79mJ/cm^2. When high laser fluence was used, the melting extends to the whole surface and much denser nucleation sites can be generated. The crystallized Si can be easily aggregated between the randomly distributed nucleation sites to form a nanocrystalline Si with large size and the area density was consequently reduced.

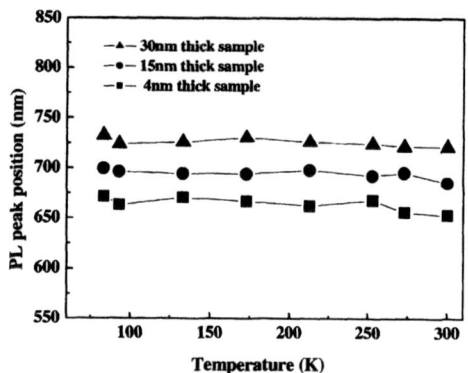

Fig. 5. Temperature-dependent photoluminescence spectra for irradiated a-Si:H samples with various film thicknesses. The excitation light source is Ar$^+$ laser with wavelength of 488nm.

The photoluminescence (PL) measurements were carried out under Ar$^+$ laser excitation with wavelength of 488nmfor all the samples. It was found that the PL signals can not be detected for samples with out subsequently thermal annealing due to the existence of a large amount of nonradiative defects states in laser-irradiated samples. However, after thermal annealing at 900°C for 30min, light emission can be obtained for samples with various film thicknesses and the PL peak energy is located at 660nm (1.87eV), 695nm (1.78eV) and 725nm (1.7eV) for samples with initial a-Si:H thickness of 4nm, 15nm and 30nm, respectively.

The presence of nanocrystalline Si due to the laser annealing plays an important role in obtaining the light emission in our case and the emission bands are dependent of the film thickness. However, it is unlikely that the emission is associated with the band-to-band transition in formed nanocrystalline Si because the mean size of nanocrystalline Si is as large as 9-11nm for the sample with the initial a-Si:H film thickness larger than 15nm as characterized by TEM measurements. In order to further clarify the luminescence mechanism, temperature-dependent PL measurements were carried out and the results are given in the inset of Fig. 5. It is shown that the PL peak position is kept the same at different measured temperature for all samples, including the 4nm-thick one. Since the energy gap of semiconductor have to be increased with decreasing the temperature as expected by Varshni Formula [11], the luminescence peak has to be blue shifted with temperature if it is related to band to band recombination in nanocrystalline Si. We tentatively attributed the visible light emission to the radiative recombination via the interface states or luminescent centers existing in the Si nanodots surface in our samples.

CONCLUSIONS

In conclusion, single layer of nanocrystalline Si with high density can be formed on insulating layer by using KrF pulsed excimer laser irradiation on ultrathin a-Si:H films (4-30nm) followed by thermal annealing. The areal density of formed nanocrystalline Si is as high as $10^{11}cm^{-2}$ and the size can be confined by the a-Si:H film thickness. The main advantages of this techniques is its low temperature process due to its short pulse length and optical absorption depth, which is very important to minimize the damage of glass substrates and to integrate devices based on temperature-sensitive electronic materials. Visible light emission was observed from the laser irradiated sample after subsequently thermal annealing and the luminescence peak is varied from 660 to 725 nm with increasing the amorphous Si film thickness. The luminescence process can be attributed to the interface state assisted radiative recombination due to the formation of nanocrystalline Si.

ACKNOWLEDGMENTS

The work was supported by NSF of China (No. 60425414, 50472066, 90301009, 10574069), NSF of Jiangsu Province (BK2006715) and the State Key Program for Basic Research of China (2006CB932202).

REFERENCES

1. K. L. Wang, J. Nanosci. and Nanotech. 2, 235 (2002) and references therein.
2. J. Heitmann, F. Muller, M. Zacharias and U. Gosele, Advanced Materials 17, 797 (2005).
3. Y. Kanemitsu, S. Okamoto, M. Otobe and S. Oda, Phys. Rev. B 55, R7375 (1997).
4. M. Yang, K. Cho, J. Jhe, S. Seo, J. H. Shin, K. J. Kim, and D. W. Mon, Appl. Phys. Lett. 85, 097401 (2004).
5. J. Mei, Y. Rui, Z. Ma, J. Xu, D.Zhu, L. Yang, X.Li, W.Li, X.Huang and K. Chen, Solid State Commun. 131, 701 (2004).
6. K. Chen, X. Huang, J. Xu and D. Feng, Appl. Phys. Lett. 61, 2069 (1992).
7. D. Fork, G. Anderson, J.Boyce, R. Johnson and P. Mei, Appl. Phys. Lett. 68, 2138 (1996).
8. J. S. Im and R. S. Sposili, MRS Bull. 21, 39 (1996).

9. M. Zacharias, J. Bläsing, P. Veit, L. Tsybeskov, K. Hirschman, and P. M. Fauchet, Appl. Phys. Lett. 74, 2614 (1999).
10. Z. Cen, J. Xu, Y. Liu, W. Li, L. Xu, Z. Ma, X. Huang and K. Chen, Appl. Phys. Lett. 89, 163107 (2006).
11. X. X. Wang, J. G. Zhang, L. Ding, B. W. Cheng, W. K. Ge, J. Z. Yu, and Q. M. Wang, Phys. Rev. B 72, 195313 (2005).

Mater. Res. Soc. Symp. Proc. Vol. 989 © 2007 Materials Research Society   0989-A13-04

Rapid Crystallization of Amorphous Silicon Utilizing the Plasma Annealing at Atmospheric Pressure

Hajime Shirai[1], Yusuke Sakurai[1], Mina Ye[1], Koji Haruta[1], Tomohiro Kobayashi[2], and Yu-ichiro Takemura[3]

[1]The Graduate School of Science and Engineering, Saitama University, 255 Shimo-Okubo, Sakura-ku, Saitama, 338-8570, Japan

[2]The Institute of Physics and Chemical Research, 2-1 Hirosawa, Wako, Saitama, 351-0198, Japan

[3]Japan Science and Technology Agency, 2-1-6 Sen-gendai, Tsukuba, 305-0047, Japan

ABSTRACT

The rapid crystallization of amorphous silicon utilizing the rf inductive coupling thermal plasma jet of argon is demonstrated. Highly crystallized Si films were fabricated on th-SiO$_2$ and textured a-Si:H:B/SnO$_2$/glass by adjusting the translational velocity of the substrate stage. The H concentration in the films decreased from 10^{21} cm^{-3} to 10^{19} cm^{-3} with no marked increases in oxygen and nitrogen impurity concentrations and defect density. The crystallization proceeds from the bottom to front surface in terms of the volume expansion during the solidification and crystallization of liquid Si.

INTRODUCTION

Low-temperature growth of polycrystalline silicon (poly-Si) films on glass is currently attracting considerable attention as a novel low-cost and stable photovoltaic and Si thin film transistors (TFTs) material. To date, extensive studies have been performed for synthesizing crystalline Si films on glass using various methods such as plasma-enhanced chemical vapor deposition (PE-CVD) of silane and pulsed-laser crystallization of amorphous silicon (a-Si) films [1-3]. The crystal size, however, is limited in several tens nanometer in the case of PE-CVD of SiH$_4$, because the surface diffusivity of deposition precursors such as SiH$_3$ determines the crystal size on the growing surface. On the other hand, a large crystal grain size of 1-5 μm has been established, whose TFT mobility reaches 100-600 cm^2/Vs. However, expensive and complex components such as a high-power laser and optical systems are needed. To date, we have employed the rapid crystallization of a-Si utilizing the thermal microplasma jet [4]. In this paper, we demonstrate the rapid crystallization of a-Si utilizing the very-high-frequency VHF thermal plasma annealing..

EXPERIMENTAL DETAILS

The experimental se up is schematized in Fig. 1. A quartz glass tube with an inner diameter of 2 mm was used as the argon gas inlet. The VHF power was supplied to the 3 turn coil through the matching circuit. The crystallization of a-Si was performed with translating velocity of the substrate stage, v_{sub}, and Ar flow rate, [Ar], as variables, by a low-pressure rf PE-CVD method, using a SiH_4-H_2 mixture at a substrate temperature of 250°C. Crystallized Si films were evaluated by Raman spectroscopy, Fourier-transform infrared reflection absorption spectroscopy (FTIR), electron-spin-resonance (ESR), spectroscopic ellipsometry (SE), and transmission electron microscopy (TEM) characterizations. The depth profiles of residual H, O, N and C impurity concentrations in the films were measured by secondary ion mass spectroscopy (SIMS) before and after the plasma annealing.

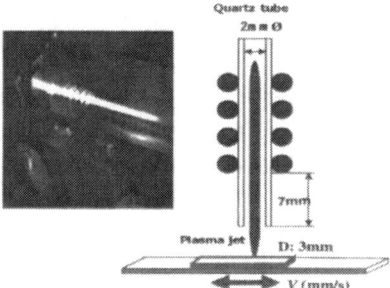

Figure 1. Schematic of experimental device and Ar emission.

RESULTS AND DISCUSSION

Figure 2 shows Raman and reflectance FTIR spectra of 300-nm-thick crystallized a-Si films annealed at different translating velocities of the substrate stage v_{sub}s. Notably, a Raman peak appeared at 518-520.5 cm^{-1}, corresponding to the c-Si transverse optical (TO) phonon mode, in the case of crystallized a-Si films annealed at a stage velocity of 0.01-350 mm/s. Thus, there exist a optimal v_{sub} value that yield high film crystallinity, although it also depends on the thickness of a-Si film precursor and VHF discharge power. In addition, the SiH$_x$ related absorption peaks at 2000-2100 cm^{-1} were upward pointing peak, whereas the SiO$_x$ absorption peak at 1100 cm^{-1} was a downward pointing peak. These results suggest that the crystallization of a-Si proceeds with an abstraction of hydrogen and the surface oxidation during the plasma exposure. In addition, the ESR study revealed that the spin density of the films increased by about one order of magnitude from 3×10^{16} cm^{-3} to $2\text{-}4\times10^{17}$ cm^{-3} for the plasma exposure time above 1 s.

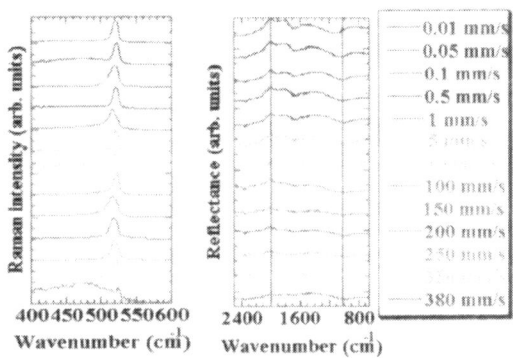

Figure 2. Raman and reflectance FTIR spectra of a 0.3-µm-thick crystallized Si film by the plasma annealing at different stage velocities of substrate v_{sub}s.

Figures 3 shows typical cross-sectional TEM image and SIMS depth profiles of H, O, N, and C impurity concentrations before and after the plasma annealing for 150-nm-thick crystallized a-Si films fabricated on *th*-SiO$_2$. The film crystallization proceeded toward the film thickness and the grain size reached at 150-200 nm.

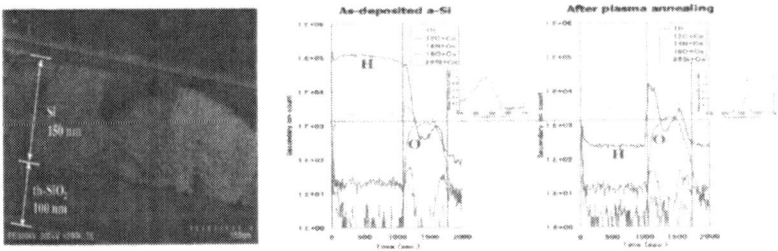

Figure 3. Cross-sectional TEM image and SIMS depth profile of crystallized a-Si film deposited on *th*-SiO$_2$ before and after the plasma annealing.

The H concentration in the films decreased markedly by 2-3 orders of magnitude after plasma annealing. The O, N and C concentrations in the films decreased slightly in the bulk and bottom interface regions, but increased slightly near the front surface.

Figure 4 shows the Raman spectra, cross-sectional TEM images, and SIMS depth profiles of crystallized Si films deposited on textured a-Si:H:B/SnO₂/glass annealed at different v_{sub}s. The crystallization was promoted uniformly with decreasing v_{sub}. Note that the crystallization proceeded from the bottom texture to front surface and the crystallization proceeded at the top surface of a-Si:H:B layer and no inter-diffusions of B and Sn elements to crystallize a-Si layer were observed before and after the plasma annealing.

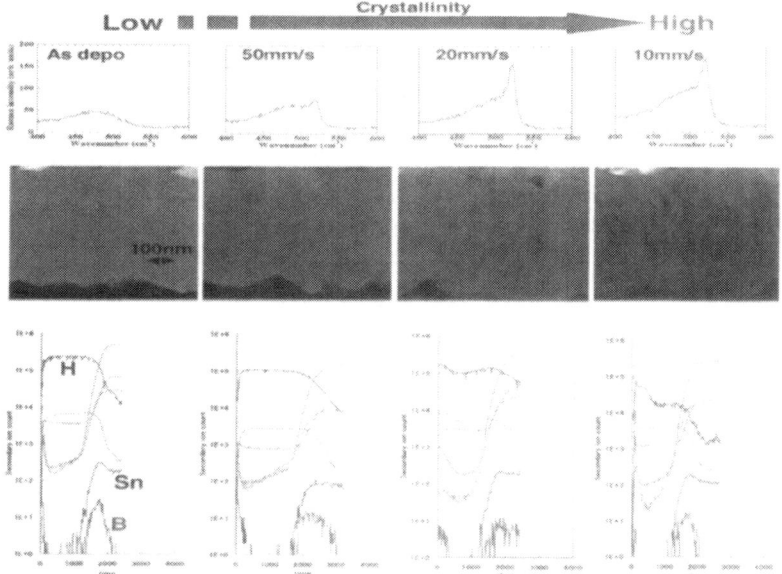

Figure 4. Raman spectra, cross-sectional TEM images, and SIMS depth profiles of crystallized aSi films on a-Si:H:B/SnO₂/glass after plasma annealed at different v_{sub}s.

Figure 5a shows the dielectric function $\langle \varepsilon_2 \rangle$ spectra of a 300-nm-thick crystallized Si films at different stage velocities. The optical model used for the spectrum analysis is also shown in the inset. The fine structures observed at 3.2 and 4.3 eV, corresponded, respectively, the optical band transition energies E_1 and E_2, for crystallized Si films.

Fitting procedure was performed by combining reference ε_2 spectra of c-Si, a-Si and void given by Jellison with Bruggeman effective medium approximation (EMA) [5]. The spectra were analyzed using several optical three-layer models composed of bulk and surface oxidation. Among several optical models used for the analysis, the optical model shown in the inset was the most plausible structure. In figure 5b, the variation of volume fractions $f'_{c\text{-}Si}$ and $f'_{a\text{-}Si}$ (i=1,2) in the ith layer and surface oxidation layer, as determined by the ε_2 spectra analysis are plotted with the thickness of the surface oxidation layer (layer 3) as a function of v_{sub}. The $f'_{c\text{-}Si}$ in layer 1 (95-100%) was almost independent of v_{sub}, and increased markedly below a v_{sub} of 50 mm/s. The $f^2_{c\text{-}Si}$ and $f^2_{a\text{-}Si}$ in layer 2 and the thickness of the surface oxidation layer depended on v_{sub}. Under the plasma condition used in this study, increasing v_{sub} above 50 mm/s suppressed crystallization because of insufficient plasma annealing period. These results also suggest that the film crystallization of a-Si proceeds from the bottom to front surface.

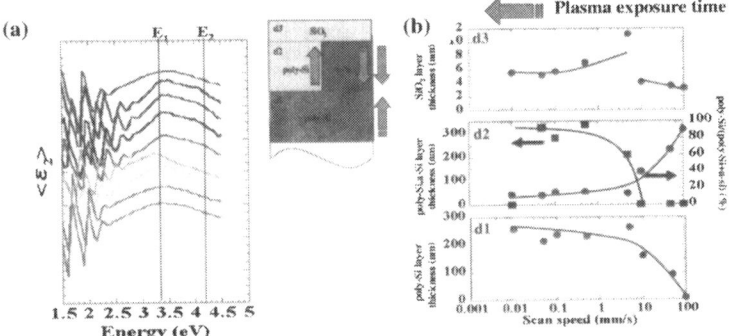

Figure 5(a) $\langle\varepsilon_2\rangle$ spectra of crystallized a-Si films annealed at different stage velocities. **(b)** The $f'_{c\text{-}Si}$ and $f'_{a\text{-}Si}$ (i=1,2) and the thickness of surface oxidation layer plotted as a function of translating velocities of the substrate stage V_{sub}s.

In addition, the surface morphology was characterized using the thickness profiler for corresponding crystallized a-Si films. The hump structure was observed within 2 mm in radius corresponding to the inner diameter of the quartz tube at v_{sub} above 50 mm/s, whereas the film ablation was dominant at v_{sub} below 50 mm/s. In general, the crystallization from liquid occurs at melting temperature accompanying with the volume reduction, when a super-cooled liquid is cooled through a first-order transition. On the other hand, the crystallization from liquid state takes place with the volume expansion in the case of Si. These results suggest that crystallization of a-Si by the plasma annealing occurs during the solidification and crystallization from the molten Si induced by the local heating. Thus, the surface morphology is determined mostly by both the volume expansion during the solidification of liquid Si and the ablation of a-Si. In addition, the transient optical reflectivity measurement revealed the crystallization occurred with a time constant of 2-4 ms, which were 4-6 orders of magnitude slower than the laser crystallization of a-Si ($\sim$100 ns). The nucleation and crystal grain growth occurs preferentially at the bottom surface because sufficient high thermal and plasma energy storage at the interface. The lateral growth of c-Si was also enhanced efficiently for thinner a-Si deposited on thicker th-SiO$_2$ substrate. The field-effect mobility and threshold voltage of 35-60 cm^2/Vs and 3-6 V, respectively, have been achieved in the crystallized Si thin-film transistors (TFTs) with a bottom gate structure. These imply that the carrier transport properties at the interface between crystallized Si and th-SiO$_2$ are improved markedly by the plasma annealing.

CONCLUSIONS

A VHF thermal plasma jet of argon can rapidly and effectively crystallize a-Si film on th-SiO$_2$ and a-Si:H:B/SnO$_2$/glass substrates. The crystallization is promoted from the bottom to front surface with crystal size of 150-200 nm. The crystallized a-Si by the plasma annealing is a promising material for TFTs and Si thin-film solar cells.

ACKNOWLEDGMENTS

This work was financially supported in part by a Grant-in-Aid for Scientific Research from the Ministry of Education, Culture, Sport, Science and Technology of Japan and Japan Science and Technology agency (JST). The authors appreciate Mr. M. Sato of Mitsubishi Material Co. Ltd for the SIMS measurement.

REFERENCES

1. A. Matsuda, *J. Non-Cryst. Solids* **338-340**, 1 (2004).

2. T. Sameshima, M. Hara and S. Usui, *Jpn. J. Appl. Phys.* **28,** 789 (1989).

3. S. Y. Yoon, S. J. Park, K. H. Kim, J. Jang, *Thin Solid Films* **383,** 34 (2001).

4. H. Shirai, Y. Sakurai, M. Ye, T. Kobayashi, and T. Ishikawa, *Eur.Phys. J. Appl. Phys.* **37**, 315 (2007).

5. G. E. Jellison, Jr. and F. A. Modine, *Appl. Phys. Lett.* **69**, 2137 (1996).

Thin Film Transistors II

Mater. Res. Soc. Symp. Proc. Vol. 989 © 2007 Materials Research Society   0989-A14-03

Drain Bias Dependent Threshold Voltage Shift of a-Si:H TFT Due to the Pulsed Stress

Sang-Geun Park, Jae-Hoon Lee, Sang-Myen Han, Sun-Jae Kim, and Min-Koo Han
Electrical Engineering and Computer Science, Seoul National University, 130-Dong, 305-Ho, San 56-1, Shillim-Dong, Gwanak-Gu, Seoul, 151-742, Korea, Republic of

ABSTRACT

We have investigated the threshold voltage shift (V_{TH}) in the a-Si:H TFTs due to the various negative pulse width stress. The drain bias dependent V_{TH} in the pulsed stress of a-Si:H TFT for AMOLED backplane was also measured and analyzed. When a positive gate and drain bias is applied to a-Si:H TFT (W/L = 200/4 μm), V_{TH} of a-Si:H TFT is increased during the stress time due to the defect state creation and charge trapping. V_{TH} of a-Si:H TFT is increased from 1.645V to 2.53V (ΔV_{TH}=0.885V) after the DC gate bias stress of V_{GS}=15V, V_{DS}=0V for 20,000sec. When the pulsed negative bias stress is applied to the gate electrode of the current driving a-Si:H TFT with the drain bias, V_{TH} shift is considerably reduced due to the hole trapping into the gate insulator during the stress. When a negative pulse width is 16msec (pulse of 60Hz), the V_{TH} is increased form 1.594V to 2.195V (ΔV_{TH}=0.601V). When a negative pulse width increases from 16msec to 5sec without drain bias (V_{DS}=0V), V_{TH} is increased from 1.615V to 2.055V (ΔV_{TH}=0.44V). When a drain bias is increased from 0V to 15V, V_{TH} is slightly decreased from 1.58V to 1.529V (ΔV_{TH}=-0.051V) due to large (-30V) V_{GD} (V_G=-15V, V_D=15V) bias, while it is increased from 1.66V to 2.078V (ΔV_{TH}=0.418V) width DC gate bias stress of V_{GS}=15V, V_{DS}=15V for 20,000sec.

INTRODUCTION

Active matrix organic light emitting diode (AMOLED) employing thin film transistor (TFT) pixels have attracted considerable attentions due to high brightness, compactness and wide viewing angle [1-3]. Recently, poly-Si TFTs are considered as the pixel element of AMOLED due to good electric characteristics such as high mobility and superior stability. However it suffers from a current non-uniformity caused by inherent fluctuation of excimer laser energy. The hydrogenated amorphous silicon (a-Si:H) TFT could be suitable for pixel elements of AMOLED due to an excellent uniformity up to large area with mature fabrication process. However it is well known that the electrical stability of a-Si:H TFT is rather poor. The stability of a-Si:H TFT needs to be improved for high quality AMOLED display circuit.

When a positive gate and drain bias is induced in the a-Si:H TFT, V_{TH} of a-Si:H TFT increases over the stress time due to the defect state creation and charge trapping. An increased V_{TH} decreases the brightness of AMOLED display. Recently, we have reported that the negative gate bias annealing can cure a degraded a-Si:H TFT due to the hole trapping into SiNx gate insulator. In this work, the gate electrode of a-Si:H TFT is subjected to the pulsed stress rather than DC negative gate bias stress in order to investigate the negative bias annealing effects on practical operation of OLED device. The pulsed stress is induced to the gate electrode of the current driving a-Si:H TFT with the drain bias. While pulsed stress for LCD application is previously reported, the drain bias has not been considered due to the small (~3) drain bias for driving liquid crystal. The purpose of our work is to report the drain bias dependent threshold voltage shift in the pulsed stress in a-Si:H TFT for AMOLED backplane. The negative pulse width dependent V_{TH} shift was also measured and analyzed.

EXPERIMENT

Fabrication of device

The a-Si:H TFTs with an inverted staggered bottom gate type were fabricated by a standard process. Fig. 1. shows the device structure of a-Si:H TFT using our experiment. The Mo/AlNd double layer was deposited on Corning 1737 glass by DC sputtering for the gate electrode. Triple layer of SiN_x (4500Å), a-Si:H (2000Å), n^+ a-Si:H (500Å) was deposited on the gate by plasma-enhanced chemical vapor deposition (PECVD). After active island patterning, Mo/AlNd/Mo triple layer (4500Å) was deposited for the source and drain electrode by sputtering. After patterning the source and drain electrode by a wet etching, the n^+ a-Si layer between the source and drain electrode was removed by a dry etching to make an etch back type channel structure. 3000Å thick SiNx was deposited for a passivation. After contact holes were formed, an indium tin oxide (ITO) electrode was deposited and patterned.

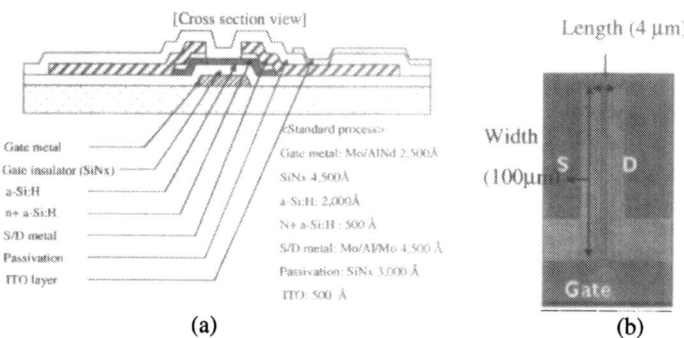

(a)   (b)

Figure 1. (a) The schematic of the fabricated a-Si:H TFT. (b) The microscopic picture of the fabricated a-Si:H TFT

Pulsed negative bias annealing with V_{DS}=0(V)

Fig. 2 shows that threshold voltage shift of a-Si:H TFT when it applied bias stress of V_{GS} = 15V, V_{DS} = 0V for 20,000sec. The V_{TH} was increased about 0.885V during bias stress. The straight lines are from the measured V_{TH} shift data. V_{TH} was defined by the well known $\sqrt{I_D} - V_G$ characteristics from a saturation mode ($V_{DS}=V_{GS}$). ΔV_{TH} is associated with the time shown in equation ($\Delta V_{TH} = A \times t^{\beta}$) [4].

$$\Delta V_{TH}(t) = 0.00138\left(V_{ST} - V_{TH\_i}\right) t^{0.368}$$

where V_{ST} is the gate bias stress voltage, V_{TH_i} is the initial V_{TH} of the TFT, and t is the bias stress time duration. This equation means that a-Si:H TFT will be severely degraded with increasing a stress duration. Gradually increased V_{TH} of a-Si:H TFT decreases an OLED current continuously. OLED is activated by a TFT used for a current source, which is operated in the saturation region.

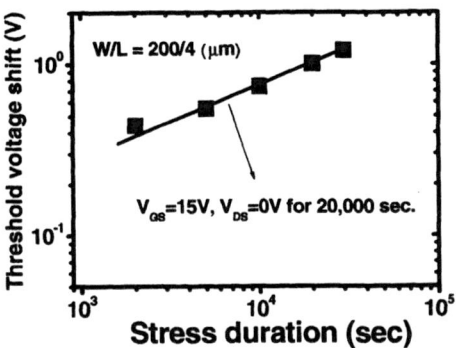

Figure. 2. The measured results for V_{TH} shift under positive bias stress width with $V_{GS}=15V$ and $V_{DS}=0V$ for 20,000 seconds. The V_{TH} was increased about 0.885V during bias stress.

Fig. 3 is the measured results for V_{TH} shift dependent on the negative bias pulse width with $V_{DS}=0V$. The duty ratio in the AC bias stress is 50% and the stress conditions for the negative bias pulse width are 16.7msec and 5sec, respectively. The stress duration in the AC bias stress considers the positive bias duration for a fairly comparison with DC bias stress. V_{TH} shift of the DC gate bias ($V_{GS}=15V$, $V_{DS}=0V$) after 20,000sec is 0.885V due to a defect state creation in a-Si layer and a charge trapping into SiN_x gate insulator. In case of AC bias stress entitled the negative bias annealing, V_{TH} shift can be considerably reduced due to the hole trapping into the gate insulator. When the negative pulse width is increased from 16msec to 5sec, V_{TH} shift after the positive bias stress duration of 20,000sec is reduced from 0.601V to 0.44V. These values are 68% and 50% of V_{TH} shift in the DC gate bias stressed sample (0.885V), respectively.

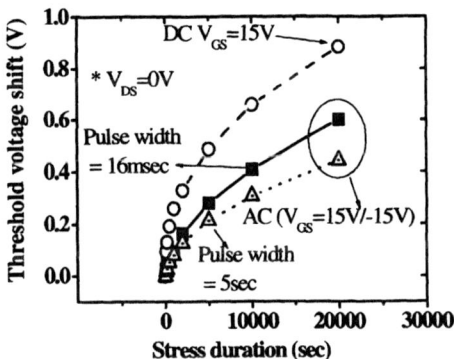

Figure. 3. The measured results for V_{TH} shift dependent on the negative bias pulse width with $V_{DS}=0V$. When the negative pulse width is increased from 16msec to 5sec, V_{TH} shift after the positive bias stress duration of 20,000sec is reduced from 0.601V to 0.44V

Pulsed negative bias annealing with V$_{DS}$=15(V)

Fig. 4 shows V$_{TH}$ shift in the negative bias pulse width with V$_{DS}$=15V. It should be noted that the drain bias of a-Si:H TFT using the current source of AMOLED is large (> 10V) to provide the saturation current for each pixel. V$_{TH}$ shift of the DC gate bias (V$_{GS}$=15V, V$_{DS}$=15V) stress for 20,000sec is 0.418V. This value is smaller than the DC gate bias with V$_{DS}$=0V (0.885V). It may be attributed that an electron concentration is decreased with increasing a drain bias. When the negative pulse width is increased from 16msec to 5sec, V$_{TH}$ shift after the positive bias stress duration of 20,000sec is reduced from 0.162V to -0.051V. These values are 39% and 0% of V$_{TH}$ shift in the DC gate bias stressed sample (0.418V), respectively. When a drain bias is increased from 0V to 15V, V$_{TH}$ shift in the negative bias annealed (pulse duration a-Si:H TFT: 5sec) can be reduced from 0.44V to -0.051V due to large (-30) V$_{GD}$ (V$_G$=-15, V$_D$=15)..

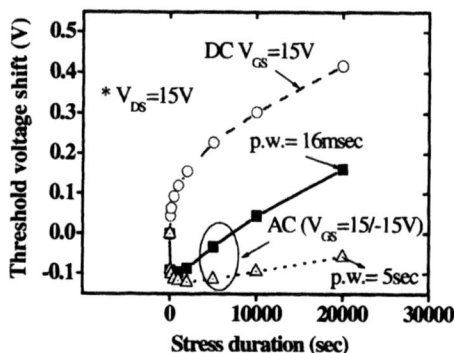

Figure. 4. V$_{TH}$ shift in the negative bias pulse width with V$_{DS}$=15V. When a drain bias is increased from 0V to 15V, V$_{TH}$ shift in the negative bias annealed (pulse duration a-Si:H TFT: 5sec) can be reduced from 0.44V to -0.051V due to large (-30) V$_{GD}$ (V$_G$=-15, V$_D$=15).

Negative bias pulse width dependence on the V$_{TH}$ shift of a-Si:H TFT

Fig. 5 shows negative bias pulse width dependence on the negatively V$_{TH}$ shift of a-Si:H TFT at a room temperature. The negative V$_{TH}$ shift effect on a-Si:H TFT is largest when the negative bias pulse width is increased to 1 sec. When the negative bias annealing is applied to the a-Si:H TFT pixel for AMOLED, the number of the current source in each pixel is determined by the negative pulse width. If the negative bias pulse width is increased to 1 sec, two current sources in each pixel are required to prevent a flicker image caused by long negative bias annealing period. Therefore the aperture ratio may be considerably decreased due to the two current sources. The trade-off between the aperture ratio in the pixel and the negative bias annealing effect on the device can be selected by the various factors, such as the size of the display and OLED emission type.

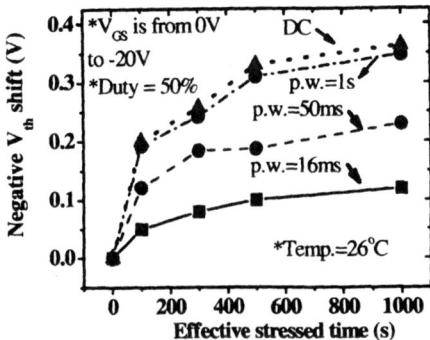

Figure. 5. The negative bias pulse width dependence on the negatively V_{TH} shift of a-Si:H TFT at a room temperature. When a negative bias pulse width is increased up to 1sec, the negatively V_{TH} shift effect on a-Si:H TFT is maximized.

DISCUSSION

Since the threshold voltage shift shows the shift directional subject to stress polarity, the shift may be due to carrier trapping at the a-Si/SiN interface. Therefore, when TFTs are stressed by pulse voltage, the voltage across the gate insulator (V_i) should be considered. V_i is calculated by the ordinary charging model of RC circuits, as follows [5].

$$Vi = Vg \times \{1 - \frac{Ci}{Cs + Ci} \bullet \exp(-\frac{t}{\tau})\}, \tau = (Cs + Ci) \bullet (\frac{1}{Rs} + \frac{1}{Ri})^{-1}$$

, where Vg is the applied gate voltage to the gate electrode, Ci and Cs are the capacitance of the insulator and silicon, respectively. Ri and Rs are the resistances of the insulator film and the a-Si film, respectively, t is the stress time, and τ is the time constant of the circuit.

The specific resistivity of a-Si is known to be 4×10^{10} ($\Omega \bullet cm$) at 25°C, and 2×10^9 ($\Omega \bullet cm$) at 60°C. When a positive pulse is applied to the gate, the resistance of a-Si film (Rs) is sufficiently low so that the time constant (τ) is the order of μsec. It means when a positive pulse stress is induced to the a-Si TFT, the effective voltage across the insulator is the same as the DC positive voltage. In case of the negative pulsed stress, that the time constant (τ) is increased up to the order of msec. Therefore, V_{TH} shift in the negative bias annealing is dependent on the negative bias pulse width. As shown above the V_{TH} shift decreases as negative bias pulse width increases.

When the V_{DS} bias is applied in the a-Si:H TFT during the bias annealing, the V_{TH} shift was smaller than that of bias annealing with $V_{DS} = 0V$. It may be attributed that an electron concentration is decreased with increasing a drain bias. In the model of defect state creation, the electron during the stress can break weak-weak silicon bonding so that V_{TH} shift with increasing a drain bias is decreased due to decreased electron concentration with increasing drain bias [4].

CONCLUSIONS

We have investigated the negative pulse width dependent V_{TH} shift. The drain bias dependent threshold voltage shift in the pulsed stress in a-Si:H TFT for AMOLED backplane is also measured and analyzed. When the positive bias stress is applied in the a-Si:H TFT, the V_{TH} was increased due to trapped charges or defect creation. The V_{TH} shift can be reduced significantly by pulsed negative bias annealing. When the DC bias stress (V_{GS}=15V, V_{DS}=0V) is applied in the a-Si:H TFT for 20,000, the V_{TH} of a-Si:H is increased from 1.645V to 2.53V (ΔV_{TH}= 0.885V). When the pulsed negative bias stress is applied to the gate electrode of a-Si:H TFT, V_{TH} is increased from 1.615V to 2.055V due to hole trapping into the gate insulator during the stress. When the pulsed negative bias annealing with V_{DS} = 15V is applied, the V_{TH} of a-Si:H TFT is decreased slightly from 1.58V to 1.523V due to the large negative V_{GD} during the negative bias stress. The shift of V_{TH} in a-Si:H TFT is decreased by the pulsed negative bias annealing.

The stability of a-Si:H TFT may be improved by the pulsed negative bias annealing. It is favorable for AMOLED application because drain bias is always applied to the current driving a-Si:H TFT.

REFERENCES

1. J.L.Sanford, and F.R.Libsch, "TFT AMOLED pixel circuits and driving methods", *SID Tech. Dig.*, vol.34, pp.10-13, 2003.
2. J-H Lee, J-H Kim, and M-K Han, "A New a-Si:H TFT Pixel Circuit Compensating the Threshold Voltage Shift of a-Si:H TFT and OLED for Active Matrix OLED", *IEEE Electron Device Letters*, vol.26, Issue12. pp.897-899, Dec, 2005.
3. J.C.Goh, J.Jang, K.S.Cho, and C.K.Kim, "A new a-Si:H thin film transistor pixel circuit for active matrix organic light emitting diodes", *IEEE Electron Device Letters*, vol.24, pp.583-585, Sep. 2003
4. K.S.Karim, A.Nathan, M.Hack, and W.I.Milne, "Drain-Bias Dependence of Threshold Voltage of Amorphous Silicon TFTs", *IEEE Electron Device Letters*, vol.25, pp.188-190, April.2004
5. C.S.Chiang, J.Kanicki and K.Takechi, "Electrical instability of hydrogenerated amorphous silicon thin-film transistors for active matrix liquid crystal displays" *Jpn.J.Appl.Phys.*, vol.37, pp.4704-4710, 1998

Mater. Res. Soc. Symp. Proc. Vol. 989 © 2007 Materials Research Society 0989-A14-05

Noise Performance of High Fill Factor Pixel Architectures for Robust Large-Area Image Sensors using Amorphous Silicon Technology

Jackson Lai[1], Yuri Vygranenko[1], Gregory Heiler[2], Nader Safavian[1], Denis Striakhilev[1], Arokia Nathan[3], and Timothy Tredwell[2]

[1]Department of Electrical and Computer Engineering, University of Waterloo, 200 University Avenue West, Waterloo, N2L 3G1, Canada
[2]Eastman Kodak Company, 1700 Dewey Avenue, Rochester, NY, 14650-1822
[3]London Centre for Nanotechnology, University College London, London, WC1H OAH, United Kingdom

ABSTRACT

Large area digital imaging made possible by amorphous silicon thin-film transistor (a-Si TFT) technology, coupled with a-Si photo-sensors, provides an excellent readout platform to form an integrated medical image capture system. Major development challenges evolve around optimization of pixel architecture for detector fill factor, signal propagation performance, and manufacturability, while suppressing noise stemming from pixel array and external electronics. This work analyzes a novel vertically integrated pixel design based on signal readout and noise performance, and compares with conventional co-planar and continuous detector architectures. In addition, the analysis will consider various substrate options including glass and robust substrates such as polymer and metal foil. Our evaluation have demonstrated state-of-the-art radiographic detector system with electronic noise under 2000 electrons at 150 μs frame time for an imaging arrays on robust substrate.

INTRODUCTION

Large-area digital image sensors are revolutionizing medical imaging by enabling electronic storage capability, immediate feedback, and possibilities to support previously unachievable applications related to computer aided image processing. Hydrogenated amorphous silicon thin-film transistor (a-Si:H TFT) technology, frequently used in liquid crystal display applications, is extended to perform back-plane readout for large-area detector system. Technological attributes of a-Si:H such as uniformity over large area, compatibility with various substrate material, and research maturity have provided an excellent development platform for high performance, low noise, and fully integrated digital detector system.

In conventional imaging array design [1][2] where the thin-film transistor (TFT) and photo-sensor in a pixel are placed side by side on the same plane, the fill-factor, signal-to-noise ratio and dynamic range are limited. An alternative approach is a vertically integrated high fill factor architecture where the photo-sensor is implemented as segmented [1] or non-segmented continuous layer [2]. Vertically integrated architectures increases the fill factor to achieve higher detected quantum efficiency (DQE), leading to better signal-to-noise ratio and is crucial to applications with stringent signal specification such as medical imaging.

Meanwhile, alternative substrates options such as conducting stainless steel for hosting a-Si:H pixelated detector array has also gained considerable interest due to substrate physical robustness and flexibility for better system portability and reliability. However, major challenge here originates from capacitive coupling between active device layers and the conductive substrate, hindering low-noise and high-speed readout operation.

The following section will present and study novel high fill factor pixel architecture specially designed for both non-conducting and conducting substrate. Comparison with conventional co-planar and continuous sensor architectures will be discussed to high light the applicability of the proposed pixel architecture.

PIXEL ARCHITECURES

Co-planar

A conventional co-planar pixel process places the sensor and TFT beside each other on the imaging panel. A cross-sectional diagram for a conventional non-overlapped pixel process using an a-Si:H p-i-n photodiode appears in Figure 1. The in-pixel readout circuit, photo sensor, and interconnecting metal lines together define the pixel pitch. While this provides consistent manufacturing yield, the photodiode competes with the TFT and metal lines for pixel space, leading to a reduced pixel fill factor. The typically low fill factor in co-planar design serves as one of the bottlenecks for both high-resolution imaging and achievable signal-to-noise ratio.

Furthermore, since the signal and address lines are placed in the same plane as the photo sensor and TFT, they are close to the substrate and are separated only by the thin CVD inter-layer dielectric. This results in high parasitic capacitive coupling and is especially pronounced in conductive substrate scenarios. Table 1 shows a 35% higher dataline capacitance in conductive substrate for co-planar architecture.

Continuous sensor

Figure 2 shows a vertically integrated design with a continuous sensor fully overlapping the pixel TFT plane. The continuous p-i-n photo sensor is used to address fill factor requirements. The overlapping architecture eliminates the pixel space sharing issues, and the sensor bottom electrode dictates the pixel resolution. This architecture can theoretically achieve fill factor over 90% and is suitable to maximize detected quantum efficiency (DQE).

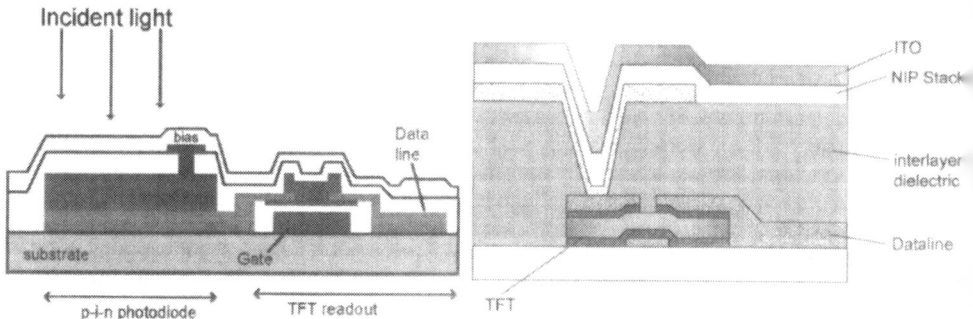

Figure 1. Co-planar pixel architecture.   **Figure 2. Continuous sensor pixel architecture.**

On the other hand, a continuous sensor for the array promotes crosstalk between adjacent pixels especially for tightly spaced pixel electrodes. Patterning the bottom n+ a-Si layer can alleviate the issue by providing pixel isolation. However, this will require an additional masking step, adding to processing complexity and consequently lower device yield. Moreover, since the i-p stack have to be separately deposited, uniform, consistent, and low defect interface with n+ layers become difficult to maintain.

Compared to the co-planar design, the dataline capacitance is higher due to the fully overlapping sensor stack. For conductive substrate scenario, the dataline coupling to the substrate will be similar to the co-planar design. As a result, the dataline capacitance for continuous sensor design is higher for both conductive and non-conductive options (see Table 1).

<u>High fill factor segmented sensor</u>

To alleviate the limitations of co-planar and continuous sensor architectures, a high fill factor (HFF) pixel design is proposed. The cross sectional diagram is shown in Figure 3, where we retain the segmented photodiode stack and integrated it above the TFT layers to form a vertically overlapping structures.

This proposed design addresses the design limitations presented in the other pixel architectures. Firstly, by building the sensor stack on top of the TFT layers, the fill factor is increased in comparison to co-planar architecture.

Secondly, the p-i-n photo sensor is segmented, thus assist in suppressing crosstalk between adjacent pixels that exist in continuous sensor structures. In addition, the entire p-i-n stack can be deposited in one processing step without intermediate patterning. Hence, it preserves the integrity of the interfaces, resulting in better performance consistency.

Thirdly, the signal and sensor bias lines are placed on top of the entire structure. This metal is typically deposited as the last step of array processing, thus allowing better flexibility to the choice of material for better substrate compatibility due to processing temperature constraints. In addition, it also allows thickness optimization for lower line resistance. The signal metal lines are also located relatively far away from the substrate without any overlap with the sensor. This results in smaller overall line capacitance, and is shown in Table 1 where C_{PD}, C_{PD_SUB}, $C_{L\_nonconduct}$, and $C_{L\_conduct}$, is the photo-sensor capacitance, photo-sensor coupling capacitance with the substrate, data line capacitance for non-conducting substrate, and data line capacitance for conducting substrate respectively.

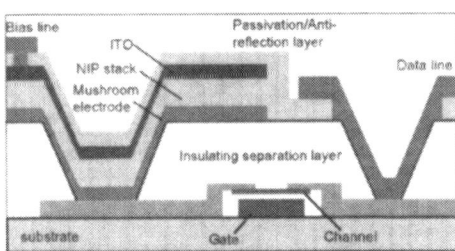

Figure 3: High fill factor segmented sensor pixel architecture

	Co-planar	HFF segmented	Continuous
C_{PD}	2.8 pF	2.98 pF	3.94 pF
C_{PD_SUB}	0.826 pF	0.877 pF	1.16 pF
$C_{L\_nonconduct}$	94.5 pF	56.9 pF	209 pF
$C_{L\_conduct}$	145.5 pF	78.4 pF	283 pF

Table 1: Equivalent lumped parameter values for pixel architectures on 2333 x 2867 matrix array.

NOISE PERFORMANCE AND SIGNAL PROPAGATION ANALYSIS

This section evaluate and compare the pixel design architectures described previously based on timing response and noise analysis due to the combined effect on detector applicability and feasibility. The passive pixel circuit design is used for all architectures for comparison without loss of generality. Figure 4 and 5 illustrate the circuit schematic for timing and noise analysis respectively, where each pixel consist a single TFT and a photodiode. The TFT gate-source overlap capacitance for the non-accessed pixels, and data line capacitive coupling together constitute the data line capacitance and are represented as the distributed data line elements in the schematic. The data line is then connected to an output amplifier to allow charge readout, followed by correlated double sampling, and anolog-to-digital converter (ADC) stage.

The schematic is compatible with both conductive and non-conductive substrates, with the difference in the capacitance values. The lumped parameter values in Table 1 are distributed in the data line and the model is simulated for signal propagation performances. Figure 6 summarizes the results, where the proposed HFF architecture on both conductive and non-conductive substrates yields the smallest time delay. The advantage of the proposed HFF design is more pronounced in conductive substrates; where it is 13% and 34% better than co-planar and continuous sensor architectures.

Noise behaviour on pixel architectures requires a comprehensive set of model because of the complex interaction between different noise sources introduced in the pixel array configuration. The same array and readout configuration is assumed and Figure 5 shows the equivalent block diagram for pixel noise modelling.

The total system additive noise can be broken down into pixel and base noise,

$$s_{add}^2 = \sqrt{s_{pix}^2 + s_{base}^2} \tag{1}$$

where s_{add}^2, s_{pix}^2, and s_{base}^2 are total additive noise, pixel noise, and base noise respectively. TFT thermal, transient, flicker, as well as photodiode shot noise are the main contributors to pixel noise, such that

$$s_{pix}^2 = \sqrt{s_{TFT\text{-}thermal}^2 + s_{tran}^2 + s_{pd\text{-}on}^2 + s_{pd\text{-}off}^2 + s_{TFT\text{-}off}^2} \text{ and } s_{base}^2 = \sqrt{s_{amp}^2 + s_{dataline}^2 + s_{ext}^2 + s_{dig}^2} \tag{2}$$

while amplifier, dataline thermal, external spurious, and digitization noise are taken consideration in readout electronics base noise as illustrated in second half of (2).

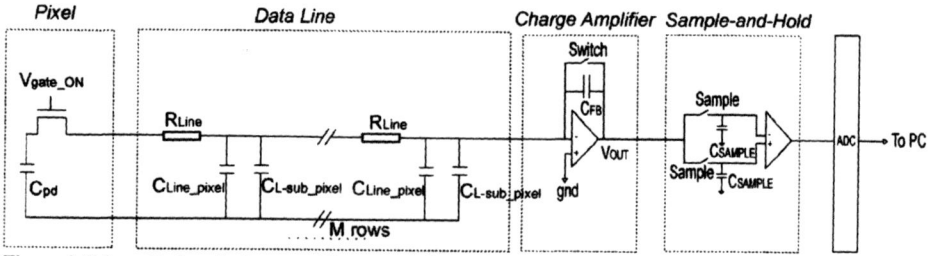

Figure 4: Schematic for pixel array with external electronics.

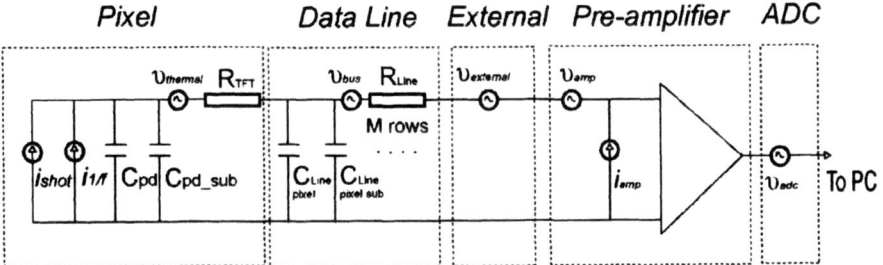

Figure 5: Noise model for pixel architectures.

Table 2: List of noise model equations for pixel architectures.

Symbols	Noise components	Definitions and estimations
$\sigma_{TFT\text{-thermal}}$	TFT thermal noise (Applies for $f_0 \gg \dfrac{1}{2\rho t_{pix}}$)	$= \sqrt{2\dfrac{C_s kT}{q^2}[1- \exp(- \dfrac{T_{INT}}{R_{ON}C_s})]^2} \quad C_s = \dfrac{C_{PIN}C_{amp}}{C_{PIN}+C_{amp}}, \quad$ [3][4]
$\sigma_{dataline}$	Data line thermal noise	$= \dfrac{1}{q}C_{dataline}\sqrt{pkTR_{data}f_0}$
σ_{tran}	TFT transient noise	$= \sqrt{\dfrac{1}{q}Q_{tran}(1+\dfrac{f_L}{f^n})}$ [4][5][6]
$\sigma_{pd\text{-off}}$	Photodiode shot noise and 1/f noise with TFT off	$= \sqrt{\dfrac{1}{q}I_{pd\text{-}leak}t_{TFT\text{-}off}(1+\dfrac{f_L}{f^n})}$
$\sigma_{pd\text{-on}}$	Photodiode shot noise and 1/f noise with TFT on	$= \sqrt{\dfrac{1}{q}I_{pd\text{-}leak}t_{TFT\text{-}on}(1+\dfrac{f_L}{f^n})}$
$\sigma_{TFT\text{-off}}$	TFT shot noise and 1/f noise with TFT off	$= \sqrt{\dfrac{1}{q}I_{TFT\text{-}leak}t_{TFT\text{-}off}(1+\dfrac{f_L}{f^n})}$
σ_{dig}	Digitization quantum noise	$= \dfrac{1}{q}\dfrac{Q_{signal}}{\sqrt{12x2^{bits}}}$

The individual noise model details are shown in Table 2. Here, f_L is a empirical parameter that defines the corner frequency at which the 1/f noise becomes equal to the shot noise, f_0 is the noise bandwidth, Q_{signal} the maximum signal charge in units of electrons to be digitized, and n the spectral slope determines the slope of 1/f noise spectral density.

Analytical calculations and simulations are performed for the noise behaviour analysis for all three architectures on both conducting and non-conducting substrates. The results are shown in Figure 6. The pixel level noise is dominated by the TFT thermal noise, which is largely determined by the TFT on-resistance and thus dependent on transistor dimension. Hence, pixel level noise is comparable among pixel architectures. Base noise, on the other hand, is dominated by line thermal noise and is directly proportional to data line capacitance. Here, we see the benefits of small capacitive coupling for the proposed design, leading to a significantly lower base noise figure. Hence, the proposed design performs better than co-planar and continuous sensor architectures especially on conductive substrates in terms of noise performance.

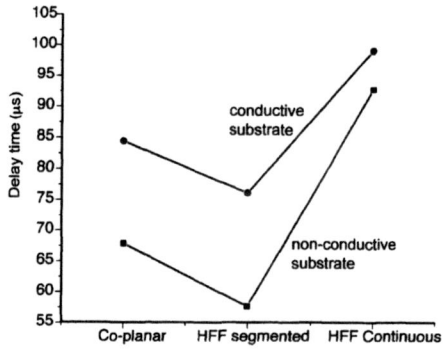

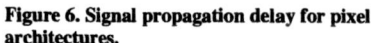

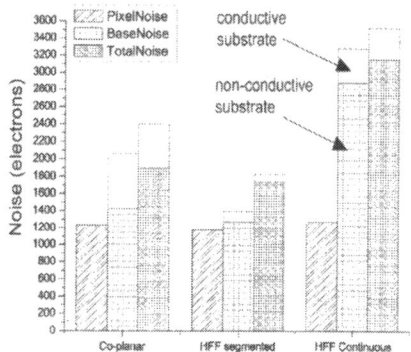

Figure 6. Signal propagation delay for pixel architectures.

Figure 7. Noise performance for pixel architectures.

CONCLUSIONS

This work investigates the feasibility of high fill factor pixel architectures for low-noise large area imaging array. High fill factor segmented pixel architecture is proposed and it is evaluated and compared with co-planar and continuous sensor alternatives based on manufacturability, readout rate, and noise performance. The proposed HFF design have demonstrated state-of-the-art radiographic detector system with electronic noise under 2000 electrons at 150 µs frame time for an imaging arrays on conductive substrate. Both readout rate and noise performances of the high fill factor pixel correspond to at least 15% benefit over alternative architectures under consideration, and the advantage is more prominent on conductive substrates.

REFERENCES

1. R. Weisfield; W. Yao; T. Speaker; K. Zhou; R.E. Colbeth; C. Proano, J. Proc. of the SPIE, Vol. 5368, pp. 338-348 (2004).
2. R.A. Street, X.D. Wu, R.L. Weisfield, S. Ready, R. Apte, M. Nguyen, and P. Nylen, MRS Symp. Proc. 377, p. 757, 1995.
3. I. Fujieda, R.A. Street, R.L. Weisfield, S. Nelson, P. Nylen, V. Perez-Mendex, and G. Cho, Jpn. J. Appl. Phys. V.32 (1993) p.198-202.
4. M. Maolinbay, Y. El-Mohri, L.E. Antonuk, K.-W. Jee, S. Nassif, X. Rong, and Q. Zhao, Med. Phys. 27(8), August 2000.
5. G. Cho, J.S. Drewery, I. Fujieda, T. Jing, S.N. Kaplan, V. Perez-Mendex, S. Qureshi, D. Wildermuth, and R.A. Street, *MRS Symposium Proceedings*, **192**, pp. 393-398, 1990.
6. N. Matsuura, W. Zhao, Z. Huang, J.A. Rowlands, *Medical Physics*, **26**(5), pp. 672-681, May 1999.

Solar Cells II

Mater. Res. Soc. Symp. Proc. Vol. 989 © 2007 Materials Research Society   0989-A15-01

Status of nc-Si:H Solar Cells at United Solar and Roadmap for Manufacturing a-Si:H and nc-Si:H Based Solar Panels

Baojie Yan, Guozhen Yue, and Subhendu Guha
United Solar Ovonic LLC, 1100 West Maple Road, Troy, MI, 48084

ABSTRACT

This paper reviews the research and development of hydrogenated nanocrystalline silicon (nc-Si:H) solar cells at United Solar Ovonic LLC. We have been studying nc-Si:H solar cells since 2001 and have made significant progress. We have achieved an initial active-area cell efficiency of 15.1% using an a-Si:H/a-SiGe:H/nc-Si:H triple-junction structure, a stable active-area cell efficiency of 13.3% using an a-Si:H/nc-Si:H/nc-Si:H triple-junction structure, and a stable aperture-area (420 cm^2) fully encapsulated module efficiency of 9.5% using an a-Si:H/nc-Si:H double-junction structure. Although the cell efficiencies with nc-Si:H in the middle and/or bottom cells have exceeded the corresponding efficiencies achieved using a-Si:H and a-SiGe:H, we still need to address several critical issues before using nc-Si:H in photovoltaic manufacturing plants. First, the cell efficiency needs to be improved further to show a clear advantage over the conventional a-Si:H/a-SiGe:H/a-SiGe:H triple-junction cell structure. Second, we need to increase the deposition rate further to make the nc-Si:H based technology more cost effective. Third, we need to develop a machine design to overcome the large-area uniformity issue, especially for very high frequency glow discharge deposition. Fourth, we need to qualify nc-Si:H based solar cell product, especially with respect to long term reliability. We have been addressing these critical issues, and will discuss the roadmap for manufacturing a-Si:H and nc-Si:H based solar panels using the roll-to-roll technology.

INTRODUCTION

Since first reported by the Neuchâtel group in 1994 [1], hydrogenated microcrystalline silicon (μc-Si:H) solar cell has attracted significant attention and has been studied widely. Because μc-Si:H materials contain nanometer-size grains and amorphous tissues, μc-Si:H is now more often called nanocrystalline silicon (nc-Si:H) to align with other nano-technologies. Compared to hydrogenated amorphous silicon (a-Si:H) and silicon germanium alloy (a-SiGe:H) solar cells, nc-Si:H solar cell has two advantages. First, its lower optical bandgap close to the value of crystalline silicon results in a high short-circuit current density (J_{sc}), because of enhanced long wavelength response. Second, optimized nc-Si:H solar cells show very little light-induced degradation. However, the major drawback of nc-Si:H cells is the indirect bandgap in the crystalline phase. The optical absorption coefficient in the short wavelength region is lower than that in a-Si:H and a-SiGe:H. Therefore, a thick nc-Si:H intrinsic layer of over one micrometer is normally needed to achieve high J_{sc} . In order to make nc-Si:H solar cell technology cost-effective for solar panel manufacturing, a high deposition rate is essential. The Neuchâtel group also pioneered the use of very high frequency (VHF) glow discharge to increase the deposition rate of a-Si:H and nc-Si:H [1-3]. The second effective method of increasing the nc-Si:H deposition rate is using high pressure and high power radio frequency (RF) glow discharge in the depletion regime [4, 5]. Lately, these two techniques have been combined for even higher deposition rate up to 20-30 Å/s [6, 7]. With efforts from many laboratories over the world, nc-Si:H solar cell efficiencies have improved significantly in the last several years with nc-Si:H single-junction cell efficiency exceeding 10% [8, 9], multi-junction cell efficiency over 15% [10, 11], and large-area module efficiency ~ 13% [11, 12].

United Solar has been studying a-Si:H and a-SiGe:H based solar cells for many years. A world record stable active-area efficiency of 13% was achieved with an a-Si:H/a-SiGe:H/a-SiGe:H triple-junction structure in 1997 [13] followed by a stable aperture-area module efficiency of 10.5% using the same structure in 1998 [14]. This triple-junction spectrum-splitting technology has been successfully transferred to mass production with a significant expansion in manufacturing capacity. We have continued to improve the current technology by optimizing the a-Si:H and a-SiGe:H quality and increasing the deposition rate. At the same time, we started nc-Si:H solar cell research in the summer of 2001 [15-17]. The research strategy is as follows: First, we optimize the deposition parameters at a low rate with small-area cells to prove the concept. Second, we investigate various methods to increase the deposition rate for small-area cells to meet the manufacturing requirements. Third, we test large-area cell deposition to ensure uniformity, and fabricate modules to conduct stability and reliability studies. Finally, once the above studies are completed, we shall build a roll-to-roll pilot deposition for a pre-manufacturing test. This strategy has been proved to be an effective approach for transferring new technology from research laboratory to mass production lines. In this paper, we review our progress in the nc-Si:H solar cell research along this line and compare with the current a-Si:H/a-SiGe:H/a-SiGe:H triple-junction technology, discuss the technical barriers to be overcome before transferring the nc-Si:H solar cell technology to mass production lines, and propose a roadmap for using nc-Si:H based technology in solar panel manufacturing with a high module efficiency at a low cost.

EXPERIMENTAL APPROACHES

Our major work has been focused on small-area nc-Si:H material and device optimizations with a limited effort on large-area module fabrication. The substrates are normally 5-mil-thick stainless steels coated with textured Ag/ZnO or Al/ZnO back reflector. The intrinsic nc-Si:H layer is deposited using various glow discharge techniques including conventional RF, high pressure/high power RF, modified VHF (MVHF), and microwave [16, 17] For small-area cells, indium-tin-oxide (ITO) dots with an active-area of 0.25 cm^2 are deposited as the top transparent contact, and metal grids are deposited on top of the ITO dots for current collection. For large-area module fabrication, we used a multi-chamber RF glow discharge batch machine to deposit the doped and intrinsic layers. Other processes of module fabrication are similar to those used in the manufacturing lines, including short/shunt passivation, wiring, and encapsulation. The small-area cells are characterized using the current density versus voltage (J-V) measurement under a solar simulator with a spectrum close to AM1.5 illumination. Quantum efficiency (QE) measurement is carried out with no optical bias at short circuit for single-junction cells and under appropriate optical and electrical biases for multi-junction cells. J_{sc} is calculated by integration of the QE curve with the AM1.5 spectrum from 300 to 1100 nm. In order to achieve high efficiency, various multi-junction cell structures have been studied, including a-Si:H/nc-Si:H double-junction, and a-Si:H/a-SiGe:H/nc-Si:H and a-Si:H/nc-Si:H/nc-Si:H triple-junction structures. Light soaking experiments are carried out under various light illumination conditions with different electrical biases.

STATUS OF nc-Si:H BASED SOLAR CELLS

Optimization of nc-Si:H single-junction cells and control of nanocrystalline evolution

In the early stage of our nc-Si:H research, we experienced and solved many issues. The first problem was an ambient degradation observed in nc-Si:H solar cells without intentional light soaking, which was caused by high porosity in the material. [15]. The porous structure of some unoptimized nc-Si:H allowed impurity diffusion into the material and degraded the cell performance. In some cases, the ambient degradation was very severe even in an a-Si:H/nc-Si:H double-junction cell; the porous structure of the nc-Si:H caused micro-cracks of the a-Si:H top cell as evidenced by the ambient degradation in the a-Si:H/nc-Si:H double-junction cells [15]. We investigated the deposition parameters and successfully reduced the porosity in the nc-Si:H and improved the cell performance.

The second problem concerns nanocrystalline evolution. It has been reported that nc-Si:H materials made under the conditions close to the nanocrystalline to amorphous transition have a compact structure and the corresponding solar cells showed high efficiencies [18]. This phenomenon implies that the nc-Si:H with a high crystalline volume fraction normally contains a high defect density, presumably resulting from poor grain boundary passivation and post-oxidation through the porous structure. The crystalline volume fraction increases with the nc-Si:H thickness [19]. An example is given in Fig. 1, where the plot on the left is a Raman spectrum from a nc-Si:H solar cell and the one on the right is the crystalline volume fraction estimated from decomposition of the Raman spectra for cells with different intrinsic layer thicknesses [20]. It clearly shows that the crystalline volume fraction increases with the film thickness. It is logical to expect that an increase in the nc-Si:H intrinsic layer thickness would result in an increase in J_{sc}. However, we found experimentally that under certain conditions, J_{sc} reaches a maximum value; further increasing the intrinsic layer thickness does not lead to an increase in J_{sc} [20]. The loss of J_{sc} in the thicker nc-Si:H cells mainly resulted from the reduction in the long wavelength response. Because this loss could be recovered by a reverse bias during the QE measurement, we concluded that the lower J_{sc} in the thicker nc-Si:H cells was not from insufficient absorption, but from poor collection. The collection problem is normally related to

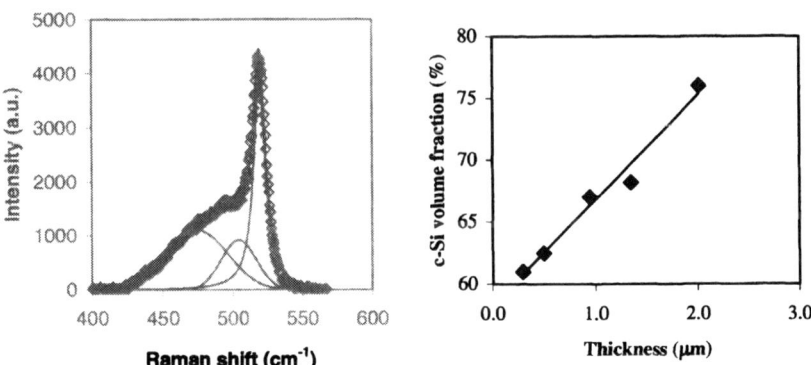

Figure 1. (left) Raman spectrum of a nc-Si:H solar cell with three components of the decomposition and (right) the crystalline volume fraction versus cell thickness.

recombination through defects. The extra thickness caused a high defect density near the *i/p* interface region. Based on a previous study, the *i/p* junction is critical for *n-i-p* (the same for *p-i-n*) solar cells, since more carriers are generated in this region, and the holes have to pass through this region to be collected [21]. The increased defect density could be related to the higher crystalline volume fraction with larger grain sizes, which caused poorer grain boundary passivation. In order to suppress the crystalline evolution, we have developed a hydrogen dilution profiling technique with a continually decreasing hydrogen dilution during the nc-Si:H deposition [20]. A very high hydrogen dilution ratio during the initial deposition reduces the amorphous incubation layer. Dynamically reducing hydrogen dilution during the deposition controls the increase of crystalline volume fraction and grain size. This technique improved the nc-Si:H cell efficiency significantly. Table I lists the J-V characteristics of nc-Si:H single-junction solar cells made with various hydrogen dilution profiles. The baseline cells with an intrinsic layer thickness ~1.2 μm showed an average efficiency of 6.6%, with an average J_{sc} of 21.8 mA/cm^2. Increasing the intrinsic layer thickness did not result in an increase in the J_{sc}, but rather reduced the value. By optimizing the hydrogen dilution profiling, the cell efficiency has been significantly improved, caused by an increase in J_{sc}. Figure 2 shows a comparison of the QE curves of one baseline cell (#14554) with a constant hydrogen dilution ratio, a thicker cell (#14559) with the same constant hydrogen dilution ratio, and two cells with hydrogen dilution profiles (#14578 and #14660). The thicker cell (#14559) did not show a higher long wavelength response, but reduced response in the middle wavelength region, as normally seen in the oxidized cell caused by ambient degradation [15]. The increases of crystalline volume fraction and grain size in the additional thickness are believed to be the origin of the reduced long wavelength response. The hydrogen dilution profiling successfully solved the problem and increased the cell efficiency, especially by enhancing J_{sc} due to the increased middle and long wavelength responses.

Table I. Summary of J-V characteristics of nc-Si:H single-junction solar cells made with various hydrogen dilution profiles. From Profile 1 to Profile 6, the slope of hydrogen dilution ratio versus time was increased.

Sample #	Eff (%)	J_{sc} (mA/cm^2)	V_{oc} (V)	FF			R_s ($\Omega.cm^2$)	Comments
				AM1.5	Blue	Red		
14554	6.74	22.58	0.495	0.603	0.652	0.615	4.4	Baseline
14568	6.48	22.15	0.488	0.599	0.648	0.599	4.0	
14592	6.78	21.32	0.488	0.642	0.690	0.658	3.1	
14594	6.61	20.79	0.486	0.654	0.687	0.648	3.1	
14596	6.61	22.05	0.482	0.622	0.656	0.605	4.0	
14559	6.54	21.48	0.482	0.632	0.678	0.637	4.0	20% thicker
14562	6.81	21.57	0.484	0.652	0.692	0.651	3.4	than baseline
14578	6.63	23.22	0.482	0.594	0.646	0.631	4.3	Profile 1
14580	7.04	22.58	0.484	0.644	0.688	0.662	3.2	Profile 2
14612	7.29	24.41	0.485	0.616	0.659	0.647	3.6	Profile 3
14619	7.81	24.63	0.492	0.645	0.683	0.641	3.4	Profile 4
14642	8.01	23.42	0.502	0.681	0.706	0.700	2.7	Profile 5
14660	8.37	25.15	0.502	0.663	0.679	0.693	3.1	Profile 6

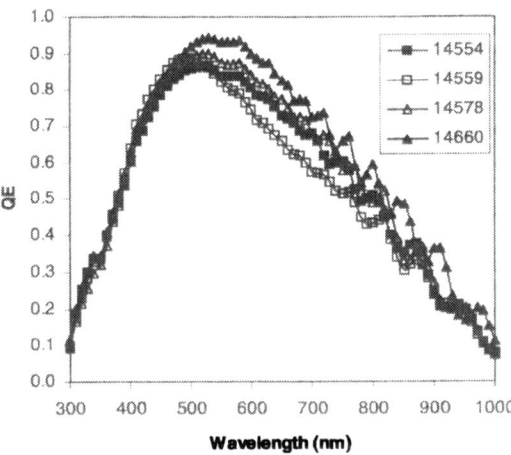

Figure 2. Quantum efficiency of nc-Si:H single-junction cells with various hydrogen dilution profiles. The sample number and the corresponding hydrogen dilution profiles are listed in Table I.

The hydrogen dilution profiling technique was initially developed during the optimization of nc-Si:H solar cells using RF glow discharge at a low rate of ~ 1 Å/s. Since the principle is the same for different deposition techniques at different rates, we applied the same technique to the MVHF deposition and increased the high rate nc-Si:H cell efficiency significantly. Combined with other optimization procedures described in our previous work [16,17], such as properly designed n/i and i/p buffer layers and optimized light trapping from the textured substrates, we have achieved an initial active-area efficiency of 9% as shown in Fig. 3 [10]. Although this cell has the highest efficiency, the J_{sc} is not sufficient for use as a bottom cell in triple-junction structures. We need to develop deposition parameters to obtain high J_{sc}, especially with an improved long wavelength response. We have optimized the deposition parameters further and obtained a nc-Si:H cell with a J_{sc} close to 27 mA/cm² (26.94 mA/cm²), V_{oc}=0.538 V, and

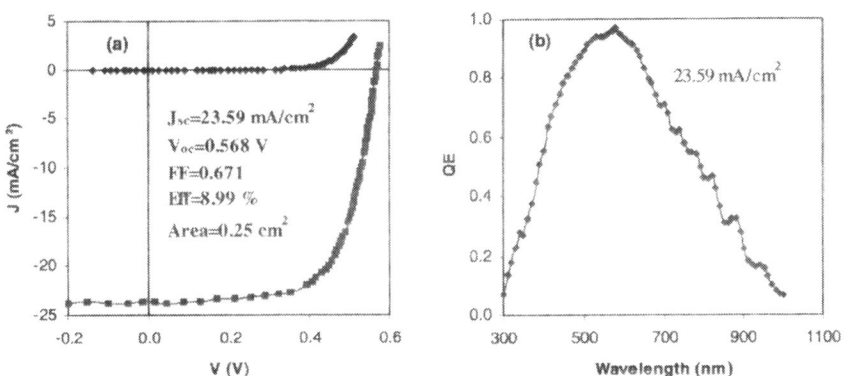

Figure 3. (a) J-V characteristics and (b) quantum efficiency of the best nc-Si:H solar cell, which is suitable for the middle cell in a-Si:H/nc-Si:H/nc-Si:H triple-junction structures. FF=0.608 for an efficiency of 8.81%.

We have also carried out metastability studies of nc-Si:H solar cells under various illumination and electrical bias conditions [22-25]. We found that nc-Si:H solar cells showed many intriguing metastability behaviors. The light-induced degradation has a range from zero to 15%, depending on the material quality and cell structure. A spectrum dependent light-soaking study found that no light-induced degradation was observed under red light illumination [22], which not only confirms that the light-induced defect generation occurred mainly in the amorphous and grain boundary regions, but also ensures that a nc-Si:H cell will not degrade after prolonged light soaking when it is used as the bottom cell in multi-junction structures. In addition, an enhanced light-induced degradation was observed in nc-Si:H cells under reverse bias during light soaking [23, 24], which is opposite to the situation of a-Si:H solar cells. We also found that an optimized hydrogen dilution profile not only improves the cell efficiency but also the stability against prolonged light soaking [25]. The optimized nc-Si:H single-junction solar cells show an average of ~5% light-induced degradation. The degradation is mainly in J_{sc}, resulting from the reduction of short wavelength response. In practice, nc-Si:H cells are mainly used as the bottom cell (or middle cell) in multi-junction structures, where the short wavelength light does not reach the nc-Si:H bottom cell, and therefore very little light-induced degradation is expected in the nc-Si:H bottom cell.

High Efficiency Multi-Junction Solar Cells with nc-Si:H in the Bottom Cell

We used the optimized component cells to fabricate triple-junction cells. Table II lists the J-V characteristics of several a-Si:H/a-SiGe:H/nc-Si:H cells in the initial and stable states, where the stable state was reached by light soaking under ~100 mW/cm² white light at 50 °C for over 1000 hours. The highest initial active-area efficiency of 15.1% is achieved with the initial J-V characteristics and QE shown in Fig. 4 [10]. After light soaking, a stable efficiency of 13.0% is obtained. From the data in Table II, one can see that the bottom cell current did not degrade after light soaking. Therefore, a bottom cell limited current mismatch resulted in a degradation of only 9.0% (#15506-34). However, the cell with the highest initial efficiency degraded by 17.1%, a large reduction of FF for a middle cell limited current mismatch condition.

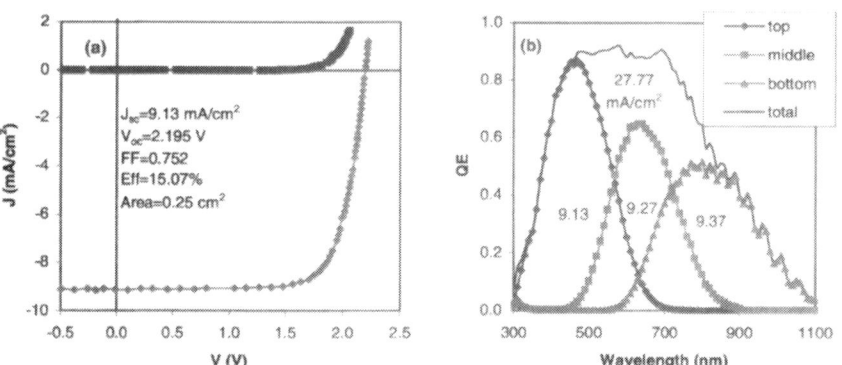

Figure 4. (a) J-V characteristics and (b) quantum efficiency of an a-Si:H/a-SiGe:H/nc-Si:H triple-junction solar cell.

Table II. J-V characteristics of high efficiency a-Si:H/a-SiGe:H/nc-Si:H triple-junction solar cells. Deg. denotes the percentage of light-induced degradation. The bold numbers are the highest efficiencies and italic numbers are the limited current densities for J_{sc}.

Sample	State	Eff (%)	J_{sc} (mA/cm^2)	QE (mA/cm^2) top	middle	bottom	V_{oc} (V)	FF
15501-	Initial	14.77	9.11	*9.11*	9.24	9.20	2.145	0.756
34	Stable	12.62	8.69	8.85	*8.69*	9.12	2.101	0.691
	Deg.	14.6%	4.6%	2.9%	6.0%	0.9%	2.1%	8.6%
15506-	Initial	**15.07**	9.13	*9.13*	9.27	9.31	2.195	0.752
33	Stable	12.49	8.63	8.74	*8.63*	9.24	2.116	0.684
	Deg.	17.1%	5.5%	4.3%	6.9%	0.8%	3.6%	9.0%
15506-	Initial	14.27	8.74	9.25	9.41	*8.74*	2.167	0.755
34	Stable	**12.98**	8.74	8.97	8.88	*8.74*	2.110	0.704
	Deg.	9.0%	0	3.0%	5.6%	0	2.6%	6.7%

Table III summarizes the J-V characteristics of several a-Si:H/nc-Si:H/nc-Si:H triple-junction solar cells in the initial and light-soaked states. The highest initial active-area efficiency of 14.1% was achieved [10,25]. Although the a-Si:H/nc-Si:H/nc-Si:H triple-junction structure did not have the highest initial efficiency, it showed a very small light-induced degradation of 4.3%, resulting from a better stability of the nc-Si:H middle cell than that of the a-SiGe:H middle cell. A stable active-area efficiency of 13.3% is achieved with this cell structure, which exceeds the stable active-area efficiency achieved using the conventional a-Si:H/a-SiGe:H/a-SiGe:H triple-junction structure [13].

Table III. J-V characteristics of high efficiency a-Si:H/nc-Si:H/nc-Si:H triple-junction solar cells. Deg. denotes the percentage of light-induced degradation. The bold numbers are the highest efficiencies and italic numbers are the limited current densities for J_{sc}.

Sample	State	Eff (%)	J_{sc} (mA/cm^2)	QE (mA/cm^2) top	middle	bottom	V_{oc} (V)	FF
13955-	Initial	**14.14**	9.11	9.11	9.72	*9.11*	1.965	0.790
33	Stable	13.19	8.79	*8.79*	9.56	9.04	1.947	0.771
	Deg.	6.7%	3.5%	3.5%	1.6%	0.8%	0.9%	2.4%
13955-	Initial	13.86	8.89	9.02	9.52	*8.89*	1.981	0.787
24	Stable	**13.26**	8.72	8.75	9.25	*8.72*	1.973	0.771
	Deg.	4.3%	1.9%	3.0%	2.8%	1.9%	0.4%	2.0%
14005-	Initial	13.67	8.99	9.44	9.54	*8.99*	1.944	0.782
33	Stable	13.24	8.92	9.04	9.42	*8.92*	1.933	0.768
	Deg.	3.1%	0.8%	4.2%	1.3%	0.8%	0.6%	1.8%

Large-Area a-Si:H/nc-Si:H Double-Junction Solar Cells

The next step towards the manufacturing of nc-Si:H based solar panels is to test the feasibility of making large-area modules [26]. We used an a-Si:H/nc-Si:H double-junction structure on Ag/ZnO BR to prove the concept, because it is easier to make than triple-junction cells. We used a large-area multi-chamber RF glow discharge system, which can hold a 14×15 in^2 substrate. The deposition time of the nc-Si:H intrinsic layer was 50 minutes at a deposition

Table IV. Initial and stable performance of a-Si:H/nc-Si:H double-junction mini-modules with an aperture area of 45 cm^2.

Sample #	State	Temp (°C)	V_{oc} (V)	FF	J_{sc} (mA/cm^2)	P_{max} (W)	Eff (%)	Eff$_{corr}$ (%)
10490F2	Initial	26.6	1.425	0.741	11.40	0.542	12.03	12.07
	Stable	25.4	1.403	0.688	11.25	0.489	10.86	10.87
10500F1	Initial	26.6	1.430	0.738	11.46	0.544	12.08	12.12
	Stable	25.6	1.409	0.665	11.43	0.482	10.71	10.72
10500F2	Initial	26.9	1.432	0.736	11.64	0.552	12.26	12.31
	Stable	25.6	1.406	0.682	11.44	0.494	10.97	10.98
10500G3	Initial	26.6	1.441	0.743	11.37	0.548	12.18	12.22
	Stable	25.9	1.419	0.686	11.12	0.487	10.82	10.83

rate ~4-5 Å/s. We first used the small-area (0.25-cm^2 active area) cells to check the uniformity of the cell efficiency distribution over the deposition area. By optimizing the deposition parameters, a reasonable uniformity was established with a standard deviation of 5.2% over an area of 645 cm^2. The highest initial active-area (0.25 cm^2) cell efficiency of 13.6%, with J_{sc}=12.15 mA/cm^2, V_{oc}=1.429 V, and FF=0.783, has been achieved. After light soaking, this cell stabilized at 12.4% with J_{sc}=12.01 mA/cm^2, V_{oc}=1.423 V, and FF=0.726. We made a-Si:H/nc-Si:H modules with different aperture areas. Table IV listed the performance data of mini-modules with an aperture area of 45 cm^2 but without encapsulation. The initial aperture-area efficiency is slightly over 12%, and the average light-induced degradation is about 11%. Table V lists the performance data of a-Si:H/nc-Si:H double-junction modules with an aperture area of ~420 cm^2, where the initial aperture-area efficiency before the encapsulation is about 11.5%. It is smaller than the efficiency of the mini-modules due to the non-uniformity of the deposition. In the final state after encapsulation and light soaking, the efficiency dropped to 9.0%-9.5%,

Table V. Performance of a-Si:H/nc-Si:H double-junction modules with an aperture area of ~420 cm^2, where T is the measurement temperature, Eff the efficiency at T, and Eff$_c$ the corrected efficiency to 25 °C. State A is the as-deposited state before encapsulation, State B is after encapsulation and light soaking under 100 mW/cm^2 of white light at 50 °C for 500 hours. The comment column indicates the measurement site at United Solar (USO) or at NREL.

Serial #	State	Temp (°C)	V_{oc} (V)	FF	J_{sc} (mA/cm^2)	P_{max} (W)	Eff (%)	Eff$_c$ (%)	Comment
10534	A	26.1	1.425	0.694	11.49	4.770	11.36	11.38	USO
(01)	B	27.6	1.383	0.624	10.6	3.834	9.13	9.18	
10536	A	26.4	1.422	0.708	11.31	4.784	11.39	11.42	USO
(03)	B	27.6	1.382	0.624	10.52	3.807	9.07	9.12	
		25.0	1.455	0.642	9.68	3.816		9.13	NREL
10582	A	25.6	1.440	0.729	11.21	4.943	11.77	11.78	USO
(04)	B	27.8	1.388	0.628	10.50	3.845	9.15	9.22	
10583	A	25.6	1.435	0.722	11.10	4.829	11.50	11.51	USO
(02)	B	27.6	1.400	0.641	10.38	3.917	9.33	9.38	
		25.0	1.461	0.648	9.74	3.797		9.22	NREL
10587	A	25.6	1.438	0.689	11.68	4.859	11.57	11.58	USO
(05)	B	27.8	1.400	0.650	10.44	3.987	9.49	9.56	

resulting from ~5-6% encapsulation loss and ~12-14% light-induced degradation. Measurements at NREL gave efficiencies similar to those measured at United Solar, although the individual parameters are different, possibly caused by the difference in the spectra of the solar simulators.

FUTURE WORK AND THE ROADMAP FOR nc-Si:H BASED SOLAR PANEL MANUFACTURING

United Solar has been manufacturing multi-junction solar panels since 1986. From a pilot plant production of about 600 kW in 1986, it has emerged as one of the largest U.S. owned manufacturers of solar panels having a current annual production capacity of 58 MW. Two additional 60 MW plants are under construction in Michigan, USA, and the total annual capacity will reach 180 MW in 2008. Our plan calls for 300 MW annual capacity by 2010. United Solar's evolution of manufacturing capacity and future expansion plan are illustrated in Fig. 5. The spectrum splitting a-Si:H/a-SiGe:H/a-SiGe:H triple-junction cell structure has been used in the two manufacturing plants today and will be used in the two 60 MW plants currently under construction. The question is which cell structure will be chosen for the future plants and whether the a-Si:H and nc-Si:H based technology is ready for manufacturing.

Several companies have been working on a-Si:H and nc-Si:H based thin film solar panel manufacturing [11, 27, 28], demonstrating that the concept has been proven and certain technical barriers have been overcome. The majority of these companies are using the *p-i-n* cell structure on glass superstrate. We make *n-i-p* cells on flexible steel substrates in a roll-to-roll operation. Currently, we still need to address the following critical issues.

The first issue is to achieve higher module efficiency than that in the current product. Figure 6 plots the cell, module, and product efficiencies achieved using the a-Si:H/a-SiGe:H/a-SiGe:H triple-junction structure at various stages. The highest initial and stable active-area efficiency of 15.2% and 13.0% were achieved using a low rate deposition of ~ 1Å/s on Ag/ZnO back reflector [13]. By increasing the deposition rate to 3 Å/s, the efficiencies became 13.0% and 11.3%; respectively. On Al/ZnO back reflectors, the efficiencies further dropped to 11.7% and 9.8%, respectively. The initial and stable aperture-area module efficiencies of 10.5% and 8.9% were achieved at ~3Å/s on Al/ZnO in the research and development group. After transferring this technology to the manufacturing lines, the current products show an average

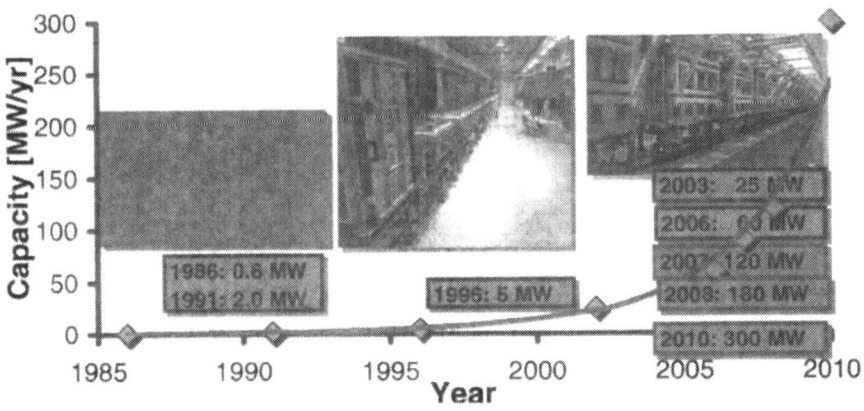

Figure 5. History and future expansion plan of United Solar's manufacturing capacity.

aperture-area efficiency of ~8%. As discussed in the previous section, our major research activities are still focused on the cell level. The champion cell performance with nc-Si:H in the bottom cell is similar to the highest efficiency achieved with the a-Si:H/a-SiGe:H/a-SiGe:H triple-junction structure. In order to make a sound business decision, the new cell structure with nc-Si:H has to show a significant advantage of at least >10% in the cell performance over the conventional structure. Therefore, a higher stable active-area efficiency of over 14% is our near term goal. The most important milestone for nc-Si:H based modules made with a similar total deposition time to the current manufacturing process on Al/ZnO back reflector is a stable aperture-area efficiency of larger than 10%. It is clear that our current status has not reached the efficiency milestone yet.

The second issue is to increase the nc-Si:H deposition rate to a level resulting in equal or shorter total deposition time. Compared to the current manufacturing machine, the deposition rate of nc-Si:H needs to be at least 20 Å/s if an a-Si:H/nc-Si:H/nc-Si:H triple-junction structure is used. Currently, the deposition rate of nc-Si:H intrinsic layer is ~5-8 Å/s. Significant efforts are needed to increase the deposition rate to meet the requirement of equal (or shorter) deposition time compared to the current manufacturing technology.

The third issue is the large-area uniformity and compatibility with the roll-to-roll deposition. At the current stage, VHF has been shown to be a possible deposition method for the intrinsic nc-Si:H deposition. However, the large-area uniformity is a great challenge for the cathode design. In principle, it is achievable by simulating the electromagnetic field distribution for various cathode geometries, such as the ladder structure used by Mitsubishi Heavy Industries, Ltd. [27]. At United Solar, we have not put, as yet, enough effort on large-area VHF cathode design. In addition, the high pressure/high power depleting regime has been used for increasing the deposition rate while maintaining high-quality nc-Si:H. One critical requirement for this regime is a small spacing between the cathode and the substrate. For a very large roll-to-roll machine with flexible substrate, keeping the same uniform spacing is also a great challenge.

The fourth issue is the reliability of the product. United Solar's a-Si:H/a-SiGe:H/a-SiGe:H triple-junction solar panels on Al/ZnO has passed various reliability tests under different harsh conditions. Any new product also must pass all the reliability tests. Currently, we have not carried out such tests. One concern is the adhesion of the solar cells to the substrate, because

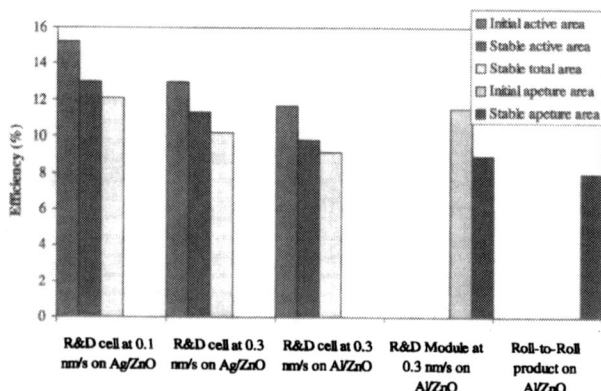

Figure 6. Cell, module, and PV product efficiencies using a-Si:H/a-SiGe:H/a-SiGe:H triple-junction technology.

the nc-Si:H intrinsic layers are much thicker than a-Si:H and a-SiGe:H intrinsic layers in multi-junction solar cells. The internal stress may accumulate with the film thickness, and then the adhesion may become an issue.

The four areas are identified as the major tasks for future research towards the development of a-Si:H and nc-Si:H based solar panel manufacturing technology. We have proposed a roadmap to achieve the goals to U.S. Department of Energy's (DOE) Solar America Initiative (SAI) Program. We shall demonstrate that by integrating a-Si:H and nc-Si:H in a multi-junction structure, an enhancement in stable module efficiencies of >10% can be achieved with the bench mark being the current stable production module efficiency of 8%. In the meantime, we shall increase the nc-Si:H deposition rate to the level such that the total deposition time is similar to the current manufacturing process. We shall design and test various VHF cathodes to achieve large-area uniform nc-Si:H deposition, and evaluate the a-Si:H and nc-Si:H deposition process in a pilot roll-to-roll machine. Finally, we shall conduct reliability testing of the a-Si:H and nc-Si:H based solar panels.

SUMMARY

We have demonstrated that by using nc-Si:H as the bottom cell in triple-junction structures, we can achieve the same initial and stable cell efficiencies as the a-Si:H/a-SiGe:H/a-SiGe:H triple-junction structure. However, there are still major issues that need to be addressed before considering the use of nc-Si:H in solar panel manufacturing plants, namely, increasing the efficiency further; increasing the deposition rate further; improving the large-area uniformity; solving any potential problem associated with the nc-Si:H deposition; and testing the reliability of the a-Si:H and nc-Si:H based panels. We have planned out technical approaches and are ready to resolve these issues under the SAI program.

ACKNOWLEDGEMENTS

This work was supported by the NREL's Thin Film Partnership Program under Subcontracts Nos. ZDJ-2-30630-19 and ZXL-6-44205-14. We have benefited from the collaborative work of the team members under the Thin Film Partnership Program. We also thank Jeff Yang and the great team effort of the R&D colleagues at United Solar.

REFERENCES

1. J. Meier, R. Flückiger, H. Keppner, and A. Shah, Appl. Phys. Lett. **65, 860** (1994).
2. A. Shah, J. Dutta, N. Wyrsch, K. Prasad, H. Curtins, F. Finger, A. Howling, and Ch. Hollenstein, Mater. Res. Soc. Symp. Proc. **258**, 15 (1992).
3. J. Meier, P. Torres, R. Platz, S. Dubail, U. Kroll, J. A. Anna Selvan, N. Pellaton Vaucher, Ch. Hof, D. Fischer, H. Keppner, A. Shah, K.-D. Ufert, P. Giannoulès, and J. Koehler, Mater. Res. Soc. Symp. Proc. **420**, 3 (1996).
4. L. Guo, M. Kondo, M. Fukawa, K. Saitoh , and A. Matsuda, Jpan. J. Appl. Phys. Part 2 **37**, L1116 (1998).
5. B. Rech, T. Roschek, J. Müller, S. Wieder, and H. Wagner, Technical Digest of 11th PVSEC, Sapporo, Japan, 1999, p. 241.
6. T. Matsui, A. Matsuda, and M. Kondo, Mater. Res. Soc. Symp. Proc. **808**, 557 (2004).
7. Y. Mai, S. Klein, R. Carius, J. Wolff, A. Lambertz, F. Finger, and X. Geng, J. Appl. Phys. **97**, 114913 (2005).

8. Y. Mai, S. Klein, R. Carius, H. Sriebig, X. Geng, and F. Finger, Appl. Phys. Lett. **87**, 073503 (2005).

9. J. K. Rath, A. Verkerk, R. Franken, C. van Bommel, K. van der Werf, A. Gordijn, and R. Schopp, Conf. Record of the 2006 IEEE 4th World Conf. on Photovoltaic Energy Conversion, Hawaii, USA, May 7-12, 2006, p. 1473.

10. B. Yan, G. Yue, J. M. Owens, J. Yang, and S. Guha, Conf. Record of the 2006 IEEE 4th World Conf. on Photovoltaic Energy Conversion, Hawaii, USA, May 7-12, 2006, p. 1477.

11. K. Yamamoto, A. Nakajima, M. Yoshimi, T. Sawada, S. Fukada, T. Suezaki, M. Ichikawa, Y. Koi, M. Goto, T. Meguro, T. Matsuda, T. Sasaki, and Y. Tawada, Conf. Record of the 2006 IEEE 4th World Conf. on Photovoltaic Energy Conversion, Hawaii, USA, May 7-12, 2006, p. 1489.

12. K. Saito, M. Sano, S. Okabe, S. Sugiyama, and K. Ogawa, Solar. Energy Matter & Solar Cells **86**, 565 (2005).

13. J. Yang, A. Banerjee, and S. Guha, Appl. Phys. Lett. **70**, 2975 (1997).

14. A. Banerjee, J. Yang, and S. Guha, Mater. Res. Soc. Symp. Proc. **557**, 743 (1999).

15. B. Yan, K. Lord, J. Yang, S. Guha, J. Smeets, and J-M. Jacquet, Mater. Res. Soc. Symp. Proc. **715**, 629 (2002).

16. B. Yan, G. Yue, J. Yang, A. Banerjee, and S. Guha, Mater. Res. Soc. Symp. Proc. **762**, 309 (2003).

17. B. Yan, G. Yue, J. Yang, K. Lord, A. Banerjee, and S. Guha, Proc. of 3rd World Conference on Photovoltaic Energy Conversion, May 11-18, 2003, Osaka, Japan, p. 2773

18. T. Roschek, T. Repmann, J. Müller, B. Rech, and H. Wagner, Proc. of 28th IEEE Photovoltaic Specialists Conference, Anchorage, AK, Sep. 15-22, 2000, (IEEE New York, 2000), p. 150.

19. J. Kočka, A. Jejfar. V. Vorlíček, H. Stuchlíkova, and J. Stuchlík, Mater. Res. Soc. Symp. Proc. **557**, 483 (1999).

20. B. Yan, G. Yue, J. Yang, S. Guha, D. L. Williamson, D. Han, and C.-S. Jiang, Appl. Phys. Lett. **85**, 1955 (2004).

21. A. Pawlikiewicz and S. Guha, Mater. Res. Soc. Symp. Proc. **118**, 599 (1988).

22. B. Yan, G. Yue, J. M. Owens, J. Yang, and S. Guha, Appl. Phys. Lett. **85**, 1925 (2004).

23. G. Yue, B. Yan, J. Yang, and S. Guha, Appl. Phys. Lett. **86**, 092103 (2005).

24. G. Yue, B. Yan, J. Yang, and S. Guha, J. Appl. Phys. **98**, 074902 (2005).

25. G. Yue, B. Yan, G. Ganguly, J. Yang, S. Guha, C. W. Teplin, and D.L. Williamson, Conf. Record of the 2006 IEEE 4th World Conf. on Photovoltaic Energy Conversion, Hawaii, USA, May 7-12, 2006, p. 1588.

26. G. Ganguly, G. Yue, B. Yan, J. Yang, S. Guha, Conf. Record of the 2006 IEEE 4th World Conf. on Photovoltaic Energy Conversion, Hawaii, USA, May 7-12, 2006, p. 1712.

27. H. Takatsuka, Y. Yamauchi, Y. Takeuchi, M. Fukagawa, S. Goya, and A. Takano, Conf. Record of the 2006 IEEE 4th World Conf. on Photovoltaic Energy Conversion, Hawaii, USA, May 7-12, 2006, p. 2028.

28. T. Repmann, S. Weider, S. Klein, H. Stiebig, and B. Rech, Conf. Record of the 2006 IEEE 4th World Conf. on Photovoltaic Energy Conversion, Hawaii, USA, May 7-12, 2006, p. 1724.

Mater. Res. Soc. Symp. Proc. Vol. 989 © 2007 Materials Research Society   0989-A15-02

Advanced Deposition Phase Diagrams for Guiding Si:H-Based Multijunction Solar Cells

Jason A. Stoke, Nikolas J. Podraza, Jian Li, Xinmin Cao, Xunming Deng, and Robert W. Collins
Department of Physics and Astronomy, University of Toledo, Toledo, OH, 43606

ABSTRACT

Phase diagrams have been established to describe very high frequency (vhf) plasma-enhanced chemical vapor deposition (PECVD) processes for intrinsic hydrogenated silicon (Si:H) and silicon-germanium alloy ($Si_{1-x}Ge_x$:H) thin films using crystalline Si substrates that have been over-deposited with n-type amorphous Si:H (a-Si:H). The Si:H and $Si_{1-x}Ge_x$:H processes are applied for the top and middle i-layers of triple-junction a-Si:H-based n-i-p solar cells fabricated at University of Toledo. Identical n/i cell structures were co-deposited on textured Ag/ZnO back-reflectors in order to correlate the phase diagram and the performance of single-junction solar cells, the latter completed through over-deposition of the p-layer and top contact. This study has reaffirmed that the highest efficiencies for a-Si:H and a-$Si_{1-x}Ge_x$:H solar cells are obtained when the i-layers are prepared under maximal H_2 dilution conditions.

INTRODUCTION

State-of-the-art solar cells based on a-Si:H prepared by PECVD employ a triple-junction design [1,2]. Optimization of the a-Si:H i-layer of the top cell has been widely successful through the concept of maximal H_2 dilution [3,4]. The benefits of H_2 dilution include enhanced surface passivation and hence diffusion of film precursors in the PECVD process, as well as enhanced relaxation of sub-surface strained Si-Si bonds [4]. The resulting "protocrystalline" nature of the i-layer provides the highest device performance and stability. In this study, vhf PECVD phase diagrams have been developed for Si:H and $Si_{1-x}Ge_x$:H i-layers deposited from Si_2H_6 + GeH_4 mixtures on a-Si:H n-layers under conditions used for high-efficiency triple-junction n-i-p cells. These diagrams have been correlated with single-junction cell performance in order to further explore and extend the applicability of the maximal H_2 dilution concept.

EXPERIMENTAL DETAILS

The Si:H and $Si_{1-x}Ge_x$:H i-layers for phase diagram development by real time spectroscopic ellipsometry (RTSE) were deposited on c-Si/(native-oxide)/(n-type a-Si:H) substrate structures using multichamber vhf (70 MHz) PECVD. Such substrates ensure a specular surface to aid in the utilization of RTSE [5] for the first time in the characterization of PECVD a-$Si_{1-x}Ge_x$:H from S_2H_6 and GeH_4. In order to apply deposition phase diagrams from RTSE for insights into n-i-p solar cell performance, additional samples ~2000 Å thick were co-deposited onto textured Ag/ZnO/(n-type a-Si:H) back-reflectors in the device configuration simultaneously with the specular c-Si/(native-oxide)/(n-type a-Si:H) substrates. For depositions performed versus the phase diagram variable R=[H_2]/{[Si_2H_6]+[GeH_4]}, all other parameters were selected as those used for the previously-optimized top and middle i-layers of a triple junction solar cell, including a vhf power of 8 W, a low source gas [Si_2H_6]+[GeH_4] partial pressure of p < 0.004 Torr, a low total gas pressure of p_{tot} ~ 0.2 Torr, and nominal substrate temperatures of T=200°C for Si:H and T=350°C for $Si_{1-x}Ge_x$:H. These nominal values correspond to calibrated values of T=107°C and 170°C, respectively, determined by RTSE [6]. In order to produce the desired optical gaps, the flow ratio G=[GeH_4]/{[Si_2H_6]+[GeH_4]} was set at G=0 for the top i-layer, resulting in growth rates ranging from ~2.3 to ~0.8 Å/s as R increases from 60 to 150, and G=0.286 for the middle i-layer, resulting in growth rates ranging from ~2.6 to ~1.1 Å/s as R increases from 45 to 150.

RESULTS

Figure 1 shows the optical band gaps plotted versus the H_2 dilution flow ratio R for the a-Si:H and a-Si$_{1-x}$Ge$_x$:H i-layers. These optical gaps were obtained from RTSE data collected at the deposition temperature after i-layer thicknesses of 150-250 Å, and were deduced both from extrapolations of $\varepsilon_2^{1/2}$ versus E (open circles) and from complete fits using a parameterized Cody-gap-modified Lorentz oscillator (solid squares) [7]. Both approaches are based on the assumptions of parabolic bands and a constant dipole matrix element. For a-Si:H films prepared over the range in R from 60 to150, gaps from 1.65 to 1.80 eV are observed. From R = 60 to 120, the band gap shows an increase with increasing R; however, from R=120 to 150 the gap is nearly constant. For a-Si$_{1-x}$Ge$_x$:H films prepared over the range in R from 45 to 150, gaps from 1.35 to 1.48 eV are observed with a continuous linear increase over the full range. The variation in E_g versus R for a-Si$_{1-x}$Ge$_x$:H may arise not only from an increase in H-content as in a-Si:H, but also from a variation in the Ge content x due to differing relative incorporation rates of Si and Ge based precursors at different H_2-dilution levels.

Figure 2 shows the surface roughness layer thickness versus the bulk layer thickness during Si:H growth with two H_2 dilution levels. Examples are provided for a Si:H film that remains amorphous throughout growth (R=60) and for one in which microcrystallites nucleate from the amorphous phase, grow preferentially, and coalesce to a single-phase microcrystalline structure (R = 150). In particular, the R=150 data reveal the amorphous-to-(mixed-phase microcrystal-line) roughening transition, denoted a → (a+μc), and the (mixed-phase)-to-(single-phase) microcrystalline smoothening transition, denoted (a+μc) → μc. The latter films can be analyzed using a four-medium (pseudo-substrate/outer-layer/surface-roughness/ambient) virtual interface technique designed to extract the volume fraction of microcrystallites in the top ~10 Å of the bulk layer and thus to generate a depth profile in the microcrystalline fraction as shown in Fig. 3.

Similar data in the evolution of the roughness thickness and microcrystalline volume fraction were compiled for all Si:H films prepared versus R in order to establish the phase diagram in Fig. 4. This diagram demonstrates that, under the vhf PECVD conditions used here, the Si:H films remain amorphous throughout ~2000 Å of bulk layer growth for R ≤ 80. At higher H_2 dilution

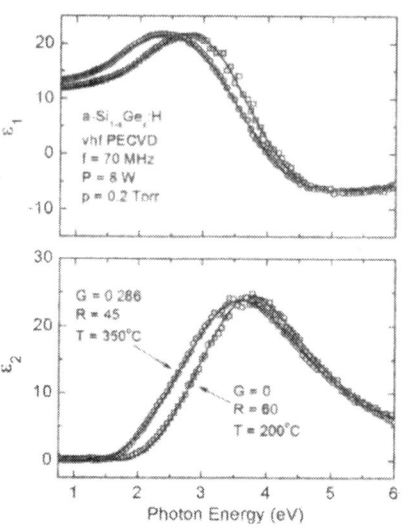

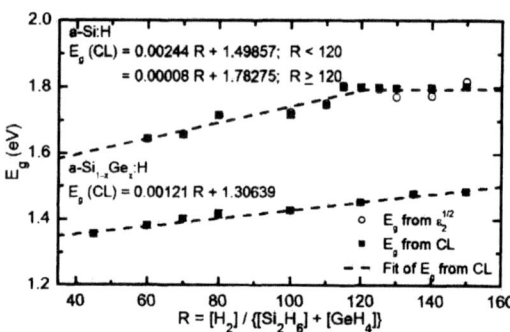

Figure 1. (above) Optical band gaps as functions of the H_2-dilution flow ratio R determined from RTSE applying both a linear extrapolation of $\varepsilon_2^{1/2}$ spectra (open circles) and a complete fit using a Cody-gap-modified Lorentz oscillator function (solid squares); two examples of the complete fits (solid lines) to data (points) are shown at the left.

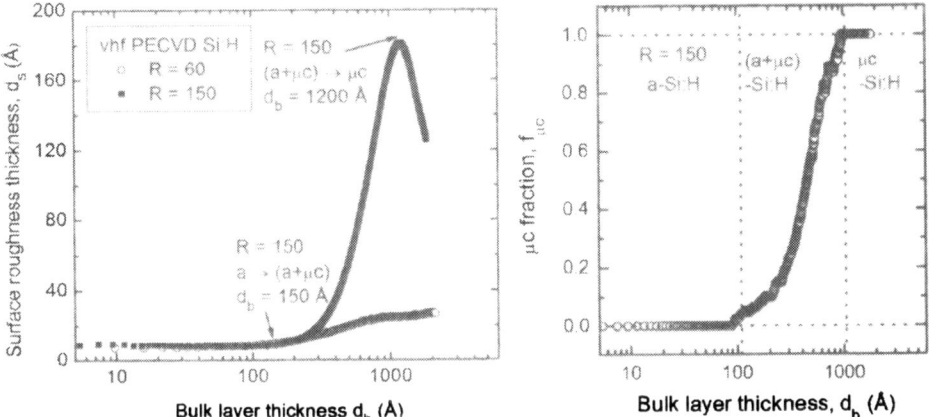

Figure 2 (left). Surface roughness evolution for a vhf PECVD Si:H film in the fully amorphous growth regime (R = 60) and one in the microcrystalline (μc) evolution regime (R = 150).
Figure 3 (right). The μc fraction in the top 10 Å of the bulk layer, determined from virtual interface analysis of RTSE data for the R = 150 Si:H film, plotted versus the accumulated bulk layer thickness.

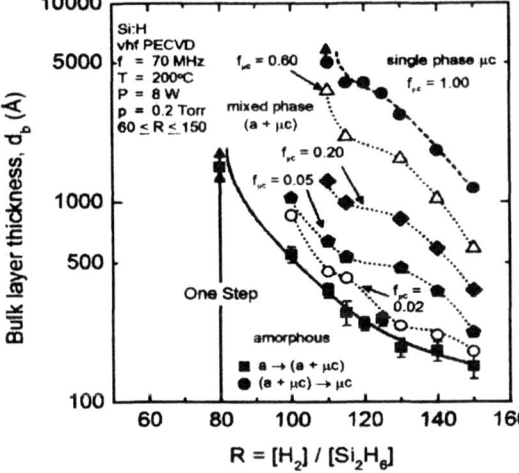

Figure 4. A deposition phase diagram for Si:H films prepared under vhf PECVD conditions. The bulk layer thicknesses at which the a → (a+μc) transition (solid line, squares), and the (a+μc) → μc transition (dashed line, circles) occur are depicted, as deduced from data such as those of Fig. 2. Up-arrows indicate that the transitions occur at thicknesses above the indicated values. Contour lines in the microcrystallite volume fraction $f_{μc}$ are plotted ranging from 0.02 to 0.60 (dotted lines). It is clear that the surface roughening effect provides sensitivity to the a → (a+μc) transition at microcrystallite volume fractions of better than 0.02 within the near-surface of the bulk layer.

(100 ≤ R ≤ 150), the Si:H films initially nucleate as a-Si:H but undergo the a → (a+μc) transition at a thickness that decreases with increasing R. Figure 4 shows that the surface roughening effect as in the data for R=150 in Fig. 2 provides sensitivity to the a → (a+μc) transition at microcrystalline volume fractions of < 0.02 within the near-surface of the bulk layer.

The structural evolution of the $Si_{1-x}Ge_x$:H exhibits behavior analogous to that of the Si:H series with respect to the appearance of the a → (a+μc) and (a+μc) → μc transitions. The surface roughness and microcrystallite evolution data acquired for depositions at varying R were used to compile the vhf PECVD phase diagram in Fig. 5. This diagram demonstrates that, under

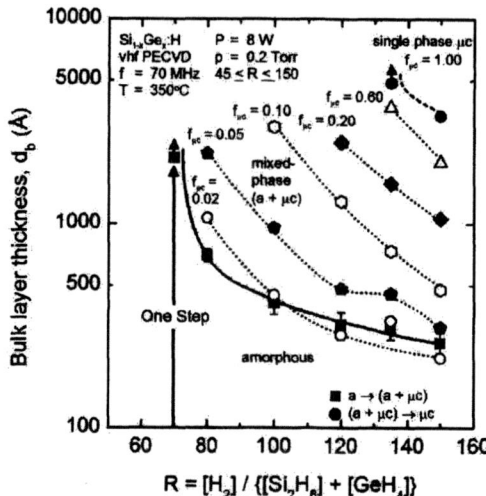

Figure 5. A deposition phase diagram for $Si_{1-x}Ge_x$:H films prepared under vhf PECVD conditions with G=0.286. The bulk layer thicknesses at which the a → (a+μc) transition (solid line, squares), and the (a+μc) → μc transition (dashed line, closed circles) occur are depicted, as deduced from the surface roughness evolution. Up-arrows indicate that the transitions occur at thicknesses above the indicated values. Contour lines in the microcrystallite fraction $f_{μc}$ are plotted ranging from 0.02 to 0.60 (dotted lines). The surface roughening effect provides sensitivity to the a → (a+μc) transition at microcrystalline volume fractions of ~0.01-0.03 within the near-surface of the bulk layer.

the deposition conditions used here, films remain amorphous throughout ~2000 Å of bulk layer growth for R ≤ 70. At higher H_2 dilutions (80 ≤ R ≤ 150), the $Si_{1-x}Ge_x$:H films initially nucleate on the underlying n-layer as a-$Si_{1-x}Ge_x$:H but, as in the case of Si:H, undergo the a → (a+μc) transition at a thickness that decreases with increasing R. Figure 5 shows that the a → (a+μc) transition as observed in the surface roughness thickness evolution is sensitive to microcrystalline volume fractions of <0.02 in the near-surface of the bulk layer at low R, but the sensitivity degrades at higher R as the a → (a+μc) transition shifts to lower thickness.

The n-i structures co-deposited on the textured Ag/ZnO back-reflectors from the two series of Figs. 1-5 (including additional films in which the depositions were terminated at ~2000 Å) were then fabricated into solar cells and characterized to extract the open circuit voltage (V_{oc}), short circuit current (J_{sc}), fill factor (FF), and efficiency (η). Results for a-Si:H with G=0 and for the alloys with G=0.286 are given in Figs. 6 and 7, respectively, and will be discussed next.

DISCUSSION

The vhf PECVD phase diagrams of Figs. 4 and 5 have been developed with the intent of optimizing systematically the top and middle i-layers of triple junction n-i-p solar cells. Toward this goal, i-layers of these two series have been incorporated into single-junction devices with bulk layer thicknesses of ~2000 Å and with 25°C Tauc gaps of ~1.8 eV and ~1.5 eV for the top and middle i-layers, respectively. The diagrams of Figs. 4 and 5 provide insights into how these Si:H and $Si_{1-x}Ge_x$:H films evolve with thickness at different H_2 dilution levels. In fact, the diagrams for vhf PECVD from Si_2H_6+GeH_4 show all the features characteristic of previous diagrams for films from SiH_4+GeH_4. Thus, by understanding the nature of microcrystallite evolution and quantifying the phase composition as in Figs. 4 and 5, it may be possible to select H_2 dilution levels in order to maximize film quality for one-step, multi-step, or graded layers.

In the case of a 2000 Å thick a-Si:H i-layer, the optimum one-step deposition process is predicted on the basis of the phase diagram to occur at the maximal value of R = 80, i.e., the largest R possible such that the i-layer remains amorphous throughout its thickness during growth. This prediction is borne out in the device performance. At R=100 the a → (a+μc) transition occurs at 600 Å, however, and the microcrystalline fraction in the near-surface of the

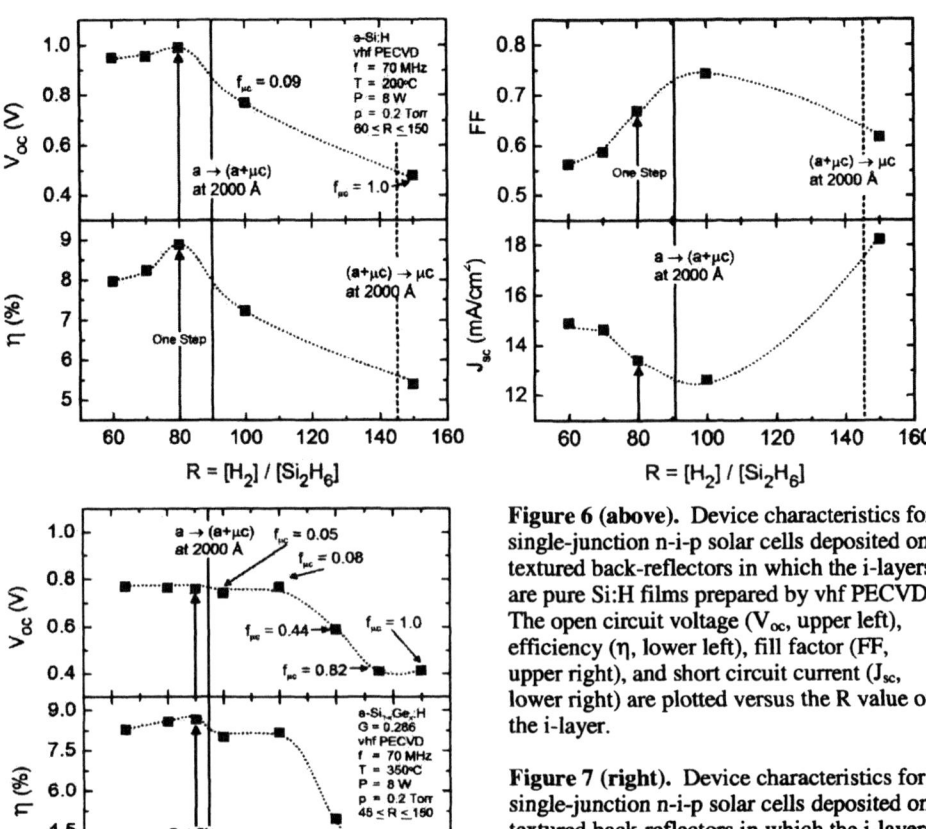

Figure 6 (above). Device characteristics for single-junction n-i-p solar cells deposited on textured back-reflectors in which the i-layers are pure Si:H films prepared by vhf PECVD. The open circuit voltage (V_{oc}, upper left), efficiency (η, lower left), fill factor (FF, upper right), and short circuit current (J_{sc}, lower right) are plotted versus the R value of the i-layer.

Figure 7 (right). Device characteristics for single-junction n-i-p solar cells deposited on textured back-reflectors in which the i-layers are $Si_{1-x}Ge_x$:H. The open circuit voltage (V_{oc}, upper) and efficiency (η, lower) are plotted versus the R value of the i-layer.

2000 Å i-layer is ~ 0.09. The Si:H crystallites appear to degrade V_{oc} significantly due to their presence at the i/p interface. The one-step optimized process for a 2000 Å thick a-$Si_{1-x}Ge_x$:H i-layer is expected at R = 70 according to the same principle, and this optimum is also borne out in the devices, but less distinctly. Above R=70 a detectable volume fraction of microcrystalline $Si_{1-x}Ge_x$:H is present in the near-surface of the film ranging from 0.05 at R = 80 to 0.08 at R=100. In contrast to Si:H, the presence of the microcrystallites at the i/p interface of the $Si_{1-x}Ge_x$:H solar cells does not significantly decrease V_{oc} and efficiency. Although the overall performance of the devices remains acceptable for 70 < R ≤ 100 in Fig. 7, this region is avoided due to irreproducibility, including a dependence on the underlying n-layer structure and poor yield. Next V_{oc} and FF will be discussed further with strategies for multistep improvements.

Considering first the variation in V_{oc} for Si:H, this parameter increases with increasing R for i-layers below the optimum at R=80 due to the increase in band gap; however, a more abrupt increase occurs that defines the optimum as the a → (a+μc) transition is approached. This is the protocrystalline regime, characterized by an improvement in ordering as well as an increase in band gap. Because V_{oc} is strongly influenced by the film properties at the very top of the i-layer, there is an abrupt decrease in its value above R=80 due to the presence of crystalline nuclei at the top of the film. When the top of the film is mixed-phase Si:H, V_{oc} lies between the values for protocrystalline Si (~ 1 eV) and μc-Si:H (~0.5 eV). For R values above the (a+μc) → μc transition for a 2000 Å thick film, a V_{oc} value appropriate for μc-Si:H is expected. In contrast, V_{oc} for $Si_{1-x}Ge_x$:H shows only a weak dependence on R throughout the amorphous growth regime and even across the a → (a+μc) transition. In this case, the effect of R on the optical band gap of $Si_{1-x}Ge_x$:H is also weaker. The lack of a strong effect of the i/p interface crystallites for $Si_{1-x}Ge_x$:H may be due to the fact that over a wide range of R the microcrystallites do not appear to grow preferentially as cone-like structures as in Si:H, but rather as isolated clusters.

The behavior of the fill factor (FF) with R for both Si:H and $Si_{1-x}Ge_x$:H suggests the possibility of an improvement in the cell performance with multistep processing. As an example, Fig. 6 shows that, although the FF increases with R, its maximum of 0.74 is reached at an R value larger than that maximizing V_{oc}. Because the FF is controlled predominantly by the bulk i-layer, the presence of a small volume fraction of crystallites at the i/p interface that reduces V_{oc} does not adversely affect the FF. The FF actually seems to benefit from the incorporation of microcrystallites in the bulk of the Si:H; however, the improvement may in fact be due to improvements in the amorphous material when it is deposited with increasing R from 80 to 100 (i.e., increases in the band gap and protocrystalline ordering). One may be able to take advantage of an optimum V_{oc} and FF simultaneously by depositing the bulk (~1800 Å) of the i-layer with R=100, then depositing a thin (< 100 Å) substrate-memory-erasing low R layer -- similar to the starting n-layer, before finally completing the cell with a second ~200 Å R=100-120 layer. The key, however, is to ensure that the memory-erasing layer first is successful at suppressing the continued growth of crystallites, and second is not detrimental to the FF.

SUMMARY

Deposition phase diagrams have been developed and augmented for vhf PECVD of thin film Si:H and its alloys with Ge by incorporating contour lines representing the crystalline fraction in the top ~10 Å of the i-layer at a given bulk layer thickness. These diagrams predict optimum one-step i-layer deposition processes for the top and middle cells of a triple-junction device that are in consistency with the performance of single-junction devices. V_{oc} and FF appear to be optimized at different values of R suggesting that a multistep PECVD process may be beneficial.

ACKNOWLEDGMENTS

This research was supported by NREL (Subcontract No. RXL-5-44205-06).

REFERENCES

[1] J. Yang, A. Banerjee, and S. Guha, *Appl. Phys. Lett.* **70**, 2975 (1997).
[2] X. Deng, X.B. Liao, S.J. Han, H. Povolny, and P. Agarwal, *Solar Energy Mater. Solar Cells* **62**, 89 (2000).
[3] D.V. Tsu, B.S. Chao, S.R. Ovshinsky, S. Guha, and J. Yang, *Appl. Phys. Lett.* **71**, 1317 (1997).
[4] R.W. Collins and A.S. Ferlauto, *Curr. Opinion Solid State Mater. Sci.* **6**, 425 (2002).
[5] I. An, J.A. Zapien, C. Chen, A.S. Ferlauto, A.S. Lawrence, and R.W. Collins, *Thin Solid Films* **455**, 132 (2004).
[6] P. Lautenschlager, M. Garriga, L. Viña, and M. Cardona, *Phys. Rev. B* **36**, 4821 (1987).
[7] A.S. Ferlauto, G.M. Ferreira, J.M. Pearce, C.R. Wronski, R.W. Collins, G. Ganguly, and X. Deng, *J. Appl. Phys.* **92**, 2424 (2002).

Mater. Res. Soc. Symp. Proc. Vol. 989 © 2007 Materials Research Society   0989-A15-03

Triple Junction n-i-p Solar Cells with Hot-Wire Deposited Protocrystalline and Microcrystalline Silicon

Ruud E.I. Schropp, Hongbo Li, Ronald H.J. Franken, Jatindra K. Rath, Karine van der Werf, Jan Willem Schüttauf, and Robert L. Stolk

Faculty of Science, Utrecht University, Department of Physics and Astronomy, SID - Physics of Devices, P.O. Box 80.000, Utrecht, 3508 TA, Netherlands

ABSTRACT

We have implemented a number of methods to improve the performance of proto-Si/proto-SiGe/μc-Si:H triple junction n-i-p solar cells in which the top and bottom cell i-layers are deposited by Hot-Wire CVD. Firstly, a significant current enhancement is obtained by using textured Ag/ZnO back contacts developed in house instead of plain stainless steel. We studied the correlation between the integrated current density in the long wavelength range (650-1000 nm) with the back reflector surface roughness and clarified that the *rms* roughness from 2D AFM images correlates well with the long wavelength response of the cell when weighted with a Power Spectral Density function. For single junction 2-μm thick μc-Si:H n-i-p cells we improved the short circuit current density from the value of 15.2 mA/cm^2 for plain stainless steel to 23.4 mA/cm^2 for stainless steel coated with a textured Ag/ZnO back reflector. Secondly, we optimized the μc-Si:H n-type doped layer on this rough back reflector, the n/i interface, and in addition we used a profiling scheme for the H$_2$/SiH$_4$ ratio during i-layer deposition. The H$_2$ dilution during growth was stepwise increased in order to prevent a transition to amorphous growth. The efficiency that was reached for a single junction μc-Si:H n-i-p cell was 8.5%, which is the highest reported value for hot-wire deposited cells of this kind, whereas the deposition rate of 2.1 Å/s is about twice as high as in record cells of this type so far. Moreover, these cells are shown to be totally stable under light-soaking tests. Combining the above techniques, a rather thin triple junction cell (total silicon thickness 2.5 μm) has been obtained with an efficiency of 10.9%. Preliminary light-soaking tests show that these triple cells degrade by less than 4%.

INTRODUCTION

Hot-wire chemical vapor deposition (HWCVD) has become a viable method for the preparation of high-quality silicon and silicon alloy materials for application in thin film transistors and solar cells. At Utrecht University, we have developed HWCVD intrinsic protocrystalline silicon (proto-Si:H), which is characterized by an enhanced medium range structural order and a higher stability against light-soaking [1] compared to amorphous silicon, and microcrystalline silicon (μc-Si:H), which is characterized by a low density of states [2] at a crystalline volume fraction of ~40% as determined by Raman spectroscopy. These materials were successfully applied in thin film solar cells on plain stainless steel [3,4].

To enhance the efficiency, multibandgap structures are required [5]. We first developed proto-Si/μc-Si/μc-Si triple junction cells, but it appeared that the intrinsic absorber layers have to be made extremely thick in order to match the currents generated in the stacked subcells. We therefore implemented three modifications to our cell design. First, the μc-Si:H absorber layer of the middle cell was replaced by a plasma-enhanced chemical vapor-deposited (PECVD) protocrystalline silicon-germanium (proto-SiGe:H) layer, which is highly stable against light-soaking [6]. This allows for more efficient spectral splitting and thus higher solar cell efficiency. The wider band gap of proto-SiGe:H increases the open-circuit voltage (V_{oc}). Second, a textured back reflector (TBR) was incorporated in the cell to improve the short-circuit current density (J_{sc}). As a part of this step, the n-layer deposition of the μc-Si:H bottom cell was re-optimized. Third, the n/i-interface and i-layer of the bottom cell were profiled to improve its performance [7].

Fig. 1. Schematic cross section of a triple junction thin film solar cell deposited onto a stainless steel substrate with a textured back reflector. The different subcell absorber materials are μc-Si:H, proto-SiGe, and proto-Si (from bottom to top). On top of the silicon layers is an indium tinoxide/gold contact.

A schematic picture of the triple junction cell structure is shown in Fig. 1. In the present paper, we report recent results for the improved single junction μc-Si:H bottom cell and the triple junction cell including the stability of these cells under light soaking.

EXPERIMENT

The silicon layers of the n-i-p structured solar cells were deposited in the PASTA multi-chamber ultra-high vacuum system. Details of the cell structures can be found in previous publications [4,8,9]. Doped layers and intrinsic proto-SiGe:H [9] were prepared using 13.56 MHz PECVD, whereas HWCVD was applied to fabricate intrinsic proto-Si:H [3] and μc-Si:H [5,10]. For the hot-wire deposition, two straight Ta filaments with a diameter of 0.5 mm were used, through which a current of 10.5 A was passed, yielding a wire temperature of 1850 °C (vacuum calibration). The calibrated substrate temperature was 250 °C. Proto-Si:H was deposited from undiluted SiH_4, whereas H_2-diluted SiH_4 was used for μc-Si:H deposition (H_2 dilution (H_2-flow/total gas flow) of around 0.95). The respective deposition rates were 10 Å/s and 2.1 Å/s. The μc-Si:H is a so-called mixed phase or transition material, consisting of nanocrystallites in an a-Si:H matrix [3,9]. The proto-SiGe:H was optimized on textured Asahi U-type SnO_2:F substrates conformally coated with Ag and ZnO to provide a constant-quality textured back contact. The source gases for proto-SiGe:H were SiH_4, GeH_4, and H_2. Typical band-gap values of a-SiGe:H are 1.5-1.6 eV [11]. The $GeH_4/(SiH_4+GeH_4)$ flow ratio was 0.45 and the $H_2/(SiH_4+GeH_4)$ ratio was 45. We used an *exponential grading* profile for the GeH_4 flow, based on the work described in [12]. Two types of substrates were used: a Ag/ZnO TBR made on stainless steel (SS) foil in our laboratory [13,14] and a SS/Ag/ZnO substrate provided by United

Solar Ovonic LLC Corporation, for comparison. Indium-tin-oxide served as an anti-reflecting TCO top window; an evaporated gold grid on top facilitated a proper charge carrier collection. Both the metal oxide layers and the textured Ag of the TBR were deposited by rf magnetron sputtering in our SALSA system [13,14]. The active area of the solar cells was 0.13 cm^2.

RESULTS

Using our in-house Ag/ZnO sputtering tool SALSA we studied the correlation between the integrated current density in the long wavelength range (650-1000 nm) and the back reflector surface roughness, for μc-Si:H n-i-p cells with an i-layer thickness of 1.5 μm. It became clear that the *rms* roughness from 2D AFM images correlates well with the long wavelength response of the cell when weighted with a Power Spectral Density (PSD) function. The nature of this PSD function is that it gives a larger weight to the lateral features with dimensions similar to the effective wavelength to be scattered [15]. The *rms* roughness σ was weighted with a weighing factor PSD(λ_{scat})/PSD$_{tot}$, where PSD(λ_{scat}) is the contribution of the features with sizes in the region 350 nm – 1400 nm.

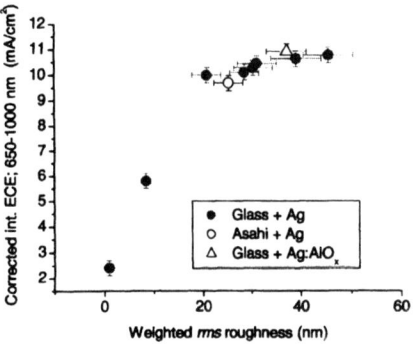 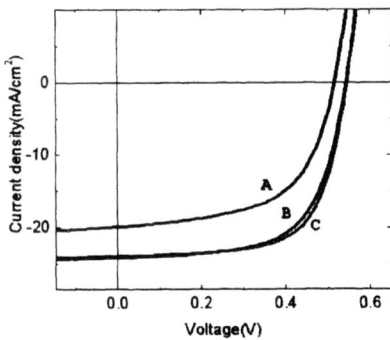

Fig. 2. The correlation between the plasmon corrected long-wavelength generated current density of μc-Si:H n-i-p cells and the weighted *rms* roughness for a large variety of back reflector morphologies.

Fig. 3: Light J-V curves for cells with a 2.0 μm HWCVD μc-Si:H i-layer: A) deposited using a constant hydrogen dilution of 0.948, B) deposited using H$_2$ profiling, and C) deposited using H$_2$ profiling and a "dynamic start".

A property that limits the optical performance of the back reflector is its surface plasmon absorption. The correlation between surface morphology and generated current improves if this is taken into account. The extended surface plasmon absorption of the rough Ag was calculated from total reflection measurements in air [15]. It is known that the presence of the intermediate refractive index interlayer of ZnO:Al between the Ag surface and μc-Si:H i-layer results in a weaker plasmon coupling. Fig. 2 shows the correlation between the external collection efficiency, integrated from 650 to 1000 nm, corrected for plasmon-induced losses, versus the *rms* roughness, weighted using PSD functions.

It can be seen in Fig. 2, that the long-wavelength ECE correlates well with the weighted *rms* roughness and that it shows saturation at high weighted *rms* values. The saturated current of

11 mA/cm^2 is the current generated by light in the 650 to 1100 nm region, and is still not as large as the optical limit given by the number of photons in this region. Nevertheless, for single junction 2-μm thick μc-Si:H n-i-p cells we improved the short circuit current density (for AM1.5 100 mW/cm^2 illumination) from the value of 15.2 mA/cm^2 for plain stainless steel to 23.4 mA/cm^2 for stainless steel coated with a textured Ag/ZnO back reflector [14].

The growth of microcrystalline silicon layers near the a-Si:H/μc-Si:H phase transition regime with HWCVD shows a different trend in the thickness direction than previous phase diagram descriptions for Plasma-Enhanced CVD deposited layers [16]. When the hydrogen dilution ratio is kept constant, the material evolves from microcrystalline into an amorphous network rather than into a microcrystalline structure with higher crystallinity. There are several possible origins for this reverse phase boundary: (i) the HWCVD growth mechanism is considerably different from PECVD in that there are no ions influencing the growth of the film, (ii) the filament surface conditions, namely an increasing degree of coverage by a silicide layer, can change the nature of growth radicals arriving at the film surface, (iii) film crystallinity is greatly substrate-dependent and the local epitaxial growth on the μc-Si:H n$^+$ layer is gradually taken over by amorphous growth [17]. Whatever the origin for the reverse phase change may be, the observation led us to stepwise increase the H$_2$ dilution during growth in order to prevent the transition to amorphous growth [7]. Fig. 3 shows the J-V curves for three μc-Si:H n-i-p cells (2.0 μm thick) on a textured back reflector using different H$_2$ dilution schemes for the i-layer.

It can be seen in the figure that J-V curve improves greatly by profiling the H$_2$ dilution such that it increases from 0.948 to 0.952. A further minor improvement is obtained by using a "dynamic start", which means that the deposition is started before the substrate reaches its equilibrium temperature. In this mode, the substrate temperature increases while the i-layer grows thicker. Presently, we are investigating the origin of the structure evolution and using filament monitoring and control schemes to dynamically adjust the filament current such that homogeneous growth over the entire thickness is maintained.

DISCUSSION

The efficiency that was reached for a single junction μc-Si:H n-i-p cell was 8.5 %. The solar cell parameters are listed in Table 1. This is the highest reported value for hot-wire deposited cells of this kind, whereas the deposition rate of 2.1 Å/s is about twice as high as in record cells of this type so far [18,19]. The performance is in line with state-of-the-art results (8.81 %) for similar bottom cells with VHF PECVD i-layers (with unknown thickness) [5]. The efficiency reached for proto-Si/proto-SiGe/μc-Si:H triple junction n-i-p solar cells is 10.9%. The parameters are listed in Table I.

Earlier, we have made triple junction cells of the type proto-Si/μc-Si:H/μc-Si:H, comprising active i-layers that were all deposited by HWCVD. The efficiency reached was 8.9%. The preparation of this type of cell is much more cumbersome, since this design leads to the use of very thick absorber layers, even when enhanced scattering back reflectors are used. For instance, if the top cell is 165 nm in order to generate ~8 mA/cm^2, the middle cell and the bottom cell have to be made as thick as ~2.4 μm and ~3.7 μm, respectively. The total thickness of over 6 μm of μc-Si:H material is very large compared to the micromorph tandem concept. If the μc-Si:H middle cell is replaced by a proto-SiGe:H cell, then not only the middle cell can be made an order of magnitude thinner, but also the bottom μc-Si:H cell can be made considerably thinner (almost half the thickness), since the SiGe:H middle cell does not absorb within exactly the same spectral region. A second important advantage is that a higher V_{oc} of the triple cell can be obtained, because SiGe:H has a higher band gap than μc-Si:H. Thirdly, in principle the achievable conversion efficiency is higher, because all three band gaps are different and less photon energy is lost as heat.

Table I: J-V parameters of the best microcrystalline single junction and triple cells.

Type of cell	V_{oc} (V)	FF	J_{sc} (mA/cm^2)	Efficiency (%)
Single junction μc-Si:H n-i-p	0.55	0.66	23.4	8.5
Triple junction proto-Si/ proto-SiGe:H/ μc-Si:H	1.98	0.66	8.35	10.9

Stability under light soaking

Single junction and triple junction cells have been exposed to prolonged illumination with 1-sun intensity under open circuit conditions and a temperature of 50 °C. Fig. 4a shows the degradation kinetics of the single junction cell and Fig. 4b that of a bottom cell limited triple cell. We found that the degradation of the μc-Si:H n-i-p cell is negligible (0.6%), while that of the triple cell is 3.5%. It is noted that while the initial efficiency of the triple cell is not at a world record level, the stability of it is excellent. There is an additional economical advantage since the thickness of the entire stack of silicon films is only 2.5 μm.

For a multitude of cells on the same substrate the degradation was in the range of 3 - 6 %. Annealing experiments at 150 °C showed that this degradation was reversible and therefore related to the Staebler-Wronski effect.

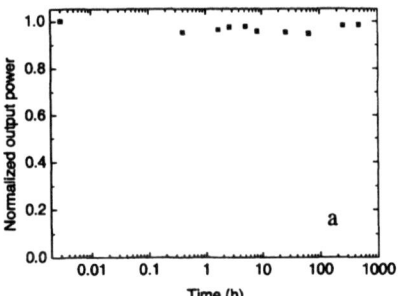 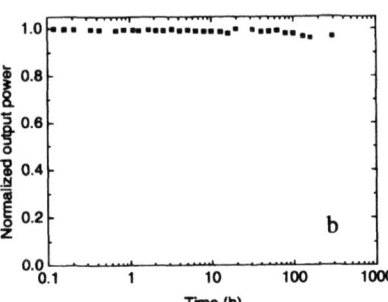

Fig. 4: a) Normalized light-induced performance evolution for a 2 μm thick μc-Si:H n-i-p cell, and b) Normalized light-induced performance evolution for a 2.5 μm thick proto-Si/proto-SiGe/μc-Si:H triple junction n-i-p solar cell.

CONCLUSIONS

We conclude that the most economically viable thin film silicon triple junction consists of a microcrystalline silicon bottom cell, a silicon-germanium middle cell, and an amorphous (preferably protocrystalline) silicon top cell. HWCVD shows to be a reliable technique for high-quality thin film silicon deposition. Triple cells with a total silicon thickness of only 2.5 μm show an efficiency of 10.9%, and the light-induced degradation stays within only 4% relative. Further improvement of the triple cell would mainly come from optimization of the middle cell.

ACKNOWLEDGMENTS

We thank United Solar Ovonic LLC for providing stainless steel substrates with textured Ag/ZnO coatings. The research was financially supported by SenterNovem.

REFERENCES

1. R.E.I. Schropp, M.K. van Veen, C.H.M. van der Werf, D.L Williamson, A.H. Mahan, Proceedings of the 19th European Photovoltaic Solar Energy Conference, Paris (France), June 2004, p. 1526.
2. J.J.H. Strengers, F.A. Rubinelli, J.K. Rath, R.E.I. Schropp, Thin Solid Films **501** (2006) 291
3. M.K. van Veen and R.E.I. Schropp, Appl. Phys. Lett. **82** (2003) 287.
4. R.L. Stolk, H. Li, R.H. Franken, J.J.H. Strengers, C.H.M. van der Werf, J.K. Rath, R.E.I. Schropp, J. Non-Cryst. Solids **352** (2006) 1933.
5. B. Yan, G. Yue, J.M. Owens, J. Yang, S. Guha, 4th WCPEC, May 2006, Waikoloa Village, Hawaii (USA).
6. A. Gordijn, R. Jimenez Zambrano, J.K. Rath, R.E.I. Schropp, IEEE Trans. Elec. Dev. **49** (2002) 949.
7. H. Li, R.H. Franken, R.L. Stolk, C.H.M. van der Werf, J.K. Rath, R.E.I. Schropp, accepted for publication in Thin Solid Films.
8. R.L. Stolk, J.J.H. Strengers, H. Li, R.H. Franken, C.H.M. van der Werf, J.K. Rath, R.E.I. Schropp, Proc. 20th PVSEC, June 2005, Barcelona (Spain), p. 1655.
9. H. Li, R.L. Stolk, C.H.M. van der Werf, R.H. Franken, J.K. Rath, R.E.I. Schropp, J. Non-Cryst. Solids **352** (2006) 1941.
10. M.K. van Veen, Ph.D. thesis, Utrecht University, the Netherlands, 2003.
11. H. Li, R.L. Stolk, C.H.M. van der Werf, R.H. Franken, J.K. Rath, R.E.I. Schropp, Journal of Non-Crystalline Solids **352** (2006), 1941-1944
12. R. Jimenez Zambrano, J.K. Rath, R.E.I. Schropp, J. Non-Cryst. Solids **299-302** (2002) 1131-1135
13. R.H. Franken, R.L. Stolk, H. Li, C.H.M. van der Werf, J.K. Rath, R.E.I. Schropp, Proceedings of the 21st European Photovoltaic Solar Energy Conference, Dresden (Germany), September 2006, p. 1565.
14. R.H. Franken, R.L. Stolk, H. Li, C.H.M. van der Werf, J.K. Rath, R.E.I. Schropp, Proceedings of the 21st European Photovoltaic Solar Energy Conference, Dresden (Germany), September 2006, p. 1744.
15. R.H. Franken, R.L. Stolk, H. Li, C.H.M. van der Werf, J.K. Rath, R.E.I. Schropp, accepted for publication in J. Appl. Phys, 2007.
16. R.W. Collins and A.S. Ferlauto, Curr. Opin. Solid State and Mater. Sci. **6** (2002) 425.
17. J. Koh, A.S. Ferlauto, P.I. Rovira, C.R. Wronski, R.W. Collins, Appl. Phys. Lett. **75** (1999) 2286.
18. S. Klein, F. Finger, R. Carius, B. Rech, L. Houben, M. Luysberg, M. Stutzmann, Mater. Res. Soc. Symp. Proc. **715** (2002) A26.2.1
19. M. Kupich, P. Kumar, R.O. Dusane, N. Schwender, D. Grunsky, B. Schröder, Proceedings of the 20th European Photovoltaic Solar Energy Conf., Barcelona (Spain), June 2005, p. 1679.

Mater. Res. Soc. Symp. Proc. Vol. 989 © 2007 Materials Research Society   0989-A15-04

High Rate Deposition of Amorphous Silicon Based Solar Cells using Modified Very High Frequency Glow Discharge

Guozhen Yue, Baojie Yan, Jeffrey Yang, and Subhendu Guha
United Solar Ovonic LLC, 1100 West Maple Road, Troy, MI, 48084

ABSTRACT

We report our recent progress on high rate deposition of hydrogenated amorphous silicon (a-Si:H) and silicon germanium (a-SiGe:H) based *nip* solar cells. The intrinsic a-Si:H and a-SiGe:H layers were deposited using modified very high frequency (MVHF) glow discharge. We found that both the initial cell performance and stability of the MVHF a-Si:H single-junction cells are independent of the deposition rate up to 15 Å/s. The average initial and stable active-area cell efficiencies of 10.0% and 8.5%, respectively, were obtained for the cells on textured Ag/ZnO coated stainless steel substrates. a-SiGe:H single-junction cells were also optimized at a rate of ~10 Å/s. The cell performance is similar to those made using conventional radio frequency technique at 3 Å/s. By combining the optimized component cells made at 10 Å/s, an a-Si:H/a-SiGe:H double-junction solar cell with an initial active-area efficiency of 11.7% was achieved.

INTRODUCTION

High rate deposition of thin film solar cells is desirable for increasing throughput and reducing cost of manufacturing solar panels. However, hydrogenated amorphous silicon (a-Si:H) solar cells made using conventional radio frequency (RF) glow discharge at high rates commonly exhibit poor quality [1]. The material contains high density of defects, microvoids, and di-hydride structures, which lead to low initial solar cell efficiencies and poor stability after prolonged light soaking. Therefore, new deposition techniques are needed for increasing the deposition rate without compromising the material quality.

Very high frequency (VHF) glow discharge has been widely used in the deposition of a-Si:H and hydrogenated nanocrystalline silicon (nc-Si:H) materials and devices [2, 3]. Under a similar excitation power density, VHF plasma has a higher ion flux intensity and lower ion energy than the conventional RF plasma, resulting in improved material quality at higher deposition rates [4]. In our laboratory, we used a modified VHF (MVHF) system with an excitation frequency range of 60-75 MHz to make a-Si:H, a-SiGe:H, and nc-Si:H solar cells. An initial efficiency of 11.2% was reported previously in an a-Si:H/a-SiGe:H double-junction solar cell with the a-Si:H intrinsic layer deposited at 8 Å/s and the a-SiGe:H intrinsic layer at 6 Å/s [5]. Recently, we have achieved an initial active-area efficiency of 9.0% for a nc-Si:H single-junction solar cell and a stabilized active-area efficiency of 13.3% for an a-Si:H/nc-Si:H/nc-Si:H triple junction solar cell [6, 7]. In this paper, we report our recent results on high rate deposition of a-Si:H and a-SiGe:H based solar cells. The growth parameters are similar to those used for nc-Si:H cells [6, 7] in the high pressure and high power regime. The initial cell performance and stability as a function of the deposition rate have been systematically studied. The results showed that for a-Si:H single-junction cells made using MVHF, both the initial cell performance and stability are independent of the deposition rate of up to 15 Å/s. This phenomenon is different from a-Si:H cells made using conventional RF glow discharge at high rates [5]. In the RF cells, the cell performance usually decreases and the light-induced degradation increases with increasing deposition rate [1].

We will also present the results of a-SiGe:H single-junction and a-Si:H/a-SiGe:H double junction solar cells made using MVHF technique.

EXPERIMENTAL

A series of a-Si:H n-i-p solar cells was made with an MVHF high rate a-Si:H intrinsic layer and RF doped layers. For a given deposition condition, two runs were made: one on specular stainless steel (SS) substrate and the other on Ag/ZnO back reflector (BR) coated SS substrate. The deposition rate of the a-Si:H intrinsic layer was changed from 5 to 15 Å/s by varying the VHF power and the pressure. The thickness of the intrinsic layer was in the range of 200-220 nm. For a-SiGe:H cells, the deposition time of the intrinsic layer was fixed at 4 minutes, which corresponds to a deposition rate of ~10 Å/s. Indium-Tin-Oxide dots with an active-area of 0.25 cm^2 were deposited on the p layer as the top contact for current density versus voltage (J-V) and quantum efficiency (QE) measurements. The J-V measurements were conducted under an AM1.5 light at 25 °C. The QE measurements were performed under the short-circuit condition at room temperature in the wavelength range from 300 to 1000 nm. The stabilized efficiency of a-Si:H solar cells was reached by light-soaking under the open-circuit condition with 100 mW/cm^2 white light at 50 °C for 1000 hours. For a-SiGe:H cells, J-V measurements were also carried out using the same light source with a 530-nm long-pass filter to simulate the condition for a bottom cell in an a-Si:H/a-SiGe:H double-junction structure.

RESULTS AND DISCUSSION

1. a-Si:H single junction cells

The deposition rate of a-Si:H depends on many parameters such as the excitation power density, gas pressure, substrate temperature, and hydrogen dilution ratio. Under a given condition, the most common way to increase the deposition rate is to increase the excitation power. Figure 1 shows the deposition rate of a-Si:H as a function of VHF power. The deposition rate increases continuously in the range of 9-15 Å/s with the VHF power, indicating a non-depleting regime. The deposition rate is also very sensitive to the gas pressure. In the high pressure regime, increasing the pressure leads to a decrease in the deposition rate. In this study, we focused on the range of deposition rate from 5 to 15 Å/s. Table I lists the J-V characteristics of a-Si:H cells on both SS and BR substrates, where the data of short-circuit current density (J_{sc}) are from the integral of the QE curves and the AM1.5 solar spectrum. We used the fill factor (FF) obtained from the measurements under a blue light (FF$_b$) and a red light FF$_b$ (FF$_r$) to

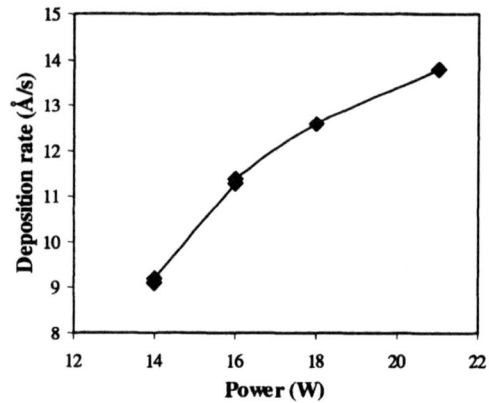

Figure 1. Deposition rate as a function of VHF power.

Table I. J-V characteristics of a-Si:H single-junction solar cells made at different rates on BR and SS substrates. FF_b and FF_r represent the FF measured under low-intensity blue and red lights, respectively. The values of J_{sc} are obtained from the QE measurement.

Sample No.	V_{oc} (V)	FF			J_{sc} (mA/cm^2)	Eff (%)	Substrate	Rate (Å/s)	i layer time (s)
		AM1.5	FF_b	FF_r					
14325	0.983	0.712	0.759	0.753	14.34	10.04	BR	5.2	420
14166	1.000	0.735	0.767	0.749	9.71	7.14	SS		
14324	1.001	0.723	0.763	0.750	13.68	9.90	BR	6.1	300
14139	0.974	0.727	0.765	0.780	9.62	6.81	SS		
14347	0.987	0.702	0.722	0.744	14.67	10.16	BR	9.2	240
14336	0.988	0.710	0.760	0.770	10.36	7.27	SS		
14323	0.993	0.709	0.741	0.741	14.69	10.34	BR	9.3	240
14318	0.978	0.697	0.754	0.763	10.48	7.14	SS		
14346	0.986	0.696	0.716	0.742	14.46	9.92	BR	11.4	222
14335	0.995	0.698	0.764	0.747	10.43	7.24	SS		
14338	0.997	0.704	0.743	0.751	13.68	9.60	BR	11.9	180
14330	0.985	0.714	0.755	0.768	9.96	7.00	SS		
14345	0.984	0.697	0.717	0.742	14.20	9.74	BR	12.6	175
14342	0.986	0.713	0.760	0.767	9.89	6.95	SS		
14348	0.982	0.710	0.717	0.743	14.11	9.84	BR	13.8	156
14333	0.995	0.713	0.769	0.763	9.64	6.84	SS		

probe the quality of the material near the i/p interface and the bulk of the intrinsic layer, respectively. One can see that the cell performance does not depend strongly on the deposition rate. Most cells show a high FF_r, reflecting high quality of the a-Si:H intrinsic layer. The average efficiency is around 7% for the cells on SS and 10% on BR. Compared to the cells on SS, the open-circuit voltage (V_{oc}) and FF do not change much for the cells on BR. However, the J_{sc} increased by 40% as a result of the light trapping on the textured Ag/ZnO BR. The intrinsic layer thickness was controlled to be in the range of 200-220 nm for each sample. By increasing the intrinsic layer thickness to 330 nm, an initial active-area efficiency of 10.7% (J_{sc}=15.94 mA/cm^2, V_{oc}=0.993 V, and FF=0.674) was obtained for an a-Si:H single-junction cell deposited on a Ag/ZnO BR substrate at 10 Å/s.

Stability experiments were carried out for all of the a-Si:H cells with different deposition rates on SS and BR substrates. The solar cells were light-soaked under 100 mW/cm^2 white light at 50 °C for 1000 hours. As a reference, two a-Si:H cells made with RF at a low rate of ~1 Å/s on SS and Ag/ZnO BR substrates were light-soaked together with the high rate MVHF cells. Figure 2 shows the initial and stable efficiencies as well as the light-induced degradation rate as a function of the deposition rate. Within the experimental variation, the light-induced degradation of the efficiency is around 15.5±1.5% for all the cells on both BR and SS substrates. It does not show a clear dependence on the deposition rate. The highest stabilized efficiency of 8.64% (J_{sc}=13.92 mA/cm^2, V_{oc}=0.956 V, and FF=0.649) was achieved with an a-Si:H single-junction cell made at 9.3 Å/s on the BR substrate.

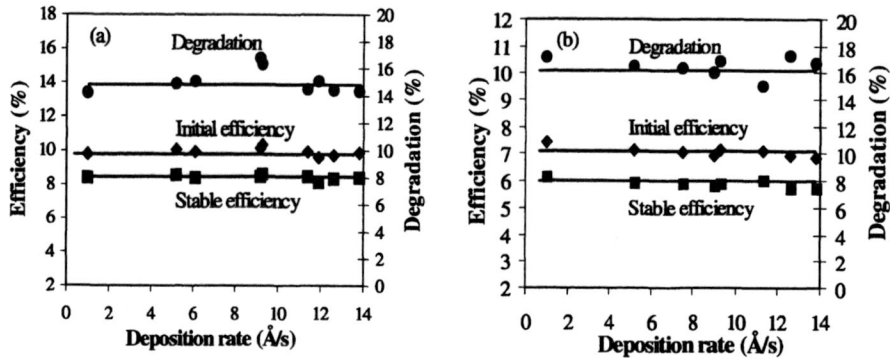

Figure 2. The initial and stable efficiencies and the degradation as a function of deposition rate of a-Si:H solar cells on (a) BR and (b) SS substrates. The cells for 1 Å/s were made by RF.

2. a-SiGe:H single and a-Si:H/a-SiGe:H double junction cells

As mentioned above, both initial performance and stability for the high rate a-Si:H solar cells made by MVHF are independent of the deposition rate of up to 15 Å/s. Encouraged by this result, we then proceeded to optimize high rate a-SiGe:H component cells with the aim of making high efficiency a-Si:H/a-SiGe:H double-junction cells. Usually, high rate a-SiGe:H cell optimization is more difficult than its a-Si:H counterpart. First, compared to the a-Si:H deposition, heavier GeH$_3$ radicals have a lower mobility on the growing surface, which requires a longer relaxation time to find energetically favorable sites to stay. Second, the Ge-H bond is weaker than the Si-H bond. For a given substrate temperature, the hydrogen content decreases with the Ge/Si ratio, which leads to an insufficient dangling bond passivation thus a high defect density. Increasing the substrate temperature can increase the surface mobility of impinging radicals, but it reduces hydrogen content further and results in a high defect density in the film. Therefore, an optimized temperature is critical for the high rate a-SiGe:H growth. Third, a proper band gap profiling is needed for enhancing the hole collection [8]. In addition, we need to optimize the n/i and i/p interfaces to reduce the shunt current and interface recombination, and the doped layers to reduce the series resistance and increase V_{oc}. Considering these factors, we chose to optimize a-SiGe:H single-junction cells at a deposition rate of ~10 Å/s, where the deposition time of the a-SiGe:H intrinsic layer was fixed at 4 minutes. Table II lists the J-V characteristics of typical a-SiGe:H cells made at high rate. For comparison, two cells made using RF at deposition rates of 1 Å/s and 3 Å/s are also included. In this study, we focused on intermediate band gap solar cells, which give an AM1.5 open-circuit voltage (V_{oc}) around 0.75 to 0.80 V and is suitable for the bottom cell in an a-Si:H/a-SiGe:H double-junction structure. In order to investigate the long wavelength response and simulate the performance of an a-SiGe:H cell in an a-Si:H/a-SiGe:H double-junction structure, we measured the cell performance under an AM1.5 solar simulator with a 530-nm long-pass filter. From our previous experience, a maximum power density (P_{max}) of 4 mW/cm^2 under the filtered light is a benchmark for good a-SiGe:H component cells on SS substrates. In this study, we achieved an active-area P_{max} of 4.4 mW/cm^2 under an AM1.5 light with the 530-nm long-pass filter. The performance is lower than the cell made by conventional RF glow discharge at 1 Å/s, but comparable to that made at 3 Å/s as shown in Table II. Figure 3 shows the J-V curves and QE spectrum of the optimized a-SiGe:H component cell.

362

Table II. J-V characteristics of typical a-SiGe:H component cells made using RF and MVHF at different rates. The results in the parentheses were measured under AM1.5 light with a 530 nm long-pass filter.

Sample No.	V_{oc} (V) (> 530 nm)	FF (> 530 nm)	J_{sc} (mA/cm^2) (> 530 nm)	P_{max} (mW/cm^2) (> 530 nm)	Comments
15333	0.763(0.744)	0.690(0.693)	16.48(10.47)	8.68(5.40)	RF 1 Å/s
5183	0.768(0.749)	0.614(0.641)	15.20(9.17)	7.18(4.40)	RF 3 Å/s
14407	0.795(0.775)	0.667(0.696)	13.49(7.82)	7.15(4.22)	MVHF 10 Å/s
14583	0.768(0.751)	0.648(0.671)	14.41(8.66)	7.17(4.36)	MVHF 10 Å/s
14584	0.768(0.750)	0.664(0.681)	14.23(8.60)	7.26(4.39)	MVHF 10 Å /s

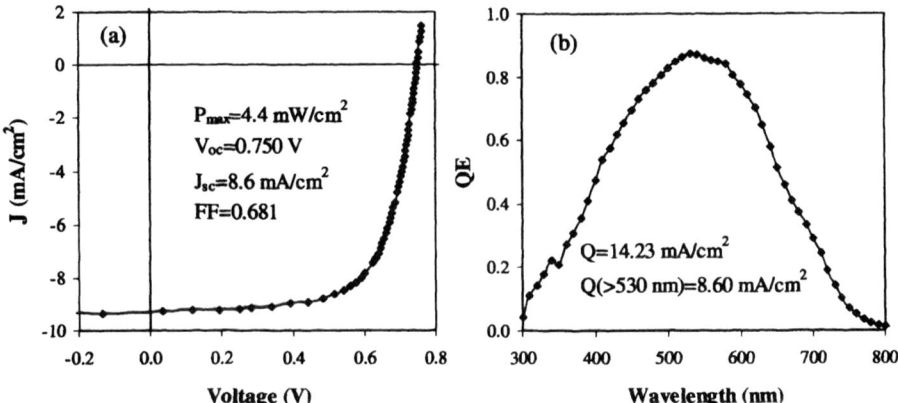

Figure 3. (a) J-V characteristics measured under AM1.5 light with a 530 nm long-pass filter and (b) QE curve of the optimized a-SiGe:H component cell made at 10 Å/s.

Having optimized the a-Si:H top and a-SiGe:H bottom cells, we made a-Si:H/a-SiGe:H double-junction solar cells on Ag/ZnO BR coated SS substrates. The deposition time for both the a-Si:H and a-SiGe:H intrinsic layers was 4 minutes. The J-V characteristics of typical double-junction cells are listed in Table III. An initial active-area efficiency of 11.7% has been achieved. The J-V curves and QE spectrum of the best a-Si:H/a-SiGe:H double-junction cell are plotted in Fig. 4. The light soaking data for the a-SiGe:H single-junction and a-Si:H/a-SiGe:H double-junction cells will be reported when they are available.

Table III. J-V characteristics of the a-Si:H/a-SiGe:H double-junction cells made at high rates. The deposition time of the intrinsic layers for both the top and bottom cells was 4 minutes.

Sample No.	V_{oc} (V)	FF	QE (mA/cm^2) Top	QE (mA/cm^2) Bottom	QE (mA/cm^2) Total	P_{max} (mW/cm^2)
14520	1.719	0.717	9.50	10.94	20.44	11.71
14530	1.696	0.664	10.23	10.16	20.39	11.44
14534	1.700	0.679	9.78	10.48	20.26	11.29

CONCLUSION

We have found that the MVHF-deposited a-Si:H solar cells showed good initial efficiency and stability. The most important result is that the cell performance and stability do not depend on the deposition rate of up to 15 Å/s. This is different from the cells made using RF at high rates. The degradation of the RF cells usually increases with the deposition rate. We have also optimized the a-SiGe:H cells at high deposition rates on SS substrates. A P_{max} of 4.4 mW/cm^2 has been achieved for a 0.25 cm^2 cell under the AM1.5 light with a 530-nm long-pass filter, which is comparable to the cells made with RF at 3 Å/s. By combining the optimized component cells, we have achieved an initial active-area efficiency of 11.7% with an a-Si:H/a-SiGe:H double-junction cell on Ag/ZnO coated SS substrates at 10 Å/s.

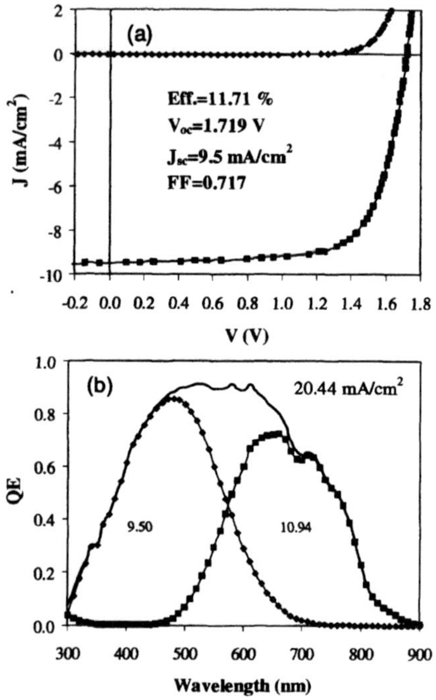

Figure 4. (a) J-V characteristics and (b) QE spectrum of an a-Si:H/a-SiGe:H double-junction cell made at 10 Å/s using MVHF.

ACKNOWLEDGMENTS

The authors thank T. Palmer, L. Sivec, and E. Chen for sample preparation and measurements, and X. Xu for fruitful discussion and critical reading. This work was supported in part by NREL under the Thin-Film Partnership Program Subcontract No. ZXL-6-44205-14.

REFERENCES

[1] S. Guha, J. Yang, S. Jones, Y. Chen, and D.L. Williamson, Appl. Phys. Lett. **61**, 1444 (1992).

[2] A. Shah, J. Dutta, N. Wyrsch, K. Prasad, H. Curtins, F. Finger, A. Howling, and Ch. Hollenstein, Mater. Res. Soc. Symp. Proc. **258**, 15 (1992).

[3] J. Meier, P. Torres, R. Platz, S. Dubail, U. Kroll, J.A. anna Selvan, N. Pellaton Vaucher, Ch. Hof, D. Fischer, H. Keppner, A. Shah, K.D. Ufert, P. Giannoulès, and J. Koehler, Mater. Res. Soc. Symp. Proc. **420**, 3 (1996).

[4] B. Yan, J. Yang, S. Guha, and A. Gallagher, Mater. Res. Soc. Symp. Proc. **557**, 115 (1999).

[5] J. Yang, B. Yan, J. Smeets, and S. Guha, Mater. Res. Soc. Symp. Proc. **664**, A11.3 (2001).

[6] G. Yue, B. Yan, G. Ganguly, J. Yang, S. Guha, C. Teplin, and D.L. Williamson, *Proc. of 4th World Conf. on Photovoltaic Energy Conversion (Hawaii, USA)*, p. 1588 (2006).

[7] G. Yue, B. Yan, G. Ganguly, J. Yang, S. Guha, and C. Teplin, Appl. Phys. Lett. **88**, 263507 (2006).

[8] S. Guha, J. Yang, A. Pawlikiewicz, T. Glatfelter, R. Ross, and S.R. Ovshinsky, Appl. Phys. Lett. **54**, 2330 (1989).

Mater. Res. Soc. Symp. Proc. Vol. 989 © 2007 Materials Research Society   0989-A15-05

Fabrication and Optimization of a-Si:H n-i-p Single-junction Solar Cells with 8 Å/s Intrinsic Layers of Protocrystalline Si:H Materials

Xinmin Cao, Wenhui Du, Y. Ishikawa, Xianbo Liao, Robert W. Collins, and Xunming Deng
Department of Physics and Astronomy, University of Toledo, Toledo, OH, 43606

ABSTRACT

At the University of Toledo (UT), we have investigated hydrogenated amorphous silicon (a-Si:H) n-i-p solar cells with intrinsic layers deposited at high rates, ~ 8 Å/s, using a multi-chamber load-locked PECVD system. a-Si:H i-layers were grown with a VHF plasma density of ~ 0.2 W/cm^2 and a frequency of 70 MHz using various hydrogen dilution levels. It is observed from the current-voltage (I-V) device performance characteristics that the open-circuit voltage (V_{oc}) increases with increasing hydrogen dilution, reaches a maximum, and then decreases. This drop in V_{oc} can be attributed to the transition region (or protocrystalline regime) from an amorphous phase into a mixed amorphous+nanocrystalline (a + nc) phase for the i-layer. An initial efficiency of $\eta = 9.99\%$ ($V_{oc} = 0.986$ V, $J_{sc} = 13.98$ mA/cm^2, FF = 72.5%) was obtained. Quantum efficiency (QE) measurement has shown that the blue light response increases as the hydrogen dilution increases. High blue light spectral response with QE values over 0.7 at $\lambda = 400$ nm has been obtained for a-Si:H cells made under specific deposition conditions in which tailored protocrystalline silicon materials were incorporated at the i/p interface region. A stabilized efficiency of ~ 8.2% has been achieved for the cells with protocrystalline or (a+nc)-mixed Si:H of a small portion being evolved at the top-most i-layer near p/i interface.

INTRODUCTION

A high V_{oc} and a high spectral response in the short wavelength region are desired for the hydrogenated amorphous silicon (a-Si:H) n-i-p solar cell, which has a wide bandgap E_g of ~1.8 eV and normally serves as a window top-cell in the a-Si based multi-junction solar cell [1].

A detailed phase diagram describing hydrogenated silicon materials, including (i) purely amorphous silicon, a-Si:H, (ii) protocrystalline Si:H, (iii) mixed amorphous+micro (or amorphous+nano) crystalline silicon, (a + μc)-Si:H [or (a + nc)-Si:H], and (iv) single-phase micro- (or nano-) crystalline silicon, μc-Si:H (or nc-Si:H) have been developed by Wronski and Collins *et al.* [2, 3]. The protocrystalline Si:H materials are more ordered than conventional a-Si:H materials and evolve with thickness from an amorphous phase into first an (a+nc)-Si:H mixed-phase and subsequently into nc-Si:H, a fully nanocrystalline phase. Such silicon materials at the transition regime between the amorphous phase and the (a+nc)-Si:H mixed-phase can be defined as protocrystalline silicon.

For the fabrication and optimization of a-Si:H solar cells, the properties and growth conditions of protocrystalline Si:H materials are of great interest. The materials made in the protocrystalline regime contain a predominant amorphous component possibly with a very small volume fraction of nano-crystalline inclusions and as a result exhibit unique optoelectronic properties. By further understanding and developing Si:H deposition phase diagrams, we can fabricate and optimize high quality a-Si materials and high performance solar cells in a more controlled way.

The VHF PECVD technique along with H_2 dilution grading processes have been applied in our laboratory in the fabrication of a-Si based solar cells at deposition rates in the range 2 ~ 40 Å/s. In the research reported here, we investigated a-Si:H n-i-p solar cells with intrinsic layers deposited at a high deposition rate of ~8 Å/s. By adjusting the i-layer deposition conditions, such as the H_2 dilution ratio and H_2 dilution grading profile, a-Si:H solar cells have incorporated protocrystalline materials near the amorphous/nanocrystalline transition. The device performance of the a-Si:H solar cells have been greatly improved as a result.

EXPERIMENT

a-Si:H n-i-p single-junction solar cells were fabricated using PECVD techniques in our UT multi-chamber load-locked deposition system. The substrates were stainless steel (SS) coated with Ag/ZnO back reflectors. The Si:H i-layers were prepared at a high deposition rate of ~ 8 Å/s using the VHF-PECVD technique with a frequency of 70 MHz, a plasma density of ~ 0.2 W/cm², and under a low deposition pressure region. As a result, the deposition time for an a-Si:H i-layer was about 5 minutes, corresponding to a thickness of ~ 240 nm. The nominal substrate temperature T_{sub} was in the range of 100 – 300 °C. A high hydrogen dilution ratio $R=[H_2]/[Si_2H_6]$ in the range of 30 – 100 has been used for i-layer deposition. The doped n and p layers were prepared using the conventional 13.56 MHz RF-PECVD technique at deposition rates of 1 Å/s under amorphous and protocrystalline growth conditions, respectively [4]. After fabrication of the n-i-p single-junction structure, the cells were completed by depositing indium tin oxide (ITO) transparent conductive layers through a mask to serve as the front contacts. Standard RF magnetron sputtering deposition techniques were used for the ITO. Each cell has an active area of 0.25 cm². For characterization of the as-deposited cells, standard dark current-voltage (I-V) and AM1.5G light I-V measurements were made. Standard QE measurements were made in the wavelength range of 350 – 1000nm. Light-soaking experiments were undertaken with ~ 100 mW/cm² white light at 50 °C.

RESULTS AND DISCUSSIONS

It is well known that hydrogen dilution can strongly influence a-Si:H based solar cell performance [5, 6]. In particular, the open-circuit voltage (V_{oc}) increases gradually with an increase in hydrogen dilution and reaches a maximum value of ~ 1.0 V, where protocrystalline material is believed to be present. Normally, for protocrystalline material there is no indication of a crystalline component in the ellipsometry and Raman spectra; however, through TEM one can detect a small crystalline component in the films deposited under conditions very close to the transition [3, 4]. If the hydrogen-dilution exceeds a certain level, (a + nc)-Si:H mixed-phase materials and solar cells can be made, and the corresponding V_{oc} values start to drop. By increasing hydrogen dilution even further, a V_{oc} value around 0.5 V can be attained -- a typical value for nc-Si:H solar cells [7].

So we can evaluate the structural phase information at the top of the as-grown i-layers simply from the V_{oc} value and its trend for the corresponding actual solar cells. We have used high hydrogen dilution ratios R in the range of 30 – 100 for a series of i-layers focusing on the protocrystalline phase in the transition region between a purely a-Si:H phase and an (a + nc)-Si:H mixed-phase. Meanwhile, a specific hydrogen dilution grading approach has been applied for the i-layer deposition, designed so that optimum protocrystalline silicon materials are incorporated at the i/p interface region.

Table I displays the AM1.5G device I-V performance characteristics in the initial state and after 100 hours of light soaking for a-Si:H n-i-p single-junction solar cells on Ag/ZnO back reflector coated stainless steel substrates. The 240 nm-thick i-layer has been made by 70 MHz VHF-PECVD with a deposition rate of 8 Å/s and a deposition time of 5 minutes. An average initial active-area efficiency of ~ 9.6% has been obtained for the a-Si:H cells of sample GD1828.

Table I. I-V and QE data in the initial state and after 100 hours of light soaking for 8 Å/s VHF a-Si:H n-i-p solar cells made on Ag/ZnO coated SS substrate.

Cell Nr.		V_{oc} (V)	J_{sc} (mA/cm^2)	QE @ λ = 400 nm	FF (%)	η (%)	comments
GD1828-2.51 (near gas inlet)	initial	0.946	14.97	0.591	65.2	9.23	upstream side
	100 hrs LS	0.922	14.73	0.542	59.3	8.05	
	degradation (%)	-2.5	-1.6	-8.3	-9.0	-12.8	
GD1828-2.31	initial	0.967	14.38	0.602	69.2	9.62	
	100 hrs LS	0.943	14.14	0.577	63.7	8.49	
	degradation (%)	-2.5	-1.7	-4.2	-7.9	-11.7	
GD1828-2.11 (center)	initial	0.986	13.98	0.600	72.5	9.99	
	100 hrs LS	0.962	13.75	0.586	66.8	8.84	
	degradation (%)	-2.4	-1.6	-2.3	-7.9	-11.6	
GD1828-4.11 (center)	initial	0.968	13.69	0.633	73.8	9.78	down-stream side
	100 hrs LS	0.956	13.50	0.627	67.8	8.75	
	degradation (%)	-1.2	-1.4	-0.9	-8.1	-10.5	
GD1828-4.12	initial	0.954	13.76	0.664	72.8	9.56	
	100 hrs LS	0.950	13.64	0.654	67.9	8.80	
	degradation (%)	-0.4	-0.9	-1.5	-6.7	-7.9	
GD1828-4.13	initial	0.938	14.12		71.8	9.51	
	100 hrs LS	0.939	14.00	0.690	67.0	8.81	
	degradation (%)	0.1	-0.8		-6.7	-7.4	
GD1828-4.14 (near pumping outlet)	initial	0.927	14.37	0.737	70.9	9.44	
	100 hrs LS	0.932	14.20	0.722	66.3	8.77	
	degradation (%)	0.5	-1.2	-2.0	-6.5	-7.1	

It should be noted that the gas feeding and pumping system has a cross-flow configuration in the present i-layer deposition chamber whereby the Si_2H_6 and H_2 gases enter from one side port of the chamber and exit from the opposite side port with gases flowing across the 4" x 4" substrate. This i-layer chamber has been used for routine fabrication of a-Si:H based cells using conventional RF PECVD at rates less than 2 Å/s and has yielded relatively good uniformity not only in the thickness and phase structure of the material but also in the device performance. However, this cross-flow configuration results in a non-uniformity of ~ ±10% in film thickness across the 4" x 4" substrate under the high rate deposition conditions of ~ 8 Å/s.

The AM1.5 light and dark I-V plots of the best a-Si cell of 9.99% initial efficiency (V_{oc}= 0.986 V, J_{sc}= 13.98 mA/cm^2, and FF=0.725) are displayed in Fig 1. This best cell is located at the upstream side near the sample center having a i-layer thickness of 220 nm with a deposition rate of 7.3 Å/s. Figure 2 plots the corresponding external quantum efficiency versus wavelength over the range of 380-1000 nm for cells whose I-V data are listed in Table I.

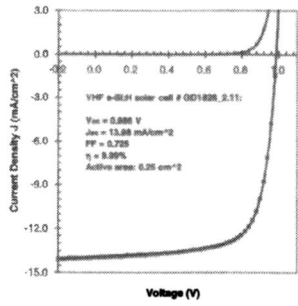

Fig. 1. AM1.5G light and dark I-V plots for a 7.3 Å/s VHF a-Si solar cell on Ag/ZnO BR.

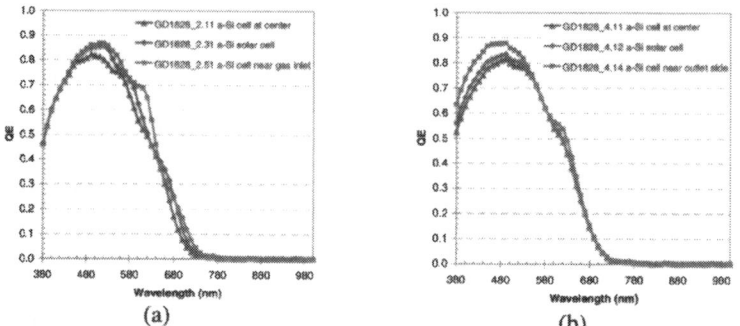

(a)   (b)

Fig. 2. External quantum efficiency plots for 8 Å/s VHF PECVD a-Si solar cells on Ag/ZnO BR: a) cells on the upstream side, and b) cells on the downstream side.

The V_{oc} values of the solar cells with i-layers made in the protocrystalline regime are very sensitive to the hydrogen dilution. For the sample GD1828 shown in Table I, the device performance characteristics show clear differences for solar cells fabricated across the substrate between the upstream side (near the gas inlet port) and the downstream side (near the outlet pumping port). These differences reflect underlying structural and thickness variations for the as-grown i-layer across the substrate. Starting at a V_{oc} value of 0.946 V for the solar cell nearest the gas inlet side, the V_{oc} values increase with cell position across the substrate, reach a maximum value of 0.986 V for the cell in the center, and then decrease to a minimum V_{oc} value of 0.927 V for the cell nearest to the outlet side. This behavior indicates that there is an obvious non-uniformity in H_2 dilution across substrate because of the high consumption of silicon atoms at high deposition rates. Thus, the structural phase of the as-grown i-layer changes from a purely amorphous phase for cells at the upstream side with relatively low H_2 dilution, to the protocrystalline regime for cells at the sample center with moderate H_2 dilution, and subsequently to an (a+nc)-Si:H mixed-phase for cells at the downstream side with higher H_2 dilution than that near the gas inlet. Based on the above analysis, we can easily interpret I-V and QE results for the cells as functions of H_2 dilution rather than cell position across the substrate. Thus, we can reveal the typical device performance behavior as a function of H_2 dilution instead of cell position for a-Si solar cells fabricated in a single run in the same way as those from a series of a-Si cell depositions with H_2 dilution variations from deposition to deposition.

For the initial device performance characteristics among the set of solar cells along gas flow direction, the cell GD1828-2.51, which is at the position nearest to the gas inlet, has the highest J_{sc} of 14.97 mA/cm^2. There are two major reasons contributing to the highest J_{sc}. One reason is that this cell has the thickest i-layer of d = 240 nm; the second reason is that the cell incorporates a pure a-Si i-layer of the lowest band gap and thus has the highest light absorption and QE in the long wavelength range of 550 – 750 nm (as shown in Fig. 2a).

The cell GD1828-4.11 in the sample center has the best initial state FF of 0.738, but the lowest J_{sc} of 13.69 mA/cm^2 due in part to the lower i-layer thickness of 220 nm. For cells on the downstream side near the pumping outlet, the i-layer thickness decreases further from 220 nm to 200 nm. In spite of this, however, the J_{sc} increases from 13.69 to 14.37 mA/cm^2. As shown in Fig. 2b, the increase in J_{sc} mainly arises from the increase in blue QE in the short wavelength range of 380 – 480 nm, whereas the spectral responses remain about same in the long wavelength range of 550 – 750 nm. The latter similarity arises because the i-layers are protocrystalline materials with about same wide optical bandgap. The cell GD1828_4.14, being the nearest to the pumping outlet side, has the lowest V_{oc} of 0.927V whereby a small portion of (a+nc)-Si:H mixed-phase is evolved at the top-most i-layer near the i/p interface. High blue light spectral responses with QE values over 0.7 at $\lambda = 400$ nm have been obtained here. There are two key reasons for the increase in blue QE for the cells on the downstream side of the center. One is the presence of improved protocrystalline material in the bulk i-layer characterized by improved mobility of the photo-generated carriers and reduced recombination. The second is the reduced absorption in the window p-layer possibly due to the presence of small nanocrystallites. When using the standard protocrystalline Si:H p-layer deposition condition for these a-Si solar cells, the as-grown p-layer becomes more transparent, especially for blue light, when it is deposited on the protocrystalline i-layer near the boundary of the transition to (a+nc)-Si:H growth.

After 100 hours of light soaking, all cells have only small drops of 1-2% in J_{sc}. For cells along the gas flow direction, with the increase of H$_2$ dilution, the light-induced degradation in FF is reduced from 9% to 6.5%, meanwhile the degradation in V_{oc} is also reduced. The cell GD1828_4.14, being nearest to the pumping outlet side, even exhibits an increase of 5 mV in V_{oc} after 100 hrs of light soaking. Consequently, cells prepared with higher H$_2$ dilution exhibit greater stability in efficiency. Moreover, the smaller thickness should also be one of the reasons for the lower degradation.

For the stability study, a prolonged light soaking of 5000 hours has also been taken to this sample. Figure 3 plots a comparison of cell efficiencies as functions of cell position for the initial state, 100 hours, and 5000 hours of light soaking, where the cell position means the distance between the measured solar cell and the sample upstream edge near the gas inlet. A most stabilized efficiency of ~ 8.2% has been achieved for the downstream cells with protocrystalline or (a+nc)-mixed Si:H of a small portion being evolved at the top-most i-layer near p/i interface. But only a 7.35% efficiency was obtained for the cell with purely amorphous Si:H bulk i-layer being the nearest to the gas inlet. The degradation ratio in efficiency decreases from 20.4% for the cell near the gas inlet to 13.2% for the cell near the pumping outlet.

Finally, it should be noted that the highest stabilized performance of V_{oc}, FF, and J_{sc} occurs at different H$_2$ dilution levels across the substrate area. If we were to ascribe the dominant contributor to the maximum V_{oc} to the i/p interface, the FF to the bulk i-layer, and J_{sc} to a combination of the bulk i-layer (enhanced hole diffusion length and reduced recombination) and p-layer transparency, then a new multistep approach with substrate memory erasing regions and crystallite enhancing regions may be proposed to achieve these maxima in a single device.

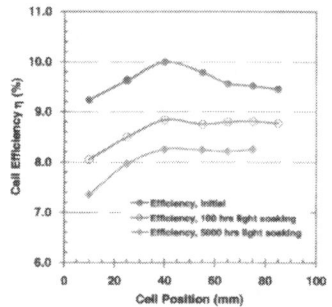

Fig. 3. A comparison of cell efficiencies as functions of cell position
for the initial state, 100 hours, and 5000 hours light soaking.

CONCLUSIONS

Using a Si:H deposition phase diagram as guidance, we have fabricated and optimized high performance single-junction a-Si:H n-i-p solar cells with 8 Å/s intrinsic layers of protocrystalline Si:H materials. We have found that the V_{oc} values of the solar cells with i-layers made near the protocrystalline regime are very sensitive to the hydrogen dilution ratio $R=[H_2]/[Si_2H_6]$. An initial efficiency of η = 9.99% was obtained for the cell with improved protocrystalline silicon material evolved at the i/p interface region. The light-induced stability in efficiency is improved for cells with protocrystalline materials or (a+nc)-mixed Si:H of a small portion being evolved at the top-most i-layer near i/p interface. High blue light spectral response with QE values over 0.7 at λ = 400 nm has been obtained here. A stabilized efficiency of ~ 8.2% has been achieved after 5000 hours of light soaking.

ACKNOWLEDGMENTS

This work was supported by the National Renewable Energy Laboratory "Thin Film Photovoltaic Partnership Program" under subcontract No. ZXL-5-44205-06.

REFERENCES

1. X. Deng and E. Schiff, a chapter in *The Handbook of Photovoltaic Science and Engineering*, edited by A. Luque & S. Hegedus, (John Willey & Sons, Ltd., 2003).
2. C. R. Wronski and R. W. Collins, *Solar Energy* 77, 877-885 (2004).
3. R. W. Collins, A. S. Ferlauto, G. M. Ferreira, C. Chen, J. Koh, R. J. Koval, Y. Lee, J. M. Pearce, and C. R. Wronski, *Solar Energy Materials & Solar Cells* 78, 143 (2003).
4. W. Du, X. Liao, X. Yang, H. Povolny, X. Xiang, X. Deng, and K. Sun, *Solar Energy Materials and Solar Cells* 90, 1098-1104 (2006).
5. B. Yan, J. Yang, G. Yue, and S. Guha, *2003 MRS Spring Meeting*, (San Francisco, CA; USA; 22-25 Apr. 2003), pp. 363-368 (2003).
6. S. Guha, J. Yang, A. Banerjee, B. Yan, and K. Lord, *Solar Energy Materials and Solar Cells* 78, 329-347 (2003).
7. X. Cao, W. Du, X. Yang, and X. Deng, *4th World Conference on Photovoltaic Energy Conversion*, May 2006, Waikoloa HI (IEEE, Piscataway NJ, 2006).

Crystallization Techniques II

Mater. Res. Soc. Symp. Proc. Vol. 989 © 2007 Materials Research Society   0989-A16-01

Laser Crystallization of Silicon for Large Area Electronics

Toshiyuki Sameshima
Graduate School of Engineering, Tokyo University of Agriculture & Technology, 2-24-16, Nakamachi, Koganei, 184-8588, Japan

ABSTRACT

Laser crystallization of silicon is discussed for forming polycrystalline silicon thin films used to fabricate polycrystalline silicon thin film transistors (poly-Si TFTs). Laser-induced rapid heating is important for crystalline film formation with a low thermal budget. Structural and electrical properties of poly-Si films are discussed. Reduction of electrical active defects located at grain boundaries is essential for achieving poly-Si TFTs with high performances. The internal film stress is attractive to increase the carrier mobility. Recent development in laser crystallization methods with pulsed and continuous wave (CW) lasers is then reviewed. Control of the heat flow results in crystalline grain growth in the lateral direction, which is essential for fabrication of large crystalline grains. We also report an annealing method using a high power infrared semiconductor laser. High power lasers will be attractive for rapid crystallization of silicon films over a large area and activation of doped regions.

INTRODUCTION

The technologies of polycrystalline silicon thin film transistors (poly-Si TFTs) have been developed in recent twenty years. TFTs have been applied to switching elements for liquid-crystal-flat-panel displays used in note book type personal computers, mobile phone and TV [1]. Poly-Si TFTs have been also been applied to logical integrated circuits because of their possibility of a high operation frequency [2,3]. Fabrication of TFT circuits on plastic films is also attractive for cheap, flexible and light devices [4]. For this purpose, fabrication processes of TFTs at a low temperature are important because of low heat resistivity of glass and plastic materials. Sameshima et al. developed excimer-laser-induced crystallization of amorphous silicon films formed on glass substrates [5]. They reported poly-Si TFTs using laser crystallization with a low processing temperature of 270°C. Many works on pulsed laser crystallization have been done to improve the low temperature fabrication processing of poly-Si TFTs [6-8]. The equipment of crystallization has been developed using the excimer laser at the present stage. Furthermore, crystallization of silicon films in the lateral direction has been widely investigated in order to fabricate poly-Si TFTs with high performances.

In this paper, laser crystallization technology is reviewed for low temperature fabrication processing. Structural and electrical properties of poly-Si films are argued. Passivation of SiO_2/Si interface and grain boundaries is also discussed for fabrication of TFTs with a high mobility and a low threshold voltage. Crystallization of silicon films in the lateral direction using pulsed laser as well as continuous wave laser is also referred for TFTs with high performances. We also introduce infrared semiconductor induced crystallization of silicon for high efficient laser crystallization.

LASER INDUCED MELTING OF SILICON FILMS

Pulsed UV-laser light is effectively absorbed in the silicon surface because of the high absorption coefficient ~10^6 S/cm of silicon at wavelength lower than 350 nm. The absorbed light simultaneously excites the electronic states of silicon. The energy of the excited states is relaxed to lattice vibration states within the time on the order of 10^{-12} seconds [9]. In the case of ns-order pulsed laser irradiation, lattice heat is, therefore, the most important interaction [10]. The heat energy generated at the surface region caused by light absorption diffuses into interior regions of silicon films. Numerical analysis of heat flow gives temperature change of silicon films caused by time-dependent laser heating [10,11]. In the case of silicon films formed on SiO_2 glass substrates, the laser energy rapidly heats the silicon films to a high temperature. Although the heat energy quickly diffuses into the underlying region because of large heat diffusivity of silicon, heating region at high temperature is limited in the surface region of SiO_2 glass substrates because of the low heat diffusivity of SiO_2 ~1.4 W/mK [12]. Heating properties of silicon films such as the threshold energy density for crystallization are therefore governed by the heat diffusivity, the density and the specific heat of underlying substrates if silicon films are thin enough compared with a heat diffusion length in silicon during laser irradiation. The low heat diffusivity at of quartz glass results in localization of heat energy at surface region at about 300 nm deep during laser irradiation. The system of silicon film/glass substrate gives an essential advantage of low required energy for crystallization in order to apply the laser heating method to electric devices.

Transient conductance measurements are useful to observe rapid melting and solidification phenomena [13,14]. Rapid increases in the electrical conductance were observed 20 to 30 ns after the irradiation of XeCl excimer laser as shown in Fig.2. This means that the melting of the silicon films was initiated by laser heating. The peak electrical conductance indicates the maximum volume of molten silicon. It increased as the laser energy density increased. The down slope of the electrical conductance indicates the average solidification rate. When the interface-controlled solidification with a single liquid/solid interface is assumed, the down slope results in the velocity of the liquid/solid interface of about 0.6 m/s. Change in the electrical conductance means rapid solidification to crystalline states in the

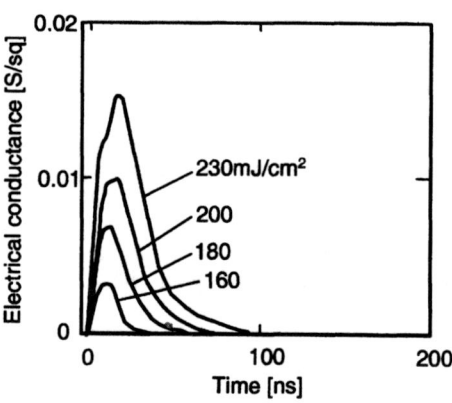

Fig.1 Transient conductance of 30-nm-thick silicon films formed on glass substrate for 30-ns-pulsed XeCl laser irradiation with different energy densities.

case of pulsed laser irradiation.

STRUCTURAL AND ELECTRICAL PROPERTIES

Measurement of transmission electron microscope (TEM) revealed that fine grain and disordered structures were observed near the crystallization threshold [15]. The average grain size increased and disordered regions were decreased as the laser energy density increased. The grain size about 0.1 μm was observed at a laser energy of 400 mJ/cm^2 for 50 nm silicon films. The preferential crystalline orientation of the laser crystallized silicon films was (111), which was observed by X-ray diffraction because the interface control growth progresses from the nucleation sites at bottom toward the top surface and the growth surface has the preferential crystalline orientation of (111) [15].

In general, thin films on foreign substrates are subjected to stress. It is well known that pulsed laser crystallized silicon films on glass substrates sustain tensile stress. We revealed that strong tensile stress was introduced by the deposited silicon films rather than by the pulsed-laser annealing. Pulsed laser crystallization maintains the existing stress at the growth sites in the bottom region of silicon film. We also confirmed that the tensile stress strongly depended on the deposition temperature of 50-nm thick amorphous silicon films, and that it increased to 9.7x10^8 Pa as the deposition temperature increased to 480°C [16]. These results show that the stress in the films changes little with the laser irradiation and significant tensile stress in silicon films is created during the film formation. Kitahara et al. precisely characterized distribution of film stress of the laser crystallized silicon films using micro-Raman spectroscopy with a highly spatial resolution [17]. The map of peak frequency shift indicated that the tensile stress was accumulated in grains and relaxed at grain boundary. The tensile stress was estimated to be 1x10^9 Pa at the central region and 7x10^8 Pa at grain boundaries. In contrast, the density of electrically active defect states was concentrated at grain boundaries and few defects existed in grains [18].

Laser crystallized silicon films have serious density of defect states at grain boundaries. The defects trap free carriers so that they affect carrier transport in polycrystalline films. Reduction of defects by post laser annealing is important for good TFT characteristics. Plasma hydrogenation has been widely investigated to terminate silicon dangling bonds with hydrogen atoms and reduce the density of electrical active defects [19-22]. Oxygen plasma treatment is also effective to reduce the density of defects [23]. The silicon dangling bonds are oxidized by oxygen atoms and then they are not sensitive for carrier trapping. High pressure H$_2$O vapor heat treatment has been also developed for defect reduction at low processing temperature [24,25].

Figure 2 shows the spin density as a function of the duration of heat

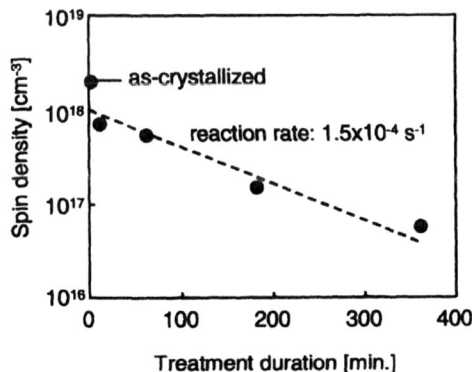

Fig.2 Spin density as a function of duration of heat treatment with 1.3x10^6 Pa H$_2$O vapor at 260°C. The dashed lines are the calculated decay of the spin density with an activation energy of 0.28 eV.

treatment with 1.3×10^6 Pa H_2O vapor at 260°C. The films as-crystallized at 305 mJ/cm² had a high spin density of 2.0×10^{18} cm⁻³ owing to the dangling bonds localized at grain boundaries [26]. The spin density was reduced to 5.0×10^{17} cm⁻³ by 10 min heat treatment. It was further reduced to 6.5×10^{16} cm⁻³ by increasing the treatment duration to 6 h. The dangling bonds were effectively passivated by heat treatment with high-pressure H_2O vapor. The spin density decreased similar to the exponential decay with a single time constant of approximately 110 min, as shown in Fig. 2. This indicates that defect reduction is governed by the pseudo-first-order reaction between the densities of dangling bonds and H_2O molecules incorporated into the silicon films. The reaction rate was the reciprocal time constant of the exponential decay of the spin density with time, as shown in Fig. 2. It was 1.5×10^{-4} s⁻¹(=1/110 [min]) for 260°C treatment with 1.3×10^6 Pa H_2O vapor. From further experimental results of the reduction in dangling bond density with different temperature between 190 and 290°C, the activation energy was estimated to be 0.26 eV. Those investigations result in that the spin density I decreased with time t as $I_0 \exp\{-0.04 \exp(-0.26/kT)^*t\}$ for heat treatment with 1.3×10^6-Pa H_2O vapor, where I_0 is the initial spin density. The low activation energy allows defect state reduction at a low temperature lower than 300°C. However, the reaction rate was low. A long time is necessary for reducing the defect density. Our recent study revealed that the oxygen concentration increased after high-pressure H_2O vapor heat treatment, and that the hydrogen concentration decreased. These results indicate a possibility of that dangling bonds were well oxidized and oxygen atoms terminated dangling bonds at grain boundaries.

Poly-Si TFT Fabricated by Excimer Laser Crystallization and Its Characteristics

In order to apply laser crystallization to fabrication of poly-Si TFTs with high performances, defect passivation is necessary for SiO_2/Si interface as well as silicon films. Because SiO_2 films formed at low temperatures are not thermal relaxed states in general, defect states can be generated at the SiO_2/Si interface. Several technologies have been developed in order to form SiO_2 with a low density of defect states at low temperatures. There have been, for example, electron cyclotron resonance plasma CVD, remote plasma CVD, plasma sputtering, molecular beam deposition. We also applied high pressure H_2O heat treatment in order to improve SiO_2/Si interface properties [24]. Through investigation of high and low frequency capacitance-voltage (C-V) characteristics for Metal-Oxide-Semiconductor (MOS) capacitor structure, we revealed that H_2O vapor heat treatment is effective to reduce defect states in SiO_2 films deposited at low temperature. Figure 3 shows the densities of interface trap states and positive fixed charges as a function of the duration of 1.3×10^6 Pa H_2O vapor heat treatment at 260°C for the MOS capacitor with 50-nm thick SiO_2 formed by the molecular beam deposition method The density of interface trap states for initial SiO_2 film was evaluated to be 3.3×10^{12} cm⁻². The H_2O vapor heat treatment reduced the

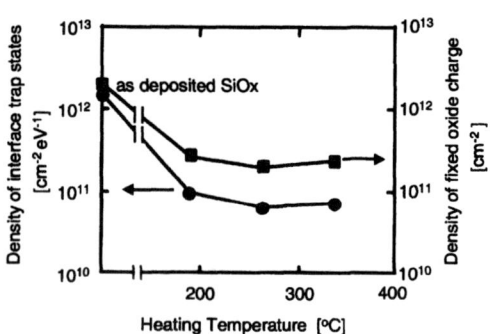

Fig.3 Densities of interface trap states and fixed oxide charge as a function of temperature of 1.3×10^6 Pa-H_2O vapor heat treatment for 3 h.

density of interface trap states with increasing the treatment duration. The interface trap state density was effectively decreased to 5.1×10^{10} $cm^{-2}eV^{-1}$ by the treatment at 260°C for 3 h. The density of fixed oxide charges was also reduced from 2.1×10^{11} (initial) to 1.3×10^{11} cm^{-2} by the treatment for 3 h, as shown in Fig. 3.

Poly-Si TFTs were fabricated by the pulsed excimer laser crystallization and the defect reduction method. 7×10^{20} cm^{-3}-phosphorus and with a thickness of 30 nm were first formed on glass substrates at 330°C using plasma enhanced chemical vapor deposition (PECVD). The doped films were removed at channel region with a length of 25 μm by etching and they were used as dopant sources for forming source and drain regions. 25-nm-thick undoped a-Si:H films were then deposited using PECVD over the whole area. The silicon layers were crystallized by irradiation of 30-ns-pulsed XeCl excimer laser at 300 mJ/cm². Undoped crystallized regions were used as the channel region. Source and drain regions were simultaneously formed through diffusion of phosphorus atoms into the overlaying silicon layer during the laser crystallization. The melt duration of silicon during laser crystallization was shorter than 100 ns so that the diffusion distance of the dopant atom was at most 60 nm in liquid silicon [27] and the 25-μm channel length hardly changed. After laser crystallization, silicon films were heated in 1.3×10^6 Pa H_2O vapor for 3 h for reduction in defects of the silicon films. The silicon films were then patterned by etching for isolation. The molecular beam deposition method was used for formation of the gate insulator. The 100-nm-thick SiO_2 films were then deposited at room temperature using evaporation of SiO powders in oxygen radical produced by induction coupled plasma. Gate, drain and source electrodes were formed with Al metals. After fabrication of the TFT structure, the samples were heated in 1.3×10^6 Pa H_2O vapor for 3 h for improving SiO_2/Si interface characteristics. Figure 4 shows the transfer characteristics of the poly-Si TFTs. The TFT fabricated with no defect reduction showed very low drain current and a high threshold voltage because of carrier trapping effect by defects. On the other hand, Heat treatment with high-pressure H_2O vapor markedly increased the drain current in the low gate voltage region because of defect reduction. The mobility increased to 600 cm^2/Vs and the threshold voltage decreased to 0.9 V. The results of Fig.

4 clearly shows that the defect reduction is in silicon as well as at the SiO_2/Si interface is essential for fabrication of TFTs with high performances. The numerical analysis of transfer characteristics resulted in that TFTs after defect reduction treatment had still serious defects [28]. Defects were mainly tail-state-type states with a width of 0.2 eV and the density of 2×10^{12} cm^{-2}, which were located in silicon. The defect states were occupied by 1×10^{11} cm^{-2} at the threshold voltage at 0.9 V because the Fermi level located at 0.9 eV measured from the valence band edge. Most of defect states stay just below the conduction band edge. Although the

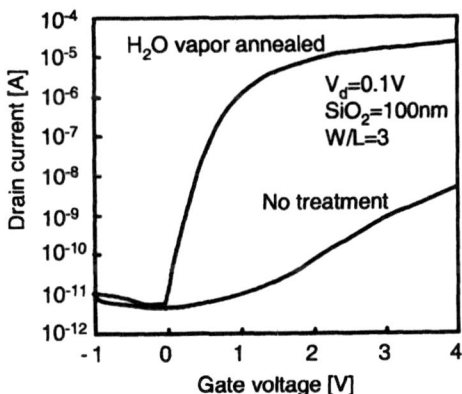

Fig.4 Transfer characteristics of poly-Si TFT fabricated in 25 nm thick silicon films crystallized by XeCl excimer laser irradiation at 275mJ/cm². Heat treatment of 1.3×10^6 Pa H_2O vapor at 260°C for 3h was carried out.

tail type states allows a low threshold voltage for the drain current increase, the numerical analysis suggests that there were still a high density of unoccupied defect states for gate voltage above the threshold voltage.

LARGE CRYSTALLINE GRAIN GROWTH

Grain growth to the lateral direction is important for formation of large crystalline grains in thin silicon films. Many technologies have been investigated for lateral grain growth in order to look for fabrication of single crystalline TFTs with no grain boundary effect. Im et al. have reported a rapid lateral grain growth ~5 μm by a single pulse irradiation of excimer laser with an energy just below the complete melting threshold energy without heating the substrate [29]. OH et al. reported crystalline grain growth ~5μm using a phase shift mask in order to make spatial distribution of excimer laser intensity for generation of temperature distribution [30]. P.C. van der Wilt et al. achieved the control of crystalline nucleation by thick silicon films and fabricated crystalline grains spatially controlled [31]. They fabricated poly-Si TFTs with an electron mobility of 345 cm^2/Vs in there silicon films with a processing temperature of 545°C. Okumura et al. developed comb shaped excimer laser beam to induce temperature gradient in the lateral direction. They achieved grain growth 5 μm long and TFT with an electron mobility of 677 cm^2/Vs with a processing temperature of 450°C [32]. Han et al. reported 5 μm long lateral grain growth by excimer laser using air gap strips below silicon films. They also fabricated TFT with an electron mobility of 452 cm^2/Vs [33].

Laser heating using CW laser beam with a Gaussian intensity profile is also attractive for melt-induced crystallization of silicon films formed on glass substrates [34,35]. The peripheral region of the molten zone is cooler than the central region and the crystalline growth will start from the molten polycrystalline silicon and proceed to the central region. Un-molten silicon at the boundary of molten zone acts as a seed for the crystallizing silicon. A high-power-diode-pumped solid state CW laser (Nd:YVO$_4$) has been recently developed. Hara et al. applied the laser to crystallization of silicon films [36]. They investigated the condition of lateral crystallization with a large grains ~20 μm. They fabricated poly-Si TFTs with an electron mobility higher 566 cm^2/Vs and a hole mobility of 200 cm^2/Vs in the silicon films with large crystalline grains. Tai et al developed a crystallization method using excimer laser for crystalline nucleation formation and CW green laser for lateral crystallization from the nucleation sites. They reported good TFTs with an electron mobility of 460 cm^2/Vs [37]. Because the intensity of CW laser is very stable, uniform crystallization will be possible.

INFRARED LASER CRYSTALLIZATION

The laser crystallization equipment has been developed with excimer laser with a power of about 300 W and it has been used for TFT production at the present stage. The low laser power is a problem increasing the tact time for crystallization of silicon films. We have proposed laser crystallization of silicon films by irradiation laser using the diamond-like-carbon (DLC) films as the photo-absorption layer [38,39]. DLC has been widely utilized for technological and industrial applications. The optical absorption property is also attractive. DLC has low refractive indices 1.3 ~ 1.9 and can have a high extinction coefficient ~0.8 for wavelength from 250 to

1000 nm. Therefore infrared laser diodes or YAG laser can be used for crystallization of silicon films by the assistance of optical absorption with DLC, while silicon films have almost no optical absorbance in the wavelength range. The power intensity of continuous wave (CW) laser diode is very stable and easily modulated by controlling the electrical current. Especially, near-infrared-laser-diode has a markedly high durability with a life time longer than 10,000 h. The laser diodes have the high energy conversion efficiency above 50 %. They can be easily assembled and give a power intensity higher than 10 kW.

An equipment for laser annealing was developed with a laser, an optics and a beam scanning mechanism. Samples were normally irradiated with a fiber coupled continuous wave (CW) laser diode with a wavelength of 940 nm in air at room temperature. A diameter of the core and a numerical aperture (NA) of the fiber was 400 μm and 0.22, respectively. The diverged beam was concentrated on the surface of samples by a combination of six aspherical lenses for 2:1 image formation. The power distribution of the beam was Gaussian like. The size of the beam spot was 180 μm at the full width at half maximum (FWHM) of the laser power distribution. Samples were mounted on an X-Y stage driven by linear motors at a constant velocity from 3 to 100 cm/s.

Figure 5 shows optical absorbance spectra for 50-nm-thick amorphous silicon (a-Si) films coated with Graphitic DLC films with thicknesses of 200 nm by the unbalanced magnetron sputtering (UBMS) method [40]. The formation 200 nm thick DLC films increased the optical absorbance to 0.64 for the wide wavelength region from 250 to 1100 nm. On the other hand, there was almost no optical absorption in the infrared region for the sample of 50-nm-thick a-Si/glass. The results of Fig. 5 clearly shows that the DLC films have a role of effective optical absorption in the infrared region.

Laser irradiation was conducted with different laser dwell times controlled by the scanning velocity of 1 m/s. The carbon layer was removed by oxygen plasma treatment. Figure 6 shows the crystalline volume ratio

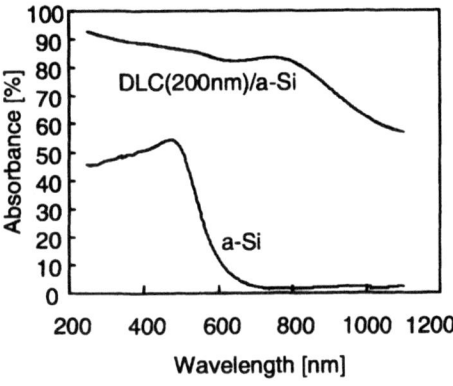

Fig.5 Optical absorbance spectra for 50-nm-thick amorphous silicon (a-Si) films coated with 200-nm-thick Graphitic DLC films and the single layer of a-Si film.

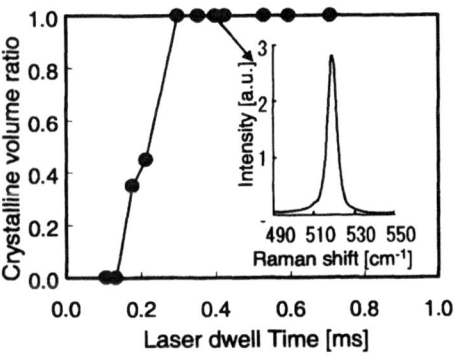

Fig.6 Crystalline volume ratio as a function of laser power. The crystalline volume ratio was obtained from analysis of optical phonon peak in Raman scattering spectra.

379

as a function of the laser dwell time at a laser power intensity of 24.7 kW/cm². The crystalline volume ratio was obtained from analysis of optical phonon peak in Raman scattering spectra. Crystallization was observed at a laser well time 0.18 ms. The crystalline volume ratio increased as the laser dwell time increased to 0.3 ms. A sharp phonon peak was observed with a high crystalline volume ratio of 1.0 for the laser dwell time longer than 0.3 ms as shown in the inset. This suggests that the DLC films effectively absorbed the 940-nm infrared laser light and were heated to enough high temperature to crystallize Si films. Electron back scattering measurement revealed that the crystalline grain growth with a size of 3 μm was achieved by laser annealing for 0.4 ms.

The present method is applied for rapid activation of impurities as well as crystallization of silicon films. Ion implantation of phosphorus atoms was carried out for p-type silicon substrates with a resistance of 10 Ωcm. The acceleration energy was 10 keV. The dose was 2×10^{15} cm⁻². DLC films with a thickness of 200 nm were formed on the silicon surface. Laser was irradiated at a power density of 70 kW/cm². The laser beam was scanned at a velocity of 7 cm/s, which gave an effective dwell time of the laser beam of 2.6 ms. Figure 7 shows the phosphorus atoms in-depth profiles 2×10^{15} cm⁻²-as-implanted at 10 keV and laser annealed. The phosphorus atoms located at the surface region and the depth of the peak phosphorus concentration was 13.7 nm for as-implanted at 10 keV. The phosphorus implantation made the silicon surface region disordered as shown by a broad phonon peak in Raman scattering spectra in Fig. 8. No serious change in the phosphorus concentration profile was observed after laser annealing at 70 kW/cm² for 2.6 ms, as shown in Fig. 7. After laser annealing, the peak phosphorus concentration located at 14.9 nm. Slight broadening of the depth profile about 5 nm for the concentration from 1×10^{20} to 4×10^{18} cm⁻³ was observed. We interpret that the silicon surface regions were heated to a very high temperature by laser annealing enough to cause recrystallization of phosphorus implanted regions. The diffusion coefficient of

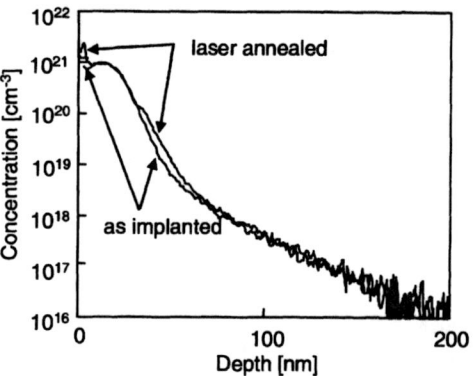

Fig. 7 Phosphorus atoms in-depth profiles 2×10^{15}-cm⁻²-as-implanted at 10 keV and laser annealed with an effective dwell time of 2.6 ms.

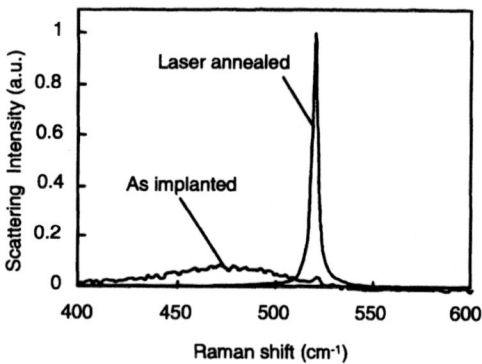

Fig. 8 Raman scattering spectra of samples phosphorus-as-implanted and laser annealed.

380

phosphorus atoms in solid silicon was about 1×10^{-11} cm^2/s at a high temperature around 1300°C [41]. The heat duration of 2.6 ms can allow a diffusion length of 3.4 nm $\{\sim 2 \times (1 \times 10^{-11} \times 2.6 \times 10^{-3})^{0.5}\}$. It is close to our maximum diffusion length of 5 nm observed by SIMS as shown in Fig.7. The implanted surface region was almost completely crystallized by laser annealing as shown by a sharp phonon peak in Raman scattering spectra in Fig. 8. The sheet resistance decreased to 102 Ω/sq by laser annealing. The results of Figs. 7 and 8 indicate that the present annealing method achieve the activation of the implanted region with keeping initial dopant concentration profiles.

CONCLUSIONS

Rapid laser crystallization of silicon films was discussed for forming polycrystalline silicon thin films used to fabricate poly-Si TFTs. Pulsed laser rapidly heats silicon films and melts them. Rapid melt-regrowth of silicon films results in poly-Si films with small crystalline grains ~0.1 μm because and a preferential orientation of (111). We revealed that the stress of laser crystallized films was mainly determined by stress at the bottom interface depending on film formation temperature. Electrical properties of poly-Si films were governed by defect located at grain boundaries. Treatments of hydrogen plasma, oxygen plasma and high-pressure H$_2$O vapor were developed in order to reduce the density of defect states. TFT fabrication was reported using high-pressure H$_2$O vapor heat treatment. Defect reduction in poly-Si silicon and gate insulator/silicon interfaces resulted in TFT characteristics with a low threshold voltage and a high mobility. Recent development in laser crystallization methods with pulsed and continuous wave (CW) lasers was reviewed. The controls of the heat flow and the crystalline nucleation have resulted in lateral crystalline grain growth in a few micro meters. Those techniques are important for fabrication of large crystalline grains and high performance TFTs. We reported an annealing method using a high power infrared semiconductor laser. Silicon films were successfully crystallized by the infrared laser using a carbon optical absorption layer. Grain growth in a few micron was achieved. The activation of the silicon surface region implanted at 10 keV with 2×10^{15}-cm^{-2}-phosphorus was also achieved by the infrared laser annealing. Laser annealing for 2.6 ms almost completely recrystallized the implanted region with keeping the initial phosphorus in-depth profile. The low resistance at 102 Ω/sq was achieved.

ACKNOWLEDGMENTS

The author thank to N. Sano, N. Andoh, Y. Andoh, Y. Matsuda for their supports.

REFERENCES

1. S. Uchikoga and N. Ibaraki, Thin Solid Films, **383** 19 (2001).
2. S. Inoue, K. Sadao, T. Ozawa, Y. Kobashi, H. Kwai, T. Kitagawa and T. Shimoda, Tech. Dig. IEDM, 2000, p197.

3. K.Shibata and H. Takahashi, *Proc Int. Workshop on Active Matrix Liquid Crystal Displays'01* (Tokyo, 2001), p219.

4. Dharam Pal Gosain, Takashi Noguchi, Akio Machida and Setsuo Usui, Proc. in Workshop on Active Matrix Liquid Displays (Tokyo, 1999) p239.

5. T. Sameshima, S. Usui and M .Sekiya, IEEE Electron Device Lett., 7 276 (1986).

6. K. Sera, F. Okumura, H. Uchida, S. Itoh, S. Kaneko and K. Hotta, IEEE Trans. Electron Devices 36 2868(1989).

7. T. Serikawa, S. Shirai, A. Okamoto and S. Suyama, Jpn. J. Appl. Phys., 28 L1871 (1989).

8. A. Kohno, T. Sameshima, N. Sano, M. Sekiya and M. Hara, IEEE Trans. Electron Devices, 42 251(1995).

9. P. L. Liu, R. Yen, N.Bloembergen and R.T.Hodson, Appl. Phys. Lett., 34, 864 (1979).

10. R. F. Wood and C. E. Giles, Rhys. Rev., B23, 2923 (1981).

11. H. S. Carslaw and J. C. Jaeger, *Conduction on Heat in Solid*, Oxford University Press, Oxford, 1959, Chapters 2 and 10.

12. A. Goldsmith, T. E. Waterman and H. J. Hirschorn, *Handbook of Thermophysical Properties of Solid Materials*, Pergamon Press, New York, 1961, Vols. 1 and 3.

13. G. J. Galvin, M. O. Thompson, J. W. Mayer, R. B. Hammond, N. Paulter and P.S. Peercy, Phys. Pev. Lett., 48, 33 (1982).

14. T. Sameshima, M.Hara and S.Usui, Jpn. J. Appl. Phys., 28, 1789 (1989).

15. S. Higashi, Ph.D. dissertation, Tokyo University of Agriculture and Technology, 2001, Chap 3, p. 34.

16. S. Higashi, N. Andoh, K. Kamisako and T. Sameshima, Jpn. J. Appl. Phys., 40, 731 (2001).

17. K. Kitahara, A. Moritani, A. Hara and M. Okabe, Jpn. J. Appl. Phys., 38, L1312 (1999).

18. T. Sameshima, K. Saitoh, N. Aoyama, S. Higashi, M. Kondo and A. Matsuda, Jpn. J. Appl. Phys., 38, 1892 (1999).

19. R. A. Ditizio, G. Liu, S. J. Fonash, B.-C. Hseih and D. W. Greve, Appl. Phys. Lett., 56, 1140 (1990).

20. I-W Wu, A.G.Lewis, T-Y, Huang, A.Chiang, IEEE Electron Device Lett., 10, 123 (1989).

21. U. Mitra, B. Rossi and B. Khan, J. Electrochem. Soc., 138, 3420 (1991).

22. D. Jousse, S. L. Delage and S. S. Iyer, "Grain-boundary states and hydrogenation of fine-grained polycrystallione silicon films deposited by molecular beams," Phil. Mag., B63, 443 (1991).

23. Y. Tsunoda, T. Sameshima and S. Higashi, Jpn. J. Appl. Phys., 39, 1656 (2000).

24. T. Sameshima and M. Satoh, Jpn. J. Appl. Phys. 36, L687 (1997).

25. K. Asada, K. Sakamoto, T. Watanabe, T. Sameshima and S. Higashi, Jpn. J. Appl. Phys., 39, 3883 (2000).

26. T. Sameshima, H. Hayasaka, M. Maki, A. Masuda, T. Matsui and M. Kondo, Proc. in Workshop on Active Matrix Flat panel Displays and Devices (Shinjuku, 2006) p143.

27. T. Sameshima, M. Hara and S. Usui, Mat. Res. Soc. Symp. Proc., 158, (PA, 1990) 255.

28. H. Watakabe and T. Sameshima, IEEE Trans. Electron device 49, 2217 (2002).

29. J.S.Im and H.J.Kim, Appl. Phys. Lett., 63 (1993) 1969.

30. Chang-Ho OH, M. Ozawa and M. Matsumura, Jpn. J. Appl. Phys, 37, L492 (1998).

31. Paul Ch. van der Wilt, B. D. van Dijk, G. J. Bertens, R. Ishihara, and C. I. M. Beenakker, Appl. Phys. Lett. 79, 1819 (2001).

32. M. Nakata, H. Okumura, H. Kanoh, and H. Hayama, H., Proc. in AsiaDisplay/IMID'04 (Tegu, Korea, 2004) p412.
33. C. H. Kim, I. H. Song, W. J. Nam, and M. K. Han, IEEE Electron Device Lett., 23 315 (2002).
34. T. J. Stultz and J. F. Gibbons, Appl. Phys. Lett. 39, 498 (1981).
35. K. F. Lee, J. F. Gibbons, K. C. Saraswat and T. I. Kamins, Appl. Phys. Lett. 35, 173 (1979).
36. A. Hara, F. Takeuchi, M. Takei, K. Suga, K. Yoshino, M. Chida, Y. Sano and N. Sasai, Jpn. J. Appl. Phys, 37, L5 (2002).
37. M. Tai, M. Hatano, S. Yamaguchi, T. Noda, P. S-Kee, T. Shiba and M. Ohkura, IEEE Trans. Electro. Dev., 51, 934 (2004).
38. T. Sameshima and N. Andoh, Jpn. J. Appl. Phys., 44, 7305(2005).
39. N. Sano, M. Maki, N. Andoh and T. Sameshima, Mater. Res. Soc. Symp. Proc. 910, (PA, 2006), A14-02.
40. G. Viera, S. Huet, L. Boufendi, J. Appl. Phys., 90 4175 (2001).
41. A. S. Grove: *Physics and Technology of Semiconductor Devices*, (John Wily & Sons, Inc. 1967) Chap. 3.

Mater. Res. Soc. Symp. Proc. Vol. 989 © 2007 Materials Research Society　　　0989-A16-03

Defect characterization of polycrystalline silicon layers obtained by aluminum-induced crystallization and epitaxy

Dries Van Gestel, Ivan Gordon, Lodewijk Carnel, Guy Beaucarne, and Jef Poortmans
IMEC, Leuven, B-3001, Belgium

ABSTRACT

In order to reduce the harmful influence of grain boundaries in thin-film polycrystalline Si solar cells we form absorber layers on foreign substrates with columnar grains with a grain width much larger than the layer thickness. Such layers with a grain size in the range of ~1-100 μm can be obtained by aluminum-induced crystallization and epitaxy. Until now however, the open-circuit voltage of solar cells made from these layers was quasi-independent of the grain size. To understand this fact, defect etching combined with electron microscopy, as well as Electron Backscattered diffraction (EBSD) measurements were performed to investigate the crystallographic defects. A very large density (~ 10^9 cm^{-2}) of intra-grain defects (IGD) was found. Room temperature Electron Beam Induced Current (EBIC) measurements were carried out to localize and investigate the electrically active defects. The intra-grain defects found with defect etching showed a strong recombination activity. These results indicate that the unexpected quasi-independence on the grain size of the open-circuit voltage of our pc-Si solar cells is due to the presence of numerous electrically active intra-grain defects.

INTRODUCTION

Thin-film polycrystalline silicon (pc-Si) solar cells (grain size ~ 0.1 – 100 μm) on foreign substrates are a promising approach for the next generation silicon solar cells. Aluminum-induced crystallization (AIC) in combination with epitaxy is a possible way to obtain pc-Si layers with columnar grains on foreign substrates [1]. Over the last few years at IMEC we have achieved with this approach an increase in absolute efficiency of ~1.5 % per year, the best efficiency being 8% at the moment [2]. To maintain this trend, further research is necessary to e.g. enhance the layer quality.

Because there is a high minority carrier recombination probability at grain boundaries, control over the grain size distribution is important. The AIC seed layer formed on the foreign substrate before epitaxial growth allows us to influence this distribution of the absorber layer [3]. However, we recently showed that solar cells made from pc-Si layers with very small grains of 0.2 μm had almost the same open-circuit values (V_{oc}) as solar cells made from AIC-based pc-Si layers with grain diameters of up to 50 μm [4,5]. Since the V_{oc} of thin-film solar cells is a measure for the electronic quality of the absorber material, the quasi-independence of the open-circuit voltage on the grain size indicates that at the moment grain boundaries and grain size distribution are not the only factors limiting the electronic quality of our pc-Si layers. It is known that intra-grain dislocations can be harmful for the electrical layer quality, and different models have already been proposed to describe their effect on the minority carrier lifetime and the V_{oc} of silicon solar cells [6,7].

In this paper we use defect etching, electron backscattering diffraction (EBSD) and electron beam induced current (EBIC) measurements to study the effect of intra-grain defects (IGD) on the electrical quality of pc-Si layers obtained by AIC and epitaxial growth.

EXPERIMENTAL DETAILS

We made pc-Si films on alumina substrates (CoorsTek ADS996R) by epitaxial thickening of AIC seed layers. The substrates were covered by a spin-on flowable oxide (FOx-25 from Dow Corning) to reduce their surface roughness, prior to the seed layer formation [8]. Next, double layers of Al and a-Si were deposited on these substrates in an electron-beam high-vacuum evaporator. In between the two depositions, the aluminum was oxidized by exposure to air for seed layers with grain sizes up to 15 μm or treaded by HNO_3 for samples with grain sizes up to 50 μm [4]. The nominal thickness of the Al and a-Si layers was fixed at 200 nm and 230-250 nm respectively. After deposition, the samples were annealed in a tube furnace under nitrogen ambient at 500°C for 4 hours. During this annealing, the a-Si crystallized into pc-Si (with an Al doping of ~2.6 10^{18} cm^{-3}) and both layers exchanged places [3]. The top Al layer was removed by selective chemical etching. Only on the HNO_3 treated samples, secondary Si crystallites were removed before epitaxial growth by a treatment with concentrated phosphoric acid at temperatures between 70 and 120 °C [4].

Absorber layers were deposited on the AIC layers by thermal CVD. The depositions were performed in a single-wafer epitaxial reactor (ASM Epsilon 2000) under atmospheric pressure, at a temperature of 1130°C. The growth rate was around 1.4 μm / min. 2 μm thick double layers of p+ and p silicon were made. The final p+ layer (both AIC seed layer as well as 2 μm thick epitaxial grown p+ layer) acts as a back surface field (BSF) and as a conductive channel for majority carriers, while the p layer is the actual absorber layer. Typical doping densities were 2 x 10^{19} cm^{-3} for the BSF layer and 10^{16}-10^{17} cm^{-3} for the absorber layers. SIMS measurements showed out diffusion of Al from the seed layer up to 1 μm deep in the epitaxial layer.

To complete the solar cell, homojunction emitters were formed at high temperatures (> 800°C) by diffusion from a phosphorous source. Defect passivation of the layers was performed by plasma hydrogenation in a PECVD system. Finally, contact structures were made by lithography, wet etching and metal evaporation.

RESULTS

In this section we first show that beside grain boundaries also high densities of intra-grain (crystallographic) defects are present in the pc-Si absorber layers. Secondly, the layers were measured by room temperature EBIC to investigate the local electrical behavior. From the combination of both results we conclude that most of the intra-grain defects are electrically active. Together with the quasi independency of the V_{oc} on grain size this is a strong indication that IGD are indeed a major concern for devices based on pc-Si, especially the ones based on AIC and epitaxial growth.

Crystallographic defects

We polished the absorber layers and used a slightly adapted Schimmel defect etch (1 (HF 49%): 1 (CrO$_3$ 0.7M)) to make the crystal defects visible [9]. Afterwards, intra-grain defects were easily distinguishable from grain boundaries. The grains showed a very high IGD density of ~ 10^9 cm^{-2} (see Figure 1a). Beside IGD appearing as points (e.g. dislocation lines crossing the surface), also square shaped structures, U shaped lines and perpendicular lines were found. To understand the origin of these lines, EBSD analysis was performed on such defect etched layers. Figure 1b is a SEM image of an area used for this. Three large grains can be recognized (example of easy distinction between grain boundaries and IGD). Figure 1c and d are the inverse pole (IPF) EBSD maps of the same area as figure 1b, showing respectively the orientation along the normal to the sample surface (IPF_Z0) and perpendicular to the normal (IPF_X0). In all measured grains where the square shaped structures, the U shaped lines and the perpendicular lines were visible after defect etching, the grains had a (100) orientation (see figure 1c). Most of the grains in our pc-Si layers have a (100) orientation as a result of the AIC process. Despite the (100) orientation of these grains, they can still have a different rotation around the [100] axis. Grains with a different rotation around this axis, which is visible in the IPF_X0 map (figure 1d), also showed a mutual different orientation of the defect etched lines (figure 1b). From these results we attributed the observed structures and lines to epitaxial stacking faults along (111) planes when looking at a (100) plane. From the length of the squares and the lines, we can therefore calculate, as illustrated in figure 1e, that most defects were formed in the seed layer or at the seed layer-epitaxial layer interface.

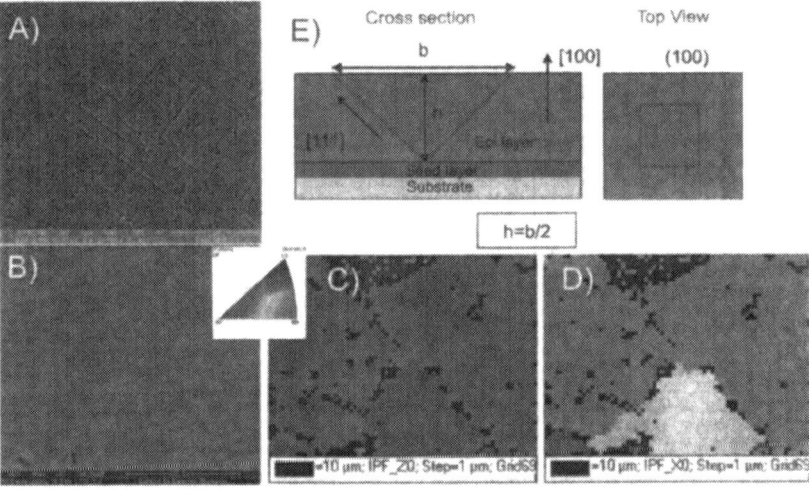

Figure 1. Defect etching and interpretation of defect etched pc-Si layers. A) Top view SEM picture of a defect etched intra-grain area (reused with permission from [13]) B) SEM image of the area used for an EBSD measurement on a defect etched layer. C) IPF_Z0 EBSD map of the area shown in part b. The represented orientation is the one normal to the surface. D) IPF_X0 EBSD map of the area shown in part b. The represented orientation is perpendicular to the normal on the surface. E) Schematic illustration of the origin of the observed lines after defect etching, showing a link between the length of the lines and the depth at which the defect is started.

Room temperature EBIC

We performed room-temperature EBIC measurements in a Philips XL 30 system to characterize local electrical defects. Although heterojunction emitters (a-Si/c-Si) give ~80mV higher V_{oc} values than homojunction emitters [10], solar cells with diffused emitter were used for the EBIC measurements. EBIC measurements on samples with a heterojunction emitter were so far not possible. This was probably due to an e-beam induced degradation process [11] of the ~15nm thick amorphous Si layer. Figures 2a and 2c show top view SEM images of solar cells respectively with grain sizes up to 15 and 50 μm and V_{oc} values of 439 and 448 mV. Figures 2b and 2d are room temperature EBIC images of the same areas using acceleration voltages of 10 kV. The topology of the surface influences the EBIC image due to the position-dependent generation volume in the pc-Si layers [12], but the EBIC images clearly show more than only a topology contrast. In both cases the same patterns as after defect etching were found, showing that the crystallographic IGD have a strong recombination activity. The different orientation of the defect lines in e.g. both grains of Figure 3d is a consequence of a different rotation around the [100] axes of both (100) grains. This is similar as the EBSD results presented in figure 1. Since the pc-Si layers show a large density of electrically active intra-grain defects and since Voc values of pc-Si solar cells are quasi-independent on the grain size, we believe that intra-grain defects are the major limiting factor for the electrical quality of pc-Si layers at the moment.

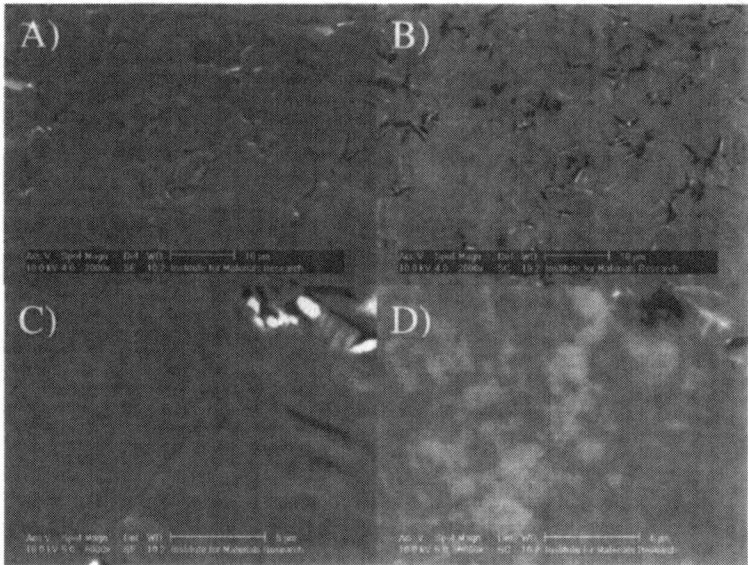

Figure 2. Top view SEM (a,c) and EBIC (b,d) images of the same area of a pc-Si solar cells with diffused emitter. The solar cells have a maximum grain size of 15 μm with a V_{oc} of 439 mV (a,b) and 50 μm with a V_{oc} of 448 mV (c,d). Images c and d are reused with permission from [13].

Grain boundaries are places of high carrier recombination activity and are therefore expected to appear as dark lines in EBIC images. However, in both EBIC images of figure 2 the grain boundaries are visible as bright lines. To further investigate this phenomenon, EBIC images of the same area as shown in figure 2 d were taken at different acceleration voltages. The results can be found in figure 3. As expected, the IGD contrast decreases with increasing acceleration voltage, since a higher acceleration voltage leads to a larger carrier generation ·volume. On the other hand, grain boundary contrast was light for low acceleration voltages (e.g. 10 kV in Figure 2d) but became dark for high (e.g 20 kV in figure 3) acceleration voltages. This is a consequence of the emitter diffusion which results in deep emitter spikes at grain boundaries due to enhanced diffusion of phosphorus along these grain boundaries [5]. At low acceleration voltage, the maximum electron-hole generation depth is around 1 µm which is comparable to the depth of the emitter spikes at the grain boundaries. The grain boundaries therefore show an increased collection compared to the intra-grain regions. At higher acceleration voltages (e.g. 20 kV), electron-hole generation extends well below the junction spikes (maximum electron-hole generation depth ~ 5 µm). In this case recombination at the grain boundaries below the emitter spikes dominates, resulting in a dark EBIC contrast at the grain boundaries.

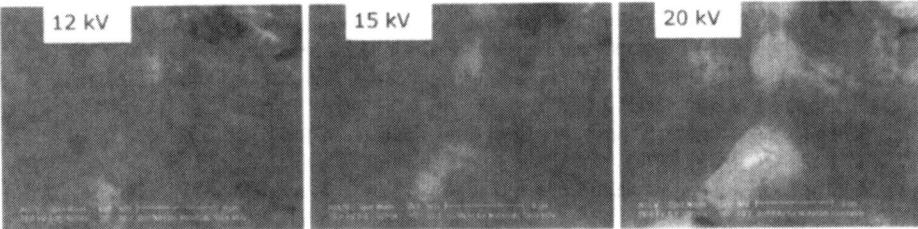

Figure 3. EBIC at different acceleration voltages. The contrast of the GB is changing with the voltage.

Due to the possibility of having IGD with different recombination activity, characterization by defect etching alone is not enough. It cannot be excluded that higher Voc values can be reached with the same IGD density, due to a change in recombination activity at the defects. Recently we showed cathodoluminescence and spectrum imaging measurements to get more insight into the origin of the electrical activity of the defects [13]. Two deep-level radiative transitions were found that have relative intensities that vary from grain to grain. Differences in metal impurity concentration seem to be a plausible reason, but this is still not completely proven. In the past it was shown that defects can have different temperature dependent EBIC contrasts, which was explained by a difference in metal impurity concentration [14]. We will therefore check whether the from grain to grain varying behavior found by cathodoluminescence can also be observed by temperature dependent EBIC measurements.

CONCLUSIONS

We conclude that the large number of electronically active intra-grain defects are found in epitaxially thickened AIC layers and that this is the major reason for the unexpected quasi-independence on grain size of the V_{oc} values of solar cells made from such layers. Besides grain size distribution control, intra-grain quality improvement is therefore very important to obtain thin-film polycrystalline layers with good electronic quality.

ACKNOWLEDGMENTS

This work was partly funded by the European Commission under contract number 019670-FP6-IST-IP ('ATHLET'). The authors thank M. J. Romero from NREL for the fruitful discussions and Jan D'Haen for the EBSD measurements and the help with the EBIC measurements. Kris Van Nieuwenhuysen is acknowledged for the epitaxial depositions.

REFERENCES

1. D. Van Gestel, I. Gordon, L. Carnel, L.R. Pinckney, A. Mayolet, J. D'Haen, G. Beaucarne, and J. Poortmans, *Mater. Res. Soc. Symp. Proc* **910**, A26-04 (2006).
2. I. Gordon, L. Carnel, D. Van Gestel, G. Beaucarne, and J. Poortmans, Paper presented at this conference
3. Olivier Nast and Stuart R. Wenham, J. Appl. Phys. 88, 124 (2000)
4. D. Van Gestel, I. Gordon, L. Carnel, K. Van Nieuwenhuysen, J D'Haen, J. Irigoyen, G. Beaucarne and J. Poortmans, Thin Solid Films, **511 – 512**, 35 (2006)
5. L. Carnel, I. Gordon, D. Van Gestel, G. Beaucarne, J. Poortmans and A. Stesmans, J. Appl. Phys. **100**, 063702 (2006)
6. Mitsuru Imaizumi, Tadashi Ito, Masafumi Yamaguchi and Kyojiro Kaneko, J. Appl. Phys. **81**, 7635 (1997)
7. Thomas Kieliba, Stephan Riepe and Wilhelm Warta, J. Appl. Phys, **100**, 063706 (2006)
8. Gordon I, Van Gestel D, Van Nieuwenhuysen K, Carnel L, Beaucarne G, Poortmans J. Thin Solid Films, **487**, 113-117 (2005).
9. D. G. Schimmel, J. Electrochem. Soc., **126**, 479 (1979)
10. L. Carnel, I. Gordon, D. Van Gestel, K. Van Nieuwenhuysen, G. Agostinelli, G. Beaucarne, J. Poortmans, Thin Solid Films, **511 – 512**, 21 (2006)
11. B.G. Yacobi, C.R. Herrington and R.J. Matson, J. Appl. Phys. **56**, 557, (1984)
12. A.B. Sproul, T. Puzzer and R.B. Bergmann, Proceedings of the 2nd World Conference on Photovoltaic Solar Energy Conversion, Vienna, 6-10 July 1998, 1355
13. D. Van Gestel, M.J. Romero, I. Gordon, L. Carnel, J. D'Haen, G. Beaucarne, M. Al-Jassim, and J. Poortmans, *Appl. Phys. Lett.* **90**, 092103 (2007).
14. V. Kveder, M. Kittler, W. Schroter, Physical Review B, **63**, 115208 (2001)

Mater. Res. Soc. Symp. Proc. Vol. 989 © 2007 Materials Research Society　　　　0989-A16-04

Comparative Study of Solid-Phase Crystallization of Amorphous Silicon Deposited by Hot-wire CVD, Plasma-Enhanced CVD, and Electron-Beam Evaporation

Paul Stradins[1], Oliver Kunz[2], David L. Young[1], Yanfa Yan[1], Kim M. Jones[1], Yueqin Xu[1], Robert C. Reedy[1], Howard M. Branz[1], Armin G. Aberle[2], and Qi Wang[1]

[1]National Renewable Energy Laboratory, Golden, CO, 80401
[2]The University of New South Wales, Sydney, Australia

ABSTRACT

Solid-phase crystallization (SPC) rates are compared in amorphous silicon films prepared by three different methods: hot-wire chemical vapor deposition (HWCVD), plasma-enhanced chemical vapor deposition (PECVD), and electron-beam physical vapor deposition (e-beam). Random SPC proceeds approximately 5 and 13 times slower in PECVD and e-beam films, respectively, as compared to HWCVD films. Doping accelerates random SPC in e-beam films but has little effect on the SPC rate of HWCVD films. In contrast, the crystalline growth front in solid-phase epitaxy experiments on (100)-oriented silicon wafer substrates propagates at similar speed in HWCVD, PECVD, and e-beam amorphous Si films. This strongly suggests that the observed large differences in random SPC rates originate from different nucleation rates in these materials while the grain growth rates are relatively similar. The larger grain sizes observed for films that exhibit slower random SPC support this suggestion.

INTRODUCTION

Solid-phase crystallization (SPC) of amorphous Si (a-Si) has become an important process in the thin-film electronics and the solar cell industry [1]. In these applications, obtaining large (at least a few microns), high-quality grains is essential. Thermal annealing of a-Si films typically produces submicron grains with a broad size distribution [2]. Laser crystallization [3] can produce uniform, larger grains but not with an inexpensive batch process. The electronic quality of thermally crystallized grains can be significantly improved by Rapid Thermal Processing (RTP) and subsequent H-passivation [4], enabling inexpensive large-scale production of devices with reasonable efficiencies [5]. Nevertheless, a deeper understanding of thermally induced nucleation and subsequent growth in solid phase, as well as the role of impurities and network microstructure is necessary for further progress. We have previously found that a-Si:H films deposited by hot-wire (HW) chemical vapor deposition (CVD) and plasma-enhanced CVD (PECVD) have different SPC rates at a given temperature [6, 7]. This indicates that the network microstructure might play an important role in SPC.

In this work, we compare SPC of a-Si prepared by HWCVD, PECVD, and electron beam physical vapor deposition. By comparing the random SPC and solid-phase epitaxy (SPE) in the same films, we demonstrate that it is most likely the nucleation that leads to the observed differences in overall random SPC rates. The role of dopants in the SPC of these films is also addressed.

EXPERIMENTAL

The HWCVD and PECVD a-Si:H films were grown at NREL on Corning 1737 glass substrates, to about 1 micron thickness, at substrate temperatures and deposition rates of 15 Å/s at 410°C (HWCVD), and 2 Å/s at 250°C (PECVD). A tungsten filament heated to ~ 2000°C was used in the HWCVD. For SPE studies, a-Si:H films were grown at 40 ° /s on (100) polished c-Si wafers at 370°C by HWCVD. The substrates were cleaned by a modified RCA process followed by a dip in 5% HF solution, and treated by HW atomic hydrogen for 60 s before film deposition to improve adhesion. Dopants were introduced in HWCVD films by adding PH_3 and TMB gases to the SiH_4 deposition gas.

The e-beam a-Si films, each about 1 micron thick, were deposited at the University of New South Wales in an e-beam system at base pressures of about 10^{-7} Torr. Dopants (phosphorus and boron) were added via effusion cells. The SPC films were deposited on Borofloat33 glass substrates coated with a ~70 nm thick SiN_x buffer layer deposited by microwave semi-remote PECVD. The SPE films were deposited in the same system as the SPC films, but onto (100) polished c-Si wafer substrates. Both SPE and SPC substrates received a 6 min H_2SO_4/H_2O_2 surface clean (piranha clean) followed by a 5% HF dip of 30 sec in order to remove the surface oxide and to obtain hydrogen-terminated surfaces. The deposition rates for SPC films were 50 nm/min except for the p$^+$ film (25 nm/min), due to the limited dopant flux from the effusion cell. The deposition rates of the SPE films (all films intrinsic) were 300 nm/min. The deposition temperature was between 210 and 230°C.

The films were thermally crystallized in N_2 ambient in a chamber equipped with in-situ spectral optical reflectance $R(\lambda,t)$ monitoring. This technique [6] allows us to study the kinetics and mode of SPC, as well as to stop the crystallization at a desired moment. The as-deposited and crystallized films were studied ex-situ by secondary-ion mass-spectrometry (SIMS) and cross-sectional transmission electron microscopy (TEM), as well as by Electron Backscatter Diffraction (EBSD).

RESULTS

Figure 1 shows the temperature dependence of the random solid-phase crystallization time τ_{cryst}. Here, τ_{cryst} denotes the time to fully complete the crystallization process, as indicated by in-situ $R(\lambda,t)$, for films on glass (HWCVD, PECVD) or SiN_x-coated glass (e-beam). Circular, diamond, and square-shaped symbols correspond to HWCVD, PECVD, and e-beam films, respectively. For each symbol shape, intrinsic a-Si is marked by the solid (i.e., filled) symbol, and n- and p-type dopings are shown by open symbols and dotted-open symbols, corres-pondingly. Dopant levels in both HWCVD and e-beam films are ~10^{19} cm^{-3} and ~10^{20} cm^{-3} for p-type and n-type doping, respectively.

According to Fig. 1, the average SPC times of intrinsic PECVD and e-beam films at 600°C are a factor of 5 and 13 longer, respectively, than that of HWCVD a-Si:H films (about 1 hour). Introducing either type of dopants reduces the SPC time of e-beam a-Si films. In contrast, p-type doping has no effect on SPC in HWCVD films while n-type doping even increases the SPC time. In-situ reflectance measurements indicate that in most of the films, SPC occurs uniformly in the bulk. However, some p-type e-beam films showed indications of delayed surface crystallization with respect to the bulk, which might be due to their more contaminated surface as indicated by SIMS. Note that Fig. 1 shows the *total* crystallization time while in Ref. 6 we present the time of the crystallization onset (corresponding to crystalline volume fraction of few %).

In addition to network microstructure, differences in bulk impurity contents might affect SPC time. We have therefore measured the former by SIMS and plot in Fig. 2 the crystallization time τ_{cryst} at 600°C as function of residual oxygen content. Oxygen is known to influence SPC [8] and its content in the films varies depending on the deposition method, base pressure, deposition rate, etc. Yet, Fig. 2 suggests that despite these variations, the τ_{cryst} seems to be more affected by the deposition method itself rather that the oxygen content. Note that all three types of films (including e-beam) also contain residual hydrogen. Typical H levels in crystallized films are about 3×10^{19} cm^{-3}, 4×10^{18} cm^{-3}, and $(4-10) \times 10^{18}$ cm^{-3} in HWCVD, PECVD, and e-beam films, respectively.

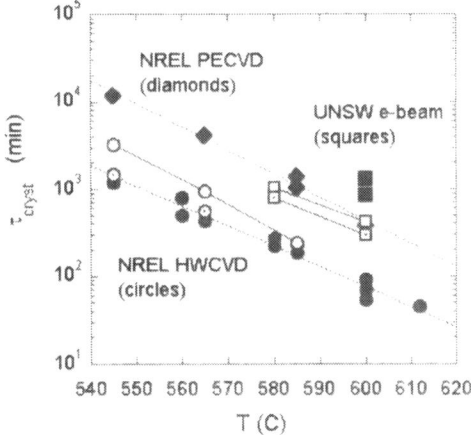 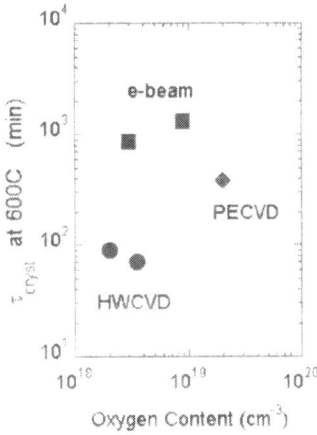

Fig. 1. Temperature dependence of the SPC crystallization time τ_{cryst} in HWCVD (circles), PECVD (diamonds), and e-beam a-Si films (squares). Solid symbols denote intrinsic films, open symbols n-type films, and dotted open symbols p-type films. The dotted lines are exponential fits.

Fig. 2. Dependence of the crystallization time at $T = 600$°C on the oxygen content measured by SIMS in the a-Si films deposited by three different methods.

Since we have films with different network structure and/or chemical composition resulting from a given deposition method, we expect different nucleation and grain growth rates for each material. However, from the random SPC rates in Fig. 1 it is difficult to distinguish which of the two is the rate limiting process for the final crystallization time. To separate them, we studied SPE in HWCVD, PECVD, and e-beam films deposited on polished (100) c-Si wafers. Since nucleation does not take place in SPE, the SPE speed is an indicator of the SPC growth rate for a particular growth direction.

Figure 3 shows the SPE thickness d as function of time in HWCVD, PECVD, and e-beam films. In this experiment, films of different thickness were produced either by several depositions, or by controlled thinning via reactive-ion etching (SF$_6$ plasma). All films were kept at 560°C and the time t of the arrival of the SPE front at the film surface was detected using in-situ $R(\lambda,t)$. Specifically, the sudden appearance of the c-Si UV signature marked the arrival of the (100)-oriented crystalline phase at the film's surface [6]. Symbol shapes are the same as in Fig. 1, however, the solid symbols here denote (100) SPE while the open symbols show for comparison the random SPC times of 1 micron thick HWCVD and PECVD films on glass.

Figure 3 demonstrates that, at 560°C, the SPE front moves approximately linearly in time at a speed of about 10 nm/min. Furthermore, the SPE speed is not reduced in PECVD and e-beam films with respect to HWCVD films, as one might expect from Fig. 1. In addition, SPE speeds in our films are close to the speed measured in ion-implanted Si (filled triangle) [9].

Figure 4 shows a cross-sectional TEM micrograph of a partially crystallized solar cell structure deposited by e-beam on SiN$_x$-coated glass. The as-deposited structure consisted of a 100 nm thick n^{++} a-Si emitter (10^{20} cm^{-3}, phosphorus-doped), followed by a 1500 nm p-type base layer (10^{17} cm^{-3}, boron-doped a-Si), topped by a 100 nm thick p$^+$ back-surface field layer (10^{19} cm^{-3}, boron-doped).

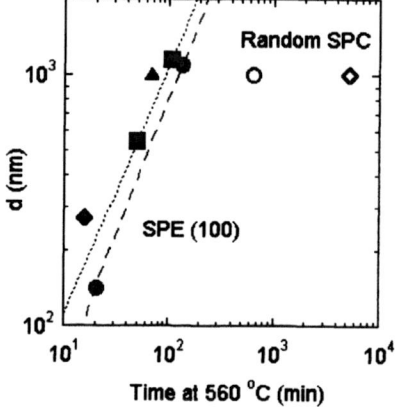

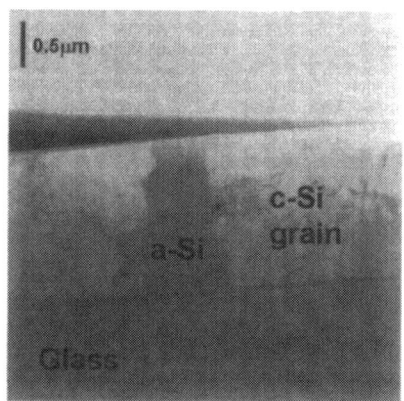

Fig. 3. Time dependence of the SPE thickness d in a-Si deposited by HWCVD (circles), PECVD (diamonds), and e-beam (squares) on (100)-oriented c-Si. Annealing temperature was 560°C. Open symbols denote random SPC times in 1 μm thick PECVD and HWCVD films at 560°C.

Fig. 4. Cross-sectional TEM micrograph of an e-beam deposited solar cell structure, partially crystallized at 600°C. The crystallization process was interrupted at the point in time when the in-situ optical reflectance $R(\lambda,t)$ interference fringe amplitude reached its minimum corresponding to a mixed-phase material.

This device structure was annealed at 600°C and monitored using in-situ $R(\lambda,t)$. The crystallization was interrupted at $t = 560$ min when the in-situ $R(\lambda)$ fringe pattern collapsed, signifying about 50% average crystallization by volume. The film was then examined by TEM (Fig. 4). TEM shows relatively large grains, above 1 micron, and a large amorphous region between them without signs of nucleation.

DISCUSSION

The observed large differences in SPC times shown on Fig. 1 are unlikely related to concentrations of common impurities in our films known to affect SPC [8]. For example, oxygen content variations are relatively small and do not correlate with SPC rate variations between the three types of films (Fig. 2). In HWCVD films, small trace amounts of filament metal atoms (W) might be present, typically at the SIMS detection limit of $\sim 10^{16}$ cm^{-3}, that could accelerate SPC. On the other hand, e-beam films crystallize about 3x slower than PECVD films at 600°C despite elevated O contents in the latter (Fig. 2). This indicates that either the SPC is strongly influenced by very small trace impurity contents or it is affected by different network microstructures produced by different deposition methods. We have suggested earlier [6, 7] that HWCVD a-Si crystallizes faster than PECVD a-Si due to its more ordered microstructure; yet, better understanding is needed about the origin and effect of this difference. It is interesting to note that in HWCVD a-Si showing fast SPC, doping has no effect on SPC (or even delays it in n-type material), while similar doping accelerates SPC in e-beam films.

The SPE results presented in Figure 3 suggest that despite large differences in random SPC times in three differently deposited films (Fig. 1), the *growth velocity* of the (100)-oriented crystalline front (at a given temperature) is approximately the same in all a-Si films studied. Since random SPC time is governed both by nucleation rate and growth velocity of the crystalline grains, we conclude that there must exist significant differences in *nucleation rate* between HWCVD, PECVD, and e-beam films. This narrows the range of possible microscopic mechanisms that could explain the different SPC times of Fig. 1. Random SPC growth proceeds mostly in growth directions other than (100), yet the above result suggests that the atomic rearrangements near the growth front are not strongly affected by the deposition method within this study.

Large variations in the nucleation rates in films prepared by different methods agree also with TEM studies of random SPC [8]. Recently, in-situ TEM revealed considerably larger differences in the nucleation rate as compared to the grain growth velocity between HWCVD films grown at different temperatures [10] and between HWCVD and PECVD films [11]. Moreover, as predicted by classical theory [8], different nucleation rates with similar crystal growth velocities (resulting in different ratios between these two parameters) are expected to result in different grain sizes in the three films studied. Figure 4 shows grains in e-beam a-Si, partially crystallized at 600°C. These grains are larger than 1 micron, separated by more than 1 micron, with a large amorphous region in between them. The amorphous zone shows no signs of crystallite nucleation. In contrast, grains 1 micron thick HWCVD films are separated by submicron distances and grow to sizes of only ~ 0.5 micron at 560°C [6] or to below 0.4 micron at 600°C [10]. In PECVD films crystallized at 600°C, grains approach 1 micron size (EBSD preliminary data, not shown). Thus, longer crystallization time results in larger grains, in agreement with the approximately identical growth rate suggested by the SPE experiment in Fig. 3. The large nucleation rate in HWCVD films results in small grains and fast crystallization, while in e-beam films the nucleation rate is likely the lowest amongst the three deposition methods studied, resulting in large grains and long crystallization times.

Also, despite the faster crystallization of the doped e-beam films in Fig. 1, there is no clear evidence in Fig. 4 that the crystallization started in the bottom n^{++} layer and propagated upwards into the weakly doped p-type layer. This is not surprising since the limited thickness of the highly doped regions considerably limits the chance of nucleation within these layers. Rather, it is the overall low nucleation density that resulted in the large grains.

From the results obtained in this work it appears important to understand how the material structure and composition of each type of amorphous silicon governs the nucleation. This is a task for further study.

SUMMARY

Solid-phase random crystallization proceeds at different rates in a-Si prepared by HWCVD, PECVD, and e-beam deposition. In contrast, solid-phase epitaxy in the same films on (100)-oriented polished Si wafer substrates proceeds at very similar speed of the crystalline growth front, for all films studied. This strongly suggests that the observed large differences in random SPC rates originate from different nucleation rates in these materials. This agrees with larger grain sizes observed for films that exhibit slower random SPC.

ACKNOWLEDGEMENTS

The authors thank E. Iwaniczko, M. Page, and L. Roybal for sample preparation, and M. Romero and F. Liu for EBSD measurements. This work is supported by the U.S. Department of Energy under Contract No. DE-AC36-99G010337. The work at the University of New South Wales is supported by the Australian Research Council.

REFERENCES

1.   M. A. Green, P. A. Basore, N. Chang, et al., Sol. Energy 77 (2004) 857.
2.   R. B. Bergmann, F. G. Shi and J. Krinke, 80 (1998) 1011.
3.   J. B. Boyce, R. T. Fulks, J. Ho, et al., Thin Solid Films 383 (2001) 137.
4.   A. G. Aberle, J. Cryst. Growth 287 (2006) 386.
5.   P. A. Basore, 31st IEEE Photovoltaic Specialists Conference (2005) 967.
6.   P. Stradins, D. L. Young, Y. Yan, et al., Appl. Phys. Lett. 89 (2006) 121921.
7.   D. L. Young, P. Stradins, Y. Xu, et al., Appl. Phys. Lett. 89 (2006) 161910.
8.   C. Spinella, S. Lombardo and F. Priolo, J. Appl. Phys. 84 (1998) 5383, and references therein.
9.   J. A. Roth, G. L. Olson, D. C. Jacobson, et al., Appl. Phys. Lett. 57 (1990) 1340.
10.   S. P. Ahrenkiel, B. Roy, A. H. Mahan, et al., Mat. Res. Soc. Proc. 910 (2006) 175.
11.   S. P. Ahrenkiel, (2007), South Dakota School of Mines and Technology, private communication.

Poster Session:
Thin Film Transistors

Mater. Res. Soc. Symp. Proc. Vol. 989 © 2007 Materials Research Society 0989-A17-02

Manufacturing of TFTs with High Deposition Rated Microcrystalline Silicon using Plasma Enhanced Chemical Vapor Deposition

Kyung-Bae Park, Ji-Sim Jung, Jong-Man Kim, Myung-kwan Ryu, Sang-Yoon Lee, and Jang-Yeon Kwon
Display Lab, Samsung Advanced Institute of Technology, Mt14-1, Nongseo-dong, Giheung-Gu, Gyeonggi-Do, Yongin-Si, 446-712, Korea, Republic of

ABSTRACT

Microcrystalline silicon was deposited on glass by standard plasma enhanced chemical vapor deposition using H_2 diluted SiH_4. Raman spectroscopy indicated a crystalline volume fraction of as high as 40% in films deposited at a substrate temperature 350°C. The deposition rate in films was as high as 10Å/sec. This process produced μc-Si TFTs with both an electron mobility of 10.9cm2/Vs, a threshold voltage of 1.2V, a subthreshold slop of 0.5V/dec at n-channel TFTs and a hole mobility of 3.2cm^2/Vs, a threshold voltage of -5V, a subthreshold slop of 0.42V/dec at p-channel TFTs without post-fabrication annealing.

INTRODUCTION

Recently, directly-deposited microcrystalline silicon(μc-Si) has been intensively introduced as attractive material for the TFT active layer [1][2]. Compared to a-Si TFTs, μc-Si TFTs have demonstrated higher electron mobility and Active Matrix display for driving devices. However disadvantages of μc-Si are inhomogeneity due to a complex crystalline structure composed of grains, amorphous silicon layer and grain boundaries, which limit the transport properties and the TFT sizes [3]. It is important to obtain higher crystalline Si layers on large size panel with high deposition rates for reducing the production costs of device. In addition, because of the large and well-established manufacturing base for a-Si TFTs, there is an overwhelming need for μc-Si TFTs fabricated using the conventional 13.56 MHz PE-CVD technique [4].Presently a deposition rate of a-Si for fabrication of TFT-LCD is as high as 15Å/sec. but a deposition rate of μc-Si is lower than deposition rate of a-Si.

In an effort to develop a high throughout process to fabricate a higher crystalline si using conventional PE-CVD with highly H_2 diluted SiH_4 plasma at 350°C. The μc-Si films were characterized using Raman scattering, UV reflectance, Scanning Electron Microscopy(SEM) and Transmission Electron Microscope(TEM).

EXPERIMENT

Microcrystalline Si films were deposited on Corning 1737 substrates at a substrate temperature 350°C in a RF Plasma Enhanced Chemical Vapor Deposition (PE-CVD) system with conventional discharge frequency 13.56MHz using the high purity SiH_4, H_2 as feed gas. In Raman scattering, SEM and TEM were used to determine the structure of the films. Both n-channel and p-channel thin film transistors are fabricated in a top-gated structure. All process temperature was done below 200°C except μc-Si deposition at 350°C. The buffer layer was silicon nitride deposited on Corning glass. Microcrystalline silicon was deposited from suitable process condition from mixture of SiH₄ and H₂ in PE-CVD reactor. The microcrystalline Si film was patterned into the island structure using mask lithography. The SiO_2 for gate insulator material and Nd incorporated Al for the gate electrode material were deposited, using ICP-CVD [12] and sputtering, respectively. Afterwards, the source and drain were doped using P^+ and B^+ ion implantation at a dose of 5×10^{15} cm^{-2}, and activated using low-energy density excimer laser irradiation. Both of the n-type and p-type TFTs characteristic were analyzed by measuring the basic transfer curve and the effective field effect mobility of the channel was calculated.

DISCUSSION

The microcrystalline silicon produced high deposition rate and high crystal volume

The microcrystalline Si films were prepared at 350°C by PE-CVD at various dilution conditions. The dilution gas and dilution ratio of each microcrystalline Si film is listed in Table 1.

Table 1. Deposition conditions, deposition rate and crystal fraction of microcrystalline Si films

Sample Number	Power(W)	Pressure (Torr)	H₂ Ratio [H₂/(H₂+SiH₄)]	Depo.rate (Å/sec)	Crystal Volume Fraction (%)
1	100	8	0.99	6.24	22
2	600	8	0.99	10.8	40
3	300	1	0.95	5.35	24
4	300	5	0.95	8.04	33
5	600	5	0.91	10.04	25
6	600	5	0.99	10.16	0.36

The observed data indicate that the plasma power as well as the hydrogen ratio are crucial parameters for microcrystalline phase formation. It is thought that the plasma power to break silicon-hydrogen bonding effectively. And both the pressure and H₂ ratio affect that microcrystalline silicon nucleation is due to enhanced surface diffusion of SiH₃ radicals due to the coverage of the surface by hydrogen. This enhanced surface mobility would then promote the incorporation of silicon into site of minimum energy and form stable crystallites. Moreover, etching and selective etching of the amorphous phase by atomic hydrogen [5][6]have also been considered, even though these processes do not account for the formation of stable nuclei [7]. Therefore the deposition rate was thought that high power and high pressure lead to maximum ion energies for deposition [8]. The condition of number 2 showed as high as 10.8Å/sec. and a crystalline volume fraction of over 40% through high power, high pressure and high H₂ ratio.

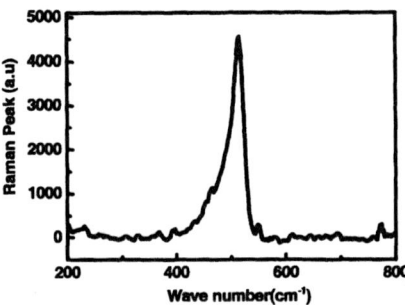

Fig. 1 Raman spectra of process condition of number 2

Fig. 1 shows Raman spectra on the same film. As indicated in the Fig.1, the sample produced direct deposition has a little amorphous contribution. Moreover, the smaller value of the width at half maximum ($15cm^{-1}$) indicates a larger grain in process condition number 2. Fig. 2 indicates 70~90nm of the silicon crystal grain in the same sample. Uniformly the silicon grain formed the whole area despite its 100nm as small as thickness.

Fig. 2 SEM image indicate a crystal silicon grain size of process condition number 2.

The crystalline growth structure and thickness of incubation layer of same microcrystalline Si film was observed by a cross-sectional transmission electron microscopy (TEM). Fig. 3 presents the TEM image of process condition number 2. In Fig.3 (a), at the bottom of the microcrystalline Si film, an amorphous phase or a mixed phase region called an incubation layer is observed. The thickness of this incubation layer measured by the TEM is 20nm. Above the incubation layer, many crystalline Si formations are observed. As shown in the Fig. 3 (a), the crystalline formation has clear Si lattice patterns which can be detected TEM image of well crystallized Si film typically. The obvious Si lattice pattern was shown in the Fig. 3 (b). It can be deduced that the amorphous Si layer was first deposited prior to the formation of crystalline components.

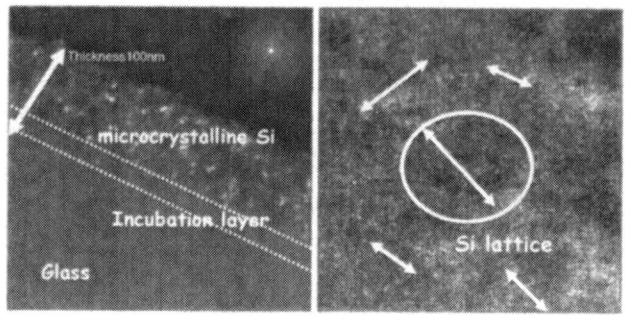

Fig. 3 (a) the cross sectional TEM image and electron diffraction. The film formed incubation layer and microcrystalline Si. (b) Crystal silicon lattice pattern in the microcrystalline Si region.

The amorphous Si, incubation layer, results from the competition between deposition rate and the supply of hydrogen to reach the crucial concentration for stable nuclei to form and grow. If the deposition rate is too high or the hydrogen flux too low, the critical concentration for nucleation will not be reached at the beginning of deposition but after some time and an amorphous Si will be left behind [9]-[11].

Fabrication of microcrystalline Si Thin film transistors

Fig. 4 shows the n-type and p-type TFT transfer characteristics (drain-current vs. gate voltage) using condition process number 2, respectively. Here, the data for the channel width(W)/ channel length (L) =20/20μm are shown. This process produced microcrystalline Si TFTs with both an electron mobility of 10.9cm^2/Vs, a threshold voltage of 1.2V, a subthreshold slop of 0.5V/dec at n-channel TFTs and a hole mobility of 3.2cm^2/Vs, a threshold voltage of -5V, a subthreshold slop of 0.42V/dec at p-channel TFTs at both V_d 0.1V without post-fabrication annealing.

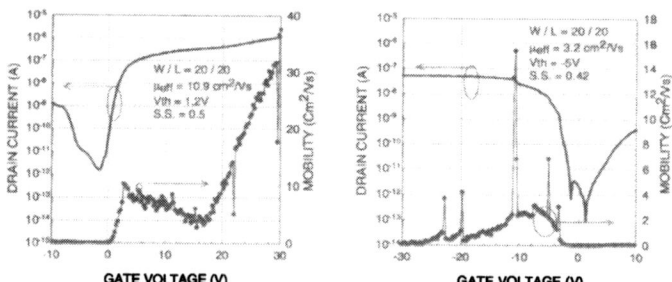

Fig. 4 (a) I_d-V_g transfer characteristic of n-type microcrystalline Si TFT (b) I_d-V_g transfer characteristic of p-type microcrystalline Si TFT.

These results demonstrated the feasibility of a directly-deposited process and silicon TFT technology for flat panel electronics. Especially AMOLED in the flat panel display will be promising technology without additional capital investment in the manufacturing.

CONCLUSIONS

In this paper, we have developed a method for the growth of high quality microcrystalline Si films on glass at 350°C in a conventional PE-CVD system. We have presented results on the effects of deposition conditions. The microcrystalline Si have a characteristic of high deposition rate over 10.8 Å/sec and crystalline volume fraction as high as 40%. Therefore the achievement of a high crystalline fraction leads to a high performance TFT characteristic, which can be explained by the about 90nm grain size of the films. Finally, we have also obtained stable TFTs with a linear mobility of 10.9cm^2/Vs and 3.2cm^2/Vs at n-type and p-type, respectively.

REFERENCES

1. P.Joubert, B.Loisel, Y. Chouan, and L. Haji, J. Electrochem. Soc. **134**, 2541 (1987)
2. S. Vepřek and V. Marecek, Solid State Electronics, **11**, 683 (1968).
3. I.D. French, S.C. Deane, P. Roca i Cabarrocas, IDW'01, Korea, 367–370, (2001)
4. F.Demichelis, G. Crovini, F. Giorgis, C.F. Pirri and E. Tresso, J. App. Phys. **79**, 1730, (1996)
5. C. C. Tsai, in: Amorphous Silicon and Related Materials, edited by H. Fritzsche (World Scientific, Singapore,1989), Vol. A, p. 123.
6. I. Solomon, H. Shirai, and N. Layadi, J. Non Cryst. Solids **164–166**, 989 (1993).
7. A. Matsuda, Thin Solid Films, **337**, 1 (1999).
8. J. E. Gerbi and J. R. Abelson, J. Appl. Phys. **89**, 1463 (2001).
9. S. Hamma, D. Colliquet, and P. Roca i Cabarrocas,. MRS Symp. Proc. Series, Vol. 507, 505 (1998).
10. J. Koh, A. S. Ferlauto, P. I. Rovira, C. R. Wronski, and R. W. Collins, Appl. Phys. Lett. 75, 2286 (1999)
11. A. Fontcuberta i Morral and P. Roca i Cabarrocas, Thin Solid Films **383**, 161 (2001).
12. J. S. Jung, J. Y. Kwon, Y. S. Park, H. S. Cho, K. B. Park, Y. X. Huaxiang, W. X. Xianyu and T. Noguchi, J. Korean Phys. Soc., **45**, S861 (2004)

Mater. Res. Soc. Symp. Proc. Vol. 989 © 2007 Materials Research Society   0989-A17-05

Effect of a Channel Length and Drain Bias on the Threshold Voltage of Field Enhanced Solid Phase Crystallization Polycrystalline Thin Film Transistor on the Glass Substrate

Won-Kyu Lee[1,2], Sang-Myeon Han[1], Sang-Geun Park[1], Young-Jin Chang[2], Kee-Chan Park[3], Chi-Woo Kim[2], and Min-Koo Han[1]

[1]School of Electrical Engineering, Seoul National University, Bldg. 130, Rm. 302, San 56-1, Sillim-dong, Gwanak-gu, Seoul, 151-742, Korea, Republic of
[2]LCD Business, Samsung Electronics Co., San-24, Nongseo-dong, Giheung-gu, Yongin, 449-711, Korea, Republic of
[3]Dept. of Electronics Engineering, Konkuk University, Seoul, 143-701, Korea, Republic of

ABSTRACT

We have fabricated a new magnetic field enhanced solid phase crystallization (FESPC) polycrystalline silicon (poly-Si) thin film transistors (TFTs), which show the excellent electrical characteristics and superior stability compared with hydrogenated amorphous silicon (a-Si:H) TFTs. The mobility (μ) and threshold voltage (V_{TH}) of p-type TFTs of which the channel width and length are 5 μm and 7 μm, respectively are 31.98 cm^2/Vs and -6.14 V, at $V_{DS} = $ -0.1 V. In the FESPC TFTs, the characteristics caused by grain boundary are remarkable due to large number of grain boundaries in the channel compared with poly-Si TFTs. The V_{TH} of the TFT which have 5 μm channel length is smaller than that of 18 μm channel length by 1.36 V, which is considerably large value. It is due to the large number of grain boundaries in the channel and the high lateral electric field. The grain boundary potential barrier height is decreased, when the large lateral electric field is applied (which is called DIGBL effect). As a result of increased mobility, the drain current is increased, and V_{TH} can be decreased. The activation energy (E_a) is strongly depended on the drain bias and the number of grain boundaries. E_a is decreased, caused by the large drain bias and/or smaller number of grain boundaries. This decreased E_a can be reduced V_{TH} due to increased the drain current. V_{TH} of p-type poly-Si TFT employing FESPC on the glass substrate is affected by channel length and V_{DS} due to energy barrier lowering effect at the grain boundary by increased lateral electrical field.

INTRODUCTION

It is well known that a-Si:H TFTs, which have a considerable attention as pixel elements for active matrix organic light-emitting diodes (AMOLEDs) displays due to an excellent uniformity up to large area [1-4]. However, the electrical performance of a-Si:H TFTs still needs to be improved in respect to mobility and threshold voltage (V_{TH}) shift [5]. Low temperature poly-Si TFTs based on excimer laser crystallization (ELC) method are stable to electrical bias and have high field-effect carrier mobility, but non-uniformity of TFT characteristics which caused by laser shot characteristics should be improved. In order to obtain uniform TFT characteristics and reliable devices without laser energy, several crystallization methods have been proposed [6-7]. Solid phase crystallization (SPC) of a-Si is a typical method of obtaining poly-Si film. And it has many advantages over ELC, such as simplicity, low cost, uniformity and large-area applicability.

However, heat treatments of SPC require high temperature above 600 °C and long annealing time (typically longer than 10h), preventing its use on thermally susceptible glass substrate [8-12]. There were a number of reports to decrease the crystallization temperature and to shorten the crystallization time [13-15].

Recently, FESPC on the glass substrate has been reported [16]. It was employing field-enhanced rapid thermal annealing where rapid thermal annealing of halogen lamps is combined with alternating magnetic fields. It could reduce the crystallization time, which made it possible to make poly-Si film employing SPC method on the glass substrate.

The purpose of our work is to report the electrical characteristics, such as V_{TH} and mobility, of p-type poly-Si TFTs employing FESPC method on the glass substrate with various channel length and V_{DS}. FESPC TFTs have small grains, (under 100 nm) due to very short crystallization time. The grain boundary area in the FESPC TFTs is considerably larger than that of other poly-Si TFTs, so that the characteristics at the grain boundary become more important. It can be contributed to understand FESPC TFTs and to enlarge low temperature poly-Si TFT applications.

EXPERIMENT

We have fabricated top gate coplanar poly-Si TFTs employing FESPC method on the glass substrate. Figure 1 shows the cross sectional view of top gate coplanar FESPC TFT. Silicon oxide (SiO_2) as buffer layer and a-Si:H layer were deposited by plasma enhanced chemical vapor deposition (PECVD) at 400 °C on the glass substrate. The a-Si film was crystallized by FESPC at 700 °C within 30 minutes. After poly-Si islands were defined, SiO_2 gate insulator was deposited and gate electrode was patterned. Next, ion doping followed and silicon nitride (SiN_x) inter-layer dielectric (ILD) was deposited. After contact holes were formed, source and drain (S/D) electrodes were patterned. The width of channel is 5 μm and channel length is varied from 5 μm to 18 μm.

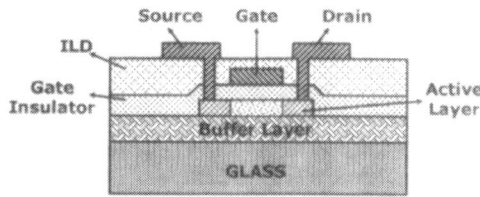

Figure 1. The cross sectional view of fabricated top gate coplanar FESPC TFT.

DISCUSSION

We have measured I-V characteristics of p-type FESPC TFTs, which have been varied with channel lengths and the several source-drain biases (V_{DS}) in order to evaluate lateral field effects. Figure 2 shows the transfer characteristics of p-type FESPC TFTs with the various channel length at $V_{DS} = -10.1$ V. The V_{TH} at $V_{DS} = -0.1$ V is defined as the gate voltage for a drive current of $(W/L) \times 10^{-8}$ A, and at $V_{DS} = -10.1$ and -20.1 V it is for a current of $(W/L) \times 10^{-7}$ A, where W and L are the channel width and length, respectively. Figure 3 shows that the channel length affects the V_{TH}. When drain bias is -10.1 V, the V_{TH} of 5 μm and 18 μm length are -5.22 V and -6.58 V, respectively.

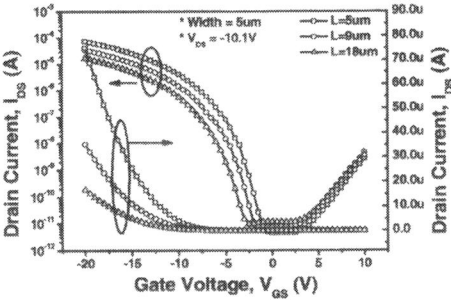

Figure 2. Transfer characteristics of fabricated p-type FESPC TFTs with various channel length.

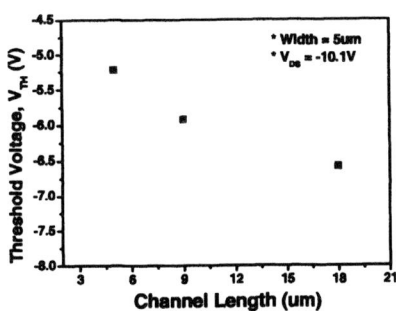

Figure 3. Dependence of V_{TH} on the channel length at $V_{DS} = -10.1$V.

The difference of V_{TH} (ΔV_{TH}) is 1.36 V, which is rather higher than in a-Si:H TFTs and poly-Si TFTs. In the short channel device, when the channel length becomes shorter, the V_{TH} decrease is caused by charge sharing effect or drain induced barrier lowering (DIBL) effect. However, though 5 μm length FESPC TFT is not a short channel device in the present TFT technology,

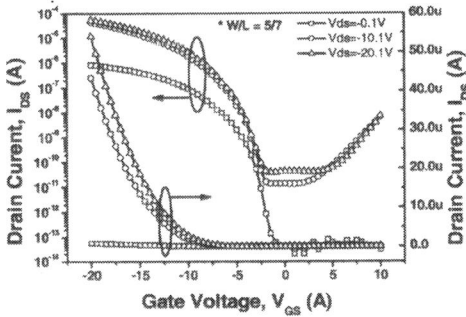

Figure 4. Transfer characteristics of fabricated p-type FESPC TFTs with various V_{DS}.

the reduction of V_{TH} at higher channel length is seen in the FESPC TFTs. In order to evaluate the lateral field effect on the FESPC TFTs, we changed V_{DS} from -0.1 V to -20.1 V. Figure 4 shows the transfer characteristics of FESPC TFTs with various V_{DS}. The channel width and length are 5 μm and 7 μm, respectively.

Figure 5 shows that as drain bias becomes more negative (p-type TFT), V_{TH} increase is reduced. The measured V_{TH} is varied from -6.14 V (at V_{DS} = -0.1 V) to -5.71 V (at V_{DS} = -20.1 V). The ΔV_{TH} is 0.43 V. These results show that V_{TH} is decreased in TFTs with short channel length and at higher V_{DS}.

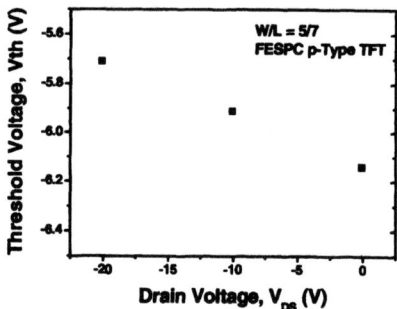

Figure 5. Dependence of V_{TH} on the V_{DS}. The channel width and length are 5 μm and 7 μm, respectively.

The mobility (μ) is 31.98 cm²/Vs, at V_{DS} = -0.1V, where the channel width and length are 5 μm and 7 μm, respectively. The small mobility compared with poly-Si TFT [17] is caused by small grain size. In the FESPC poly-Si film, its grain size is small (under 100 nm) due to very short crystallization time, so that number of grain boundaries is much larger than that of low temperature poly-Si. Therefore, the characteristics are due to large number of grain boundaries in the channel compared with poly-Si TFTs. Potential barrier at each grain boundary can be lowered with increasing drain bias, which is called drain induced grain barrier lowering (DIGBL) effect [18]. The barrier height $\Psi_B(V_i)$ on the lowered side of boundary can be obtained by solving Poisson's equation

$$\Psi_B(V_i) = \Psi_{B0}(V_i) - \frac{N_t^* \overline{\varepsilon_L}}{2n_{inv}} \qquad (1)$$

where $\Psi_{B0}(V_i)$ is the ith grain boundary potential barrier when the drain voltage is low, N_t^* is the ionized grain boundary trap density, $\overline{\varepsilon_L}$ is the average lateral electric field emanating from the drain, and n_{inv} is the inversion carrier concentration [18]. As the channel length is decreased and V_{DS} is increased, the lateral electric field is increased. The energy barrier lowering effect caused by lateral electric field in the FESPC poly-Si film is distinguished from other poly-Si TFTs (see equation 1).

The carrier mobility (μ) in a polycrystalline material is given by the following,

$$\frac{1}{\mu} = \frac{1}{\mu_G} + \frac{1}{\mu_{eff}} \qquad (2)$$

408

where μ_G is the mobility in the grains similar to the mobility in single crystal, and μ_{eff} is the effective mobility that has the following expression,

$$\mu_{eff} = \mu_o \exp\left(-\frac{E_B}{kT}\right) \qquad (3)$$

where μ_o is linearly increasing with grain size, and E_B is the energy barrier height [19]. Typically $\mu_G \gg \mu_{eff}$, then the carrier mobility (μ) strongly depend on the μ_{eff}. From equations 1 to 3, as the lateral electric field is increased, the barrier height is decreased, so that μ_{eff} and carrier mobility (μ) are increased. We have defined the V_{TH} as the gate voltage for a drive current of (W/L) × 10^{-8} A at low drain bias, and for a drive current of (W/L) × 10^{-7} A at high drain bias. Therefore, the drain current is increased according to increasing carrier mobility, so that V_{TH} can be decreased.

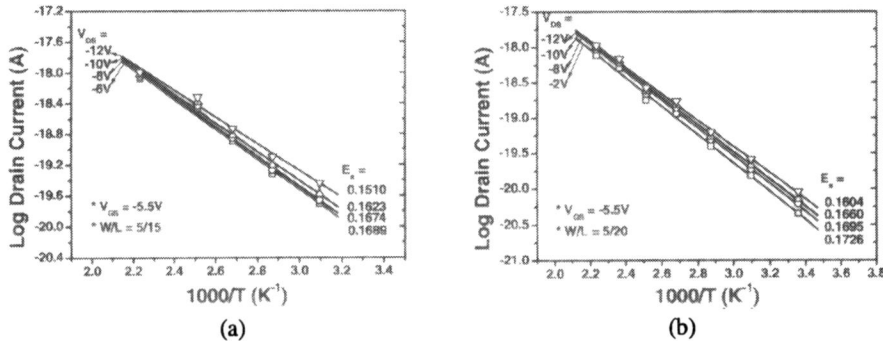

Figure 6. Arrhenius plot of the source-drain current versus the inverse of the absolute temperature for the fabricated FESPC TFTs with (a) W/L = 5/15 and (b) W/L = 5/20. The absolute temperature T varies from 293 K to 448 K.

Figure 6 shows the temperature dependence of the output characteristics for the p-type FESPC TFTs with different channel length at (V_{GS}-V_{TH}) ≈ 1 V. Malhi *et al.* [20] reported that the activation energy is dependent on the drain bias and the number of grain boundaries (see equations 4):

$$E_a = E_B - \frac{qV_D}{2N_g} \qquad (4)$$

where E_a is the activation energy, and N_g is the number of grains in the channel. Our measured data show that the increasing V_{DS} and short channel decreased the activation energy. That is well coincided with equations 4. For FESPC TFTs, the grain size is so small, that N_g dependence of E_a is considerably larger than that of other poly-Si TFTs. This energy barrier lowering in the whole channel results in the V_{TH} reduction at short channel and high V_{DS}.

CONCLUSIONS

We have fabricated and measured the p-type magnetic field enhanced SPC TFTs. Our results show that the V_{TH} of the TFTs which have 5 µm channel length (-5.22 V) is smaller than that of 18 μm channel length (-6.58 V) by 1.36 V. And the V_{TH} is varied from -6.14 V (at V_{DS} = -0.1 V) to -5.71 V (at V_{DS} = -20.1 V). The ΔV_{TH} is 0.43 V. Those differences are considerably large, which is due to the large number of grain boundaries in the TFT channel and the DIGBL effect. Due to the DIGBL effect, the carrier mobility and the drain current are increased, and V_{TH} is decreased. The number of grain boundaries has influence on the activation energy. In the case of the small number of grain boundaries, the activation energy is reduced. This decreased activation energy can reduce V_{TH} due to higher drain current. Therefore, the reduction of V_{TH} in the p-type FESPC TFTs on the glass substrate is affected by the number of grain boundaries, and the lateral electric field. This report could contribute to understand FESPC TFTs and to advance low temperature poly-Si TFT applications.

REFERENCES

1. A. K. Saafir, J. K. Chung, I. S. Joo, J. M. Huh, J. S. Rhee, S. K. Park, B. R. Choi, C. S. Ko, B. S. Koh, J. H. Hung, J. H. Choi, N. D. Kim, and K. H. Chung, *SID '05 Digest* 968 (2005).
2. J. J. Lih, C. F. Sung, C. H. Li, T. H. Hsiano, and H. H. Lee, *SID' 04 Digest* 1504 (2004).
3. J.-H. Lee, W.-J. Nam, K.-S. Shin, and M.-K. Han, *J. Non-Cryst. Solids* 352, 1719 (2006).
4. J.-H. Lee, J.-H. Kim, and M.-K. Han, *IEEE Electron Device Lett.* 26, 897 (2005).
5. V. D. Bui, Y. Bonnassieux, J. Y. Parey, Y. Djeridane, A. Abramov, P. R. Cabarrocas, and H. J. Kim, *SID '06 Digest* 204 (2006).
6. S.-W. Lee, and S.-K. Joo, *IEEE Electron Device Lett.* 17, 160 (1996).
7. S. Y. Yoon, K. H. Kim, C. O. Kim, J. Y. Oh, and J. Jang, *J. Appl. Phys.* 82, 5865 (1997).
8. T. Sameshima, *J. Non-Cryst. Solids* 227-230, 1196 (1998).
9. Y. Zhao, W. Wang, F. Yun, Y. Xu, X. Liao, Z. Ma, G. Yue, and G. Kang, *Sol. Energy Mater. Sol. Cells* 62, 43 (2000).
10. S. Y. Yoon, S. J. Park, K. H. Kim, and J. Jang, *Thin Solid Films* 383, 34 (2001).
11. J. H. Ahn, J. N. Lee, Y. C. Kim, and B. T. Ahn, *Curr. Appl. Phys.* 2, 135 (2002).
12. S. H. Park, H. J. Kim, K. H. Kang, J. S. Lee, Y. K. Choi, and O. M. Kwon, *J. Phys. D: Appl. Phys.* 38, 1511 (2005).
13. Y. Kawazu, H. Kudo, S. Onari, and T. Arai, *Jpn. J. Appl. Phys.* 129, 729 (1990).
14. C. Hayzelden, J. L. Bastone, and R. C. Cammarata, *Appl. Phys. Lett.* 60, 225 (1992).
15. Y. J. Choi, W. K. Kwak, S. J. Park, S. J. Yoon, C. O. Kim, and J. Jang, *SID' 99 Digest* 508 (1999).
16. B. S. So, Y. H. You, H. J. Kim, Y. H. Kim, J. H. Hwang, D. H. Shin, S. R. Ryu, K. Choi, and Y. C. Kim in *Application of Field-Enhanced Rapid Thermal Annealing to Activation of Doped Polycrystalline Si Thin Films*, (Mater. Res. Soc. Proc. 862, 2005) pp. 275-280.
17. I.-H. Song, S.-H. Kang, W.-J. Nam, and M.-K. Han, *IEEE Electron Device Lett.* 24, 580 (2003).
18. G.-Y. Yang, S.-H. Hur, and C.-H. Han, *IEEE Trans. Electron Devices* 46, 165 (1999).
19. O. Bonnaud, T. Moammed-Brahim, and D. G. Ast in *Thin Film Transistors – Materials and Processes Volume 2 – Polycrystalline Silicon Thin Film Transistors*, edited by Y. Kuo (Kluwer Academic Publishers, New York, 2004), p. 37.
20. S. D. S. Malhi, H. Shichijo, S. K. Banerjee, R. Sundaresan, M. Elahy, G. P. Pollack, W. F. Richardson, A. H. Shah, L. R. Hite, R. H. Womack, P. K. Chatterjee, and H. W. Lam, *IEEE J. Solid-State Circuits* SC-20, 178 (1985).

Poster Session:
Solar Cells

Mater. Res. Soc. Symp. Proc. Vol. 989 © 2007 Materials Research Society   0989-A18-02

Influence of Amorphous Layers on Performance of Nanocrystalline/Amorphous Superlattice Si Solar Cells

Atul Madhavan[1], Debju Ghosh[1], Max Noack[2], and Vikram Dalal[1]

[1]Electrical and Computer Engineering, Iowa State University, Ames, IA, 50011
[2]Microelectronics Research Center, Iowa State University, Ames, IA, 50011

ABSTRACT
    Nanocrystalline Si:H is an important material for solar cells. The electronic properties of the material depend critically upon the degree of crystallinity, the efficacy of passivation of the grain boundaries and concentration of impurities, particularly oxygen, in the material. In this paper, we examine different degrees of passivation of grain boundaries by amorphous Si, by deliberately introducing various thicknesses of amorphous tissue layers at the grain boundaries. The device structure consisted of a p+nn+ cell on stainless steel. The base n layer was fabricated using alternating layers of amorphous (a-Si) and crystalline (nc-Si) phases, creating a superlattice structure. The thicknesses of the amorphous and crystalline phases were varied to study their influence on structural and electrical properties such as grain size and diffusion length. We find that <111> grains continued to be nucleate independently of the thickness of the amorphous layer, but the size of <220> grains decreased when the thickness of the tissue layer became very large. We also find that as the thickness of the amorphous tissue layer increased, the quantum efficiency at 800 nm decreased and the open circuit voltage increased. For significant thickness of amorphous layer (>10nm), the transport properties degrade dramatically, causing an inflection in the I-V curve, probably because of difficulty of holes tunneling across the barriers.

INTRODUCTION
    Nanocrystalline Si:H is an increasingly important material for solar cells, thin film transistors, integrated sensors etc. [1-6]. It consists of small nano-sized grains of Si surrounded by a thin amorphous Si:H tissue. Numerical simulations show that H is bonded at the grain boundary [7], and there may also be amorphous tissue surrounding the grain. The electronic properties can be expected to depend critically upon recombination at the grain boundaries and transport through the surrounding amorphous tissue. In particular, since there is expected to be a considerable mismatch between the amorphous and crystalline valence bands, holes photo-generated within the crystalline grain may have difficulty tunneling through the amorphous phase if it is too thick. In addition to this effect, one may expect that the crystalline grain structure itself may change as we increase the thickness of the amorphous tissue layer. This effect may arise because the growth of thermodynamically preferred grains depends upon the template effect related to thickness and crystallinity of the preceding layers, whereas those grains which nucleate randomly do not need a template. In this paper, we explore the crystalline structure and transport of holes through a mixed phase nanocrystalline device, where the thickness of the amorphous tissue is deliberately varied by using an alternating layer growth technique, creating a superlattice of alternating amorphous and nanocrystalline layers.

GROWTH TECHNIQUE

Both amorphous and nanocrystalline layers were grown using a VHF parallel plate plasma at frequencies in the range of 45-50 MHz. The materials and devices were both grown on stainless steel substrates. Both materials and devices consisted of alternating layers of amorphous and nanocrystalline materials. The deposition pressure was 100 mTorr and the temperature during growth was 250 °C. For amorphous layers, the power was small (~2.5 Watts) and for nanocrystalline layers, the VHF power was ~25 Watts. The power could be switched between the two values using a modulated HP 8116 function generator as the input source. Experiments showed that switching of power levels changed the crystallinity of the layers and that we did not need to change the hydrogen dilution. The thickness of the nanocrystalline layer was varied between 3 and 10 nm, and that of amorphous layers, between 1 and 10 nm. The total thickness of the film , and of the base n layer in the p^+nn^+ cell was about 0.7 micrometer.

STRUCTURAL PROPERTIES

The structural properties were measured using Raman spectroscopy, with a Renishaw instrument. The wavelength of the laser was 488 nm. In addition, x-ray diffraction measurements were also done on the films.

In Fig. 1, we show the results of the x-ray diffraction measurements on one of the films. It is seen that both <111> and <220> grains are present in the film. As we vary the thickness of the amorphous layer, the x-ray diffraction changes, as shown in Fig. 2. As seen from that figure, the <111> grains remain essentially of the same size, but the <220> grains decrease in size. From these data, we postulate that the <111> grain is the randomly nucleated grain, which does not need a template, whereas the thermodynamically preferred grain is the <220>, which needs a template to grow. Having a thick amorphous Si layer interrupts the growth of that grain. This result confirms our earlier interpretation, obtained from continuously grown films, where we showed that the grain size of <111> remained relatively independent of growth temperature, whereas the <220> grain size increased at higher temperatures [8], thereby suggesting that <111> was randomly nucleated whereas <220> was the thermodynamically preferred grain.

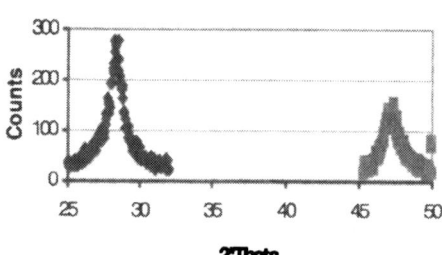

Fig. 1 x-ray diffraction spectrum of a superlattice film when the amorphous Si layer was thin (~1 nm). Both <111> and <220> grains are present.

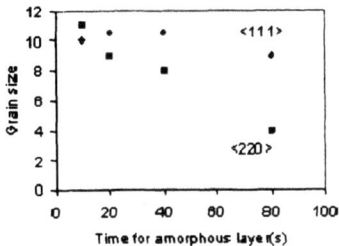

Fig. 2 Changes in grain size as the time for deposition of the a-Si superlattice layer is varied. <111> grain size does not change, whereas the <220> grain size decreases systematically.

The corresponding Raman spectrum is shown in Fig. 3, where we show what happens to the ratio of crystalline 520 cm^{-1} peak to the amorphous shoulder at ~485 cm^{-1} as a function of thickness of the amorphous layer for a fixed thickness of the nanocrystalline layer. Very clearly, from this figure, we can deduce that when the time for amorphous layer becomes large (60 s), the 520 cm^{-1} peak is reduced, again suggesting that the film is becoming less crystalline.

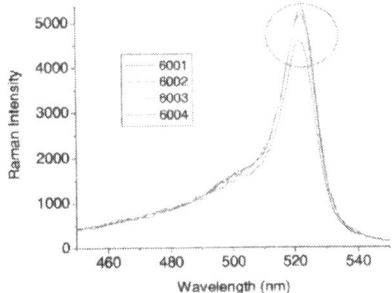

Fig. 3 Raman spectrum of the superlattice films as the thickness of a-Si layer changes. 6001 represents the thinnest layer (~1nm), and 6004 the thickest (~8-9 nm).

DEVICE RESULTS

Standard p$^+$nn$^+$ devices were fabricated on stainless steel substrates. An ITO contact was deposited on the p+ layer to act as a transparent electrode. The n+ layer was generally amorphous, followed by the base n layer which was the superlattice layer composed of alternating amorphous and nanocrystalline layers. The top p+ layer was nanocrystalline topped with a thin amorphous p+ layer to prevent oxygen penetration. Both p+ and n+ layers were deposited in another reactor to prevent cross-contamination of the base n layer. The thickness of the amorphous layer in the superlattice was deliberately varied by varying the deposition time. The total thickness of the nano-Si part of the superlattice layer was again kept similar to what was utilized for structural measurements, so that one can make a meaningful comparison of the long wavelength quantum efficiencies of the different devices.

In Fig. 4, we show the I-V curve for the device when the amorphous layer is kept thin (2 nm) and for the device where the amorphous layer is thick (8 nm). It is clear that a standard I-V curve results without any inflection points when the a-Si layer in the superlattice is thin (sample 10024),whereas the device with the thick a-Si layer (# 9947) shows a distinct inflection point. The inflection point arises in the curve because there is a hole transport problem in this device. This transport problem arises because both a-Si and nc-Si have approximately the same electron affinity [9], which means that the valence band of the larger bandgap a-Si is significantly lower in energy than the valence band of nc-Si. Therefore, the hole generated in the nc-Si layer gets trapped in that well. Under zero or reverse bias, the electric field in the device is large enough to allow the transport of holes through tunneling, but under forward bias, near open circuit voltage conditions, the electric field is reduced and the hole cannot escape from the well. When the thickness of the a-Si layer is low, the hole can tunnel through that amorphous layer, but not when it is high.

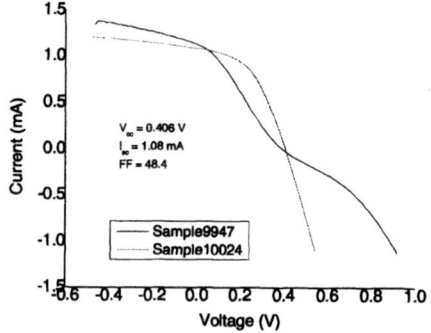

Fig. 4 I-V curves for two devices, one with thick a-Si layer in the superlattice (#9947) and one with thin a-Si layer(#10024).

In Fig. 5, we show the quantum efficiency and open circuit voltage at 800 nm of the devices as the thickness of a-Si layer is increased. It is sent that there is a progressive decrease in QE at 800 nm, particularly when the thickness of the amorphous layer begins to exceed ~5 nm, implying that the collection of holes generated within the nano Si layers is becoming difficult. As the amorphous phase becomes more dominant, one expects an increase in open-circuit voltage, particularly for large a-Si superlattice layer thicknesses, which is seen in Fig. 5.

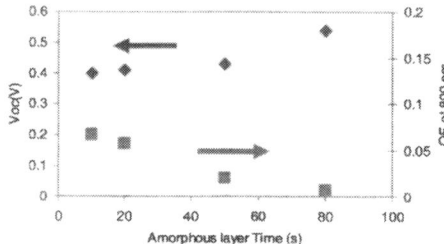

Fig. 5 Influence of varying the thickness of a-Si layer of superlattice solar cells on QE at 800 nm (squares)and open circuit voltage (diamonds)

CONCLUSIONS

From our results, we can conclude that the structural properties of the film depend upon the template effect of underlying layers. The randomly nucleated grain, <111> remains relatively independent of the template, whereas the thermodynamically preferred grain, <220> needs an underlying crystalline template to grow. A thick amorphous interlayer suppresses the growth of <220> grain. The device results show that for relatively thin amorphous layers, the holes can tunnel through, whereas for thicker layers, they get trapped with in the nanocrystalline quantum well and need a field assist to get out. Further experiments are in progress to understand the nature of recombination and hole transport properties in such superlattice structures.

ACKNOWLEDGEMENTS

This work was partially supported by NSF. We also thank our colleague Kay Han for her help. One of us, Atul Madhavan, was supported by a Catron Fellowship.

REFERENCES

1. Kenji Yamamoto, Masashi Yoshimi, Yuko Tawada, Susumu Fukuda, Toru Sawada, Tomomi Meguro, Hiroki Takata, Takashi Suezaki, Yohei Koi, Katsuhiko Hayashi Solar Energy Mater. and Solar Cells, $\underline{74}$, 449 (2002)
2. A. V. Shah, J. Meier, E. Vallat-Sauvain, N. Wyrsch, U. Kroll, C. Droz and U. Graf, Solar Energy Mater. and Solar Cells, $\underline{78}$, 469 (2003)
3. B. Rech, O. Kluth, T. Repmann, T. Roschek, J. Springer, J. Müller, F. Finger, H. Stiebig and H. Wagner, Solar Energy Mater. and Solar Cells, $\underline{74}$, 439 (2002)
4. K. Yamamoto, A. Nakajima, M. Yoshimi, T. Sawada, S. Fukuda, K. Hayashi, M. Ichikawa, Y. Tawada, Proc. 29th. IEEE Photovolt. Spec. Conf.(2002), p.1110
5. A. Sazonov, D. Striakhilev, C-H Lee, and A. Nathan: Proceedings of the IEEE, $\underline{93,}$ 1420 (2005).
6. I-C Chen and S. Wagner: IEE Proc.- Circuits, Devices Syst., $\underline{150}$, 339(2003).
7. B. C. Pan and R. Biswas, J. Appl. Phys. $\underline{96}$, 6247(2004)
8. Vikram L. Dalal, Kamal Muthukrishnan, Satya Saripalli, Dan Stieler and Max Noack, Proc. MRS, $\underline{910}$, 293(2006)
9. S. R. P. Silva, R. D. Forrest, J. M. Shannon, and B. J. Sealy, J. Vac.Sci. and Tech. B, $\underline{17}$,596(1999)

Mater. Res. Soc. Symp. Proc. Vol. 989 © 2007 Materials Research Society   0989-A18-03

Improved Efficiency of Single Junction Microcrystalline Silicon n-i-p Solar Cells with an i-Layer Made by Hot-Wire CVD

Hongbo Li, Ronald H.J. Franken, Robert L. Stolk, C. H.M. van der Werf, Jan-Willem A. Schuttauf, Jatin K. Rath, and Ruud E.I. Schropp

Surfaces, Interfaces and Devices, Faculty of Science, Utrecht University, P.O.Box 80.000, Utrecht, 3508TA, Netherlands

ABSTRACT

The influence of the surface roughness of Ag/ZnO coated substrates on the AM1.5 J-V characteristics of microcrystalline silicon (μc-Si:H) solar cells with an i-layer made by the hot-wire chemical vapour deposition (HWCVD) technique is discussed. Cells deposited on substrates with an intermediate *rms* roughness show the highest efficiency. When using *reverse* hydrogen profiling during i-layer deposition, an efficiency of 8.5% was reached for single junction μc-Si:H n-i-p cells, which is by far the highest for μc-Si:H n-i-p cells with a hot-wire i-layer.

INTRODUCTION

Intrinsic hydrogenated microcrystalline silicon (μc-Si:H) have been proven to be suitable for use in thin film silicon cells. For single junction μc-Si:H solar cells, independently confirmed efficiencies over 10% have been reported [1, 2], whereas multi-junction cells with a μc-Si absorber in the bottom cell have yielded an efficiency of around 12% [1,3]. Although most of the groups use plasma enhanced chemical vapour deposition (PECVD) for the deposition of intrinsic μc-Si:H [4], it is of interest to study the hot-wire CVD technique [5], because of its potential to be faster and more cost-effective than conventional PECVD, while scaling-up for industrial production is expected to be straightforward.

Almost 10 years ago, we initially developed highly crystalline silicon material at a high substrate temperature (~500 °C). Samples made on Corning glass showed a compact, singly oriented poly-Si structure with purely intrinsic electrical transport properties [6, 7]. Single junction solar cells made on bare stainless steel foil with 1.2 μm of this material as the absorber layer showed a remarkably high short-circuit current density (J_{sc}) of over 19 mA/cm^2 [8], considering no textured back reflector was used at that time. However, the cell open circuit voltages (V_{oc}) and fill factors (FF) obtained at that time are low compared to those of μc-Si:H cells developed nowadays. These differences can be attributed to the high density of electrical defects present in the i-layer [6]. Though this poly-Si material did not yet result in high efficiency solar cells, crucial knowledge of silicon growth with the hot-wire technique was obtained [8 -11].

Lowering the substrate temperature, the Jülich group obtained significantly higher solar cell efficiencies with hot-wire deposited μc-Si:H i-layers: an initial efficiency of 7.5% was reported for an n-i-p cell with an i-layer deposited at a substrate temperature around 260 °C [12]. Successive study further improved the efficiency of p-i-n cells to 9.4% [13], which is the highest published efficiency for single junction hot-wire μc-Si:H thin film solar cells so far. At Utrecht

University, we have been concentrating on developing hot-wire μc-Si:H solar cells with a n-i-p structure. Due to the lack of a proper back reflector, the solar cell efficiency was relatively low in the beginning [14]. Upon using ZnO coated Asahi-U type TCO glass as textured substrates, the efficiency of n-i-p μc-Si:H cell reached 8.4%, with a J_{sc} of 23.60 mA/cm^2 [15]. By further improvement on the i-layer quality by introducing a reverse H_2 profiling technique [16] and implementing an in-house developed Ag/ZnO rough substrate [17], efficiency of 8.5% has been reached, with a J_{sc}, V_{oc} and FF of 23.4 mA/cm^2, 0.545 V and 0.668, respectively [16]. This is the highest efficiency so far for a single junction μc-Si:H n-i-p cell with a hot-wire CVD deposited i-layer.

In this contribution, we discuss a few key factors to produce high quality μc-Si:H n-i-p cells with a hot-wire deposited i-layer.

EXPERIMENT

The solar cell structure discussed in this contribution is as the following: substrate (stainless steel or Corning 1737 glass) /rough Ag/ZnO /n-type μc-Si:H /intrinsic μc-Si:H /buffer /p-type μc-Si:H /ITO /Au (gridlines). Ag layers were deposited by rf magnetron sputtering at an elevated temperature to form a rough surface, ZnO and ITO by sputtering at room temperature. The silicon layers were deposited in a multi-chamber UHV system. The intrinsic layers were deposited by means of HWCVD, whereas for the doped layers in the cell PECVD at 13.56 MHz was used. Two straight tantalum wires (0.5 mm in diameter) were used for the hot-wire decomposition of the source gasses. They were mounted in parallel and were about 4 cm away from the substrate. Throughout the deposition, a constant current of 10.5 A was used, yielding a wire temperature around 1850 °C, as measured with a pyrometer in vacuum without source gas. No extra substrate heating was used, which results in a substrate temperature of around 250 °C when the system is in an equilibrium condition. Au grids are evaporated in vacuum. The active cell areas were defined by the geometry of the ITO top contact and the Au-grid, and were around 0.13 cm^2.

The substrate surface *rms* roughness values were estimated from Atomic Force Microscope (AFM) pictures of Ag/ZnO samples deposited on Corning 1737 glass. The thickness of each silicon layer was measured on accompanying glass substrates using a surface step profiler (Dektak) and calibrated with cross-sectional transmission electron microscope (XTEM) pictures. The average deposition rate for intrinsic μc-Si:H layers was calculated accordingly and was typically around 2 Å/s. The current density – voltage (J-V) characteristics of the solar cells were measured at 25 ℃ under AM1.5 100 mW/cm^2 white light generated by a dual beam solar simulator (WACOM). A 0.3 mm thick SS mask used for depositing the ITO top contact was used during the measurement to have a precise definition of the cell area. Due to the finite thickness of the mask and the accompanying shadow effect, the thus measured J_{sc} is slightly underestimated.

RESULTS AND DISCUSSION

Influence of rough substrate on solar cell output performance

The key to increasing the efficiency of a μc-Si:H solar cell is an increase of J_{sc} by more efficient light-trapping, which currently means the use of a properly designed rough substrate. Figure 1a) to d) show the correlation between the surface root mean square (rms) roughness of a Ag/ZnO coated Corning 1737 glass substrate and AM1.5 J-V characteristics of a series of single junction μc-Si:H n-i-p cells. Except for the difference in substrate roughness, all deposition parameters were identical, with a constant H_2 dilution ratio R_H of 0.953 (R_H is defined as the gas flow ratio $H_2/(SiH_4 + H_2)$), and with an i-layer thickness of around 1.3 μm. From Figure 1d, it is clear that with the use of a rough substrate, the cell efficiency greatly increases, from less then 4% on a nearly flat substrate till around 7% on a substrate with an *rms* roughness of ~ 70 nm. This is obviously because of the increase of J_{sc} (Figure 1a) due to the enhanced light-scattering effect brought about by the substrate roughness. The efficiency reaches its highest value and saturates at *rms* values of 60 nm to 90 nm. With a further increase of the substrate roughness, the cell efficiency gradually goes down (Figure 1 d), similar to the trend of J_{sc} (Figure 1a). The FF for this set of samples shows a scattered distribution, but its maximum appears for cells made on a medium-rough substrate (Figure 1b, *rms* between 60 nm - 90 nm).

The limitation in J_{sc} can be explained by plasmon absorption of the rough Ag substrate, which increases with increasing surface roughness [17], and by the formation of structural defects in the i-layer of the cells [18]. The presence of electronic defects in the μc-Si:H i-layer is evident from the dark J-V properties, showing an increase in both the diode quality factor and the dark reverse saturation current density with increasing *rms* roughness [18]. As a result, a continuous decrease in cell V_{oc} is observed (Figure 1c). The increased density of electrical defects in the i-layer of the cells deposited on rougher substrates can also result in a decrease in FF, especially when the density of photogenerated carriers does not further increase with increasing substrate roughness (Figure 1d, *rms* > around 90 nm).

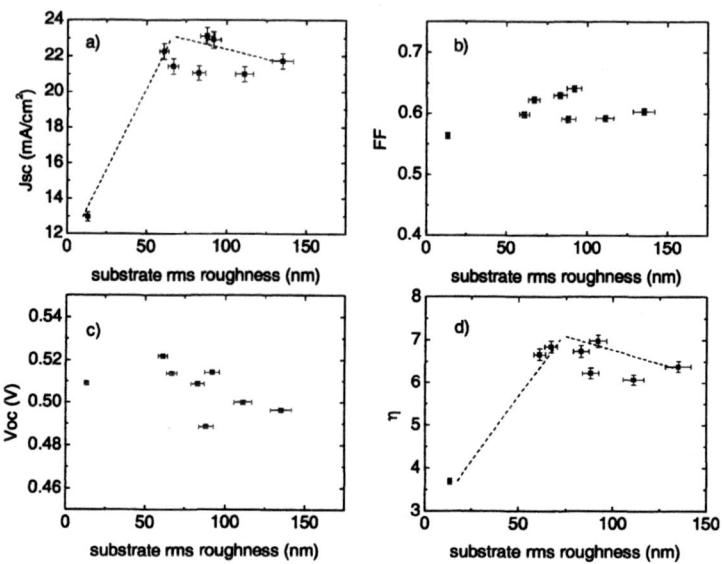

Figure 1 AM1.5 J-V characteristics of μc-Si:H n-i-p solar cells with hot-wire i-layers and PECVD doped layers. The cells are made on Ag/ZnO coated Corning 1737 glass substrate.

Increase of cell efficiency by using a reversed hydrogen profiling technique

One of the key parameters for a high efficiency μc-Si:H cell is the i-layer crystallinity. It has been shown that the maximum cell efficiency always appears when the intrinsic μc-Si:H is made near the phase transition regime between amorphous and microcrystalline structure [19, 20]. Since it was found that the crystallinity of μc-Si:H thin films made by PECVD tends to increase with layer thickness [21], to have an accurate control on the i-layer crystallinity, a hydrogen profiling technique was introduced, viz. decreasing the hydrogen dilution ratio during the i-layer deposition in order to keep the i-layer crystallinity unchanged. This has resulted in a large improvement on the efficiencies of cells made by PECVD [22]. For μc-Si:H n-i-p cells with an i-layer made by hot-wire CVD, we showed that the crystallinity may evolve in the other direction, i.e. it can *decrease* with the layer thickness [23], especially when the filament current is held constant during the i-layer deposition [16]. This is probably related to the progressing filament aging during the deposition of μc-Si:H i-layer. We therefore introduced a technique in which the hydrogen dilution ratio is *increased* instead of decreased. The preliminary deposition shows a promising result. For cells containing a ~ 2 μm thick μc-Si:H i-layer, which includes a two step reverse H_2 profiling, the first 1 μm at $R_H = 0.948$ and the second 1μm at $R_H = 0.953$, a well-calibrated AM1.5 efficiency of 8.5% has been obtained. The J-V parameters of this cell and two reference cells (at a constant R_H of 0.953 and 0.948, respectively) are shown in Table 1 (sample 1, 2, and 3). Since the cell structure and the substrate roughness for this cell are not optimal yet, we expect an increase of the efficiency after further optimization work.

Table 1 AM1.5 J-V characteristics of single junction n-i-p solar cells with i-layers deposited by hot-wire CVD.

Sample No.	remarks	Efficiency [%]	J_{sc} [mA/cm^2]	V_{oc} [V]	FF
1	~2 μm i-layer, constant R_H =0.953	7.5	23.5	0.498	0.641
2	~2 μm i-layer, constant R_H =0.948	6.5	21.2	0.517	0.594
3	~2 μm i-layer, two step H$_2$ profiling	8.5	23.4	0.545	0.668

SUMMARY AND CONCLUSIONS

Two key factors influencing the efficiency of μc-Si:H n-i-p solar cells with a hot-wire i-layer are discussed. To obtain a maximum AM1.5 J-V performance, the substrate roughness has to be carefully chosen. On substrates with an intermediate *rms* roughness, single junction μc-Si:H n-i-p cells showed the maximum efficiency. Upon using a reverse hydrogen profiling technique in the deposition of i-layer, efficiency of 8.5% has been reached, which is by far the highest for hot-wire μc-Si:H n-i-p cells.

ACKNOWLEDGMENTS

The research was financially supported by Netherlands Agency for Energy and Environment (NOVEM, The Netherlands).

REFERENCES

1. Solar cell Efficiency Tables (version 29). Prog. Photovolt.: Res. Appl. 2007; 15: 35-40.
2. Yamamoto K, Toshimi M, Suzuki T, Tawada Y, Okamoto T, Nakajima A., MRS Spring Meeting, April 1998; San Francisco.
3. Yoshimi M, Sasaki T, Sawada T, Suezaki T, Meguro T, Matsuda T, Santo K, Wadano K, Ichikawa M, Nakajima A, Yamamoto K. Conference Record, 3rd World Conference on Photovoltaic Energy Conversion, Osaka, May 2003; 1566-1569.
4. See, for example, 4th World Conferences on Photovoltaic Energy Conversion, or 31st IEEE Photovoltaic Specialists Conferences and the conferences hold before.
5. R. E.I. Schropp, Thin Solid Films 451 –452 (2004) 455–465.
6. J.K. Rath, F.D. Tichelaar, H.Meiling and R.E.I. Schropp, Rat. Res. Soc. Symp. Proc. 507 (1998) 879.
7. Schropp, R.E.I., Feenstra, K.F., Molenbroek, E.C., Meiling, H., Rath, J.K. , Philosophical Magazine B: Physics of Condensed Matter; Statistical Mechanics, Electronic, Optical and Magnetic Properties, 76 (1997)309-321.
8. J.K. Rath, F.D. Tichelaar, R.E.I. Schropp, Solid State Phenomena 67-68 (1999) 465.
9. P. A. T. T. van Veenendaal, O. L. J. Gijzeman, J. K. Rath, and R. E. I. Schropp, Thin Solid Films 395 (2001) 194-197.

10. C. H. M. van der Werf, P. A. T. T. van Veenendaal, M. K. van Veen, A. J. Hardeman, M. Y. S. Rusche, J. K. Rath and R. E. I. Schropp, Thin Solid Films 430 (2003) 46-49.

11. Li, H., Stolk, R.L., Van Der Werf, C.H.M., Rusche, M.Y.S., Rath, J.K., Schropp, R.E.I., Thin Solid Films 501(2006) 276-279.

12. Stefan Klein, Johannes Wolff, Friedhelm Finger, Reinhard Carius, Heribert Wagner and Martin Stutzmann, Jpn. J. Appl. Phys. 41 (2002) L10 – L12.

13. S. Klein, T. Repmann and T. Brammer, Solar Energy 77 (2004) 893-908.

14. M. K. van Veen, C. H. M. van der Werf, J. K. Rath and R. E. I. Schropp, Thin Solid Films 430 (2003) 216-219.

15. Robert L. Stolk, Hongbo Li, Ronald H. Franken, Karine H.M. van der Werf, Jatindra K. Rath and Ruud E.I. Schropp, MRS Symp. Proc. 910, A.26.03, (2006).

16. H. Li, R.H. Franken, R.L. Stolk, C.H.M. van der Werf, J.K. Rath, R.E.I. Schropp, 4th international Conference on Hot-Wire CVD (Cat-CVD) process, October 4-8, 2006, Takayama, Gifu, Japan.

17. R.H. Franken , Phd thesis, Utrecht University, 2006.

18. Hongbo Li, Ronald Franken, Robert Stolk, Jatin Rath and Ruud E. I. Schropp, to be published.

19. O. Vetterl, F. Finger, , R. Carius, P. Hapke, L. Houben, O. Kluth, A. Lambertz, A. Mück, B. Rech and H. Wagner, Sol. Energ Mat. and Sol. C. 62 (2000) 97-108.

20. Y. Mai, S. Klein, R. Carius, J. Wolff, A. Lambertz, F. Finger and X. Geng, J. Appl. Phys. 97 (2005) 114913.

21. R.W. Collins and A. S. Ferlauto, Curr. Op. Sol. Stat. & Mat. Sci. 6 (2002) 425, and R.W. Collins, A.S. Ferlauto, G.M. Ferreira, Chi Chen, Joohyun Koh, R.J. Koval, Yeeheng Lee, J.M. Pearce, C.R. Wronski, Sol. Energ. Mat. Sol. Cell. 78 (2003) 143.

22. B. Yan, G. Yue, J. Yang, and S Guha, D. L. Williamson, D. Han, C. Jiang, App. Phys. Lett. 85 (2004) 1955.

23. M. van Veen, PhD thesis, Utrecht University, The Netherlands, 2003.

Mater. Res. Soc. Symp. Proc. Vol. 989 © 2007 Materials Research Society   0989-A18-04

Resistive Losses at c-Si/a-Si:H/ZnO Contacts for Heterojunction Solar Cells

Florian Einsele[1], Phillip Johannes Rostan[1], and Uwe Rau[2]
[1]Institut für Physikalische Elektronik, Universität Stuttgart, Pfaffenwaldring 47, Stuttgart, 70569, Germany
[2]IEF-5, Photovoltaics, Forschungszentrum Juelich, Jülich, 52425, Germany

ABSTRACT

We study resistive losses at (p)c-Si/(p)Si:H/(n)ZnO heterojunction back contacts for high efficiency silicon solar cells. We find that a low tunnelling resistance for the (p)a-Si:H/(n)ZnO part of the junction requires deposition of Si:H with a high hydrogen dilution $R_H > 40$ resulting in a highly doped µc-Si:H layer. Such a µc-Si:H layer if deposited directly on a Si wafer yields a surface recombination velocity of $S \approx 180$ cm/s. Using the same layer as part of a (p)c-Si/(p)Si:H/(n)ZnO back contact in a solar cell results in an open circuit voltage $V_{OC} = 640$ mV and a fill factor $FF = 80$ %. Insertion of an (i)a-Si-layer between the µc-Si:H and the wafer leads to a further decrease of S and, for the solar cells to an increase of V_{OC}. However, if the thickness of this intrinsic layer exceeds a threshold of 4-5 nm, resistive losses lead to a degradation of the fill factor of the solar cells. These resistive losses result from a valence band offset ΔE_V between a-Si:H and c-Si of about 600 meV. The fill factor losses overcompensate the V_{OC} gain such that there is no benefit of the (i)a-Si:H interlayer for the overall solar cell performance when using an (i)a-Si:H/(p)µc-Si:H double layer.

INTRODUCTION

Heterojunction solar cells from crystalline Si (c-Si) and amorphous hydrogenated Si (a-Si:H) are a promising alternative to conventional solar cells made from c-Si with diffused emitters. This is mainly due to the much lower process temperatures required for the deposition of the a-Si:H layers as compared to those required for dopant in-diffusion. In addition, the thin intrinsic (i)a-Si:H layer introduced between the doped a-Si:H layers of the front and the back contacts and the c-Si absorber leads to a very low interface recombination velocity for the photo-generated charge carriers at both interfaces [1].

When using a-Si:H/c-Si heterojunctions for the contacts to c-Si solar cell absorbers, the band alignment between the two materials plays a crucial role. There is a wide agreement in the literature on the fact that the a-Si:H/c-Si band offset mainly shows up in the valence band. Using different characterization methods, Refs. [25] report on valence band offsets ΔE_V between 0.44 eV and 0.71 eV whereas the conduction band offset ΔE_C is found to be $\Delta E_C < 0.1$ eV in all cases. Because of this band offset asymmetry, it is obviously much easier to extract electrons from the c-Si absorber into an a-Si:H contact than it is to extract holes. Sanyo's successful HIT solar cell [1] is made up of an n-type doped c-Si absorber, an emitter consisting of an intrinsic and a p-type doped a-Si layer, and a corresponding i/n type structure as the back contact. All these layers are deposited using Plasma Enhanced Chemical Vapour Deposition (PECVD). In this structure, extraction of electrons, i.e. the majority carriers, into the back contact poses no problem because of the small value of ΔE_C. In contrast, upon using p-type c-Si absorbers, the extraction of holes into the p-type a-Si:H back contact becomes a challenge due to the large band offset ΔE_V in the valence band. Despite of this drawback highly efficient solar cells can be produced using p-type c-Si absorbers combined with a four-layer sequence (i)a-Si:H/(p)a-Si:H/(p$^+$)μc-Si/ZnO as the back contact [6].

This paper presents a detailed study on resistive losses that occur during hole transport across the (p)c-Si/(i)a-Si:H heterojunction contacts and during transport across the (p$^+$)μc-Si:H/(n-type) ZnO:Al tunnel junction. We demonstrate that a low tunnelling resistance for the latter interface is achieved if the silicon is deposited by PECVD with a hydrogen dilution above a certain threshold. We argue that this threshold coincides with the threshold for microcrystalline growth [7]. We show further that the conductance of (p)c-Si/(i)a-Si:H heterojunction critically depends on the (i)a-Si:H thickness. Since a certain thickness d_i of the (i)a-Si:H layer is necessary for the minimization of the interface recombination velocity for electrons, optimization of d_i for solar cells requires compromise between the increase of series resistance and the decrease of recombination. This effect is finally demonstrated by an investigation of the fill factors FF and the open circuit voltages V_{OC} of c-Si solar cells with various d_i within their back contact structure.

TUNNEL JUNCTION ZnO/a-Si:H

This section concentrates on the tunnelling resistance between the PECVD-deposited (p-type) Si and the n-type ZnO:Al. The test structures (shown in the inset of Fig. 2) consist of 40 nm layer of boron doped a-Si:H directly deposited onto the surface of highly doped ($\rho \approx 10...20$ mΩcm) c-Si wafers. For this deposition step, we use a gas mixture of 2 % B$_2$H$_6$ in SiH$_4$ and vary the hydrogen dilution ratio R_H = H$_2$/SiH$_4$. The plasma frequency during PECVD deposition was v_P = 80 MHz. The test devices are then finished by sputter deposition of a 40 nm ZnO:Al layer and by evaporation of Al contacts on both sides of the wafer. The Al contact on the back of the wafer is a full area contact, whereas the Al contact on top of the PECVD Si is evaporated through a shadow mask onto areas of 0.1 cm^2. Subsequently, we remove ZnO between the contact pads by a short etch in 1 % HCl solution.

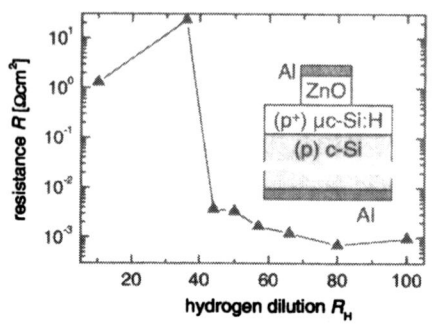

Figure 1: *Resistance R of the test device (see inset) depending on the hydrogen dilution ratio R_H used for the deposition of the Si:H layer. We ascribe the transition of R > 1 Ωcm^2 to R < 10^{-2} Ωcm^2 around $R_H \approx 40$ to the transition between amorphous and microcrystalline growth of Si:H.*

Figure 1 shows the functional dependence of the zero-bias resistance R of the test device on the hydrogen dilution ratio R_H. We observe a significant step by more than two orders of magnitude when the hydrogen dilution exceeds a threshold at $R_H \approx 40$. We ascribe this dramatic change to a transition from amorphous to microcrystalline growth due to the increasing hydrogen dilution [7]. For dilution ratios $R_H > 40$ the resistance decreases further and reaches values up to $R < 10^{-3}$ Ωcm^2. This value is by far low enough to avoid any significant resistive losses in heterojunction solar cells. In contrast, for dilution ratios $R_H < 40$, the contact resistance is pronouncedly too high for usage in a solar cell.

HETEROJUNCTION c-Si / (i) a-Si

The insertion of an (i)a-Si:H interlayer between the highly doped μc-Si:H and the c-Si absorber is desirable to reduce interface recombination [1,8,9]. However, this layer also introduces additional resistive losses that finally lead to a trade-off between low recombination and low contact resistance. In order to study the effect of interlayer thickness on the contact resistance of the c-Si/μc-Si:H/ZnO heterojunction, we fabricate samples with various (i) a-Si:H deposition times. The test structures as shown in the inset of Fig. 2 are essentially the same as those for analyzing the Si:H/ZnO tunnel resistance except for the (i)a-Si:H interlayer.

For the deposition of the (i)a-Si:H layers, we use either no H$_2$ added to the process gas (type I) or a hydrogen dilution ratio of $R_H = 2.5$ (type II) and a plasma frequency $\nu_P = 13.56$ MHz. The deposition rates measured from approximately 100 nm thick films grown on a glass substrate are $r_I \approx 7.5$ nm/min for type I and $r_{II} \approx 5.5$ nm/min for type II films. In all these test structures, a 40 nm thick μc-Si:H layer grown with high hydrogen dilution warrants a low tunnelling resistance to the ZnO back contact.

Figure 2 shows the contact resistance R as a function of the deposition time t_{dep} for the two different films. After a deposition time of $t_{dep} = 30$ s we obtain for both types of contacts resistances $R < 10^{-2}$ Ωcm^2, i.e., values that are very well suited for a low-ohmic back contact in a solar cell. At $t_{dep} = 35$ s (type I) and 50 s (type II), the resistance remains below 10^{-1} Ωcm^2 before rather abruptly exceeding 10 Ωcm^2 and, thus, leaving the range of a reasonably low series resistance for a solar cell.

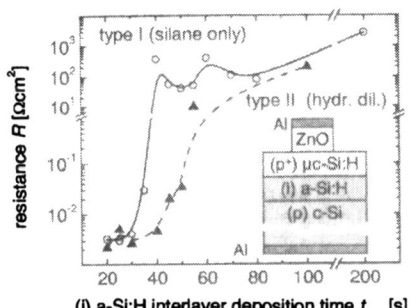

(i) a-Si:H interlayer deposition time t_{dep} [s]

Figure 2: *Resistance R of the heterojunction back contact in dependence of the deposition time t of the passivating (i) a-Si:H interlayer deposited without additional H$_2$ (type I) and with hydrogen dilution (type II). The lines are guides to the eye. The inset sketches the used device geometry.*

Accepting the growth rates extrapolated from the growth of thicker films, we judge the critical thickness to be about 4-5 nm for both types I and II of (i)a-Si:H deposition. The rather abrupt increase of the resistance by more than four orders of magnitude when exceeding this critical value might be a result of tunneling possibly combined with a fluctuation of the thickness and/or the barrier height of the thin (i)a-Si:H film [10]. We note that the temperature dependence of the resistance of the contacts with t_{dep} > 50 s unveils thermally activated behaviour with an activation energy E_a in a range of 490 - 650 meV [10]. If we identify E_a with the valence band offset ΔE_V between the c-Si wafer and the (i) a-Si:H interlayer this result corresponds reasonably well to the band offset data reported earlier [2-5]. In view of such a high band offset it appears impossible to achieve a low-ohmic contact without the support of additional transport paths provided by tunnelling, band gap and thickness fluctuations.

LIFETIME MEASUREMENTS

This section studies the influence of the (i)a-Si:H deposition time t_{dep} on the surface recombination velocity of 250 μm thick, 100 mm diameter, p-type float-zone Si wafers with resistivity ρ ≈ 1 Ωcm. The inset of Fig. 3 sketches the cross-section of these samples being passivated one side with 30 nm thick (i)a-Si:H and on the other one with (i)a-Si:H layers of various deposition times followed by the highly doped μc-Si:H layer (R_H = 100, v_P = 80 MHz).

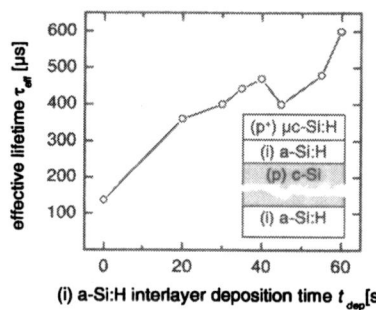

Figure 3: *Effective lifetime τ_{eff} with respect to deposition time t of the (i) a-Si:H passivating interlayer between the crystalline wafer and the (p^+) μc-Si layer . The inset sketches a cross-section of our samples with 30 nm of passivating (i) a-Si:H layer on the rear side. The lines are given as guide to the eye.*

The effective minority carrier lifetime τ_{eff} as determinded by the Quasi Steady-State Photo Conductance method [11] for reference wafers covered with (i)a-Si:H layers of 30 nm thickness on each side is τ_{eff} = 1 ± 0.1 ms. Assuming negligible bulk recombination, we derive a surface recombination velocity S ≈ 13 cm/s using the model of Sproul [12]. For the asymmetric structures depicted in the inset of Fig. 3, we find that the effective lifetimes increases from 100 μs (for no (i)a-Si:H interlayer) up to 600 μs (after a deposition time t_{dep} = 60 s). Accepting S ≈ 13 cm/s for the side passivated with 30 nm (i)a-Si:H, we find values between S = 180 cm/s and S = 30 cm/s for the side passivated with the (i)a-Si:H/(p)μc-Si:H double layer. First of all, there seems to be a reasonable surface passivation quality for any thickness of the (i)a-Si:H layer and even when omitting it. This result is in contrast to the work of de Wolf and Beaucarne [13] who found that the passivation quality of 3nm thin (i)a-Si:H layers is destroyed by deposition of an additional (p)a-Si:H layer grown without hydrogen dilution. The high hydrogen dilution that we use for our doped layers obviously makes this difference. From the present results we see that increasing the deposition time for the (i)a-Si:H interlayer decreases the surface recombination velocity. Whether or not the simultaneous increase of the contact resistance compensates the expected efficiency gain is investigated in the next section.

SOLAR CELLS

To study the influence of the (i)a-Si:H thickness on the performance of a-Si:H back contacts for solar cells, we use test devices made from 250 μm thick p-type Si wafers (resistivity $\rho \approx 1$ Ωcm) with a diffused emitter and a random pyramid texture at the front surface. For the (i) a-Si:H layer we use no hydrogen dilution (type I in Fig. 2), and we again vary the thickness of this interlayer.

Figure 4 depicts the open circuit voltages V_{OC} and fill factors FF of the completed cells as well as the product $FF \times V_{OC}$. Firstly, we see from Fig. 4 that already without an (i)a-Si:H interlayer, we obtain a $V_{OC} = 640$ mV. Insertion of the (i)a-Si-layer between the μc-Si:H and the wafer leads to an increase of V_{OC} with increasing interlayer thickness (expressed as deposition time t_{dep} in Fig. 4). A maximum $V_{OC} = 687$ mV is achieved after a deposition time $t_{dep} = 40$ s. However, the intrinsic layer leads to a decrease of the fill factor FF with increasing thickness. As can be seen from Fig. 4, the fill factor drops dramatically from 78.7 % to below 60 % between $t_{dep} = 20$ s and $t_{dep} = 30$ s. This loss clearly overcompensates the simultaneous gain of few mV in V_{OC}. We use the product $FF \times V_{OC}$ as a figure of merit because the variations of the short circuit current density J_{SC} due to slight differences in the optical properties of the different cells are too large to allow us the use of the output power $P = J_{SC} \times FF \times V_{OC}$ for the same purpose. The product $FF \times V_{OC}$ does barely change from $t_{dep} = 0$ to $t_{dep} = 20$ s and then drops steeply because of the decline of the fill factor.

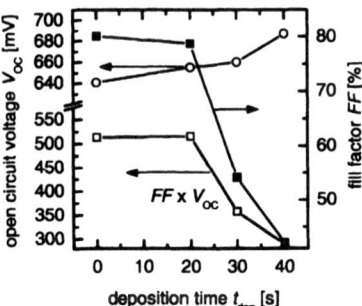

Figure 4: *Open circuit voltages V_{OC}, fill factors FF, and the product $V_{OC} \times FF$ of solar cells with a (i)a-Si:H/(p)μc-Si:H double layer heterostructure back contact with varying (i) a-Si:H deposition time t_{dep}.*

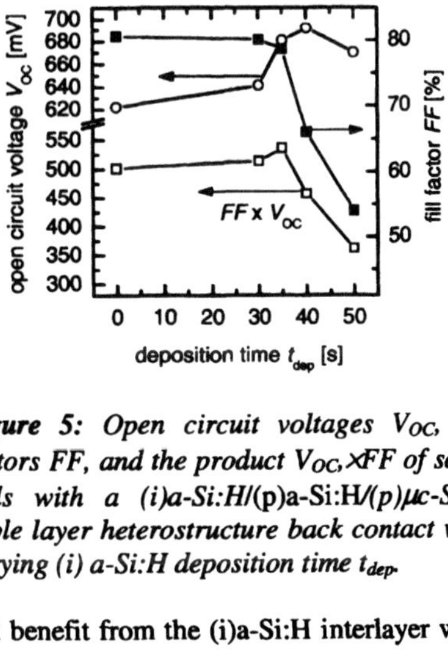

Figure 5: *Open circuit voltages V_{OC}, fill factors FF, and the product $V_{OC} \times FF$ of solar cells with a (i)a-Si:H/(p)a-Si:H/(p)μc-Si:H triple layer heterostructure back contact with varying (i) a-Si:H deposition time t_{dep}.*

From Fig. 4, we conclude that there is no real benefit from the (i)a-Si:H interlayer when directly combined with the μc-Si:H contact layer. In contrast, our best back contacts use an additional (p)a-Si:H layer prepared at $R_H = 80$ and $v_P = 13.56$ MHz [6] in between the intrinsic passivation layer and the μc-Si:H contact layer. Figure 5 depicts the dependence of V_{OC}, FF, and

FF×V_{OC} on the intrinsic layer deposition time. Here, the increase of V_{OC} with increasing t_{dep} is more pronounced than in Fig. 4 whereas the decline of FF at $t_{dep} > 20$ s is less dramatic as for the two layer system. For the (i)a-Si:H/(p)a-Si:H/(p)µc-Si:H three layer system of Fig. 5, we actually find a maximum for FF×V_{OC} around $t_{dep} = 35$ s that combines a high $V_{OC} = 680$ mV with a high fill factor FF = 78.7 % corresponding to a confirmed efficiency η = 21 % as reported in Ref. [6].

CONCLUSIONS

The present study on resistive losses at (p)c-Si/(p)Si:H/(n)ZnO heterojunction back contacts for high efficiency silicon solar cells has demonstrated that a low tunnelling resistance for the (p)a-Si:H/(n)ZnO part of the junction requires deposition of Si:H with a high hydrogen dilution $R_H > 40$ resulting in a highly doped µc-Si:H layer. Such a µc-Si:H layer if deposited directly on a Si wafer yields already a relatively low surface recombination velocity of $S \approx 180$ cm/s. Thus, using the same layer as part of a (p)c-Si/(p)Si:H/(n)ZnO back contact in a solar cell results in an open circuit voltage $V_{OC} = 640$ mV and a fill factor $FF = 80$ %. Insertion of an additional (i)a-Si-layer between the µc-Si:H and the wafer leads to a further decrease of S and, for the solar cells to an increase of V_{OC}. However, if the thickness of this intrinsic layer exceeds a threshold of nominally 4-5 nm resistive losses lead to a degradation of the fill factor of the solar cells. These fill factor losses overcompensate the V_{OC} gain such that we are unable to find a benefit of the (i)a-Si:H interlayer for the overall solar cell performance when using an (i)a-Si:H/(p)µc-Si:H double layer. In contrast, the triple-layer system (i)a-Si:H/(p)a-Si:H/(p)µc-Si:H used for our earlier solar cells [6] exhibits a maximum performance at a non-zero (i)a-Si:H interlayer thickness. Thus, this relatively complicated structure appears necessary to combine a good surface passivation with relatively low series resistance.

References

1. M. Taguchi, K. Kawamoto, S. Tsuge, T. Baba, H. Sakata, M. Morizane, K. Uchihashi, N. Nakamura, S. Kiyama, and O. Oota, *Prog. Photov.: Res. Appl.* **8**, 503 (2000).
2. H. Mimura and Y. Hatanaka, *Appl. Phys. Lett.* **50** (6), 326 (1987).
3. J. M. Essick and J. D. Cohen, *Appl. Phys. Lett.* **55** (12), 1232 (1989).
4. J. M. Essick, Z. Nobel, Y.-M. Li, and M. S. Bennett, *Phys. Rev. B* **54** (7), 4885 (1996).
5. M. Sebastiani, L. Di Gaspare, G. Capellini, C. Bittencourt, and F. Evangelisti, *Phys. Rev. Lett.* **75**, 3352 (1995).
6. P. J. Rostan, U. Rau, V. X. Nguyen, T. Kirchartz, M. B. Schubert, and J. H. Werner, *Sol. En. Mat. Sol. Cells* **90**, 1345 (2006).
7. H. Keppner, J. Meier, P. Torres, D. Fischer, and A. Shah, *Appl. Phys. A* **69**, 169 (1999).
8. H. Plagwitz, M. Nerding, N. Ott, H. P. Strunk, and R. Brendel, *Prog. Photov.: Res. Appl.* **13**, 381 (2005).
9. N. Jensen, R. M. Hausner, R. B. Bergmann, J. H. Werner, and U. Rau, *Prog. Photov.: Res. Appl.* **10**, 1 (2002).
10. F. Einsele, P. J. Rostan, and U. Rau (unpublished).
11. R. A. Sinton and A. Cuevas, *Appl. Phys. Lett.* **69**, 2510 (1996).
12. A. B. Sproul, *J. Appl. Phys.* **76** (5), 2851 (1994).
13. S. De Wolf and G. Beaucarne, *Appl. Phys. Lett.* **88**, 022104 (2006).

Mater. Res. Soc. Symp. Proc. Vol. 989 © 2007 Materials Research Society

Hydrofluoric Acid Treatment of Amorphous Silicon Films for Photovoltaic Processing

M. Burrows[1,2], U. Das[1], M. Lu[1], S. Bowden[1], R. Opila[2], and R. Birkmire[1]

[1]Institute of Energy Conversion, University of Delaware, 451 Wyoming Rd., Newark, DE, 19716
[2]Materials Science and Engineering, University of Delaware, 201 Dupont Hall, Newark, DE, 19716

ABSTRACT

Hydrofluoric acid (HF) is commonly used in Si wafer processing as a surface treatment to remove surface oxide and provide a H-terminated surface passivation that resists contamination within short time scales. During silicon heterojunction (SHJ) device fabrication a similar oxide removal and surface passivation is desired for doped and intrinsic hydrogenated amorphous silicon (aSi) films and therefore studied as follows. X-ray photoelectron spectroscopy is employed to evaluate surface chemical composition, especially with regard to oxygen removal, resistance to hydrocarbon adsorption, and fluorine incorporation post HF treatment. Variable angle spectroscopic ellipsometry is used to determine the growth rate of native oxide and terminal oxide thickness. The electrical effects of aSi native oxide at contact interfaces of SHJ cells are evaluated with current-voltage measurements. HF treatment is effective for oxide removal and provides surface passivation of aSi similar to the crystalline counter-part. Further, fluorine bonding is enhanced for p-type aSi films and control of native oxide thickness below 20Å is not essential for typical electrical contacts of the SHJ photovoltaic cell.

INTRODUCTION

Silicon heterojunction devices which involve aSi films deposited on crystalline Si wafer (cSi) are being produced industrially claiming 6% of the total photovoltaic market and studied by a number of groups internationally [1]. Excellent cSi wafer surface passivation can be achieved by thin intrinsic aSi layers and doped aSi films provide the emitter and back contact. Device contacts typically consist of a transparent conducting oxide for the illuminated junction and metallization for the rear. Due to the numerous deposition steps required it is difficult to realize this completed device structure without exposure to atmosphere. The work herein is designed to answer the question of what is on the a-Si surface after a simple wet chemical H-termination procedure and thus compliments the body of literature on cSi surface treatment and furthers that of SHJ processing.

EXPERIMENT

Two types of double-side polished float zone (111) silicon wafer have been used in this work: phosphorous doped silicon (n-cSi) of 2.5-3Ωcm resistivity and 300μm thickness and boron doped silicon (p-cSi) of identical resistivity and thickness. Four different aSi film conditions deposited by plasma enhance chemical vapor deposition are analyzed. These include 2% gas phase concentration PH_3/SiH_4 (n-aSi), 2% gas phase concentration B_2H_6/SiH_4 (p-aSi), undoped or intrinsic films (i-aSi) deposited by DC plasma and a i-aSi film deposited by 13.56MHz RF plasma. Other deposition conditions include substrate temperature of 200°C, pressure of 1.25Torr, .078mA/cm² discharge current for DC plasma and 19mW/cm² power for RF plasma. The silane gas flow rate of 20sccm is constant throughout and H_2 to SiH_4 gas flow rate ratio of ~6 is also maintained.

Atomic force microscopy (AFM) measurements were performed on a Digital Instruments Dimension 3100 Scanning Probe Microscope. X-ray Photoelectron Spectroscopy (XPS) was performed using a Physical Electronics PHI 5600 using Al K_α radiation with a base pressure in the low 10^{-10} range. The electron take-off angle was approximately 30° from the surface normal and the high resolution quantitative scan pass energy was set to 23.5eV. A variable angle spectroscopic ellipsometer (VASE) from J. A. Woollam Co. was used and corresponding optical models were built and analyzed in the WVASE32 software. The ellipsometer was fit with a N_2 purge source directed at the sample stage and cover to protect from room turbulence during the 12min scan time.

Two cleaning processes were used immediately prior to deposition or analysis. A three step approach including a pre-etch ultrasonic clean to remove debris and dissolve organic contamination; a wet-chemical oxidation that both dissolve impurities including ionic, metallic, and surface hydrocarbon as well as causing further oxidization into the Si bulk; finally this surface oxide is etched and a new H-terminated Si surface is created using an HF solution. Table 1 contains the cleaning procedures' details.

Table 1. Cleaning procedure prior to deposition or analysis.

		Deposition					Analysis		
	Step	Time (min)	Temp (°C)	Description	Step	Time (min)	Temp (°C)	Description	
Pre-etch Ultrasonic Clean	1	2	80	ultrasonic soap solution clean	1	4	RT	ultrasonic acetone clean	
	2	3	70	ultrasonic DIH₂O‡ rinse	2	4	RT	ultrasonic methanol clean	
	3	10	85	hot flowing air dry	3	4	RT	ultrasonic 18MΩ DIH₂O clean	
					4	3	RT	flowing 18MΩ DIH₂O rinse	
Wet -Chemical Oxidation	4	5	60-70	piranha† etch	5	5	73	piranha etch	
	5	5	RT*	flowing 18MΩ DIH₂O rinse	6	5	RT	flowing 18MΩ DIH₂O rinse	
Oxide etch and H-termination	6	1	RT	10% HF etch	7	1	RT	10% HF etch	
	7	0.33	RT	N₂ jet dry	8	0.05	RT	18MΩ DIH₂O rinse	
					9	0.33	RT	N₂ jet dry	

† Piranha etch is 4:1 H_2SO_4 to H_2O_2 allowed to self heat to 60-70°C or held at 73°C in a hot water bath
‡ DIH₂O is de-ionized water
* RT denotes room temperature

RESULTS AND DISCUSSION

Sample Roughness

The root mean square surface roughness (R_q) of a sample set of 200nm DC i-aSi / n-cSi and bare n-cSi in four stages of extended clean processing were evaluated: after pre-etch ultrasonic clean and after 1, 2, and 3 series of oxidation in piranha solution followed by HF dip. For both the deposited film (R_q = 4.4-6.0Å) and wafer (R_q = 1.3-2.0Å) there is no detectable roughening upon single or repeated etching. Also it was observed that at the above deposition conditions surface roughness is independent of film thickness up to a few hundred nanometers.

Atomic Surface Concentration and Bonding

XPS was used to quantify oxidation, hydrocarbon adsorption, and residual fluorine for five identically treated samples at various stages of the cleaning process. Importantly the time between the last N_2 jet dry step and evacuation of the XPS load-lock to the 10^{-5} Torr range is 4-5 minutes and matches the typical time window required for substrate loading and evacuation of the PECVD deposition system.

Figure 1(a) is a summary of the results; 'nox' is short for native oxide, 'pox' for piranha etch oxide, and 'HF' is HF dipped and loaded as described above. In all cases the piranha etch oxide step is seen to effectively decrease the carbon signal and increase the oxygen. The O fraction increases from 35.8±1.3% to 45.4±1.1% (a 27% increase) on average for the four aSi films and from 35.6% to 40.2% for the n-cSi wafer. Conversely the C fraction decreases from 12.0±2.3% to 4.1±0.6% on average for the four aSi films and 10.7% to 5.4% for the n-cSi wafer. The increase in oxygen signal is believed to be from two sources; the more complete oxidation (greater fraction of SiO_2 relative to SiO_z, $z < 2$) and further oxidation into the Si bulk. Considering the aSi films and assuming a fixed sampling depth we can estimate that the average increase of SiO_2 to SiO ratio upon piranha etch (about 0.67 to 0.76) to account for a 5% gain in oxygen signal, the remaining 22% must come from further oxidation of Si bulk. A similar argument holds for the n-cSi wafer. Figure 1(b) illustrates the Si2p region of the n-aSi sample and is a good illustration of the variation in intensity of oxide state from native to piranha etched oxide to finally the HF treated where no shifted components are detectable. A single Gaussian is applied to fit the elemental, sub-oxide, and dioxide peaks. It is not reasonable to deconvolute the Si region into +1 to +4 oxidation states in our experimental set-up.

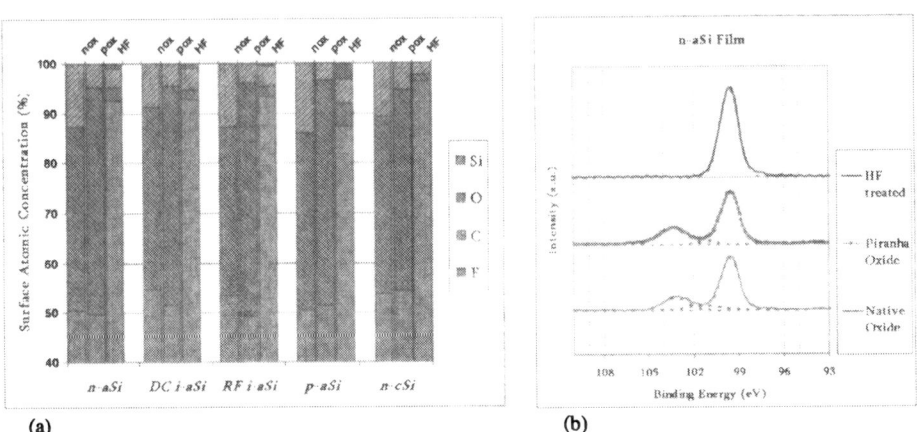

(a)   (b)

Figure 1. (a) Graph of surface atomic concentration at various conditions. Note that for higher resolution in the region of interest the plot is truncated below 40% with the remainder understood to be Si. 'nox' is short for native oxide, 'pox' for piranha etch oxide, and HF is post HF dip. (b) Si2p region of 200nm 2% n-aSi film with Shirley background correction then deconvolution into elemental, monoxide, and dioxide regions for native oxide, piranha etch oxide, and HF treated.

Three trends can be distinguished post HF dip. n-aSi, DC i-aSi, and RF i-aSi can be considered to have identical results within our experimental resolution, 2.2% O, 4.0% C, and 0.8% F. The n-cSi wafer was measured to have extremely low contamination with 1.3% O, 2.3% C and F below detection (less than 0.5%). Comparable XPS wafer clean studies reported in literature often contain greater than 5% O, 10% C and at least 1%F [1-3]. It is proposed that two important controls employed are the sources of this improvement. The final three second DIH$_2$O dip has been shown to reduce F contamination by several percent absolute [3]. Also the final post HF carbon concentration is significantly reduced and O to a lesser extent when the piranha etch temperature is maintained at 73°C versus a self-heating strategy. Lastly it is interesting to note that the p-aSi film with 4.6% O, 4.9% C, and 3.1%F contains twice the O and over three times the F of the n- and i-aSi films. In all cases the F is preferentially bonded to Si (binding energy of 685.9eV) versus adsorbed F (binding energy of 687.4eV) [3].

Resistance to Oxide Growth

In VASE analysis by ellipsometric modeling a number of important features are used as fitting parameters including surface roughness, oxide thickness, aSi film thickness and optical constants. However in this study the focus is not the aSi film bulk optical properties that are of interest but rather the changing surface upon HF treatment and subsequent exposure to air.

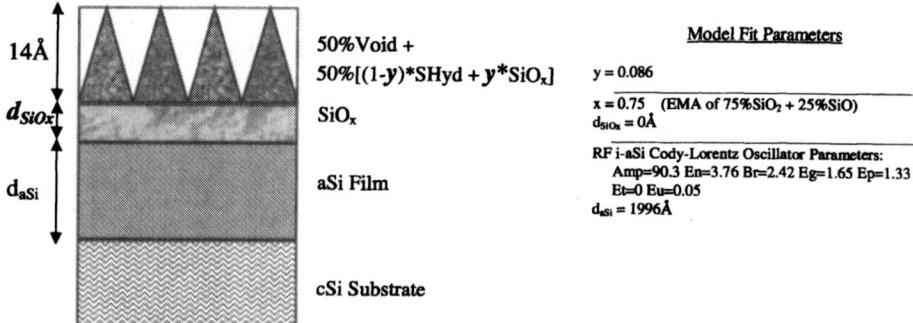

Figure 2. Schematic of model structure used in ellipsometry analysis. Also shown is best fit of RF i-aSi 5min post HF treatment resulting in MSE of 1.386. Note that during oxide fitting only the italicized d_{SiOx} and y, $0 < y < 1$, are varied. SHyd is an aSi like layer whose optical parameters are established when the oxide thickness is known from XPS and is generally similar to the aSi film underneath with a slightly wider band gap, Eg.

Figure 2 shows the pertinent details of our modeling strategy. The modeling of the aSi films is performed as follows: the roughness of aSi surface is set to 14Å (peak to peak) as estimated from AFM measurements; the monoxide to dioxide fraction from XPS measurements is built into a Bruggeman effective medium approximation (EMA) oxide layer composed of SiO and SiO$_2$; the native oxide and piranha etched oxide cases are thoroughly analyzed to establish the optical constants and thickness of the aSi films modeled with a single Cody-Lorentz oscillator and are similar to other literature references [4]. With this information the HF etched sample is repeatedly analyzed over the duration of several weeks in order to estimate the oxide layer growth as a function of time. From XPS measurements the oxide thickness for each aSi film plus (111) and (100) n-cSi is estimated to be between 1.4-0.3Å at 5min after HF dip from the ratio of the I[O1s]/I[Si2p]=0.26 for 1 monolayer of oxide (3Å). Thus, a well defined model can be established for this time interval. For other times, including a refit of native and piranha oxide only two fit parameters are adjusted: the fraction of oxide, y, in the surface roughness

EMA and the thickness of oxide below the surface roughness, d_{SiOx}. Also note that the surface roughness is held *constant* at 14Å and the only fit variables pertain to oxide thickness. The oxide thicknesses reported in Figure 3 are the sum of d_{SiOx} and equivalent thickness in the roughness layer.

Figure 3 shows the oxide growth with time four five different samples. In each case except p-aSi there appears to be an 'incubation' time in which the oxide growth is slow and the thickness estimate is below 1Å. Following this stage there is an approximately logarithmic growth up to the final data point which is ~100 days after deposition, taken to be a good estimate of the terminal oxide thickness. Further support is given by the excellent agreement with a similar clean treatment that reports 2Å of oxide after 70minutes atmospheric exposure on a cSi(111) surface matching our estimate [5]. The important observation for the current work is that the removal and control of surface oxide on aSi films is confirmed on the time scales required for substrate loading into deposition systems. More precisely, less than 1Å of surface oxide after 5min of atmospheric exposure for both cSi and aSi films exist. In the p-aSi case for which the HF treatment is less effective in removing oxide and resisting growth; the oxide thickness 5min after HF dip is estimated to be 1.4Å for a p-aSi film.

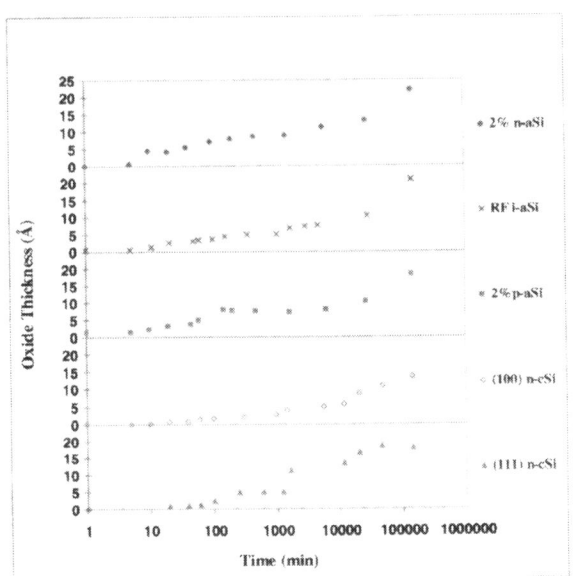

Figure 3. Oxide growth with time in atmosphere for various aSi and cSi samples.

Effect of Oxide Thickness on Electrical Contacting

The effect of the oxide on SHJ devices was evaluated on symmetric test structures where 500nm of electron-beam evaporated Al and 70nm of sputtered indium-doped tin oxide (ITO) were deposited on 12nm n-aSi / n-cSi / 12nm n-aSi and similarly 12nm p-aSi / p-cSi / 12nm p-aSi. Half the sample set were processed according to the cleaning procedure in Table 1 up to step 6 to produce a controlled 15-18Å 'piranha' oxide. The remaining samples were further processed through the HF dip and loaded under vacuum within 5 minutes for contact deposition.

There are two important trends: first that Al makes a good ohmic contact with a low measured resistance at open circuit (R_{oc}) of 0.6-$2.2\Omega cm^2$. This demonstrates that there is low series resistance through silicon films and wafer section of both the n- and p-type structures. ITO contacted areas, on the other hand, had $R_{oc} = 9.9$-$12.4\Omega cm^2$ and thus would effectively lower fill factors on SHJ devices. The effect due to variation of oxide thickness is minimal compared to the contact material itself. It is proposed that Al reacts with a portion of the aSi and oxide during deposition to create a conductive interface of Al, aSi, and oxides of each. ITO sputtering likely entails significant high energy particle bombardment of the aSi surface and, thus, the presence of a thin native oxide does not inhibit the aSi/ITO contact significantly.

CONCLUSIONS

A simple wet-chemical procedure developed to clean and H-terminate cSi surfaces has been applied to aSi. AFM determined roughness is not affected by this clean and a low surface roughness of $R_q = 5\text{Å}$ is established. From XPS results the effect of the wet-chemical oxidation is to remove surface carbon, increase the silicon dioxide fraction over sub-oxides and further oxidize into the aSi or cSi bulk. Also the effective removal of this oxide by a 60sec 10%HF dip followed by a 3sec DIH$_2$O rinse is shown. C and F surface contaminants are well controlled in the n- and i-aSi as well as n-cSi cases, but less so for p-aSi. It is also established that less than 0.8Å of oxide exists on the n- and i-aSi films and n-cSi wafer 5 minutes after HF treatment, whereas up to 1.4Å exists on the p-aSi surface in the same time frame. Thus, good control of oxide thickness is possible for SHJ processing. Contact resistance was unaffected by the thickness of surface oxide of aSi up to 20Å.

ACKNOWLEDGMENTS

The author would like to thank Shannon Fields and Kevin Hart for assistance with depositions and Vicky DiNetta for AFM measurements, all of whom are from the Institute of Energy Conversion at the University of Delaware. From J.A. Woollam Co. Ron Synowicki's input on the ellipsometry modeling was also very helpful. This work was supported by Department of Energy and the National Renewable Energy Laboratory subcontract #ADJ-1-30630-12.

REFERENCES

1. R. Barrio, C. Maffiotte, J.J. Gandía, and J. Cárabe, *Journal of Non-Crystalline Solids* 352, 945-949 (2006).
2. J. Alay, S. Verhaverbeke, W. Vandervorst, and M. Heyns, *Jpn. J. Appl. Phys.* 32, 358-361 (1993).
3. C.H. Bjorkman, H. Nishimura, T. Yamazaki, J.L. Alay, M. Fukuda, and M. Hirose, *Mat. Res. Soc. Symp. Proc.* 386, 177-182 (1995).
4. D.H. Levi, C.W. Teplin, E. Iwaniczko, R.K. Ahrenkiel, H.M. Branz, M.R. Page, Y. Yan, Q. Wang, and T.H. Wang, *Mat. Res. Soc. Symp. Proc.* 808, 239-244 (2004).
5. H. Angermann, W. Henrion, M. Rebien, and A. Röseler, *Solar Energy Materials and Solar Cells* 83, 331-346 (2004).

Mater. Res. Soc. Symp. Proc. Vol. 989 © 2007 Materials Research Society   0989-A18-08

High-Performance, Tandem-Type Amorphous Silicon Solar Cell

Porponth Sichanugrist, Nirut Pingate, and Channarong Piromjit
National Science and Technology Development Agency, 111 Paholyothin Rd., Klong Luang, Pathumthani, 12120, Thailand

ABSTRACT

Microcrystalline silicon oxide (μc-SiO) deposited by VHF plasma from the gas mixture of silane and carbondioxide was reported by authors to be the promising material for thin-film silicon solar cell fabricated on glass substrate, comparing with the conventional amorphous silicon oxide or microcrystalline silicon p-layer. High-performance, tandem-type solar cell with amorphous and microcrystalline cells (double cells) has been achieved by applying this μc-SiO to the p-layer of both cells. Further work has been extended to triple-junction type solar cell using amorphous silicon oxide (a-SiO), amorphous silicon germanium (a-SiGe) and microcrystalline Si (μc-Si) films as i-layers of the top, middle and bottom cells, respectively. Up to now, cell efficiency of 15.7 % has been achieved using this novel μc-SiO.

INTRODUCTION

Microcrystalline silicon oxide (μc-SiO) deposited from the gas mixture of silane and carbondioxide using VHF plasma was reported by authors to be the promising material for thin-film silicon solar cell fabricated on glass substrate, comparing with the conventional amorphous silicon oxide (a-SiO) or microcrystalline silicon (μc-Si) p-layer [1]. High-performance, tandem-type solar cell with amorphous and microcrystalline cells (double cells) has been achieved by applying this μc-SiO to the p-layer of both cells. On the other hand, several research laboratories are working with triple-junction type solar cells such as Kaneka Corp. [2] and United Solar Ovonic Corp. [3]. Both of them use a-Si, a-SiGe and μc-Si cells and obtained the efficiency as high as 15%. Since we found that higher performance can be obtained with μc-SiO p-layer and both of them did not use it in their triple cells, we can expect to achieve higher cell efficiency if we apply this μc-SiO p-layer into the triple cells. So our R&D about μc-SiO p-layer has been extended to triple-junction type solar cell.

EXPERIMENT

A cluster-type, multi-chamber system has been used to deposit various films such as Ag, ZnO, a-Si, a-SiO, a-SiGe, μc-Si and μc-SiO films on 30 cm x 40 cm area without breaking the pressure. The detailed cell structure of our triple cell is shown in Fig.1. VHF (Very High Frequency) of 60 MHz, carbon dioxide gas and higher hydrogen dilution are used for μc-SiO deposition. Tin-oxide coated, low-Fe glass is used as the substrate. Thin-film ZnO is deposited by DC sputtering on it in order to promote the crystallization of the μc-SiO p-layer. The i-layer of the top cell is normally deposited at lower substrate temperature in order to increase the open circuit voltage (Voc) of the cell. However, this lower temperature method can not be used with single chamber configuration. Since our work will be extended to one single chamber in the future, we used a-SiO i-layer deposited by VHF plasma from SiH_4 , CO_2 and H_2 in order to increase Voc instead. Hydrogen dilution with profiling has been used in the deposition of a-SiGe and μc-Si i-layer in order to increase the cell performance.

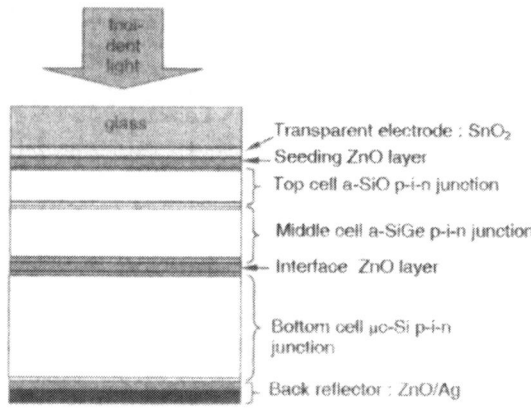

Figure 1. Schematic view of the cell structure for triple cell

     N-type μc-SiO has also been used in order to minimize the optical loss due to the absorption. Interface ZnO layer has been deposited between the a-SiGe middle and μc-Si bottom cell by DC sputtering in order to increase the current of the a-SiGe middle cell. Some treatments have been done to improve the cell interface properties. Finally, cell performance was measured under AM1.5, 100 mW/cm² at 25 °C. The typical deposition conditions are shown in Table I and II.

Table I. Typical deposition conditions

Layer	$CO_2/(CO_2+SiH_4)$	Gas ratio	H_2/SiH_4	Pressure (mTorr)	Power density (mW/cm²)	Thickness (nm)
p(μc-SiO)	0.28	TMB/(TMB+SiH₄) = 0.90	75	500	67	15 – 25
n(μc-Si)	-	PH₃/(PH₃+SiH₄) = 0.90	25	500	67	25 – 35
n(a-Si)	-	PH₃/(PH₃+SiH₄)= 0.75	3	500	25	35 – 45
i(a-SiO)	-	-	10	1000	50	200 – 300
i(a-SiGe)	-	GeH₄/(GeH₄+SiH₄) = 0 - 1.3	10	1000	25	400 – 800
i(μc-Si)	-	-	11	1000	100	1800 – 2500

Table II. Typical deposition conditions

Layer	Ar (sccm)	Substrate temperature (°C)	Pressure (mTorr)	Power density (mW/cm²)	Deposition time (sec)
Front ZnO	6	25	4 - 20	1125	570
Interface ZnO	6	25	4 - 20	625	800
Back reflector : ZnO	6	25	4	750	1170
Back reflector : Ag	15	25	4	1125	360

The effect of CO_2 /SiH_4 in the top cell has been investigated. Furthermore, the effect of the sputtering pressure of ZnO deposited before the p-layer of the top cell and ZnO inserted between the middle and bottom cells has also been investigated.

RESULTS

Effect of CO_2 in the i-layer of top cell

Figure 2 shows Voc of the triple cells for various CO_2/SiH_4 ratios. Here, the hydrogen dilution ratio (H_2/SiH_4) is about 10 times. The thickness of i-layer has been adjusted in order to obtain the same current. As seen in Fig. 2, Voc of 2.11 V has been achieved at the CO_2/SiH_4 ratio of about 0.18. Lower Voc and cell performance are obtained with higher CO_2/SiH_4 due to the fact that thicker i-layer is needed with wider bandgap i-layer and will decrease Voc and cell performance.

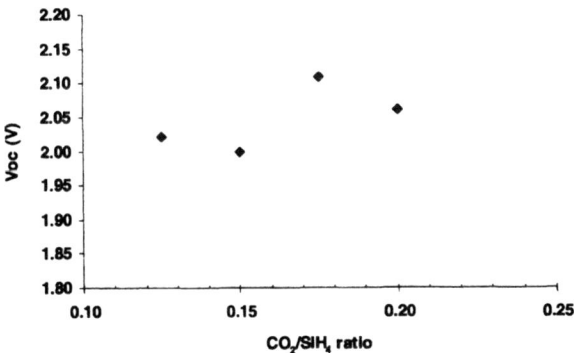

Figure 2. Voc of tandem cells vs CO_2/SiH_4 gas ratio in the top i-layer

Effect of interface ZnO layer

Thicker a-SiGe i-layer can not be used due to its poor film quality. So the current of the triple cell will be limited by the current of middle a-SiGe cell. In order to increase the current of middle cell, interface ZnO layer has been added and deposited by DC sputtering at low and high pressure before the bottom cell fabrication.

Table III shows the effect of interface ZnO layer on the cell performance. Higher current has been obtained with ZnO as expected. Furthermore, for the cells with interface ZnO layer, higher current can be obtained with the one deposited at high processing pressure This may come from the fact that the transmission of ZnO layer made at higher pressure is better [4].

Table III. Detailed cell parameters of the triple cells vs the effect of interface ZnO layer and its s sputtering pressure

	Voc (V)	Jsc (mA/cm^2)	FF	Eff (%)
Cell with no interface ZnO	2.18	6.55	0.65	9.37
Cell with interface ZnO deposited at low pressure	2.20	8.11	0.72	12.9
Cell with interface ZnO deposited at high pressure	2.20	9.16	0.71	14.3

Effect of front seeding ZnO layer

Front seeding ZnO layer is deposited on SnO$_2$-coated glass in order to promote the crystallization of the μc-SiO p-layer in the top cell. We have investigated the effect of sputtering pressure by using low and high pressure. The results of its effect is shown in Table IV. The cell efficiency of 15.7% (Figure 3) has been obtained with the one deposited at high pressure. This may be similar to the effect described above.

Table IV. Detailed cell parameters of the triple cells for the effect of front ZnO layer deposited at low and high pressure

	Voc (V)	Jsc (mA/cm^2)	FF	Eff (%)
Cell with front ZnO deposited at low pressure	2.20	9.16	0.71	14.3
Cell with front ZnO deposited at high presure	2.21	10.4	0.68	15.7

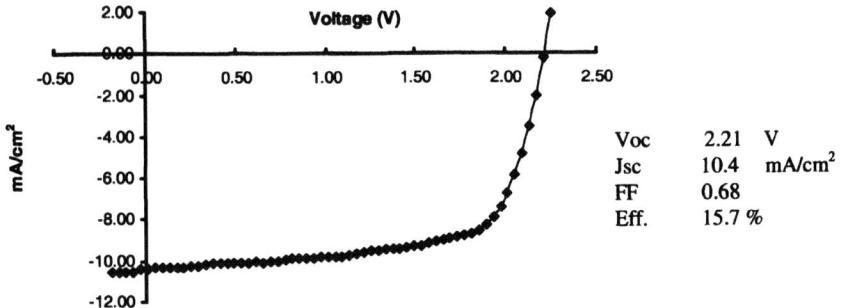

Figure 3. I-V curve of the triple cells

CONCLUSION

Novel p(μc-SiO) has been developed by using VHF plasma with the gas mixture of SiH_4 and CO_2. It has been applied as the p-layer of a-SiO top cell, a-SiGe middel cell and μc-Si bottom cell in the triple cell configuration.

It was found that interface ZnO layer between the middle and bottom cells has to be deposited at higher sputtering pressure due to its better transmission. Furthermore, the same result has also found for the front ZnO layer using as the seeding layer.

Up to now, an efficiency of 15.7 % has been achieved for a-SiO/a-SiGe/μc-Si triple cell. The results show that μc-SiO fabricated by VHF plasma is the promising material for thin-film Si solar cell fabricated on the glass substrate.

REFERENCES

1. P. Sichanugrist et al., Proc. of MRS Fall Meeting, Boston, 2006.
2. K. Yamanoto et al., Proc. Of 3rd Photovoltaic Energy Conversion, 2789 (2003)
3. J. Yang et al., Thin Solid Films 487, 162 – 169 (2005).
4. O. Kluth et al., Proc. of the 26th PVSC, Anaheim, 1997.

Mater. Res. Soc. Symp. Proc. Vol. 989 © 2007 Materials Research Society

Hydrogen Passivation of Thin-Film Polysilicon Solar Cells

Lode Carnel, Ivan Gordon, Dries Van Gestel, Guy Beaucarne, and Jef Poortmans
MCP SSC, IMEC, Kapeldreef, Leuven, 3001, Belgium

ABSTRACT

In this work we characterized fine-grained polysilicon layers with a grain size of only 0.2 µm before and after passivation. Plasma hydrogenation led to a higher hydrogen concentration in the first micron of the layer than nitride passivation. The highest efficiency of 5.0 % was reached when nitride passivation was followed by plasma passivation.

INTRODUCTION

Over the last decade, the production of photovoltaic modules has increased with an impressive annual growth rate of more than 30 %. This resulted in a significant shortage of solar grade Si material and a sharp increase of the Si feedstock price. To reduce the cost price of photovoltaic energy, other technologies need to be developed that use less silicon. The ultimate silicon technology could be a thin ($\sim$ 5 µm) polycrystalline silicon layer deposited from the gas phase onto a foreign carrier substrate such as an alumina ceramic or a glass-ceramic substrate [1]. Such a technology could lead to a sharp cost reduction while maintaining the efficiency and the stability of a fully crystalline Si absorber. However, current efficiencies are still well below those achieved for bulk silicon solar cells, mainly due to the large number of defects both at the grain boundaries and in the grain itself [2]. To passivate these defects, hydrogen is introduced into the layers.

EXPERIMENTAL DETAILS

In this work we studied the hydrogen passivation of defects using fine-grained ($\sim$ 0.2 µm grains) polysilicon. The p-type doped fine-grained polysilicon layers were obtained by direct deposition of silicon on top of an oxidized silicon wafer. The deposition is done by thermal chemical vapour deposition at temperatures above 1000 °C and at atmospheric pressure. The passivation is carried out using two different methods: a hydrogen plasma treatment and a high-temperature anneal (firing) of a hydrogenated silicon-nitride (a-SiNx:H) layer. Plasma hydrogenation offers an infinite amount of hydrogen, while the a-SiNx:H is a solid source with only a finite amount of hydrogen. The hydrogen incorporation in the layers was measured with secondary ion mass spectroscopy (SIMS) using deuterium as an easily traceable isotope of hydrogen. The passivated layers were studied electronically by mobility, temperature-dependent resistivity and spreading resistance profiling (SRP) measurements. Finally, solar cells were made to compare both passivation methods at device level.

RESULTS

p-type layer characterization

Figure 1 plots the D-profiles of the first two microns of identical polysilicon layers after both the plasma and the a-SiN:D passivation. The influence of the presence of a diffused emitter on the D-profiles is also shown for both passivation techniques.

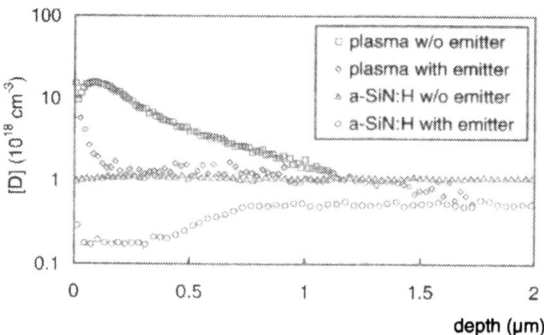

Figure 1. Comparison between remote plasma and a-SiN:D passivation of fine-grained polysilicon layers (Boron concentration of 10^{17} cm^{-3}) and influence of the presence of a diffused emitter.

The remote plasma leads to an exponentially decaying D-profile. The a-SiN:D passivation on the other hand leads to a flat profile throughout the whole layer. In the first micron the D-concentration is larger after plasma passivation than after nitride passivation. After the first micron, both methods result in the same D-content with the D-concentration after plasma passivation decreasing further exponentially. Unfortunately the measurements were stopped after 1 μm for the plasma passivation. The decay for the remote plasma passivated sample is not observed after the a-SiN:D passivation. This is due to the much higher temperatures reached during the firing (> 700 °C) of the nitride than during the remote plasma treatment (~ 380°C). These higher temperatures result into a deeper penetration of hydrogen in the Si layer. The presence of a highly n$^+$ doped emitter leads to a worse D-penetration for both passivation techniques. This was described in detail in [3].

For the remote plasma passivation the D-content is largely influenced by the sample temperature during the plasma treatment. SIMS measurements after plasma passivation at different temperatures are shown in Figure 2.

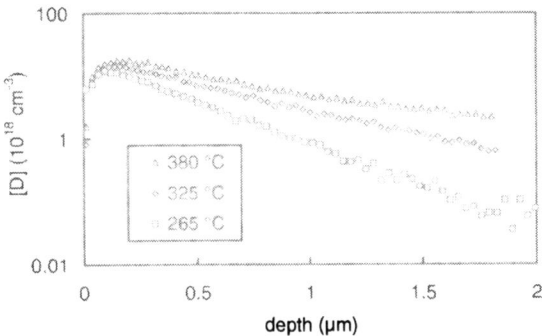

Figure 2. Influence of the sample temperature during the plasma treatment on the D-content of fine-grained polysilicon layers (Boron concentration of 10^{17} cm^{-3}).

Higher temperatures result in a higher D incorporation into the layers and in a flatter D-profile. This results from the activated behavior of the deuterium diffusion which leads to a higher diffusion coefficient for higher temperatures. If we assume that the D-diffusion is limited by defect trapping, we obtain an activation energy of 0.44 eV. This value is in good agreement with previous publications on polycrystalline silicon [4].

In general we observe that the D-concentration in the layer is larger than the defect density of our layers which is around 10^{18} cm^{-3} [5]. Johnson et al. also showed a much larger D (factor of thousand difference) concentration than the number of dangling bonds for polysilicon [6]. Despite this higher concentration they found that the D-concentration was a good indicator of the number of dangling bonds, with a higher D concentration leading to a lower spin density. The higher H-concentration was described as due to the breakage of weak Si-Si bonds by the H atoms in the neighborhood of the grain boundaries. Calculations by Van de Walle et al. indeed confirmed that an increase in bond length of Si-Si bonds (for instance around the grain boundaries) may lead to a lower formation energy of the Si-H-Si configuration [7].

To investigate the influence of the higher D-content with a higher hydrogenation temperature, we characterized the passivated samples with mobility and resistivity measurements. Figure 3 shows the mobility and the activation energy of the resistivity (E_{act}) for various hydrogenation temperatures. E_{act} is a measure for the potential energy barrier at the grain boundaries and for the number of ionized defects at the grain boundaries.

These measurements indeed show an improved majority carrier transport (i.e. a higher mobility and a lower E_{act}) for a higher hydrogenation temperature. The number of defects decreased roughly by a factor of three as deduced from E_{act}. With nitride passivation, we also obtained an increase of the mobility to 18 cm^2/Vs and a decrease of the number of defects by a factor 3.

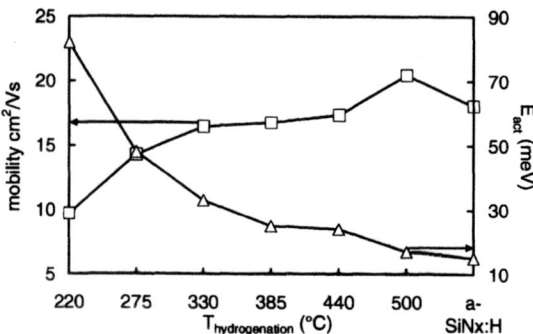

Figure 3. Influence of hydrogenation temperature on the mobility and the activation energy of fine-grained polysilicon layers (Boron concentration of 10^{17} cm^{-3}).

To examine the influence of the reduced hydrogen penetration when an emitter is present on the passivation of the layer, we did spreading resistance profile (SRP) measurements. We measured an as-grown sample, a passivated n$^+$p junction created by P diffusion and a passivated p-type sample (Figure 4).

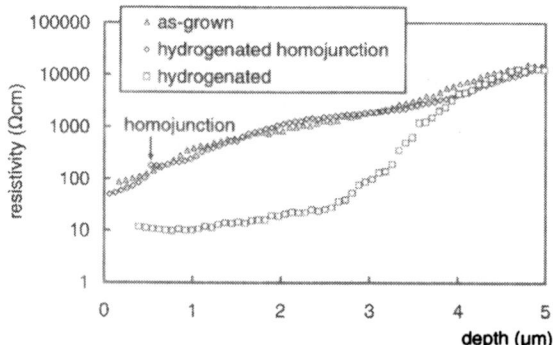

Figure 4. Influence of hydrogen plasma treatment on the resistivity profiles of p and n$^+$p-type devices (Boron concentration of 10^{16} cm^{-3}).

The SRP-profile of the three samples shows an increasing resistivity towards the Si-SiO$_2$ interface, which is because of the reduced material quality in this region. The hydrogenation of the p-type layer results in a reduced resistivity due to the smaller potential energy barriers at the grain boundaries. The electrical junction for the diffused emitter is observed at 0.6 μm by the peak in the resistivity. The resistivity after hydrogenation of the homojunction is as high as for the as-grown polysilicon sample. One possible reason for this high resistivity is indeed that the n$^+$ layer acts as a blocking layer for the hydrogen atoms resulting in a worse passivation. However, this effect alone cannot explain why the change in resistivity of the homojunction after passivation is so small. As mentioned in previous papers, the diffused n$^+$ emitter also leads to preferential dopant spikes along the grain boundaries [8]. These highly doped n$^+$ regions result in depletion regions in the grains itself and potential energy barriers for the holes, which might partly explain the high resistivity of the layers.

Solar cell results

An alternative to a diffused homojunction emitter is an amorphous silicon emitter on top of the polycrystalline base. In that case, emitter formation can occur *after* passivation due to the low deposition temperature of the amorphous silicon layer, and therefore does not hinder the passivation. These heterojunction emitters typically yield an increase of the open-circuit voltage (V_{oc}) of 80 mV compared to homojunction emitters [9]. Table I shows the impact of the hydrogen plasma on the IV-measurements of fine-grained polysilicon cells with a heterojunction.

Table I. Influence of hydrogen passivation of fine-grained polysilicon solar cells with a heterojunction emitter (1 μm p thickness).

Passivation	J_{sc} mA/cm^2	V_{oc} mV	Fill factor %	Efficiency %	R_{series} mΩcm^2	R_{shunt} Ωcm^2
No	6.6	417.0	57.7	1.6	8351.8	807.9
Yes	11.3	511.5	66.1	3.8	2425.3	1610.6

The V_{oc} of 417 mV before hydrogenation is quite impressive for fine-grained polysilicon, and again confirms the independency of the open-circuit voltage on the grain size of the layer, since it is even larger than the V_{oc} of coarse-grained material before hydrogenation [9]. The hydrogenation results in a large improvement of both the short-circuit current density (J_{sc}) and the V_{oc}, while the efficiency increases by a factor of more than 2. EQE-measurements showed an improved long-wavelength (> 400 nm) response due to a longer diffusion length. Using the equation proposed by Taretto et al. [10], we calculated the diffusion length of our cells from the J_{sc} and the V_{oc}, and obtained an increase from 0.01 μm to 2.2 μm after hydrogenation.

Despite the higher D-content and the improved majority carrier properties, no influence of the hydrogenation temperature was found on cell level. A hydrogenation at 265 °C resulted in the same passivation on cell level as a hydrogenation at 440 °C. This is unexpected and contrary to SIMS and majority carrier measurements.

Finally, we also investigated nitride passivation on cell level and compared it with plasma hydrogenation (Table 2). Before deposition of the amorphous silicon heterojunction, the nitride layer was removed in an HF solution (5 %). We also investigated nitride passivation followed by plasma hydrogenation.

Table 2. Influence of different passivation techniques on the cell efficiency of fine-grained heterojunction solar cells (3 μm p thickness).

Passivation	J_{sc} mA/cm^2	V_{oc} mV	FF %	Efficiency %
Plasma	12.2	448.7	61.8	3.4
a-SiNx:H	9.8	369.6	56.4	2.1
a-SiNx:H + plasma	12.3	472.8	69.9	4.1

With the heterojunction emitter, the plasma hydrogenation clearly showed better passivation than the a-SiNx:H hydrogenation alone, resulting in a higher J_{sc} and V_{oc}. This probably means that the reduced H-concentration in the case of a-SiNx:H hydrogenation leads to a reduced passivation of the grain boundaries. The best results at cell level were obtained when a-SiNx:H passivation was followed by a hydrogen plasma treatment. The sequence was found to be important since the reverse order showed an as bad passivation as the nitride alone. When we lowered the p-type thickness from 3 μm to 2 μm, we obtained the highest efficiency so far of 5 % on polysilicon material with a grain size of only 0.2 μm.

CONCLUSIONS

In this work we have shown the impact of hydrogen passivation on majority and minority carrier transport. The number of defects decreased by a factor of three after both the nitride and the plasma passivation of the fine-grained polysilicon films. The hydrogen concentration in the polysilicon layers depended on the hydrogenation temperature and the presence of the diffused emitter. On cell level, the hydrogen plasma passivation outperformed the nitride passivation. Best passivation of the fine-grained polysilicon was obtained if a combination was made of a nitride and a hydrogen plasma passivation, leading to the highest efficiency of 5.0%.

ACKNOWLEDGMENTS

This work was partly funded by the European Commission under contract number 019670-FP6-IST-IP ('ATHLET'). The authors thank Kris Van Nieuwenhuysen for the CVD depositions.

REFERENCES

1. G. Beaucarne and A. Slaoui in *Thin-film Solar Cells: Fabrication, Characterization and Applications,* edited by J. Poortmans and V. Arkhipov (Wiley, New York, 2006) pp. 97-131
2. D. Van Gestel, et al., Appl. Phys. Lett. **90**, 092103 (2007)
3. L. Carnel, et al., Electron Device Lett. **27**, 163 (2006)
4. N. H. Nickel, et al., Phys. Rev. B **66**, 075211 (2002)
5. L. Carnel, et al., J. of Appl. Phys. **100**, 063702 (2006)
6. N. M. Johnson, et al., Appl. Phys. Lett. **40**, 882 (1982)
7. C. G. Van de Walle, et al., Phys. Rev. B **51**, 2636 (1995)
8. G. Beaucarne, et al., Solid State Phenom. **67 – 68**, 577 (1999)
9. L. Carnel, et al., Thin Solid Films **511-512**, 21 (2006)
10. K. Taretto, et al., J. Appl. Phys. **93**, 5445 (2003)

Mater. Res. Soc. Symp. Proc. Vol. 989 © 2007 Materials Research Society   0989-A18-12

Photocurrent Profile in a-SiC:H Monolithic Tandem Pinpin and Pinip Photodiodes

Alessandro Fantoni, Manuela Vieira, and Yuri Vygranenko
DEETC, ISEL, Rua Conselheiro Emidio Navarro, Lisbon, 1949-014, Portugal

ABSTRACT

We present in this paper results about the analysis of photocurrent and spectral response in a-SiC:H/ a-Si:H pinpin and pinip structures. Our experiments and analysis reveal the photocurrent profile to have a strong nonlinear dependence on the externally applied bias and on the light absorption profile, i.e. on the incident light wavelength and intensity. Our interpretation points out the cause of such effect to a self biasing of the junctions under certain unbalanced light generation of carriers and to an asymmetric reaction of the internal electric fields to the externally imposed bias. The possibility to relate such a behavior to the light intensity and wavelength indicates realistic hypothesis of using these structures and this effect for color recognition sensors.

We present results about the experimental characterization of the structures and numerical simulations obtained with the program ASCA. Considerations about electrical field profiles and inversion layers will be taken into account to explain the optical and voltage bias dependence of the spectral response. Our results show that in both structures the application of an external electrical bias (forward or reverse) mainly influences the field distribution within the less photo excited sub-cell.

INTRODUCTION

Optical properties of amorphous silicon (a-Si:H, a-SiC:H) p-i-n structures allow to identify and develop good sensors for image pattern recognition and color filtering. This has been an important topic of research in the field of sensing applications [1, 2, 3]. Work on characterization and modeling of image sensor electrical behavior [4] led the way to the development of multilayer structures of interest for image and color sensor application: p-i-n structures for pattern detection of B/W images through the LSP (Laser Scanned Photodiode) thechnique [5]; and tandem structures for additional light wavelength discrimination [6, 7].

Light filtering properties, namely light wavelength discrimination, depend on the structure of the sensor, mainly carbon doping, thickness of each pin cell absorbing layer, and the selected sequence of the cells in the multilayer structure.

A numerical ASCA simulation (Amorphous Silicon solar Cell Analysis [8]) procedure was performed for different sensor configurations. Adjusting the values of some parameters or the type of the simulation analysis, it was possible to explain some of the observed singularities of the sensor, aiming to explain color detection, light-to-dark sensitivity and image pattern recognition. The ASCA simulations show a strong dependence of the device effective functioning on the optical characteristic (absorption coefficient and optical gap) of the materials used, leading to a introduction of a-SiC:H layers into the stacked structure in order to improve RGB color screening.

EXPERIMENTAL

Figures 1 illustrates the two different sensor configurations that were tested by an experimental optical-readout procedure. The first one (Figure 1a), the pinpin sensor, consists of a glass/ITO/p-i-n a-SiC:H multilayer structure which faces the incident illumination (front diode) followed by a a-SiC:H(p)/a-Si:H(i)/a-Si:H(n) /ITO heterostructure (back diode) that allows the optically addressed readout. The second one (Figure 1b), the pinip sensor, is formed by two pin/nip diodes built one after the other. The a-SiC:H(p)/ a-Si:H(i)/ a-SiC:H(n)/ a-Si:H(i)/ a-SiC:H(p) structure consists of two back to back diodes that share the same n-type doped layer.

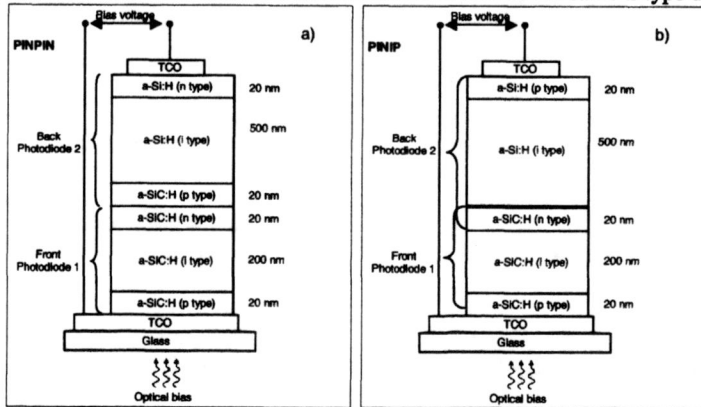

Figure 1. Schematics layer sequence of the analyzed srtructures

In both configurations, color detection is attempted based on spatially separated absorption of the different wavelengths. A trade-off between intrinsic layers thickness and light depth penetration was established to spatially decouple the generated carriers. The blue sensitivity, the red transmittance and the green complete absorption were optimized in the pinpin structure, respectively, through an a-SiC:H front absorber (200 nm) and a thick a-Si:H back absorber (500 nm). In the pinip structure the thickness of the intrinsic layers was chosen to decouple the blue from the red absorptions. Both structures were characterized by DC and AC current-voltage measurements in dark and under different optical (λ_L=450 nm, 550 nm, 650 nm) and electrical (-6 V<V<6V) bias conditions. The steady state optical bias is applied through the front diode. The DC behavior of both structures has been simulated with the ASCA program, obtaining results in coherence with the measurements. The simulation permits an insight of the internal distribution of potential and electric field and leads to an overall description of the device physics and functioning.

DISCUSSION

The a-SiC:H diode

The principal action to control light penetration and absorption, in order to obtain color separation and detection, is the introduction of an a-SiC:H diode within the structures. We have measured and simulated the electrical behavior of the a-SiC:H pin under different wavelength of the incident light and under different externally applied bias. Figure 2 shows the absorption coefficient (α) of the intrinsic a-SiC:H film as a function of the light wavelength. The high values of α for low energy radiation is an indication of a high density of localized states within the gap, as already well known in published literature about structural properties of a-SiC:H films [9]. A fit to this plot together with a choice of an high value of density of defect states (5×10^{17} cm^{-3}) has been taken as input for the ASCA simulations. Figure 3 reports the spectral response of a a-SiC:H pin diode, revealing a strong dependence of the responsivity on the applied bias when reversely polarized. Not only the peak value of the responsivity changes with the applied bias, but also the peak position move from the green toward the blue region with the increasing bias. Such dependence has not been observed under forward bias conditions. Figure 4 shows the simulated J-V characteristic of structure illuminated with different light wavelengths in the blue-green range while Figure 5 reports the simulated spectral response of the same a-SiC:H pin structure. The measured bias dependence of the responsivity is also reported by the simulation, and this effect appears in the simulated J-V curve as a very strong dependence of the shunt resistance on the applied bias and on the light wavelength and can be explained by the high density of localized states within the absorbing intrinsic layer

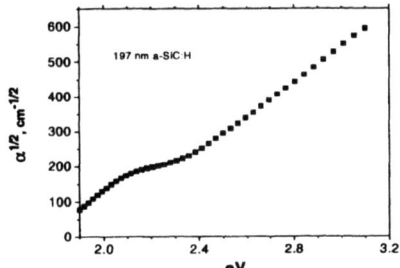

Figure 2. Absorption coefficient of an intrinsic a-SiC:H film with thickness of 197 nm

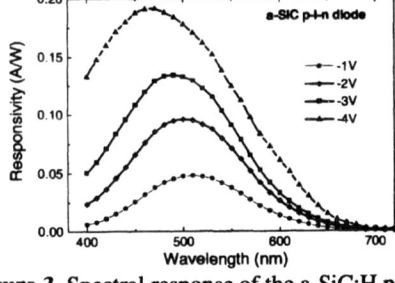

Figure 3. Spectral response of the a-SiC:H pin structure

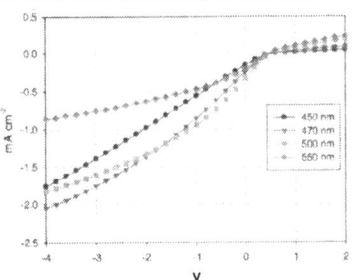

Figure 4. Simulated J-V characteristic of the a-SiC:H pin structure

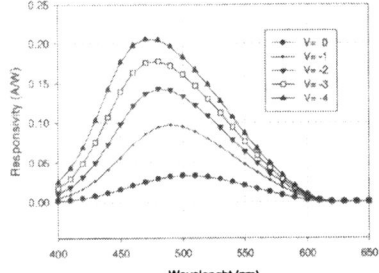

Figure 5. Simulated spectral response of the a-SiC:H pin structure

The a-SiC:H / a-Si:H pin/pin structure

The a-SiC:H / a-Si:H pin/pin structure has been intensively studied and characterized for color sensing application [6] and as an enhanced short wavelength light sensor [10]. Figure 6 reports the simulated spectral response of the a-SiC:H/ a-Si:H pin/pin structure showing a peak under reverse bias condition in the green region which corresponds to a well balanced photo-generation profile along the overall structure. The bias dependence of the responsivity under reverse bias is to be ascribed to the top a-SiC:H diode, as previously observed in Figure 4.

Figure 7 show the simulated data for recognition of blue radiation by using a DC emulation of the LSP technique obtained by comparing the spectral responses calculated with and without the superposition of a red (650 nm) light (the LSP scanner). The scanner is tuned to compensate the self biasing effect in the bottom cell caused by the asymmetry in the photo-generation profile that does not reach the a-Si:H diode under blue irradiation and the pin/pin structure shows to be very sensitive under reverse bias to the blue (450 nm) wavelength. Good separation of blue and red colors can be obtained by increasing the thickness of the a-Si:H intrinsic layer of the bottom cell up to 1500 nm and by properly tuning the intensity of the red scanner. As small variation with the applied bias are observed in the LSP signal produced under a green irradiation, process for green color extraction is presently investigated by the application of a proper conversion algorithm.

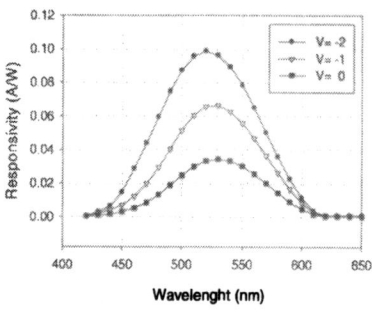

Figure 6. Simulated spectral response of the a-SiC:H/ a-Si:H pin/pin structure

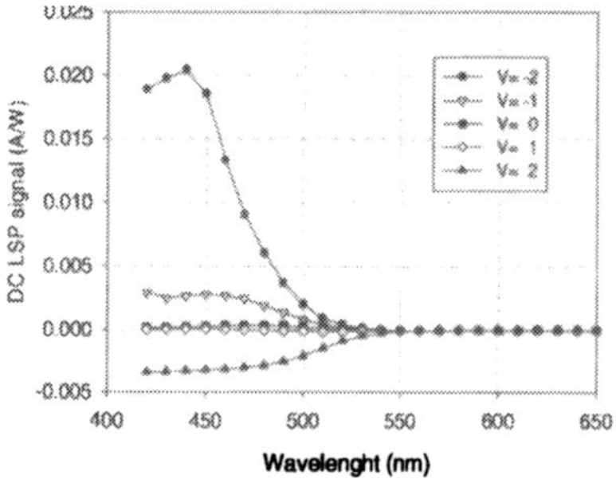

Figure 7. DC emulation of the signal extracted by the LSP technique

The a-SiC:H / a-Si:H pinip structure

In order to improve color detection, we have recently started to analyze stacked pinip structures under different condition of illumination and of applied bias. As a first approach, and as a term of comparison, a fully a-Si:H pinip structure (200 nm thickness for the intrinsic layer of the front diode, 500 nm the back one) has been considered. Figure 8 and 9 report, respectively, the simulated J-V characteristic and the spectral response. The current saturates for small values of both forward and reverse bias (we consider here forward bias when the voltage is applied to the p-layer of the front diode) showing some asymmetry only in the blue region which is due to the unbalanced photogeneration profile within the two sub-cells. Also, the spectral response does not show any defined peak, being the responsivity low when no bias is applied. We can conclude that no clear color separation can be extracted directly from these data by application of the LSP technique. In order to improve the possibility of color recognition, the top photodiode has been substituted by an a-SiC:H one (see Figure 1b). The thickness of the absorber layers is adjusted to the light absorption profile with the front a-SiC:H diode photo excited under blue light conditions and the back one under red light. Figure 10 shows the measured spectral response of the a-SiC:H/a-Si:H pinip structure. The responsivity plot shows a clear asymmetry on the polarity of the applied bias in respect to the wavelength of the incident light. The structure gives a good response under reverse bias when illuminated with blue light (absorbed mainly in the front

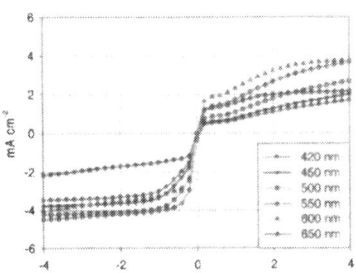

Figure 8. Simulated J-V characteristic of the illuminated a-Si:H pinip structure

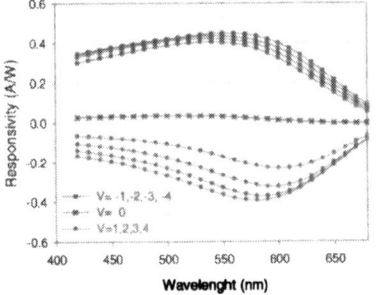

Figure 9. Simulated spectral response of the a-Si:H pinip structure

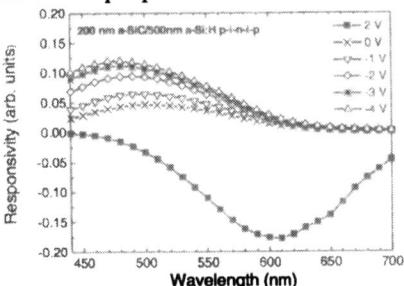

Figure 10. Measured spectral response of the a-SiC:H / a-Si:H pinip structure separation of blue and red colors

diode) and under forward bias when illuminated with yellow-red light (absorbed mainly in the back diode).

Figure 11 and 12 report, respectively, the spectral response and the J-V characteristic obtained by the ASCA simulations. The simulation results are in agreement with the experimental ones and show the same asymmetry in the responsivity plot while the symmetry in the J-V characteristic observed in the full a-Si:H structure remain still observable only under green illumination. The effect of the different wavelength of the incident light on the internal electric configuration can be observed in Figure 13, where it shown the potential distribution within the a-SiC:H / a-Si:H pinip structure for different light wavelengths. The applied reverse bias drops almost entirely on the front diode, so that, if it is photo-excited (blue light), it produces its correspondent reverse photocurrent while the back diode becomes polarized on its correspondent quiescent point (forward bias correspondent to a current value equal to the one photo-generated by the front diode) that permits the current flow along the structure. A similar mechanism is observed for the forward bias condition up to wavelengths in the yellow region, the exception is that under red light illumination the quiescent point of the (not photo-excited) forward biased front diode must be much higher to permit the flow of the photocurrent produced within the back diode.

CONCLUSION

We have presented an analysis based on ASCA numerical simulation and experimental results about the physics of a-SiC:H/a-Si:H pinip and pin/pin structures for color sensing application. It has been shown that, by properly control the layer thickness in regard to the absorption properties of the material, the insertion of a-SiC:H layers lead to a well defined

ACKNOWLEDGMENT
This work has been founded by the Portuguese FCT-LAXOR and IPL 5822/2004 projects which are gratefully acknowledged.

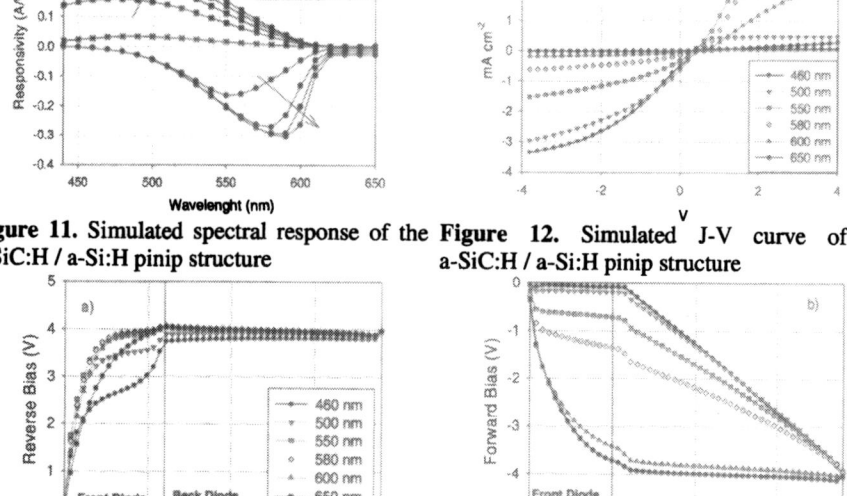

Figure 11. Simulated spectral response of the a-SiC:H / a-Si:H pinip structure

Figure 12. Simulated J-V curve of the a-SiC:H / a-Si:H pinip structure

Figure 13. Potential distribution within the a-SiC:H / a-Si:H pinip structure for different light wavelengths under 4 V reverse (a) and forward (b) bias

REFERENCES

1. G. de Cesare, F. Irrera, F. Lemmi, F. Palma, *IEEE Trans. on Electron Devices*, **42**, pp. 835-840 (1995)
2. A. Zhu, S. Coors, B. Schneider, P. Rieve, M. Bohm, *IEEE Trans. on Electron Devices*, **45**, pp. 1393-1398 (1998).
3. M. Topic, H. Stiebig, D. Knipp, F. Smole, J. Furlan, H. Wagner, *J. Non Cryst. Solids* **266-269** pp. 1178-1182 (2000).
4. M. Vieira, M. Fernandes, J. Martins, P. Louro, A. Maçarico, R. Schwarz, M. Schubert, *IEEE Sensor Journal*, 1, no.2 (August, 2001) pp. 158-167.
5. Fernandes M, Vieira M, Martins R, *Journal of Non-Crystalline Solids* 352 (9-20): 1801-1804.
6. P. Louro , M. Vieira , A. Fantoni ,M. Fernandes, C. N. Carvalho , G. Lavareda, *Sensors and Actuators* A, **123–124**, pp. 326–330 (2005)
7. J. Martins, M. Fernandes, A. Fantoni, and M. Vieira, J. Non Cryst. Solids, (2006).
8. A.Fantoni,M.Vieira,R.Martins,Math.Comput.Simul.49 (1999)
9. G. Ambrosone, U. Coscia, S. Ferrero, F. Giorgis, P. Mandracci and C. F. Pirri, Philosophical Magazine B, **82**, pp. 35-46 (2002)
10. A. Fantoni, P. Louro, M. Fernandes, M.Vieira, G. Lavareda, C.N. Carvalho, *Sensors & Actuators: A. Physical*, **123-124**, pp.343-348 (2006)

Poster Session:
Imagers, Sensors and
Novel Applications

Mater. Res. Soc. Symp. Proc. Vol. 989 © 2007 Materials Research Society   0989-A19-01

Study of a Fabrication Process and Characterization of One Dimensional Array of Uncooled Micro-bolometers Based on Germanium Films Deposited by Plasma

Mario Moreno, Andrey Kosarev, Alfonso Torres, and Roberto Ambrosio

Electronics, Institute National of Astrophysics, Optics and Electronics, L.E. No. 1 Tonanzintla., Puebla, 72840, Mexico

ABSTRACT

In our previous works, we have studied the fabrication process and characterization of single cell micro-bolometers based on germanium thin films deposited by low frequency (LF) PECVD technique at low temperature and fully compatible with the IC fabrication technology. We have demonstrated promising properties of those devices for further development of IR imaging systems [1-2].

In this work, we report the study and characterization of the fabrication process of a lineal array of 32 un-cooled micro-bolometers. We have used surface micro-machining techniques for the array fabrication onto a silicon wafer. The micro-bolometers in the array have a "bridge type" configuration. In this case, a SiN_x supporting film is suspended 2.5 µm from the substrate by two legs, which form the bridge and provide sufficient thermo-isolation to the thermo-sensing layer. The thermo-sensing layer was deposited on the bridge by using LF PECVD. The a-Ge_xSi_y:H film used in this devices showed high activation energy $E_a = 0.34$ eV, providing high thermal coefficient of resistance, TCR=α=0.043 K^{-1} and improved but still high resistance. We studied the effect of the addition of boron to the a-Ge_xSi_y:H film deposition process, for reducing its undesirable high resistance and, the resulting layer (a-$Ge_xB_ySi_z$:H) is used as thermo-sensing film in the micro-bolometers arrays. The active area of the cell in the array is A_b=70x66 µm^2 and the area of the array including interconnection lines and pads is A_A=1600x3120 µm^2. The temperature dependence of conductivity $\sigma(T)$, current-voltage characteristics I(U), and spectral noise density have been measured in the array and the main figures of merit such as, responsivity and detectivity have been obtained.

INTRODUCTION

The maturity of the MEMS structures fabrication process through the surface micro-machining techniques, has allowed the development of low cost and reliable night vision systems based on thermal detectors [3-4]. Among the thermal detectors used for IR arrays, the micro-bolometer is one of them. The main requirements for the thermo-sensing materials used in micro-bolometers are: high value of the temperature coefficient of resistance, TCR ($\alpha(T)$), defined as $\alpha(T) = (1/R)|dR/dT| = E_a/kT^2$, where E_a is the activation energy, moderated resistivity, low noise and compatibility with standard IC fabrication processes. Currently, it have been developed large arrays of un-cooled micro-bolometers that use different thermo-sensing materials. But none of them can be consider as the optimum one. Among the materials preferably used as thermo-sensing films are vanadium oxide, amorphous and polycrystalline semiconductors, and some metals [5-6]. Those materials have shown advantages and disadvantages. Vanadium oxide has a high value of TCR but it is not a standard material in IC technology, resulting in a more complex fabrication process with special installations.

Metals are compatible with the standard IC fabrication technology, have low resistance but low values of TCR, which is transformed in a low responsivity.

Amorphous silicon (a-Si:H) has showed high values of TCR and is fully compatible with the silicon technology. However, intrinsic amorphous semiconductors have very high resistance, which often cause a mismatch with the read out circuits. In order to reduce the high resistance of amorphous materials, boron doping has been employed.

In our previous work intrinsic a-Ge$_x$Si$_y$:H films obtained by PE CVD were used as thermo-sensing layers in micro-bolometers, providing high activation energy and improved, but still high resistance. We demonstrated promising properties of those devices for further development of IR imaging systems [1-2]. In this work we have added boron during the deposition process of the a-Ge$_x$Si$_y$:H thermo-sensing film for reducing its high resistance. The resulting film (a-Ge$_x$B$_y$Si$_z$:H) was also used as thermo-sensing film in micro-bolometers. With these films, we have fabricated one dimensional arrays (1-D) of 32 elements with the both types of thermo-sensing films: a-Ge$_x$Si$_y$:H and a-Ge$_x$B$_y$Si$_z$:H. We selected one cell of each array and such characteristics as responsivity, spectral density of noise, and detectivity were compared.

EXPERIMENT

The micro-bolometer arrays fabrication process is as follows. A 200 nm-thick SiO$_2$ layer was thermally grown on a c-Si wafer and a 2.5 μm-thick aluminum layer was deposited by e-beam evaporation over it, the aluminum layer is used as sacrificial film. A lithographic step and wet etching is carried out in order to pattern the aluminum layer. A 0.8 μm-thick SiN film was deposited at low temperature (350°C) by low frequency PE CVD over the aluminum pattern. The SiN film is patterned by reactive ion etching (RIE) in order to form a SiN bridge over the aluminum pattern. A 200 nm-thick titanium layer was deposited by e-beam evaporation over the SiN bridge, and it was patterned in order to form the contacts. A 0.5 μm-thick thermo-sensing a-Ge$_x$Si$_y$:H film was deposited over the Ti contacts by low frequency LF PECVD technique at a rf frequency f=110 kHz, temperature T=300 °C, power W=350 W and pressure P=0.6 Torr. The a-Ge$_x$Si$_y$:H film was obtained from a SiH$_4$ (100%)+ GeF$_4$ (100%) + H$_2$ (100%) mixture with gas, at the following gas flows: Q$_{SiH4}$=25sccm, Q$_{GeF4}$ =25 sccm, Q$_{H2}$=1000 sccm, resulting in a Ge content in solid phase Y=0.88 and a Si content in solid phase Y=0.11, SIMS was used for the content determination. The thermo-sensing film was covered with a 0.2 μm-thick SiN film deposited by PE CVD. The active area was patterned by RIE and finally, the aluminum sacrificial layer was removed by wet etching.

The boron alloy thermo-sensing film (a-Ge$_x$B$_y$Si$_z$:H), was deposited with the same conditions as those for of the a-Ge$_x$Si$_y$:H film, but with different gas mixture. The film was deposited from a SiH$_4$ (100%) + GeF$_4$ (100%) + B$_2$H$_6$ (1%) mixture with the following gas flows: Q$_{SiH4}$=50sccm, Q$_{GeF4}$ =50 sccm, Q$_{B2H6}$=5 sccm. This results in a Ge content in solid phase Y=0.675, B content in solid phase Y=0.262 and Si content in solid phase Y=0.055.

The active area of the thermo-sensing layer is A$_b$=70x66μm^2 and the area of the array including interconnection lines and pads is A$_A$=1600x3120 μm^2. A SEM picture of a top view of one micro-bolometer in the array is shown in Figure 1 a) and a SEM picture of a fragment of the 1-D array is shown in Figure 1b). The performance of the micro-bolometers in the arrays was studied through the measurement of I(U) characteristics in dark and under IR illumination conditions. The source of IR light is a "Globar", which provides an intensity I$_0$=5.3x10^{-2} W/cm^2 on the surface of the sample. The samples were placed in a vacuum thermostat and illuminated through a zinc selenide window (ZnSe). The window has a 70% of transmission in the range of λ=0.6 – 20 μm. The I(U) measurements were performed at pressure P≈10 mTorr, at room temperature, current was measured with an electrometer ("Keithley"- 6517-A) and the applied voltage was varied from U = 0 to 7 V. The responsivity was calculated from the I(U) measurements. Noise measurements in the micro-bolometers were performed with a lock-in amplifier ("Stanford Research Systems" - SR530). The noise of the system and the total noise (system + cell noise) were measured separately, and a subtraction of the system noise allowed us to obtain the noise of the device. The detectivity was calculated from the I(U) characteristics and noise measurements.

 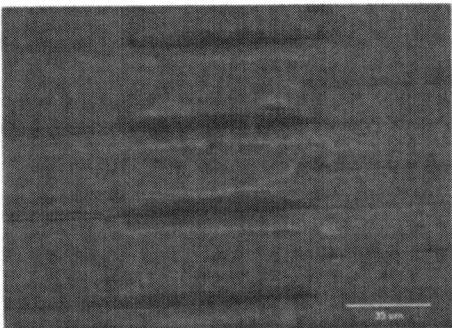

Figure 1. a) Top optical view of one micro-bolometer in the array, b) SEM picture of a fragment of 1-D 32 elements array.

DISCUSSION

We have measured the temperature dependence of conductivity with increasing and decreasing temperature, in the range of 300K to 400K on both the a-Ge_xSi_y:H, and the a-$Ge_xB_ySi_z$:H thermo-sensing films. The activation energy for the a-Ge_xSi_y:H thermo-sensing film measured in a test structure is E_a=0.34 eV, which provides a TCR α=0.043 K^{-1} and a measured conductivity at room temperature σ_{RT}=6x10^{-5} $Ohm^{-1}cm^{-1}$. The activation energy of the a-$Ge_xB_ySi_z$:H film measured in a test structure is E_a=0.21 eV, which results in a TCR α=0.027 K^{-1} and the conductivity at room temperature σ_{RT}=1.3x10^{-2} $Ohm^{-1}cm^{-1}$.

For a complete characterization and comparison of the devices with the different thermo-sensing films, a device from the array with a-Ge_xSi_y:H film, and other from the array with a-$Ge_xB_ySi_z$:H film were selected. The current-voltage I(U) characteristics in dark and under IR illumination are shown in Figure 2 for both type of micro-bolometers in the arrays. Linear I(U) characteristics are observed for both type of devices, as it is demonstrated in the inserts in Figures 2 a) and 2 b).

Figure 3 shows the responsivity for the micro-bolometers with the a-Ge_xSi_y:H (3a) and a-$Ge_xB_ySi_z$:H thermo-sensing films (3b). The responsivity R_I, is defined as R_I=(I_{IR}-I_d)/I_0, where, I_{IR} is the current under IR illumination, I_d is the dark current, and I_0 is the incident IR intensity. From Figure 3 it can be seen that R_I for the a-Ge_xSi_y:H thermo-sensing film micro-bolometer, is one order of magnitude smaller than that of the a-$Ge_xB_ySi_z$:H thermo-sensing film micro-bolometer for the same bias voltage and IR illumination conditions. However, the relative responsivity (I_{IR}/I_d) is larger for the a-Ge_xSi_y:H film micro-bolometer.

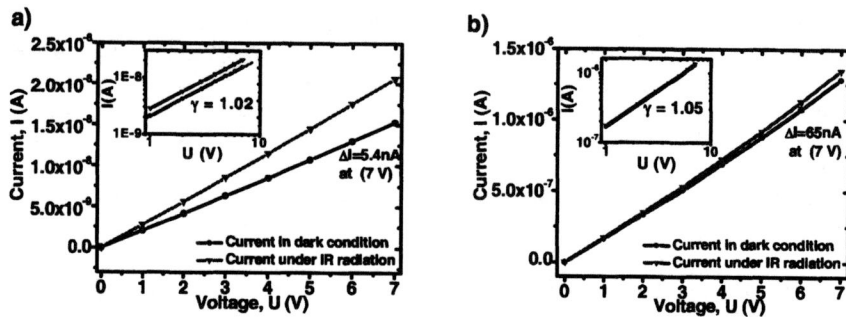

Figure 2. I(U) characteristics of micro-bolometers for: a) a-Ge$_x$Si$_y$:H and b) a-Ge$_x$B$_y$Si$_z$:H thermo-sensing films. In the inserts the I(U) curves are shown in double log scales.

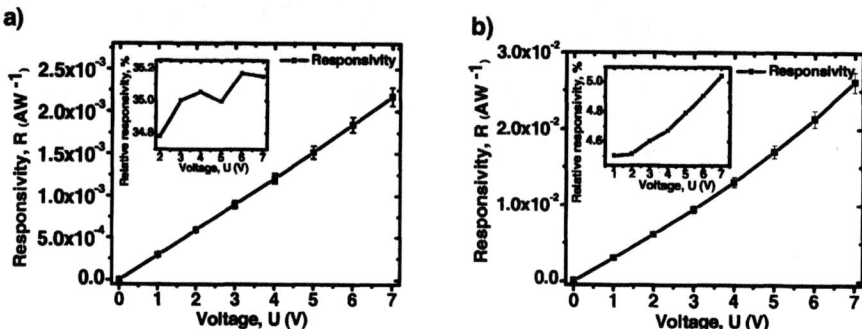

Figure 3. Responsivity of micro-bolometers for: a) a-Ge$_x$Si$_y$:H thermo-sensing film and b) a-Ge$_x$B$_y$Si$_z$:H thermo-sensing film. The inserts show relative responsivity versus bias voltage.

The current noise spectral density (NSD) $I_{cell\ noise}$ (f) for both micro-bolometers is shown in Figure 4. Where $I_{cell\ noise}(f) = I_{cell\ +\ system\ noise}(f) - I_{system\ noise}(f)$, $I_{cell\ +\ system\ noise}(f)$ is the NSD measured at the micro-bolometer cell in with the measuring system. The $I_{system\ noise}(f)$ is the NSD measured in the system without the micro-bolometer cell. The $I_{cell\ noise}$ (f) values observed for a-Ge$_x$Si$_y$:H thermo-sensing film micro-bolometer are around one order of magnitude smaller than that of the a-Ge$_x$B$_y$Si$_z$:H film micro-bolometer as it is presented in Figure 4.

From the measured responsivity and noise we have calculated detectivity D* for both devices. For The a-Ge$_x$Si$_y$:H thermo-sensing film device we obtained D*= 7x10^9 cmHz$^{1/2}$W^{-1} while for the a-Ge$_x$B$_y$Si$_z$:H micro-bolometer D*= 5.9x10^9 cmHz$^{1/2}$W^{-1}. It is important to note that both values of detectivity are very similar. This is because the a-Ge$_x$B$_y$Si$_z$:H micro-bolometer show both higher responsivity and higher noise with respect to those in a-Ge$_x$Si$_y$:H film device. The measured performance characteristics of the devices here studied are listed in Table 1, and are compared with data recently reported in literature.

Additionally, we have studied the yield of the fabrication process here presented. It is found that the 1-D arrays with a-Ge$_x$Si$_y$:H film resulted in an average yield of 50% (on 15 measured arrays containing 476 cells), while the 1-D arrays using a-Ge$_x$B$_y$Si$_z$:H film had an average yield of 60% (19 measured arrays containing 608 cells). In order to compare the average responsivity of the fabricated arrays with the films studied, we have selected one array of each type. After measurements of each one of the cells conforming the array it was found the following: The 1-D array with the a-Ge$_x$Si$_y$:H micro-bolometers resulted in a yield of 60%, with an average responsivity of 1.2 x10^{-3} AW^{-1}. While the 1-D array with the a-Ge$_x$B$_y$Si$_z$:H micro-bolometers showed a yield of 85% with an average responsivity of 1.5 x10^{-2} AW^{-1}. Figure 5 shows the normalized responsivity (R / R$_{max}$) for both micro-bolometers arrays.

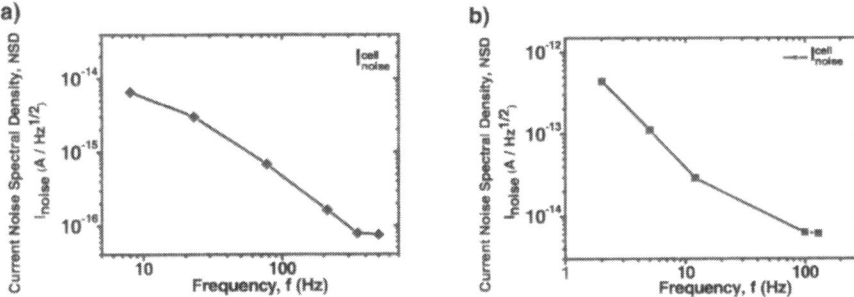

Figure 4. Current noise spectral density I$_{noise}$, as a function of frequency for the micro-bolometers with a) a-Ge$_x$Si$_y$:H thermo-sensing film and b) a-Ge$_x$B$_y$Si$_z$:H thermo-sensing film.

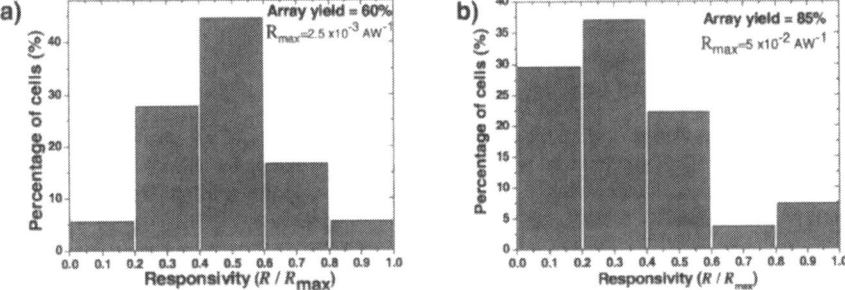

Figure 5. Yield of the cells as a function of normalized responsivity of two 1-D arrays: a) a-Ge$_x$Si$_y$:H thermo-sensing film micro-bolometers and b) a-Ge$_x$B$_y$Si$_z$:H thermo-sensing film micro-bolometers.

CONCLUSIONS

1-D arrays of 32 elements with two different thermo-sensing films: intrinsic thermo-sensing film (a-Ge$_x$Si$_y$:H) and boron alloy thermo-sensing film (a-Ge$_x$B$_y$Si$_z$:H) have been fabricated and characterized. From the characterization we can conclude: The intrinsic thermo-sensing film has larger activation energy, E$_a$ = 0.34 eV and consequently higher TCR than the boron alloy thermo-sensing film, E$_a$ = 0.21 eV. However, the intrinsic film resistivity is 3 orders of magnitude larger than the boron alloy film resistivity. The responsivity of the boron alloy film micro-bolometer is one order of magnitude larger than that of the intrinsic film micro-bolometer. The boron alloy film micro-bolometer resulted in a noise density of 1 order of magnitude larger than that of the intrinsic film micro-bolometer. As a consequence, the detectivity of both devices is very similar, ($D^* = 7 \times 10^9$ cmHz$^{1/2}$W^{-1} for the intrinsic film micro-bolometer and $D^* = 5.9 \times 10^9$ cmHz$^{1/2}$W^{-1} for the boron alloy film device). In spite of that, both devices showed similar detectivity and the boron alloy film micro-bolometer showed 2 orders of magnitude less resistance than that of the intrinsic material. The comparison in the average responsivity of both type of arrays resulted in the following: The 1-D array with the intrinsic film micro-bolometers had an average responsivity of 1.2×10^{-3} AW^{-1} while the 1-D array with the boron alloy film micro-bolometers had an average responsivity of 1.5×10^{-2} AW^{-1}.

Table 1. Comparison of the two types of micro-bolometers with literature.

Thermo sensing layer	E_a, eV	TCR,α K^{-1}	Pixel area, A_b, μm²	Pixel resistance, R_b, Ohm	Voltage responsivity $\mathfrak{R}_U$, VW^{-1}	Current responsivity $\mathfrak{R}_I$, AW^{-1}	Spectral Response, μm	Detecti vity, D^*, cmHz$^{1/2}$W^{-1}	References
Ge$_x$Si$_{1-x}$ O$_y$	-	0.048	50 x 50	-	1x10^5	-	10	6.7x10^8	[5]
a-Si:H,B	-	0.028-0.039	48 x 48	3 x10^7	10^6	-	5 - 14	-	[6]
a-Ge$_x$Si$_y$:H	0.34*	0.043	70 x 66	5x10^8	7.2x10^5	2x10^{-3}	2 - 14	7x10^9	a-Ge$_x$Si$_y$:H micro-bolometer
a-Ge$_x$B$_y$Si$_z$:H	0.21*	0.027	70 x 66	6x10^6	2.8x10^5	2.6x10^{-2}	2 - 14	5.9x10^9	a-Ge$_x$B$_y$Si$_z$:H micro-bolometer

* Measured in a test structure.

ACKNOWLEDGMENTS

The authors acknowledge the support of this research by CONACyT project No. 48454 and greatly appreciate Dr. Yu. Kudriavtsev, CINVESTAV, Mexico, for SIMS measurements and analysis. M. Moreno acknowledges CONACyT for support granted through scholarship # 166011.

REFERENCES

1. R. Ambrosio, A. Torres, A. Kosarev, A. Illinski, C. Zúñiga, A. S. Abramov, *J. Non-cryst. Solids*, 338-340, 91-96 (2004).
2. A. Torres, A. Kosarev, M. L. García Cruz, R. Ambrosio, *J. Non-cryst. Solids*, 329, 179 - 183 (2003).
3. R. Ambrosio, A. Torres, A. Kosarev, M. Moreno, *Book of Abstracts ICANS 21 – Science and Technology*, WO4.4, 215 (2005).
4. T. Adrega, M. Almeida, D.M.F. Prazeres, V. Chu, J.P. Conde, *Journal of Non-Crystalline Solids*, Volume 352, Issues 9-20, Pages 1999-2003, June 2006.
5. A. H. Z. Ahmed, R. N. Tait, *IEEE Trans Electr.Dev.*, v52(8) 1900-1906 (2005).
6. A. J. Syllaios, T. R. Schimert, R. W. Gooch, W. L. Mc.Cardel, B. A. Ritchey, J. H. Tregilgas, *Mat. Res. Soc. Symp. Proc.* 609 A14.4.1 (2000).

Mater. Res. Soc. Symp. Proc. Vol. 989 © 2007 Materials Research Society   0989-A19-02

Preliminary Results on Large Area X-ray a-SiC:H Multilayer Detectors with Optically Addressed Readout

Manuela Vieira[1], Yuri Vygranenko[1,2], Miguel Fernandes[1], Paula Louro[1], Pedro Sanguino[1], Alessandro Fantoni[1], and Reinhard Schwarz[3]

[1]DEETC, ISEL, Rua Conselheiro Emidio Navarro, Lisbon, 1949-014, Portugal
[2]Electrical and Computer Engineering, University of Waterloo, Waterloo, Ontario, N2L3G1, Canada
[3]Fisica, IST, Lisbon, 1049-001, Portugal

ABSTRACT

This paper investigates the feasibility of using a large area image sensor with an optically addressed readout for medical X-ray diagnostic imaging. The device prototype comprises a multilayer glass/ZnO:Al/p (a-SiC:H)/i (a-Si:H)/ n (a-SiC:H)/ i(a-Si:H)/p (a-SiC:H)/ a-SiN$_x$/ITO structure coupled to a scintillator layer. Here, the p-i-n-i-p structure works in both sensing and switching modes depending on the biasing conditions. A numerical simulation is used to optimize the semiconductor layer thicknesses in order to achieve a photocurrent matching between back-to-back diodes in switching mode. The charge carrier transport within the p-i-n-i-p structure is also analyzed under different electric and optical biasing conditions. A physical model supports the results.

INTRODUCTION

Flat-panel digital radiography systems have recently been introduced as a new digital radiography technology [1, 2]. Modern hydrogenated amorphous silicon (a-Si:H) based radiographic imagers have active areas up to 17x17 sq. inch with the pixel pitch ranging from 127 to160 μm [3]. 3D pixel architecture was developed to integrate a switch use TFT with photodiode coupled to scintillator [4]. Specially designed charge amplifiers with extreme performances are used to attain a system noise in the range 1000-2000 electrons. The complexity of the array and associated electronics make these devices very costly. While these systems meet the requirements of the technology in terms of size and imaging performance, numerous innovations can be expected to increase the sensitivity and resolution, improve the device yield, as well as to reduce device cost.

Typical pixel architecture consists of an a-Si:H n-i-p sensing diode and a TFT switch. The signal charge is stored on the sensor capacitance during the integration period, and is subsequently transferred to an external charge-sensitive amplifier during readout phase via the TFT switch. An alternative pixel design consists on the integration of the sensing element with an optoelectronic switch in order to enable an optically addressed readout.

This paper reports on design, fabrication, characterization and modeling of an X-ray image sensor with an optically addressed readout. The implemented concept is the conversion of the X-ray image into an optical image by a scintillator and the conversion of the optical image into an electric one by a large area Laser Scanned Photodiode (LSP). We discuss the device operating principles, along with device characterization and numerical simulation results.

EXPERIMENT

Figure 1 shows a schematic diagram of an X-ray image sensor. The device comprises a glass/ZnO:Al/a-Si:H p-i-n-i-p/a-SiN$_x$:H/ITO structure optically coupled to a phosphor layer. The pixels are defined by patterning the top p-layer. The fabricated imager prototype has 42x42 pixels with a pixel pitch of 250 um. The area of the top ITO electrode is 1.2x1.2 cm^2. In order to decrease the lateral currents that could lead to image smearing, the conductivity of the 30-nm-

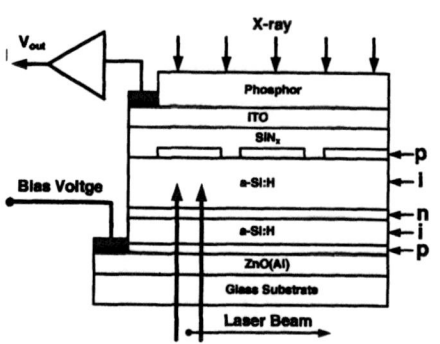

Figure 1. Schematic diagram of an X-ray image sensor.

thick n-layer was reduced by introducing methane during the deposition process [5]. A lightly doped p-type a-SiC:H with an optical bandgap of 1.95 eV was used in order to reduce the absorption loss in the 25-nm-thick p-layers. To avoid the defect creation at the p-i interface, which can cause an increase in the leakage current, a graded 4-nm-thick a-SiC:H layer is inserted between the i- and p-layers. The i-layer thicknesses in the top and bottom diodes are 600 nm and 250 nm, respectively. The a-Si:H, a-SiC:H and SiN$_x$ films were deposited on a parallel-plate PECVD reactor at 220 °C. The deposition conditions have been described elsewhere [6].

Figure 2 shows an equivalent circuit and timing diagrams of the bias voltage, light and output signals. The pixel circuit includes a capacitor for charge storage C$_s$ and back-to-back photodiodes. The capacitor comprises a segment of the p-layer, and the SiN$_x$ and ITO layers. Depending on the biasing conditions the p-i-n-i-p structure is used for signal detection in photoelectric conversion mode and as a switching element in readout mode.

During the conversion period positive voltage is applied to the bottom electrode causing a potential redistribution between the charge storage capacitor and p-i-n-i-p structure, defined by the ratio of their geometrical capacitances. Since the SiN$_x$ layer is thinner than the a-Si:H stack, the voltage drop across p-i-n-i-p structure is larger than the one across the SiN$_x$ layer. The light emitted by the phosphor layer is absorbed at the top n-i-p diode, which is reverse-biased, and the generated photo-charge is accumulated on the storage capacitor.

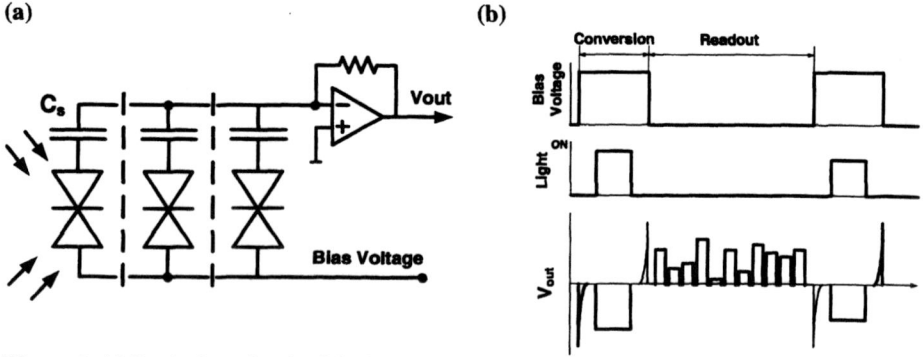

Figure 2. (a) Equivalent circuit of the image sensor and (b) signal timing diagrams.

During the readout period the bias voltage is set to zero, and the p-i-n-i-p switches are consecutively biased by the scanning laser beam (λ=633 nm) to transfer the signal charge from pixels to an external charge amplifier. The scanner is based on a two-axis deflection system capable of high-speed scan, where two additional photodiodes generate the synchronization signals used for the image restoration process. The output signal is converted to digital format by a signal acquisition card installed on a computer. Further processing algorithms like fixed pattern noise suppression are performed by software. The developed setup is described elsewhere [7].

RESULTS AND DISCUSSION

Figure 3 shows a typical current-voltage characteristic of the test glass/metal/p-i-n-i-p/ITO structure under a photon flux of 10^{15} cm^{-2}s^{-1} at a wavelength of 633 nm. The light is incident trough the bottom electrode where the bias voltage applied. The insert in Figure 3 shows the equivalent circuit of the test structure comprising two back-to-back diodes D_1 and D_2, two photocurrent sources I_1 and I_2, and a series resistor Rs. For the p-i-n-i-p structure under quasi-monochromatic illumination the amplitude of the current in the saturation region, which is I_1 or I_2 depending on the polarity of the applied bias voltage, is determined by wavelength of the incident light and i-layer thicknesses. For the measured J-V characteristics the photocurrents are about the same at both positive and negative polarities because the i-layer thicknesses have been optimized for a scanning beam wavelength of 633 nm.

In the transition region under low bias voltages the current-voltage dependence is linear and the switching structure can be characterized by its on-state resistance R_{ON} and open circuit voltage V_{OC}. For back-to-back diodes with the same diode ideality factor n the on-state resistance is

$$R_{ON} = \frac{4nkT}{e(I_1 + I_2)} + R_s,$$ (1)

where Rs is the series resistance. For the optimized p-i-n-i-p structure I_1=I_2 at a scanner wavelength λ_s yielding

$$R_{ON} = \frac{2nkT}{e \cdot R(\lambda_s) \cdot P_\lambda} + R_s,$$ (2)

where $R(\lambda_s)$ is the spectral responsivity of the switching structure, P_λ the incident light power. The power of the scanning beam should be adjusted to achieve an efficient charge transfer from the pixel to the charge amplifier for a readout time T_r=5C_sR_{ON}. For the fabricated device, C_s=23 pF, R=0.16 A/W at λ_s=630 nm, n=1.5, and assuming $R_{ON} \gg$ Rs, the light power is determined to be 1.1 µW at T_r=50 µs.

Figure 4 shows a processed line signal at an amplitude of the bias pulse of 5 V. The image pattern was formed using a shadow mask having three slits. The line scan frequency was 400 Hz. The observed image smearing is caused by an optically induced pixel cross-talk. The laser beam should be accurately focused to form a light spot much smaller than the pixel. The smearing effect can be suppressed by improving the optical system and implementing a grating metal layer between the glass substrate and bottom electrode.

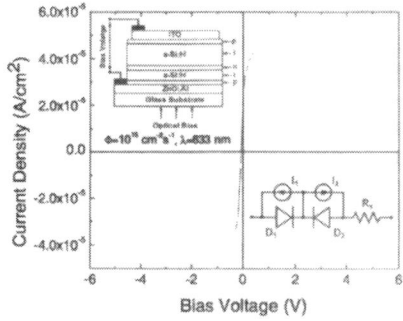
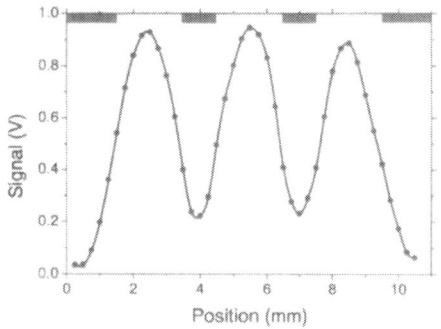

Figure 3. J-V characteristic of the p-i-n-i-p structure under a photon flux of 10^{15} cm^{-2}s^{-1} at a wavelength of 633 nm.

Figure 4. Measured line signal. The image pattern is formed using a shadow mask having three slits.

A device simulation program AMPS-1D was used to analyze the behavior of the a-Si:H p-i-n-i-p structure under illumination [8]. The device has been simulated for the same layer thicknesses used in the actual device. The charge carrier density profile and an energy band diagram of the p-i-n-i-p structure in thermodynamic equilibrium are shown in Figure 5a and 5b, respectively. Electron and hole densities in the n- and p-layers are 2×10^{14} cm^{-3} and 4×10^{14} cm^{-3}, respectively. The model predicts a charge carrier accumulation at both p-i and n-i interfaces due to the conduction and valance band offsets.

Figure 6 shows the calculated current-voltage characteristic of the p-i-n-i-p structure under a photon flux of 10^{15} photons·cm^{-2}·s^{-1} (λ=630 nm). The photocurrent slowly saturates with increasing bias, i. e. the p-i-n diode with 600-nm-thick i-layer is reverse-biased. To explain this behavior we analyzed the charge carrier transport under different biasing conditions. Figure 7 shows the energy band diagrams of the p-i-n-i-p structure under bias voltages of +3 V and -3 V.

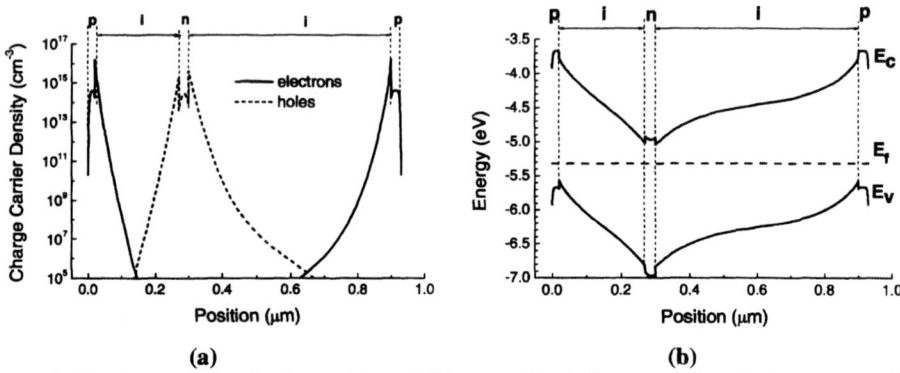

(a)   (b)

Figure 5. (a) Charge carrier density profile and (b) energy band diagram of the p-i-n-i-p structure in thermodynamic equilibrium.

466

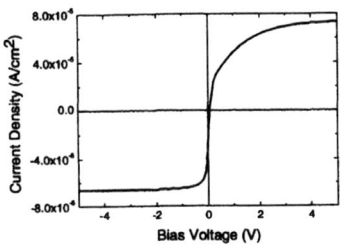

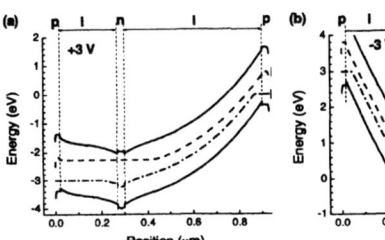

Figure 6. Simulated I-V characteristic of the p-i-n-i-p structure under a photon flux of 10^{15} photons·cm^{-2}·s^{-1}, λ=630 nm.

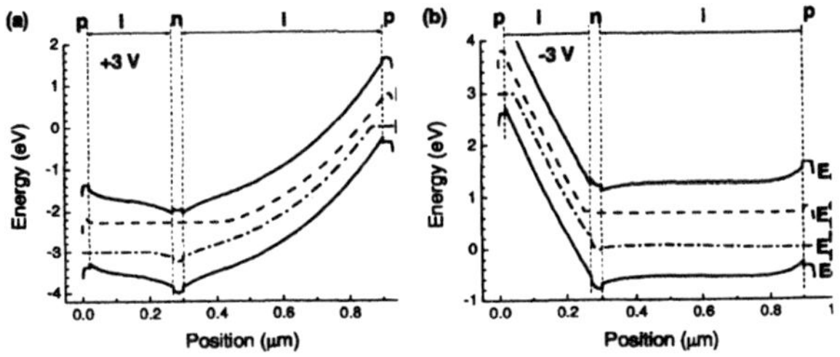

Figure 7. Energy band diagrams of the p-i-n-i-p structure at different bias voltages: (a) +3 V and (b) -3 V. The light bias is 10^{15} photons·cm^{-2}·s^{-1} at a wavelength of 630 nm.

If a negative voltage is applied to the bottom electrode, the potential drop occurs mostly across the thin i-layer of the top n-i-p diode, the electric field strength is high yielding an efficient charge collection. Under positive polarity the band bending is not uniform across the i-layer in the thick p-i-n diode. Figure 8 shows the generation-recombination profile within the p-i-n-i-p structure for this case. It reveals that in the reverse-biased diode the recombination rate is highest within a position range of 100-200 nm from the n-i interface. The dominant recombination mechanism is determined to be through acceptor-like Gaussian states. Figure 9 shows the simulated electric field profile within the p-i-n-i-p structure under illumination for different bias voltages. The electric field strength is also low in the same position range. Thus, the effective carrier transport through the 600-nm-thick i-layer requires strong electric fields.

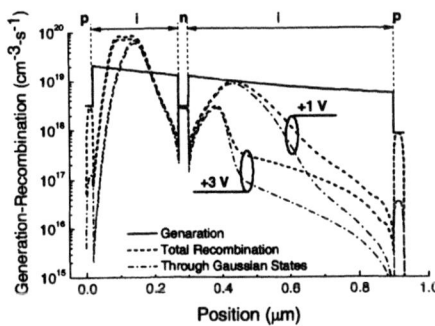

Figure 8. Generation-Recombination profile within the p-i-n-i-p structure under a photon flux of 10^{15} photons·cm^{-2}·s^{-1}, λ=630 nm

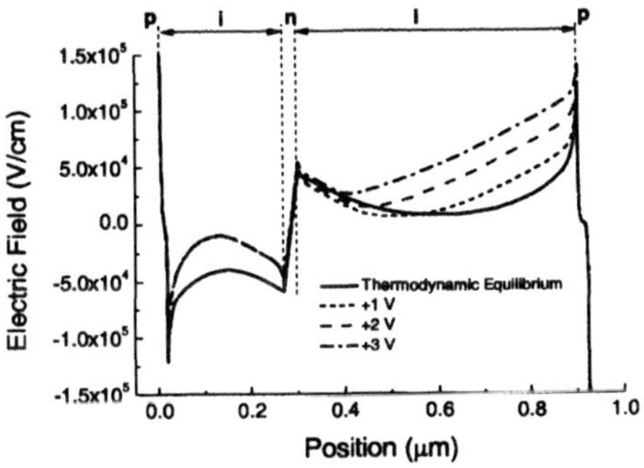

Figure 9. Electric field profile within the p-i-n-i-p structure under illumination at different bias voltages.

CONCLUSIONS

We have presented a novel X-ray image sensor with an optically addressed readout. It is demonstrated that the p-i-n-i-p/insulator/TCO structure can be used as an efficient imager backplane. The device design and readout conditions have been optimised using a physical model. Further optimization of the device design has to be done to improve the image performance.

ACKNOWLEDGMENTS

This work has been financially supported by POCI/CTM/56078/2004, research grant SFRH/BPD/20264/2004 and by Fundação Calouste Gulbenkian.

REFERENCES

1. R. A. Street (Ed.), in *Technology and Applications of Amorphous Silicon* (Springer, Berlin, 2000) pp.147-221.
2. 1. J. Beutel, and H. L. Kundel, Handbook of Medical Imaging, SPIE Press, 2000.
3. Xiujiang J. Rong et al. *Med. Phys.* 28 (11) 2328 (2001).
4. J.-P. Moy, *Nuclear Instruments and Methods in Physics Research,* A 442, (2000) 26.
5. M. Vieira, M. Fernandes, P. Louro, R. Schwarz, M. Schubert, *J. Non-Cryst. Solids,* 299–302 (2002) 1245
6. Y. Vygranenko, P. Louro, M. Vieira, J. Chang, and A. Nathan, *J. Non-Cryst. Solids,* 352 (2006)1837.
7. M. Fernandes, M. Vieira and R. Martins, *Sensors and Actuators A-Physical,* 115 (2004) 357.
8. http://www.cneu.psu.edu/amps/default.htm.

Mater. Res. Soc. Symp. Proc. Vol. 989 © 2007 Materials Research Society   0989-A19-03

Modeling the Laser Scanned Photodiode S-Shaped J-V Characteristic

Miguel Fernandes[1], Manuela Vieira[1], and Rodrigo Martins[2]
[1]DEETC, ISEL, Rua Conselheiro Emidio Navarro, 1, Lisbon, 1949-014, Portugal
[2]Materials Science Dept., FCT-UNL, Campus da FCT-UNL, Quinta da torre, Lisbon, 2825 Monte da Caparica, Portugal

ABSTRACT

The devices analyzed in this work present an S-shape J-V characteristic when illuminated. By changing the light flux a non linear dependence of the photocurrent with illumination is observed. Thus a low intensity light beam can be used to probe the local illumination conditions, since a relationship exists between the probe beam photocurrent and the steady state illumination. Numerical simulation studies showed that the origin of this S-shape lies in a reduced electric field across the intrinsic region, which causes an increase in the recombination losses. Based on this, we present a model for the device consisting of a modulated barrier recombination junction in addition to the p-i-n junction. The simulated results are in good agreement with the experimental data.

Using the presented model a good estimative of the LSP signal under different illumination conditions can be obtained, thus simplifying the development of applications using the LSP as an image sensor, with advantages over the existing imaging systems in the large area sensor fields with the low cost associated to the amorphous silicon technology.

INTRODUCTION

The Laser Scanned Photodiode (LSP) concept enables the fabrication of large area image sensors with application in fields where low cost, and design simplicity are of major importance [1,2]. The device is a large area p-i-n amorphous silicon based structure deposited over a glass substrate by a PECVD technique with two semitransparent electrical contacts. In order to decrease the conductivity of the doped layers methane is introduced in the reactor during the growth of the n and p layers. This step is of major importance since the lateral currents in the doped layers degrade the sensor performance.

The LSP technique relies on the fact that the photocurrent of the device does not change linearly with the light intensity, and thus the local illumination conditions over the active area of the structure can be evaluated by measuring the current generated by a low power light beam. In order to extract the image information the low power modulated light beam scans the active area of the device, in raster mode, and the photocurrent generated in each position is measured at the electrical contacts by a lock-in amplifier. By recording the photocurrent measured on each point of the sensor an image of the light pattern captured by the sensor is obtained without the need of any additional signal processing.

In this paper a model for p-i-n heterojuntion photodiodes based on the modulated barrier photodiode is presented, and the results from simulation are compared with the experimental data.

EXPERIMENT

The devices investigated in this work are large area (4 x 4 cm²) amorphous silicon p-i-n structures in the assembly glass/ZnO:Al/p-(Si:H)/i-(Si:H)/n-(Si:H)/Al. The semiconductor layers were fabricated by Plasma Enhanced Chemical Vapor Deposition, at 13.56 MHz radio frequency, the deposition conditions are presented elsewhere [3] (sample #m06301). The front transparent contact (ZnO:Al) was produced by rf-sputtering and the metal back contact (Al) by thermal evaporation. In order to decrease the conductivity of the doped layers methane is introduced in the reactor during the growth of both layers. This step is of major importance since the lateral currents in the doped layers degrade the sensor performance, decreasing the spatial resolution.

The current to voltage characteristics of the samples was evaluated with the device illuminated by the p side under different illumination conditions. The illumination came from a 150W tungsten-halogen lamp with a collimating/homogenizing optics followed by interchangeable color and neutral density filters. The color filters were broadband interference filters with 50 nm bandwidth with wavelengths centered in 450 nm, 550 nm and 650 nm. Fig. 1 shows the measured J-V characteristics under different photon fluxes under steady state illumination with filtered light ($\lambda_L =450$ nm).

Results from the analyzed samples show that, for low electrical bias the curve deviates from the standard diode curve, showing a significant decrease of the short circuit current. This S-shaped J-V characteristic is similar to the one observed in a-Si/c-Si heterojunctions [4].

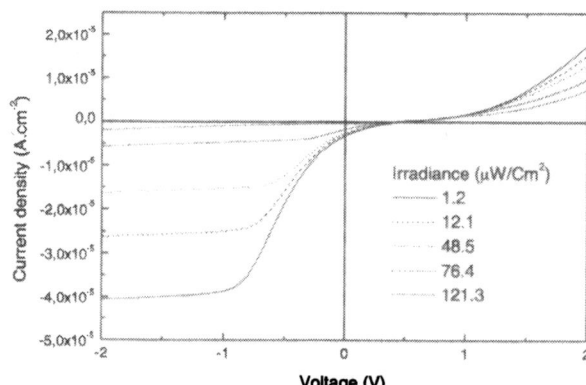

Figure 1. Experimental J-V characteristics obtained under different illumination conditions with filtered light (450 nm).

DISCUSSION

The use of relatively simple electrical models for describing the complex physical operation of electronic devices is a powerful tool in research. Frequently, the variations in the device structure cause a change in the terminal characteristics of the device making the existent model inaccurate, requiring a further enhancement of the model in order to take into account the new physical processes. The low doping concentration and incorporation of carbon in a p-i-n structure leads to an S-shaped J-V characteristic, as seen in figure 1, in this way, the standard diode model [5] fails to predict the behavior of the device in the low bias range. The analytical model is given by Eq. (1) and is schematically depicted in figure. 2. Here, the photo induced current I_{ph} is added as a current source to the current through the diode.

$$I = -I_{ph} + I_0\left(\exp\frac{q(V - R_s I)}{\eta kT} - 1\right) + \frac{V - R_s I}{R_p} \tag{1}$$

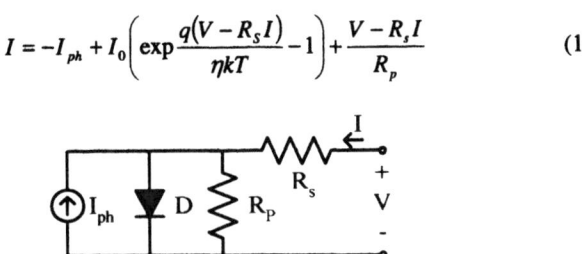

Figure 2. Equivalent circuit diagram of the one diode macroscopic model describing a photodiode.

Where q is the electron charge, I_0 represents the diode saturation current, k is the Boltzman constant, T the temperature in Kelvin, V the external applied bias, η is the ideality factor of the diode, R_s its serial resistance, and R_p the parallel resistance.

This model accurately describes the behaviour of typical solar cells and photodiodes by choosing the convenient set of parameters and it can be still used to model the present devices in the large bias regime. From the forward bias portion of the data, an estimate of the series resistance can be obtained, while the shunt resistor is determined from the analysis of the reverse bias curve, after the saturation of the current. The aim of this work is to adapt the standard model in order to achieve a best agreement with the experimental data.

To gain insight on the physical processes inside the device we used the AMPS device simulation tool to analyse the carrier transport under AM1.5 at zero voltage bias [6]. The typical band tail and gap state parameters of amorphous materials were taken as a basis, while the doping level was adjusted in order to obtain approximately the same layer conductivity as in the tested samples.

Figure 3. shows the simulated band diagrams for samples with and without carbon on the n -layer. In the heterojunction the potential drop occurs mainly across the low conductivity and wide gap doped layer, and the electric field strength decreases significantly inside the i-layer, while in the homojunction the electric field remains high across the entire absorber layer. Due to the band offset observed in the heterojunctions the photogenerated electrons and holes tend to accumulate at the i-(Si:H)/n-(SiC:H) interface where they are trapped or recombine, giving rise to a potential barrier limiting the charge transport. The SiC:H doped layers act as blocking layers and prevent, under illumination, the electrons and holes to be injected into the i-layer. Here, the field strength is not enough to sweep the photocarriers to the contacts before they recombine resulting in a low collection. The field-dependent collection reduces the photocurrent at low applied bias, while for higher bias the lowering of the barrier height leads to a full collection of the carriers.

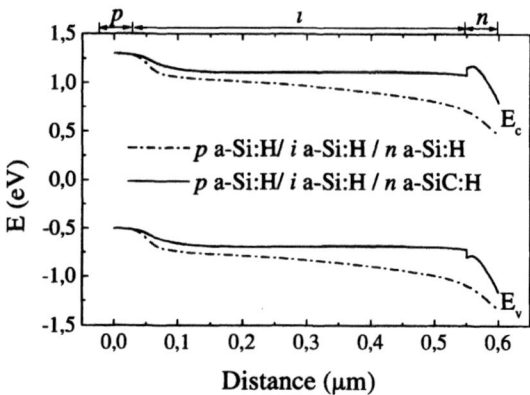

Figure 3. Simulated band diagram for a p-i-n photodiode with a a-SiC:H low doped n-layer (solid line), and for a a-Si:H with the same dimensions (dashed line).

Other models, regarding the influence of direct recombination processes have been discussed for a-Si devices [7]. We tested these models to fit the data presented in Figure 1, and it turned out, that direct recombination processes are not relevant for the present data. The previous discussion suggests extending the one diode model by adding a modulated barrier photodiode (MBP) in series [8]). The new model is presented in figure 4, with the I-V characteristic of the MBP given by equation 2 and depicted in figure 5 (dotted line).

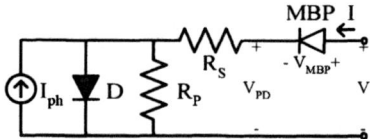

Figure 4. Equivalent circuit diagram of the two diode macroscopic model for describing a p-i-n photodiode with a a-SiC:H low doped n-layer.

$$I = A^{*}T^{2}\left(\exp\frac{-q\phi_{b}}{kT}\right)\left(\exp\frac{qV_{MBP}}{n_{f}kT} - \exp\frac{-qV_{MBP}}{n_{r}kT}\right) \qquad (2)$$

where A^{*} represents the effective Richardson constant, T the temperature in Kelvin, V the voltage across the diode, ϕ_{b} is the barrier height, and n_{r} and n_{f} are the ideality factors of the I-V characteristic of the MBP under forward and reverse bias respectively. The Richardson constant was obtained from the literature [6], while the barrier height was found to be equal to the band offset in the interface a-SiC:H/a-Si:H, the other parameters where adjusted in order to fit the measured data.

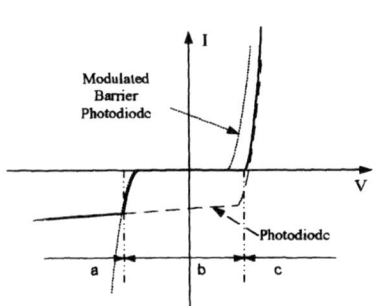

Figure 5. Typical I-V characteristics for a MBP (doted line) and for a photodiode (dashed line).

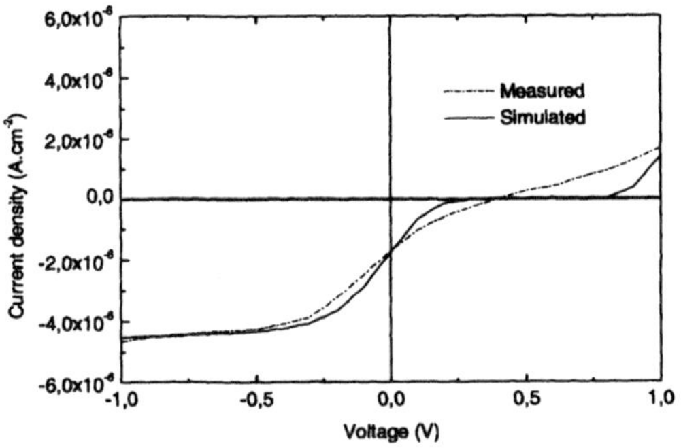

Figure 6. Measured and simulated J-V characteristic under illumination with filtered light (λ=450 nm) with a light flux of 12.1 μW/cm^2.

Taking into account Figures 4 and 5. the physical meaning of this model is straightforward, since the two diodes are in series and so, the overall voltage V will be distributed across the two diodes and both share the same current. The current through the device is then limited differently depending on the magnitude of the applied bias. In the low voltage bias regime (region a) the MBP is the limiting factor, while in the high bias (forward or reverse) the standard photodiode is responsible by the net current.

The MBP device is used to model the internal recombination process due to the low electrical field in the absorber near short circuit conditions, as seen in Figure 3. In the high bias regime (region c), the diode D is the responsible by the I-V characteristic trends. In this regime the MBP is "on", and acts as a constant voltage source either for reverse or forward bias. For negative applied voltage (region a) the current saturates and is given by the sum of I_{ph}, I_D and I_{Rp}. Under forward bias the direct current across the standard photodiode defines the behavior of the curve.

Figure 6 represents the fitting of the curve relative to the 12.1 μW/cm^2 irradiance. A simulation of the proposed model was performed using the circuit simulator program pSpice, in order to extract the I-V characteristics for different illumination conditions. The obtained simulated results are in fair agreement with the measured ones, namely for the range of interest (below the open circuit voltage). In the forward bias regime the simulation deviates from the measured data, due to the non optimal set of model parameters.

CONCLUSIONS

To model the experimental I-V characteristics observed in the LSP devices we have expanded the standard photodiode model by including a Modulated Barrier Photodiode to model the internal recombination process occurring under low applied voltages. The simulated results where compared with the experimental data, showing a fair agreement. The developed model enables the circuit based simulation of LSP sensors, leading to an easier design and optimization of large area image sensors based on this technology.

Despite of the agreement between measured and simulated data, further work is needed in order to extract the ideal set of parameters by numerical simulation.

ACKNOWLEDGMENTS

This work has been financially supported by POCI/CTM/56078/2004, and by Fundação Calouste Gulbenkian.

REFERENCES

[1]   M. Vieira, M. Fernandes, J. Martins, P. Louro, A. Maçarico, R. Schwarz and M. Schubert, IEEE Sensors Journal, vol.1, no.2 (2001)158-67.
[2]   M. Fernandes, M. Vieira, I. Rodrigues, and R. Martins, Sensors and Actuators A physical, (2004) 360-364.
[3]   P. Louro, M. Vieira, Yu. Vygranenko, M. Fernandes, R. Schwarz, M. Schubert, Applied Surface Science, 184 (2001) 144-149.
[4]   M.W.M. van Cleef, F.A. Rubinelli, J.K. Rath, R.E.I. Schropp, W.F. van der Weg, R. Rizzoli, C. Summonte, R. Pinghini, E. Centurioni, R. Galloni, J. Non-Crystalline Solids, 227-230 (1998) 1291-1294.
[5]   S. M. Sze, Semiconductor Devices -Wiley, New York, (1985).
[6]   P. J. McElheny, J. K. Arch, H. S. Lin, S. J. Fonash, J. Appl. Phys., 64 (1988) 1254.
[7]   J. Merten, J. M. Asensi, C. Voz, A. V. Shah, R. Platz, and J. Andreu, IEEE Trans. Electron Devices 45, 423 (1998).
[8]   Kwok K. Ng, The Complete Guide to Semiconductor Devices, Mc Graw Hill, New York, (1995).

Mater. Res. Soc. Symp. Proc. Vol. 989 © 2007 Materials Research Society   0989-A19-04

Temperature Dependence of Leakage Current in Segmented a-Si:H *n-i-p* Photodiodes

Jeff Hsin Chang[1], Tsu Chiang Chuang[1], Yuri Vygranenko[1], Denis Striakhilev[1], Kyung Ho Kim[1], Arokia Nathan[2], Gregory Heiler[3], and Timothy Tredwell[3]

[1]Electrical and Computer Engineering, University of Waterloo, 200 University Avenue West, Waterloo, N2L 3G1, Canada

[2]London Centre for Nanotechnology, University College, London, WC1H OAH, United Kingdom

[3]Eastman Kodak Company, Rochester, NY, 14652-3487

ABSTRACT

Hydrogenated amorphous silicon (a–Si:H) *n–i–p* photodiodes may be used as the pixel sensor element in large-area array imagers for medical diagnostics applications. The dark current level is an important parameter that dictates the performances of these types of pixelated imaging devices. Through measurements performed at different ambient temperatures, the leakage current components of segmented a–Si:H *n–i–p* photodiodes were extracted and analyzed. It was found that the central component of the reverse current depends exponentially on bias and temperature. The activation energy of this component is independent of bias. The peripheral component of reverse current exhibits linear bias dependence at temperatures up to 50°C, while the contribution of this component diminishes at high temperatures. The dependence of dark current components on bias and temperature could be described by compact analytical equations. The model of forward and reverse dark current characteristics in temperature range was implemented in Verilog-A hardware description language.

INTRODUCTION

Hydrogenated amorphous silicon (a–Si:H) *n–i–p* photodiodes are commonly used in large area flat-panel imagers (FPI) for medical imaging applications [1]. The leakage current level of these devices has important implications as it dictates the performances of the imaging systems. While a major portion of the total leakage current in large a–Si:H *n–i–p* photodiodes may be attributed to central leakage, the peripheral component, affected by the details of the sensor structure and fabrication process, may very well dominate the total leakage current level in small photodiodes, thereby limiting the practical dimensions of such devices. The majority of studies of the leakage components were performed at room temperature for various photodiode sizes, deposition temperature, and etching processes to separate the central and peripheral components [2, 3]. Even though temperature dependence of the reverse dark current (I_D) for large area photodiodes were previously analyzed [4], the temperature dependences of various current components are still to be investigated. This work presents a systematic study of the temperature dependence of dark current characteristics for segmented a–Si:H *n–i–p* photodiodes, allowing the separation of different leakage components at various temperatures.

EXPERIMENT

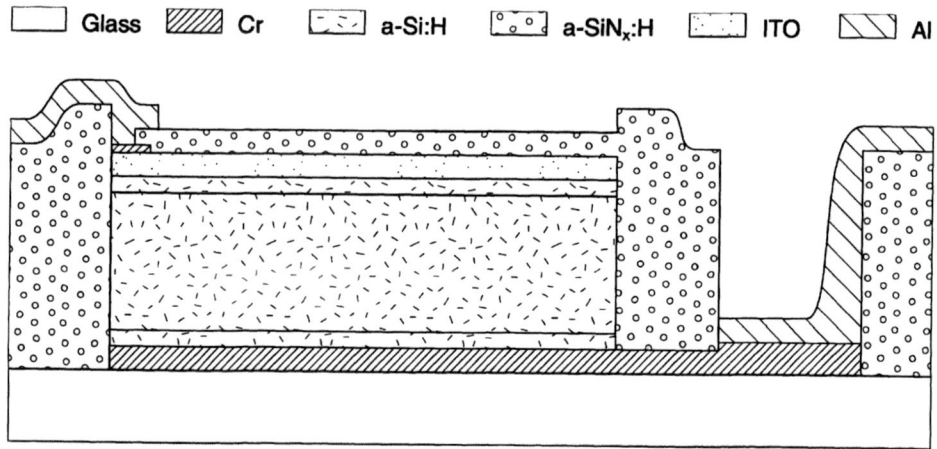

Fig. 1: Cross-sectional diagram of the fabricated a–Si:H *n–i–p* photodiode.

Square segmented a–Si:H *n-i-p* photodiodes of various sizes were prepared using standard plasma-enhanced chemical vapor deposition (PECVD) technique at 220°C; photosensitive areas of the photodiodes range from 125 ° 125 μm^2 to 2 × 2 mm^2. The device cross section is shown in Fig. 1. The thickness of *i–*, *n$^+$–* and *p$^+$–*layer are 500 nm, 40 nm and 20 nm, respectively. The detailed fabrication process is presented elsewhere [5]. The dark current-voltage (I_D–V) characteristics were measured in the range of 40–80°C using a Cascade Microtech Summit probe station in conjunction with Keithley 4200 semiconductor characterization system. The forward and reverse dark current characteristics are measured separately in the range of 0 V to –5 V and 0 V to 1 V at 50 mV steps. The leakage current level of the measurement system is ~10 fA. Small photodiodes are connected in parallel for measurement under the assumption that each has very similar characteristics. To ensure that none of the photodiodes have shunts, the forward current densities of the smaller photodiodes are compared with data collected from other photodiode sizes.

RESULTS AND DISCUSSIONS

Fig. 2 shows the dark current density–voltage (J_D–V) characteristics of square photodiodes measured for various side lengths (L) at different ambient temperatures (T). For clarity, only data for $T = 40$°C (main figure) and $L = 1$ mm (figure inset) are shown. The forward bias characteristics follow an exponential behaviour in the bias range of 0.1 – 0.6 V, and the dark current densities (J_D) match very well for photodiodes with different L. The exponential region can be fitted using the diode equation

$$J_D = J_{Do}(T)\left[\exp\left(\frac{qV}{n(T)kT}\right) - 1\right] \qquad (1)$$

where $J_{Do}(T)$ is the saturation current density, q is the elementary charge, $n(T)$ is the photodiode ideality factor, and k is Boltzmann's constant. The temperature dependence of J_{Do} and n obtained from measurement data are shown in Fig. 3. The parameter $J_{Do}(T)$ follows the Arrhenius relationship

$$J_{Do}(T) = A_o \exp\left(-\frac{E_a}{kT}\right) \qquad (2)$$

where A_o is the pre-exponential factor dependent on the processing conditions, and E_a is the forward bias activation energy. The activation energy, averaged to be ~0.85 eV from the extraction, is shown in Fig. 3 inset; it is similar to the previous reported values [6]. The diode ideality factor at 40°C is ~1.4 and decreases linearly with temperature at the rate of ~0.003K^{-1}.

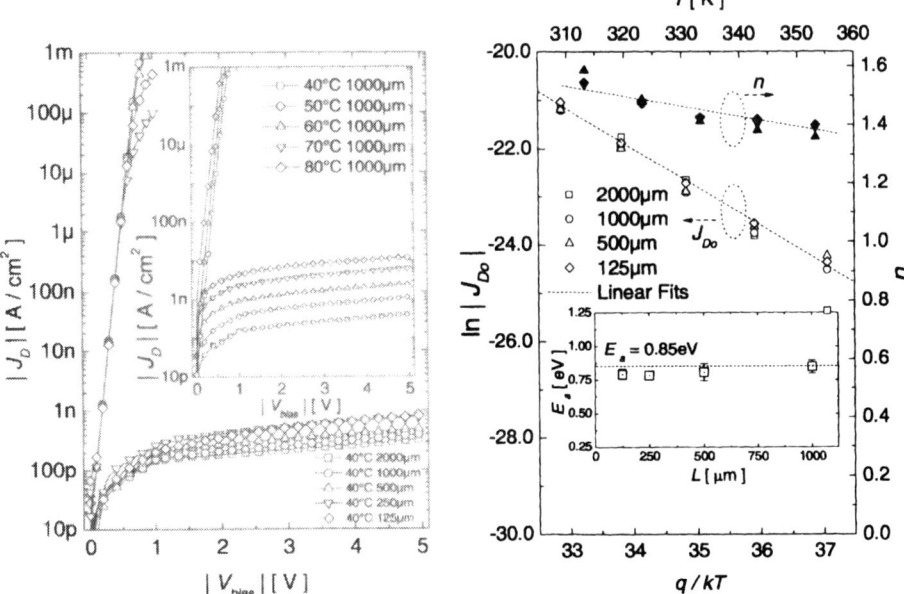

Fig. 2: Measured J_D–V characteristics of the fabricated photodiodes plotted at 40°C for various sizes, and plotted for L = 1000μm at various temperatures (inset). The J_D doubles every ~10°C.

Fig. 3: Extracted parameters $J_{Do}(T)$ and $n(T)$ from measured forward J_D–V characteristics. Inset shows the extracted forward activation energy which is averaged to be ~0.85 eV.

From the measured data of photodiode samples baring the same L, the observed reverse photodiode leakage current level is fairly consistent. It is evident that segmented photodiodes have two distinct contributions referred to as central and peripheral components [1]; additional leakages from the photodiode corners have also been inferred [3]. The total current can be expressed by

$$I_D = J_B L^2 + 4 J_P L + I_o \tag{3}$$

where J_B is the central current density and J_P is the peripheral current density. The parameter Io accounts for the current contributions from photodiode corners and instrumentation leakage. Fig. 4 shows the I_D/L and I_D/L^2 plots with respect to L at 40°C for different voltage biases. Since I_D/L varies linearly with L for $L > 500$ µm, J_B and J_P can be obtained from the slope and intercept values, respectively, via linear regression. The magnitude of Io is comparable with the total leakage current for photodiodes with $L < 250$µm.

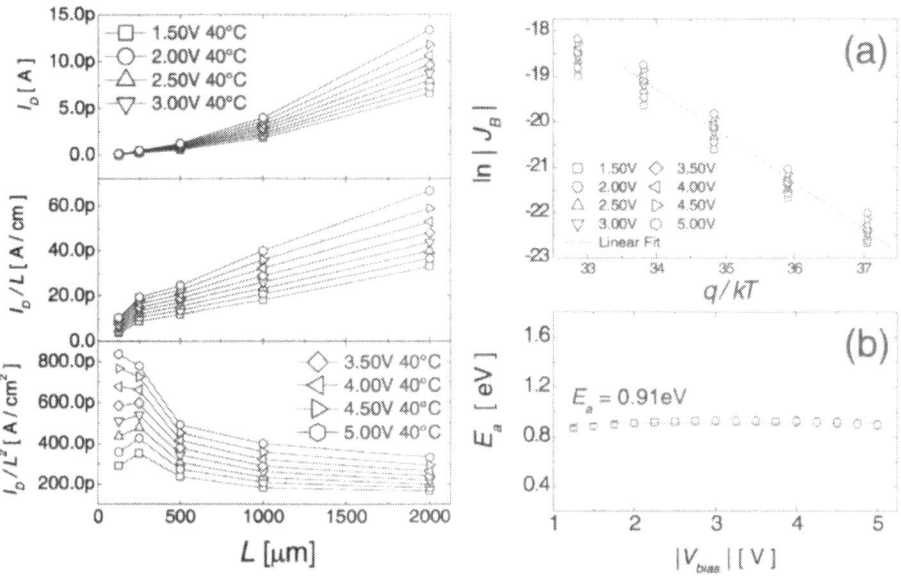

Fig. 4: I_D/L and I_D/L^2 as functions of the diode side length (L) measured at different voltage biases for 40°C.

Fig. 5 Temperature dependence of extracted central leakage current density (a) and activation energy for different reverse bias voltages (b).

The Arrhenius plot of J_B shown in Fig. 5a illustrates that the central component of the leakage current increases exponentially with temperature. Activation energies obtained at different reverse biases are shown in Fig. 5b; there is no noticeable bias dependence of E_a. The average value of E_a is 0.91 eV ± 0.03 eV, which is approximately half of the mobility gap of device quality a–Si:H [7]. Since the dependence of E_a on the bias voltage is weak, the field and thermal component of J_B can be decoupled and modeled separately. J_B at various temperatures and bias voltages can be described by

$$J_B(V,T) = J_{Bo} \exp\left(-\beta_t \frac{V}{t_i}\right) \exp\left(-\frac{E_a}{kT}\right). \tag{4}$$

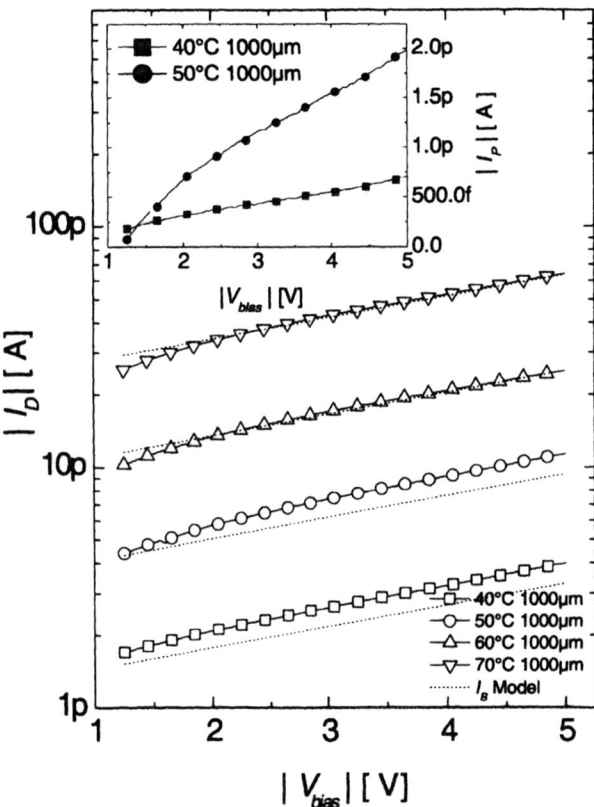

Fig. 6: Reverse I_D–V plot of $L = 1000\mu m$ a–Si:H n–i–p photodiode at various temperatures. The measurement results are shown with hollowed symbols. Dotted lines are the modeled I_B using the extracted activation energy.

From measured data, β_t is estimated to be ~1.03 × 10^{-5} cm/V assuming the field is uniform across the i–layer with thickness t_i, and J_{Bo} ~5.8 × 10^4 A/cm². Fig. 6 shows the modeled I_B plotted with measured I_D for an $L = 1$ mm photodiode. It is evident from this plot that as temperature increases, I_P contribution towards I_D diminishes. The noticeable deviation of the experimental curves from values calculated via Equation (4) is owed to the incomplete depletion of the i-layer at low biases. Fig. 6 inset shows the plot of peripheral current component extracted for 40°C and 50°C. At higher temperatures, the I_P component is much smaller than I_B, which prevents accurate extraction of I_P. The perimeter current increases linearly with voltage, which gives the opportunity for this leakage component to be modeled as a shunt resistor.

The peripheral unit conductance $G = (dI_P/dV)/4L$ are estimated to be 0.3 pS/cm and 1 pS/cm at 40°C and 50°C, respectively. This mounts to a peripheral shunt resistance of 64 TΩ for an $L = 125\mu m$ photodiode at 40°C. The calculations of the perimeter and central components have been implemented as a part of the Verilog–A photodiode model. The input parameters include the device dimensions, layer thicknesses, material properties, bias conditions and temperature.

CONCLUSIONS

Dark leakage current in segmented a–Si:H n–i–p photodiodes are measured and analyzed at different temperatures. The central and peripheral components are successfully extracted from the collected data. The activation energies for forward and reverse characteristics are extracted to be ~0.85 eV and ~0.91 eV, respectively. The reverse bias activation energy for the central current component is independent of bias, thus allowing the decoupling of the temperature and bias effects. The contribution of the peripheral current component towards the total current diminishes with elevating temperature. An empirical model is formulated for the leakage current characteristics of segmented a–Si:H n–i–p photodiodes at different temperatures and biases. The model is implemented in Verilog-A, and simulated under Cadence analog design environment.

ACKNOWLEDGMENTS

This research is funded by Eastman Kodak Company.

REFERENCES

1. R. A. Street, in *Technology and Applications of Amorphous Silicon*, edited by R. A. Street, (Springer-Verlag, Heidelberg, 2000), p. 204.
2. E. A. Schiff, R. A. Street and R. L. Weisfield, *J. Non-Cryst. Solids* **192–220**, 1155 (1996).
3. M. Mulato, C. M. Hong and S. Wagner, *J. Electrochem. Soc.* **150**, G735 (2003).
4. N. Kramer and C. van Berkel, *Appl. Phys. Lett.* **64**, 1129 (1994).
5. Yu. Vygranenko, J. H. Chang and A. Nathan, *Mater. Res. Soc. Symp. Proc.* **862**, A9.4.1 (2005).
6. H. Matsuura, A. Matsuda, H. Okushi and K. Tanaka, *J. Appl. Phys.* **58**, 1580 (1985).
7. R. A. Street, *Hydrogenated Amorphous Silicon*, (Cambridge University Press, Cambridge, 1991), p. 82.

Mater. Res. Soc. Symp. Proc. Vol. 989 © 2007 Materials Research Society   0989-A19-05

High Performance Hydrogenated Amorphous Silicon n-i-p Photo-diodes on Glass and Plastic Substrates by Low-Temperature Fabrication Process

Kyung Ho Kim[1], Yuriy Vygranenko[1], Mark Bedzyk[2], Jeff Hsin Chang[1], Tsu Chiang Chuang[1], Denis Striakhilev[1], Arokia Nathan[3], Gregory Heiler[2], and Timothy Tredwell[2]

[1]Electrical and Computer Engineering, University of Waterloo, 200 University Ave. W, Waterloo, N2L 3G1, Canada

[2]Eastman Kodak Company, Rochester, NY, 14652-3487

[3]London Centre for Nanotechnology, University College, London, WC1H OAH, United Kingdom

ABSTRACT

We report on the fabrication and characterization of hydrogenated amorphous silicon (a-Si:H) films and n-i-p photodiodes on glass and PEN plastic substrates using low-temperature (150 °C) plasma-enhanced chemical vapor deposition. Process conditions were optimized for the i-a-Si:H material which had a band gap of ~1.73 eV and low density of states (of the order 10^{15} cm^{-3}). Diodes with 0.5 µm i-layer demonstrate quantum efficiency ~70%. The reverse dark current of the diodes on glass and PEN plastic substrate is $\sim 10^{-11}$ and below 10^{-10} A/cm^2, respectively. We discuss the difference in electrical characteristics of n-i-p diodes on glass and PEN in terms of bulk- and interface-state generation currents.

INTRODUCTION

There is strong commercial interest at present on the development of thin-film electronics on flexible plastic substrates. Substantial research has been carried out to adopt a-Si:H technology for the fabrication of thin-film semiconductors on plastic. The a-Si:H technology is widely used [1] for industrial production of flat-panel displays, large area imaging devices, and solar energy conversion systems. It conventionally relies on plasma-enhanced chemical vapor deposition (PECVD) of thin film silicon and related alloys on glass substrates with typical process temperatures of 250 ~ 300 °C. For the fabrication on plastic substrates, the process temperature often needs to be reduced down to 100 ~ 200 °C, and process parameters need to be re-optimized to avoid serious deterioration of material properties. This re-optimization comes at the cost of deposition rate. On the other hand, processing on flexible substrates at reduced temperatures induces less thermal stress in fabricated structures, potentially allowing for thicker films to be deposited and better registration accuracy [2] between device layers.

It has been reported that a-Si solar cells [3], thin-film transistors [4], and transistor circuits [5] with good performance characteristics can be fabricated on plastic substrates by plasma deposition at 110 ~ 150 °C. In this paper we demonstrate that it is feasible to produce high performance a-Si:H n-i-p photodiodes on plastic substrates using low temperature PECVD.

EXPERIMENTAL PROCEDURE

A-Si:H films and n-i-p diode structures were grown at 150 °C in a multi-chamber thin-film deposition system by MVSystems. Undoped a-Si:H layers were deposited from a mixture of silane and hydrogen. To optimize deposition conditions, we varied the silane gas fraction from process to process and evaluated film composition, optical and electronic properties, and mechanical stress. In particular, infrared spectra were measured to identify bonding configurations of the hydrogen in the films, and to estimate the hydrogen concentration. Optical transmission data in the spectral range from 400 to 1000 nm were used to calculate the absorption coefficient and the optical bandgap. Mechanical stress in the films was evaluated from the measurements of substrate curvature before and after deposition.

Corning glass wafers and 150 µm-thick Teonex® PEN plastic foils were used as substrates for the n-i-p a-Si:H photodiodes. Prior to device fabrication, PEN foils were annealed in vacuum, adhesive-laminated onto glass carriers and coated with a silicon nitride passivation layer. From this point on wards, plastic and glass substrates were subjected to the same processing steps. A schematic cross section of a test n-i-p photodiode structure is shown in Fig. 1. A continuous 60 nm-thick Mo layer was sputter-deposited for the bottom contact followed by deposition of the n-i-p stack. Doped n- and p- type a-Si:H materials were prepared by adding small amounts of phosphine or tri-methyl boron, respectively, to the silane-hydrogen process mixture. The thicknesses of n- , i-, and p-type layers were 30, 500, and 20 nm, respectively. The active diode area was defined by patterned top electrodes (ITO or Al) ranging from 500×500 μm^2 to 1.5×6.5 mm^2.

Current-voltage and transient current characteristics of n-i-p diodes were measured at room temperature using a Cascade Summit probe station and a Keithley 4200 semiconductor characterization system. Spectral response characteristics of the photodiodes were measured in wavelength range of 400 - 900 nm. The characterization set-up is described elsewhere [6].

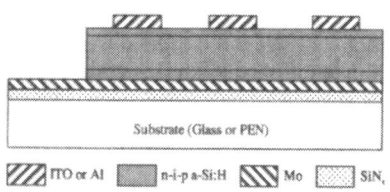

Figure 1. Structure of a-Si:H n-i-p photodiodes.

RESULTS AND DISCUSSIONS

The effect of the silane fraction in the gas mixture on deposition rate, and the properties of the intrinsic a-Si:H films are presented in Fig. 2 (a-d). The decrease in silane fraction from 1.0 to 0.2 causes a three fold decrease in the deposition rate (Fig. 2(a)). Simultaneously, the hydrogen content (Fig. 2(c)) and relative fraction of SiH_2 groups in the film (inset of Fig. 2c) decrease. The compressive stress in the film increases, approaching 600 MPa for the film deposited with a silane fraction of 0.2 (Fig. 2(b)). For the films deposited with silane fraction between 0.2 and 0.5, the optical bandgap has the same value of ~1.73 eV, while a slightly higher value of 1.77 eV was obtained for films from pure silane. The correlation between SiH_2 fraction, hydrogen content, and mechanical stress follows a previously reported trend [7] for relatively low deposition rates. Denser films with better microstructure and optimal hydrogen content are characterized by higher compressive stress, ranging from a few hundred MPa to 1 GPa. This combination of structural and mechanical properties appears to correlate well with the electronic properties desired for device applications [8]. For the low temperature PECVD, which is used in this work, the desired improvement in material structure is achieved at higher dilution ratios and low (1 ~ 1.5 Å/s) deposition rates. The a-Si:H material deposited with silane fraction of 0.2 is characterized by a hydrogen content of ~13%, a small fraction of SiH_2 groups relative to the total hydrogen concentration, $I_{SiH2}/(I_{SiH} + I_{SiH2}) \sim 0.1$, and optical bandgap of 1.73 eV. These material parameters closely approach that of "device grade" a-Si:H [1,9].

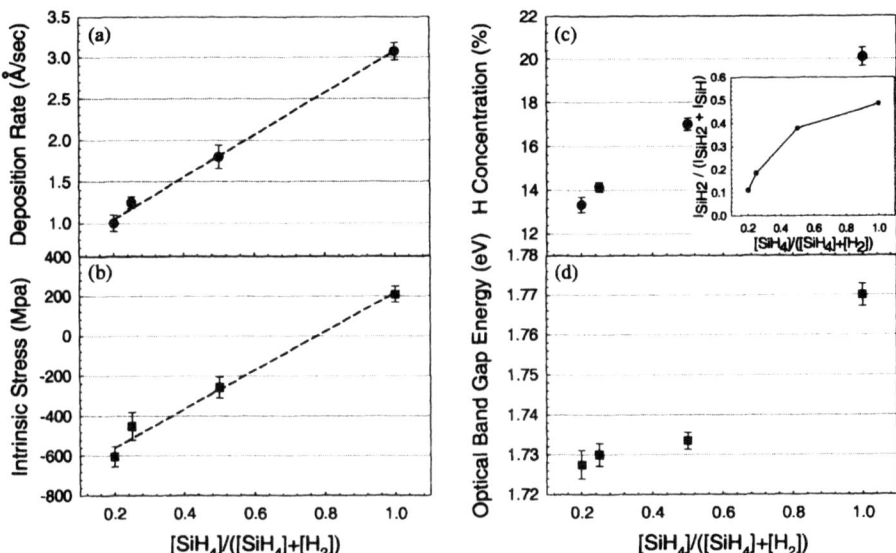

Figure 2. Deposition rate (a), intrinsic stress (b), hydrogen content (c), and optical bandgap (d) as functions of silane gas fraction in process gas. Insert of Fig. 2 (c) shows the relative infrared absorption fraction of the SiH_2 group as a function of silane fraction.

The electronic quality of intrinsic layers deposited with a different silane fraction was examined by fabricating n-i-p diodes on glass substrates and measuring their dark current-voltage characteristics. Results for 500×500 μm^2 diodes with Al top contact are presented in Fig. 3(a). It can be seen, that lowering silane fraction for the i-layer deposition from 1.0 to 0.2 corresponds to a reduction in reverse current by an order of magnitude due to the large reduction in defect state density in the i-a-Si:H layer.

The transient current characteristics of the photodiode prepared with a silane fraction of 0.2 are presented in Fig. 4(a). Following application of the reverse bias, the current exhibits a monotonic decay for the entire duration of the test, and approaches a steady-state value, $J_{ss}(V)$. The observed steady-state current of 12 pA/cm² at -1 V bias (Fig. 4(b)) is very close to the low-field leakage current in optimized a-Si:H n-i-p diodes [10,11]. The transient current behavior (see Fig. 4), suggests that contact injection current is negligible for this device and that the reverse current is determined by the thermal generation mechanism [10]. Assuming that the current is pre-dominantly generated from traps in the bulk of i-layer, we can estimate that the density of trap states. This value turns out to be approximately is 10^{15} cm⁻³.

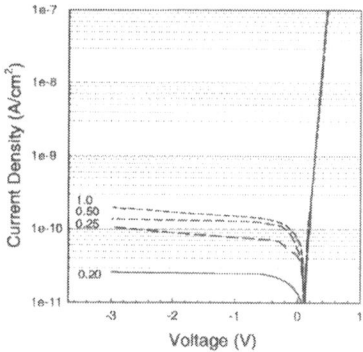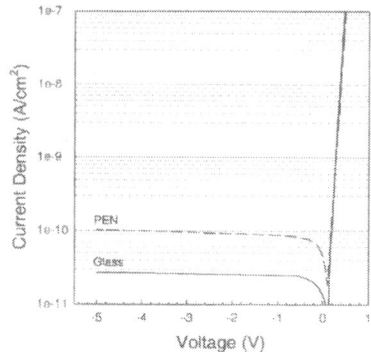

Figure 3. Current-voltage characteristics of a-Si:H n-i-p photodiodes for the indicated silane fractions for the i-layer deposition (a) and current voltage characteristics of the photodiodes on glass and PEN plastics substrate (b).

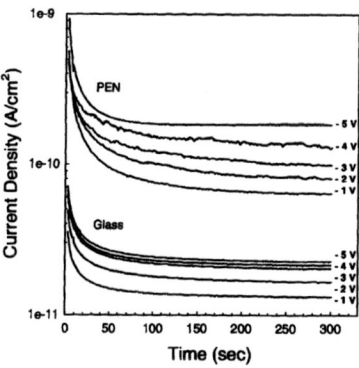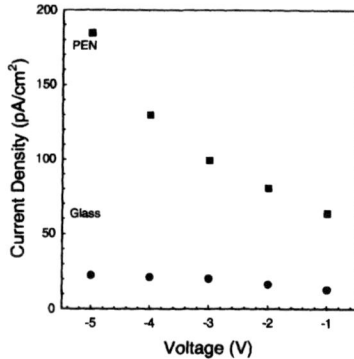

Figure 4. Dark current decay of a-Si:H n-i-p photodiodes (a) and voltage dependence of the steady-state current (b) on glass and PEN plastic substrates.

The spectral response of the optimized photodiode is shown in Fig. 5. The quantum efficiency for our device reaches a peak value of 68%. The wavelength at which the maximum quantum efficiency is observed tends to shift to lower values for diodes with a thinner intrinsic layer. In diodes that have a 0.5 μm-thick i-layer and use no optical interference coating, the maximum quantum efficiency occurs at 560 nm wavelength. The spectral response is lower at shorter wavelength due to absorption in the p-layer and absorption/reflection from ITO contact, while the

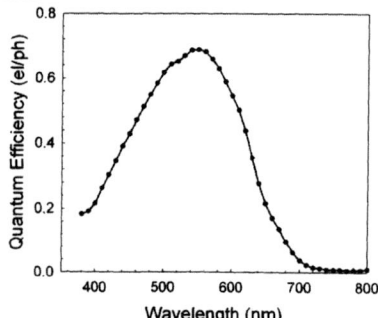

Figure 5. Spectral response characteristic of a-Si:H n-i-p photodiode.

484

decay in the red region is caused by a decrease in absorption in the intrinsic layer. The observed behavior of spectral response is typical for a-Si:H photodiodes.

The deposition conditions that yielded the lowest diode leakage current were used to fabricate the n-i-p photodiodes on PEN substrates. Current-voltage and transient characteristics of 500×500 μm^2 diode on PEN are presented in Fig. 3 (b) and Fig. 4(a), respectively. The plot of steady-state current versus bias is shown in Fig. 4(b). For a reverse bias of 1 V and 5 V, the steady-state currents are ~65 and 185 pA/cm^2, respectively. This level of diode leakage current is low enough to meet the requirements of digital X-ray radiography. The transient current behavior (Fig. 4(a)) is similar to the diode on glass, suggesting that the dominant current mechanism is thermal generation. However the magnitude of the transient- and steady-state currents is noticeably higher for each bias value. For example, at low reverse bias voltages, the current in the diode on PEN is a factor of ~5 higher compared to the diode on glass. We estimated the depletion charge for both samples at low biases by integrating I(t)-I$_{ss}$ from Fig. 4(a). At a reverse bias of 2V, the depletion charge is higher for the samples on PEN by an order of magnitude. This qualitative agreement between steady-state currents and depletion charge values is responsible to confirm that the current in both samples is due to thermal generation. However, the relative contributions of the bulk and interface traps in thermal generation current can be different for these devices. Indeed, an increase in bulk defects alone would cause a uniform shift of current-voltage characteristic to higher values [12,13]. Furthermore, from numerical modeling studies [13,14], field-enhanced generation from deep states in the bulk systematically underestimates the typical experimentally measured currents and their increase with applied field. It was also shown, that a good match measured values can be achieved if the generation from interface states is added to the model [13].

What we observe in Fig. 3(b) and 4(b) is that the reverse current of the diode on PEN is higher than that for the diode on glass in the entire bias range. Besides, the exponential field dependence of steady-state current for the diode on plastic is stronger than that of the diode on glass, which may result from a higher density of interface traps. It also worth to mention that the diode on PEN qualitatively reproduces the field dependence of the reverse dark current reported by several groups for diodes fabricated on glass at higher temperatures [10,11,15]. In contrast, a weak field dependence of the reverse current for the diode on glass may result from dominance of bulk generation in this device, while the relative contribution of interface states is lower compared to the diode on PEN and previously reported data.

Alternatively, a presence of high local fields near the interface can also cause a stronger bias dependence of the leakage current. An increase of local electric fields may come from interface roughness. The roughness of PEN substrates in this study is ~10 nm as measured by AFM on the top of the silicon nitride-passivated plastic foil. This is a much higher value compared to the roughness of glass substrate (~0.5 nm). High surface roughness originating from the plastic substrate will also likely appear at the device interface, which can cause both an increase of electric field near the interface and higher density of trap states.

In summary, a relatively weak voltage dependence of the current for our low-temperature diode on glass, as compared our sample on PEN, and published data [10,11,14], can be explained by a reduced density of interface defects and the dominance of bulk generation in the device. The density of trap states for in the bulk is of the order of 10^{15} cm^{-3}. The higher leakage current and stronger exponential field dependence of the reverse dark current for the diode on PEN are most likely caused by a higher density of defects in the bulk i-layer and at the diode interfaces.

CONCLUSIONS

High performance a-Si:H n-i-p photodiode structures were fabricated on glass and on plastic substrates. Plasma-enhanced chemical vapor deposition at low process temperature (150° C) was used for compatibility with inexpensive polymer substrates. The deposition conditions of a-Si:H films were optimized to preserve high electronic integrity. Electrical characteristics of fabricated photodiodes revealed a low density of the electronic defects in the undoped layer and the interfaces. The reverse leakage current of $500 \times 500 \ \mu m^2$ 0.5 μm-thick n-i-p photodiodes on glass and plastic substrate was 25 and 185 pA/cm^2, respectively, at -5 V bias. A quantum efficiency of 68% was achieved for a 0.5 μm-thick photodiode. The characteristics of the photodiodes presented here are of high enough quality to satisfy the requirements of medical X-ray imaging. The results of the work demonstrate that high-performance imaging devices can be fabricated on unconventional substrates by low temperature PECVD.

ACKNOWLEDGMENTS

The authors are grateful to Eastman Kodak Company and Natural Science and Engineering Research Council of Canada (NSERC) for financial support of this research.

REFERENCE

1. *Technology and Applications of Amorphous Silicon*, Ed. R. A. Street, (Springer, 2000).
2. H. Gleskova , I. C. Cheng, A. Kattamis, S. Wagner, Z. Suo, ECS Transactions 3, 249 (2006).
3. Y. Ishikawa, M. Schubert, Jpn. J. Appl. Phys. **45**, 6812 (2006).
4. H. Gleskova, S. Wagner, Z. Suo, MRS Symp. Proc. **508**, 73 (1998).
5. A. Nathan et al., MRS Symp. Proc. **814**, 61 (2004).
6. Y. Vygranenko, J.H. Chang, A. Nathan, IEEE J. Quant. Electronics 41. 697 (2005).
7. S.Nanomura et al., J. Non-Crystalline Solids **266-269**, 474 (2000).
8. E. Spanakis et al., J. Appl. Phys., **89**, 4294 (2001).
9. R.E.I. Schropp, M. Zeman *Amorphous and Microcrystalline Silicon Solar Cells: Modeling, Materials and Device Technology* (Kluwer 1998).
10. R.A. Street, Phil. Mag. B 63, 1343 (1991).
11. J.A. Theil, MRS Symp. Proc. **762** publ.#A21.4.1 (2003).
12. R.A. Street, Appl. Phys. Lett, **57**, 1334 (1990).
13. J.K. Arch, S.J. Fonash, Appl. Phys. Lett. **60**, 757 (1992); J. Appl. Phys. **72**, 4483 (1992).
14. A. Ilie, B. Equer, Phys. Rev. B **57**, 15349 (1998).
15. Y. Vygranenko, R. Kerr, J.H. Chang, D. Striakhilev, A. Nathan, G. Heiler, T. Tredwell, MRS Spring meeting 2007, A12.2.

Poster Session:
Electronics and
Flexible Substrates

Mater. Res. Soc. Symp. Proc. Vol. 989 © 2007 Materials Research Society   0989-A20-04

Hot-Wire Deposited Nanocrystalline Silicon TFTs on Plastic Substrates

Farhad Taghibakhsh, Michael M. Adachi, and Karim S. Karim
School of Engineering Science, Simon Fraser University, 8888 University Drive,
Burnaby, V5A 1S6, Canada

ABSTRACT

Hot-wire chemical vapor deposition (HWCVD) was used to deposit nanocrystalline silicon (nc-Si) thin film transistors (TFT) on thin polyimide sheets. Two straight tantalum filaments at 1850°C with a substrate to filament distance of 4 cm was used to deposit HWCVD nc-Si with no thermal damage to plastic sheet. Top-gate staggered TFTs were fabricated at 150 °C and 250 °C using a HWCVD nc-Si channel, PECVD silicon nitride gate dielectric, and microcrystalline n^+ drain/source contacts. A Leakage current of 3.3×10^{-12} A, switching current ratio of 3×10^6, and sub threshold swing of 0.51 V/decade were obtained for TFTs with aspect ratio of 1400 µm / 100 µm fabricated at 150 °C. The highest electron field effect mobility was found to be 0.3 cm^2/Vs observed for TFTs deposited at lower substrate temperature. Measurements showed superior threshold voltage stability of HW nc-Si TFTs over their amorphous silicon (a-Si) counterparts.

INTRODUCTION

Rugged plastic sheets can provide lightweight inexpensive substrates for large area thin film displays. Hot-wire chemical vapor deposition (HWCVD) technology is gaining popularity for depositing high quality materials because of its efficient use of source gases, and its simple and inexpensive setup which is easily scaleable for large area applications [1, 2]. Lack of ion bombardment and powder formation during deposition are the other advantages of the HWCVD technique. HWCVD is a physical deposition process, and unlike plasma enhanced deposition (PECVD) technique, a HWCVD system is entirely independent of the substrate (substrate does not need to be electrically grounded) which is a major advantage especially in roll-to-roll deposition systems for plastic sheets. However, adapting HWCVD for deposition on plastic substrates is a challenge due to the low temperature requirement of plastic substrates. Compared to state-of-the-art hydrogenated amorphous silicon thin film transistors (a-Si TFTs), nanocrystalline silicon (nc-Si) TFTs offer superior threshold voltage stability, as well as higher carrier mobility for both electrons and holes [3], providing the possibility of realizing an electronic system on plastic using thin film complementary metal oxide semiconductor (CMOS) transistors.

The structure of the transistors plays an important role in performance of nonocrystalline devices. Because of conical growth of crystallites in a nc-Si thin film, top gate structures show better performance than bottom gate configurations due to higher crystallinity of silicon film at the top of the layer. Also, the formation of an incubation bottom layer, which is amorphous, increases drain/source access resistance, and reduces the effective carrier mobility for top gate staggered TFT structures. This problem can be avoided in coplanar top gate configurations. For HWCVD deposited devices on glass substrates, electron field effect mobilities of 22 cm²/Vs and 5 cm²/Vs has been reported for top-gate coplanar and staggered structures (with sputtered SiO₂ dielectric) respectively. [3, 1].

We previously reported high quality TFTs by HWCVD of amorphous silicon on plastic substrates at a filament temperature of 1500°C [4]. In this study, we investigate device performances by substituting amorphous silicon with nanocrystalline silicon directly deposited using HWCVD at higher filament temperature (1850 °C) and high hydrogen dilution on thin plastic substrates. In addition to minimizing thermal damage to thin plastic substrates due to heat radiation from the filament, other challenges such as depositing device quality nanocrystalline silicon, minimizing mechanical stress in the film, and maintaining a high deposition rate exist. Details of the fabrication process of HWCVD nc-Si TFTs in top gate staggered and coplanar configuration are presented and results are discussed in the following sections.

EXPERIMENT

Square sheets of thin polyamide (50μm thick, Kapton E5) were first out gassed in an oven in air at 250°C, and then passivated on both sides with 300 nm PECVD silicon nitride to minimize inflation or shrinkage because of gas intake and outgassing during the fabrication process. Plastic sheets were then mechanically fixed on stainless steel substrate holders; care was taken to make sure that the plastic sheet is in direct contact with the thick metal back plate of the substrate holder. High hydrogen dilution and high filament temperature was used to deposit nc-Si films. Both on polyimide and glass substrates, films were deposited for process characterization [5], and the film at the border of nanocrystalline and amorphous phase that had a relatively high photo to dark current conduction ratio was selected for the TFT active layer. An MVSystem® cluster tool was used for PECVD and HWCVD, as well as aluminum sputtering. Details of the deposition conditions for various films used in this research are listed in table 1.

Table 1. Deposition conditions for different silicon films. For all depositions, substrate temperature was 150 °C, however, for HW nc-Si deposition, the substrate temperature increased to almost 200 °C because of high radiation from glowing filaments.

Type of layer	Gas mixture	Flow rate (sccm)	Pressure (mTorr)	Filament temp or rf power density
HWCVD nc-Si	SiH₄:H₂	2:33	40	1850 °C
PECVD SiNₓ	SiH₄:NH₃:H₂	5:50:200	500	20 mW/cm²
PECVD n⁺ μc-Si	SiH₄:PH₃:H₂	1:0.05:200	1900	200 mW/cm²

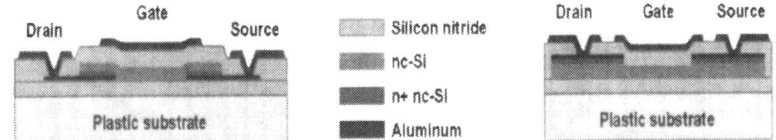

Figure 1. Schematic diagram of two different top-gate structures used in this research: top gate staggered (left) and top gate coplanar (right).

For top-gate staggered configuration (figure 1), substrates were first coated with 70 nm aluminum, and then 25 nm microcrystalline PECVD n^+ silicon. Dry and wet etching was used to pattern the n^+ μc-Si and aluminum layer to form drain/source contacts. Two straight tantalum filaments (0.5 mm in diameter, 10 cm long, 2.5 cm apart, and 4.5 cm away from substrate) were heated up to 1850 °C for HWCVD deposition of nc-Si , followed by a thin passivation layer of PECVD silicon nitride layer. Active islands were patterned using dry etching and a second PECVD silicon layer was deposited as gate dielectric. Contact vias were opened and a thick aluminum layer was deposited and patterned to form gate electrode and drain/source access pads.

Top-gate coplanar devices were fabricated by depositing bi-layer of HWCVD nc-Si and PECVD n+ μc-Si followed by sputtering a thin aluminum layer. Active islands were patterned first, and drain source contacts were defined by removing aluminum from channel area. The n+ layer was etched off the channel by dray etching, and silicon nitride was deposited, followed by opening vias and metallization to form gate drain and source contacts (Figure 1). All devices were annealed in 150 °C for two hours before any electrical measurement. Lower drain and source contact resistance, and therefore better output characteristics is expected from top gate coplanar design, while a lower leakage and better transfer characteristics is expected from the top gate staggered configuration. Both types of devices were fabricated using four masks.

RESULTS AND DISCUSSION

Dark conductivity and crystalline volume fraction of hot wire deposited nanocrystalline silicon film with deposition condition listed in table 1 was measured to be 2×10^{-8} S/cm, and 30% respectively. Properties of other films such as n^+ μc-Si and silicon nitride have been previously reported [4].

Transfer characteristics of a top gate staggered TFT with 150 nm thick nc-Si film and aspect ratio of 10 is shown in figure 2a. For $V_{DS} = 10V$, the leakage current and the sub-threshold swing is 3.3 pA, and 0.51 V/decade respectively. Field effect mobility, extracted from gated four probe TFT test structures [6], was measured to be close to 0.3 cm^2/Vs, with threshold voltage of 4.0 V. Compared to previously fabricated hot wire a-Si TFTs on plastic, electron mobility, leakage current and sub-threshold voltage swing have degraded, all of which are indications of increased trap density in deep and tail states.

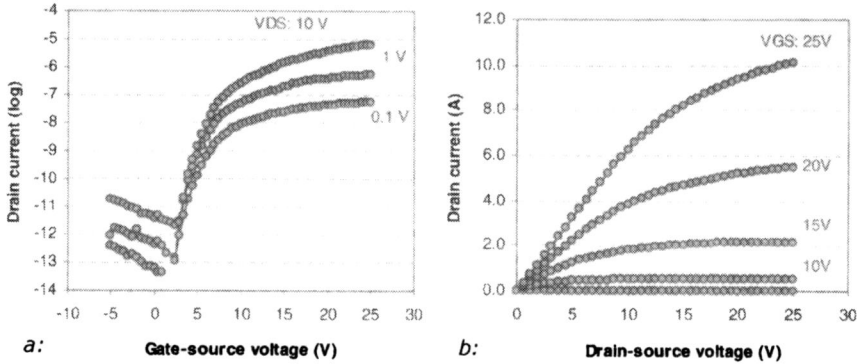

Figure 2. Characteristics of hot wire nc-Si TFT on plastic substrate with W/L of 10. (a) transfer characteristics showing low leakage and high switching current ratio, (b) the output characteristics.

For negative gate-source voltages, a significant increase in leakage current is observed as the drain-source voltage increases. The problem is less obvious for thinner deposited silicon films, however, measurements showed that a reduction of silicon film thickness resulted in lower ON current and mobility probably because of deposition of a totally amorphous silicon film (i.e. the incubation layer prior to nanocrystalline growth phase). As the output characteristics show, no severe current crowding effect was observed at low drain-source voltages, which indicates the formation of a good quality interface between the intrinsic and n^+ doped nc-Si film.

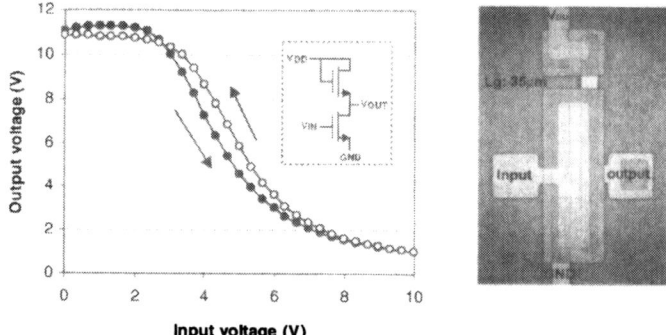

Figure 3. Transfer characteristics, schematic diagram (inset) and picture of a fabricated NMOS thin film inverter gate using top gate staggered TFTs on plastic. The circuit can operate by supply voltage of as low as 15 V.

Transfer characteristics of a fabricated HW nc-Si top gate coplanar TFT showed very large sub-threshold voltage swing of ~1.5 V/decade, high OFF voltage of -15 V or less, as well as smaller electron mobility, 0.2 cm^2/Vs. We believe the above problems in coplanar TFTs stem from the fact that the channel area of such transistors are formed by removing the n$^+$ silicon layer using RIE, which can potentially creates lots of surface defects at the nc-Si and a-SiN$_x$ interface, as well as plasma damage to the silicon film. However, the output characteristics showed better contact resistance than that found in the staggered configuration, which can be attributed to a better Al / n$^+$ / nc-Si interface since they were deposited in successive steps in the cluster tool.

Figure 3 shows transfer characteristics of an NMOS inverter gate fabricated using top gate staggered TFTs on plastic, biased with a supply voltage as low as V_{DD} = 15 V. The input voltage was swept from 0.0 to 10.0 V and back to 0.0 V to show the small hysteresis of the fabricated gate as well as the excellent output level restoration of the inverter.

To further investigate the effect of nanocrystallinity on the stability of threshold voltage, constant stress voltage of 15V was applied to the gate of a HW nc-Si TFTs with W/L of 5 for 4 hours while source and drain terminals were kept grounded and threshold voltage was measured in 5 minutes intervals as described earlier [4, 7]. Time variations of the change in threshold voltage, $\Delta V_T(t)$, is expressed as $\Delta VT(t) = A (V_{ST}-V_{TI}) t^\beta$ in which, A nd β are temperature dependent parameters, V_{ST} is the gate bias stress voltage, V_{TI} is the initial threshold voltage, and t is time [7]. Room temperature measurements resulted β of 0.19 for HW nc-Si TFTs (V_{Ti}: 4.0V). Compared to HW a-Si TFTs fabricated on plastic [4], nanocrystalline silicon TFTs appeared to have a smaller threshold voltage shift (figure 4).

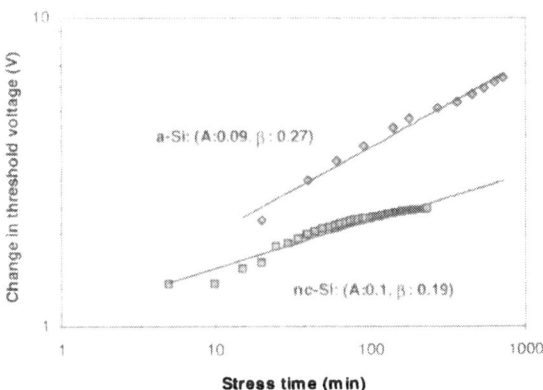

Figure 4. Comparison of time bias stress measurement of HW nanocrystalline and amorphous silicon TFTs fabricated on plastic substrates for gate stress voltage of V_{ST} = 15 V. The nc-Si TFT shows better stability of threshold voltage.

CONCLUSIONS

Hot wire CVD with filament temperature of 1850 °C was used to deposit nanocrystalline silicon for fabrication of thin film transistors and circuits on plastic substrates at 150 °C. Results show the applicability of this technique to fabrication of low operating voltage thin film circuits on plastic. Compared to HW a-Si TFTs fabricated at filament temperature of 1500°C, nanocrystalline devices showed much better threshold voltage stability, but in terms of mobility and sub-threshold swing, nc-Si TFTs appeared to be slightly degraded which may be due to higher filament temperature, nc-Si to a-SiN$_x$ interface quality, or possible doping via outgassed material from the heated plastic during deposition. Generally, a lower temperature for both the filament and substrate resulted in higher mobility.

ACKNOWLEDGMENTS

Part of this research was funded by NSERC (Natural Sciences and Engineering Research Council of Canada), CFI (Canada Foundation for Innovation) and SFU (Simon Fraser University). Also, authors would like to thank Bill Woods and Dr. X. Wong for their valuable helps in fabrication.

REFERENCES:

1: I. Schropp, Japanese Journal of Applied Physics **45**, no. 5B, 4309-4312, (2006).

2: K. Ishibashi, M. Karasawaa, G. Xua, N. Yokokawaa, M. Ikemotoa, A. Masudab, H. Matsumurab, Thin Solid Films, **430**, issues 1-2, 58-62, (2003).

3: M. Fonrodona, J, Escarre, F. Villar, D. Soler, J. Bertomeu, J. Andreu, A. Saboundji, N. Coulon, N. Mohammed-Brahim, 2005 Spanish Conference on Electron Devices, 183-186, (2005).

4: F. Taghibakhsh and K. S. Karim, Mater. Res. Soc. Symp. Proc., **910**, 429 (2006).

5: M.M. Adachi, F. Taghibakhsh, K. Kavanagh, K.S. Karim, Mater. Res. Soc. Symp. Proc., **989**, (2007) (in press).

6: C. Chun-Ying, J. Kanicki, Electron Device Letters, IEEE, **18**, issue 7, 340-342, (1997).

7. K.S. Karim, A.Nathan, M.Hack, W.I. Milne, IEEE Electron Device Letters, **25**, issue 4, 188-190, (2004).

Poster Session:
Electronic Properties
and Metastability

Mater. Res. Soc. Symp. Proc. Vol. 989 © 2007 Materials Research Society   0989-A21-06

Growth and Electronic Properties in Hot Wire Deposited Nanocrystalline Si Solar Cells

Kamal Muthukrishnan[1,2], Vikram Dalal[1], and Max Noack[2]

[1]Electrical and Computer Engr., Iowa State University, Coover Hall, Ames, IA, 50011

[2]Microelectronics Res. Ctr., Iowa State University, 1925 Scholl Rd., Ames, IA, 50011

ABSTRACT

We report on the growth and properties of nanocrystalline Si:H grown using a remote hot wire deposition system. Unlike previous results, the temperature of the substrate is not significantly affected by the hot filament in our system. The crystallinity of the growing film and the type of grain structure was systematically varied by changing the filament temperature and the degree of hydrogen dilution. It was found that high hydrogen dilution gave rise to random nucleation and <111> grain growth, whereas lower hydrogen dilution led to preferable growth of <220> grains. Similarly, a high filament temperature gave rise to preferential <111> growth compared to lower filament temperature. The electronic properties such as defect density and minority carrier diffusion length were studied as a function of the degree of crystallinity. It was found that the lowest defect density was obtained for a material which had an intermediate range of crystallnity, as determined from the Raman spectrum. Both highly amorphous and highly crystalline materials gave higher defect densities. The diffusion lengths were measured using a quantum efficiency technique, and were found to be the highest for the mid-range crystalline material. The results suggest that having an amorphous tissue surrounding the crystalline grain helps in passivating the grain boundaries.

INTRODUCTION

Nanocrystalline Si:H is an important electronic material with wide applications in solar cells [1-4] and thin film transistors [5-7]. The material is generally deposited using a plasma process, though it can also be deposited using a hot wire process [8-10]. High quality solar cells have been fabricated with both processes, producing solar cells with conversion efficiencies of ~ 10%. The hot wire process is interesting because, in principle, one can get higher growth rates, higher grain size and higher mobilities in hot wire films [11]. Most of the previous work on hot wire systems has been done using a proximity type hot wire, where the filament is only a few cm away from the substrate. For such cases, it is reported that the substrate temperature can be increased significantly because of the radiation for the proximate hot filament. Klein et al have reported that the substrate temperature increased by 185 °C when the hot filament was turned on in their system[8]. In this paper, we use a remote hot wire system so that we can better control the substrate temperature, with temperature rise being limited to ~10 °C . We report on the relationship between hydrogen dilution and filament temperature and grain crystallinity, and then investigate fundamental properties such as defect densities and hole diffusion lengths in hot wire deposited materials.

EXPERIMENTAL GROWTH TECHNIQUE

The remote hot wire system used three Tantalum filaments which were 11 cm away from the substrate. In Fig. 1, we show the change in substrate temperature as a function of filament temperature. The substrate temperature was monitored using a very thin thermocouple mounted inside a cavity within a stainless steel substrate which was clamped to a heater block. It is clear from the figure that for most films reported here, which correspond to filament temperatures of <1800 C (11 A current), the substrate temperature only increased by 10-15 C, and for the most extreme case, 1920 C filament temperature, it increased by 25 C. Thus, we have shown that a remote filament system significantly reduces the accidental heating of the substrate.

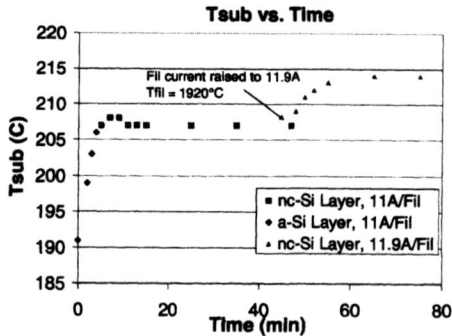

Fig. 1 Influence of turning on the filament on substrate temperature. The set point was 200 °C. 11 A filament corresponds to a filament temperature of 1800 °C.

EXPERIMENTS ON HDYROGEN DILUTION AND CRYSTALLINITY

The films were grown mostly on stainless steel substrates using varying ratios of hydrogen to silane at different filament temperature and chamber pressure. The degree of crystallinity was estimated from the measurement of Raman spectrum a Renishaw Raman microscope, using the 488 nm Ar laser line. In Fig. 2, we show the typical Raman spectrum for a film with a reasonably high degree of crystallinity (80%). The approximate degree of crystallinity was estimated from the ratio of the peak at 520 cm^{-1} to the broad peak at 485 cm^{-1}. The grain size was measured using x-ray diffraction and estimated using Scherer's formula. Changing the deposition conditions led to changes in the degree of crystallinity.

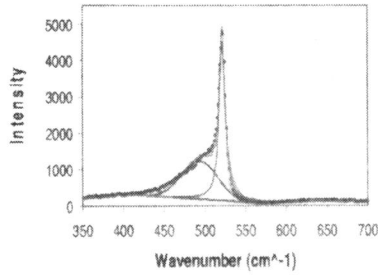

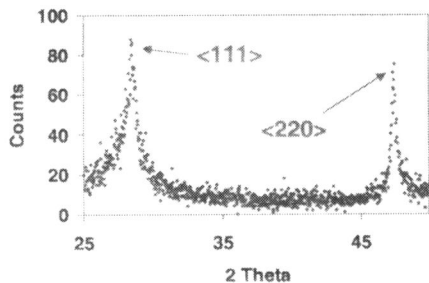

Fig. 2 Raman spectrum for a hot wire
nanocrystalline Si film.

Fig. 3 x-ray spectrum for a hot wire deposited
nanocrystalline Si film

In Fig. 3, we show the typical x-ray spectrum for the films. Both <111> and <220> peaks
are evident. In Fig. 4, we show the ratio of <111> to <220> intensity as a function of hydrogen
dilution. As the hydrogen dilution increases, the <111> peak becomes more prominent. The
grain sizes are plotted in Fig.5 as a function of hydrogen dilution ratio, again showing that <220>
grain size reduces as hydrogen dilution increases.

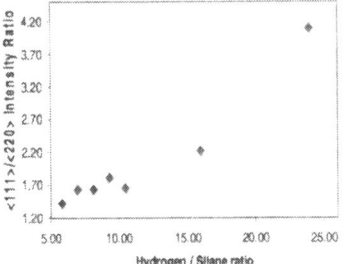

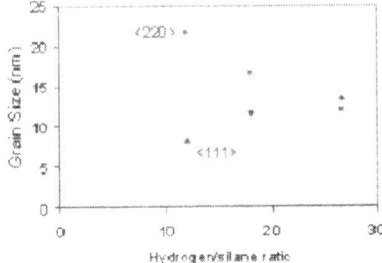

Fig.4 Ratio of <111> to <220> peak intensities
as a function of hydrogen/silane ratio

Fig. 5 Grain size as a function of hydrogen/
silane ratio.

A very similar result is obtained when the filament temperature is increased for a given
hydrogen dilution. It is useful to remember that hydrogen molecule has a higher bond strength
than silane molecule, and therefore, a higher temperature is more effective in decomposing
hydrogen, which then results in a higher ratio of H atoms to silane radicals. Therefore, increasing
filament temperature is similar to increasing hydrogen dilution in terms of controlling the
hydrogen atom/silane radical ratio at the growing surface. Therefore, from these sets of
experiments, we can conclude that excess of hydrogen atoms at the surface leads to excessive
<111> growth and a reduction in <220> grain size, results which we ascribe to random
nucleation being preferred when excessive H atoms are present at the surface. We have
previously shown that the preferred growth direction in both Si and Ge is the <220>, and that
<111> arises because of random nucleation [12].

In Fig. 6, we show the influence of chamber pressure on the grain size. As the pressure increases, <220> grain size increases up to a point, whereas <111> seems to remain the same. We do not understand this result. It clearly has some relationship to the collisions between various radicals produced by the hot filament, causing changes in which radical arrives preferentially at the surface to cause growth, but in the absence of any experimental data, we cannot speculate on what is happening.

Fig. 6. Influence of chamber pressure on grain size

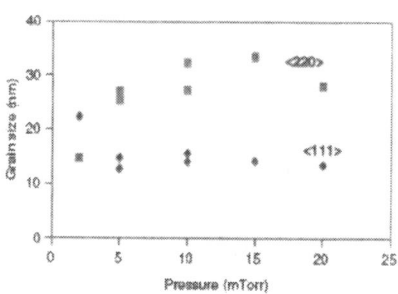

MEASUREMENT OF ELECTRONIC PROPERTIES

To measure electronic properties of minority carriers, we made p+nn+ devices on stainless steel, with the n+ layer next to steel. The top contact was sputtered ITO. Only the base n layer was made using the hot wire technique. The p+ and n+ layers were fabricated using a PECVD process in a different reactor so as to avoid contaminating the base n layer.

We focused on two sets of measurements, mid-gap defect density using the capacitance-voltage-frequency technique described in an earlier paper[13], and diffusion length using the quantum efficiency vs. voltage technique [13]. A series of devices with varying crystalline ratio were prepared and defect densities and diffusion lengths were measured in these devices. In Fig. 7, we show the changes in defect density as a function of crystalline ratio for devices prepared at 250 C substrate temperature. It is seen that the least defect density occurs at a crystalline/amorphous peak ratios of about 1.7-3. Much lower or higher ratios lead to increases in defect density.

In Fig. 7, we also show the corresponding hole diffusion length, also measured as a function of crystalline ratio. As expected, in accordance with the changes in defect density, the highest diffusion lengths are obtained when the crystalline ratio is between 1.7 and 2.5. Surprisingly, higher crystalline ratio leads to higher defect density and lower diffusion lengths. There can be two reasons for such a behavior. One, a highly crystalline material may have very porous grain boundary, where oxygen can diffuse in and change the material properties. Alternatively, having a thin amorphous tissue surrounding the grain may passivate the grain more effectively than just H. We cannot definitively say which one of these mechanisms is responsible for better electronic properties for material with crystalline/amorphous peak ratios in the 2:1 range.

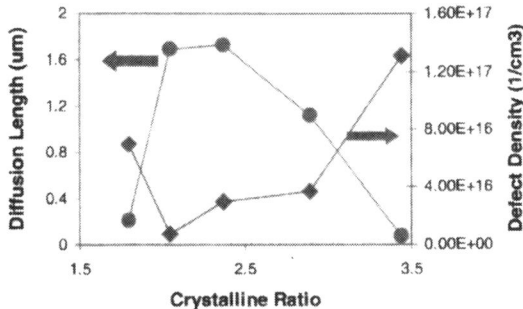

<u>Fig. 7</u> The influence of crystallinity on the midgap defect density and the diffusion length of holes. A moderate degree of crystallinity appears to be best for photovoltaic devices.

CONCLUSIONS

In conclusion, we have shown that a remote hot wire system does not cause the temperature of the substrate to vary significantly compared to the set point. We have also shown that high hydrogen dilutions or more H at the growing surface lead to more random <111> grain nucleation as opposed to the thermodynamically preferred <220> grain. We have also shown that the defect density increases significantly when the Raman ratio is either very low or very high, with the lowest defect densities being obtained in the range of Raman ratio of 1.5-3. The defect density measurements correlated very well with the minority carrier diffusion length measurements, with the highest diffusion lengths being obtained in the same range of Raman ratios as the lowest defect densities.

ACKNOWLEDGMENTS

This work was partially supported by NSF and NREL.

REFERENCES

1. Kenji Yamamoto, Masashi Yoshimi, Yuko Tawada, Susumu Fukuda, Toru Sawada, Tomomi Meguro, Hiroki Takata, Takashi Suezaki, Yohei Koi, Katsuhiko Hayashi Solar Energy Mater. And Solar Cells, <u>74</u>, 449 (2002)
2. A. V. Shah, J. Meier, E. Vallat-Sauvain, N. Wyrsch, U. Kroll, C. Droz and U. Graf, Solar Energy Mater. And Solar cells, <u>78</u>, 469 (2003)
3. B. Rech, O. Kluth, T. Repmann, T. Roschek, J. Springer, J. Müller, F. Finger, H. Stiebig and H. Wagner, Solar Energy Mater. And Solar Cells, <u>74</u>, 439 (2002)
4. K. Yamamoto, A. Nakajima, M. Yoshimi, T. Sawada, S. Fukuda, K. Hayashi, M. Ichikawa, Y. Tawada, Proc. Of 29th. IEEE Photovolt. Spec. Conf.(2002), p.1110
5. A. Sazonov, D. Striakhilev, C-H Lee, and A. Nathan: Proceedings of the IEEE,. <u>93,</u> No. 8, (2005).

6.   I-C Chen and S. Wagner: IEE Proc.- Circuits, Devices Syst., Vol. 150, No. 4, 2003

7.   Durga Panda and Vikram Dalal, Proc. Of MRS, Vol. 910,615(2006)

8.   S. Klein, F. Finger and R. Carius, J. Appl. Phys., 98, 024905(2005)

9.   H. Matsumura, A. Masuda and H. Umemoto, Thin Solid Films, 501,58(2006)

10.   R.E.I. Schropp, Thin Solid Films, 395, 17(2001)

11.   V. L. Dalal, K. Muthukrishnan, S. Saripalli, D. Stieler and M. Noack, Proc. Of MRS, Vol. 910, 293(2006)

12.   Xuejun Niu and Vikram L. Dalal , J. Appl. Physics, 98, 096103 (2005)

13.   Vikram Dalal, Puneet Sharma , Appl. Phys. Lett. 86, 103510 (2005)

Mater. Res. Soc. Symp. Proc. Vol. 989 © 2007 Materials Research Society 0989-A21-07

Evolution of Structural and Electronic Properties in Boron-Doped Nanocrystalline Silicon Thin Films

Hyun Jung Lee[1], Andrei Sazonov[1], and Arokia Nathan[2]

[1]Department of Electrical and Computer Engineering, University of Waterloo, Waterloo, Ontario, N2L 3G1, Canada

[2]London Centre for Nanotechnology, University College London, London, WC1H 0AH, United Kingdom

ABSTRACT

We report on the boron-doping dependence of the structural and electronic properties in nanocrystalline silicon (nc-Si:H) films directly deposited by plasma-enhanced chemical vapor deposition (PECVD). The crystallinity, micro-structure, and dark conductivity (σ_d) of the films were investigated by gradually varying a ratio of trimethylboron [$B(CH_3)_3$ or TMB] to silane (SiH_4) from 0.1 to 2 %. It was found that the low level of boron doping (≤ 0.2 %) first compensated the nc-Si:H material which demonstrates slightly n-type properties. As the doping increased up to 0.5 %, the maximum σ_d of 1.11 S/cm was obtained while high crystalline fraction (X_c) of the films (over 70 %) was maintained. However, further increase in the TMB-to-SiH_4 ratio reduced σ_d to the order of 10^{-7} S/cm due to a phase transition of the films from nanocrystalline to amorphous, which was indicated by Raman spectra measurements.

P-channel nc-Si:H thin film transistors (TFTs) with top gate and staggered source/drain contacts were fabricated using the developed p^+ nc-Si:H layer. The fabricated TFT exhibits a threshold voltage (V_{Tp}) of -26.2 V and field effective mobility of holes (μ_p) of 0.24 cm^2/V·s.

INTRODUCTION

Low-power complementary (n- and p-channel) operation of TFT circuits is of great importance for highly functional and ultra-compact system-on-panel. While laser-annealed polycrystalline silicon (poly-Si) TFTs, one of the promising candidates for this purpose, still suffer from relatively high manufacturing cost, complex processing, and non-uniform electrical characteristics over large area [1], there has been considerable interest in nc-Si:H TFTs as a high-performance and low-cost alternative [2]. However, achieving heavily doped p-type layers remains an issue for the high-performance p-channel nc-Si:H TFTs [3], [4].

In this work, we investigate the effect of boron doping on the structural and electronic properties in nc-Si:H thin films and its application to TFTs by using TMB as a doping gas, which has the increased thermal stability over diborane (B_2H_6) currently used for the TFT fabrication [5].

EXPERIMENTS

Film Deposition and Characterization

The films were deposited on Corning 1737 glass substrates using a conventional (13.56 MHz) PECVD system (PlasmaTherm 790 Series). The substrate temperature, RF power density, pressure, and hydrogen dilution ratio, $[H_2] / ([H_2] + [SiH_4]) \times 100 \%$, were fixed at 260 °C, 70 mW/cm^2, 900 mTorr, and 99 %, respectively. The structural and electronic properties of the boron-doped films were studied as a function of a TMB-to-SiH$_4$ ratio, $[TMB] / [SiH_4]$, which was gradually varied from 0.1 to 2 %.

The growth rate (r_d) was calculated from the film thickness measured with a Dektak 8 surface profiler. The film crystallinity was determined from Raman spectra measured in the backscattering geometry using a Renishaw micro-Raman 1000 spectrometer with a 633 nm He-Ne laser source. The power of the incident beam was kept below 50 mW to avoid the film recrystallization. X_c was deduced from the integrated Raman intensity ratio, $X_c = I_{520} / (I_{520} + \eta L_{480})$, where I_{520} and L_{480} are the deconvoluted intensities of the Raman spectra in crystalline silicon (c-Si) transverse optical (TO) (~ 520 cm^{-1}) and amorphous silicon (a-Si:H) TO (~ 480 cm^{-1}) peaks, respectively, and η is the ratio of the back scattering cross-sections, being set to be 0.8 here [6]. X-ray diffraction (XRD) peaks were measured with a PANalytical X'Pert PRO x-ray diffractometer using Cu-Kα_1 line ($\lambda = 1.54056$ Å) at a grazing incidence (0.5°). The grain size (d_g) was calculated by the Scherrer formula, $d_g = k\lambda / B\cos\theta_B$, where k ~ 0.9, B is the FWHM of the peaks (in units of 2θ), and θ_B is the angular position of the peak. For σ_d measurements, 18 mm long, 1 mm apart coplanar Cr contacts were deposited through a shadow mask using an Edwards E306A RF sputtering system. The activation energy (E_a) was calculated from the temperature dependence of σ_d using $\sigma_d = \sigma_0 \exp[-E_a/(k_B T)]$, where k_B is the Boltzmann constant. For all measurements, the film thickness was ~ 50 nm, which was the same thickness as that of source/drain contacts in the fabricated TFTs.

TFT Fabrication

P-channel top-gate TFTs were fabricated using the p$^+$ nc-Si:H film developed in this work as a source/drain contact layer, and their schematic cross-section is depicted in Figure 1. The process flow is as follows. First, 50 nm-thick Cr was sputtered on a Corning 1737 glass substrate, and then the p$^+$ contact layer in the same thickness was deposited by PECVD. After the source/drain area was patterned by wet chemical etching of Cr followed by reactive ion etching (RIE) of the p$^+$ nc-Si:H, 100 nm-thick nc-Si:H channel and 200 nm-thick gate SiO$_2$ were successively deposited by PECVD without breaking vacuum to provide better interface quality. Finally, contact holes were opened to the bottom Cr layer, and 300 nm-thick Al electrodes were sputtered and patterned by wet chemical etching.

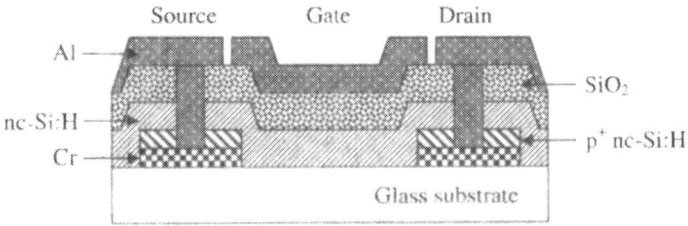

Figure 1. Schematic cross-section of the p-channel nc-Si:H TFT.

RESULTS AND DISCUSSION

<u>Film Properties</u>

Table 1 summaries the structural and electronic properties of selected p-doped nc-Si:H films. r_d of the doped films increases by a factor of 1.4 as [TMB] / [SiH$_4$] increases from 0.1 to 2 %, which is in agreement with that previously reported for p-type nc-Si:H films grown by using B$_2$H$_6$ as a doping gas [7]. The increase in r_d of a-Si:H films upon addition of B$_2$H$_6$ has been explained by the catalytic effect of BH$_3$ radicals that abstract hydrogen atoms from the film growing surface and thus enhancing the sticking probability of SiH$_3$ precursor radicals [8]. From similar trend in r_d observed in our case, CH$_3$ radicals are attributable to the hydrogen removal process from the surface and hence promoting the growth process.

Table 1. Structural and electronic properties of selected p-doped nc-Si:H films.

[TMB] / [SiH$_4$] [%]	r_d [nm/min]	X_c [%]	d_g [nm]	σ_d [S/cm]	E_a [meV]
0.2	3.50	72	8.15	9.45×10^{-9}	540
0.4	3.75	73	9.56	0.24	70.4
0.5	3.97	71	7.4	1.11	36.7
0.7	4.51	51	5.75	0.47	45
2	4.83	Amorphous	N/A	1.94×10^{-7}	530

With respect to the crystallinity of the doped films, high X_c over 70 % was observed when [TMB] / [SiH$_4$] is below 0.5 %. However, X_c significantly drops at [TMB] / [SiH$_4$] of 0.7 %, and the films undergo a phase transition from nanocrystalline to amorphous as [TMB] / [SiH$_4$] increases further, which is in agreement with the previously reported data [9]. In Figure 2, Raman spectra of the p-doped films at [TMB] / [SiH$_4$] of 0.2 and 0.7 % are compared. It is seen that X_c decreases by a factor of 1.4 as [TMB] / [SiH$_4$] increases from 0.2 to 0.7 %. We also investigated detailed micro-structural properties of the films using XRD. Figure 3 shows that the grains are oriented at (111), (220), and (311) planes. d_g calculated from the (111) peak decreases by a factor of 1.7 as [TMB] / [SiH$_4$] increases from 0.4 to 0.7%. This also implies that the films are amorphized by increased doping.

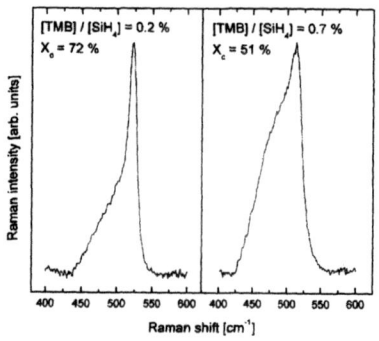
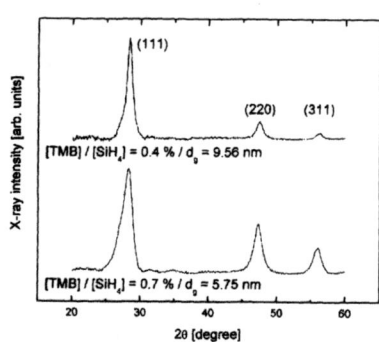

Figure 2. Raman spectra of p-doped nc-Si:H films at [TMB] / [SiH₄] of 0.2 and 0.7 %.

Figure 3. XRD spectra of p-doped nc-Si:H films at [TMB] / [SiH₄] of 0.4 and 0.7 %.

The temperature dependence of σ_d shown in Figure 4 illustrates that when [TMB] / [SiH₄] is below 0.2 %, boron compensates the nc-Si:H material which is known to be typically n-type due to oxygen impurity incorporated during deposition [10], and σ_d was reduced from 6.31×10^{-8} to 9.45×10^{-9} S/cm ($E_a = 0.54$ eV) as [TMB] / [SiH₄] increases from 0.1 to 0.2 %. Up to [TMB] / [SiH₄] of 0.5 %, the films form good ohmic contacts with the maximum σ_d of 1.11 S/cm ($E_a = 36.7$ meV). However, further increase in [TMB] / [SiH₄] to 2 % reduces σ_d to 1.94×10^{-7} S/cm (a corresponding increase in E_a to 0.53 eV) due to the amorphization of nc-Si:H films.

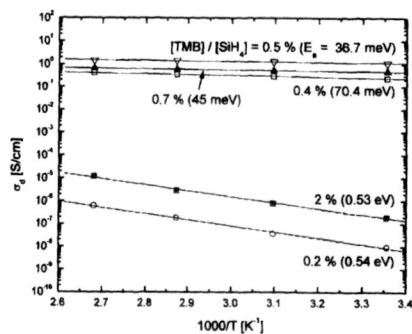
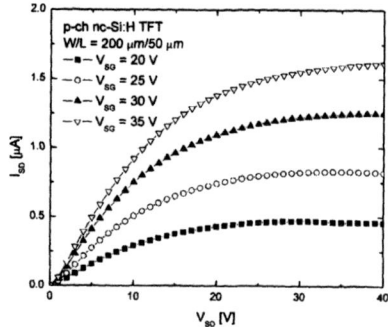

Figure 4. Temperature dependence of σ_d for p-doped nc-Si:H films.

Figure 5. Output characteristics of the fabricated p-channel nc-Si:H TFT.

TFT Characteristics

The TFT characteristics were measured using a Keithley S4200 semiconductor parameter analyzer. The TFT output characteristics are shown in Figure 5. As seen here, no distinct source-to-drain current (I_{SD}) crowding was observed at low source-to-drain bias (V_{SD}), which means the TFT has low source/drain contact resistance. V_{Tp} and μ_p

were extracted based on conventional MOSFET theory as shown in Figure 6, and are -26.2 V and 0.24 cm²/V·s, respectively. The transfer charactersistics are presented in Figure 7. For the small reverse bias (V_{SD} = 5 V), it is seen that the leakage current scarcely increases with increasing gate-to-source bias (V_{GS}), which indicates that the p⁺ nc-Si:H contact layer is capable of blocking the injection of electrons efficiently.

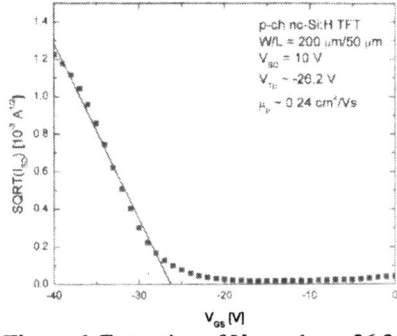

Figure 6. Extraction of V_{Tp} and μ_p: -26.2 V and 0.24 cm²/V·s, respectively.

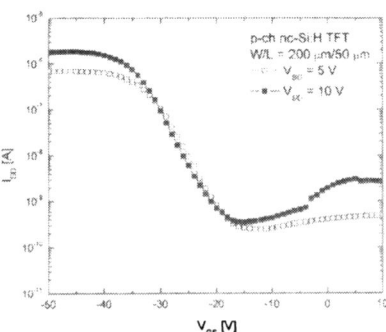

Figure 7. Transfer characteristics of the fabricated TFT.

CONCLUSION

The influence of boron doping on the structural and electronic properties in nc-Si:H thin films grown by PECVD has been systematically studied. The crystallinity, micro-structure, and σ_d of the p-type films were found to be tailored by controlling [TMB] / [SiH₄], and the maximum σ_d over 1 S/cm was achieved. By employing the p-doped thin film as a source/drain contact layer, p-channel nc-Si:H TFTs were fabricated. The developed p⁺ nc-Si:H contact layer has demonstrated increased efficiency in blocking the injection of electrons and thus good capability of suppressing the leakage current.

ACKNOWLEDGEMENTS

This work was supported by the Natural Sciences and Engineering Research Council of Canada (NSERC). The authors thank Dr. Liyan Zhao of the Department of Chemistry at the University of Waterloo for the help with XRD measurements.

REFERENCES

1. A. W. Wang and K. C. Saraswat, IEEE Trans. Elec. Dev., **47**, 1035 (2000).
2. I.-C. Cheng and S. Wagner, MRS Symp. Proc., **808**, A4.6 (2004).
3. P. Alpuim, V. Chu, and J. P. Conde, J. Vac. Sci. Technol. A, **19**. 2328 (2001).

4. D. P. Panda and V. Dalal, MRS Symp. Proc., **910**, A8.3 (2006).
5. H. Tarui, T. Matsuyama, S. Okamoto, H. Dohjoh, Y. Hishikawa, N. Nakamura, S. Tsuda, S. Nakano, M. Ohnishi, and Y. Kuwano, Jpn. J. Appl. Phys., **28**, 2436 (1989).
6. T. Okada, T. Iwaki, H. Karasawa, and K. Yamamoto, Jpn. J. Appl. Phys., **24**, 161 (1985).
7. T. Matsui, M. Kondo, and A. Matsuda, J. Non-Cryst. Sol. **338-340**, 646 (2004).
8. J. Perrin, Y. Takeda, N. Hirano, Y. Takeuchi, and A. Matsuda, Surf. Sci., **210**, 114 (1989).
9. P. Kumar, M. Kupich, D. Grunsky, and B. Schroeder, Thin Solid Films, **501**, 260 (2006).
10. R. Flückiger, J. Meier, M. Goetz, and A. Shah, J. Appl. Phys., **77**, 712 (1995).

Poster Session:
Structural Properties

Mater. Res. Soc. Symp. Proc. Vol. 989 © 2007 Materials Research Society

0989-A22-01

Elastic Properties of Several Silicon Nitride Films

Xiao Liu[1], Thomas H. Metcalf[1], Qi Wang[2], and Douglas M. Photiadis[1]
[1]Naval Research Laboratory, Washington, DC, 20375
[2]National Renewable Energy Laboratory, Golden, CO, 80401

ABSTRACT

We have measured the internal friction (Q^{-1}) of amorphous silicon nitride (a-SiN$_x$) films prepared by a variety of methods, including low-pressure chemical-vapor deposition (LPCVD), plasma-enhanced chemical-vapor deposition (PECVD), and hot-wire chemical-vapor deposition (HWCVD) from 0.5 K to room temperature. The measurements are made by depositing the films onto extremely high-Q silicon double paddle oscillator substrates with a resonant frequency of ~5500 Hz. We find the elastic properties of these a-SiN$_x$ films resemble those of amorphous silicon (a-Si) films, demonstrating considerable variation which depends on the film growth methods and post deposition annealing. The internal friction for most of the films shows a broad temperature-independent plateau below 30 K, characteristic of amorphous solids. The values of Q^{-1}, however, vary from film to film in this plateau region by more than one order of magnitude. This has been observed in tetrahedrally covalent-bonded amorphous thin films, like a-Si, a-Ge, and a-C. The PECVD films have the highest Q^{-1} just like a normal amorphous solid, while LPCVD films have an internal friction more than one order of magnitude lower. All the films show a reduction of Q^{-1} after annealing at 800°C, even for the LPCVD films which were prepared at 850°C. This can be viewed as a reduction of structural disorder.

INTRODUCTION

Amorphous silicon nitride (a-SiN$_x$) thin films, as commonly deposited by chemical-vapor deposition techniques have a variety of appealing properties, such as, high hardness, chemical and thermal stability, and low permeability, making it one of the most widely used material in microelectromechanical systems as main building material, diffusion barrier, passivation layer, and dielectric insulator [1]. More recently, it is used as thermal isolator in bolometric and microcalorimetric applications at low temperatures [2]. Studies show that the elastic properties of a-SiN$_x$ films are sensitive to way a film is deposited and post-deposition heat treatment [3,4].

In the last few years, we have systematically studies a variety of amorphous silicon (a-Si), amorphous germanium (a-Ge), and amorphous carbon (a-C) films. We have reached the conclusion that covalent tetrahedral bonding is an important factor in reducing atomic tunnelling states in these elementary amorphous thin films at low temperatures [5,6]. Structurally, amorphous form of stoichiometric silicon nitride Si$_3$N$_4$ has the fundamental tetrahedrally bonded SiN$_4$ and NSi$_3$ basic units of its crystalline counterpart. It is one of the few compound amorphous solids with similar tetrahedral unit microstructure. It would be interesting to see whether the conclusion that we reached earlier on tetrahedrally covalent-bonded elementary amorphous thin films also applies to compound ones.

In this work, we study the internal friction (Q^{-1}) of a-SiN$_x$ films prepared by a variety of deposition methods, including low-pressure chemical-vapor deposition (LPCVD), plasma-enhanced chemical-vapor deposition (PECVD), and hot-wire chemical-vapor deposition (HWCVD) from T = 0.5K to room temperature. The effect of post-deposition heat treatment is also studied by annealing of these films to 800°C.

EXPERIMENT

Measurements of Q^{-1} were performed using the double-paddle oscillator (DPO) technique [7]. The DPOs were fabricated out of high purity P doped silicon wafers, which were <100> oriented and had resistivity >5 kΩcm. The overall dimension of a DPO is 28 mm high, 20mm wide, and 0.3 mm thick; see inset of figure 1. The DPO consists of a head, a neck, two wings, a leg, and a foot. The main axes are along the <110> orientation. On the back of the DPO a metal film (30Å Cr and 500Å Au) was deposited from the foot up to the wings but not on the neck and the head. The DPO was then clamped to an invar block using invar screws and a precision torque wrench. This minimized the effect of thermal contraction during cool down and ensured reproducibility after repeated remounting of the same DPO. Two electrodes were coupled to the wings from the back side so that the DPO could be driven and detected capacitively. For our internal friction measurements, we used the so-called second antisymmetric mode oscillating at ~5500 Hz. It has an exceptionally small background internal friction ($Q^{-1} \approx 2 \times 10^{-8}$) at low temperatures (T<10 K) which is reproducible within ±10% for different DPOs. The small Q^{-1} is attributed to its unique design and mode shape. During oscillation, the head and the wings vibrate against each other, which lead to a torsional oscillation of the neck while leaving the leg and the foot with little vibration, minimizing the external loss. The internal friction results presented in this work were obtained exclusively using this mode for maximum detection sensitivity. For this purpose, the films to be studied should at least cover the whole neck area.

The LPCVD a-SiN$_x$ films were deposited on a whole wafer at 850°C using a mixture of dichlorosilane (SiH$_2$Cl$_2$) and ammonia (NH$_3$) in a flow rate of 47 and 11 SCCM (standard cubic centimeter per minute), respectively. The DPOs were fabricated after deposition. The thickness of the LPCVD on each side of the wafer was determined by Ellipsometer to be 536Å. A small piece of the wafer was further checked by Rutherford backscattering. The LPCVD coating was found to be 516Å thick, followed by a native oxide layer of 10Å thick. Since the thickness of all the other a-SiN$_x$ films were measured by Ellipsometer, we used that value for consistency, and ignored the contribution from the native oxide layer, which should exist in all DPOs and have minimal impact on the Q^{-1} of the films. Two PECVD a-SiN$_x$ films were studied. They were deposited on one side of the DPOs in two different chambers, with the same substrate temperature of 250°C, but with silane (SiH$_4$) and NH$_3$ flow rate of 20 (280) SCCM and 7 (9) SCCM, respectively. The HWCVD a-SiN$_x$ film was deposited only on the neck area at the National Renewable Energy Laboratory, with substrate temperature of 300°C. The SiH$_4$, H$_2$, and NH$_3$ flow rates were 2.5, 45, and 5 SCCM, respectively. One film out of each deposition technique was annealed at 800°C for 20 minutes in a vacuum of 10^{-5} Torr. To avoid the impact of molecular hydrogen generation during annealing, the temperature increase was carefully controlled. Still, the annealed PECVD film showed sign of burst of hydrogen bubbles and lose of weight of the film. More detailed sample parameters are given in Table 1.

The results of the internal friction of some of the DPOs after deposition, Q^{-1}_{osc}, are shown in figure 1, together with the internal friction of a bare DPO as the background. Deposition of a thin film onto a DPO changes its Q^{-1}_{osc}, as well as its resonance frequency, f_{osc}, from those of a

Table 1. Summary of amorphous silicon nitride thin film samples studied in this work.

Type of thin films	T_{sub} (°C)	Reactive gases	Flow rate (SCCM)	Thickness (Å)	G_{film} (GPa)
LPCVD	850	SiH_2Cl_2/NH_3	47/11	$2 \times 5.4 \times 10^2$	110
PECVD1	250	SiH_4/NH_3	20/7	2.6×10^3	40.8
PECVD2	250	SiH_4/NH_3	280/9	2.6×10^3	40.8
HWCVD	300	$SiH_4/H_2/NH_3$	2.5/45/5	1.0×10^4	40.8

bare DPO, Q_{sub}^{-1}, and f_{sub}, respectively. From the difference, the shear modulus, G_{film}, and the internal friction of the film, Q_{film}^{-1}, can be calculated by

$$Q_{film}^{-1} = \frac{G_{sub} t_{sub}}{3 G_{film} t_{film}} \left(Q_{osc}^{-1} - Q_{sub}^{-1} \right) + Q_{osc}^{-1}, \tag{1}$$

$$\frac{f_{osc} - f_{sub}}{f_{sub}} = \frac{t_{film}}{2 t_{sub}} \left[\frac{3 G_{film}}{G_{sub}} - \frac{\rho_{film}}{\rho_{sub}} (1 + \eta)^{-1} \right], \tag{2}$$

where t, ρ, and G are thicknesses, mass densities, and shear moduli of substrate and film, respectively; η is the ratio of moments of inertia of the uncoated versus the coated part of head and neck. For the PECVD films, we used equation 2 and determined $G_{film} = 40.7$ GPa by measuring the frequency shift and mass change of the DPOs after removal of the films by a HF

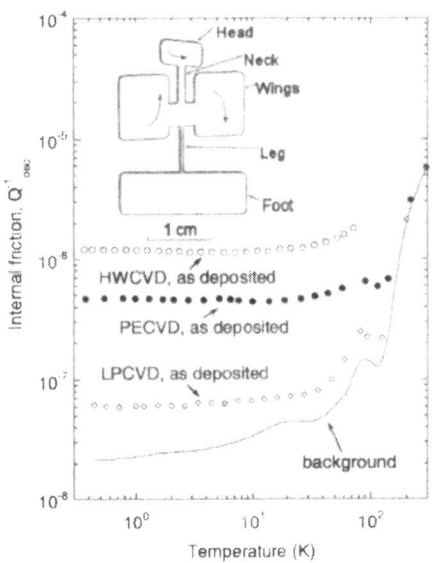

Figure 1. The internal friction of double-paddle oscillators (DPOs) carrying three different types of amorphous silicon nitride films. The background internal friction of a bare DPO is shown as a solid line. The inset: a schematic drawing of DPO with the 2nd-antysymmetric mode indicated.

dip. At present, we have not been able to measure G_{film} of HWCVD films. We assume the same G_{film} as for PECVD. For the LPCVD films, although we have not found any direct shear modulus measurements in literature, the Young's modulus (E) and Poisson's ratio (ν) have been reported a few times [8-10]. Using $G = E/2(1+\nu)$, the calculated G value varies from 100 to 120 GPa. We use $G_{film} = 110\,GPa$.

DISCUSSION

Using equation 1, we calculate the Q_{film}^{-1} of LPCVD a-SiN$_x$ films. The results are shown in figure 2 for both the as-deposited and 800°C annealed films. The temperature independent internal friction at low temperature is a characteristic feature of amorphous solids, and is a direct consequence of elastic energy dissipation by tunneling states due to the broad distribution of their density of states, see [11] for a review. The flat internal friction is called as the internal friction plateau, Q_0^{-1}. While this quantity has previously been shown to be of a nearly universal magnitude from 1.5×10^{-4} to 1.5×10^{-3} for all amorphous solids, called 'glassy range', it has recently been found that it can be reduced several orders of magnitude [12]. The variation of Q_0^{-1} outside the glassy range only occurs in a-Si, a-Ge [5], and a-C [6], which share the same tetrahedral bonding. In figure 2, the internal friction plateaus of the LPCVD films are clearly smaller than the glassy range found in all other amorphous systems, indicated by the double-arrow and the internal friction of a prototypical glass, amorphous SiO$_2$ (a-SiO$_2$), measured at 4500 Hz [13]. Although the film was deposited at 850°C, annealing at 800°C still further lowers its internal friction. Note that 800°C anneal would not lead to crystallization of LPCVD a-SiN$_x$

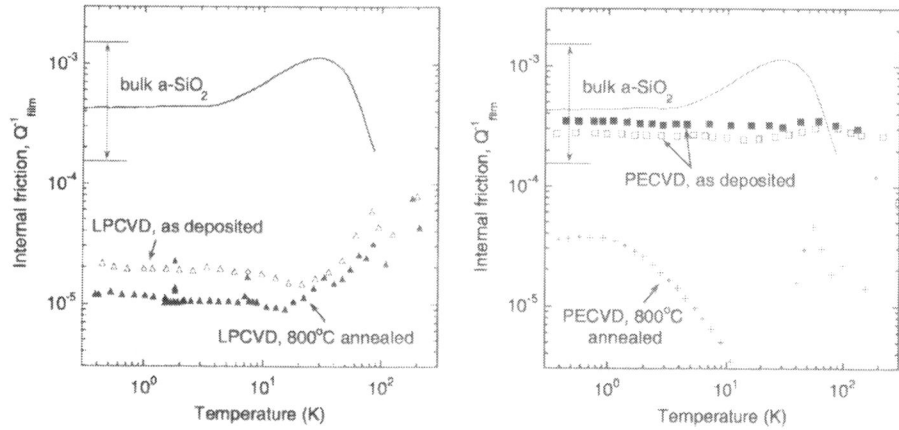

Figure 2. (left) The internal friction of as-deposited and annealed LPCVD *silicon nitride* films. The internal friction of bulk a-SiO$_2$, is shown for comparison. The double arrow denotes the "glassy range" explained in the text.
Figure 3. (right) The internal friction of two as-deposited and one annealed PECVD *silicon nitride* films.

[14].

Figure 3 shows the Q_{film}^{-1} of PECVD a-SiN$_x$ films. The internal frictions of the as-deposited films are as high as the a-SiO$_2$, also shown in the figure. Although the internal frictions of the two as-deposited films are quite identical, the one shown as solid squares has much lower SiH$_4$ flow rate than the one shown as open squares, making it a silicon-rich a-SiN$_x$. The silicon-rich film was annealed at 800°C. The internal friction of the annealed film show strong temperature dependence and the data are noisy at above 30 K. It is possible that the excess a-Si in the silicon-rich a-SiN$_x$ starts to crystallize at 800°C [15].

Figure 4 shows that the Q_{film}^{-1} of as-deposited HWCVD a-SiN$_x$ film locates at the low edge of the glassy range. Both the as-deposited and annealed films have the typical temperature independent internal friction plateau as other amorphous solids. This indicates that the film has not been crystallized after annealing. Annealing reduces the Q_{film}^{-1} by more than a factor of two, making it lower than the glassy range.

Figure 5 compares the internal friction of some of the a-SiN$_x$ films presented in this work with those of a-Si films of various types that we measured previously [12]. Both as-deposited PECVD and HWCVD a-SiN$_x$ films have higher internal frictions than e-beam a-Si — the highest internal friction in a-Si. In contrast, the LPCVD a-SiN$_x$ films have amazingly low internal friction, lower than the glassy range by more than one order of magnitude. The comparison shows that a-SiN$_x$ films have similar variation of Q_0^{-1} as in a-Si.

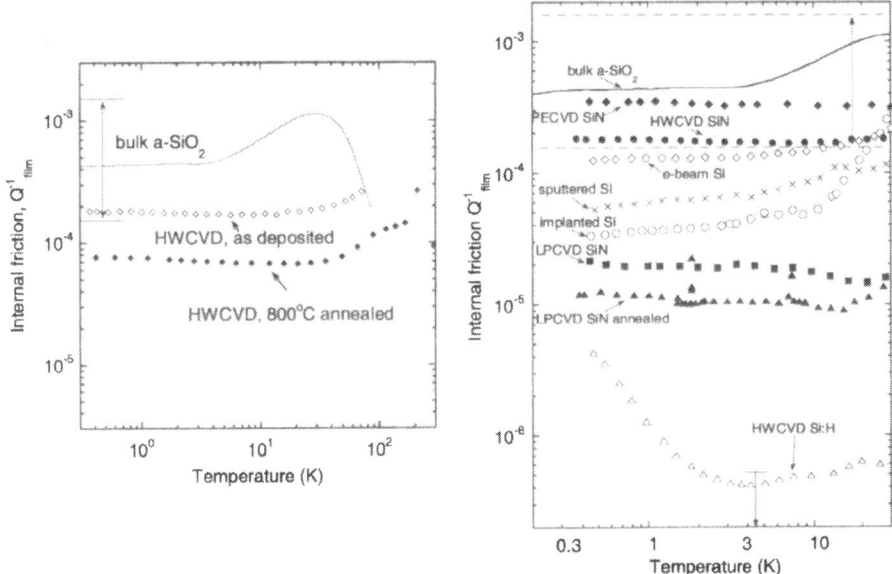

Figure 4. (left) The internal friction of as-deposited and annealed HWCVD amorphous silicon nitride films.

Figure 5. (right) The internal friction of some of the amorphous silicon nitride films studied in this work together with some of the representative a-Si films from [12].

CONCLUSIONS

The low temperature internal friction of a-SiN$_x$ films below 30 K is temperature independent and has a wide range of variation, depending strongly on the methods of deposition and post deposition heat treatment, just like the case in a-Si, a-Ge, and a-C films. Although the lowest internal friction is not as low as that of a-Si, it is better than those of a-Ge and a-C films studied so far. From its microstructure point of view, the PECVD a-SiN$_x$ films are the least ordered form of a-SiN$_x$ studied in this work, which is as disordered as any normal amorphous solids. The LPCVD a-SiN$_x$ film, on the other hand, is a well ordered amorphous solid. Since a-SiN$_x$ films has the similar tetrahedral covalent bonding as in a-Si, a-Ge, and a-C except for having a compound, and hence more complicated structure, our study demonstrates once again tetrahedral covalent bonding, or high average coordination number is critical in reducing the atomic tunneling states, and this conclusion applies to compound tetrahedrally bonded amorphous system.

ACKNOWLEDGMENTS

This work is supported by the Office of Naval Research. We thank Christoph L. Spiel for the Rutherford backscattering measurement.

REFERENCES

1. F.. L. Riley, J. Am. Ceram. Soc. **83**, 245 (2000).
2. F. Giazotto, T. T. Heikkila, A. Luukanen, A. M. Savin, and J. P. Pekola, Rev. Mod. Phys. **78**, 217 (2006).
3. B. A. Walmsley, Y. Liu, X. Z. Hu, M. B. Bush, K. J. Winchester, M. Martyniuk, J. M. Dell, and L. Faraone, J. Appl. Phys. **98**, 044904 (2005).
4. D. M. Profunser, J. Vollmann, and J. Dual, Ultrasonics, **42**, 641 (2004).
5. X. Liu, D. M. Photiadis, H. D. Wu, D. B. Chrisey, R. O. Pohl, and R. S. Crandall, Philos. Mag. B **82**, 185 (2002).
6. X. Liu, T.H. Metcalf, P. Mosaner, and A. Miotello, Appl. Surf. Sci. (in press).
7. B. E. White, Jr., R. O. Pohl in *Thin Films: Stresses and Mechanical Properties V*, edited by S. P. Baker, C. A. Ross, P. H. Townsend, C. A. Volkert, and P. Borgesen, (Mat. Res. Soc. Symp. Proc. **356**, Pittsburgh, PA, 1995) pp. 567-573.
8. G. Carlotti, P. Colpani , D. Piccolo, S. Santucci, V. Senez, G. Socino, and L. Verdini, Thin Solid Films, **414**, 99 (2002).
9. A. Khan, J. Philip, and P. Hess, J. Appl. Phys. **95**, 1667 (2004).
10. W. Chuang, T. Luger, R. K. Fettig, and R. Ghodssi, J. Microelectromech. Syst., **13**, 870 (2004).
11. R. O. Pohl, X. Liu, and E. Thompson, Rev. Mod. Phys. **74**, 991 (2002).
12. X. Liu, B. E. White, Jr., R. O. Pohl, E. Iwanizcko, K. M. Jones, A. H. Mahan, B. N. Nelson, R. S. Crandall, and S. Veprek, Phys. Rev. Lett. **78**, 4418 (1997).
13. J. E. Van Cleve, Ph.D. thesis, Cornell University, 1991.
14. G. Beshkov, D.B. Dimitrov, N. Velchev, P. Petrov, B. Ivanov, L. Zambov, and T. Dimitrova, Vacuum, **58**, 509 (2000).
15. C. Kaya, T. P. Ma, T. C. Chen, R. C. Barker, J. Appl. Phys. 64, 3949 (1988).

Mater. Res. Soc. Symp. Proc. Vol. 989 © 2007 Materials Research Society 0989-A22-03

Structural Analysis of Nanocrystalline Silicon Prepared by Hot-wire Chemical Vapor Deposition on Polymer Substrates

Michael M. Adachi[1], Farhad Taghibakhsh[1], Karen L. Kavanagh[2], and Karim S. Karim[1]

[1]School of Engineering Science, Simon Fraser University, 8888 University Drive, Burnaby, V5A 1S6, Canada

[2]Department of Physics, Simon Fraser University, 8888 University Drive, Burnaby, V5A 1S6, Canada

ABSTRACT

Nanocrystalline silicon (nc-Si:H) films were deposited by hot-wire chemical vapor deposition (HWCVD) directly onto Corning glass and polyimide (Kapton E) substrates. The effect of silane concentration (in hydrogen carrier gas) on film crystallinity and conductivity were studied for a constant substrate growth temperature of 220 °C. Raman spectroscopy, X-ray diffraction and cross-sectional transmission electron microscopy (XTEM) showed that nc-Si:H (grain-size 20-65 nm) was observed for silane concentrations below 5.8 %. Similar to previous reports, closer inspection using XTEM found that there was an initial growth of an amorphous interfacial layer which then crystallized into a randomly-oriented polycrystalline material after 10 – 100 nm of growth. However, unlike previous reports, there was no detectable difference in the structure or conductivity for films grown on the two types of substrates. In both cases, the dark conductivity decreased with increasing silane concentration while the photo-conductivity was uniform for all films at values between 2 and 4×10^{-5} S/cm.

INTRODUCTION

Nanocrystalline silicon has gained considerable attention for use as the absorber layer in thin-film silicon solar cells. This is due to stability against light induced degradation compared to hydrogenated amorphous Si (a-Si:H), and sensitivity in the near-infra red wavelengths. Films prepared by hot-wire chemical vapor deposition (HWCVD) [1], also known as catalytic chemical vapor deposition (CAT-CVD) [2], has gained interest due to the potential for high deposition rates and simple, large-area scale-up [3]. In addition, HWCVD facilitates low temperature deposition allowing films to be deposited onto flexible substrates such as polymer foils. Flexible substrates in turn facilitates roll-to-roll manufacturing, a means to reducing manufacturing costs for large area applications.

Films grown by ECR plasma deposition have been reported to require an a-Si or Moly interlayer to grow crystalline films when deposited directly only polyimide substrate [4]. In this study, nc-Si:H films with high crystalline volume fractions were deposited by HWCVD on both Corning glass and polyimide substrates at silane concentrations near the known transition point between nanocrystalline and amorphous phases. Films were characterized by electrical conductivity measurements as well as structural analysis by x-ray diffraction (XRD), Raman spectroscopy and cross-sectional transmission electron microscopy (XTEM).

EXPERIMENT

Amorphous and nanocrystalline films were deposited in a commercially available HWCVD chamber [5]. Two straight 10 cm Ta filaments were heated up to a filament temperature of 1850°C measured using an optical thermometer. In addition, a heater well located behind the substrate set to a temperature of 210°C was used for preheating the substrate. With gases flowing and the filament turned on the substrate temperature was measured to stabilize at 220°C using a thermocouple connected to a test Si wafer. The substrate to filament distance was kept at a constant value of d_{f-s}= 4.5 cm. Total gas flow rates were held constant at a value 35 sccm and silane concentrations diluted in hydrogen were calculated as SC = [SiH$_4$]/([SiH$_4$]+[H$_2$]). Deposition pressure was 5 Pa for all depositions, automatically controlled via a throttle valve. The system base pressure was 1×10^{-8} Torr. Films were deposited on 0.7 mm thick Corning glass and 50 μm thick polyimide Kapton E substrates for electrical conductivity, Raman spectroscopy, and XRD measurements and on single crystalline (100) silicon and polyimide Kapton E for TEM measurements. Single crystalline silicon substrate was used to facilitate an ion milling step needed to prepare the cross sectional samples for TEM measurements. Films were deposited directly onto polyimide substrate without use of an interlayer. Before deposition, polyimide substrates were degassed by baking in air at 220°C for 1 hour. Thicknesses were measured on Corning glass substrates using a stylus-based surface profiler with sub-angstrom precision.

Electrical conductivities were measured using two strips of sputtered aluminum contacts 20 mm in length and 1 mm apart. Photoconductivity was measured using a Halogen lamp at an irradiance of 100 mW/cm^2. Raman measurements were performed in a backscattering set-up with a 488 nm laser excitation focused to a spot diameter of 1 μm. A semi-quantitative measure of crystalline volume fraction was calculated as the ratio of Raman intensity of the crystalline silicon peaks to that of the combined crystalline and amorphous peaks defined as $R_c = (I_{500} + I_{520})/(I_{480} + I_{500} + I_{520})$ where I_x is the integrated de-convoluted Gaussian curve corresponding to mode x [6,7]. X-ray diffraction measurements were taken using a Phillips Diffractometer equipped with a fixed Cu tube source and scintillator detector with a curved crystal monochromator situated between the detector and sample. TEM measurements were performed on a TEM with a LaB$_6$ thermal source operating at 200 keV.

RESULTS AND DISCUSSION

Electrical Properties

Figure 1 shows the dark conductivity (σ_d) and photoconductivity (σ_{ph}) of films grown on glass and polyimide as a function of silane concentration (SC). The decrease in dark conductivity at SC's above 5.8% indicates a transition from nanocrystalline to amorphous growth and occurs at about the same SC on both substrate types. The photoconductivity for films grown on both glass and polyimide ranged between 2 and 4×10^{-5} S/cm, a value similar to those of device quality films reported by other authors [8]. Note that both dark and photo conductivities of films grown on polyimide match very well with those grown on Corning glass, suggesting that no interlayer was necessary to improve electrical characteristics of films when grown on these polyimide substrates. The deposition rates of films grown on Corning glass, up to 2.9 A/s for nc-Si:H films are shown in Figure 2. Film thicknesses ranged from 300 to 1200 nm depending on the silane concentration used during deposition.

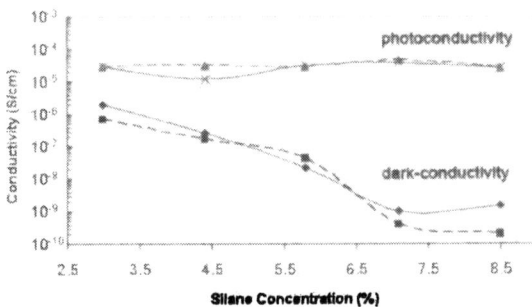

Figure 1: Dark and photoconductivity as a function of silane concentration for films grown on glass (solid line) and on polyimide (dashed line) substrate.

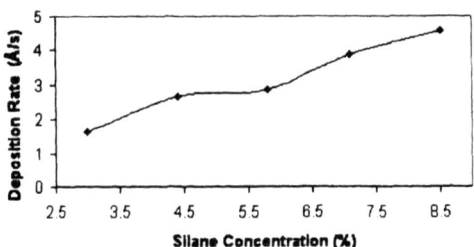

Figure 2: Deposition rate on Corning glass substrate as a function silane concentration.

Structural Properties

The crystalline volume fraction of nc-Si:H over a range of silane concentrations is shown in Figure 3. For films grown on both Corning glass and polyimide substrates, the amorphous to crystalline transition point occurred between 5.8 % and 7.1 % SC. Films grown on Corning glass substrate consistently show higher crystalline volume fraction than those grown on polyimide substrate although the difference is within error of ± 0.1. The small difference in crystal volume fraction could be due to small differences in substrate temperatures during deposition originating from different substrate thermal conductance and transparency.

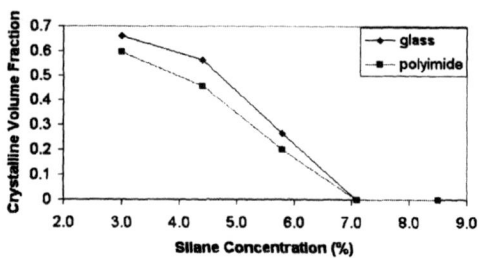

Figure 3: Crystalline Volume Fraction of nc-Si:H films as a function of silane concentration.

Figure 4 shows the x-ray diffraction spectra for nc-Si:H films grown at 3.0%, 4.4% and 5.8% SC on both glass and polyimide substrates. Other than a very small (111) crystalline peak at 2θ = 28.4° the film grown at 5.8% SC is mostly amorphous on both substrates. As the silane concentration is lowered to 4.4 and 3.0 % we notice much larger peaks corresponding to the (111), (220) and (311) reflecting planes. Note that the large signal below 2θ = 30° in Figure 4b originates from the polyimide substrate itself and covers up most of the (111) silicon reflection. Using Scherrer's equation, at a silane concentration of 4.4 %, the grain size was calculated to be ~ 20 nm for both the film grown on Corning glass and that on polyimide. Similarly, at 3.0%, the grain size was found to be ~ 65 nm for films grown on both substrates. These results suggest that no interlayer was necessary to grow highly crystalline films on polyimide substrates by HWCVD. In comparison, films prepared by remote ECR plasma deposition required a buffer layer or conductive coating to grow crystalline films directly on polyimide [4], possibly because energetic ion bombardment, which is absent in HWCVD, can break polyimide bonds creating compounds which possibility inhibits crystalline growth.

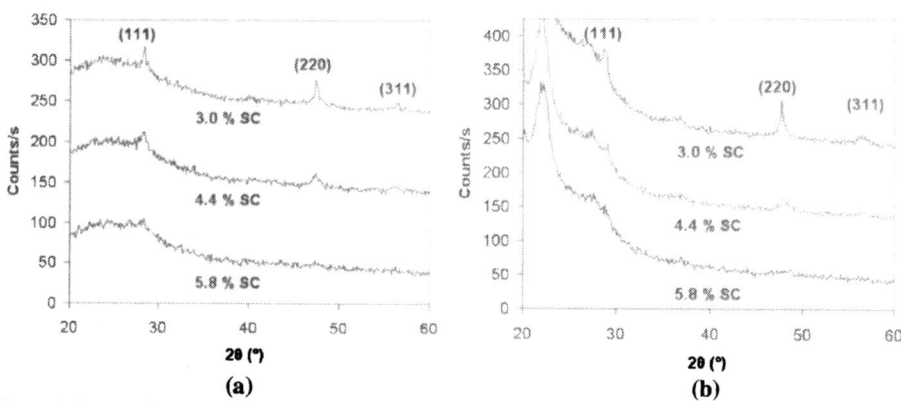

Figure 4: X-ray Diffraction spectra of nc-Si:H films grown at different silane concentrations on Corning glass substrates (a) and polyimide substrates (b).

XTEM bright field (BF) images and corresponding selected area diffraction (SAD) patterns for nc-Si:H grown at 4.4 % SC on a (100) single crystalline silicon wafer or polyimide substrates are shown in Figure 5 (a). From this figure, we observe that the film starts its growth from an amorphous silicon layer which has a thickness between 10 and 150 nm. Although this incubation layer is rather thick it is expected to be much smaller for lower silane concentrations [9]. Note that the native oxide layer was not stripped from the surface of the c-Si wafer which is one explanation why the film starts its growth as amorphous silicon rather than epitaxial. Figure 5(a) shows conical crystallites grow upwards at angles ranging from 55 to 75° and then form compact columns that grow vertically, similar to that found in previous reports [9,10]. The ring diameters and intensities in the SAD are consistent with the XRD results and the known diamond cubic structure of polycrystalline Si [11].

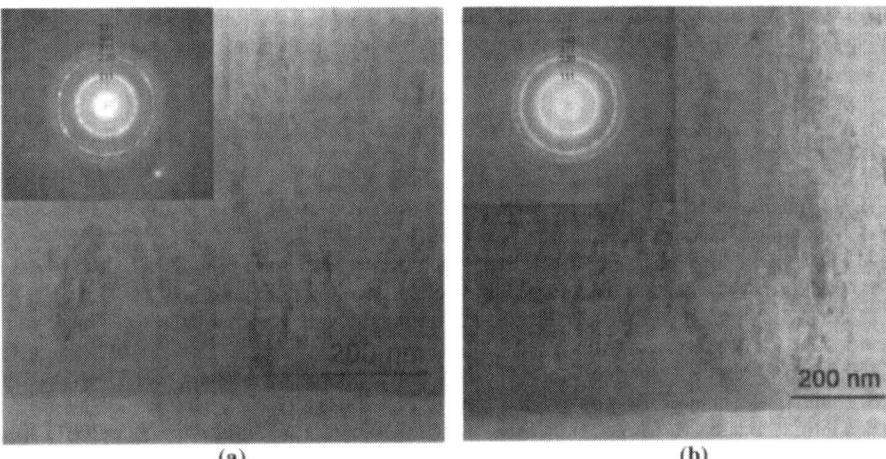

(a)   (b)

Figure 5: TEM cross section of nc-Si:H film grown on (a) single crystalline silicon wafer and (b) polyimide substrates.

The same cone-shaped crystallites that grow into vertical columns which were observed in the film grown on c-Si substrate are found again in the film grown on polyimide, shown in Figure 5 (b). Again, there is a thin amorphous layer from which the film starts its growth, this time with a thickness between 10 and 100 nm. In terms of film strain, a minor tensile strain causing substrate curvature but no cracking or peeling was noticed in all films grown on polyimide substrate.

CONCLUSIONS

Nanocrystalline silicon films grown on polyimide showed very similar dark and photo conductivities as those grown on Corning glass both in the amorphous and crystalline phases. Raman spectroscopy showed that films grown on Corning glass were slightly higher than those grown on polyimide substrate but silane concentration had the largest impact on crystalline volume fraction. XRD measurements showed that nc-Si:H films grown both on Corning glass and polyimide had the expected (111) and (220) preferred crystal orientations and that grain sizes ranged between 20-65 nm. Finally, TEM images showed that crystalline growth started from a thin amorphous region and grew into vertical columns both when grown on c-Si and polyimide substrate. Based on the Raman and XRD data, we assume that the same crystalline morphology occurred with the Corning glass substrates. Polyimide showed no disadvantage compared to Corning glass both in terms of σ_{ph}/σ_d ratio and crystallinity.

ACKNOWLEDGMENTS

The authors would like to thank Dr. Xinyu Wang for technical support and Dr. Michael Chen for access to Raman spectroscopy measurement apparatus. This project was funded by NSERC (Natural Science and Engineering Council of Canada), CFI (Canada Foundation for Innovation), BCKDF (British Columbia Knowledge Development Fund) and Simon Fraser University.

REFERENCES

1. H. Wiesmann, A. K. Ghosh, T. McMahon, M. Strongin, J. Appl. Phys. 50 (1979) 3752-3754.
2. H. Matsumura, Japanese J. Appl. Phys. 37 (1998) 3175-87.
3. A. Ledermann, U. Weber, C. Mukherjee, B. Schroeder, Thin Solid Films 395 (2001) 61-65.
4. K. Erickson, V. L. Dalal, J. Non-crystalline Solids, 266-269 (2000) 685-688.
5. M. M. Adachi, W. F. L. Tse, G. Cluff, K. L. Kavanagh, K. S. Karim, Mater. Res. Soc. Symp. Proc. 910 (2006) 0910-A08-05.
6. Van Veen, Marieke Katherine, Ph.D. Thesis, Utrecht University, Netherlands, 23-46 2003.
7. S. Klein, T.Repmann,T.Brammar, Solar Energy 77, (2004) 893-908.
8. S. Klein, F. Finger, R. Carius, H. Wagner, M. Strutzmann, Thin Solid Films 395 (2001) 305-309.
9. H.R. Moutinho, C.S. Jiang, J. Perkins, Y. Xu, B.P. Nelson, K.M. Jones, M.J. Romero, M.M. Al-Jassim, Thin Solid Films 430 (2003) 135–140.
10. J. Bailat, E. Vallet-Sauvain, L. Feitknecht, C. Droz, A. Shah, J. Appl. Phys. 93 (2003) 5727-5732.
11. D.B.Williams, C.B.Carter, *Transmission Electron Microscopy*, (Plenum Press, New York, 1996), pp. 265-286.

Manipulating the Hydrogen-Bonding Configuration in ETP-CVD a-Si:H

M. A. Wank[1], R. A. C. M. M. van Swaaij[1], and M. C. M. van de Sanden[2]

[1]DIMES-ECTM, Delft University of Technology, P. O. Box 5053, Delft, 2600 GB, Netherlands

[2]Department of Applied Physics, Eindhoven University of Technology, P.O.Box 513, Eindhoven, 5600 MB, Netherlands

ABSTRACT

The effect of ion bombardment on the relationship between the critical hydrogen concentration c_{crit} and the reactor pressure has been investigated for hydrogenated amorphous silicon (a-Si:H) deposited with the expanding thermal plasma-CVD (ETP-CVD) method. By increasing the reactor pressure, in particular above 0.24 mbar, the ionic cluster formation in the plasma can be increased, resulting in a decrease of c_{crit}. It is observed that this decrease of c_{crit} with increasing reactor pressure can not be compensated by ion bombardment at 14V biasing. Biasing with 20V however increases c_{crit} nearly up to the value obtained at low pressures. This observation indicates that the incorporation of ionic clusters formed at elevated reactor pressures can be reduced by substrate biasing, possibly due to break-up upon impact on the substrate surface or due to processes occurring in the secondary plasma close to the substrate. The onset of void formation in the film is found to depend on reactor pressure and substrate biasing, indicating that the maximum hydrogen solubility for ETP material might be affected by these deposition parameters.

INTRODUCTION

Hydrogen plays an important role in a-Si:H. It passivates dangling bonds in the a-Si:H network, reducing the defect density significantly compared to a-Si and permits the application of a-Si:H as material in, e.g., solar-cell devices. However, hydrogen is also involved in metastability mechanisms and a study of the hydrogen bonding configuration is therefore important to assess the material quality. Hydrogen bonding configuration is usually studied by Fourier transform infrared (FTIR) absorption spectroscopy. Three vibration modes of SiH_x hydrides can be observed: a wagging mode at 640 cm^{-1}, bending modes in the range of 840-890 cm^{-1}, and two stretching modes at 1980-2030 cm^{-1} (low stretching mode, LSM) and 2060-2160 cm^{-1} (high stretching mode, HSM). The LSM absorption is attributed to monohydrides in mono- and di-vacancies, whereas the HSM is mainly ascribed to Si-H bonds located at void surfaces [1].

From film density measurements, it was recently concluded that when the total hydrogen concentration, c_H, is above a critical value, c_{crit}, of 14 at.% the material is regarded as void-dominated and below 14 at.% as vacancy-dominated [1]. In the vacancy dominated region the integrated absorption of the LSM mode, I_{LSM}, is larger than that of the HSM mode, I_{HSM}, and vice versa in the void dominated region. Therefore, the critical hydrogen concentration, c_{crit}, can be determined as that c_H for which $I_{LSM} = I_{HSM}$. Recently, Petit [4] has investigated the dependence of c_{crit} on the reactor pressure during deposition. It was shown that increasing the reactor pressure above ~ 0.20 mbar leads to a reduction of c_{crit} [4] while it remains essentially independent of pressure below 0.20 mbar. The dependence of c_{crit} on pressure was attributed to an increased contribution of ionic clusters to the growth flux at increasing pressures. The production of these clusters depends on the silane density in the plasma beam [3] and the reactor pressure can be used to increase the silane density, thus increasing cluster ion production.

Alternatively the material structure can be related to the hydrogen solubility limit (HSL), which is generally regarded as the hydrogen concentration that can be incorporated into the amorphous network without introducing hydrogen-rich voids into the material. While the HSL for a-Si, in which hydrogen was implanted, was found to be around 4%, higher hydrogen concentrations can be obtained in directly deposited a-Si:H (up to 11%) [14,15] while maintaining a void-free structure. However these films are considered to be thermally unstable and void formation can be observed upon annealing above 450°C.

Ion bombardment in remote plasmas can be achieved by applying a sinusoidal RF signal to the substrate. As a result of the ion bombardment several processes can occur, like enhanced surface mobility, surface and bulk atom displacement, and sputtering at very high ion energies. Additionally, substrate biasing might also lead to breaking up of ionic clusters in the sheath region around the substrate due to interactions with the secondary plasma or upon impact on the substrate surface. Ion energies of a few electron volts are already sufficient to affect material properties [2]. Several studies on the effect of ion bombardment on a-Si:H film growth have been carried out [7-10], generally describing a densification of the films upon ion bombardment.

In this work, we investigate how the relationship between c_{crit} and reactor pressure is affected by RF biasing. By changing the reactor pressure the ionic cluster formation in the plasma can be varied and the influence of these clusters on the material properties can be studied. The effect of the pressure on c_{crit} was first determined for our gas-flow conditions and then the influence of the RF biasing was investigated. FTIR spectroscopy will be used to analyze the hydrogen bonding configuration.

EXPERIMENT

The ETP-CVD technique is a deposition method based on a remote plasma created in a cascaded arc and has been described in more detail elsewhere [6]. Silyl-radicals (SiH$_3$) are the dominant growth radical responsible for 90% of the film growth [11]. Kessels et al. [3] (Check the referencing!) have detected ionic clusters with up to 10 silicon atoms, which have a detrimental effect on the material density, especially at substrate temperatures below 300°C [1]. An externally induced ion bombardment is created by applying a 13.56 MHz RF signal to the substrate. Depending on the applied RF power a DC voltage on the substrate between 0 and 70 V can be obtained. Additionally, RF biasing generates a secondary plasma around the substrate holder.

For all depositions reported here, identical gas flows were used: 570 sccm Ar and 190 sccm H$_2$ in the arc, 150 sccm H$_2$ in the nozzle, and 230 sccm SiH$_4$ in the injection ring. The current in the arc is 40 A. The substrate temperature was varied between 100 and 400°C for each deposition series. The reactor pressure between the different series was varied by partially closing the valve to the roots pumps. Depositions were carried out with and without RF substrate biasing at reactor pressures of 0.18 mbar, 0.24 mbar and 0.3 mbar. For 0.3 mbar, two biased series were deposited with 14 V and 20 V, at an RF power of ~ 20 W and ~ 90 W, respectively. The deposition rate varied from 1.5 to 2.5 nm/s, depending on the pressure.

The a-Si:H films have been deposited on c-Si wafers (prime wafer, 500-550 μm, native oxide). From the FTIR spectra the integrated absorption of the wagging mode, and the low and high stretching mode (LSM at 2000 cm^{-1} and HSM at 2100 cm^{-1}, respectively) have been determined, from which the total hydrogen concentration, c_H, the hydrogen concentration incorporated in vacancies (LSM), c_{LSM}, and voids (HSM), c_{HSM}, respectively, is calculated [1]. The proportionality constants we used were obtained from Smets et al. [1]: $A_{640} = 1.6 \times 10^{19}$ cm^{-2} and $A_{2000} = A_{2100} = 9.1 \times 10^{19}$ cm^{-2}.

RESULTS AND DISCUSSION

Figure 1 shows the concentration of hydrogen obtained from LSM (left-hand side) and HSM (right-hand side) plotted versus the substrate temperature for the three pressure series (top to bottom), each with and without RF substrate biasing. For all pressures, LSM increases with biasing. Additionally, for 0.3 mbar we observe that LSM increases further for biasing at a higher voltage of 20 V. HSM is not affected by 14-V biasing for all investigated pressures, however, for 20 V we observe a decrease of HSM by 2-3% at temperatures below 300°C. For increasing reactor pressures, c_{LSM} shifts to lower concentrations at all temperatures, while c_{HSM} shifts up slightly and mainly at low temperatures.

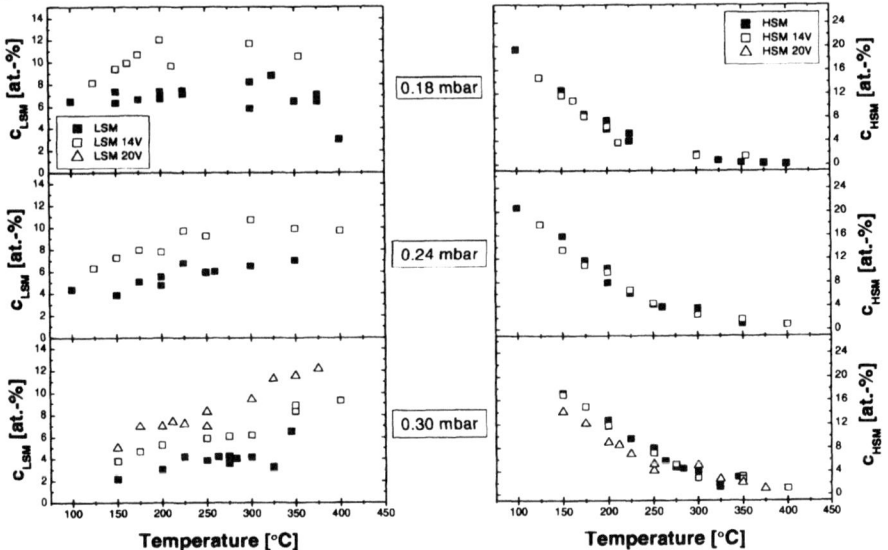

Figure 1. c_{LSM} and c_{HSM} for different substrate temperatures of all seven series. Full symbols are from unbiased and empty symbols from biased depositions.

In Figure 2 both c_{LSM} and c_{HSM} have been plotted versus the total hydrogen concentration, c_H, for the seven different depositions series. In this figure c_{crit} is marked with vertical lines. The graphs on the left-hand side show the three unbiased series. While c_{crit} does not change much when increasing the reactor pressure from 0.18 to 0.24 mbar (around 14% for both pressures), there is a substantial change when the pressure is increased further to 0.3 mbar. In the latter case we observe a shift of c_{crit} from 14% down to 9.5%, mainly due to an absolute drop of c_{LSM} by roughly 3%, but also a slight shift of the c_{HSM} to higher concentrations. These changes are observed mainly at higher total hydrogen concentrations (which corresponds to deposition at low temperatures). Petit [4] observed a *stronger* dependence of c_{crit} on the reactor pressure, with a drop from 14% to 7% when decreasing the pressure from 0.23 mbar to 0.25 mbar. However, her depositions were carried out at higher growth rates (3.2 to 6 nm/s compared to 1.5 to 2.5 nm/s for our series). An increased contribution of cluster ions to the growth flux can explain both the drop of c_{crit} with increasing reactor pressure and the stronger dependence of c_{crit} on reactor pressure at higher growth rates. With increasing pressure, the beam diameter is reduced, the silane density increased and thus the probability of ion-molecule (i.e. silane) interaction increases. Since ion-molecule reaction in the plasma occurs at near-collision rates, the increased interaction probability results in more and larger cluster ions formed in the beam [5,6]. For the same reason, more ionic cluster formation occurs at high growth rates, corresponding to higher gas flows of the growth precursor. These cluster ions and dust particles have higher sticking coefficients than SiH_3 radicals and contribution of these ionic clusters to the growth flux can result in poor material quality [12].

A similar trend can be observed for the depositions biased with 14 V as shown on the right-hand side of Figure 2. When increasing the pressure from 0.18 mbar to 0.24 mbar, c_{crit} changes slightly from 15.5% to 14.5%. At 0.3 mbar, a c_{crit} of 11% is found, which again we can attribute mainly to a decrease of c_{LSM} by approximately 3%. So for both biased and unbiased depositions, c_{crit} mainly changes due to lower c_{LSM}, corresponding to less hydrogen in vacancies, while the c_{HSM} remains largely unaffected. Compared with the unbiased series, biasing at 14-V hardly affects c_{crit}.

As shown above, we do not see a significant change in c_{crit} at 14-V biasing for the series at 0.3 mbar. However, at 20-V biasing we find a c_{crit} of 13% (dotted line), which is close to the 14% we obtain for the depositions at lower pressure. This high value for c_{crit} is due to an increase in LSM and a decrease in HSM. The increase in LSM and the decrease in HSM can both be explained by a reduced incorporation of cluster ions into the film. Cluster ions might break up upon impact on the surface, or cluster ions might break up in the

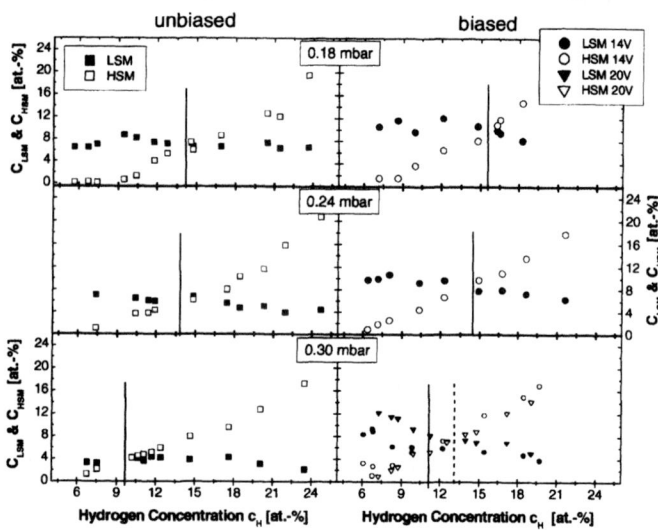

Figure 2: Plot of c_{LSM} and c_{HMS} versus c_H for the seven different series. The critical hydrogen concentration is marked by the vertical solid lines for 14-V biasing, and by the vertical dashed line for 20-V biasing.

secondary plasma around the substrate holder. The reduced cluster incorporation is not necessarily a result of the increased substrate voltage alone, also a higher ion current at increased RF power might contribute as it increases the ion-radical flux ratio and thus the energy per deposited cluster molecule. The ion-energy distribution of RF biasing is broad and bimodal [13], which complicates the interpretation of the observed trends further as both high and low energy ions are present in the plasma.

The LSM integrated absorption at the onset of HSM absorption varies as a function of reactor pressure and substrate biasing, as can be seen in Figure 2. This observation might imply that the HSL is influenced by these parameters. Since annealing at high temperatures was not carried out, it can not be concluded if the films are thermally stable.

Figure 3 shows the total hydrogen concentration, c_H, at different substrate temperatures. The typical decrease in c_H with increasing temperature can be seen for all series. An increase in pressure leads to an increased c_H for temperatures below 350°C, while virtually no changes can be observed above 350°C. The solid lines indicate the temperature, $T(c_{crit})$, at which c_{crit} is obtained from the unbiased samples in Figure 2. Above this $T(c_{crit})$, vacancy dominated material is obtained, whereas the material is void dominated material below $T(c_{crit})$. $T(c_{crit})$ shifts to higher temperatures with increasing reactor pressure, which demonstrates how the temperature range in which we obtain vacancy-dominated

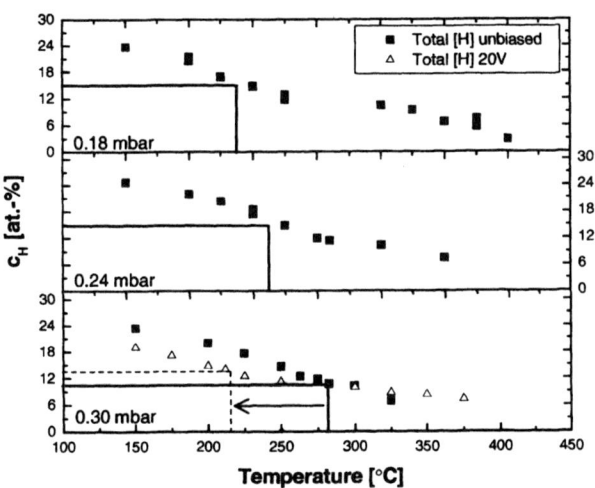

Figure 3: The total hydrogen concentration versus the substrate temperature for all series. Full symbols are from unbiased and empty symbols from biased depositions. The solid lines indicate the substrate temperature at which the critical hydrogen concentration is obtained, the dashed line does the same for the series at 0.3 mbar reactor pressure and 20 V substrate biasing.

material is reduced with increasing reactor pressures, thus reducing the range in which solar-grade material can possibly be obtained. The dashed line in the lowest graph in Figure 3 indicates $T(c_{crit})$ for the series biased with 20 V at 0.3 mbar reactor pressure. We can clearly see how, due to the increased c_{crit} and the reduced hydrogen concentration at lower temperatures, $T(c_{crit})$ is shifted downwards, thus increasing the temperature range in which vacancy-dominated material can be obtained.

CONCLUSIONS

The influence of RF substrate biasing on the relationship between the critical hydrogen concentration c_{crit} and the reactor pressure has been investigated. An increase in reactor pressure leads to a decrease in c_{crit}, mainly above 0.24 mbar, which is attributed to an increased contribution of cluster ions and polymers to the growth flux. Ion bombardment at 14 V has hardly an effect on c_{crit}, yet an increase in LSM could be observed for all temperatures and pressures. Ion bombardment at 20 V for the high pressure series resulted in an increase of c_{crit} from 9% unbiased to 13% biased, along with a decrease of HSM and an increase of LSM. This is attributed to a reduced incorporation of ionic clusters and polymers into the film. A dependence of hydrogen solubility on reactor pressure and substrate biasing could be observed, but the thermal stability of these films has not been investigated.

ACKNOWLEDGMENTS

M. Tijssen and K. Zwetsloot are acknowledged for their skillful technical assistance. This research was financially supported by SenterNovem within the framework of EOS-LT.

REFERENCES

1. A. H. M. Smets, W. M. M. Kessels, and M. C. M. van de Sanden, *Appl. Phys. Lett.* **82**, 1547 (2003).
2. H. R. Kaufman and J. M. E. Harper, *J. Vac. Sci. Technol. A*, **22**, 221 (2004).
3. W. M. M. Kessels, M. C. M. van de Sanden, and D. C. Schram, *Appl. Phys. Lett.* **72**, 2397 (1998).
4. A. M. H. N. Petit, Ph. D. thesis, Delft University of Technology (2006).
5. M. H. Brodsky, Thin Solid Films **40**, L23 (1977).
6. W. M. M. Kessels, C. M. Leewis. A. Leroux, M. C. M. van de Sanden, and D. C. Schram, *J. Vac. Sci. Technol.* **A17**, 1531 (1999).
7. M. A. Ring, V. L. Dalal, and K. K. Muthukrishnan, *J. Non-Cryst. Solids* **338-340**, 61-64 (2004).
8. B. Drevillon, J. Perrin, J. M. Siefert, J. Huc, A. Lioret, G. de Rosny, and J. P. M. Schmitt, *Appl. Phys. Lett.* **42**, 801 (1983).
9. E. A. G. Hamers, W. G. J. H. M. van Sark, J. Bezemer, H. Meiling, and W. F. Van der Weg, *J. Non-Cryst. Solids* **226**, 205 (1998).
10. T. V. Herak, T. T. Chau, S. R. Mejia, P. K. Shufflebotham, J. J. Schellenberg, H. C. Card, K. C. Kao, and R. D. McLeod, *J. Non-Cryst. Solids* **97-98**, 277 (1987).
11. W. M. M. Kessels, A. Leroux, M. G. H. Boogaarts, J. P. M. Hoefnagels, M. C. M. van de Sanden, and D. C. Schram, *J. Vac. Sci. Technol. A*. **19**, 467 (2001).
12. W. M. M. Kessels, M. C. M. van de Sanden, R. J. Severens, and D. C. Schram, *J. Appl. Phys.* **87**, 3313 (2000).
13. E. Kawamura, V. Vahedi, M. A. Liebermann, and C. K. Birdsall, *Plasma Sources Sci. Tech.* **8**, R45 (1999).
14. S. Acco, D. L. Williamson, W. G. J. H. M. van Sark, W. C. Sinke, W. F. van der Weg, A. Polman, and S. Roorda, *Phys. Rev. B* **58**, 12853 (1998).
15. W. Beyer and U. Zastrow, *J. Non-Cryst. Solids* **227-230**, 880 (1998).

Poster Session:
Nanocrystals, Nanoclusters
and Nanowires

Mater. Res. Soc. Symp. Proc. Vol. 989 © 2007 Materials Research Society

Tailored Deposition by LPCVD of Non-stoichiometric Si Oxides and their Application in the Formation of Si Nanocrystals Embedded in SiO₂ by Thermal Annealing

Bruno Morana[1], Juan Carlos G. de Sande[2], Andrés Rodríguez[1], Jesús Sangrador[1], Tomás Rodríguez[1], Manuel Avella[3], Ángel Carmelo Prieto[3], and Juan Jiménez[3]

[1]Tecnología Electrónica, Universidad Politécnica de Madrid, E.T.S.I.T., Madrid, 28040, Spain
[2]I. Circuitos y Sistemas, Universidad Politécnica de Madrid, E.U.I.T.T., Madrid, 28031, Spain
[3]Física de la Materia Condensada, Univ. de Valladolid, E.T.S.I.I., Madrid, 47011, Spain

ABSTRACT

Silicon oxide films with excess of Si were deposited by Low Pressure Chemical Vapor Deposition. The growth rate of the films and the excess of silicon in them have been modeled using a Face-centered Central Composite Design experiment. Samples annealed at 1100 °C show luminescence (665 nm) at 80 K and at room temperature associated to Si nanocrystals.

INTRODUCTION

Si nanocrystals embedded in SiO₂ have been used in non-volatile memories and optoelectronic devices compatible with the CMOS technology. A simple method to fabricate this kind of structures consists of the precipitation by thermal annealing of the excess of Si incorporated in non-stoichiometric silicon oxide films, which may be obtained using different techniques [1, 2]. In this work, the non-stoichiometric oxide layers have been deposited using a commercial Low Pressure Chemical Vapor Deposition (LPCVD) reactor. To find the experimental conditions that allow us to achieve a sufficiently slow growth rate to be able to control the thickness of the films and to incorporate the desired amount of Si in excess, the system has been characterized using a Face-centered Central Composite Design (FCCD) experiment [3]. The as-deposited samples were characterized by FTIR spectroscopy and spectroscopic ellipsometry to determine their composition, thickness and refraction index. The deposition process was characterized by the growth rate and the excess of Si incorporated in the films. The effect of the post-growth annealing process to form Si nanocrystals has been studied using Raman spectroscopy and Cathodoluminescence spectroscopy.

EXPERIMENT

Samples preparation

The deposition was carried out on Si wafers using a LPCVD deposition system with Si₂H₆ and O₂ as precursor gases and N₂ as carrier gas. The process parameters were selected based on our previous results on the deposition of stoichiometric oxides [4]. The pressure was varied between 185 and 300 mTorr and the temperature between 250 and 450 °C. The Si₂H₆ / O₂ gas flow ratio was varied between 2 and 5 to ensure that the gas is rich enough in the precursor of Si. The total flow was 102 sccm, being 90 sccm the flow of N_2 in all the experiments and 12 sccm the total flow of the reactant species, which was kept constant. The deposition time was two hours except in some samples in which it was three hours.

The volume fraction occupied by the excess (E) of Si incorporated into the deposited layers and the growth rate (V) were modeled as a function of the process temperature (T), pressure (P) and Si_2H_6 / O_2 flow ratio (R) using a Face-centered Central Composite Design experiment [3, 5]. The coded values of the main variables ranging from -1 (lowest value) to +1 (highest value) were used for the analysis. The experiments necessary for the full characterization of the system are the eight vertices (±1, ±1, ±1), the central points of the six faces (±1, 0, 0), (0, ±1, 0) and (0, 0, ±1) as well as three repetitions of the central point of the cube (0, 0, 0). The significance of the effects (mains, linear interactions and quadratic terms) was determined by the p-value (prob > F) obtained from the F distribution. The F values (ratio between the mean square of each effect and the mean square of the noise of the experiment) for each effect were determined from the experimental results. The data were analyzed using the Design-Expert® 7.1.1 Software [6].

The as-deposited samples were annealed in a furnace at a temperature of 1100 °C for 1 h to segregate the Si in excess. Additional treatments were performed in H_2-N_2 at 450 °C for 1 h.

Characterization techniques

The composition of the as-deposited and annealed oxides was qualitatively analyzed by Fourier transform infrared spectroscopy (FTIR). The spectra were acquired using a Perkin Elmer Spectrum 100 spectrometer in the 200 to 7800 cm^{-1} wavenumber range. The bare Si substrates obtained by etching the deposited layers were used as references to acquire the background in each case, in such way that the spectra of the samples account only for the absorption of the layers. The absorbance spectra were normalized to the layer thickness yielding the absorption coefficient spectra.

The thickness, the refraction index (n) and the volume fraction of the excess of Si in the layers were determined by spectroscopic ellipsometry (SE) measurements carried out in the 270 to 1200 nm wavelength range for incidence angles of 70°, 75°, and 80° using a variable angle rotating analyzer ellipsometer (J. A. Woollam Co., Inc.). The growth rate was obtained as the ratio between the layer thickness and the deposition time. To estimate the volume fraction of Si in excess, the as-deposited samples were modeled as a mixture of SiO_2 and amorphous Si on crystalline Si [7]. A model of the sample structure formed by one layer with optical properties represented by an effective medium approximation (EMA) on top of the Si substrate was used. For small values of the excess of Si (below 5 % in volume) Maxwell-Garnett and Bruggeman EMA give almost the same result, but for large values, Bruggeman EMA gives a more appropriate description of the optical properties of the films [7,8].

The annealed samples were characterized by Raman spectroscopy and cathodoluminescence (CL). The Raman spectra were acquired with UV (325 nm line from a HeCd laser) excitation, to enhance the sensitivity of the technique to small amounts of crystalline material, using a HR UV Labram Jobin-Yvon Raman spectrometer. The CL emission was measured in a XiCLOne system (Gatan) attached to a SEM (JEOL 820). The spectra were measured at liquid nitrogen temperature and at room temperature using an e-beam acceleration voltage of 5 kV and a beam current of 10 nA.

RESULTS AND DISCUSSION

As-deposited samples

Figure 1 shows a selected region of the absorption coefficient spectra obtained by FTIR spectroscopy of several as-deposited layers of different volume fraction of Si in excess. The spectra of the layers with a very small excess of Si show the three bands characteristic of SiO_2 located around 450 (not included in the figure), 810 and 1080 cm^{-1}, although they are slightly shifted towards lower wavenumbers due to the non-stoichiometric composition of the as-deposited oxides. The absorption bands which appear at 700, 850 and 875 cm^{-1} in the spectra of samples with a higher excess of Si have been attributed to the bending mode of Si-H bonds with one, two and three oxygen atoms backbonded to the Si atom respectively [9]. The 870 cm^{-1} band has also been attributed to the SiH_2 scissors vibration [10]. The presence of these bands in the spectra confirms the non-stoichiometry of the oxides. The intensity of these peaks decreases as the deposition temperature increases, due to the enhancement of the hydrogen desorption with the temperature.

Figure 2 shows the volume fraction of Si in excess as a function of the refraction index at two different wavelengths (620 and 300 nm). The experimental values of the volume occupied by the excess of Si ranges from 0 to around 50 %. The obtained growth rates, ranging from 15 to 60 nm/h, are appropriate for the controlled deposition of layers with thickness from 10 nm to several tens of nm in a reasonable process time.

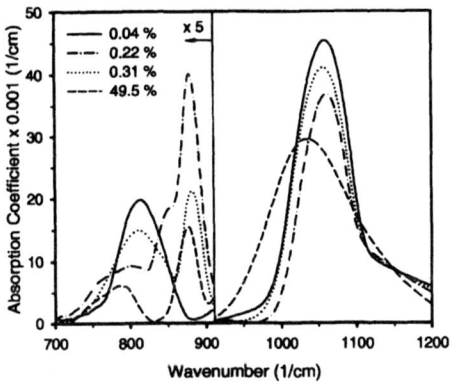

Figure 1. Absorption coefficient spectra of selected as-deposited samples with different Si contents.

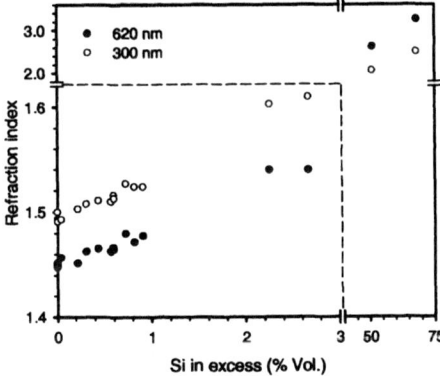

Figure 2. Refractive index at 620 and 300 nm of the as-deposited samples as a function of the Si content.

Figures 3 and 4 summarize the main results of the FCCD experiment. The excess (E) of incorporated Si depends significantly (small p values) on the temperature and on the gas flow ratio, being noticeable the importance of the effect of the interaction between them, which reinforces the incorporation of Si to the films when both variables take high values. Increasing the pressure causes the Si in excess to increase slightly. However, the p value for the pressure main effect is 0.62, indicating that its influence is not significant from the statistical point of view. The F-value obtained for the whole analysis is 11.76, which implies that the model is significant, with only a 0.05 % (p value) chance that an F-value this large could occur due to noise. The obtained expression, valid for E > 0, is

$$E(\% \text{ Si Vol.}) = 11.5 + 15.9 \cdot T + 14.8 \cdot R + 8.4 \cdot T \cdot R,$$

where the coded values of the variables T and R (ranging from -1 to +1) must be used.

The significant effects (small p values) of the growth rate are the temperature, the temperature squared, the gas flow ratio and the interaction between the temperature and the gas flow ratio. The p value for the pressure main effect is 0.63 in this case, so the effect of this variable is not significant. The F-value is now 13.47, which implies that model results for the growth rate are also significant. There is only a 0.02 % (p value) chance that a F-value this large could occur due to noise. The obtained expression is

$$V(\text{nm/h}) = 52.2 - 5.5 \cdot T - 10.0 \cdot R - 6.0 \cdot T \cdot R - 16.4 \cdot T^2,$$

where the coded values of T and R must be used. For low values of the temperature and the flow ratio, the amount of Si in excess is very small, so the deposited material is almost pure SiO_2. If the gas flow ratio is increased keeping constant the temperature, the growth rate decreases due to the progressive incorporation of Si in excess to the films. On the other hand, if the deposition temperature is increased keeping constant the flow ratio, the oxide growth rate first increases due to the effect of the temperature and then decreases since the deposited layer is incorporating an increasing amount of Si in excess. The main effect of the interaction between the temperature and the gas flow ratio is to reinforce the decrease of the growth rate at high values of both variables.

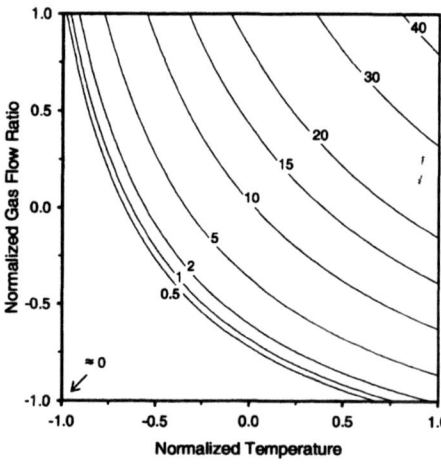 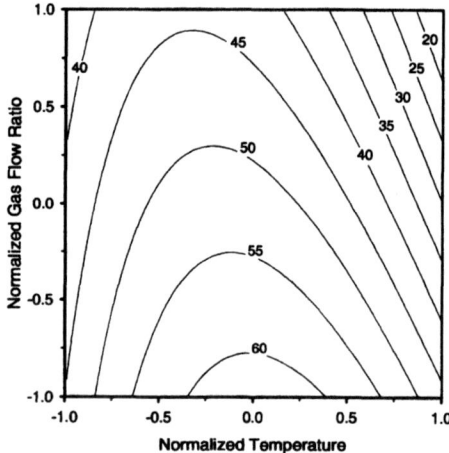

Figure 3. Excess of Si, E (% Si Vol.), as a function of the normalized values of T and R at P = +1.

Figure 4. Growth rate, V(nm/h), as a function of the normalized values of T and R at P = +1.

Annealed samples

The Raman and CL spectra of some annealed samples are shown in figures 5 and 6 respectively. The broadening of the optical phonon first order Raman band of Si observed in the spectra of the annealed samples can be explained by the presence of Si nanocrystals, with roughly estimated diameters ranging from 7 to 10 nm, in which the phonon correlation length is reduced due to finite size effects. Note that this band in the spectrum of the Si substrate is already broadened and shifted due to disorder induced by polishing, which becomes observable when excited with 325 nm light. In the samples with a high excess of Si the spectrum shows, in addition to the presence of nanocrystals, a broad band around 485 cm^{-1}, which can only be accounted for the presence of a highly disordered Si phase.

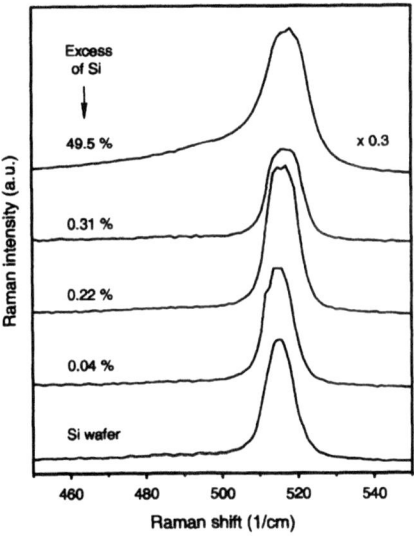

Figure 5. Raman spectra of samples with layers of different Si contents annealed at 1100 °C for 1 h and a spectrum of a bare Si wafer for comparison. The spectra have been shifted for a sake of clarity.

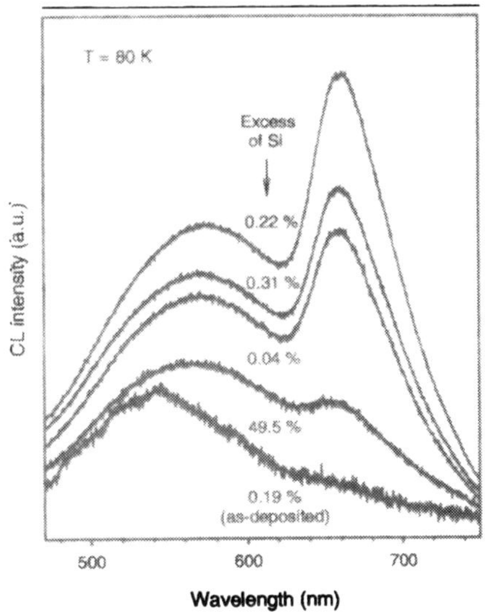

Figure 6. CL spectra of samples with layers of different Si contents annealed at 1100 °C for 1 h and spectrum of an as-deposited sample. The spectra have been divided by the layer thickness.

The CL spectra show two broad luminescence bands. One of them at around 570 nm, that is also present in the as-deposited samples, and the other one at 665 nm, present only in the annealed samples. The first one is associated in the literature to the SiO_2 matrix [11]. The second band is associated to the presence of the Si nanoparticles formed during the annealing process and detected by Raman spectroscopy. This association is supported by the increase of around a 30 % in the intensity of the CL emission which was detected when the annealed samples are additionally treated in forming gas (not shown). The intensity of this band increases with the excess of Si and decreases again when this excess is too high. The presence of a highly disordered Si phase, detected by Raman spectroscopy, in the samples with a high excess of silicon could explain the poor luminescence emission of these samples. The CL band position does not depend on the deposition parameters, and therefore on the size of the nanoparticles formed during the annealing process, which suggests that the luminescence emission is related to recombination at deep levels, without any contribution from quantum confinement effects. The CL spectra of the samples measured at RT show only the 665 nm band, with an intensity around ten times lower than at 80 K.

CONCLUSIONS

The growth rate and the excess of silicon of non-stoichiometric silicon oxides deposited by LPCVD have been modeled as a function of the process parameters, identifying the main effects and the interactions between them. These models allow us to design the deposition processes to obtain non-stoichiometric silicon oxide films with the desired thickness and excess of silicon to be used in the synthesis of luminescent structures with Si nanocrystals.

ACKNOWLEDGMENTS

This work was funded by the Spanish Government through CICYT Project MAT2004-04580-C02. The authors would like to thank Rosalía Serna (Instituto de Óptica, CSIC, Madrid) for allowing them to carry out the ellipsometric measurements in her laboratory.

REFERENCES

1. J. R. Aguilar, G. Monroy, M. Cárdenas, G. S. Contreras. Materials Sci. Eng. C (2007). In press (available online).
2. G. G. Ross, D. Barba, C. Dahmoune, Y. Q. Wang, F. Martin. Nucl. Instrum. and Meth. B **256**, 211 (2007).
3. D. C. Montgomery. *Design and Analysis of Experiments* (John Wiley & Sons, New York, 2001).
4. M. I. Ortiz, J. Sangrador, A. Rodríguez, T. Rodríguez, A. Kling, N. Franco, N. P. Barradas, C. Ballesteros. Phys. Stat. Sol. (a) **203**, 1284 (2006) and references therein.
5. M. J. Anderson, P. J. Whitcomb. *RSM Simplified* (Productivity Press, New York, 2004).
6. http://www.statease.com/
7. *Handbook of Optical Constants of Solids*, edited by E. D. Palik (Academic, San Diego, CA, 1998).
8. J. C. G. de Sande, R. Serna, J. Gonzalo, C. N. Afonso, D. E. Hole, and A. Naudon. J. Appl. Phys. **91**, 1536 (2002).
9. H. Rinnert, M. Vergnat, G. Marchal, A. Burneau. Appl. Phys. Lett. **72**, 3157 (1998).
10. X. Wu, Ch. Ossadnik, Ch. Eggs, S. Veprek, F. Phillipp. J. Vac. Sci. Technol B **20**, 1368 (2002).
11. H. Nishikawa, E. Watanabe, D. Ito, Y. Sakurai, K. Nagasawa, Y. Ohki. J. Appl. Phys. **80**, 3513 (1996).

Mater. Res. Soc. Symp. Proc. Vol. 989 © 2007 Materials Research Society   0989-A23-03

Silicon Nanowires: Growth Studies Using Pulsed PECVD

David Parlevliet and John C.L. Cornish

Physics, Murdoch University, South Street, Murdoch, WA 6150, Australia

ABSTRACT

Silicon nanowires with high aspect ratios have been grown at high density using a variation of Plasma Enhanced Chemical Deposition (PECVD) known as Pulsed PECVD (PPECVD). Growth rate and morphology were investigated for a range of catalysts: gold, silver, aluminum, copper, indium and tin. The thickness of the catalyst layer was 100nm. Deposition was carried out in a parallel plate PECVD chamber at substrate temperatures up to 350°C, from undiluted semiconductor grade Silane. A 1 kHz square wave was used to modulate the 13.56 MHz RF power. Samples were analyzed using either a Phillips XL20 SEM or a ZEISS 1555 VP FESEM. The average diameter for nanowires grown using a gold catalyst layer was 150nm and the average length was 4μm although some nanowires were observed with lengths up to 20μm. Back-scattered-electron images clearly show gold present at the tips of the silicon nanowires grown using gold as a catalyst, confirming their growth by the vapor liquid solid (VLS) mechanism. Sporadic growth of nanowires was detected when using copper as a catalyst. Although gold performed best as catalyst for nanowire growth it was, however, closely followed by tin. The other catalysts produced nanowires with properties between these extremes.

INTRODUCTION

Silicon nanowires have potential uses in the semiconductor industry including possibly the next generation of photovoltaic solar cells. The most common method of fabrication involves the Chemical Vapor Deposition (CVD) of a gas containing silicon and the subsequent growth of silicon nanowires by the Vapor Liquid Solid (VLS) mechanism first proposed by Wagner and Ellis [1]. The presence of a metal catalyst droplet on one end of a nanowire is indicative of the VLS mechanism. Plasma Enhanced Chemical Vapor Deposition (PECVD) is a technique widely used in the production of amorphous and nanocrystalline silicon thin films. When used with substrates covered with a metal catalyst, PECVD has been used to produce silicon nanowires and is known to improve their deposition rate. [2]. A modification of PECVD is pulsed PECVD which uses a modulated plasma to aid the deposition process. We have previously shown [3] that PPECVD can be used to produce silicon nanowires with a greater area density than conventional PECVD. In the VLS mechanism the role of the catalyst is to encourage the growth of single crystal silicon nanowires. For silicon to be absorbed by the catalyst it needs to be highly soluble in the chosen metal. The choice of the metal catalyst is also known to affect the electrical properties of resulting nanowires [4].

A number of catalysts are commonly used in the growth of silicon nanowires. The most common is gold which does not form a silicide and has a bulk Au/Si eutectic temperature that is fairly low (363°C), this results in low temperature growth [2]. Gallium is another catalyst used mainly in the growth of bulk Si nanowires. The main advantage of using Gallium is the extremely low Ga/Si eutectic temperature (30°C) which should permit low temperature growth of silicon nanowires [5]. Titanium has been used to grow silicon nanowires; however a mechanism other than VLS is required to explain the growth of these nanowires as the growth temperature is well below the melting point of the catalyst. The advantage of Ti is that it does not form deep mid-gap levels and the solubility and diffusion coefficients of Ti are low in silicon [6]. Iron has been used to grow silicon based nanowires, however, the deposition method used involves physical evaporation or laser ablation of an Au/Fe target [7,8]. Cobalt has been used to grow silicon nanostructures via the VLS mechanism, the resulting growth described as a "flower" [9]. Zinc has also been used to grow nanowires via the VLS mechanism, the resulting nanowires were observed to have different electrical properties than those grown with a gold catalyst under the same conditions [4]. VLS growth of silicon whiskers has also been shown to occur for catalysts including Pt, Ag, Pd, Cu and Ni [1]. The aim of this work was to use as catalysts gold, tin, aluminum, indium, silver and copper with pulsed PECVD in an endeavor to improve the yield of silicon nanowires under the conditions imposed by the available equipment.

EXPERIMENT

Both glass and polished n-type Si (100) substrates were used for the deposition of silicon nanowires in this study. The substrates were cleaned in an ultrasonic bath in several steps using decon-90, ultra-pure water and propanol. After a final ultra-pure water rinse the substrates were dried using high purity nitrogen before being transferred into a vacuum system for deposition of the catalyst layer. The native oxide layer was left intact on the crystalline silicon substrates to ensure non-epitaxial growth. Catalyst layers with an average thickness of 100nm were deposited onto the substrates by thermal evaporation of high purity catalyst metal from a tungsten wire or boat under vacuum. The thickness of the catalyst layer was measured in-situ using a quartz crystal microbalance. Silicon nanowires were grown in a separate chamber using pulsed PECVD which has been previously shown to promote improved growth. The PECVD system has parallel plate electrodes and uses a 13.56 MHz RF signal to generate the plasma. The addition of square wave modulation at 1 kHz produces pulsed PECVD. The deposition was performed in 2.7 to 3.0 Torr of silane and at a substrate temperature of 335°C. Plasma duration ranged from 10 to 40 seconds. A control sample using gold as a catalyst was present during each deposition. The samples were analyzed using either a Philips XL20 SEM or ZEISS 1555 VP FESEM.

DISCUSSION

Each of the catalysts used produced nanowires, however, the results ranged from a high density of nanowires over a substantial area of the substrate to others where few nanostructures grew on localized regions of the substrate. The morphology of the nanowires grown as part of this study was found to be dependant on the catalyst that was used. Nanowires produced using Silver are shown in figure 1. They were found to have a range of diameters and lengths. These nanowires exhibit some growth defects such as kinking but generally tended to grow as long straight wires. Present on the end of some of the nanowires is a bright tip that is similar in appearance to the droplet seen on gold nanowires grown by the VLS mechanism. Aluminum was a largely unproductive catalyst yet did produce some structures similar in appearance to the silver catalyzed nanowires shown in figure 1. They were much larger, however, having diameters up to ten times as large as nanowires grown by other catalysts. The diameters of the wire-like structures were fairly uniform and many exhibited a droplet at one end, indicative of VLS growth. Nanowires grown by using indium were similar in appearance to silver and aluminum. These also exhibited a range of diameters.

Figure 1. Silver-catalyzed nanowires growing out of more complex structures.

Nanowires grown using gold have largely uniform diameters as seen in figure 2.

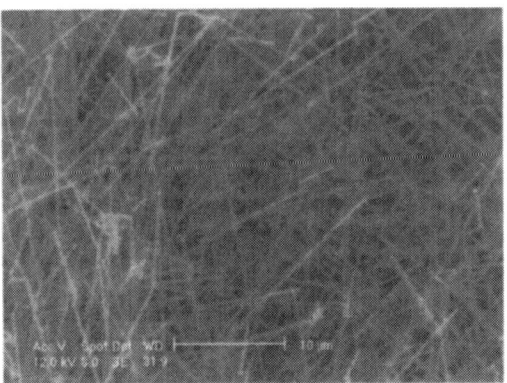

Figure 2. The appearance of a high density of Silicon nanowires grown using Au as a catalyst.

These show some growth defects such as kinking and commonly have a droplet visible on the tip of the nanowire. Although the nanowires are mainly straight and of high aspect ratio some curled or worm-like structure have also been observed. Nanowires grown using gold have a much higher growth density (NW/μm^2) than other catalysts. Using copper yielded very few nanowires under the growth conditions used. Copper mainly produced droplet-like structures on the surface of the substrate. Some of these may exhibit signs of growth. Scattered amongst these were the occasional nanowire, with largely uniform diameters and few growth defects. Indium produced very few nanowires under the growth conditions used. The nanowires had both crystalline like growth defects and amorphous worm-like structures, often in the same wire. A bright tip was visible on the end of a number of the nanowires. Tin produced a high density of very fine, uniform and long nanowires which are ribbon like in appearance as can be seen in figure 3. These nanowires have few growth defects and tend to bend rather than kink. Their sinuous form is in contrast to the ramrod straightness of nanowires initiated by the other catalysts.

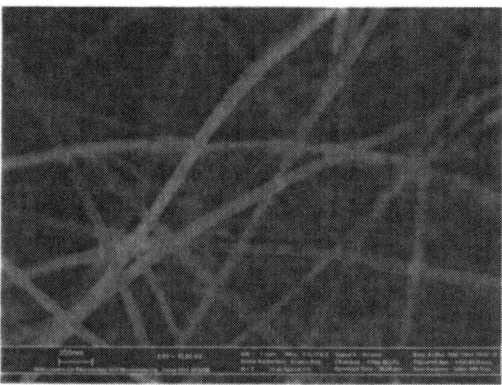

Figure 3. Long ribbon like nanowires grown using Sn as a catalyst.

The average diameters of the nanowires are compared in figure 4a. It can be seen from this plot that aluminum catalyzed wires have the greatest diameter, tin nanowires have the smallest and the other catalysts have very similar diameters. To compare how well the nanowires are produced by the different catalysts the "coverage" has been plotted in figure 4b. The coverage is expressed as a percentage of the sample covered by the nanowires. It can be seen that gold produces the greatest coverage of nanowires at 84.5% followed by tin at 72.3%. The other catalysts have coverage below 12% indicating that tin and gold produce a greater density of nanowires for the growth conditions used.

The physical properties of the catalyst used to grow the silicon nanowires are likely to have an impact on the growth and morphology of the silicon nanowires. For example the melting point, or eutectic point of Au and Si determine the minimum temperature required for the growth of silicon nanowires via the VLS mechanism.

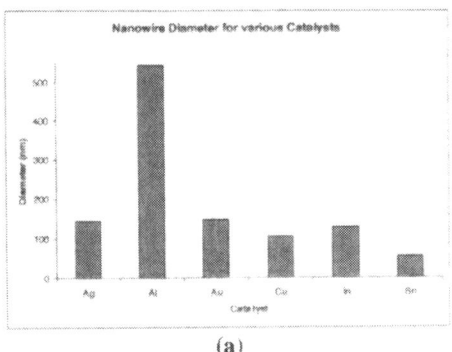
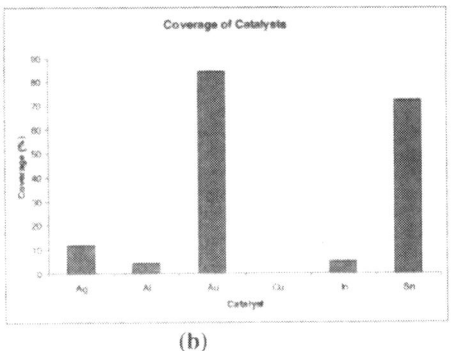

(a)   (b)

Figure 4. **(a)** Nanowire diameters for various catalysts.
             (b) The coverage of nanowires for various catalysts.

The melting and eutectic points of the various catalysts used in this study are listed in table I.

Table I. Melting points and eutectic points of the six catalysts used in this study [2, 10, 11, 12]

Catalyst	Melting Point °C	Si Eutectic Point °C
Ag	961.78	845
Al	660.32	577
Au	1064.18	363
Cu	1084.62	802
In	156.6	156.63
Sn	231.93	231.9

      The nanowires grown using gold as the catalyst have shown the gold tip on the nanowires that is indicative of the VLS growth mechanism. This confirms that the VLS mechanism applies to the growth of silicon nanowires using PPECVD. The VLS mechanism is the most probable growth mechanism for the nanowires catalyzed by a thin film of tin, as the deposition temperature was above the melting point of tin. The diameter of the nanowires produced in this study was on average 54nm. The long nanowires curve gently and are loosely bound and appear to be ribbon like. There are very few kinks or growth defects that would indicate crystallinity. They may be amorphous but this has not been confirmed. It was found in this study that the use of both silver and copper as catalysts produced silicon nanowires. The silver catalyzed nanowires often showed a bright tip under examination by the SEM which indicates the VLS mechanism is responsible for the growth. The Cu catalyzed nanowires did not show any droplets on the end of the nanowires. This may be the result of either the catalyst being consumed by the silicon nanowire to the point where the diameter and composition of the tip is indeterminate from the bulk, or a different nanowire growth mechanism. The higher temperature of the eutectic points of

Al-Si, Ag-Si and Cu-Si provides an explanation as to why the growth using these catalysts was so ineffective. The substrate temperature employed in this study was too low for the VLS mechanism to be operative, resulting in sporadic growth due to some other mechanism.

CONCLUSIONS

Silicon nanowires have been grown by PPECVD using Ag, Al, Au, Cu, In and Sn as catalysts. It was found that gold was the most effective catalysts from this group at the growth temperatures and conditions used producing nanowires with substrate coverage of seven to eight times greater than the other catalysts. Tin was almost as effective and produced nanowires with an average diameter less than the other catalysts. The tin-catalyzed nanowires showed a morphology which appeared to possibly be amorphous. For high density growth of silicon nanowires at ~340°C via pulsed PECVD tin and gold are the preferred catalysts.

ACKNOWLEDGMENTS

The analysis of samples using the ZEISS 1555 VP FESEM was carried out using facilities at the Centre for Microscopy and Microanalysis and Biomedical Image and Analysis Facility, The University of Western Australia, which is supported by University, State and Federal Government funding. The electron microscope studies were funded by a REGS grant from Murdoch University

REFERENCES

1. R. S. Wagner and W. C. Ellis, Applied Physics Letters **4(5)**, 89-90 (1964).
2. S. Hofmann, C. Ducati et al., Journal of Applied Physics **94(9)**, 6005-6012 (2003).
3. D. Parlevliet and J. C. L. Cornish, Proceedings of the International Conference on Nanoscience and Nanotechnology, Brisbane, Australia 35-38 (2006).
4. J.-Y. Yu, S.-W. Chung et al., Journal of Physical Chemistry B **104(50)**, 11864-70 (2000).
5. S. Sharma and M. K. Sunkara, Nanotechnology **15(1)**, 130-134 (2004).
6. T. I. Kamins, , R. S. Williams, et al., Applied Physics Letters **76(5)**, 562-4 (2000).
7. D. P. Yu, Z. G. Bai, et al., Applied Physics Letters **72(26)**, 3458-3460 (1998).
8. D. P. Yu, Q. L. Hang, et al., Applied Physics Letters **73(21)**, 3076-3078 (1998).
9. Z. Y. Qui, H. W. Bing, et al., Advanced Materials **11(10)**, 844-7 (1999).
10. T. B. Massalski, J. L. Murray, et al., Eds. Binary alloy phase diagrams. Metals Park, Ohio, American Society for Metals. (1986).
11. D. R. Lide, Ed. Handbook of Chemistry and Physics. CRC. Boca Raton, Taylor & Francis Group. (2005).
12. Y. Wang, V. Schmidt, et al., **1(3)**, 186-189 (2006).

Solar Cells III

Mater. Res. Soc. Symp. Proc. Vol. 989 © 2007 Materials Research Society
0989-A24-01

Recent Progress in Up-Scaling of Amorphous and Micromorph Thin Film Silicon Solar Cells to 1.4 m² Modules

Johannes Meier[1], Ulrich Kroll[1], Stefano Benagli[1], Tobias Roschek[2], Andreas Huegli[2], Joel Spitznagel[1], Oliver Kluth[2], Daniel Borello[1], Michael Mohr[2], Dmitri Zimin[2], Giovanni Monteduro[1], Jiri Springer[2], Christoph Ellert[2], Girogios Androutsopoulos[1], Gerold Buechel[2], Arno Zindel[2], Franz Baumgartner[3], and Detlev Koch-Ospelt[2]

[1]Oerlikon Solar-Lab S.A., Puits-Godet 6a, Neuchâtel, CH-2000, Switzerland
[2]OC Oerlikon Balzers A.G., Balzers, LI-9496, Liechtenstein
[3]University of Applied Science Buchs, Werdenbergstrasse 4, Buchs, CH-9471, Switzerland

ABSTRACT

In this paper an overview of our developments towards industrialization of thin film silicon PV modules is presented. Amorphous silicon p-i-n solar cells have been developed in medium size single-chamber R&D KAI-M PECVD reactors. High initial efficiencies of 10.6 % and stabilized of 8.6 % could be achieved for a 1 cm² a-Si:H p-i-n solar cell of 0.20 μm thick i-layer deposited on TCO from Asahi U type (SnO₂). On our in-house developed LPCVD ZnO we could further improve the stabilized a-Si:H p-i-n efficiency to a similar level of 8.5 %. Incorporating such cells in commercial available front TCO of lower quality still leads to high initial mini-module aperture efficiencies (10 x 10 cm²) of 9.1% and stabilized ones of 7.46% (independently measured by ESTI JRC-Ispra).

Transferring the processes from the KAI-M to the industrial size 1.1x1.25 m² KAI-1200 R&D reactors resulted in a-Si:H modules of 110.6 W using commercial TCO, respectively 112.4 W when applying in-house developed LPCVD front ZnO. Both initial module performances have been independently measured by ESTI laboratories of JRC Ispra. A typical temperature coefficient for the module power of -0.22 %/°C (relative loss) has been deduced from temperature dependent I-V characteristics at ESTI laboratories of JRC Ispra. Finally, micromorph mini-modules of 10 % initial aperture efficiency have been fabricated.

INTRODUCTION

The shortage in crystalline silicon for the PV industry calls for alternative solar cell concepts like thin film solar cells. Various studies have led to the conclusion that thin film PV has a higher potential for cost reduction compared to conventional wafer-based silicon PV [1-3]. Due its considerably reduced quantity of the absorber material, herein, thin film silicon has a huge potential as its technology is based on non-toxic and highly abundant materials involved. From all types of thin film solar cells amorphous silicon technology has shown to be the most advanced concept having already the potential for GW-mass production [3]. One main cause why thin film solar cells have not achieved the growth of crystalline PV is the lack of large-area production equipment. In addition, the built-up of a thin film silicon PV manufacturing line requires still a higher initial investment for the equipment than for c-Si lines.

Since the early 90ies the development of large-area PECVD reactors for amorphous silicon technology was driven by the flat panel display industry. This industry is much larger than all thin film silicon PV activity worldwide together [4] and, hence, can be used as driving force of new thin film silicon solar cell production tool equipment. Therefore, Oerlikon Balzers renamed from formerly Unaxis in September 2006, enters as equipment manufacturer for thin film silicon PV with its large-area KAI PECVD reactors developed for the amorphous silicon display technology. Our KAI PECVD reactors are now adapted to the deposition of thin film silicon solar cells [4-8]. In addition we develop and qualify necessary equipments like glass cleaning, laser-scribing, back contacts, quality assurance, etc for thin film silicon fabrication facilities based on mass production approved KAI-1200 PECVD reactors (substrate area 1.1x1.3 m^2). The strategy of Oerlikon Solar is to offer full thin film manufacturing lines for amorphous and micromorph silicon modules. The high productivity of Oerlikon Solar KAI production equipment is very attractive for a new generation of cost-effective thin film PV modules based on amorphous and promising new micromorph tandem cells in challenging the reduction of PV costs.

Very important in a-Si:H based PV is the quality of the intrinsic absorber layer to keep low the light-induced degradation (Staebler-Wronski effect). We have carefully optimized our amorphous silicon absorber layer in the p-i-n solar cell with respect to interfaces, stability, deposition rate and throughput. In this paper an overview of the activities in the field of thin film silicon solar cells and modules reports about the actual status and results of Oerlikon Solar.

EXPERIMENT

To enhance the deposition rate for amorphous, and especially microcrystalline silicon, the display-type reactors were modified to run at a higher excitation frequency of 40.68 MHz. Developments are currently carried out in R&D reactors of different substrate size, like the KAI-S (350x450 mm), the KAI-M (520x410 mm) and the KAI-1200 (substrate size 1250/1300x1100 mm). In the medium size reactors we first explore the fabrication processes both for the amorphous silicon [9] and microcrystalline silicon cells [10]. The developed recipes are then transferred to the industrial size KAI-1200 of 1.4 m^2 module area.

In order to evaluate the stability performance cells and mini-modules were light-soaked under 1 sun illumination at 50°C for 1000 hours.

Attracted by record stabilized efficiencies of 9.5 % confirmed by NREL for an amorphous silicon single- junction p-i-n solar cell using LPCVD (low pressure chemical vapor deposition) grown ZnO [11] (based on IMT's modified LPCVD process [12, 13]) we developed as well large-area R&D equipment (1.4 m^2) for the deposition of ZnO to open the full efficiency potential of thin film silicon solar cells. LPCVD ZnO layers have been developed with excellent transmission, haze and low sheet resistance. Standard front TCO like Asahi U-type SnO_2 and commercial SnO_2 were compared with our in-house ZnO. It has to be noted that the commercial TCO is not a high-quality TCO but commercially available for large areas and quantities. Due to the low process temperature involved, LPCVD ZnO is also suited for the back contact of cells and modules. ZnO back contacts in combination with a white reflector reveal excellent light-trapping properties [4, 7-9, 14] and have been applied in all cells and modules presented here. The test cells were scribed to areas of well-defined 1 cm^2.

Laser-scribing equipment for large-area modules (1.4 m^2) has been developed for all three pattern steps for each type of TCO. Our laser processes are running at industrial relevant high speed, throughput and precision. Module stringing and encapsulation have been developed for 1.4 m^2 substrate area. Such modules were sent to ESTI laboratories of the JRC (Ispra) for independent I-V characteristics confirmation and to the German TÜV (Rheinland) for outdoor relevant testing, like damp heat, thermal cycling and high voltage tests.

RESULTS AND DISCUSSION

Amorphous silicon cell and module technology

By careful optimization of the device in our KAI-M reactor using Asahi U-type SnO$_2$ the initial and stabilized efficiencies could be further improved by reducing the i-layer thickness from 250 nm to 200 nm [14]. Figure 1 represents our best present result of single-junction cells in the light-soaked state of 8.6 % efficiency on Asahi U. This device reveals that our single-chamber KAI process is state-of-the-art and fully equivalent to a multi-chamber p-i-n fabrication approach.

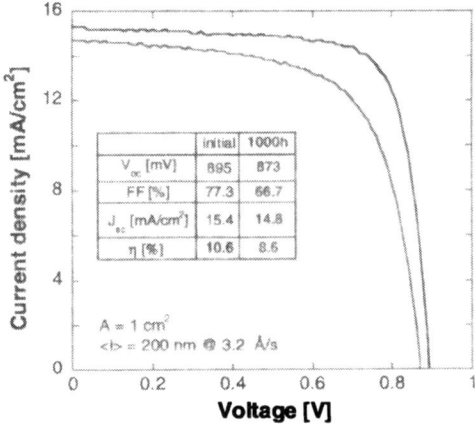

	initial	1000h
V_{oc} [mV]	895	873
FF [%]	77.3	66.7
J_{sc} [mA/cm^2]	15.4	14.8
η [%]	10.6	8.6

A = 1 cm^2
<b> = 200 nm @ 3.2 Å/s

Figure 1: Best amorphous p-i-n test cells in the initial and light-soaked state processed in a medium size R&D single-chamber KAI-M reactor. As front TCO, Asahi U-type SnO$_2$ was applied.

The deposition of high-quality front LPCVD ZnO as well as the PECVD process for the amorphous silicon solar cells deposited on ZnO has been further optimized, too. Figure 2 reveals our recent a-Si:H single-junction p-i-n device obtained in a KAI-M reactor. The absorber layer has been also reduced to 200 nm thickness leading to an improved stable solar cell of 8.5 %, coming close to the one obtained on Asahi U.

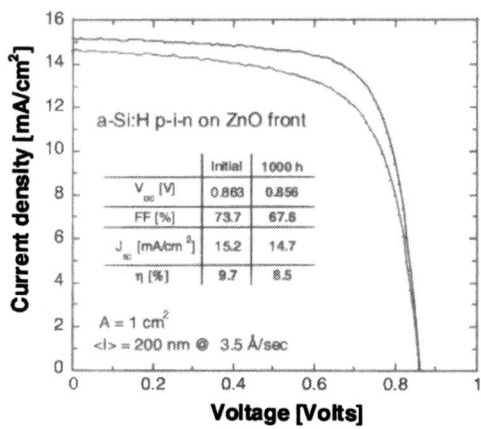

Figure 2: Recently achieved result on in-house developed LPCVD ZnO. The a-Si:H single-junction p-i-n cell (1 cm² area) has been deposited in a KAI-M R&D reactor.

The analysis of the short-circuit current density by QE-measurements in Figure 3 reveal for both cells on Asahi U and ZnO (Figures 1 and 2) a comparable photo current.

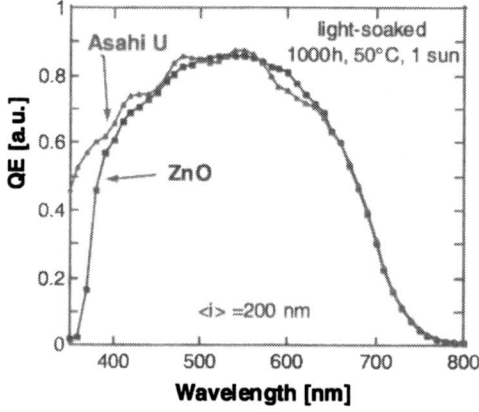

Figure 3: Comparison of the stabilized QE-characteristics of the a-Si:H p-i-n cell optimized on Asahi U SnO₂ and the one recently obtained on in-house LPCVD ZnO.

Using Asahi U substrates we scaled the solar cell up to 10x10 cm² mini-modules applying laser-processed monolithic series connection. Figure 4 represents the AM1.5 I-V characteristics of our best module in the initial and light-soaked state resulting in a stable aperture efficiency of 8.0 %. In agreement with the cell result a relative degradation of 18% has been evaluated for this type of cell system.

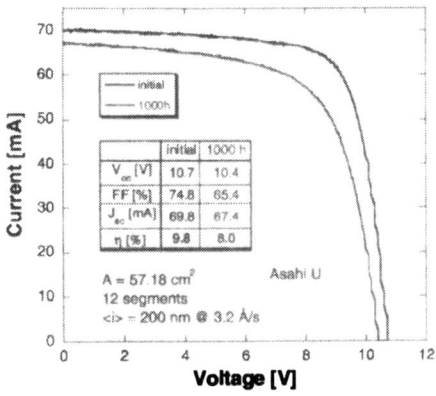

Figure 4: Best a-Si:H p-i-n 10x10 cm² mini-module on Asahi U prepared with in-house laser-patterning technique. The module was deposited in the KAI-M.

Our a-Si:H single-junction solar cell processes have been optimized to use commercial available large-area front TCO. In spite of this lower-quality TCO we found for 1 cm² test cells maximal stabilized efficiency close to 8 %, applying hereby i-layer thicknesses in the range of 200 to 250 nm [14]. The replacement of the high quality front SnO_2 from Asahi (U type) results to a loss in efficiencies of around half a percent absolute in both the initial and degraded state [14]. However, even taking into account this loss still excellent stabilized mini-module efficiencies (10x10cm²) of 7.46±0.19% (independently measured by ESTI Ispra, see Figure 5) could be achieved for this type of TCO. The high efficiency level even in case of this reduced quality front TCO might be attributed to the high light-trapping potential of our diffusive LPCVD ZnO/white back reflector contact concept [4] that has different reflectance properties compared to conventional sputtered metallic contacts.

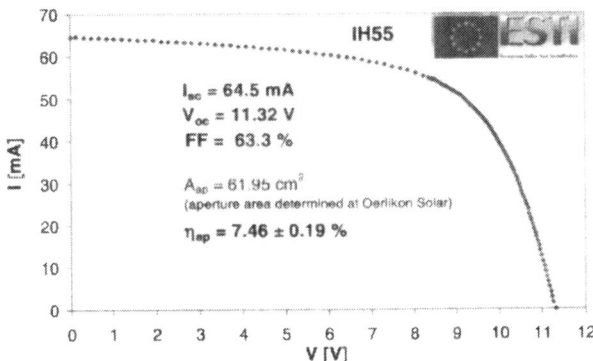

Figure 5: J-V characteristic measured by Ispra ESTI of a stabilized 10x10 cm² single-junction a-Si:H module on commercial available large-area TCO. The module was light-soaked for 1000h and the i-layer was prepared at a thickness of 250 nm and a rate of 3.2 Å/s. The module was deposited in the KAI-M.

The transfer of KAI-M processes to the industrial-size R&D KAI 1200 reactors to realize single-junction a-Si:H p-i-n modules on 1.4 m² substrate area is ongoing and further developed. Figure 6 shows such large-area modules on the outdoor testing bench based on commercial available SnO₂ and LPCVD ZnO as front TCO.

Using commercial available large-area TCO the best module (1.1x1.25m²) I-V characteristics measured at ESTI is represented in Figure 7. The AM1.5 module power of 110.6 W demonstrates that modules using such kind of front TCOs are already mature for mass production. The FF could be improved to 70 % and the V_{oc}-value of 890 mV per segment shows the well-developed p-i interface over the whole 1.4 m² substrate area. Assuming a relative degradation loss of around 25 % for a 280nm thick i-layer the stabilized output power of such modules is to be expected above 80 W.

Figure 6: a-Si:H p-i-n modules (1.4 m²) deposited on LPCVD ZnO (on the left) and deposited on commercial low-grade SnO₂ (on the right). Note, the darker appearance of the ZnO reflects the enhanced light-trapping compared to commercial TCO.

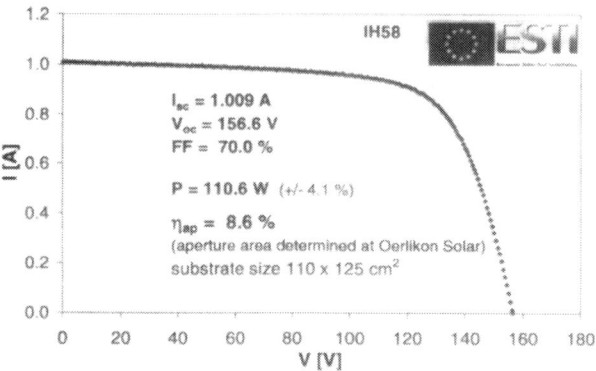

Figure 7: AM1.5 I-V characteristics of an industrial size a-Si:H p-i-n solar cell module deposited on commercial available large-area front SnO₂ (measurements by ESTI of the JRC Ispra). The device is deposited at a rate of 3.4 Å/sec to an average i-layer thickness of 280 nm.

The German TÜV, Rheinland, checked encapsulated 1.4 m^2 a-Si:H p-i-n modules with LPCVD ZnO back contacts with respect to thermal cycling (-40 ° C to + 85 ° C, 200 times), damp-heat (DH) stability and high voltage tests. Figure 8 shows the results of the former DH experiments updated with a recent improved 1.4 m^2 module having an initial output power of around 105 W.

Our damp heat and thermal cycling experiments of 1.4m^2 encapsulated a-Si:H p-i-n modules at TUEV Rheinland confirm clearly that our cell contacts and encapsulation technique are well adapted for outdoor application. Furthermore, the experiments demonstrate that LPCVD ZnO has, a proper adequate encapsulation provided, a long-term outdoor reliability.

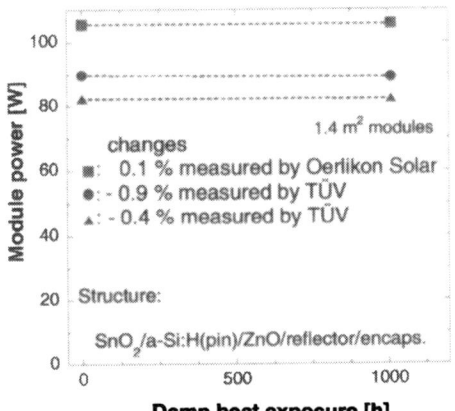

Figure 8: Damp heat experiments of 1.4 m^2 a-Si:H modules performed and characterized by TÜV Rheinland (Germany) and Oerlikon Solar (however, DH of all three modules were executed by TÜV). All modules show a perfect stability over the 1000 h of damp heat exposure.

The performance of modules at higher operation temperatures than at STC is very important and of great interest. The efficiencies evaluated at 25 °C do not fully reflect the outdoor conditions as the modules can easily reach 50 - 60 °C operation temperatures. Therefore, we investigated at ESTI the temperature-dependency of the I-V characteristics of a 1.4m^2 a-Si:H module, as given in Figure 7. The I-V curves have been taken in the temperature range of 25 °C and 50 °C. In Figure 9 the dependency of the maximal module power P_{max}, the open circuit voltage V_{oc}, the short-circuit current I_{sc} and the fill factor FF normalized to the measurements at 25°C are given. The linear fit reveals for our 1.4m^2 a-Si:H p-i-n module a temperature coefficient of -0.22 %/°C which is in good agreement with previously published studies on amorphous modules and cells of -0.20 to -0.29 %/°C [17-20].

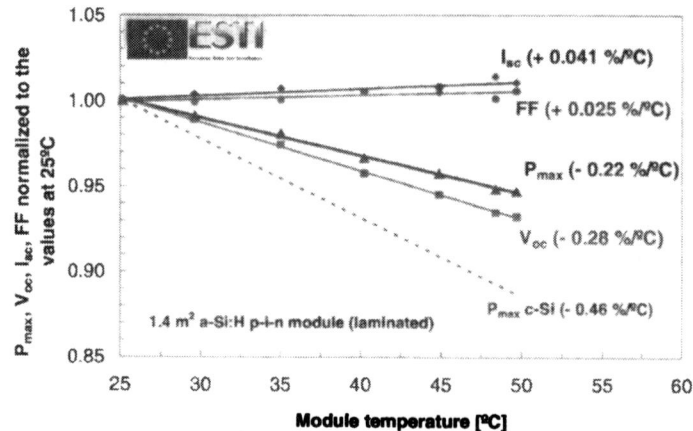

Figure 9: Solar module parameters in function of the module temperature normalized to the values measured at 25 °C. For comparison the typical crystalline module behavior as evaluated from the long-term experience of ESTI Ispra [16] is added.

Recently, a test production run of totally 32 amorphous modules of 1.4 m² using our installed R&D equipments was carried out at Oerlikon Solar. Subsequently 32 modules have been manufactured and the distribution of the modules as a function of the initial output power is represented in Figure 10. Commercial low-grade SnO_2 has been used as front TCO. It has to be noted that none of the manufactured modules has been skipped, all fabricated modules are working and all have significant output powers.

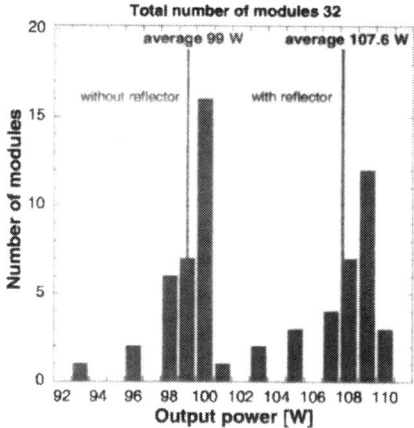

Figure 10: Output power distribution of 32 unlaminated amorphous single-junction modules of 1.4 m² with and without white back reflector subsequently manufactured in our R&D production systems.

The result of the different module powers obtained in Figure 10 demonstrates that a high yield and a narrow distribution can be achieved using our developed processes and systems. An average module output power of 99 W has been obtained for modules without white back reflector. Except for three modules (one at 93 W and two at 96 W) the output power of all other modules lays ± 1 W around the average power of 99 W reflecting a very narrow variation of our processes. Most of the modules, namely 16, reach the maximal power of 100 W and are 1 W above the average power of 99 W. The addition of the white back reflector generally shifts the distribution to higher powers. Maximal values of 110 W and an average power of 107.6 W have been achieved for the 32 modules. The slight increased number of outlays at powers between 101 W and 105 W as compared to the modules without reflector indicates that this process step still can be improved. Taking only modules into account with deviation of less than 5% from the average power as "acceptable modules", a production yield of over 96 % could be obtained.

Recently, we started up-scaling of the a-Si:H p-i-n cell on LPCVD front ZnO as next generation of improved a-Si:H modules. Our present result is presented in Figure 11 showing the AM1.5 I-V characteristics of the 1.4 m^2 module. The short-circuit current is enhanced compared to modules on commercial available SnO$_2$ due to the enhanced light-trapping and the slightly thicker cell. However, further optimization of the p-layer is necessary to obtain the full potential of a-Si:H single-junction modules using this improved front TCO.

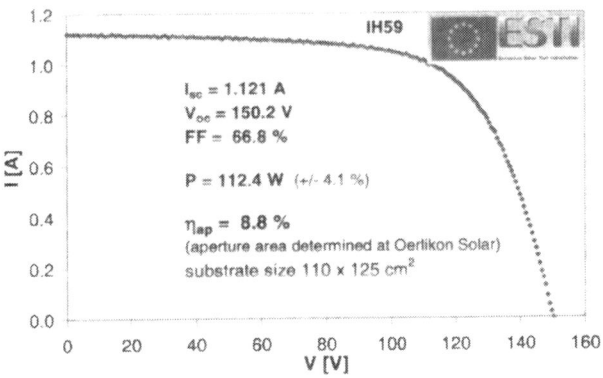

Figure 11: Initial I-V characteristics of an industrial size a-Si:H p-i-n module prepared on LPCVD front-ZnO. The i-layer thickness is 320 nm. The I-V measurements were performed by the ESTI laboratories of the JRC (Ispra).

Large-area modules of our first generation have been installed on the outdoor test bench at NTB (University of Applied Science Buchs, in the Swiss Rhein valley). These modules here have been deposited without buffer layer at the p-i interface. In Figure 12 the decay of the FF is monitored over the first year of outdoor conditions (Dec. 15, 2005 to Dec 20, 2006). One can clearly observe a seasonal effect on the FF-value. There is a slight annealing during summer time, while colder seasons reduce the FF slightly again. Nevertheless, the module fill factor stabilizes depending on the seasonal time around 58% to 60%. Recently, additional modules including buffer layers at the p-i interface (to improve stability) have been installed on the outdoor test bench.

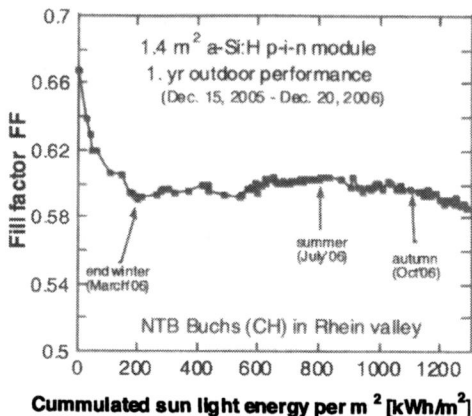

Cummulated sun light energy per m ^{2} [kWh/m^2]

Figure 12: First year outdoor experience of a 1.4 m^2 a-Si:H p-i-n module on the outdoor bench of NTB (University of Applied Science Buchs) in the Swiss Rhein valley. Note the p-i-n device has no buffer incorporated, for the i-layer a thickness of 270 nm has been applied. Fill factors have been taken at irradiation intensities between 700 to 750 W/m^2.

Micromorph tandem developments

Microcrystalline silicon growth has been investigated in the different KAI reactors. In the small KAI-S reactor high µc-Si:H deposition rates of up to 15 Å/sec at homogeneous layer thickness and crystallinity could be found [10]. In a deposition regime of around 7 Å/sec we achieved in devices V_{oc}-values of 500 mV and more. The IMT team in Neuchâtel (Institut de Microtechnique) recently boosted in their KAI-S system the µc-Si:H solar cell to remarkable cell efficiencies of 8.4 % [10].

Micromorph tandem test cells have then been prepared with µc-Si:H bottom cells at rates of around 7 Å/s and using a-Si:H top cell processes as reported above. Figure 13 represents the initial and degraded I-V characteristics of a micromorph tandem cell. As this tandem cell is bottom limited, we do not observe a loss in the J_{sc}-value after light-soaking. The relative efficiency loss of 9 % from initial 10.38 % to degraded 9.45 % is therefore relatively low, and hence, as expected, well below the degradation rates typically found for amorphous single-junction devices alone.

Finally, a 10x10 cm^2 micromorph mini-module with an initial aperture efficiency of 10.0% has been fabricated (see Figure 14).

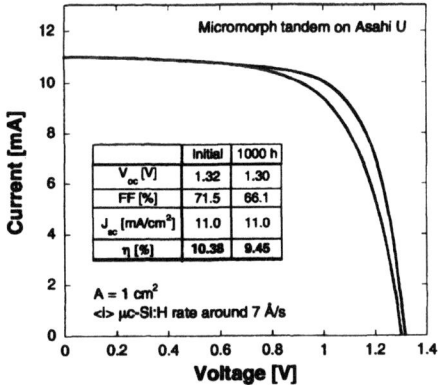

Figure 13: AM1.5 I-V characteristics of an initial and degraded micromorph (a-Si:H/μc-Si:H) tandem solar cell deposited in small KAI reactors (top cell deposited in KAI-M and bottom cell deposited in KAI-S at IMT).

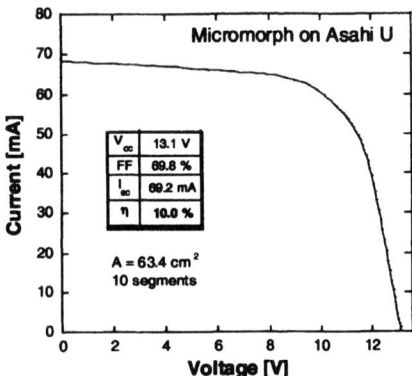

Figure 14: 10x10 cm^2 micromorph tandem mini-module of 10 segments achieving 10.0 % initial aperture efficiency.

SUMMARY

The state-of-the-art efficiency potentials for Asahi U-type and commercially available large-area SnO$_2$ front TCO have been reached for both single-junction amorphous silicon p-i-n cells and mini modules deposited in our single chamber KAI-M reactor. Stabilized a-Si:H p-i-n test cells of 8.6 % at a deposition rate of over 3 A/sec for the intrinsic layer of 200 nm thickness could be obtained. The process transfer to the industrial large-area 1.4 m^2 equipments has succeeded in a-Si:H p-i-n modules being mature for mass production. Applying commercial front glass TCO initial module powers of 110.6 W and 112.4 W when using our in-house developed LPCVD front ZnO have been independently confirmed by the ESTI laboratories of the JRC (Ispra). This corresponds to initial aperture module efficiencies of 8.6% using commercial SnO_2 as front TCO, respectively 8.8 % in case of our LPCVD ZnO. Further optimization of the cells and modules on LPCVD will be carried out.

Oerlikon Solar demonstrated to able to control all process steps for industrial amorphous silicon solar module fabrication. This includes glass cleaning, ZnO coating by LPCVD, thin film Si PECVD deposition, laser-scribing for monolithic series connection, stringing and effective encapsulation passing the DH-tests using LPCVD ZnO back contacts. The different fabrication steps are industrially advanced and developed to the level of mass fabrication of amorphous silicon modules using the equipment from Oerlikon Solar.

Micromorph tandem cells and mini-modules by medium size R&D KAI reactors have achieved initial efficiencies of 10 % at μc-Si:H deposiaround 7 A/s. Further developments are underway to improve micromorph tandems processes which will be soon transferred and implemented in the large-area 1.4 m^2 systems.

ACKNOWLEDGMENTS

The authors thank Prof. C. Ballif and the IMT PV research group for scientific and technical support. Furthermore, we wish to thank Dr. Hollenstein and his group at CRPP Lausanne for the help in developing the large-area depositions. Moreover, we gratefully acknowledge the module measurements of ESTI Laboratories of the JRC (Ispra) and discussions with Dr. H. Müllejans.

REFERENCES

1. J.M. Woodcock et al., Proc. 14th EU-PVSEC 1997, p. 857.
2. M.A. Green, Proc. 3rd WCPEC (Osaka 2003), paper OPL-02.
3. Photon International (Oct. 2004), p. 48.
4. J. Meier et al., Proc. of 19th EU-PVSEC (Paris 2004), p. 1328.
5. U. Kroll et al., Proc. of 19th EU-PVSEC (Paris 2004), p.1374.
6. U. Kroll et al., Thin Solid Films **451-452** (2004), p. 525.
7. J. Meier et al., Proc. of 31st IEEE PVSC (Orlando 2005), p. 1464.
8. U. Kroll et al., Proc. of 21st EU-PVSEC (Dresden 2006), 3DP.1.5, p. 1546.
9. S. Benagli et al., Proc. of 20th EU-PVSEC (Barcelona 2005), 3DV.3.42, p. 1671.
10. L. Feitknecht et al., Proc. 21st EU-PVSEC (Dresden 2006), 3DV.3.10, p. 1634.
11. J. Meier et al., Proc. of 3rd WCPEC (Osaka May 2003), session S2.
12. O. Kluth et al., Proc. of 20th EU-PVSEC (Barcelona 2005), 3DV.3.38, p. 1671.
13. U. Kroll et al, 3rd FVS TCO-workshop 2005, www.FV-Sonnenenergie.de/Publikationen
14. S. Benagli et al., Proc. of 21st EU-PVSEC (Dresden 2006), 3DV.3.42, p. 1719.
15. J. Meier et al., Thin Solid Films **451-452** (2004), p. 518.
16. H. Müllejans, private communication.
17. Y. Ichikawa et al., Proc. of 23rd IEEE PVSC (1993), p. 27.
18. Y.Kuwano, Solar Cells and Their Applications , Power Co. (Tokyo 1985) p. 42.
19. J. Meier et al., Proc. of Mat. Res. Soc. Symp. 507 (1998), p. 139.
20. R.R. Arya, 19th EU-PVSEC (Paris 2004), 5DP_1_05, p. 2024.

Mater. Res. Soc. Symp. Proc. Vol. 989 © 2007 Materials Research Society 0989-A24-02

Temperature Dependence of Silicon-based Thin Film Solar Cells on Their Intrinsic Absorber

Kobsak Sriprapha[1], Ihsanul Afdi Yunaz[1], Shuichi Hiza[1], Kun Ho Ahn[1], Seung Yeop Myong[1], Akira Yamada[2], and Makoto Konagai[1]

[1]Department of Physical Electronics, Tokyo Institute of Technology, 2-12-1, S9-9, O-okayama, Meguro-ku, Tokyo, 152-8552, Japan
[2]Quantum Nanoelectronics Research Center, Tokyo Institute of Technology, 2-12-1, S9-9, O-okayama, Meguro-ku, Tokyo, 152-8552, Japan

ABSTRACT

The temperature dependence of Si-based thin film single junction solar cells on the phase of the intrinsic absorber is investigated in order to find the optimal absorber at high operating temperatures. For comparison, hydrogenated amorphous, protocrystalline and microcrystalline silicon solar cells are fabricated by plasma-enhanced chemical vapor deposition (PECVD) and hot-wired chemical vapor deposition (HWCVD) techniques. Photo J-V characteristics are measured using a solar simulator at an ambient temperature in the range of 25-75 °C. We found that the protocrystalline silicon solar cells provided the lowest temperature coefficient for the efficiency, while the microcrystalline silicon solar cells were highly sensitive to the temperature. Experimental results indicated that protocrystalline silicon is a promising material for using as an intrinsic absorber of Si-based thin film solar cells which operating in high temperature regions.

INTRODUCTION

In general, the solar cell performance is measured under the standard test condition (STC) at room temperature (25 °C). Under outdoor installation, in fact, the operating temperature of solar cells considerably changes according to circumstances, i.e., the climate in the installed area. In a tropical climate region, the operating temperature often reaches more than 70 °C. The increase in the operating temperature leads to the decline in solar cell efficiency (η) mainly due to the drop in the open-circuit voltage (V_{oc}) [1-3]. In the case of silicon (Si)-based solar cells, bulk crystalline Si solar cells including single crystalline-Si (c-Si) and polycrystalline-Si (poly-Si) solar cells show higher η than thin film solar cells at room temperature. However, η of c-Si and poly-Si solar cells seriously decreases with an increase in the operating temperature, compared to hydrogenated amorphous Si (a-Si:H)-based thin film solar cells [5-8]. The main reason for the lower temperature coefficient (TC) for a-Si:H-based solar cells is due to their wide band gap intrinsic absorber. Taking real output power affected by the operating temperature and production cost into account [3], a-Si:H-based thin film solar cells have advantages over bulk crystalline-Si solar cells for use in high temperature area such as a tropical climate region.

It is well known that a-Si:H-based thin film solar cells exhibit light-induced degradation so-called the Staebler-Wronski effect (SWE). The SWE in Si-based thin film solar cells is also a very important factor that needs to be concerned for outdoor installation. During the past 30 years, extensive research works have been done to suppress the SWE. As a result, two-kinds of edge-materials near the phase boundary have been developed as stable intrinsic absorbers; one is the wide band gap hydrogenated protocrystalline Si (pc-Si:H) existing just below the a-Si:H-to-microcrystalline Si (µc-Si:H) transition. The other is the narrow band gap µc-Si:H near the onset of the phase transition. These two kinds of materials are attractive to be applied to Si-base thin film solar cells due to their low SWE [9-13].

Since the temperature dependence of pc-Si:H solar cell has not been clarified yet, in this work we investigated the temperature dependence of pc-Si:H solar cell comparing with a-Si:H and µc-Si:H solar cells in order to find the optimal absorber at high operating temperatures. We also compared the temperature dependence of a-Si:H and µc-Si:H solar cells fabricated by different methods.

EXPERIMENTAL DETAILS

Various kinds of p-i-n-type single junction solar cells were fabricated on Asahi U-type glass substrates by varying the hydrogen dilution ratio using PECVD technique. For comparison, a-Si:H and µc-Si:H solar cells were also fabricated by HWCVD technique. The detailed structure of the fabricated a-Si:H, pc-Si:H, and µc-Si:H solar cells are summarized in Table I.

Table I: Detailed structures of the solar cells under consideration. TCO stands for the transparent conducting oxide film.

Types of solar cells	Structure
a-Si:H (PECVD)	Glass/TCO/p-a-SiC:H/buffer/i-a-Si:H/n-a-Si:H/ZnO/Ag/Al
pc-Si:H (PECVD)	Glass/TCO/p-a-SiC:H/buffer/i-pc-Si:H/ n-a-Si:H/ZnO/Ag/Al
µc-Si:H (PECVD)	Glass/TCO/p-µc-Si:H/i-µc-Si:H/n-a-Si:H/ZnO/Ag/Al
a-Si:H (HWCVD)	Glass/TCO/p-a-SiC:H/buffer/i-a-Si:H/n-µc-Si:H/ZnO/Ag/Al
µc-Si:H (HWCVD)	Glass/TCO/p-µc-Si:H/i-µc-Si:H/n-µc-Si:H/ZnO/Ag/Al

The thicknesses of a-Si:H, pc-Si:H, and µc-Si:H absorbers were about 300, 340 and 1000 nm, and the optical band gap of these absorbers were approximately 1.8, 1.9, and 1.1 eV, respectively.

Photo J-V characteristics were measured using a solar simulator in a climate chamber at the ambient temperature (T) in the range of 25-75 °C with a step increment of 10 °C under 1-sun (AM 1.5, 100 mW/cm^2) irradiation. The temperature dependence of solar cells was obtained from the photo J-V measurements. The value of TC for solar cell parameters (Z) can be expressed as:

$$TC(part/°C) = \frac{1}{z}\frac{\delta Z}{\delta T}\bigg|_{T_n = 25°C} \tag{1}$$

where the normalized temperature T_n was chosen to be 25 °C regarding the standard reference condition for the solar cell measurement.

RESULTS AND DISCUSSIONS

Table II: Typical initial solar cell performances and values of TC for a-Si:H, pc-Si:H, and μc-Si:H solar cells fabricated by the PECVD and HWCVD techniques.

Type	T (°C)	Initial performance				TC (%/°C)			
		J_{sc} (mA/cm^2)	V_{oc} (V)	FF	η (%)	J_{sc}	V_{oc}	FF	η
a-Si:H (PECVD)	25	14.3	0.91	0.72	9.29	0.039	-0.30	-0.052	-0.28
	75	14.5	0.77	0.70	7.91				
pc-Si:H (PECVD)	25	12.9	0.86	0.71	7.95	0.093	-0.28	-0.018	-0.22
	75	13.5	0.74	0.71	7.03				
μc-Si:H (PECVD)	25	22.2	0.46	0.64	6.59	0.074	-0.47	-0.156	-0.54
	75	23.1	0.36	0.59	4.89				
a-Si:H (HWCVD)	25	14.4	0.84	0.58	6.99	0.053	-0.33	0.113	-0.19
	75	14.9	0.70	0.60	6.26				
μc-Si:H (HWCVD)	25	14.9	0.48	0.60	4.29	0.079	-0.47	-0.095	-0.48
	75	15.5	0.37	0.57	3.24				

The initial performances and TC for a typical of Si-based thin film solar cells under consideration are illustrated in Table II. It was found that J_{sc} and FF were less sensitive to T than V_{oc}. The drop in V_{oc} led to the reduction of η with increasing T. The decline in V_{oc} at higher temperatures can be attributed to an increase in the reverse saturation current and a decrease in the optical band gap [2, 4]. The short-circuit current-density (J_{sc}) increased when T became higher. The fill factor (FF) of any solar cells slightly decreased with increasing temperature except for the case of a-Si:H solar cell fabricated by the HWCVD technique. Normally, FF of Si-based solar cells decrease when temperature become higher but a-Si:H solar cells with low FF (< 0.6) at room-temperature measurement were reported to show the increase in FF [15]. This is probably due to the decrease in the resistance or the increase in the mobility-carrier lifetime product within the collection region in the a-Si:H solar cells [4, 15]. The same trend was found in a-Si:H solar cells fabricated by the PECVD technique. It should be note that, although a-Si:H solar cell that fabricated by HWCVD technique showed the lowest TC for η mainly due to the positive value of TC for FF, the initial η and FF of this cell were low. So, we conclude that this low TC for η should not be considered.

In case of the solar cells that fabricated by PECVD technique, pc-Si:H solar cell exhibited lower V_{oc} than a-Si:H solar cell, Since the band gap of p-layer in pc-Si:H solar cell was a little lower than that of a-Si:H solar cell, this probably caused a difference in V_{oc}. However, TC for η of pc-Si:H solar cell is smaller than that of a-Si:H solar cell due to the wider band gap and lower TC for FF and V_{oc}.

It was found that the typical TC for η of a-Si:H, pc-Si:H, and μc-Si:H solar cells fabricated by PECVD were -0.28, -0.22, and -0.54%/°C, respectively. The solar cells prepared by HWCVD technique were more stable against T than those prepared by the PECVD technique. However, the solar cells prepared by HWCVD technique provided significantly poor performances.

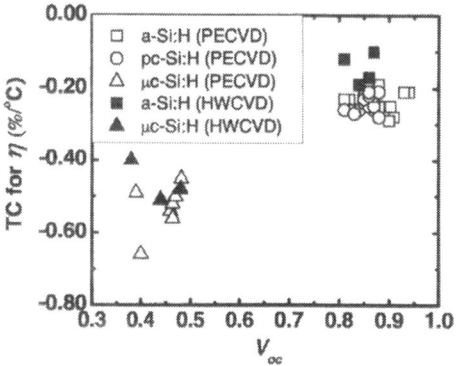

Figure 1. TC for η of various Si-based thin film solar cells as a function of initial V_{oc}.

Figure 1 shows TC for η of various Si-based thin films as a function of initial V_{oc} (V_{oc} measured at room temperature). It was found that the higher initial V_{oc} became, the lower TC for η showed. Thus, μc-Si:H solar cells with low initial V_{oc} showed high values of TC for η of -0.50 to -0.40%/°C. On the contrary, a-Si:H and pc-Si:H solar cells with higher V_{oc} showed lower temperature dependent behaviors; TC for η of -0.30 to -0.20%/°C. These results agreed well with theoretical analysis for Si-based thin-film solar cells [4, 14]. It can be concluded that the values of TC are inversely proportional to V_{oc}.

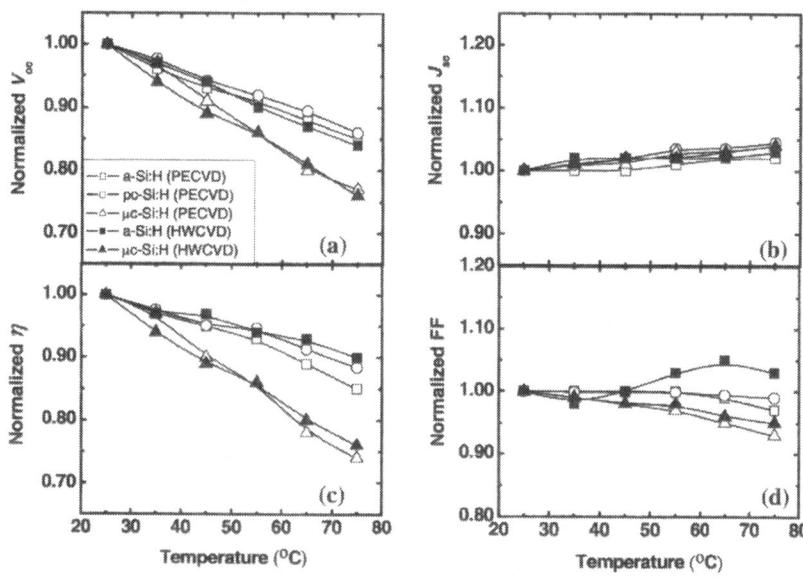

Figure 2. Normalized solar cell parameters as a function of the operating temperature (a) V_{oc}, (b) J_{sc}, (c) FF and (d) η.

Fig. 2 illustrates the normalized solar cell parameters of the Si-based thin-film single junction solar cells as shown in Table II. V_{oc} and η of the solar cells almost linearly decreased while J_{sc} progressively increased with the increase in T. Since the decrease in the optical band gap of the thin-film absorbers were not taken into consideration, the increase in J_{sc} was mainly due to the reduced recombination velocity or unfilled localized states [1]. In the case of FF, as mentioned above, it slightly decreased with increasing T. However, the contrary rather happens when a reduction of contact resistance or an increase in the collection length is caused by the T elevation. Hence, the a-Si:H solar cells fabricated by the HWCVD technique displayed the largest increment of the collection length in its absorber with the increase in T, which means a poor carrier extraction property. This may be ascribed for the increased product of the mobility and the carrier lifetime with increased T [15]. It was found that a-Si:H and μc-Si:H solar cells fabricated by the PECVD and HWCVD techniques showed similar temperature-dependent trends for the photovoltaic parameters. According to the highest increase in J_{sc}, lowest decrease in V_{oc}, and nearly invariant FF with the increase in T, the pc-Si:H solar cell prepared by the PECVD technique displayed the lowest TC for η.

CONCLUSIONS

We have investigated the temperature dependence of Si-based thin-film solar cells with different phase of intrinsic absorbers. Our experiment results indicated that the values of TC for η are oppositely proportional to initial V_{oc}. We found similar temperature-dependent trends for the photovoltaic parameters of a-Si:H and μc-Si:H solar cells fabricated by the PECVD and HWCVD techniques. The changes in TC for solar cells parameters did not depend on deposition technique. The pc-Si:H solar cell fabricated by the PECVD technique revealed low TC for V_{oc}, FF and η compared to a-Si:H and μc-Si:H solar cells. Since the pc-Si:H solar cell exhibit a good light-induced stability and low temperature dependence, it is beneficial for use in a tropical climate region.

REFERENCES

1. H. Stiebig, Th. Eickhoff, J. Zimmer, C. Beneking and H. Wagner, *Proc. of MRS* **420**, 855 (1996).
2. H. Stiebig, C. Zahren, T. Repmann, B. Rech and T. Brammer, *Proc. of the 19th EU-PVSEC*, Paris, France, 1583 (2004).
3. M. Shima, M. Isomura, K. Wakisaka, K. Murata and M. Tanaka, *Solar Energy Material and Solar Cells* **85**, 167 (2005).
4. M. A. Green, *Progress in Photovoltaics: Research and Application* **11**, 333 (2003).
5. M. Kameda, S. Sakai, M. Isomura, K. Sayama, Y. Hishikawa, S. Matsumi, H. Haku, K. Wakisaka, M. Tanaka, S. Kiyama, S. Tsuda and S. Nakano, *Proc. of the 25th IEEE PVSC*, Washington, USA. 1049-1052 (1996).
6. K. Akhmad, A. Kitmura, F. Yamamoto, H. Okamoto, H. Takakura and Y. Hamakawa, *Solar Energy Materials and Solar Cells* **46**, 209 (1997).
7. K. Fukae, C. C. Lim, M. Tamechika, N. Takehara, K. Saito, I. Kajita and E. Kondo, *Proc. of the 25th IEEE PVSC*, Washington, USA. 1227-1230 (1996).

8. M. Kondo, H. Nishio, S. Kurata, K. Hayashi, A. Takenaka, A. Ishikawa, K. Nishimura, H. Yamagishi and Y. Tawada, *Solar Energv Materials and Solar Cells* 49, 1-6 (1997).

9. J. Koh, Y. Lee, H. Fujiwara, C.R. Wronski and R.W. Collins, *Appl. Phys. Lett.* 73, 1526 (1998).

10. R. W. Collins, A. S. Ferlauto, G. M. Ferreira, C. Chen, J. Koh, R. J. Koval, Y. Lee, J. M. Pearce and C. R. Wronski, *Solar energy Materials & Solar cells* 78 143-180 (2003).

11. C. R. Wronski, J. M. Pearce, R. J. Koval, X. Niu, A. S. Ferlauto and R. W. Collins, *Proc. of MRS* 715, pp. A13.4.1-12 (2002).

12. S. Y. Myong, S. W. Kwon, K. S. Lim and M. Konagai, *Solar energy Materials & Solar cells* 85 133 (2005).

13. S. Y. Myong, S. W. Kwon, K. S. Lim, M. Kondo and M. Konagai, *Appl. Phys. Lett.* 88, 083118 (2006).

14. I. A. Yunaz, K. Sriprapha, S. Hiza, A. Yamada and M. Konagai, *Jpn. J. Appl. Phys.* (2006) (*in press*).

15. D.E. Carlson, G. Lin and G. Ganguly, *Proc. of the 28th IEEE PVSC*, Alaska, USA. 707-712 (2000).

Mater. Res. Soc. Symp. Proc. Vol. 989 © 2007 Materials Research Society 0989-A24-03

Efficient Thin-Film Polycrystalline-Silicon Solar Cells Based on Aluminium-Induced Crystallization

Ivan Gordon, Lode Carnel, Dries Van Gestel, Guy Beaucarne, and Jef Poortmans
Solar Cell Technology, IMEC, Kapeldreef 75, Leuven, B-3001, Belgium

ABSTRACT

Efficient thin-film polycrystalline-silicon (pc-Si) solar cells on inexpensive substrates could lower the price of photovoltaic electricity substantially. At the MRS conference in 2006, we presented a pc-Si solar cell with an efficiency of 5.9% that had an absorber layer made by aluminum-induced crystallization (AIC) of amorphous silicon followed by high-temperature epitaxial thickening. The efficiency of this cell was mainly limited by the current density. To obtain higher efficiencies, we therefore need to implement an effective light trapping scheme in our pc-Si solar cell process. In this work, we describe how we recently enhanced the current density and efficiency of our cells. We achieved a cell efficiency of 8.0% for pc-Si cells in substrate configuration. Our cell process is based on pc-Si layers made by AIC and thermal CVD on smoothened alumina substrates. The cells are in substrate configuration with deposited a-Si heterojunction emitters and interdigitated top contacts. The front surface of the cells is plasma textured which leads to an increase in current density. The current density is further enhanced by minimizing the back surface field thickness of the cells to reduce the light loss in this layer. Our present pc-Si solar cell efficiency together with the fast progression that we have made over the last few years indicate the large potential of pc-Si solar cells based on the AIC seed layer approach.

INTRODUCTION

The current high price of photovoltaic electricity could be lowered substantially if efficient solar cells could be made from polycrystalline-silicon thin films on inexpensive substrates. Due to the recent silicon feedstock shortage, prices of silicon solar modules have increased. A silicon thin-film technology could lead to cheaper modules by the use of less silicon material and by the implementation of monolithic module processes. A technology based on polycrystalline-silicon thin films with a grain size between 1 μm and 1 mm (pc-Si), seems particularly promising since it combines the low-cost potential of a thin-film technology with the high efficiency potential of crystalline silicon [1]. State-of-the-art pc-Si mini-modules based on solid-phase crystallization (SPC) with efficiencies close to 10% and open-circuit voltages (V_{oc}) around 500 mV per cell have recently been reported by CSG Solar AG [2].

At the MRS conference in 2006, we presented promising solar cell results that were obtained on pc-Si films made by aluminum-induced crystallization (AIC) of amorphous silicon (a-Si) followed by high-temperature epitaxial thickening [3]. The AIC process leads to very thin pc-Si seed layers with a typical grain size in the range of 5-20 μm [4]. Absorber layers for solar cells can be made by epitaxial thickening of these AIC layers [3, 5]. For this, we use chemical vapor deposition (CVD) at temperatures above 1000°C. The advantage of the AIC seed layer approach is that substantially larger grains can be obtained compared to the SPC approach that leads to grains with a typical size around 1-2 µm [5]. The best cell efficiency we presented last year on ceramic alumina substrates was 5.9%, while the highest open-circuit voltage (V_{oc}) was 533 mV [3]. The main limiting factor of our cells was the short-circuit current density (J_{sc}) which was typically around 17 mA cm^{-2}.

To obtain higher efficiencies, we need to implement an effective light trapping scheme in our pc-Si solar cell process. In this work, we describe how we enhanced the current density of our pc-Si cells by texturing the silicon front side of the cells using plasma texturing and by optimizing the pc-Si layer thickness.

EXPERIMENTAL DETAILS

We made pc-Si films on alumina substrates (CoorsTek ADS996R) by epitaxial thickening of AIC seed layers. The substrates were covered by a spin-on flowable oxide (FOx-25 from Dow Corning) to reduce their surface roughness, prior to the seed layer formation [3, 6]. Next, double layers of Al and a-Si were deposited on these substrates in an electron-beam high-vacuum evaporator. In between the two depositions, the aluminum was oxidized by exposure to air for two minutes. The nominal thickness of the Al and a-Si layers was fixed at 200 nm and 230-250 nm respectively. After deposition, the samples were annealed in a tube furnace under nitrogen ambient at 500°C for 4 hours. During this annealing, the a-Si crystallized into pc-Si and both layers exchanged places [4]. Finally, the top Al layer was removed by selective wet chemical etching.

Absorber layers were deposited on the AIC layers by thermal CVD. The depositions were performed in a single-wafer epitaxial reactor (ASM Epsilon2000) under atmospheric pressure, at a temperature of 1130°C. The growth rate was around 1.4 µm / min. Double layers of p+ and p silicon with variable thickness ratios were made. The p+ layer acts as a back surface field (BSF) and as a conductive channel for majority carriers, while the p layer is the actual absorber layer. Typical doping densities were 2 x 10^{19} cm^{-3} for the BSF layer and 10^{16}-10^{17} cm^{-3} for the absorber layers. The total epitaxial layer thickness was always between 2 and 6 µm.

After epitaxial deposition, some of the pc-Si layers were textured using plasma texturing. The plasma texturing was done in a prototype reactor from Secon using micro-wave antennas positioned above the substrates, with SF$_6$ and N$_2$O as precursor gases [7]. Typically around 1 µm of silicon was removed during texturing.

Heterojunction emitters were formed by deposition of thin double layers of undoped and phosphorus-doped a-Si using plasma-enhanced chemical vapor deposition (PECVD) at 180°C. The total thickness of this emitter was around 16 nm. Before emitter formation, defect passivation of the layers was performed by plasma hydrogenation in a PECVD system at 400°C.

To complete the solar cells, indium tin oxide (ITO) was deposited by rf-sputtering as anti-reflective coating (ARC) and metal contacts were formed. Conductive ITO is used to avoid excessive resistive losses in the thin a-Si emitter. The contacts were formed by photolithography and wet chemical etching in combination with metal evaporation. Both base and emitter contacts are on top of the cell in interdigitated finger patterns. All cells were measured under AM1.5 (1000 W/m^2) illumination with an aperture area of 1 cm^2.

RESULTS

We developed a process that yields pc-Si solar cells in substrate configuration on alumina substrates with efficiencies so far of 8.0%. Figure 1 shows a schematic cross-section of such a pc-Si solar cell. The alumina substrate is smoothened by a spin-on oxide. This increases the electronic quality of the AIC seed layers [3, 6]. The cells have heterojunction emitters consisting of thin a-Si layers that lead to much higher V_{oc} values than diffused homojunction emitters due to better hydrogen bulk passivation and due to the absence of preferential phosphorus diffusion along grain boundaries. [3, 8]. Base and emitter contacts are on top of the cell in an interdigitated finger pattern that leads to good fill factors even when thin BSF layers are used [9]. Before emitter formation the pc-Si absorber layers are plasma textured (not shown in Figure 1).

In this section we show how the current density of our pc-Si cells can be enhanced drastically by plasma texturing the front side of the cells and by minimizing the BSF layer thickness.

Plasma texturing

We developed a plasma texturing process to enhance the short-circuit current of our pc-Si solar cells by lowering the front surface reflection and obtaining an oblique coupling of incident light into the cells. Although the alumina substrates act as diffuse back reflectors, the current density of our thin pc-Si solar cells without textured front surface is relatively low with J_{sc} values around 17 mA cm^{-2} [3]. The plasma texturing process removes around 1 µm of silicon. The as-grown surface of our layers is quite rough and has a reflectance without ARC of around 35% in the visible light spectrum. After plasma texturing, the surface of the pc-Si layers shows much smaller features and has a reflectance without ARC of only 15%. Furthermore, the reflectance is nearly completely diffuse after plasma texturing while there is a large specular component in the reflectance of as-grown layers [10].

At cell level, plasma texturing the front side of our pc-Si layers on alumina led to an increase in J_{sc} by roughly 15% from 17.2 mA cm^{-2} for an untextured cell to 19.7 mA cm^{-2} for the textured cell (see Table I). Both samples had a BSF thickness of 0.5 µm and an original absorber thickness of 3 µm. The cells had heterojunction emitters and interdigitated top contacts. Besides

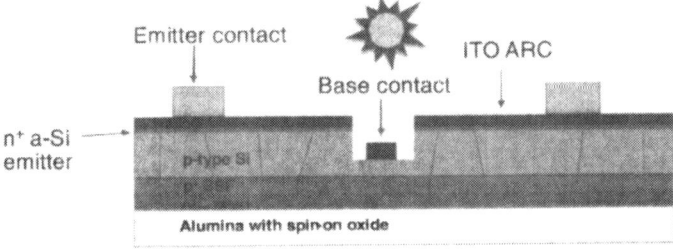

Figure 1. Schematic cross-section of our pc-Si solar cells (not in scale).

Table I. Comparison of the illuminated IV parameters of two pc-Si solar cells with and without plasma texturing.

	J_{sc} (mA cm^{-2})	V_{oc} (mV)	Fill factor (%)	Efficiency (%)
As-grown	17.2	483	69	5.7
Plasma-textured	19.7	506	71	7.0

the current increase, plasma texturing also led to a small increase in V_{oc}. This is the result of the lower total thickness of the plasma textured cell. As a result of the current and voltage increase, plasma texturing led to an increase in efficiency from 5.7% to 7.0%. The large current increase results from a lower reflectance and from an oblique coupling of the light into the cell. The external quantum efficiency of the plasma textured cell is enhanced compared to that of the untextured cell mainly at short wavelengths. Due to the reduced front surface reflection and the oblique light coupling a much larger portion of the light gets absorbed near the space charge region.

When plasma texturing is applied on cells with BSF layers thicker than 0.5 μm, the current density increase is only very small, and sometimes even lower current densities are obtained [10]. The reason for this is that the long wavelength response of our cells is largely determined by the relative thickness of the absorber layer to the BSF layer. Light that is not absorbed in the absorber layer has to pass twice through the BSF layer before it can enter the absorber layer again. Light that gets absorbed in the BSF layer does not contribute to the current density since the diffusion length in this highly doped layer is very small. After plasma texturing, cells have a thinner absorber layer than before and hence a smaller absorber to BSF thickness ratio. Moreover, the influence of the BSF layer on the external quantum efficiency is even stronger for textured cells than for untextured cells since light that does not get absorbed in the absorber layer has to pass twice through the BSF layer under a shallow angle due to the oblique light coupling. Especially for textured cells, the BSF thickness should therefore be minimized.

8% efficient cells

Our best pc-Si solar cell efficiency so far is 8.0%. Figure 2 shows the illuminated current-voltage graph of this cell. The cell had a BSF thickness of 0.25 μm and an original absorber layer thickness of 3 μm. The absorber doping concentration was 3×10^{16} cm^{-3}. The cell had a heterojunction emitter and interdigitated top contacts, and was plasma textured. The short-circuit current density is 1 mA cm^{-2} higher than that of the plasma textured cell in Table 1. This higher current arises mainly from the thinner BSF and from a slightly modified metal finger pattern. By lowering the BSF thickness, less light gets absorbed in this highly doped region. The BSF thickness of 0.25 μm is a compromise between the need to minimize the thickness of this layer to prevent light loss and the need to have a minimal lateral conductivity for the majority carriers. The fill factor of the cell was 73%, indicating that there were no series resistance problems. When going to even thinner BSF layers, the fill factor starts to decrease due to an increasing series resistance. We note that the highest J_{sc} we obtained so far is 21.2 mA cm^{-2}. By using plasma texturing in combination with a thin BSF layer, we have been able to increase the J_{sc} by 3 to 4 mA cm^{-2} compared to untextured cells while still retaining good fill factors and high V_{oc} values above 530 mV [3].

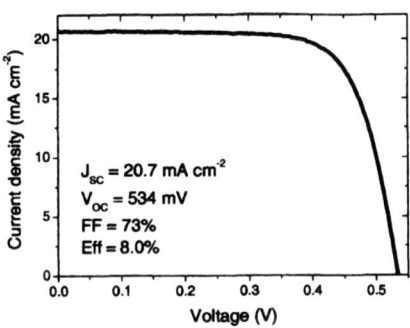

Figure 2. Illuminated current-voltage curve of our best pc-Si solar cell to date.

CONCLUSIONS

In this work we presented a cell process that enabled us to obtain polycrystalline-silicon solar cells with an efficiency of 8%. The process is based on pc-Si layers made by aluminum-induced crystallization of amorphous silicon and thermal CVD on smoothened alumina substrates. The cells are in substrate configuration, they have deposited a-Si heterojunction emitters, and they have both base and emitter contacts on top in an interdigitated finger pattern. The front surface of the cells is plasma textured to lower the front side reflection and to obtain oblique coupling of incident light into the cells, which leads to higher current densities. The current density is further enhanced by minimizing the BSF layer thickness of the cells to reduce light loss in this highly doped layer.

Our present pc-Si solar cell efficiency together with the fast progression that we have made over the last few years indicate the large potential of pc-Si solar cells based on the AIC seed layer approach. Figure 3 shows the evolution in time of our pc-Si cell efficiency. We had an

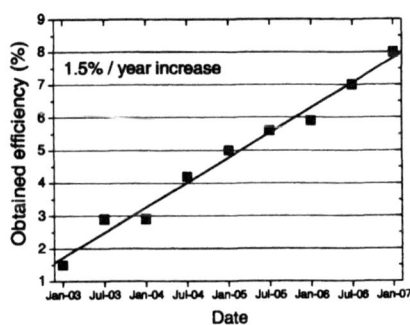

Figure 3. Evolution in time of the efficiency of our pc-Si solar cells.

average increase in absolute efficiency of around 1.5% per year over the last four years. To continue this trend we will have to introduce new features in our cell concept and to enhance the electronic quality of our pc-Si layers. By using transparent glass-ceramic substrates we will be able to make cells in a superstrate configuration in the near future [11]. Recently we showed that our absorber layers contain a large number of electronically active intra-grain defects [12, 13]. Intra-grain quality improvement will therefore be very important to further increase our pc-Si cell efficiency.

ACKNOWLEDGMENTS

This work was partly funded by the European Commission under contract number 019670-FP6-IST-IP ('ATHLET'). The authors thank Kris Van Nieuwenhuysen for the epitaxial depositions and Harold Dekkers and Filip Duerinckx for help with the plasma texturing.

REFERENCES

1. G. Beaucarne and A. Slaoui in *Thin-film Solar Cells: Fabrication, Characterization and Applications,* edited by J. Poortmans and V. Arkhipov (Wiley, New York, 2006) pp. 97-131.
2. P. Basore, *Proceedings of the 21st European Photovoltaic Solar Energy Conference,* pp. 544-548 (2006).
3. I. Gordon, D. Van Gestel, L. Carnel, K. Van Nieuwenhuysen, G. Beaucarne, and J. Poortmans, *Mater. Res. Soc. Symp. Proc* **910**, A23-04 (2006).
4. O. Nast, T. Puzzer, L.M. Koschier, A.B. Sproul, and S.R. Wenham, *Appl. Phys. Lett.* **73**, 3214 (1998).
5. A.G. Aberle, *Proceedings of the 21st European Photovoltaic Solar Energy Conference,* pp. 738-741 (2006).
6. I. Gordon, D. Van Gestel, K. Van Nieuwenhuysen, L. Carnel, G. Beaucarne, and J. Poortmans, *Thin Solid Films* **487**, 113 (2005).
7. H.F.W. Dekkers, G. Agostinelli, D. Dehertoghe, and G. Beaucarne, *Proceedings of the 19th European Photovoltaic Solar Energy Conference,* pp. 412-415 (2004).
8. L. Carnel, I. Gordon, D. Van gestel, G. Beaucarne, J. Poortmans, and A. Stesmans, *J. Appl. Phys.* **100**, 063702 (2006).
9. I. Gordon, K. Van Nieuwenhuysen, L. Carnel, D. Van Gestel, G. Beaucarne, and J. Poortmans, *Thin Solid Films* **511-512**, 608 (2006).
10. L. Carnel, I. Gordon, H. Dekkers, F. Duerinckx, D. Van gestel, G. Beaucarne, and J. Poortmans, *Proceedings of the 21st European Photovoltaic Solar Energy Conference,* pp. 830-833 (2006).
11. D. Van Gestel, I. Gordon, L. Carnel, L.R. Pinckney, A. Mayolet, J. D'Haen, G. Beaucarne, and J. Poortmans, *Mater. Res. Soc. Symp. Proc* **910**, A26-04 (2006).
12. D. Van Gestel, M.J. Romero, I. Gordon, L. Carnel, J. D'Haen, G. Beaucarne, M. Al-Jassim, and J. Poortmans, *Appl. Phys. Lett.* **90**, 092103 (2007).
13. D. Van Gestel, I. Gordon, L. Carnel, G. Beaucarne, and J. Poortmans, Paper presented at this conference.

Mater. Res. Soc. Symp. Proc. Vol. 989 © 2007 Materials Research Society   0989-A24-04

Materials Optimization for Silicon Heterojunction Solar Cells using Spectroscopic Ellipsometry

Dean Levi, Eugene Iwanizcko, Steve Johnston, Qi Wang, and Howard M. Branz

National Renewable Energy Lab, 1617 Cole Blvd., Golden, CO, 80401

ABSTRACT

We have used hot wire chemical vapor deposition (HWCVD) to fabricate silicon heterojunction (SHJ) solar cells on p-type FZ silicon substrates with efficiencies as high as 18.2%. The best cells are deposited on anisotropically-textured (100) silicon substrates where an etching process creates pyramidal facets with (111) crystal faces. Texturing increases J_{sc} through enhanced light trapping, yet our highest V_{oc} devices are deposited on un-textured (100) substrates. One of the key factors in maximizing the efficiency of our SHJ devices is the process of optimization of the material properties of the 3 – 5 nm thick hydrogenated amorphous silicon (a-Si:H) layers used to create the junction and back contact in these cells. Such optimization is technically challenging because of the difficulty in measuring the properties of extremely thin layers. In this study, we have utilized spectroscopic ellipsometry (SE) and photoconductivity decay to conclude that a-Si:H films grown on (111) substrates are substantially similar to films grown on (100) substrates. In addition, analysis of the substrate temperature dependence of surface roughness evolution reveals a substrate-independent mechanism of surface smoothening with an activation energy of 0.28 eV. Analysis of the substrate temperature dependence of surface passivation reveals a passivation mechanism with an activation energy of 0.63 eV.

INTRODUCTION

Silicon heterojunction solar cells (SHJ), first called HIT (heterojunction with intrinsic thin layer) cells by Sanyo Corp., incorporate amorphous silicon contacts on crystal silicon (c-Si) to produce high efficiency c-Si cells with relatively simple, low temperature processing [1]. Extremely thin layers of intrinsic and doped amorphous hydrogenated silicon (a-Si:H) are deposited onto a c-Si wafer to passivate the wafer surfaces while also creating an electrical junction for carrier collection. These devices present unique challenges for characterization and optimization because of the extremely thin layers used in their fabrication. The SHJ devices under development at NREL use 40Å intrinsic layers and 60Å doped layers deposited by hot wire chemical vapor deposition (HWCVD) [2].

Our research team has demonstrated increasing V_{oc} for SHJ solar cells as progress has been made in both materials optimization and substrate cleaning [3]. The increase in V_{oc} has been correlated to reduced density of interface states at the a-Si:H – c-Si interface and reduced surface recombination (SRV) at that interface [4]. Nearly all of our materials optimization studies have focused on a-Si:H grown on (100) c-Si surfaces. Recently our team has begun to fabricate SHJ devices on textured wafers in order to increase the short circuit current (J_{sc}) through reduced reflection. This has produced a new record p-type SHJ efficiency of 18.2% [5]. This anisotropic texture-etch process results in (111) face facets on a (100) c-Si wafer. Hence, the a-Si:H layers for our best SHJ cells are now grown on (111) surfaces, yet most of our materials studies have been conducted on (100) wafers. In this work we compare a-Si:H growth on (111) and (100) c-Si surfaces.

EXPERIMENT

Textured wafers present challenges for characterization, particularly for optical techniques such as ellipsometry that require a specular reflection from the surface to make a measurement. In order to gain insight into the properties of a-Si:H layers grown on textured wafers, we have performed a comparative study of the growth dynamics and surface passivation properties of a-Si:H on un-textured (100) and (111) c-Si wafers. We have used HWCVD to grow a series of 100 Å thick intrinsic layers at substrate temperatures ranging from 75°C to 175°C on both (100) and (111) wafers. Prior to deposition the c-Si substrates were thoroughly cleaned [3], with a final HF etch immediately prior to insertion into the deposition chamber. Hence film growth takes place on hydrogen-terminated c-Si surfaces. We have used *in situ* and *ex situ* spectroscopic ellipsometry to characterize the film properties during and after growth. Identical films were grown on both sides of each substrate to facilitate determination of the SRV through photoconductive decay measurements of photoexcited carrier lifetimes.

Intrinsic a-Si:H layers were grown by HWCVD using a tungsten filament and pure SiH_4 at a growth rate of ~ 2.5 Å/sec. The substrate temperature was monitored using a thermocouple in contact with the back of the substrate. Substrate temperature was also monitored using the c-Si optical constants measured by *in situ* spectroscopic ellipsometry (SE). *In situ* and *ex situ* SE measurements were performed using a JA Woollam M2000 rotating compensator spectroscopic ellipsometer with a wavelength range of 1.7 μm to 250 nm. Photoexcited carrier lifetimes were determined using transient and quasi-static photoconductivity decay measurements performed using a Sinton apparatus [6].

DISCUSSION

In situ measurements of surface roughness evolution

Film deposition was monitored using *in situ* real time spectroscopic ellipsometry (RTSE). RTSE provides real time measurement of three essential pieces of information; bulk film thickness $d_b(t)$ vs. time, surface roughness $R_s(t)$ vs. time, and the complex dielectric function ε. The dielectric function (DF) is related to the joint density of states between occupied states in the valence band and empty states in the conduction band. As such, it provides insight into the optical and electronic properties of the material. The evolution of the surface roughness R_s with film thickness d_b provides insight into the film nucleation and growth mechanisms [7]. By monitoring how the transition between increasing R_s and constant or decreasing R_s depends on film thickness and substrate temperature, we can gain insight into the nature of the mechanisms involved in film growth. Figure 1 illustrates how R_s varies with d_b and substrate temperature T_s for films grown on (100) substrates.

Prior to initiation of HWCVD deposition, RTSE indicates approximately 8 Å of surface roughness on the substrate. Because the dielectric properties of roughness and the hydrogen termination layer are very similar [8], the initial 8 Å of roughness is understood to be a combination of hydrogen termination and roughness of the c-Si substrate. Once deposition begins we note a rapid increase in surface roughness for all T_s in the figure. This corresponds to the initial nucleation phase as adatoms or radicals attach at random nucleation sites and additional adatoms attach at those sites, increasing the surface roughness prior to complete coverage and detection of non-zero film thickness. All films reach a roughness of ~ 12 A during this nucleation phase prior to measurement of non-zero film thickness.

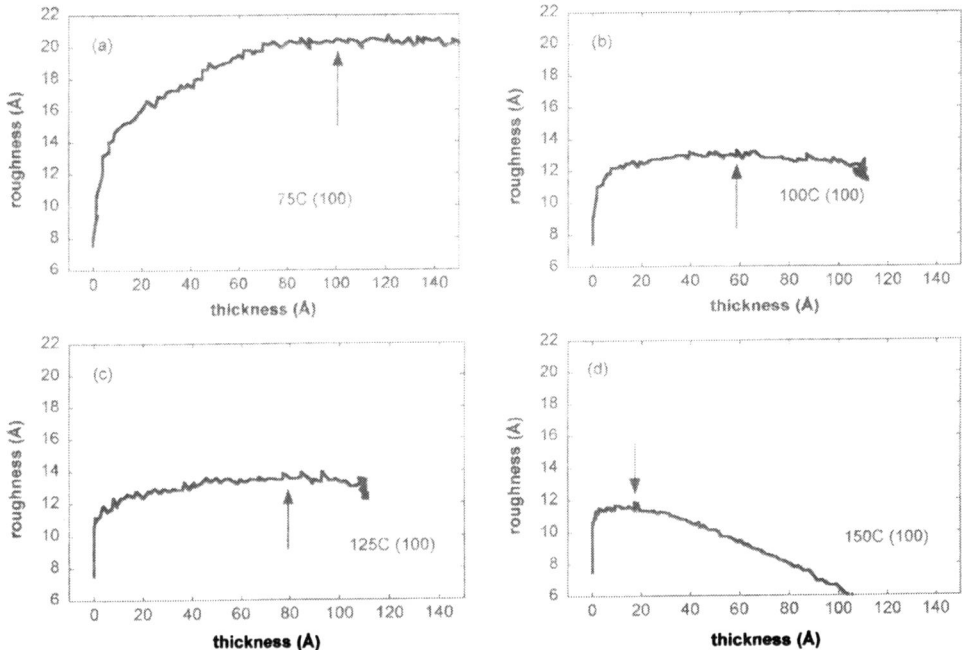

Figure 1. Evolution of surface roughness vs. film thickness at 4 substrate temperatures for films deposited on (100) c-Si measured by *in situ* RTSE. Figures (a) – (d) illustrate how the roughness evolution changes with substrate temperature for $75°C \leq T_s \leq 150°C$. The arrow in each figure indicates the film thickness where the surface roughness is no longer increasing.

After initial nucleation, R_s continues to increase in the $T_s=75°C$ film until ~ 100 Å thickness, as indicated by the arrow in figure 1(a). Such an inflection point in the $R_s(d_b)$ curve can be interpreted as the film thickness where competing mechanisms responsible for increasing and decreasing the roughness are in balance. For $T_s \leq 125°C$, R_s is constant after this point, indicating that these two mechanisms continue in balance up to the maximum thickness observed in these experiments. For $T_s = 150°C$, the roughness immediately decreases after reaching its maximum at $d_b=20$ Å, indicating that at this deposition temperature the smoothening mechanism is dominant after this film thickness. We have not plotted $R_s(d_b)$ for the $T_s = 175°C$ deposition on (100) c-Si because ellipsometry indicates that the film growth is epitaxial, hence we are unable to track $d_b(t)$ because of lack of optical contrast between the film and the substrate.

We have performed equivalent measurements and analysis of the a-Si:H film growth on (111) c-Si substrates vs. substrate temperature. The general behavior and trends in surface roughness are similar, with the films grown on (111) generally having slightly greater roughness, while there is no evidence of epitaxy even at the highest substrate temperature of 175°C. We have evaluated the film thickness where the surface roughness ceases to increase for these films as well. As discussed above, this thickness corresponds to a transition from growth conditions dominated by increasing roughness to growth conditions wherein roughening is balanced or dominated by smoothening mechanisms. The transition thickness vs. inverse substrate

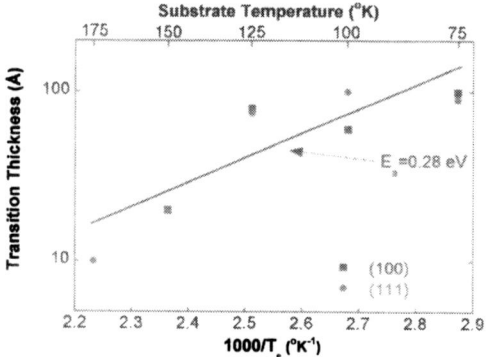

Figure 2. Arrheniusplot of the film thickness at the transition from increasing to constant or decreasing surface roughness for a-Si:H films deposited on (100) c-Si – squares, and (111) c-Si – circles. The line indicates the best-fit exponential with an activation energy of 0.28 eV.

temperature for both (100) and (111) substrates is plotted in figure 2. The line shows an exponential fit with an activation energy of 0.28 eV. Although there is some scatter in the data, the two substrate orientations show very similar trends with substrate temperature, indicating that the mechanisms responsible for the transition in surface roughness are not strongly dependent on the substrate orientation.

The surface smoothening mechanism in a-Si:H growth has been the subject of a great deal of experimental [9] and theoretical [10] investigation. Aydil and coworkers have used molecular dynamics simulations to conclude that surface smoothening is due to preferential incorporation of SiH_3 radicals into surface valleys, with an activation energy of $0.3 - 0.4$ eV [11]. This value appears to be relatively consistent with the activation energy we have determined from the transition thickness as plotted in figure 2.

Passivation and surface recombination velocity

One of the primary advantages of SHJ solar cells is the high open circuit voltages obtained due to the superior surface passivation of the silicon wafer provided by the a-Si:H layers. In order to investigate the surface passivation properties of the intrinsic a-Si:H layers deposited in this study we have utilized photoconductive decay measurements of photoexcited carrier lifetime to infer the surface recombination velocity (SRV) for each substrate orientation and deposition temperature. As discussed above, identical films were grown on each side of the substrate to facilitate accurate measurements of the SRV. Assuming that one knows the bulk lifetime in the wafer, one can calculate the SRV from the photoconductivity decay lifetime and equation (1) below, with τ_{pc} the measured photoconductivity lifetime, τ_b the bulk lifetime, d the wafer thickness, and S the surface recombination velocity. The (100) substrates used in this study were 1.2 Ω-cm FZ wafers, 300 μm thick. The (111) substrates were 75 Ω-cm CZ wafers, 540 μm

$$\frac{1}{\tau_{pc}} = \frac{1}{\tau_b} + \frac{2S}{d} \tag{1}$$

thick. Bulk lifetimes were determined using dilute HF etch to remove the native oxide followed by soaking in iodine-methanol solution to passivate the surfaces. Assuming SRV=0 in the iodine methanol solution, we obtained bulk lifetimes of 325 μs for the (100) substrates and 2.3 ms for the (111) substrates. Using these values for the bulk lifetime and equation (1), we have determined SRV values for each of the samples in this study. The results are shown vs. inverse deposition temperature in the Arrhenius plot of figure 3. We find that films grown on both (100) and (111) substrates follow a very similar substrate temperature dependence, and that this dependence is fit quite well by an exponential dependence on temperature. Fitting to all of the data points except for the 175°C (100) film that RTSE has confirmed is epitaxial, we derive an activation energy of 0.63 eV.

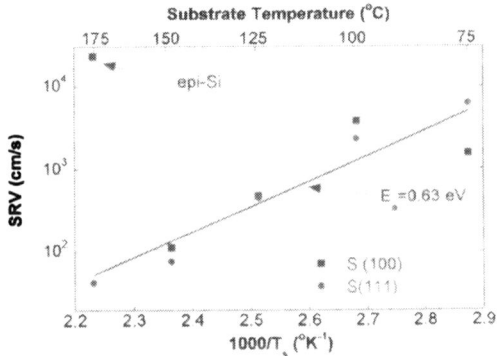

Figure 3. Arrhenius plot of surface recombination velocity for (100)-squares and (111)-circles c-Si substrates. The line is an exponential fit with an activation energy of 0.63 eV. Note the out-lying point for the (100) substrate at T$_s$=175°C. The very high SRV is associated with epitaxial deposition.

Based on the difference in the values of activation energies determined for the surface roughness transition and the surface passivation, it appears that different physical mechanisms are responsible for a-Si:H surface morphology and c-Si surface passivation. At this time it is not clear exactly what these mechanisms are, yet this study provides an insight into several possibilities. As discussed in the preceding section, the surface roughness transition appears to be consistent with a theoretical model of valley filling due to SiH$_3$ incorporation into valleys in the a-Si:H surface. It is natural to speculate that atomic hydrogen is associated with passivation of the c-Si surface in these devices. The activation energy associated with reductions in SRV could be related to diffusion of atomic hydrogen in c-Si, or perhaps at the c-Si / a-Si:H interface.

CONCLUSIONS

We have explored the substrate temperature dependence for deposition of very thin a-Si:H films on (100) and (111) c-Si substrates in the context of their use as the intrinsic layers in SHJ solar cells. Our primary conclusion in this regard is that the substrate orientation does not have a major impact on the material properties of the a-Si:H layer. The (111) substrate orientation does provide greater process latitude in that epitaxial growth, which is deleterious to the surface passivation, occurs at significantly higher temperatures than on (100) substrates.

Somewhat more interesting from a more fundamental point of view are the temperature dependences of the surface roughness transition point and the surface recombination velocity. We find that these properties are relatively independent of substrate orientation, and that they evidence activation energies of 0.28 and 0.63 eV, respectively. The surface roughness transition appears likely to be related to a surface smoothening mechanism related to preferential adsorption of SiH_3 radicals in surface valleys, while the surface passivation is likely related to diffusion of atomic hydrogen.

ACKNOWLEDGMENTS

The authors thank Matthew Page of NREL for extensive work preparing substrates for deposition, and Peter Rupnowski of NREL for HF etching and iodine-methanol treatments of substrates for bulk lifetime determinations. This work was supported by the US DOE under contract DE-AC36-99-GO10337.

REFERENCES

[1] M. Tanaka, M. Taguchi, T. Matsuyama, T. Sawada, S. Tsuda, S. Nakano, H. Hanafusa' and Y. Kuwano, *Jpn. J. Appl. Phys.*, **31** 3518 (1992).

[2] T.H. Wang, E. Iwaniczko, M.R. Page, D.H. Levi, Y. Yan, V. Yelundur, H.M. Branz, A. Rohatgi, and Q. Wang. *Proc. 31st IEEE Photovoltaic Specialists Conference* 955 (2005).

[3] M.R. Page, E. Iwanizcko, Y. Xu, Q. Wang, Y. Yan, L. Roybal, Howard M. Branz, and T. Wang., *Proc. 4th IEEE World Conf. on Photovoltaic Energy Conversion* 1485 (2006).

[4] D.H. Levi, C.W. Teplin, E. Iwanizcko, R.K. Ahrenkiel, H.M. Branz, M.R. Page, Y. Yan, and Q. Wang, *Mat. Res. Soc. Symp. Proc.* **808**, 239 (2004).

[5] T.H. Wang, E. Iwaniczko, M.R. Page, Qi Wang, Y. Xu,Y. Yan, D. Levi, L. Roybal, R. Bauer, H.M. Branz, *Proc. 4th IEEE World Conf. on Photovoltaic Energy Conversion* 1439 (2006).

[6] R.A. Sinton and A. Cuevas, *Appl. Phys. Lett.* **69**, 2510 (1996).

[7] A.S. Ferlauto, R.J. Koval, C.R. Wronski, R.W. Collins, *Mat. Res. Soc. Symp. Proc* **664** A5.4.1 (2001).

[8] H. Yao, J.A. Woollam, S.A. Alterovitz *Appl. Phys. Lett.* **62** (1993) 3324.

[9] A.H.M. Smets, W.M.M. Kessels, and M.C.M. van de Sanden, *Appl. Phys. Lett.* **82**, 865-856 (2003).

[10] M. S. Valipa, T. Bakos, E. S. Aydil, and D. Maroudas, *Phys. Rev. Lett.*, **95** 216102 (2005).

[11] Aydil, E.S.; Agarwal, S.; Valipa, M.S.; Hoex, B.; van de Sanden, M.C.M.; Maroudas, D. *Surface Science* **598**, 35 (2005).

Mater. Res. Soc. Symp. Proc. Vol. 989 © 2007 Materials Research Society   0989-A24-05

Interdigitated Back Contact Silicon Heterojunction (IBC-SHJ) Solar Cell

Meijun Lu[1,2], Stuart Bowden[1], Ujjwal Das[1], Michael Burrows[1], and Robert Birkmire[1,2]
[1]Institute of Energy Conversion, University of Delaware, Newark, DE, 19716
[2]Department of Physics and Astronomy, University of Delaware, Newark, DE, 19716

ABSTRACT

Interdigitated back contact silicon heterojunction (IBC-SHJ) solar cells have been developed. This structure has interdigitated p/n amorphous silicon (a-Si:H) films deposited by plasma enhanced chemical vapor deposition (PECVD) on the backside of crystalline silicon (c-Si) wafers, with light irradiating the front surface. IBC-SHJ cells possess advantages over front junction a-Si:H/c-Si heterojunction cells due to minimized current losses in the illuminating side, and over traditional diffused back-junction cells due to low temperature processing combined with the potential of high voltages for the heterojunction. Current-voltage curves, spectral response and laser beam induced current maps have been used to characterize the IBC-SHJ cells. It was found that the IBC-SHJ cell has non-linear illumination level dependence that correlates with measured minority-carrier lifetime. As the performance of these cells is very sensitive to the quality of passivation on front surface, they are ideally suited as a diagnostic tool for detailed characterization of surface passivation. Initial cell structures have achieved independently confirmed cell efficiencies of 11.8% under AM1.5 illumination. Device simulation shows that an efficiency of higher than 20% can be expected after optimizing the IBC-SHJ cells.

INTRODUCTION

Presently, 95 percent of solar cells are fabricated on crystalline silicon wafers and the rapid market growth has resulted in a shortage of silicon feedstock. Further cost reductions in photovoltaics require a thin, high efficiency cell to reduce the cost per watt generated. The "Interdigitated Back Contact Si HeteroJunction" (IBC-SHJ) solar cell introduced in this paper is aimed to achieve these requirements.

The IBC-SHJ solar cell is based on amorphous-silicon (a-Si) / crystalline-silicon (c-Si) heterojunction (SHJ) [1], rather than the high temperature (~800°C) diffusions used to form conventional homojunction cells. By using deposited amorphous silicon (a-Si:H) as an emitter layer, SHJ solar cells can be produced at low temperature (~200°C). The SHJ also achieves a high open-circuit voltage (V_{oc}) due to the excellent surface passivation by a-Si:H layer on both sides of Si wafer. The high V_{oc} together with non-detrimental low-temperature processes allows SHJ solar cells to achieve high efficiency. Efficiencies of 21% have been reported [2], and Sanyo has started mass production of SHJ solar cells.

Based on SHJ structure, the uniqueness of IBC-SHJ solar cells is that it combines the performance advantages of the SHJ with an IBC technique. IBC was first developed for concentrator systems about 30 years ago [3]. Recently, its application at one sun has been introduced to diffused junction cells by many groups [4], including SunPower who has reported ~22% efficiency all back-contact solar cells [5]. The IBC structure eliminates the front contact shading in conventional cells by interdigitating both contacts on the rear of the cell. Having both contacts on the rear of the cell allows for large metal electrodes and simplifies module interconnection. However, the high temperature processing for diffused back-junction cells is not easy to manufacture due to the need to separate the n and p-type regions, and the devices are often shunted. Using deposited layers of amorphous silicon largely solves this problem. The IBC structure has further advantages for SHJ cells since the emitter is now on the rear of the cell and the parasitic absorption of blue light in the emitter is avoided.

In the paper, we will report our progress in developing IBC-SHJ solar cells, including the design of structure, processing steps and the initial solar cell results. It will also show the non-linear illumination level dependence of the IBC-SHJ structure and its potential application as a diagnostic tool to characterize the surface passivation.

EXPERIMENT

The designed structure of IBC-SHJ solar cells is shown in Figure 1, where (a) is the interdigitated backside, and (b) is the cross section. Devices were fabricated on 300 μm thick, polished, n-type float-zone silicon wafers with a resistivity of 2.5 Ωcm. The front surface is presently passivated with an intrinsic a-Si:H layer of 20 nm thick deposited by DC plasma enhanced chemical vapor deposition (PECVD) system at 200°C. It is covered with a dual-layer anti-reflection coating composed of Indium Tin Oxide (ITO) and MgF_2, deposited by sputtering and electron beam evaporation respectively. At the back side of the wafer, the emitter and contacts are provided by alternating strips of p and n-type interdigitated a-Si:H layers respectively at a thickness of 20 nm. Both types of a-Si:H layers are deposited in the PECVD system at 200 °C, using separate chambers for n and p-type dopants. The bottom metal contact layer is formed by 200nm thick aluminum deposited by electron beam evaporation. The interdigitated pattern was created by two-step photolithography as described later. The finger-like p-region has lateral dimension of 1.2 mm, while n-region is 0.5 mm wide. These dimensions are tunable and can be optimized for device performance. The separation between p- and n-regions is ~2 μm, and is formed naturally by undercutting during the etching process, which is sufficient to avoid shunting. The process sequence to make the IBC-SHJ structure is described below:

1. Front surface: intrinsic a-Si deposition.
2. Back surface: p-type a-Si deposition, followed by Al (p-contact) deposition.
3. Back surface: photolithography to define p-region.
4. Back surface: Al + a-Si etching for non-p-region.
5. Back surface: n-type a-Si deposition, followed by Al (n-contact) deposition.
6. Back surface: lift-off to remove photoresist and attached n-type a-Si & n-contact on p-region.
7. Back surface: photolithography to define cell area.
8. Back surface: Al + a-Si etching outside of cell area.
9. Back surface: remove photoresist.
10. Front surface: AR coating.

Presently, photolithography is used to define the p and n strips but the large dimensions allow for the use of simpler techniques, like lasers and self-aligning masks.

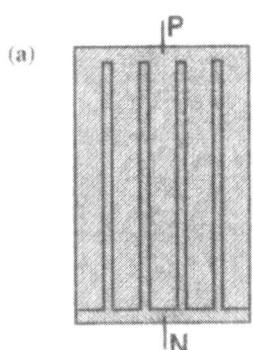

(a)

P

N

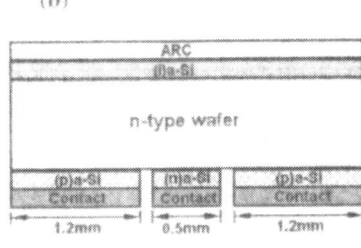

(b)

ARC
(i)a-Si
n-type wafer
(p)a-Si Contact | (n)a-Si Contact | (p)a-Si Contact
1.2mm | 0.5mm | 1.2mm

Figure 1. Schematics of IBC-SHJ structure:(a) bottom view and (b) cross section view, where the cell thickness is exaggerated.

RESULTS AND DISCUSSION

Initial IBC-SHJ cells have been fabricated according to the above sequence, and Figure 2 shows the current-voltage (I-V) curve for a cell with area of 1.32 cm² as measured by NREL under AM1.5 illumination, demonstrating an efficiency of 11.78%. The open circuit voltage (V_{oc}) of 602mV matches similar front-junction SHJ solar cells fabricated without intrinsic a-Si buffer layers. By inserting an extra passivation layer (intrinsic a-Si) between c-Si wafer and doped a-Si at back side, IBC-SHJ solar cells achieved a V_{oc} of 690 mV, but with a low fill factor (S-shape curve). Similar S-shape JV curve has been observed for front-junction SHJ as well at IEC [6], but it was resolved through process optimization, and 18% efficiency was obtained. The fill factor (FF) without i-layer in Figure 2 is 73.3%, and it can be further improved once the doped a-Si layers are optimized and/or Al layers are thickened. There is no evidence of shunting between the p and n type regions. Since no special effort was made to separate the two regions, the result demonstrates the robust nature of the process.

The short circuit current density (J_{sc}) of 26.7mA/cm² is somewhat lower than optimized IBC devices where J_{sc} is in the range 35 – 40 mA/cm². The front structure was adapted from our front junction devices and consists of a-Si to provide surface passivation and dual layers of ITO and MgF₂ as anti-reflection coatings. These layers all absorb short wavelength light (see Figure 5) and can be replaced by a single layer of silicon nitride that has much lower absorption and slightly better surface passivation. An advantage of the IBC structure is that the front surface can

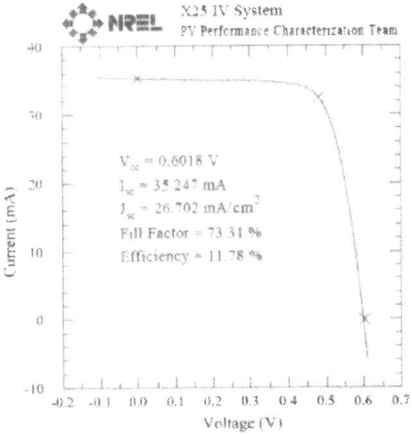

X25 IV System
PV Performance Characterization Team

V_{oc} = 0.6018 V
I_{sc} = 35.247 mA
J_{sc} = 26.702 mA/cm²
Fill Factor = 73.31 %
Efficiency = 11.78 %

Current (mA)

Voltage (V)

Figure 2. J-V curve for initial IBC-SHJ cell measured at NREL.

be optimized independent of the need to provide electrical contact. Combining the identified improvements to V_{oc}, FF and J_{sc}, an IBC-SHJ device with efficiency over 20% is feasible.

The spectral response of the IBC-SHJ cell was measured at room temperature. The reflection-corrected internal quantum efficiency (IQE) is shown in Figure 3, where the solid curve is IQE with approximately one-sun light bias, and dashed one is IQE without light bias. The difference between them is due to the changes in front surface recombination velocity under different carrier density with or without light bias. Further confirmation of this effect was provided by non-linear cell short-circuit current, which falls off dramatically at low light intensities.

To verify that the effect is due to the change of front surface recombination, a symmetrical test sample with both sides passivated by same deposition used in the front surface of IBC-SHJ cell was prepared to perform lifetime measurements. Minority carrier lifetime was measured using quasi-steady-state photoconductivity. The measured *effective minority lifetime* [7] is the net result of summing up bulk and surface recombination losses. Since the substrates are high quality float-zone with lifetimes of several milliseconds, the effective lifetime is a measure of the recombination at the surfaces. The result for the sample after annealing is shown in Figure 4. It can be seen that the lifetime decreases dramatically at lower carrier density. It drops from ~1ms at a carrier density of 10^{15} cm^{-3} to ~30μs at 10^{12} cm^{-3}, corresponding to one-sun and no light bias condition respectively in the IQE measurements. The a-Si passivates the surface of the crystalline silicon through field passivation and surface band bending, in addition to the reduction of interface states. Theoretical modeling of the interface to include effects of band bending induced by the amorphous layer [8] reproduces the curve of Figure 4 with a low lifetime at low carrier densities and a lifetime maximum in the range 10^{15} to 10^{16} cm^{-3}. This non-linear effect can be reduced by decreasing the interface state density. Similar non-linear effects have also been observed in diffused-junction IBC cells [9].

Since the IBC-SHJ cell performance sensitively depends on the quality of front surface passivation, it can also be used as a diagnostic tool for surface recombination during the process sequence. The IQE for the same IBC-SHJ cell are characterized at three stages in the process sequence and shown in Figure 5: before anti-reflection coating (ARC), after ARC, and after

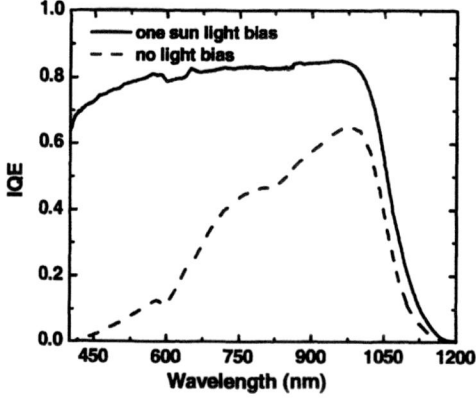

Figure 3. Internal Quantum Efficiency (IQE) curves for IBC-SHJ cell with and without light bias.

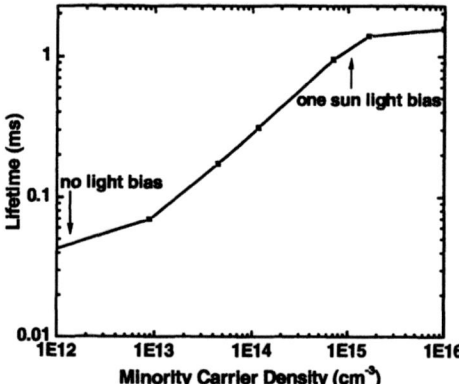

Figure 4. Effective lifetime measured on a wafer passivated on both sides with intrinsic a-Si.

annealing with ARC. The cell's performance decreases after application of the ARC, but improves after annealing to exceed the initial device performance. The modeling software PC1D was used to fit the IQE curves, and extract the front surface recombination velocity, shown as dashed lines in Figure 5. A one-dimensional modeling program such as PC1D [10] can be used because most of the light with a wavelength less than 1000 nm is absorbed in the wafer before arriving back-surface pattern. The modeling closely matched the measured data except at wavelengths shorter than 600 nm, where there is parasitic absorption by front surface a-Si layer. The surface recombination velocity, S, increases from 152 cm/s to 299 cm/s after AR-coating, while decreasing to 82 cm/s after annealing. The increase in surface recombination after AR-coating suggests the application of the AR has detrimental effect on the front surface passivation due to the ion damage during sputtering. The damage due to AR application can be removed by annealing the sample in air and exceeds the initial device performance. The result indicates that annealing not only eliminates the sputter damage but also improves the front surface passivation quality.

To verify the effect of AR coating and annealing effect, lifetime measurement were also performed on the symmetric test structure. It was found that the original lifetime 500μs decreases to 200μs after AR-coating, but increases to a higher value of 1ms after annealing, which is consistent with recombination velocity result estimation by PC1D. This consistency confirms that IBC-SHJ structure can be used as a diagnostic tool to complement the lifetime study and investigate surface passivation in an actual device structure.

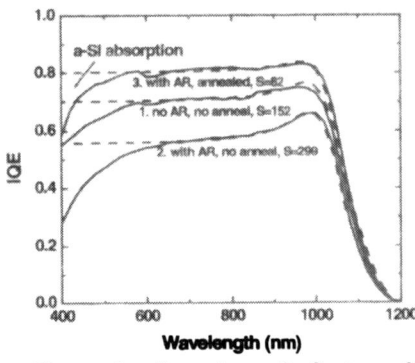

Figure 5. Extraction of front surface recombination velocity from IQE curves for a sample at three stages in the process sequence.

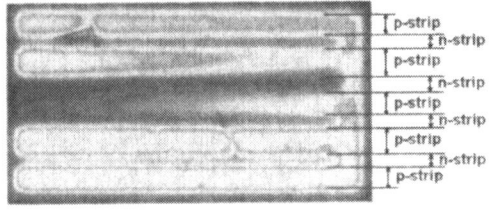

Figure 6. LBIC scan of an IBC-SHJ solar cell whose middle p-strip is poorly contacted.

Figure 6 shows a laser beam induced current (LBIC) scan of an adjacent IBC-SHJ cell on same wafer. White regions correspond to high current response and the dark areas are regions of reduced carrier collection. For this cell, the middle p-strip is poorly contacted, resulting in lower current and the dark region in the center of the LBIC scan. Another interesting feature is that n-region always provides less current than p-region, even for well-contacted strips. Minority carriers generated above the n-regions need to travel not only vertically to the back of the cell, but also laterally up to 0.5mm to reach the collecting junction at the p-region. Several measures can be taken to reduce this effect: i) improve surface passivation by adding an intrinsic buffer layer as described above, so as to reduce recombination of minority carriers during diffusing to the junction; ii) enlarge the p-region area (emitter) and reduce n-region (base contact), so minority carriers need to travel much shorter distance to reach emitter.

CONCLUSIONS

An interdigitated back-contact silicon heterojunction (IBC-SHJ) solar cell has been developed. The structure combines the advantages of both back-contact and heterojunction cells. An initial cell has achieved an independently confirmed efficiency of 11.8%. Optimizations have been proposed to improve V_{oc}, J_{sc} and FF, and simulation shows efficiency higher than 20% is achievable. It was also found that the IBC-SHJ cell has non-linear illumination level dependence that correlates with measured minority-carrier lifetime, and the sensitivity of IBC-SHJ cell's performance to front surface recombination makes the cell suitable as a diagnostic tool for detail characterization of surface passivation.

ACKNOWLEDGMENTS

The authors would like to thank Kevin Hart and Shannon Fields for film depositions and Steven Hegedus for helpful discussion. This work was partly supported by BP Solar and National Renewable Energy Laboratory under subcontract #ADJ-1-30630-12.

REFERENCES

[1] M. Taguchi, K. Kawamoto, S. Tsuge, T. Baba, H. Sakata, M. Morizane, K. Uchihashi, N. Nakamura, S. Kiyama and O Oota, *Progress in Photovoltaics* 8 (5), 503 (2000).
[2] E. Maruyama, A. Terakawa, M. Taguchi, Y. Yoshimine, D. Ide, T. Baba, M. Shima, H. Sakata and M. Tanaka, *4th World Conference on Photovoltaic Energy Conversion*, Hawaii, USA (2006).
[3] M.D. Lammert and R.J. Schwartz, *IEEE Transactions on Electron Devices* 24 (4), 337 (1977).
[4] E. Van Kerschaver and G. Beaucarne, *Progress in Photovoltaics* 14 (2), 107 (2006).
[5] K.R. McIntosh, M.J. Cudzinovic, D.D. Smith, W.P. Mulligan and R.M. Swanson, *Proceedings of the Third World Conference on Photovoltaic Energy Conversion*, Osaka, Japan, 971 (2003).
[6] U. Das, S. Bowden, M. Burrows, S. Hegedus and R.Birkmire, *4th World Conference on Photovoltaic Energy Conversion*, Hawaii, USA, 1283 (2006).
[7] T. Markvart and L. Castaner (editors), *Practical Handbook of Photovoltaics: Fundamentals and Applications*, Elsevier Science Inc., UK, (2003), Page 234
[8] I. Martin, M. Vetter, A. Orpella, J. Puigdollers, and A. Cuevas, *Applied Physics Letters*, 79 (14), 2199 (2001).
[9] P-J. Ribeyron, E. Rolland, and M. Pirot, *Proceeding of 19th European Photovoltaic Solar Energy Conference*, Paris, France, 1315 (2004).
[10] D.A. Clugston, and P.A. Basore, *Conference Record of the Twenty Sixth IEEE Photovoltaic Specialists Conference*, 207-10, (1997).

Film Growth

Mater. Res. Soc. Symp. Proc. Vol. 989 © 2007 Materials Research Society   0989-A25-02

Time-resolved Cavity Ringdown Spectroscopy as a Monitoring Technique of Nanoparticles in Pulsed VHF Plasmas

Takehiko Nagai, Arno H. M. Smets, and Michio Kondo
Research Center for Photovoltaics, National Institute of Advanced Industrial Science and Technology (AIST), Central 2, 1-1-1 Umezono, Tsukuba, Ibaraki 305-8568, Japan

ABSTRACT

Time-resolved cavity ringdown (τ-CRD) spectroscopy has been applied to monitor the silyl (SiH_3) radicals and nano-particles in pulsed very high frequency (VHF) silane (SiH_4)/hydrogen (H_2) plasmas under microcrystalline silicon (μc-Si:H) deposition conditions. After the plasma ignition, a small constant cavity loss (~100 ppm) on timescales smaller than ~1 s has been observed, whereas on time scales larger than ~1 s after plasma ignition, an additional cavity loss is observed. By variation of the wavelength of the CRD laser pulse, we demonstrate that the cavity loss on time scales smaller than ~1 s reflects the SiH_3 absorption. On time scales larger than ~1 s, the additional cavity loss corresponds to the loss of light due to mainly scattering at the nano-particles. Under the conditions studied, the light scattering at nano-particles can be described by Rayleigh scattering during its initial growth. After ~ 2.5 s, the cavity loss reflects the transition of the scattering mechanism from dominant Rayleigh to dominant Mie-scattering. These results are discussed in terms of nano-particles growing in time and further confirmed by additional scanning electron microscopy analyses on the nano-particles created in the plasma pulse.

INTRODUCTION

In the last decade, μc-Si:H thin films have been deposited under the interesting high pressure and high input power conditions, as these conditions have access to high deposition rates (2-4 nm/s). An unwelcome side effect of these typical conditions is the fact that nano-particles are easily created, which can have an effect on the film growth, the deposition chamber wall conditions and the pumping system of the deposition set-up. To improve the insights in the ratio of radicals and nano-particles created in the plasma and their effect on the μc-Si:H deposition, a measurement technique, which simultaneously monitors nano-particles and radicals, is desirable. A powerful technique to study the nucleation, growth and evolution of the nano-particles in the plasmas is the so-called laser-light-scattering (LLS) [1-9] technique, which are based upon the detection of a small fraction of the scattered probe light at the nano-particles under a spatial angle. However, this technique is insensitive for radical molecules. In contrast, Cavity ringdown (CRD) spectroscopy is a very powerful technique to detect the small fractional absorbing species, like for example the Si, SiH and SiH_3 radicals in silane plasmas [10-14]. In this paper, we will show that the CRD technique can also be used to detect UV light scattering at nano-sized particles created in the plasma. An important difference to LLS optical techniques is the fact that CRDS detects the total sphere integrated scatter loss instead of a fraction of the scattered light under a spatial angle. Here, we will monitor the creation of nano-particles in a pulsed SiH_4/H_2 VHF plasmas using time resolved (τ)-CRD measurements. We will discuss the growth of the nano-particle in time after plasma ignition. Furthermore, the size-distribution of the nano-particles has been studied by additional SEM analyses.

EXPERIMENT

The CRD technique has been demonstrated to be an useful method for the determination of small fractional absorptions down to sub-ppm levels per cavity pass. The CRD technique is based upon the measurement of the decay of the intensity of a light pulse trapped in an optical cavity, formed by two highly reflective plano-concave mirrors. The characteristic decay time depends on the reflectivity R of the mirrors, the density and the absorption cross-section of the absorbing species and light scattering particles within the optical cavity. Since the typical cavity ring down decay time is in the order of several hundreds of nanoseconds up to several microseconds, the technique is very suitable for time-resolved measurements. More details of the principle of the τ-CRD technique can be found in Refs. [15, 16].

The optical cavity consists of two high reflectivity plano-concave mirrors (Laser Optik), whose radius of curvature and diameter are -100 cm, and 2.5 cm, respectively. The cavity mirrors are attached to the VHF deposition chamber using flexible bellows. Two sets of narrow bandwidth high reflective mirrors have been used, one set optimized for the 220 nm range with reflectivity of $R=98.9\%$ and one for the 280 nm range with reflectivity of $R=99.2\%$. The cavity length (l) is equal to 125 cm. The diameter of VHF deposition chamber is 30 cm, the diameter of the electrode plates (d) is 8 cm and the distances between the grounded electrode and the powered showerhead-electrode is fixed at 8 mm for the 1.0 Torr pressure conditions. The position of the optical cavity axis has been fixed parallel at distance of 1 mm from the powered showerhead-electrode. To avoid any film deposition on the cavity mirrors, we flow H_2 gas (50 sccm) from the flexible bellows down to the deposition chamber. The SiH_4/H_2 gas flow is injected into the deposition chamber via the powered showerhead-electrode at 2/170 sccm. The total SiH_4/H_2 gas flow injected in the chamber is therefore 2/220 sccm. The temperature of the grounded electrode is kept at 473 K during the CRD measurements. The VHF (50 MHz) input power densities is 0.3 W/cm^2 at the 1.0 Torr conditions. The time period of plasma burning (t_{on}) and plasma off are 4 s and 11 s.

The CRD laser light pulse at 220 nm and 280 nm is obtained by dye excitation using the frequency double output of of a Nd:YAG-laser (Nihon laser TLD80). The pulse duration and repetition rate is ~5 ns and 10 Hz, respectively. The light leaking out of the cavity at the back mirror is dispersed by a 25 cm monochromator before it is detected by a photomultiplier (Hamamatsu, R928). The cavity ringdown transients is recorded with a digital oscilloscope (Agilent technology, Infinium). A delay generator (Stanford Inc, DG535) and pulse generator (Ascom Inc, Time98) have been used to tune the delay time between the plasma ignition and generation of the laser pulse.

The nano-particles created in the plasma have been analyzed by means of the SEM technique. The cleaned electrodes (and deposition chamber) are exposed to the SiH_4/H_2 plasma for 40 pulses. The particles are trapped on a cupper substrate by scrapping the electrodes with the cupper plates after the plasma exposure. The nano-particle size distribution is obtained from 200 nano-particles per deposition conditions to obtain an acceptable statistics. This procedure has been perform for $t_{on}=2.5$ s and $t_{on} = 3.0$ s for the 1.0 Torr conditions.

BACKGROUND THEORY

The light intensity leaking out of the cavity I_t, decays like a single exponential in time with: Eq. (1) $I_t = I_0 \exp(-t/\tau)$ where the ringdown time τ is given by: Eq. (2) $\tau = (l/c)/L_{tot} = (l/c)/(L_0 + L_{abs} + L_{sca})$ with l the cavity length, c the light velocity, L_{tot} the total cavity loss, L_0 the primary cavity loss, L_{abs} the cavity loss due to light absorption in the cavity and L_{sca} the cavity loss due to light scattering at species in the cavity. In the absence of any absorbing or scattering species between the cavity mirrors, the light intensity leaking out of the cavity (I_t) is determined by the primary cavity loss L_0, which in turn is determined by only the small transmittance of the mirrors and diffraction loss [15], i.e. Eq. (3) $L_0 = (1 - R)$ with $(1-R)$ the effective transmittance of the mirrors. The L_{abs} and L_{sca} are given by: Eq. (4) $L_{abs} = \sigma_{abs} N_{abs} d$ and Eq. (5) $L_{sca} = \sigma_{sca} N_{sca} d$ respectively, with N_{abs} the density of the absorbing species, σ_{abs} the light absorption cross section of the absorbing species, N_{sca} the density of the scattering species and σ_{sca} the cross section of the light scattering at particles. The cavity loss presented in the figures in this paper reflect the cavity loss due to absorption and scattering of the species, i.e. $L_{tot} - L_0 = L_{abs} + L_{sca}$.

RESULTS AND DISCUSSION

Figures 1 shows the cavity loss ($L_{abs} + L_{sca}$) at a wavelength of 220 nm and 280 nm as a function of t_{on} under the 1.0 Torr condition. We can distinguish four time intervals (I up to IV) in the measured cavity loss evolution, as shown in Fig. 1(a). Figure 1(b) is a close-up of the cavity loss in Fig. 1(a) at a time of –0.1 s up to 1.0 s. Interval I (0 s $<t_{on}<$ ~1 s) is characterized by an instantaneous increase of the cavity loss to a level (~1.0×10^{-4}), which remains constant during interval I at 220 nm, while the cavity loss at 280 nm is within the lower limit of this system, as shown in Fig. 1(b).

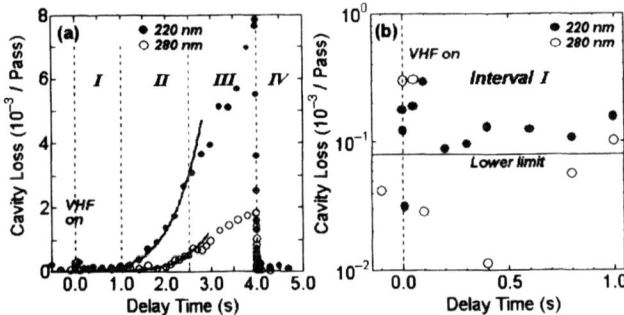

Figure 1. (a) Cavity loss versus the plasma ignition time (t_{on}) in the SiH_4/H_2 mixture gas flow rate of 2/220 sccm at the total pressure of 1.0 Torr. Solid circles and open circles indicate the cavity loss at the wavelength of 220 nm and 280 nm, respectively. (b) Close-up of the cavity loss during interval I.

We have speculated in Ref. [14] that these instantaneous present small cavity losses correspond to the absorption of SiH_3 radicals. The $A^2A'_1 \leftarrow X^2A_1$ transition of SiH_3 radicals has a broadband featureless absorption spectrum ranging from ~200 to ~260 nm [16-19]. Figure 1(b) shows indeed a cavity loss above the detection limit of ~8×10^{-5} for measurements at 220 nm corresponding to $A^2A'_1 \leftarrow X^2A_1$ transition of SiH$_3$ radicals, which is absent for measurement at 280 nm. Moreover, this transaction is not active at 280 nm, the cavity loss at 280 nm has no SiH$_3$ absorption contribution and only reflects the scatter or absorption losses at nano-particles. Not shown in this paper is the fact that the cavity loss scales linearly with the SiH$_4$/H$_2$ concentration in plasmas in interval I.

At interval II (1 s < t_{on} <~2.5 s), the cavity loss starts to increase with t_{on}, where the additional cavity loss is much larger at 220 nm compared to 280 nm. At 2.5 s a transition to interval III (2 s < t_{on} <~4 s) can be observed: in interval III the cavity loss is still increasing with t_{on}, however it does not increase as strongly as in interval II. In view of the strong frequency dependence and the timescales in which this additional cavity loss shows up in interval II and III, we argue that this additional light loss is a result of scattering at or absorption by growing nano-particles. Since the absorption and scattering at nano-particles strongly depends on radius (r_c) and the wavelength, the additional cavity loss will be determined by the nano-particle density and the size distribution of nano-particles [20, 21].

The particle distribution has been measured by SEM measurements on the captured nano-particles at t_{on} of 2.5 s and 3.0 s, respectively. Figure 2 (a) shows the SEM images of the captured particles. Figures 2 (b) shows the size distribution of particles determined from the SEM images, at t_{on} of 2.5 s and 3.0 s. In general, these a-symmetric distributions for particles obey a log-normal distribution functions [22, 23]: particle count = A x exp[-{ln(x/r_c)/γ_p}^{2}], where A is an empirical constant, r_c is the averaged particle size and γ_p is a dispersions of generated particles. Using this equation, we have deduced an r_c of ~30 nm and ~55 nm at t_{on} of 2.5 s and 3.0 s, respectively. This reflects that the nano-particles are growing in time.

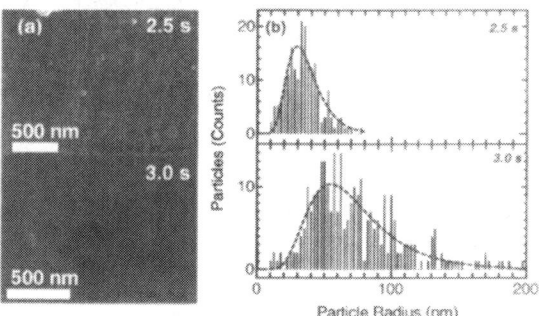

Figure 2. (a) SEM observation of particles at t_{on} of 2.5 s and 3.0 s, respectively. (b) Size distribution of growing particle at t_{on} of 2.5 s and 3 s. The dotted lines were the fitting curve using the log-normal functions.

First we will consider the contribution of light scattering at the growing nano-particles to the additional cavity loss. Since at the initial growth the nano-particle size (r_c) r_c<λ and r_c $\varepsilon^{1/2}$<λ, the light scattering has to be Rayleigh like. The cross section σ_{sca} in equation (5) is described by sphere integrated Rayleigh scattering cross section [20, 24]: Eq. (7) $\sigma_{sca} = 128/3\pi^5\{(\varepsilon-1)/(\varepsilon+2)\}^2 r_c^6/\lambda^4$ with ε the dielectric constant of the cluster material. Kujundzid et al. [8] have shown that in hydrogen diluted silane plasma the cluster radius grows linear with t_{on} in line with the results on various other gaseous plasmas [8, 9]. These results reflect a nano-particle growth from an isotropic and constant radical or ion flux (Γ). Therefore we think that it is rather acceptable to use the assumption that in our conditions r_c is also increasing linearly in time. The increase of the particle volume $V(t)$ is linear with the surface $A(t)$: Eq. (8) $dV/dt = A(t)\Gamma$. This gives for a growing particle radius $r_c = A \times t^n$, a linear increasing particle radius in time ($n=1$) as a solution. The evolution of the nano-particle density (Eq. 5) will determine the observed cavity loss evolution. Many studies have shown that the nano-particle density is slightly decreasing with time and therefore we assume Eq. (9) $N_{par} = B \times t^{-\chi}$. Substitution of equations (7) and (9) in Eq. (5) results in a cavity loss due to scattering which scales like: Eq. (10) $L_{sca} = t^{6-\chi}$.

The experimental data in interval II has been fitted using Eq. (10), as shown by the solid line in Fig. 2 and we find $6-\chi = 6.0$. The $\chi \sim 0$ result suggests that the nano-particle density is nearly constant. Furthermore, in agreement with the strong dependence of the scatter cross section with the wavelength $\sigma_{sca} \sim \lambda^{-4}$ the cavity loss at 220 nm is larger than at 280 nm. These observations make Rayleigh scattering at nano-particles the most likely explanation for the increasing cavity loss. This is even further founded if we consider the averaged particle size of ~30 nm at the transition from interval II to III and 55 nm in interval III. The light scattering can be considered Rayleigh-like up to $r_c \sim 0.1 \times \lambda$, corresponding to 22-28 nm at the used wavelength range. Above these values the scattering becomes Mie-like. Therefore we claim that at the transition from interval II to III ($r_c \sim 30$ nm) the dominant scattering mechanism shifts from Rayleigh to Mie-like. Note, that the sphere-integrated Mie scatter cross-section dependence on particle radius is an oscillating function which on averaged does not increase as strongly with r_c as for Rayleigh scattering.

Here, we can estimate the N_c at t_{on} of 2.5 s by using the assumption of $r_c \sim 30$ nm, respectively, and from Eq. (5). At t_{on} of 2.5 s, cavity loss at 220 and 280 nm are equal to 3.07 x 10^{-3} and 7.37 x 10^{-4} /pass, as shown in Fig. 1(a). Consequently, we can obtain the particle densities of 1.5 x 10^7 and 9.5 x 10^6 cm^{-3} at 220 and 280 nm, respectively.

Finally, we would like to make a remark that for nano-particles with a radius of only a few nm's, the absorption cross section is more dominant than the cross section for Rayleigh scattering. However, the density of these small nano-particles in the early nano-particle growth is too small to detect using CRDS in the presented conditions. However, in future reports we will show that under certain conditions, at certain locations in the plasma and at certain time intervals the absorption at nano-particles can also have a significant contribution to the cavity loss.

CONCLUSIONS

We have applied the τ-CRD technique for the detection of SiH$_3$ radicals and nano-particles in a SiH$_4$/H$_2$ VHF plasma. The cavity loss evolution is characterized by 4 time intervals. Interval I the cavity loss reflects the absorption at the SiH$_3$ radical, in interval II the additional cavity loss reflects Rayleigh scattering at nano-particles and in interval III the additional cavity loss reflects Mie-scattering at the nano-particles. SEM analyses on the particles size distribution are in line with the observed upper limit for Rayleigh scattering. We have shown that τ-CRD method is very helpful technique to detect not only radicals but also to monitor the nano-particle evolution after plasma ignition.

ACKNOWLEDGMENTS

The authors would like to thank Dr. Akihisa Matsuda for useful discussions.

REFERENCES

1. K. G. Spears. T. J. Roth, IEEE Trans. Plasma Sci. **PS14**, 179 (1986).
2. G. S. Selwyn, J. Singh, and R. S. Bennett, J. Vac. Sci. Technol. A7, 2758 (1989).
3. G. M. Jellum, J. E. Daugherty, and D. B. Graves, J. Appl. Phys. **69**, 6923 (1991).
4. M. Shiratani, S. Maeda, K. Koga, and Y. Watanabe, Jpn. J. Appl. Phys. **39**, 287 (2000).
5. K. Koga, Y. Matsuoka, K. Tanaka, M. Shiratani, and Y. Watanabe, Appl. Phys. Lett. **77**, 196 (2000).
6. M. A. Childs and A. Gallagher, J. Appl. Phys. **87**, 1076 (2000); **87**, 1086 (2000).
7. G. Bano, P. Horvath, K. Rozsa, and A. Gallagher, J. Appl. Phys. 98, 013304 (2005).
8. D. Kujundzic and A. Gallagher, J. Appl. Phys. **99**, 033301-1 (2006).
9. S. Nunomura, M. Kita, K. Koga, M. Shiratani, and Y. Watanabe, J. Appl. Phys. **99**, 083302-1 (2006).
10. W. M. M. Kessels, A. Leroux, M. G. H. Boogaarts, J. P. H. Hoefnagels, M. C. M. van de Sanden, and D. C. Schram, J. Vac. Sci. Technol. A **19**, 467 (2001).
11. W. M. M. Kessels, J. P. M. Hoefnagels, M. G. H. Boogaarts, D. C. Schram, and M. C. M. van de Sanden, J. Appl. Phys. **89**, 2065 (2001).
12. Y. Nozaki, M. Kitazoe, K. Horii, H. Umemoto, A. Masuda, H. Matsumura, Thin Solid Films. **395**, 475 (2001).
13. J. P. M. Hoefnagels, Y. Barrell, W. M. M. Kessels, and M. C. M. van de Sanden, J. Appl. Phys. **96**, 4094 (2004).
14. T. Nagai, A. H. M. Smets, and M. Kondo, Jpn. J. Appl. Phys. **45**, 8095 (2006).
15. K. W. Busch and M. A. Busch, *Cavity-Ringdown Spectroscopy* (Oxford University Press, 1999) Chap. 2.
16. P. D. Lightfood, R. Becerra, A. A. Jemi-Alade, R. Lesclaux, Chem. Phys. Lett. **180**, 441 (1991).
17. G. Olbrich, Chem. Phys. **101**. 381 (1986).
18. M. G. H. Boogaarts, P. J. Böcker, W. M. M. Kessels, D. C. Schram, and M. C. M. van de Sanden, Chem. Phys. Lett. **326**, 400 (2000).
19. A. V. Baklanov and L. N. Krasnoperov, J. Phys. Chem. **105**, 4917 (2001).
20. C. F. Bohren and D. R. Huffman, *Absorption and Scattering of Light by small particles* (Wiley, New York, 1983).
21. S. H. Hong and J. Winter, J. Appl. Phys. **100**, 064303 (2006).
22. S. L. Chan and S. R. Elliott, Phys. Rev. B **43**, 4423 (1991).
23. A. M. Law and W. D. Kelton, *Simulation Modeling and Analysis* (McGraw-Hill, New York, 1986), p164.
24. M. Kerker, *The scattering of light and other electromagnetic radiation*, (Academic Press, 1969). If the scatter cross-section is known (i.e. the radius and the dielectric constant of the nano-particles is known) the CRD technique provides a direct measure of the nano-particle density.

AUTHOR INDEX

Aberle, Armin G., 391
Adachi, Michael M., 489, 517
Ahn, Kun Ho, 557
Aktik, Cetin, 87
Alford, T.L., 145
Al-Jassim, M.M., 15
Ambrosio, Roberto, 457
Androutsopoulos, Girogios, 545
Apte, Raj B., 199
Astakhov, Oleksandr, 3
Avella, Manuel, 531

Ballif, Christophe, 281
Barankov, Dmitry, 3
Bauer, Russell, 41
Bauer, Todd, 293
Baumgartner, Franz, 545
Beaucarne, Guy, 385, 443, 563
Bedzyk, Mark, 481
Beenakker, C.I.M., 211
Benagli, Stefano, 545
Beyer, Wolfhard, 171
Birkmire, Robert, 431, 575
Biswas, Rana, 35
Böhm, Markus, 231
Borello, Daniel, 545
Borysenko, Valeriy, 3
Bowden, Stuart, 431, 575
Branz, Howard M., 41, 55, 121,
    133, 391, 569
Bronsveld, Paula C.P., 159
Buechel, Gerold, 545
Bunte, Eerke, 261
Burrows, Michael, 431, 575

Cao, Xinmin, 347, 365
Carius, Reinhard, 3
Carnel, Lodewijk, 385, 443, 563
Carroll, Malcolm, 293
Cen, Zanhong, 307
Chabinyc, Michael L., 199
Chan, I., 185
Chan, Kah Yoong, 261
Chang, Jeff Hsin, 275, 299, 475,
    481
Chang, Young-Jin, 405

Chen, Jian-Zhang, 205
Chen, Kunji, 307
Chen, T., 211
Cheng, Cherry Y., 151
Cheng, I-Chun, 205, 255
Childs, Kent, 293
Cho, N.I., 185
Choong, Gregory, 281
Chu, Virginia, 219
Chuang, Tsu Chiang, 475, 481
Cohen, J. David, 55
Collins, Robert W., 347, 365
Conde, Joao Pedro, 219
Cornish, John C.L., 537

Dalal, Vikram L., 101, 269, 413,
    497
Das, Ujjwal, 431, 575
Datta, Shouvik, 55
Dauskardt, Reinhold H., 109
Decker, J., 145
Deng, Xunming, 347, 365
de Sande, Juan Carlos G., 531
Dostie, Starr, 87
Du, Wenhui, 365
Duengen, Wolfgang, 75

Einsele, Florian, 425
El-Gohary, Hassan G., 151
Ellert, Christoph, 545
Esfandyarpour, Behzad, 95

Fahrner, Wolfgang, 75
Fantoni, Alessandro, 449, 463
Farjas, Jordi, 139
Farrokh Baroughi, Mahdi, 151
Fejfar, Antonin, 159
Fernandes, Miguel, 287, 299, 463,
    469
Finger, Friedhelm, 3
Franken, Ronald H.J., 353, 419

Gall, Stefan, 133
Garção, A., 287
Ghosh, Debju, 269, 413
Goldbach, Hanno D, 67

Gordon, Ivan, 385, 443, 563
Guha, Subhendu, 15, 335, 359
Gujrathi, Subhash, 87

Han, Min-Koo, 245, 321, 405
Han, Sang-Myeon, 245, 321, 405
Harrison, Walter A., 127
Haruta, Koji, 313
Hashemi, Pouya, 95
Hasoon, Falah, 127
He, Ming, 211
Heiler, Gregory, 275, 299, 327,
    475, 481
Hekmat-Shoar, Bahman, 95
Hiza, Shuichi, 557
Houweling, Zomer Silvester, 67
Hsu, Kendrick S., 115
Huang, Xinfan, 307
Huegli, Andreas, 545

Ishihara, R., 211
Ishikawa, Y., 365
Iwaniczko, Eugene, 41, 121, 569

Jackson, Warren, 205
Jaju, Vishwas, 101
Jarecki, Robert, 293
Jiang, C.-S., 15
Jiménez, Juan, 531
Job, Reinhart, 75
Johnson, Erik V., 61
Johnston, Steve, 569
Jones, Kim M., 121, 133, 391
Ju, Tong, 9, 127
Jung, Ji-Sim, 399

Karim, Karim S., 489, 517
Kaufmann, Rolf, 281
Kavanagh, Karen L., 517
Kerr, R., 275
Kherani, Nazir, 9
Kim, Chi-Woo, 405
Kim, Jong-Man, 399
Kim, Kyung Ho, 275, 299, 475,
    481
Kim, Sun-Jae, 321

Kluth, Oliver, 545
Knipp, Dietmar, 261
Kobayashi, Tomohiro, 313
Koch-Ospelt, Detlev, 545
Konagai, Makoto, 49, 81, 557
Kondo, Michio, 583
Kosarev, Andrey, 457
Kouketsu, Seiichi, 81
Krc, Janez, 23
Kroll, Ulrich, 545
Kunz, Oliver, 391
Kuo, Yue, 225
Kwon, Jang-Yeon, 399

Lai, Jackson, 299, 327
Lee, Hyun Jung, 503
Lee, Jae-Hoon, 321
Lee, Sang-Yoon, 399
Lee, Won-Kyu, 405
Levi, Dean, 41, 569
Li, Hongbo, 353, 419
Li, Jian, 347
Li, Wei, 307
Liao, Xianbo, 365
Limb, Scott, 199
Limmanee, Amornrat, 49
Lin, Chun-Jung, 237
Lin, Gong-Ru, 237
Liu, Xiao, 511
Louro, Paula, 287, 463
Lu, Meijun, 431, 575
Lujan, Rene A., 199
Lustenberger, Felix, 281

Madhavan, Atul, 413
Mahan, A.H., 55
Martins, Rodrigo, 469
Mates, Tomas, 159
Mayer, J.W., 145
Meier, Johannes, 545
Metcalf, Thomas H., 511
Metselaar, J.W., 211
Miyajima, Shinsuke, 49, 81
Mohajerzadeh, Shams, 95
Mohr, Michael, 545
Monteduro, Giovanni, 545

Moradi, M., 185
Morana, Bruno, 531
Moreno, Mario, 457
Moutinho, H.R., 15
Mueller, Thomas, 75
Muthukrishnan, Kamal, 497
Myong, Seung Yeop, 557

Nagai, Takehiko, 583
Nam, H.G., 185
Nathan, Arokia, 185, 275, 299,
   327, 475, 481, 503
Ng, TseNga, 199
Niessen, Lars, 171
Noack, Max, 413, 497
Nominanda, Helinda, 225

Ogino, Masaaki, 191
Ong, Markus D., 109
Opila, R., 431
Otsubo, Michio, 49
Oulachgar, El Hassane, 87
Ou-Yang, Jeremy, 115

Page, Matt R., 41
Pan, Grant Z., 115
Park, Kee-Chan, 405
Park, Kyung-Bae, 399
Park, Sang-Geun, 321, 405
Parlevliet, David, 537
Patil, Samadhan, 219
Pennartz, Frank, 171
Perlov, Craig, 205
Petrusenko, Yuri, 3
Photiadis, Douglas M., 511
Pingate, Nirut, 437
Piromjit, Channarong, 437
Podraza, Nikolas J., 347
Poortmans, Jef, 385, 443, 563
Prieto, Ángel Carmelo, 531

Rath, Jatindra K., 159, 353, 419
Rau, Uwe, 425
Reedy, Robert C., 391
Ren, Li P., 115

Robertson, Michael D., 95
Roca i Cabarrocas, Pere, 61, 139
Rodríguez, Andrés, 531
Rodríguez, Tomás, 531
Romero, Manuel J., 133
Roschek, Tobias, 545
Rostan, Phillip Johannes, 425
Rouhi, Nima, 95
Roura, Pere, 139
Roybal, Lorenzo, 41
Ryu, Myung-kwan, 399

Safavian, Nader, 299, 327
Saiz, Kevin, 293
Sakai, Shigeru, 191
Sakurai, Yusuke, 313
Sambandan, Sanjiv, 199
Sameshima, Toshiyuki, 373
Sangrador, Jesús, 531
Sanguino, Pedro, 463
Sato, Takehiko, 49
Sazonov, Andrei, 503
Scarlete, Mihai, 87
Schäfer, Heiko, 231
Scherff, Maximilian, 75
Schöler, Lars, 231
Schropp, Ruud E.I., 67, 159, 353,
   419
Schüttauf, Jan-Willem A., 353, 419
Schwarz, Reinhard, 463
Seibel, Konstantin, 231
Serkland, Darwin, 293
Shimizu, Ryosuke, 191
Shimogaki, Yukihiro, 191
Shinar, Joseph, 269
Shinar, Ruth, 269
Shirai, Hajime, 313
Sichanugrist, Porponth, 437
Sivoththaman, Siva, 151
Smets, Arno H.M., 583
Spitznagel, Joel, 545
Springer, Jiri, 545
Sriprapha, Kobsak, 557
Stiebig, Helmut, 261
Stoke, Jason A., 347

Stolk, Robert L., 353, 419
Stradins, Paul, 9, 121, 127, 133, 391
Street, Robert A., 199
Striakhilev, Denis, 185, 275, 299, 327, 475, 481
Su, Tining, 127
Sugiura, Tsutomu, 49
Šutta, Pavol, 179

Taghibakhsh, Farhad, 489, 517
Takemura, Yu-ichiro, 313
Taussig, Carl, 205
Taylor, P. Craig, 9, 127
Teplin, Charles W., 121, 133
Theodore, N. David, 145
Thompson, D.C., 145
Tichelaar, Frans D., 179
To, Bobby, 121
Torres, Alfonso, 457
Tredwell, Timothy, 275, 299, 327, 475, 481

van de Sanden, M.C.M., 165, 523
van der Werf, C.H.M., 419
van der Werf, Karine, 67, 353
van Elzakker, Gijs, 179
Van Gestel, Dries, 385, 443, 563
van Swaaij, R.A.C.M.M., 165, 523
Verkerk, Arjan, 159
Verlaan, Vasco, 67
Vieira, Manuela, 287, 449, 463, 469

Vygranenko, Yuriy, 275, 287, 299, 327, 449, 463, 475, 481

Wagner, Sigurd, 205, 255
Wang, Qi, 41, 127, 391, 511, 569
Wang, Tihu, 41
Wank, M.A., 523
Whitaker, Janica, 9
Wong, William S., 199
Wyrsch, Nicolas, 281

Xu, Jun, 307
Xu, Yueqin, 41, 55, 127, 391

Yamada, Akira, 49, 81, 557
Yan, Baojie, 15, 335, 359
Yan, Yanfa, 41, 391
Yang, Jeffrey, 15, 359
Yashiki, Yasutoshi, 81
Ye, Mina, 313
Young, David L., 391
Yue, Guozhen, 335, 359
Yunaz, Ihsanul Afdi, 557

Zambrano, Jimenez R., 165
Zeman, Miro, 23, 179
Zhou, Dayu, 35
Zhou, Jiang, 307
Zhou, Zhaoqun, 269
Zimin, Dmitri, 545
Zindel, Arno, 545
Zukotynski, Stefan, 9

SUBJECT INDEX

absorption, 583
amorphous, 9, 15, 23, 35, 55, 139,
       179, 205, 225, 275, 281,
       299, 321, 359, 431, 437,
       449, 475, 481, 511, 523,
       557, 583

bonding, 109

chemical vapor deposition (CVD)
    (chemical reaction), 191
    (deposition), 67, 87, 121, 191,
       237, 255, 353, 419, 517,
       523, 531
crystal growth, 139, 145, 313
crystalline, 517, 537

defects, 9, 55, 127, 385
devices, 23, 261, 287
dielectric, 101
    properties, 67, 185
diffusion, 115
dislocations, 293
display, 245, 321, 405
dopant, 145, 171

electrical properties, 81, 151, 159,
    385
electron
    irradiation, 3
    spin resonance, 3, 9
electronic
    material, 255, 497
    structure, 503
epitaxy, 121, 133, 151

film, 191

Ge, 61, 81, 293, 457
grain boundaries, 159

H, 443
hydrogenation, 95

infrared (IR) spectroscopy, 165,
    523

internal friction, 511
Ir, 457

laser, 211
    annealing, 307, 373
liquid, 231
luminescence, 237, 531

microelectro-mechanical (MEMS),
    219, 231
microelectronics, 211
microstructure, 399
microwave heating, 145

nanostructure, 61, 307, 335, 353,
    537
nitride, 49, 67, 185, 511
nucleation and growth, 391

optical, 35, 269
    properties, 49, 569
optoelectronic, 231, 469, 475, 545
oxidation, 95, 101
oxide, 75

passivation, 49, 75
photovoltaic, 15, 41, 75, 81, 127,
    133, 179, 299, 335, 347,
    353, 365, 391, 413, 419,
    425, 431, 481, 497, 545,
    557, 563, 575
plasma(-)
    deposition, 165, 313, 359
    enhanced CVD (PECVD)
       (chemical reaction), 569,
       583
       (deposition), 61, 171, 261,
       269, 275, 293, 335,
       347, 365, 425, 437,
       463, 489, 503, 545,
       575
polycrystal, 211, 385, 405, 413,
    443, 563
polymer, 87, 219

radiation effects, 299

semiconducting, 287, 327
sensor, 199, 269, 275, 281, 287,
　　327, 449, 457, 463, 469,
　　475, 481
Si, 3, 15, 23, 35, 41, 55, 95, 101,
　　109, 115, 121, 127, 133,
　　139, 151, 159, 165, 171,
　　199, 205, 225, 237, 245,
　　255, 281, 307, 313, 321,
　　347, 359, 373, 391, 399,
　　405, 413, 419, 425, 431,
　　437, 443, 489, 497, 517,
　　531, 537, 563, 575

simulation, 449, 463, 469
structural, 179
surface chemistry, 569

thin film, 41, 87, 109, 115, 185,
　　199, 205, 219, 225, 245,
　　261, 327, 365, 373, 399,
　　489, 503, 557

CPSIA information can be obtained at www.ICGtesting.com
Printed in the USA
LVOW06s0749210514

386631LV00009B/297/P